Physical Properties and Thermodynamic Behaviour of Minerals

NATO ASI Series

Advanced Science Institutes Series

A Series presenting the results of activities sponsored by the NATO Science Committee, which aims at the dissemination of advanced scientific and technological knowledge, with a view to strengthening links between scientific communities.

The series is published by an international board of publishers in conjunction with the NATO Scientific Affairs Division

A Life Sciences B Physics	Plenum Publishing Corporation London and New York
C Mathematical and Physical Sciences	D. Reidel Publishing Company Dordrecht, Boston, Lancaster and Tokyo
D Behavioural and Social Sciences E Applied Sciences	Martinus Nijhoff Publishers Dordrecht, Boston and Lancaster
F Computer and Systems Sciences G Ecological Sciences H Cell Biology	Springer-Verlag Berlin, Heidelberg, New York, London, Paris, and Tokyo

Series C: Mathematical and Physical Sciences Vol. 225

Physical Properties and Thermodynamic Behaviour of Minerals

edited by

Ekhard K. H. Salje
Department of Earth Sciences,
University of Cambridge, U.K.

D. Reidel Publishing Company

Dordrecht / Boston / Lancaster / Tokyo

Published in cooperation with NATO Scientific Affairs Division

Proceedings of the NATO Advanced Study Institute on
Physical Properties and Thermodynamic Behaviour of Minerals
Cambridge, U.K.
July 27 - August 8, 1987

Library of Congress Cataloging in Publication Data

NATO Advanced Study Institute on Physical Properties and Thermodynamic Behaviour of Minerals (1987: Cambridge, Cambridgeshire)
Physical properties and thermodynamic behaviour of minerals / edited by Ekhard K. H. Salje.
p. cm. — (NATO ASI series. Series C, Mathematical and physical sciences, vol. 225)
"Proceedings of the NATO Advanced Study Institute on Physical Properties and Thermodynamic Behaviour of Minerals, Cambridge, U.K., July 27–August 8, 1987"—T.p. verso.
"Published in cooperation with NATO Scientific Affairs Division."
Includes bibliographies and index.
ISBN 90–277–2656–6
1. Mineralogical chemistry—Congresses. 2. Thermodynamics—Congresses.
I. Salje, Ekhard K. H. II. North Atlantic Treaty Organization. Scientific Affairs Division.
III. Title. IV. Series: NATO ASI series. Series C, Mathematical and physical sciences; no. 225.
QE371.N38 1987
549′.12—dc 19 87–28892
CIP

Published by D. Reidel Publishing Company
P.O. Box 17, 3300 AA Dordrecht, Holland

Sold and distributed in the U.S.A. and Canada
by Kluwer Academic Publishers,
101 Philip Drive, Norwell, MA 02061, U.S.A.

In all other countries, sold and distributed
by Kluwer Academic Publishers Group,
P.O. Box 322, 3300 AH Dordrecht, Holland

D. Reidel Publishing Company is a member of the Kluwer Academic Publishers Group

Printed in The Netherlands

TABLE OF CONTENTS

Preface

The role played by earth sciences in the scientific community has changed considerably during this century. Since the revolutionary discoveries of global processes such as plate tectonics, there has been an increasing awareness of just how fundamental many of the mechanisms which dominate in these processes depend on the physical properties of the materials of which the earth is made. One of the prime objectives of mineral sciences is now to understand and predict these properties in a truly quantitative manner. The macroscopic properties which are of most immediate interest in this context fall within the conventional definitions of thermodynamics, magnetism, elasticity, dielectric susceptibilities, conductivity etc. These properties reflect the microscopic contributions, at an atomistic level, of harmonic and anharmonic lattice vibrations, ionic and electronic transport as well as a great variety of ordering and clustering phenomena.

The advances made by solid state physicists and chemists in defining the underlying phenomena involved in the thermal evolution of materials have stimulated major new research initiatives within the Earth Sciences. Earth Scientists have combined to form active groups within the wider community of solid state and materials scientists working towards a better understanding of those physical processes which govern not only the behaviour of simple model compounds but also that of complex materials like minerals. Concomitant with this change in direction has come an increasing awareness of the need to use the typical working tools of other disciplines. The experimental facilities in common use now range from neutron sources, spectroscopy and high resolution TEM to newly developed calorimetry and X-ray scattering techniques. Interdisciplinary research without doubt holds many exciting prospects and the ASI-meeting in Cambridge is a visible expression of the growing together of Earth Sciences with Physics and Chemistry in the fields of thermodynamics, phase transitions, kinetics, simulation-modelling, TEM and spectroscopy, including NMR, Mössbauer technique, EXFS, Raman-, infrared- and optical spectroscopy. This volume is a distillation of the meeting and, we hope, will provide fresh stimulation for scientific interaction in the future.

It is a great pleasure for me to use this opportunity of acknowledging the help given by members of the Departments of Earth Sciences and Physics in Cambridge, in the organisation of the meeting. I wish to thank, in particular, the local organising committee consisting of V. Heine, M.G. Bown, M.A.Carpenter, M.T. Dove, M.I. Johnston, J.D.C. McConnell, A. Putnis.

The meeting was generously funded by NATO and travel grants have been made available by NERC, SERC, NSF and NATO. On behalf of all the participants I thank these institutions for their support.

Cambridge

Ekhard Salje

R. J. Reeder A. Meike J. N. Boland P. H. Turner M. Rubbo P. F. Zanazzi H. Feuer E. Stuckenschmidt G. Cressey A. J. Brearley

C. Giampaolo G. L. Hovis J. Roux J. E. M. Allan S. T. Bramwell C. R. Ross, II L. Fernández-Díaz M. Tribaudino A. S. McKnight A. M. Buckley M. Gregorkiewitz D. M. Hatch A. P. Önen V. L. Escobar D. E. Laughlin

M. C. Magalhaes M. T. Dove M. J. L. Percival V. Tazzoli F. C. Hawthorne G. Nord J. Sykes-Nord H. P. Fritzer B. Palosz A. M. Glazer M. C. Domeneghetti B. Bonello M. Fischer C. Ruescher J. Copreaux

E. Baum J. Sanz I. Sobrados H. Jhans J. Honig N. J. Hess A. Pawley Shi Ping B. Güttler U. Bismayer M. J. Tello H. Dachs H. P. Weber J. M. Kennedy W. Depmeier I. Gavrieli C. Capobianco

M. Pieroth L. L. Boyle A. G. Christy B. L. Phillips R. J. Angel C. A. McCammon N. Le Breton N. E. Brown M. Hucher F. Denoyer J. Pouget H. Küppers S. Redfern W. C-K Poon D. C. Palmer B. Wruck

A. Remsberg C. Boberski F. Seifert C. Prewitt I. Jones Alian Wang J. Dubessy D. Price N. Ross K. Langer W. L. Brown W. E. Jackson G. E. Brown, Jr. S. Ghose R. C. Peterson S. Gould

J. M. D. Coey M. Madon A. Le Cleac'h P. H. Gillet O. Ballet G. Amthauer M. G. Bown A. Putnis E. Salje J. D. C. McConnell M. I. Johnston M. A. Carpenter A. Navrotsky J. M. Yeomans R. J. Kirkpatrick R. C. Liebermann W. W. Schmahe

LIST OF PARTICIPANTS

ABS-WURMBACH I.	Institut fur Mineralogie, Bochum University, Postfach 10 21 48, D-4630 Bochum, F.R. Germany
ALLAN J.	Department of Pure and Applied Physics, Trinity College, Dublin University, Dublin, Ireland
ALVES A.C.P.	Departamento de Quimica, Coimbra University, 3039 Coimbra, Portugal
AMTHAUER G.	Institut fur Geowissenschaften, Mineralogie, Hellbrunnerstrasse 34, A-5020 Salzburg, Austria
ANGEL R.	Geophysical Laboratory, 2801 Upton Street, Washington DC 20008, U.S.A.
BALLET O.	DRF/SPL-MDN C.E.N.G., 38041 Grenoble, France
BARBER D.	Essex University, Wivenhoe Park, Colchester, Essex, CO4 3SQ U.K.
BAUM E.	Institut fur Mineralogie, Marburg University, Hans Meerwein Strasse, D-3550 Marburg, FR Germany
BELLUSO E.	Dipartimento di Scienze della Terra, Universita degli Studi di Torino, 10123 Torino, Italy
BISMAYER U.	Institut fur Kristallographie und Petrologie, Hannover University, D-3000 Hannover, FR. Germany
BOBERSKI C.	Institut f. Mineralogie, Ruhr-Universitat Bochum, D-4630 Bochum, F.R. Germany
BOLAND J.	Dept. of Earth & Space Sciences, SUNY at Stony Brook, Stony Brook, NY 11794, U.S.A.
BONELLO B.	Pierre et Marie Curie University, 4 place Jussieu 75005 Paris, France
BOROWITZ J.L.	SOREQ Nuclear Research Centre, 70600 Yavne, Israel
BOWN M.	Department of Earth Sciences, University of Cambridge, Downing Street, Cambridge CB2 3EQ U.K.
BRAMWELL, S.	Inorganic Chemistry Laboratory, Oxford University, South Parks Road, Oxford OX1 3QR U.K.

BREARLEY, A.	Department of Geology, New Mexico University, Albuquerque, NM 87131, U.S.A.
BROWN G.	Department of Geology, Stanford University, Stanford, CA 94305, U.S.A.
BROWN N.	Department of Geology, Princeton University, Guyot Hall, Princeton, NY 08544, U.S.A.
BROWN W.	CNRS-CRPG, BP 20, 54501 Vandoeuvre, France
BUCKLEY A.	Inorganic Chemistry Laboratory, Oxford University South Parks Road, Oxford OX1 3QR, U.K.
CAPOBIANCO C.	Department of Earth Sciences, University of Cambridge, Downing Street, Cambridge, CB2 3EQ U.K.
CARPENTER M.A.	Department of Earth Sciences, University of Cambridge, Downing Street, Cambridge, CB2 3EQ U.K.
CATLOW C.R.A.	Department of Chemistry, Keele University, Keele, ST5 5BG, U.K.
CATTI M.	Dipartamento di Chimica Fisica ed Elettrochimica, Milan University, 20133 Milan, Italy
CHRISTY A.	Department of Earth Sciences, University of Cambridge, Downing Street, Cambridge CB2 3EQ U.K.
COEY J.M.D.	Department of Pure and Applied Physics, Trinity College, Dublin, Ireland
COPREAUX J.	Laboratoire de Mineralogie & Cristallographie, 4 place Jussieu, 75005 Paris, France
CRESSEY G.	Dept of Mineralogy, British Museum (Natural History), Cromwell Road, London SW7 5BD, U.K.
DACHS H.	Hahn-Meitner-Institut, Glienicker Str. 100, D-1000 Berlin, F.R. Germany
DENOYER F.	Laboratoire de Physique des Solides, Paris Sud University, Bat 510, 91450 Orsay, France
DEPMEIER W.	Institut fur Kristallographie, Karlsruhe University, D-7500 Karlsruhe, F.R. Germany
DOMENEGHETTI C	CNR-Centro di Studio Cristallografia Strutturale, Via Bassi n.4, 27100 Pavia, Italy
DOVE M.T.	Department of Earth Sciences, University of Cambridge, Downing Street, Cambridge CB2 3EQ U.K.

DUBESSY J.	CREGU, 3 rue du Bois de la Champelle, 54501 Vandoeuvre, France
ESCOBAR V.	Department of Mineral Resouces Engineering, Imperial College, London SW7 2AZ, U.K.
FERNANDEZ-DIAZ L.	Dpto. Cristalografia y Mineralogia, Complutense de Madrid University, 28040 Madrid, Spain
FEUER H.	Institut fur Kristallographie, Frankfurt University, D-6000 Frankfurt, F.R. Germany
FISCHER M.	Pierre et Marie Curie University, 4 place Jussieu 75005 Paris, France
FRITZER H.P.	Institut of Physical & Theoretical Chemistry, Graz Univ. of Technology, A-8010 Graz, Austria
GALLIOS G.	Aristotelian Univeristy of Thessaloniki, 54006 Thessaloniki, Greece
GAVRIELI I.	Department of Geology, The Hebrew University, Givat Ram, 91904 Jerusalem, Israel
GHOSE S.	Department of Geological Sciences, Washington University, Seattle, WA 98195, U.S.A.
GIAMPAOLO C.	Dip. Scienze della Terra, Univ. di Roma "La Sapienza", P. le Aldo Moro 5, 00100 Rome, Italy
GILLET P.	Laboratoire de Mineralogie Physique, CAESS, Rennes University, 35042 Rennes, France
GLAZER M.A.	Department of Physics, Oxford University, Oxford OX1 2JD, U.K.
GOULD S.	Beevers Miniature Models Unit, Dept. Chemistry, Edinburgh University, Edinburgh EH9 3HS, U.K.
GREGORKIEWITZ M.	Instituto de Ciencia de Materiales, CSIC, Serrano 115 bis, 28006 Madrid, Spain
GUTTLER B	Department of Earth Science, University of Cambridge, Downing Street, Cambridge CB2 3EQ U.K.
HATCH D.M.	Department of Physics & Astronomy, Brigham Young University, Provo, Utah 84602, U.S.A.
HAWTHORNE F.	Geological Sciences, Manitoba University, Winnipeg, R3T 2N2, Canada

HEINE V. Department of Physics, Univeristy of Cambridge, Cavendish Laboratory, Cambridge, CB3 OHE, U.K.

HESS N. Department of Geological Sciences, Washington University, Seattle, WA 98112, U.S.A.

HONIG J. Department of Chemistry, Purdue University, West Lafayette, IL 47907, U.S.A.

HOVIS G.L. Department of Geology, Lafayette College, Easton, PA 18042, U.S.A.

HUCHER M. Departement des Sciences de la Terre, Orleans, University, 45067 Orleans, France

JACKSON W. Department of Geology, Stanford University, Stanford, CA 94305, U.S.A.

JONES I. Cavendish Laboratory, University of Cambridge, Madingley Road, Cambridge CB3 OHE, U.K.

KARATAS C. Materials Engineering, Middle East Technical University, 06531 Ankara, Turkey

KENNEDY J. University of Kent at Canterbury, Canterbury, Kent, U.K.

KENNEDY S.W. Adelaide University, Adelaide, Australia

KIRKPATRICK J. Department of Geology, Illinois University, 1301 W. Green Street, Urbana IL 61801, U.S.A.

KRAUSE C. Kernforschungszentrum Karlsruhe, Postfach 3640, D-7500 Karlsruhe, F.R. Germany

KUPPERS H. Mineralogisches Institut, Kiel University, Olshausenstr. 40, D-2300 Kiel, F.R. Germany

LANGER K. Institut fur Mineralogie und Kristallographie, Technische University, D-1000 Berlin, F R Germany

LAUGHLIN D.E. Dept. of Metallurgical Engineering & Materials Science, Carnegie Mellon University, Pittsburgh 15213, U.S.A.

LE BRETON N. Institut fur Mineralogie, Ruhr University, D-4630 Berlin, F.R. Germany

LE CLEAC'H A. Lab. de Mineralogie Physique, CAESS, Rennes University, 35042 Rennes, France

LIEBERMANN R.	Department of Earth & Space Sciences, State Univ of New York, Stony Brook, NY 11794, U.S.A.
MADON, M.	Department of Geological Sciences, University College London, London, WC1E 6BT, U.K.
MAGALHAES C.	Department of Chemistry, Aveiro University, 3800 Aveiro, Portugal
MANGHNANI M.	Department of Geophysics, Hawaii University, 2525 Correa Road, Honolulu, Hawaii 96822 U.S.A.
McCAMMON C.	Department of Geological Sciences, British Columbia University, Vancouver BC, V6T 2B4 Canada
McCONNELL J.D.C.	Department of Earth Sciences, Oxford University, Parks Road, Oxford OX1 3PR, U.K.
McKNIGHT A.S.	Department of Geology, Hull University, Cottingham Road, Hull, HU6 8DF, U.K.
MEIKE A.	Earth Sciences Division, Lawrence Berkeley Lab., 1 Cyclotron Road, Berkeley, CA 94720, U.S.A.
MOTT N.	Department of Physics, University of Cambridge, Cavendish Laboratory, Cambridge CB3 0HE, U.K.
MULLER K.A.	IBM Research Division, Zurich Research Laboratory 8803 Ruschlikon, Switzerland
MURTY K.S.	Department of Geology, The University, Law College Compound, 440 001 Nagpur, India
NAVROTSKY A.	Department of Geological Sciences, Princeton University, Princeton, NJ 08544, U.S.A.
NORD G.	United States Geological Survey, 959 National Center, Reston, VA 22092, U.S.A.
ONEN A.P.	Geological Engineering Department, Middle East Technical University, 06531 Ankara, Turkey
PALMER D.	Department of Earth Sciences, University of Cambridge, Downing Street, Cambridge CB2 3EQ U.K.
PALOSZ B.	Institut fur Kristallographie und Mineralogie, Munchen University, D-8000 Munchen, F.R. Germany
PAWLEY A.	Grant Institute of Geology, Edinburgh University West Mains Road, Edinburgh EH9 3JW, U.K.

PERCIVAL M.J.L.	Department of Earth Sciences, University of Cambridge, Downing Street, Cambridge CB2 3EQ U.K.
PERTLIK F.	Institut fur Mineralogie und Kristographie, Wien University, A-1010 Vienna, Austria
PETERSON R.C.	Department of Geological Sciences, Queen's University, Kingston, Ontario K7L 3N6, Canada
PHILLIPS B.L.	Department of Geology, Illinois University, 1301 W. Green Street, Urbana, Illinois 61801, U.S.A.
PIEROTH M.	Fak. f. Physik, Konstanz University, Bucklestr. 13, D-7750 Konstanz, F.R. Germany
POON, W.C-K.	Cavendish Laboratory, Madingley Road, Cambridge CB3 0HE, U.K.
POUGET J.	Mechanique Theorique, Univ Pierre et Marie Curie, 4 place Jussieu, 75230 Paris, France
PREWITT C.	Geophysical Laboratory, 2801 Upton Street, N.W., Washington DC, 20008, U.S.A.
PRICE G.D.	Department of Geological Sciences, University College London, London, WC1E 6BT, U.K.
PUTNIS A.	Department of Earth Sciences, University of Cambridge, Downing Street, Cambridge CB2 3EQ U.K.
REDFERN S.A.T.	Department of Earth Sciences, University of Cambridge, Downing Street, Cambridge CB2 3EQ U.K.
REEDER R.	State University of New York at Stony Brook, Stony Brook, NY 11794, U.S.A.
REMSBERG A.	Department of Earth & Space Sciences, State Univ of New York, Stony Brook, NY 11794, U.S.A.
ROSS II C.R.	Bayerisches Geoinstitut, Bayreuth University, D-8580 Bayreuth, F.R. Germany
ROSS N.L.	Geophysical Laboratory, 2801 Upton Street NW, Washington DC 20008, U.S.A.
ROUX J.	CNRS-SCM, IA rue de la Ferollerie, Orleans, France
RUBBO M.	Dipartimento di Scienze della Terra, Via S. Massimo, 10123 Torino, Italy

SALJE E.	Department of Earth Sciences, University of Cambridge, Downing Street, Cambridge CB2 3EQ U.K.
SANTAMARIA C.M.	Dpto. Fisica, Facultad de Ciencias, Pais Vasco University, 48080 Bilbao, Spain
SANZ LAZARO J.	Instituto de Ciencia de Materials, CSIC, Serrano 115 dpdo., 28006 Madrid, Spain
SCHMAHL W.W.	Department of Earth Sciences, University of Cambridge, Downing Street, Cambridge CB2 3EQ U.K.
SCHMIDBAUER E.	Institut fur Geophysik, Munchen University, Theresienstr. 41, D-8000 Munchen, F.R. Germany
SEBALD A.	Department of Chemistry, Durham University, South Road, Durham, DH1 3LE, U.K.
SEIFERT F.	Bayerisches Geoinstitut, Bayreuth University, D-8580 Bayreuth, F.R. Germany
SHI PING	Grant Institute of Geology, Edinburgh University, West Mains Road, Edinburgh EH9 3JW, U.K.
SOBRADOS I.	Instituto di Ciencia de Materiales, CSIC, Serrano 115 bis, 28006 Madrid, Spain
SOFFA W.A.	Dept of Metallurgical Engineering & Material Science, Carnegie-Mellon Univ, Pittsburg PA 15213
STOBBS W.M.	Dept. of Materials Science & Metallurgy, Cambridge University, Cambridge CB2 3QZ, U.K.
STUCKENSCHMIDT E.	Institut fur Kristallographie und Mineralogie, Frankfurt Univ., D-6000 Frankfurt, F.R. Germany
SYKES-NORD J.	Department of Earth & Environmental Science, CUNY New York, NY 10036, U.S.A.
TAZZOLI V.	Dip. Scienze della Terra, Sez. Mineralogia, Pavia University, Via Bassi n.4, 27100 Pavia, Italy
TELLO M.J.	Departamento de Fisica, Facultad de Ciencias, Pais Vasco University, Bilbao, Spain
TRIBAUDINO M.	Dipartimento di Scienze della Terra, Torino University, 10123 Torino, Italy
TURNER P.	Bruker Spectrospin Limited, Unit 3, 209 Torrington Avenue, Coventry, CV4 9HN, U.K.

WALL A. — Department of Geological Sciences, University College London, London, WC1E 6BT, U.K.

WANG A. — Lab. Spectrochimie Infrarouge et Raman, Lille I University, 59655 Lille, France

WEBER H-P. — Institut de Cristallographie, Lausanne University B.S.P., CH-1015 Lausanne-Dorigny, Switzerland

WRIGHT K. — Department of Geological Sciences, University College London, London, WC1E 6BT, U.K.

WRUCK B. — Department of Earth Sciences, University of Cambridge, Downing Street, Cambridge CB2 3EQ U.K.

YEOMANS J. — Department of Physics, Oxford University, Oxford, OX1 2JD, U.K.

ZANAZZI P.F. — Dipartimento di Scienze della Terra, Pizza University, 06100 Perugia, Italy

ZOUBOULIS A. — Department of Chemistry, Aristotelian University, 54006 Thessaloniki, Greece

THE MICROSCOPIC MECHANISMS OF COMPLEX STRUCTURAL PHASE TRANSITIONS

Volker Heine
Cavendish Laboratory
Madingley Road
Cambridge CB3 OHE
United Kingdom

ABSTRACT. The last twenty years have seen progress in identifying quantitatively and unambiguously the mechanisms responsible for many structural phase transitions. By this we mean identifying the interatomic forces involved and showing that they do in fact lead to the phase transitions observed. Nowadays that means much more than a qualitative description. Naturally such progress implies that increasingly complex transitions can be unravelled by experimental and computational study. This theme will be illustrated by some examples particularly from the field of incommensurate structures.

1. INTRODUCTORY REMARKS

In adding my welcome to Cambridge and to this NATO Advanced Studies Institute, I am pleased to see so many countries represented from NATO and beyond. We usually read of NATO in our newspapers in a military context and it is good to be reminded of its wider role in the cooperation between Europe and North America. After World War II, NATO helped in the rebuilding of western Europe alongside the economic Marshall Plan and the Fulbright Fellowship programme. I can remember operating in my science as an Atlantic off-shore satellite of the USA until the mid 1970's. Now all that has changed and it is good to see an equal share of the scientific leadership in the programme of this ASI coming from Europe. However, we must remember that science is not equally well developed in quality and quantity in all NATO countries, and there is still work to be done to achieve this. It is right that one of the rules for NATO Advanced Studies Institutes is that an effort should be made to achieve active participation from as many NATO countries as possible. However, there is no wish to be parochial: so let me repeat my welcome to all, including equally the participants from beyond the NATO area.

To set the stage for our scientific discussions I want to refer at the outset to the important role of structural phase transitions in the wider context of earth sciences. Gravity is an amazingly strong force. When I

E. K. H. Salje (ed.), Physical Properties and Thermodynamic Behaviour of Minerals, 1–15.

drop an object it accelerates much faster than I can in my car, and when I put a pot of water to heat on the stove a lively convection is maintained by tiny differences in density. So it is with the convection in the earth's mantle that drives continental drift and the earth's magnetic field. In this context note the substantial changes in density associated with structural phase transitions compared with normal thermal expansion. For example, at the α to β phase transition in quartz induces a fractional density change $\delta\rho/\rho$ of the order of 5%, spread over a range of temperature. This is comparable to normal thermal expansion over about 400°C. Thus the study of structural phase transitions at high temperature and pressure is very relevant to geological processes. Another aspect besides density change is the energy of transformation, sometimes large and sometimes surprisingly small, as seen in specific heat anomalies. These can control the rate of cooling in some geological situations, or so my friends tell me.

By way of opening remarks, I also want to tell you of my great discovery in the last few years: I have found that there are earth scientists, chemists and materials scientists just as fascinated by structural phase transitions as I and other physicists are! It is currently fashionable to stress the importance of interdisciplinary research. But that very expression implies that you have previously in your mind divided science up into tidy separate disciplines. It isn't like that. Science is a continuous web. Of course, we each view it from our own perspective, and it is a pleasure to note the diversity of background of people here. From mineralogy, geology, physics, chemistry, materials science,...... it should make for lively discussion. In the following remarks I shall therefore not try to hide the particular intellectual baggage which I can bring to this meeting from my own particular experience as a theoretical and computational physicist.

2. MICROSCOPIC MECHANISMS

In science knowledge is organised on different levels. Much has been done by mineralogists to work out a broad and coherent understanding of mineral structures and the phase transformations observed. I have aleady mentioned its connection with another level, that of plate tectonics and mantle convection.

We are currently seeing the beginning of a new stage in this study, involving the transition to a new level, namely that of microscopic mechanisms. In the last twenty years a detailed microscopic understanding has been achieved in some simple materials, thanks largely to neutron scattering, computer modelling and other technical developments. These techniques and the insights that simple materials have already provided, are now being applied to the more complex phase transitions in important mineral structures.

Before discussing specific examples I want to add a more general remark. In some ways to talk of different 'levels' is unfortunate. Of course a more microscopic understanding can and should feed into a study of more macroscopic phenomena: conversely the existence of plate tectonics and mantle convection adds an extra relevance and spur to the study of mineral phase transitions. But it is quite wrong to suggest that levels of organising knowledge somehow imply levels of importance or scientific worth. For example it would be quite absurd to maintain that mineralogy is a 'basic' science while plate tectonics is 'merely' 'applied' mineralogy! Similarly no status differentiation is implied in the search for microscopic driving mechanisms for phase transitions. Their role is to help complete the picture. Perhaps this is obvious and you may wonder why I am flogging a dead horse. Well, unfortunately this point is not always correctly expressed. You may remember Rutherford surveying all of science, saying "there is physics and all the rest is stamp collecting". Personally, I think it says more about Rutherford than about science. The same kind of put-down was repeated by an eminent professor of this University at his inaugural lecture a few years ago. While I have great admiration for him as a scientist and a person, I think this analysis of the nature of science is simply wrong. Even within physics, we have for example macroscopic thermadynamics as a coherent branch of science with its own laws and principles which were not, and could not possible have been, discovered by 'merely' applying quantum mechanics, still less by applying quarks and unified field theory. It is important to be clear about this point at an interdisciplinary meeting such as this one.

Let me now turn to science and illustrate my theme of microscopic mechanisms with an example. Professor Vallade and co-workers have recently given a very nice picture of the incommensurate phase of quartz which exists over a narrow temperature interval at the α/β phase transition (Vallade 1987, Berge et al 1986, 1987). As is well known, the high temperature β form consists of SiO_4 regular tetrahedra joined by the sharing of vertex oxygen atoms into hexagonal spirals. At low temperature the spirals can distort slightly in a right handed or left handed sense to give the α_+ and α_- forms, presumably to optimise the Si-O-Si bond angle. The β phase can then be viewed as resulting from local oscillation between the α_+ and α_- extremes due to thermal agitation. The point is that the SiO_4 tetrahedra rotate more or less as <u>rigid</u> wholes during this transformation, and Vallade has shown that this also accounts for the incommensurate phase. It is not at all obvious that any transformations are possible either at $q=0$ or $q\neq0$ consisting only of rigid motions of the tetrahedra. There are 6 co-ordinates per tetrahedon (3 translations and 3 rotations) and 6 constraints = (1/2 x 4 x 3) from the common oxygen atoms, which leaves open the possibility of allowed modes. Vallade finds that in fact certain phonon branches are possible, including a mixed optic (α_+ to α_-) plus transverse acoustic mode in the Σ direction in the Brillouin zone which corresponds to the observed modulation in the incommensurate phase.

The second ingredient of Vallarde's theory is a coupling of the α_+/α_- optic mode to a σ_{66} shear or crystal. Why is this relevant? Basically because it kills two birds with one stone. One valid way to look at it is to say that the more free parameters you have to adjust, the more you can lower the energy. Let us write the free energy as

$$G = A\,\Psi^2 + 2\,H\Psi\emptyset + B\emptyset^2 \qquad (1)$$

where Ψ and $\emptyset$ refer to the amplitude of the optic mode and the shear respectively. The system minimises its free energy by choosing $\emptyset$ to have the value

$$Ø = - H\Psi/B \qquad (2)$$

and substituting this in (1) gives

$$G = A_{eff}\,\Psi^2 = (A - H^2/B)\Psi^2 \qquad (3)$$

This is clearly less than the value of (1) if we had neglected Ø completely, namely $G = A\Psi^2$. The full analysis is necessarily more complicated but our argument here contains the essential physical and mathematical point (Heine and McConnell 1984). Another view is to consider the fluctuations in the high temperature disordered ß phase. As the temperature is reduced, the fluctuations grow in intensity, especially those of α_+/α_- type as a precursor of the phase transition. Indeed the phase transition may be viewed as a condensation of local uncorrelated fluctuations of α_+ or α_- type into a macroscopic coherent new phase. Correlating the fluctuations into a static coherent whole clearly lowers the entropy.

Our point here is that as well as the α_+/α_- fluctuations there is the normal thermal agitation of the atoms including local σ_{66} shear fluctuations. The entropy of the system is lowered further by incorporating these shears with the precise value (2) into the coherent new phase. We conclude as before that it is favourable for the low temperature phase to incorporate both the ß to α distortion and the shear in a coupled way. The combined condensation of the two types of fluctuation can be seen very beautifully in some molecular dynamic simulations of $NaNO_2$ where the same story applies (Heine et al 1984). In fact the above remarks relate to any kind of coupling.

A particular feature of the coupling results in the incommensurate structure in quartz, $NaNO_2$ and other materials. Briefly, the coupling (1) should really be decomposed into fluctuations of all wave vectors $\underline{q}$ in the Brillouin zone, when one finds $H(\underline{q})$ is zero for $\underline{q} = 0$ and grows with $\underline{q}$ away from the origin. Thus the incommensurate (IC) phase can take advantage of the energy lowering (3) whereas the uniform α phase cannot. (The IC structure is displaced by the α phase as the temperature is lowered further because the IC phase does not lower the energy enough where the modulation wave goes through zero: but that is another story.) The reason for this particular variation with $\underline{q}$ is because we have what is called a gradient coupling. If we consider Ψ and Ø varying with position we can write the coupling as

$$H(Ø\nabla\Psi - \Psi\nabla Ø) \qquad (4)$$

Clearly, a uniform Ψ and Ø is no help, but the modulation wave of an IC structure has plenty of gradients of Ψ (and Ø) in it for the coupling (4) to make use of. There is nothing mysterious about gradient couplings. They arise from the normal interatomic forces (Heine and McConnell 1984). It is just a matter of symmetry how they are organised. In atomic terms we would write (4) as

$$H\,Ø(\underline{\ell})\,[\Psi(\underline{\ell} - \underline{a}) - \Psi(\underline{\ell} + \underline{a})], \qquad (5)$$

the square bracket being proportional to the gradient of Ψ. To lower the energy at lattice site $\underline{\ell}$ we want (for positive H and Ø ($\underline{\ell}$)) a negative Ψ in the cell one lattice spacing $\underline{a}$ to the left and a positive Ψ one lattice spacing to the right. Such a 'pull' on the left and 'push' on the right is exactly what one would expect to create a local shear σ_{66} as in quartz.

The effect is depicted for quartz in Figure 1. The antiphase boundary, being an abrupt change from positive to negative distortion of the structure, is effectively a very large gradient. In order to maintain the integrity of the regular tetrahedra at the boundary the α_+ and α_- regions have had to be shifted sideways relative to each other which constitutes a local shear there. Again the full theory by Vallade is more complicated but Figure 1 expresses the essence of it. The most important point here is that it shows how a gradient type of interaction with the symmetry of (4), (5) can arise in terms of normal interatomic forces. Incidentally, the combined modulation in the incommensurate structure can be written (McConnell and Heine 1984)

$$C_1 \text{ cos-like } (\underline{q}_{IC}\cdot\underline{r}) + C_2 \text{ sin-like } (\underline{q}_{IC}\cdot\underline{r}) \qquad (6)$$

where C_1 is, say, the pure α_+ distortion and C_2 a uniform σ_{66} shear. The modulating functions start as pure cosine and sine at the onset temperature of the IC structure, and then square up in shape at lower temperatures.

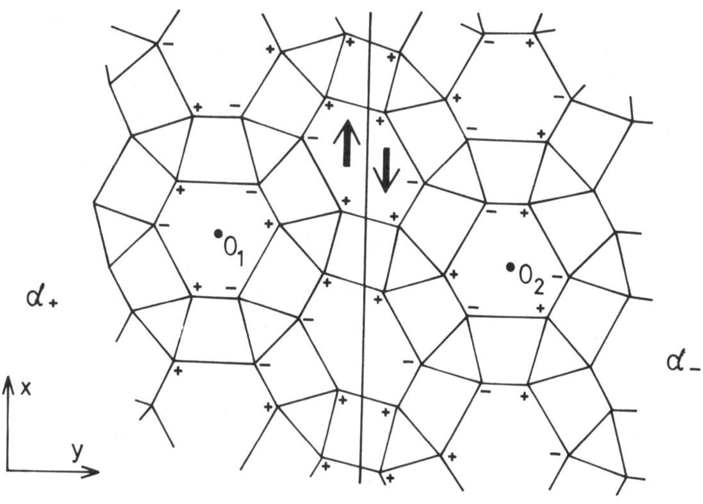

Figure 1. The low temperataure enantimorphous structures α_+ and α_- of quartz, joined at an antiphase boundary. Note the relatiave displacements of the two structures at the boundary constituting a local σ_{66} shear there. (After Vallade, private communication.)

There are several aspects which I like of Professor Vallade's work. First of all there is the elegance of the simple physical picture which, if true, can be validly transferred to a wider range of situations. Already while writing this it has suggested a new approach to a research problem that has been puzzling me. When I was a raw undergraduate I thought the best theory was the most detailed and accurate one, but that is very naive. I purposely shocked some students once by saying the best theory was the least accurate one! When we say we want to understand the weather, we do not mean that we want a set of equations that will predict where every raindrop will fall. To understand something means knowing what are the dominant effects and how they interact. Of course to propose an elegant simple model is all very well, but is it really true or just wishful thinking? Here physicists try to set very high standards for themselves. The model ideally should meet two criteria. One, of course, is to be consistent with as wide a range of data as possible. Vallade's theory explains qualitatively the origin of the IC structure, the direction of the modulation wave vector $\underline{q}_{IC}$, the different character of the soft mode dispersion relation in different directions in the Brillouin zone, the fact that there must be a set of three modulations giving a triangular pattern, and that the axes of these triangles rotate slightly with temperature. With a handful of constants he obtains a quantitative fit (Vallade 1987, Berge et al 1986, 1987). I find this very convincing, but it also explains why physicists like incommensurate structures. If I had a very simple phase transition in some cubic material, there would not be much to measure beyond the transition temperature T_c, which is just a measure of the average strength of interaction. If my theory gets that wrong, there is no indication why; and if the theory gets it right, that could equally well be good luck or good management. The result is that we are not much the wiser. Whereas incommensurate structures show such a wealth of phenomena that a coherent theoretical understanding also yields a wealth of information about how the interatomic effects work.

My other criterion is that a simple model must be mathematically justifiable. This is the hallmark of good theoretical physics. Many simple models have a validity much wider than the situation or assumption that gave rise to them. Take, for example, simple harmonic motion. It started with Galileo observing the swinging of the lights in the church (or so we are told) and yet today describes rigorously the propagation of radio waves and a great deal else in physics. In the case of quartz one could start with a very complicated set of equations describing the cores and the valence shells of the atoms to yield the eighteen phonon bands of the β phase. By suitable choice of a very large number of parameters one could no doubt describe them adequately including their temperature dependence due to anharmonicity and the softening of the optic mode plus its interaction with the other modes to give the IC phase transition. But the problem would be that the number of parameters would be so large that the fit would not be unique. The choice would therefore not represent actual atomic interactions which could be transferred to other silicates. Even aside from that, the equations for the eighteen phonon bands would not be very useful. They would be like the equations describing individual rain drops. However - and this is the crunch - we can start conceptually with those equations for the phonons and do a rigorous mathematical transformation on them, based on just the supposition of fairly rigid tetrahedra. We can eliminate almost all the unwanted internal coordinates of the tetrahedra: we do not ignore them but incorporate their effect implicitly in much the same way as we have eliminated the subsidiary order parameter Ø in going from (1) to (3) and expressed its effect in the second term in (3). The result is a simpler set of equations, as in Vallade's theory: rigorous equations. A bonus would be that these new equations would tell us the minimal number of parameters and their nature that would be needed to describe the small degree of squishyness of the tetrahedra. Again an analogous example can be drawn from eq (3). It can be shown that if H represents a nearest neighbour interaction, then $-H^2/B$ in (3) would appear as an effective second neighbour interaction (mediated by the intermediate atom and the eliminated order parameter Ø). Our mathematics implies a physical picture.

The point is that mathematics is a way of dealing with complexity and imposing a conceptual framework upon it. The equations can tell you what aspects of a model are exact, or what precise approximations are involved in getting to it. This is very important in theoretical models. If I have some model with equations and parameters for one material, I would call that a description of the material. It only becomes a proper theory for me if I can say that the symbols and numbers rigorously represent something precisely definable and real, so that I could lift those numbers and apply them with confidence when I meet the same atoms in a different material. That way our web of scientific understanding is extended from simpler materials to more complex ones. It is a very stringent test of a theory.

This is not the place to embark on a review of the different microscopic mechanisms that can give rise to incommensurate structures. The reader is referred to elsewhere for that (Heine 1987, 1988). Suffice it to say that there are several different mechanisms, but the one of gradient interaction between two order parameters is probably the most common one in the case of minerals, eg mullite and the intermediate plagioclase feldspars. There is a notion that IC structures are somehow unsatisfactory structures resulting from 'making the best of a bad job' between two competing interactions. That is so in some cases, but for quartz and the other examples just mentioned we have seen that a more accurate description would be 'making the best of a good job', insofar as the material taps into two ordering modes as we showed in connection with (1) to (3). Indeed mullite is extremely stable.

3. COMPUTER MODELLING

Over recent years computers have come to be used increasingly in a new mode. Having a theory and putting some numbers into it is conventional theoretical physics, even if it involves substantial computing. But in computer modelling one is effectively putting the laws of physics into the computer and letting the computer say what happens. In that sense it is akin to an experiment. We make 'observations' and 'measure' energies etc. Indeed they are the kinds of observations and measurements which our experimental colleagues would dearly love to make in the laboratory. The trouble is that the experimental tools are in most cases just far far too blunt at the level of individual atoms and their interactions. The same is often true of computers too, but nevertheless there are many situations where computers are able to provide important and highly detailed information.

The work in Cambridge to understand the origin of SiC polytypes is a good example. Started by K Kunc and N Churcher, most of it has been done by C Cheng and R Needs (Cheng et al 1987, 1988). The polytypes represent different regular stacking sequences of identical double layers of Si and C atoms. The i'th layer can be placed in two different orientations designated by $\sigma_i = \pm 1$ onto the layer below. It is suggested that the energy of the crystal can be expressed a

$$-\Sigma_i \Sigma_n J_n \sigma_i \sigma_{i+n} \qquad (7)$$

where J_n is the interaction between n'th neighbour double layers. Different polytypes can be created in the computer and their energies at $T = 0^oK$ 'measured' by highly sophisticated ab initio quantum mechanical calculations. These have to be done highly accurately in order to extract the energy differences between polytypes as represented by the J_n. The main point is that an indefinite degeneracy in the energies of polytypes occurs when

$$J_1 + 2J_2 - 3J_3 = 0 \qquad (8)$$

Then the energies of all structures are equal, which consist of bands of + and - layers where all bands have widths of two or three double layers. Such structures include almost all the observed polytypes. Calculations on four polytypes allowed J_1, J_2 and J_3 to be evaluated, and one finds that they do very nearly satisfy the condition (8). One can extend the idea of (7) by interpreting the J_n as free energies varying slightly with temperature T due to different phonon frequencies in the various polytypes. It turns out that tiny differences of order 0.04% in average phonon frequencies would suffice to bring the $J_n(T)$ onto the degeneracy line (8). The suggestion therefore is that the polytypes could form as equilibrium phases, or very nearly so, at a suitably high temperature.

There is more to the theory of polytypes but the initial step which we have described suffices to illustrate some points about computer modelling. First of all the only theoretical input to the calculations is the Schrodinger equation: not a theory of polytypes. The purpose is to 'measure' some energies which could in principle be measured in the laboratory except for serious experimental difficulties. The parallel between 'computer experiments' and laboratory experiments is clear. Of course, computer investigations have to be designed just as carefully and intelligently as laboratory ones, which is our second point. One does not usually go in blind. In the present instance the original idea was to test the relation (8), or rather the simpler version $J_1 + 2J_2 = 0$ omitting J_3. Indeed the unexpected discovery was that J_3, although small, is not negligible. This raises the question of why the interaction is so surprisingly long range, extending to third neighbouring double layers, ie to sixth neighbour bonds. Incidentally the suggestion is that the polytypes are stabilised as distinct phases at appropriate temperatures by small longitudinal displacements of the layers, and this is the next aspect being investigated.

By way of a third point we note that one of the basic principles of the so-called scientific method is to control some of the many variables. This the computer can do superbly. In our case we really have perfect infinite crystals at 0^oK with no impurities or defects: an experimentalist's dream! I always think of a colleague studying the initial stages of corrosion. Does he throw some nails out into the weather? Oh no. He has single crystals of iron with faces as perfectly flat as he can make them, kept clean in ultra high vacuum. Then he sees what happens when he lets in traces of oxygen, almost (it seems) a molecule at a time. Well, why not go one step further and do it all in the computer? Joking aside, we also calculated a fifth polytype with a very different type of stacking sequence from the others and from the observed ones. It verified that the representation (7) of the energy really seemed to be a very good approximation, even for this rather different structure, which is not at all obvious a priori. Incidentally, because such a stacking sequence is an unstable one, the polytype could not be produced in the laboratory and its energy measured. This is an example where the computer opens up domains of a variable not accessible experimentally. Finally an important but trivial point. Every computationalist and experimentalist has to test and calibrate his tools. The quantum mechanical calculations have to take various short cuts which all have to be validated either in the course of this work or by previous experience. There is the local approximation to the density functional formalism, the use of norm-conserving pseudopotentials and how they are derived, convergence in the basis set, the approach to self-consistency and the sampling in k-space.

This is the stage to mention different types of computer modelling. We have already discussed the quantum mechanical calculations, necessary for SiC because there are no models of the interatomic forces which are reliable at the level of the small energy differences we are dealing with in polytypes. Of course, if we could derive such interatomic forces, for example by fitting to the quantum calculations, we could take the next step and model either the lattice vibrations with their contribution to the free energy, or we could insert inpurities which are known to influence the stabilities of the polytypes. This illustrates how modelling at one level climbs onto the back of the more basic level. In this way one can obtain the dynamic phonons or deal with the large systems needed to model impurity effects. Molecular dynamic simulation (MDS) depicts the thermal agitation of the atoms and their presumed approach to whatever the equilibrium phase is. You throw the atoms into the computer with Newton's laws of motion, and give them a bit of random kinetic energy to start them off. It helps to determine the equilibrium phase, and shows the fluctuations occurring as precursors above the transition temperature. I have already referred to this process being displayed by MDS on $NaNO_2$.

Of course, the MDS relies on interatomic forces which have first been tested statically to see whether they give the right lattice constants, elastic constants and ground state structure at 0^oK: another example of the inter-relation between different types of modelling.

4. THE LANDAU FREE ENERGY FUNCTION

I want to point out a couple of developments connected with the Landau free energy function used to describe a phase transition. Originally a general theory of second order phase transitions, it was then elaborated to describe multiple order parameters and coupling to strains etc. In fact there is quite an industry to work out the appropriate form of free energy function for each particular case, and if possible to fit the parameters quantitatively. That is a good way of summing up or encapsulating the various phenomena observed and their connections with one another.

How does the Landau free energy function relate to our theme of microscopic mechanisms? The first way is through molecular dynamic simulation (MDS). The Landau function describes the fluctuations above the transition temperature as well as the phase change below, and it is the fluctuations which are generated in the MDS. Now in the MDS one can tinker with the microscopic model. One can therefore trace how observed aspects of the free energy function relate to specific features of the microscopic model. The beauty of this route is that there are no additional approximations between the microscopic model and the generated free energy expression: one is simply solving Newton's laws of motion for the atoms and applying a rigorous analysis of fluctuations.

The other connection between Landau theory and microscopic models is through a technical theoretical development. It has been known for some years but slow to permeate the literature. In fact it is just a modern way of developing mean field theory, alias Bragg Williams' theory, alias Currie Weiss' theory, but also extending it in new ways (Heine et al 1988). In the past, mean field theory has revolved around calculating the susceptibility, which gives one the transition temperature but not much more. The newer route calculates the free energy directly from the Hamiltonian. Everything else can then be calculated from the free energy, or the latter can be compared directly with the observed Landau expression. The method depends on the Bogoliubov inequality for the free energy

$$G < G_o + \langle\langle H - H_o \rangle\rangle. \tag{9}$$

Here H is the Hamiltonian and H_o some simple part of it for which the free energy G_o can be calculated exactly. The second term is the ensemble average of $H-H_o$ calculated with the distribution corresponding to G_o. The trick is to include in H_o a fictitious external field or set of fields h, and subtract it again in $H-H_o$. This is why it is a mean field theory. The h is then eliminated by minimising G(h) with respect to h, so that one has the benefit of a variational formulation. All in all, the new route is more coherent and flexible than the old mean field theory. For example in a quasi one dimensional system (with a dominant interaction in one direction) one can take H_o to be the one dimensional chain, nor is h restricted to being a small field. Another application is to an anharmonic double well potential (Heine 1988). We can approximate it in H_o by a harmonic well whose parameters are optimised by the variational pocedure. This yields the conventional theory of soft phonon models in the single site approximation. At low temperatures one may approximate the double well by a pair of parabolae to generate a different approximation.

Of course a mean field theory represents an approximation so that this route to G does not have the accuracy of MDS. But I do not see that as much of an issue. One can, for example, correct the transition temperature by a factor of around 0.7 known from simple systems. There is a curious notion in some circles that theoretical calculations are supposed to give exact agreement with experiment. But I say if you already know the answer from experiment, why are you wasting hours of Cray time calculating it. Of course if you are modelling a design of nuclear reactor before building it, you want to include every bit of realism possible. However, we are not doing that. We are doing theoretical physics to achieve understanding. Our prime consideration is to isolate the most important factors and their inter-relationships. And the logic must be absolutely rigorous. The beauty of mean field theory is that the various factors can be carried through to the end in an analytic or nearly analytic way.

We can sum up this discussion by saying that the Landau free energy function is the central meeting point of experiment, theory and computing. Free energy is a precisely defined, measurable and relevant quantity, not dependent on any theoretical notions. Symmetry determines what kind of terms are allowable, experiment can fit the constants to the observed phenonema, while computing and mean field theory establish the connection with microscopic models and interatomic forces.

5. CRITICAL FLUCTUATIONS

This brief survey would be incomplete without some mention of critical fluctuations, the subject of an enormous amount of study by physicists over the last twenty years. Near a second order phase transition at T_c various physical quantities diverge as some power of $| T - T_c |$. These powers are called critical exponents, conventionally labelled α for the divergence of the specific heat, β for the growth of the order parameter fluctuations, and γ for the susceptibility, and they are of course measurable by experiment. The point is that these critical exponents cannot have arbitrary values. They have to belong more or less to one or another 'universality class' , which depends on the <u>microscopic</u> nature of the ordering process. This is what has fascinated physicists. They love universalities like Maxwell's equations or Newton's laws. However, by the same token it is not immediately obvious what relevance critical exponents have for materials science which is essentially the study of particularities; what makes one material different from the next one.

The answer is that critical exponents are not very important but they have been used to identify what is the correct microscopic order parameter in some materials. Consider for example an atom (or SiO_4 tetrahedron or other entity) in a double well potential with central hump of height k_BT_0 (Figure 2). If T_c is much larger than T_0 the atom oscillates across the whole potential well whose double-well nature is hardly apparent or relevant. We have a purely displacive phase transition. On the other hand if the double well is very deep with T_c much less than T_0 then the atom has to sit at any time in either one side or the other which one can describe by an Ising spin. Unfortunately, in one dimension they belong to the same universility class so that measuring the critical exponents cannot distinguish between the displacive and the Ising model. But in two and three dimensions

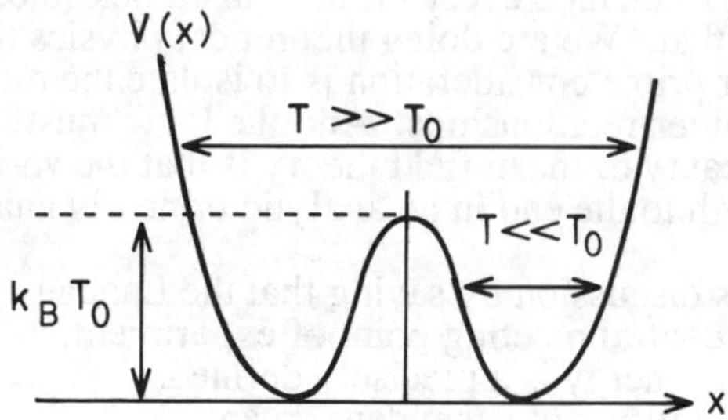

Figure 2. Fluctuations in a double potential well V(x) at temperatures T greater and less than the height of the central hump. The phase transition temperature T_c is a measure of the cooperative coupling between neighbours and can correspond to either of the two cases.

they can distinguish. For example in a trigonal system one may have continuous displacements in the x, y plane, or jumping between three equal potential wells. The critical exponents will conform to those of the two component displacive model or the three component Potts model respectively, independent of the detailed nature of the coupling between the atoms. However, this is a case where the critical exponents can take up intermediate values according to where the potential lies between the two extremes.

I want to give a brief physical picture of the main features of critical fluctuations. Their occurrence represents a breakdown of the principle of equipartition of energy. With equipartition one has $1/2k_BT$ of energy per degree of freedom and there is no divergence of the specific heat at T_c. This is also true of mean field theories which seem to be equivalent to equipartition. In reality as one approaches T_c from above, the fluctuations build up as a precursor of the phase transition. They not only build up in amplitude as one has already in mean field theory: they also build up in spatial extent. That is the true precursor. A phase transition is a cooperatiave phenomenon involving mutual reinforcement among <u>all</u> the atoms and the critical fluctuations are doing this over increasingly large regions. Because of the mutual cooperation the fluctuating modes are no longer independent so that equipartition cannot be applied.

The important fact for our purposes is that the critical fluctuations retain a memory of the microscopic nature of the ordering process, even as the fluctuations become macroscopic in extent. That is at first surprising but Figure 3 illustrates how it happens. On the left we have the boundary between an up and a down fluctuation in an Ising model. Since each site has to be in one or the other discrete state, the boundary will also be rather abrupt. On the other hand Figure 3b represents a system with a continuous displacement as order parameter. As we move from an 'up' to a 'down' region the order parameter changes smoothly and gradually over a broad transition region. Thus one can see some tell-tale signs of the microscopic mechanism in a snapshot of the fluctuations and hence in the critical exponents.

Another very important factor determining the critical exponents is the spatial dimensionality of the system, ie one for a chain, two for a layer or three for a bulk material. Here one can have quasi two dimensional behaviour in the bulk solid if the ordering interaction and hence T_c within each layer is much stronger than the coupling between planes. The reason why dimensionality is so important is illustrated in Figure 4. Let us suppose we have nearest neighbour interactions only. In a cooperative phenomenon each atom influences every other atom, but with nearest neighbour interactions it has to do so along a path of influence, each atom tending to line up the next one along the path. Indeed one can express the theory of phase transitions mathematically in these terms. In one dimension there is only one path, the direct route, connecting two atoms (Figure 4a). It turns out this does not give a strong enough influence: there can be no phase transition in a genuinely one dimensional system. Instead the fluctuations build up massively to a range of order of magnitude exp (J/k_BT) where J is the nearest neighbour coupling, but they remain fluctuations. In two dimensions there are many more possible paths (Figure 4b) and a phase transition does occur in almost all models though not in all. In some cases a 'floating' phase results. The more floppy the ordering is, the lower T_c is compared with the mean field value. Finally, in three dimensions there are again many more paths than in two dimensions and one gets sharply defined phase transitions with well ordered phases.

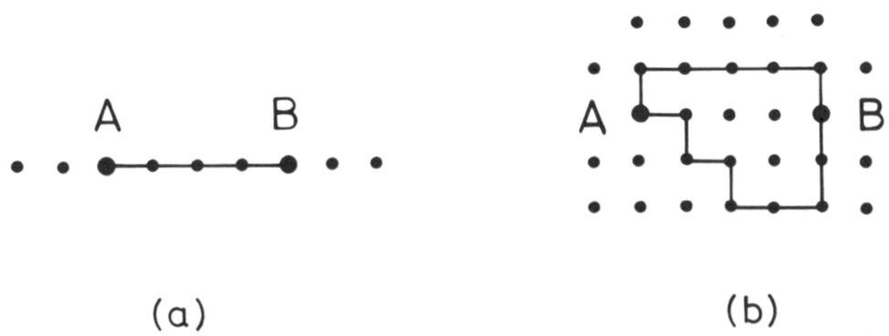

Figure 3. Instantaneous view of the boundary between positive and negative (or left and right) fluctuations just above the transition temperature. (a) For an Ising system where each site can only have the values $\sigma = \pm$. (b) For a displacive phase transition where the amount of displacement can vary continuously from zero to some maximum. Note how the boundary is much more diffuse in (b) than in (a).

Another factor influencing critical exponents is the number of components in the order parameter. For example in a trigonal or tetragonal crystal, displacive fluctuations in the x and y directions will occur together, the two being related by symmetry. This increases the amount of phase space available to them and therefore affects the critical exponents. It is also reflected in the kind of Hamiltonian one writes down. An Ising system can only fluctuate between two states and the interaction between two atoms i and j is usually written $-J\sigma_{iz}\sigma_{jz}$. However, for two or three dimensional displacements one would write $-J\underline{\sigma}_i . \underline{\sigma}_j$ in terms of vectors.

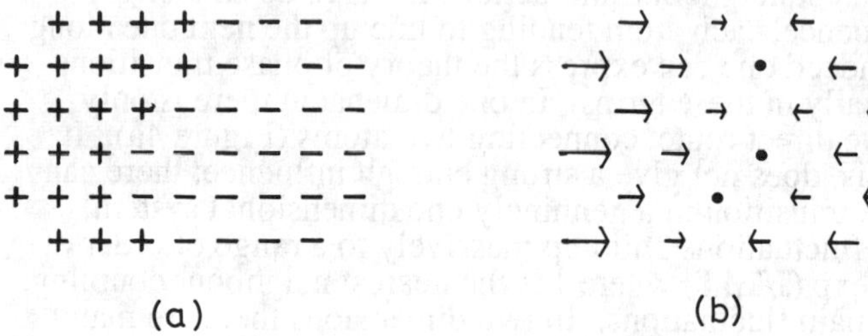

Figure 4. Nearest neighbour paths between two sites A and B, (a) in one dimension and (b) in two dimensions. Note the much greater number of possible paths in (b).

In conclusion, therefore, we have seen that critical fluctuations represent the growing spatial extent of the precursory ordering fluctuations. The precise form of the fluctuation, particularly at a boundary of an ordered region, is determined by the microscopic ordering process. This influence is reflected in turn in the critical exponents which can be measured and can give an indication of the microscopic mechanism in a given material.

ACKNOWLEDGEMENTS

It is a pleasure to acknowledge enlightening discussions with many people, and collaborations particularly with Professor JDC McConnell, Dr R Lynden-Bell, Dr R Needs and Miss C Cheng on the work mentioned here. The research has been supported financially by the Science and Engineering Research Council of the UK.

REFERENCES

Berge, B, Backheimer, J P Dolino, G and Vallade, M., 1986, Ferroelectrics, 66, 73.

Berge, B, Dolino, G and Vallade, M, 1987, to be published.

Cheng, C M, Needs, R J, Heine, V and Churcher, N, 1987, Europhys. letters 3, 475.

Cheng, C M, Needs, R J and Heine, V 1988, J Phys. C: Solid State Phys. to be published.

Heine, V, 1987, in "Competing Interactions and Microstructures: Statics and Dynamics". Proceedings of the Los Alamos workshop, May 1987. R LeSar, ed. (Springer).

Heine, V, 1988, Solid State Phys. to be published.

Heine, V, Lynden-Bell, R M, McConnell, J D C and McDonald, I R, 1984, Zeit. Phys. B: Condensed Matter 56, 229.

Heine, V, Lynden-Bell, R M and Simmons, E H, 1988, to be published.

Heine, V and McConnell, J D C, 1984, J Phys. C: Solid State Phys. 17, 1199.

McConnell, J D C and Heine, V, 1984, Acta Cryst. A40, 473.

Vallade, M, 1987, Private communication.

THE THERMODYNAMICS OF SHORT RANGE ORDER

J.D.C. McConnell,
Department of Earth Sciences,
University of Oxford,
Parks Road,
Oxford. OX1 3PR.

ABSTRACT. In many important phase transformations in minerals transformation occurs by way of modulated structures. This is true in the case of the plagioclase feldspars and potassium feldspar, and an excellent example of modulated structures is found in the transformation of cordierite. Knowledge of the general characteristics of these microstructures has long been available from electron microscopy but a proper understanding of their origins in terms of the interaction between different possible ordering schemes, and between ordering and displacement modes is only now becoming available. Current advances in understanding in this field are due primarily to the application of symmetry principles as contained in group theory. In this chapter these principles will be discussed in some detail and then used to elucidate a number of the more important examples of short range interactions and ordering in modulated mineral structures. The analysis begins with an account of the origin of simple incommensurate structures in minerals. These can best be understood in terms of interaction of two different structures in a short range distribution. From this point the analysis proceeds to a consideration of the interaction of three or more components and deals in detail with the interaction between local ordering and two displacement waves. This effect, which may be described as a triple product interaction, or three component interaction, gives rise to a very important third order term in the Landau free energy expansion, and results in a phase transformation of first order character. The triple product interaction is observed in both potassium feldspar and cordierite and both examples will be discussed in some detail in this chapter. It is relevant to note that the theory which is discussed in this chapter is not simply relevant in the case of modulated mineral structures. Some of the best examples of the triple product interaction occur in martensitic transformations in alloys, and examples of two component interactions leading to incommensurate structures are legion in both organic and inorganic materials.

E. K. H. Salje (ed.), Physical Properties and Thermodynamic Behaviour of Minerals, 17–48.

1. INTRODUCTION

The present article is concerned with the study of ordering interactions at short range which lead to the formation of stable or metastable modulated structures in minerals. Such effects are relatively common and have been recognised for a long time from electron microscopy but it has only recently been possible to establish in detail the nature of the interactions which lead to their stability. Current advances in our understanding of the thermodynamics of these ordering interactions has come largely from the use of group theory and it is important to appreciate from the outset why group theory provides such an important tool in the understanding of the interactions which we propose to discuss. In the physics of interactions in systems involving symmetry it is usual to find that symmetry rules govern the possibility of interaction and that, from symmetry arguments alone, it is frequently possible to establish whether a particular interaction does, or does not, occur. A familiar example is provided by optical spectroscopy where such rules are described as selection rules. Use is made of group theoretical arguments to establish these rules and the possibility of direct interaction between incident radiation and some vibration mode of the crystal or molecule. In the present chapter we will be concerned with the derivation of similar selection rules for interaction between different possible components of a modulation in a single crystal. By far the most difficult part of this task is to provide a brief but satisfactory introduction to the relevant group theoretical principles.

The most important practical implication which emerges from a study of the selection rules for modulated structures is the fact that it is possible to classify interactions in terms of the number of symmetry components which combine to produce the final modulated structure. In the simplest case one finds that interaction occurs between only two components and certain very simple selection rules obtain. Examples of this class of interaction in modulated mineral structures are found in the incommensurate structures of mullite and the intermediate plagioclase feldspars, and many other examples can be found in the literature. A second class of modulated structures involves the interaction of more than two components and in this case the selection rules are more complicated. It will be convenient to use this classification scheme for interactions involving short range order phenomena throughout the remainder of this chapter.

The section which follows this introduction will deal with general problems in defining selection rules for the origin of interactions in modulated structures in minerals. In Section 3 the principles of group theory will be applied to order-disorder in crystals. Section 4 will deal with ordering interactions in incommensurate structures such as mullite and the intermediate plagioclases. Section 5 will deal with interactions between ordering and phonon waves leading to a discussion of the modulated structures observed in potassium feldspar and cordierite and associated with martensitic transformation mechanisms. In the final section, Section 6, the thermodynamic implications of the theory will be used to discuss the general mechanisms of ordering transformations under both stable and metastable conditions.

2. SYMMETRY PRINCIPLES FOR INTERACTIONS IN MODULATED STRUCTURES

As we anticipated in the introduction to this chapter, symmetry principles, and the related application of group theory, are of particular importance in defining the selection rules for interactions in modulated structures. The development of the relevant theory can be considered in two parts. The first of these is concerned with the mathematical description of the symmetry properties of the chosen system and this involves the use of group representations. The second part, which is physical in origin, is concerned with the formulation of selection rules in terms of the direct products of these representations. The important physical principle here is simply that, for an interaction to be possible the direct product of representations should contain the identity representation of the full symmetry group of the crystal, which is simply the mathematical description of its full symmetry. This condition relating to the presence of the identity representation is universal in the study of all manner of physical interactions in molecules and crystals, Cotton, 1971.

2.1 Symmetry groups and their irreducible representations

This topic is very well covered in the existing literature but it is necessary to introduce the reader to suitable source books, and to provide an introduction to some of the more complex representations needed in the study of modulated mineral structures. Introductory theory on groups and group representations is provided by Cracknell,1968 and the theory of space groups and their representations is contained in Bradley and Cracknell, 1972. For readers who are completely unfamiliar with group theory a brief introduction to the relevant concepts will be given here on the understanding that facility with the methods will involve the reader spending further time with one or other of the textbooks on the subject.

We consider first the nature of a symmetry group representation. In general this comprises a set of n by n matrices, one for each element of the group, which behave on matrix multiplication exactly as the original symmetry elements did. An important theorem relating to such general sets of matrices states that on taking any arbitrary set of such matrices they may be reduced (by suitable matrix transformations) to a limited number of sets of matrices of minimal dimension. The resulting reduced sets are described as the irreducible representations of the group concerned. Such representations have another important property. In any group they are all mutually orthogonal, and if we write down the characters (the sum of diagonal terms in each matrix) for the matrices of a pair of irreducible representations for a chosen group, and take the direct product, i.e. the sum of the products of all pairs of characters for each of the group elements, taken in turn, then this sum will be equal to zero if the representations are not identical.

2.2 The origin of selection rules for interactions

We may now consider the role of irreducible representations in developing selection rules for interactions in a modulated structure. First we note that each aspect of the behaviour of the single crystal involved in the modulation may be assigned to a specific irreducible representation of the single crystal. For the moment we will assume that this has been done. The selection rules in the presence of two such components are extremely simple since the product of the two chosen irreducible representations must yield the identity representation of the group. Examination of the circumstances under which this condition holds indicates that it is only possible, Cotton, 1971, if the two representations are identical or contain the same irreducible representation of the group. We may conclude quite rightly from this observation that the conditions associated with the development of a modulated structure comprising only two components are particularly rigorous, and further that a knowledge of the symmetry of one of the components immediately defines that of the other since they are required to be the same. We will qualify this statement later in relation to space group symmetry where we must deal with both the point group and translational group aspects of the symmetry in the case of a lattice modulation function.

Next we consider the development of selection rules for modulations where three or more symmetry components are involved. In this case the appearance of the identity representation in the product of irreducible representations can occur in many ways and it is permissible for example to select three entirely different irreducible representations. The identity representation will appear in this triple product if the product of two of the representations yields the representation of the third, Cotton,1971. One obvious feature of the presence of three or more symmetry components in the modulation structure is the fact that it will in general be much more difficult to determine with confidence the symmetry components which are present.

It is important at this stage, having established the nature of the interactions of different symmetry components in a modulation, to define their thermodynamic implications. Briefly the number of symmetry components present and interacting in the modulation defines the order of the corresponding term in the Landau expansion of the free energy of the system. Thus, in the simplest case where there are only two symmetry components, these yield a Landau term of second order and in general this will lead to transformation behaviour of second order in the Landau sense. The presence of three components in interaction will yield a third order term and therefore in general produce first order transformation behaviour. It will be apparent from this analysis that the order of the interaction, i.e. the number of interacting components, is the most important single feature of such transformations and that the number of symmetry components provides a simple and effective method of classification. Since these general aspects of the manipulation of irreducible representations, and related selection rules, are of fundamental importance in the study of modulated structures the reader would be well advised to utilise the point group representation tables available in the literature to gain familiarity with the use of the product concept for representations, and then test the statements given here on the selection rules which give the identity representation.

2.3 The nature of irreducible representations of the space group.

From the last section it will be obvious that the use of the selection rules for interaction in modulated structures is quite straightforward once the irreducible representation for each of the symmetry components has been defined. In this section we discuss the nature of the space group irreducible representations needed in the analysis of a perfectly general lattice modulation.

Certain of the irreducible representations associated with space groups are particularly simple. Thus zone centre and zone boundary representations are usually equivalent to simple point group representations, or the product of point group representations with a small group which is a factor group of the complete translation group. This simplicity results from the fact that in these cases the lattice points are either all equivalent (which is true at the zone centre when the factor group is the group of order one) or are inequivalent in relation to the development of a simple superlattice in the case of a zone boundary reciprocal vector. The irreducible representations of all the zone centre and zone boundary points for all space groups have been catalogued by Bradley and Cracknell, 1972.

The derivation of irreducible representations for arbitrary points P_0 in the Brillouin zone are more complex since such points are not unique and must be considered together with all equivalent points in the Brillouin zone. This means that the dimensions of the full space group irreducible representation (i.e. the value of n for the n by n matrices of the representation) must be equal to the number of equivalent points in the Brillouin zone or some simple multiple R of this number depending on the dimension R of the representation chosen for point P_0 alone.

It is usual in the physics of interaction in k-space to deal with the point group symmetry of the vector P_0 itself and it is true that the possibility of interaction between two different functions with the same reciprocal wave vector P_0 is determined simply by the condition that they transform according to the same irreducible representation of the point group of P_0. Where one is interested simply in the possibility of interaction, this viewpoint is satisfactory but, in dealing with modulated structures of more than one symmetry component, we also need to know the phase relationships between functions for the same reciprocal wave vector P_0. In order to do this we must consider all the functions in the set (star) of vectors associated with P which are related by the point group symmetry elements which take P_0 into $-P_0$ etc. In other words we must use the full space group symmetry. The theory of the development of full space group irreducible representations has been dealt with in considerable detail by Bradley and Cracknell, 1972.

The question of the full symmetry of modulation components, which we now discuss, is of crucial importance in the present development of the theory of modulations. That part of the irreducible representation for the reciprocal vector P_0, which is concerned only with translations is very simple and has the form, 1, $\exp -iP_0 . R_1$, $\exp -iP_0 . R_2$ .. This representation component has the effect of simply multiplying any chosen basis function by a phase factor as we translate through the lattice translation vectors R_1, R_2, R_3 etc. Where two or more different functions with wave vector P_0 have the same point group irreducible representation in the group of P_0 the phase relationships between them are determined by the full space group irreducible representation and in particular the point group symmetry elements which take P_0 into $-P_0$, and into the other vectors in the star of P_0. The operation of the rules which determine the phase relationships between functions associated with the vector P_0 can be illustrated most easily for a two component modulation. In this case the star comprises P_0 and $-P_0$, a situation which corresponds to a simple incommensurate structure. This type of modulation has been analysed in detail in McConnell and Heine, 1984. There it was shown that, in describing the interaction, it is necessary to deal with the full symmetry of each of the two component structures present in the modulation. This amounts, in effect, to dealing separately with the two representations of the full space group irreducible representation. Since the theory of irreducible representations requires that these two representations should be orthogonal, the matrices for symmetry elements which take P_0 into $-P_0$ must have character zero. This condition ensures that the component functions in the modulation must differ by ± 90° in phase. (The choice of phase observed in any given incommensurate structure is dictated by by the fact that interaction involves a splitting of the energy level and loss of degeneracy. In practice the incommensurate structure is stabilised by the choice of the lower energy alternative, a point which will be discussed in some detail later in this chapter).

The definition of the phase relationships in the two component modulation, discussed above, makes it quite clear that, in dealing with the individual components of a lattice modulation in general, it is necessary to use the full symmetry of each component of the modulation. To be precise we require, in addition to the characters of the chosen irreducible representation of the point group of P_0, the characters for all point group symmetry elements which take P_0 into P_1 and into the other vectors in the star of P_0. This information is contained in the full space group irreducible representation. Throughout the remainder of the present chapter this symmetry will be described as the augmented symmetry of the modulation function. This symmetry must be used in considering the possible interaction of modulation components.

2.4 Symmetry criteria for the interaction of modulation components

We are now in a position to summarise the symmetry aspects of individual modulation waves in the context of their interaction to produce a stable or metastable modulation structure.

First we note that we may deal separately with translation and point group operations. In the case of the translation components it is necessary that the sum of the set of all interacting vectors P_1, P_2 etc. must be equal to a translation of the reciprocal lattice. This ensures that the product of representations yields the identity representation of the translation group. In the simplest case where there are two symmetry components the associated vectors are P_0 and $-P_0$ and their sum is zero as required. Where three symmetry components combine to form a modulation structure it follows that $P_1 + P_2 + P_3 = 0$. In this case the three vectors P_1, P_2 and P_3 need not be related by symmetry and may be contained in three entirely independent irreducible representations of the space group of the crystal. In practice it will usually be possible to establish the component vectors P_1 etc in a complex modulation system from data on the distribution of diffracted intensity associated with the modulation. This will occur in the diffraction pattern at points other than the reciprocal lattice points. Since in this sense the vectors P_1, P_2 and P_3 are observables it will normally be possible to define the translation components of the irreducible representations of the symmetry components present in the modulation from a careful study of available diffraction data.

The second part of the problem is more difficult since we wish to establish the specific space group representations associated with the stars of P_1, P_2 and P_3 which, in the direct product operation, yield the identity representation. The relevant theory has been presented by Birman, 1962, 1974, who used the direct product of complete irreducible representations for the full stars of P_1 and P_2 in defining selection rules for interactions in a crystal. Since in all cases such irreducible representations are multi-dimensional, and therefore contain a number of different component representations, a unique assignment of representations to a three component modulation is normally likely to be either very difficult or impossible if no additional symmetry information on the components of the modulation is available.

A considerable simplification of the problem of determining the appropriate representations associated with P_1, P_2 and P_3 occurs where it is possible to characterise some aspect of one or other of the components by other methods. Such facilities are available, for example, in electron microscopy. In the case of adularia, McConnell 1965, it was possible to establish that the components of the modulation comprised orthogonal transverse displacement waves associated with two vectors P_1 and P_2. By using the groups of P_1 and P_2, and the symmetry of the displacement vectors for both components of the modulation associated with P_1 and P_2, it is possible in terms of the theory developed here, first to establish the relevant irreducible representations of P_1 and P_2 uniquely, and then to build up the full space group irreducible representations of the stars of P_1 and P_2. Since the vectors P_1 and P_2 do not add to give a vector of the reciprocal lattice it follows that they cannot interact directly. We deduce that a third irreducible representation must be involved in the modulation complex. By using the augmented symmetry, i.e. the individual representations from the full space group irreducible representations for the stars of P_1 and P_2 it is possible to find acceptable representations of the star of $P_1 + P_2$ ($=P_3$) and hence ascertain what the third symmetry component of the modulation must be. This technique has been employed in a number of examples of phonon-order interactions in minerals to be discussed later.

2.5 Summary of the theory of short range interactions in crystals

Examination of the symmetry aspects of interactions at short range in crystals indicates that it is best to classify such interactions in terms of the number of symmetrically distinct components which are present in the modulation. The thermodynamic justification for this procedure is simply that the number of interacting components defines the order of the interaction term in the Landau expansion of the free energy as we have already noted earlier.

The simplest interaction at short range involves two symmetry components and gives rise to a class of compounds which have been described as incommensurate structures. The two component condition requires that we associate the vector P_0 with its negative $-P_0$ and in this case, since the star of P_0 contains two vectors, we need only examine a two-dimensional irreducible representation of the space group. The two representations which combine to form this two-dimensional representation are precisely related by symmetry since they share the same group characters for elements which exist in the point group of P_0, and have characters of opposite sign for each of the symmetry elements which turn P_0 into $-P_0$. It follows that if we are able to define one of the representations then we can automatically fix the symmetry of the other. Finally, since we know that the symmetry of the full irreducible representation requires that the components are in quadrature, i.e. have phase relationships of either $\pm$ 90°, we can immediately introduce physical arguments which determine the choice of phase between the two alternatives, and the resulting decrease in the free energy of the system associated with the loss of degeneracy.

In the corresponding analysis of the three component modulation two separate conditions must be fulfilled. Firstly it is necessary to ensure that the sum over the three vectors, which we describe as P_1, P_2 and P_3, must be equal to zero or a reciprocal lattice repeat. Secondly we must select a one-dimensional representation from each of the three full irreducible representations of the stars of P_1, P_2 and P_3 (the augmented symmetry groups) such that the triple direct product of these three representations yields the identity representation. The mechanism for the reduction of free energy in this case will be discussed in some detail later in the chapter.

Similar rules may be derived for modulations which contain higher order interactions associated with more than three primary interacting symmetry components but, as yet, examples of these have not been recognised. Analysis in terms of the relevant representations is likely to be extremely difficult. Such effects may be important in incommensurate structures of high symmetry where the vectors $\pm P_0$ are repeated by rotational symmetry elements in the Brillouin zone, as studied in incommensurate phase transformations in certain cubic nitride phases, McConnell and Heine, 1982.

3. THE GROUP THEORETICAL DESCRIPTION OF SHORT RANGE ORDER IN CRYSTALS

So far we have been concerned with setting up a formal method for the general analysis of short range interactions in crystals which is relevant in defining the origin of modulated structures. This theory pre-supposes that it is possible to describe order modulations in a single crystal in terms of the irreducible representations of the space group of the crystal. This question of an appropriate description of short range order is an issue of pivotal importance to which we must now turn. In dealing with this issue we will utilise a great deal of the general theory of representations which has been presented above.

3.1 Long range order and the importance of the special point vectors

In order to derive a theory for short range order phenomena in crystals in group theoretical terms it is necessary to start by considering the long range ordering possibilities in crystals. This topic has already been dealt with in detail by Landau and Lifshitz, 1958, who first demonstrated that permissible long range ordering schemes may be defined in terms of the irreducible representations for the special points of the Brillouin zone. Special symmetry points are points where some element of symmetry takes the point into another which is equivalent to it. Such points are necessarily restricted to the zone centre or the zone boundary. The argument used by Landau and Lifshitz runs as follows. If one chooses an ordering function associated with some general point P_0 in k-space, within the Brillouin zone, then in general it is necessary, since this function is complex, to associate with it its complex conjugate function with vector $-P_0$. Together these functions form a two-dimensional irreducible representation of the space group, as we have already discussed, and the combined functions transform as a vector on translation, meaning that the amplitude of the ordering function remains constant on translation, and the system remains trans-1 lationally invariant. By contrast, at a Landau type special point, there is no second function since the special point vector goes into itself under the symmetry elements of the special point and, as a result, it is impossible to build up a function which transforms as a vector under translation. It follows that translational symmetry is lost and the crystal undergoes a discrete change in symmetry as soon as the ordering function based on a symmetry point representation first appears.

As we have already noted earlier the irreducible representations of these special symmetry points at the origin of the Brillouin zone 000 and on the zone boundary are particularly simple since in general they relate to a group which is the outer direct product, or the semi-direct product of a normal point group, and a factor group for translations which is usually of small order. Landau and Lifshitz also pointed out that relevant irreducible representations must be capable of describing a density modulation in the single crystal. These authors were also particularly interested in determining the characteristics of the phase transformations associated with different symmetry point irreducible representations and in defining the conditions under which it is possible for the transformation to have second order character.

3.2. The general characteristics of short-range order

It is possible to provide a group theoretical description of short-range order effects since in general any function associated with an arbitrary vector P_0 must have the Bloch form;

$$U_r \exp(iP_0 \cdot R)$$

where U_r is a function which is periodic under the lattice translations and transforms according to some chosen irreducible representation of the point group of P_0. As already noted, with P_0 we must also associate a representation for $-P_0$, as its complex conjugate function. Remembering that the functions which we wish to derive must be real we now establish the correct linear combinations of the these functions associated with vectors P_0 and $-P_0$ by taking the sum and difference of the complex representations of the translation subgroup for these vectors,

$$\begin{matrix} 1 & \exp(iP\cdot R_1) & \exp(iP\cdot R_2) & \exp(iP\cdot R_3) & \exp(iP\cdot R_4)\dots \\ 1 & \exp(-iP\cdot R_1) & \exp(-iP\cdot R_2) & \exp(-iP\cdot R_4) & \exp(-iP\cdot R_4)\dots \end{matrix}$$

to give the functions

$$\begin{matrix} 2 & 2\,\mathrm{Cos}\,P\cdot R_1 & 2\,\mathrm{Cos}\,P\cdot R_2 & 2\,\mathrm{Cos}\,P\cdot R_3 & 2\,\mathrm{Cos}\,P\cdot R_4\dots \\ i\{0 & 2\,\mathrm{Sin}\,P\cdot R_1 & 2\,\mathrm{Sin}\,P\cdot R_2 & 2\,\mathrm{Sin}\,P\cdot R_3 & 2\,\mathrm{Sin}\,P\cdot R_4\dots\} \end{matrix}$$

which are both real when we remove the constant factor i from the second. These new functions operate to modulate real periodic functions U_r, on translation through lattice vectors R_1, R_2 etc, and are orthogonal as required. We now turn to an examination of the nature of the functions U_r which are periodic and hence must relate to specific ordering schemes within a single unit cell, or within a simple superlattice, as discussed below.

In the context of the periodic functions U_r the role of the Landau special points is pivotal. These points, at the Brillouin zone centre or on the Brillouin zone boundary, have symmetry such that the vector P_0 and $-P_0$ are one and the same and hence, in terms of a potential modulation, the wavelength is infinite and the ordering function necessarily fully periodic. This condition allows one to define appropriate basis functions U_r for all possible modulations. Clearly they comprise all possible real ordering schemes selected on the basis of the irreducible representations of the Landau type special points. Note that it is clearly permissible to select Landau special points, either at the origin or boundary of the Brillouin zone, as an origin for modulation vectors $\pm P_0$. This implies that we may modulate on the basis of the original unit cell, or on a simple superlattice associated with any of the special points on the Brillouin zone boundary.

In summary we may note that group theory imposes considerable constraints on the general form of short range order modulations in a single crystal. Each component of the order modulation must transform on translation according to real sine or cosine functions, and each modulation component must be based on an irreducible representation.

3.3. The development of an energy band theory for short range order

The constraints imposed by group theory on ordering behaviour in a single crystal lead directly to the concept of a simple band theory model for the energetics of disorder. This model may be derived by taking each of the irreducible representations of the special points in turn and modulating each one with all N of the k vectors in the Brillouin zone simultaneously. Here N is the total number of lattice points in the crystal. In as far as it is possible to isolate the energetics of this single band component of the total disorder we may define an energy in the single band which varies with k (Ek) throughout the Brillouin zone. It should be obvious that this energy functional Ek must be an even function of k at the special points leading to the existence of extrema and saddle points at these points throughout the zone, as is true in the case of phonon bands. It is also obvious that a single band must be continuous, and in general must link specific irreducible representations for ordering at each of the special points throughout the zone. Essential degeneracies at the special points must also be considered due to high symmetry, or the existence of glide and screw operations in the space group which lead to degeneracies at the zone boundary in non-symmorphic space groups.

So far we have assumed that each of the energy bands Ek, which we have constructed from the complete set of irreducible representations of the special points, is independent of the remainder and that there is no interaction between them. We now turn to considering the rules which must apply in the case of interband interactions.

3.4. The nature of interband interactions in order-disorder

The theory of interband interactions for ordering bands follows easily from the general theory of irreducible representations and their symmetry throughout the Brillouin zone. While it is self evident that there can be no interaction between the bands at individual special points in the Brillouin zone (where the relevant irreducible representations are all orthogonal) this need not be so at other points within the zone. Where two bands have the same symmetry, i.e. subduce the same representation in the group of the chosen vector P, it follows that there can be interaction and mixing between the bands. In general this process of mixing will only be important where the energy Ek is closely similar in the two bands in question. Ultimately the energy level for disorder associated with this mixing process must be determined by solving a secular equation as is usual in the case of other types of interband interactions.

There are two very important aspects of this band theory of ordering which we must now address. The first of these we have already anticipated in discussing the nature of the most general irreducible representations of the space group where the chosen vector P is in a completely general position. Energy bands at such points may be particularly inhomogeneous since they all have the same symmetry, namely that of the identity E alone, and hence the lowest energy band may well contain something from all possible ordering schemes. We may note here that the full space group irreducible representations for such points with dimension n (equal to the order n of the full point group of the crystal) must contain n degenerate representations and a similar number of basis functions U_r weighted as determined by interaction, and present in linear combination. The important point to note in relation to this part of the theory is that, close to an ordering phase transformation, it is inevitable that the relevant ordering band associated with the phase transformation must be close in energy to bands which we may associate with disorder, leading to the possibility of accidental degeneracy and concomitant anharmonic interband interactions. Among the representations of the special point 000 there is always at least one which cannot be used to describe substitutional ordering within the unit cell. This is the identity representation for the special point.

If we choose a set of general positions in the unit cell which are occupied at random by atoms of type A and B we may order them in turn on the basis of appropriate irreducible representations of the special point. If, however, we chose the identity representation of the group this requires that each of these general positions is equivalent and we must assume that each item is equal to an average atom, i.e., is statistically part A and part B. This particular special point representation therefore describes a situation involving complete disorder within a single unit cell. In general a full description of the state of disorder in the single crystal requires that we define the average situation in terms of the identity representation by defining a symmetrical function of the form $\{A + B\}/2$, and then use ordering representations to define difference functions of the general form $\pm\{A - B\}/2$. The average function and a specific difference function are both required to describe a state of long range order throughout the single crystal. This condition, which holds for the representations of the special point 000, also holds for the ordering behaviour based on the k vectors throughout the Brillouin zone.

Among the general irreducible representations of the space group based on vectors other than those for the special points it is usually possible to find some which, on symmetry grounds, admit of the possibility of a potential degeneracy between representations associated with ordering, and representations where ordering is prohibited. This situation frequently leads, in practice, to the occurrence of modulated, or incommensurate, structures in the temperature range close to an order-disorder phase transformation. An excellent example of this type of anharmonic interband interaction is found in sodium nitrite. This phase develops a modulated incommensurate structure close to the temperature associated with a zone centre para- to ferro-electric phase transformation.

The example which we have just discussed is one of the simplest cases of ordering interband interactions and provides a powerful justification for the band theory of order-disorder developed here. Another important example of short range order modulation occurs in the mineral yoderite, Fleet and Megaw, 1962. This is associated with ordering interband interactions near an invariant point associated with a screw axis and located at the Brillouin zone boundary. An analysis of the main features of this type of short range order interaction follows.

3.5. Ordering modulations associated with an invariant point

In the previous section we were concerned with the simplest possible interaction between ordering bands at a point of potential accidental degeneracy in the Brillouin zone. The theory which we used was based on the simple assumption that, where interaction is possible because the symmetry is the same for two different ordering schemes at some point other than the special points, then anharmonic interaction will result in a decrease in the free energy, and the resulting energy minimum will correspond to the preferred vector for ordering. This local minimum then leads to the order-modulated structure.

The ordering possibilities close to certain special points at the boundary of the Brillouin zone may lead to quite a different type of ordering modulation in the presence of an essential screw or glide operation, (with translation normal to the boundary), for the following reason. Suppose that, within the unit cell, atoms A and B are related by a screw diad axis. In the disordered state the operation of the screw diad is necessarily even in character since A and B must be identified with each other. For an ordered state A and B may be distinguished by assuming that the character of the screw axis is odd. At the zone origin 000 this condition will lead to a normal order-disorder transformation, or the presence of a modulated structure with vectors $\pm P_0$ positioned close to 000 if there is at some temperature an accidental degeneracy. However it is possible that the minimum of free energy in the band structure for both bands may occur near the zone boundary (near to $0\frac{1}{2}0$ for a monoclinic crystal) rather than at 000. If this should be so an essential degeneracy must be present and this degeneracy between the ordered and completely disordered components is necessarily a permanent feature of the crystal while it remains monoclinic in symmetry. The nature of the irreducible representations for such special points has been discussed in detail by Bradley and Cracknell, 1972. In the monoclinic example, which we have assumed above, the representations at the zone boundary $0\frac{1}{2}0$ are two-dimensional, and mix ordered and disordered component structures as we have already established in the case of a simple modulated structure.

Whereas the mixing of component structures in the case of a normal incommensurate structure often lead to enhanced stability and the formation of an incommensurate structure this is not the case in general for a zone boundary invariant point. The reason is not difficult to understand and relates to the fact that an unnatural symmetry constraint is imposed in the ordering structure associated with the invariant point. It assumes that there is a degeneracy between ordered and disordered components over an indefinite temperature range, for fixed content and for fixed wavelength, ($\lambda = 2b$ for a monoclinic crystal). In the absence of such an unlikely degeneracy between ordered and disordered states the minimum of free energy must occur at some small distance away from the invariant point. Off the invariant point unequal absolute weights can be given to the ordered and disordered component structures, and the wavelength may be determined by the optimum interaction between them. The nature of the associated band structure for ordering at an invariant point has been illustrated in Figure 1.

3.6. The interaction between ordering and phonon bands.

In the previous section we dealt with possible interaction between different ordering schemes at points in the Brillouin zone other than the special points where we found that the application of group theoretical principles leads to precise interband interaction rules. We now turn to discussing the nature of possible interaction between individual substitutional ordering bands and phonon bands with the object of exploring the possibility of stable modulated structures associated with such interactions.

Attention in this section will be devoted explicitly to the study of possible interactions between ordering bands and acoustical modes. We will also limit the analysis to the case of zone centre ordering transformations. Zone centre order-phonon interactions will normally be forbidden since at 000 the acoustical mode degenerates to give pure lattice displacements alone. At the same time interaction may well be possible in the case of non-zero k vectors where the acoustical wave must be represented by means of a two-dimensional irreducible representation. Here component representations correspond to both displacement and shear, the latter appearing as the gradient of the displacement wave. In considering possible mechanisms for the coupling of ordering bands and acoustical phonons in general it is obvious that shear strain is particularly important since it appears automatically where the ordering is accompanied by a reduction in symmetry. In principle, therefore, shear strain coupling between the acoustical phonons and ordering modes will be forbidden by symmetry at the zone centre, and permitted at non-zero k values. This simple symmetry consideration leads directly to the definition of a new class of stable modulated phases. The mechanism, in that it relates to non-zero k space interaction only, is closely analogous to that first introduced by the present author for order-order interband interactions in 1978.

Although the basic symmetry conditions relating to order-order and order-phonon interactions are closely comparable there remain certain differences which must be examined in more detail. It is clear from the discussion above that the coupling between ordering and acoustical phonons must involve the existence of a spontaneous strain field with the correct symmetry within the crystal. Further, we note that the full symmetry of the ordering function U_r, as defined by its special point irreducible representation, has a very precise symmetry in which each and all of the point group symmetry elements are involved. Remembering the final strain field irreducible representation must itself have exactly the same symmetry as U_r, the ordering function, it becomes apparent that this same symmetry must appear in the direct product of participating phonon waves. Hence the direct product of the full set of phonon irreducible representations, which define the symmetry of the shear strain field, must have the full symmetry of the ordering function U_r. Clearly phonon modes with k vectors in general positions in the Brillouin zone will not be candidates for inclusion in such a highly symmetrical phonon-phonon interaction complex. In brief, our analysis indicates that only certain acoustical modes can be candidates for inclusion in the strain field complex which couples to the ordering.

Thus in the context of phonon-order interactions it would seem that coupling with a single phonon will occur only through the gradient effect and not directly with the displacements. On the other hand in the case of two suitably chosen phonon vectors direct coupling to the displacements is possible since we must consider inclusion of irreducible representations for displacements for both phonon waves in the direct product. This principle will be illustrated in Section 5 in the context of certain specific examples of this phenomenon.

One final theoretical aspect of the postulated phonon-order interactions remains to be discussed. In the case of the order-order interband interactions discussed earlier it is usual to find that the associated diffracted intensity is sharply peaked at a single pair of vectors $\pm P_0$ implying that the original interband degeneracy was similarly precisely located as required by the theory of accidental degeneracy. No such precise location need exist for phonon-order interactions since it is not the individual phonon waves which interact but the phonon-phonon complex which gives rise to the modulated strain field in the crystal. In short we do not demand that the single phonon band should be in any sense soft prior to participation in the phonon complex. Hence the location of short range order-phonon scattering is likely to be dominated by criteria associated with the symmetry of the participating phonon waves rather than by the energetics of these waves prior to interaction. This analysis is endorsed by the available data on order-phonon modulated structures as discussed in detail in Section 5.

3.7. Summary of stable order modulations in crystals

The theory as developed above indicates that at least two classes of interaction may lead to the formation of thermodynamically stable modulated structures. The first relates to the situation where different ordering modes are permitted to interact by symmetry at points in k-space other than the special points through the fact that they can there have the same point group symmetry in the group of the associated wave vector P_0. The phase relationships of both structures in the modulation are fully determined from the characters of the space group irreducible representation. The phenomena can be understood in terms of an interband degeneracy which is likely to occur close to a normal ordering phase transformation where the disordered state and an appropriate ordered state are close in free energy. The theory of such modulated (incommensurate) structures is now well understood, Heine and McConnell, 1984. A subclass of order modulated structures related to ordering behaviour close to an invariant special point also exists.

Finally a new class of stable modulated structure has been recognised. In these structures an ordering transformation, associated with the Brillouin zone origin (000), couples with a strain field generated by acoustical modes with non-zero k vectors. The conditions for interaction are dictated primarily by symmetry: the symmetry of the modulation strain field associated with the phonon complex must match the symmetry of the chosen ordering mode. In practice, as will be discussed in Section 5, the resulting modulation intensity appears as sharp streaks along high symmetry directions in the Brillouin zone.

4. SHORT RANGE ORDER INTERACTIONS IN INCOMMENSURATE STRUCTURES

It is now some time since it was recognised that modulated structures in certain minerals could be related directly to interactions between different ordering schemes, McConnell, 1978. Since that time the theory as presented then has been tested and amplified in the case of a large number of modulated mineral structures. In this Section reference will be made briefly to modulated mineral structures which have been the subject of detailed study elsewhere. More attention will be devoted to some of the the examples which have not been written up in detail so far.

4.1. By far the most important example of the theory of short range ordering interactions is that of the plagioclase feldspars where an incommensurate structure exists between approximately 25 and 75 mole% anorthite. In this case vectors $\pm P_0$ are associated with the Brillouin zone boundary special point $00\frac{1}{2}$. Since the structure is triclinic with a centre of symmetry the irreducible representations of the special point are only two in number. Both correspond to simple centro-symmetric structures and they differ only in relation to the fact that one irreducible representation has a centre where the other has an anticentre. Originally, McConnell, 1978, one of the two structural components was identified with the end member structure for anorthite, implying that a Si/Al ordering scheme as observed in this end-member was appropriate. The second component structure contains an anticentre in a position such that it requires ordering of the large cations Na and Ca. Although there has been little or no experimental work done on confirming this model there is no reason to doubt its validity at the present time. It is likely, however, that the model holds best for the structure of the intermediate plagioclases in the composition range above 50% anorthite. The existence of the Boggild loop in the composition range around 50% anorthite implies that the incommensurate structure of the intermediate plagioclase feldspars in the composition range 25-50% anorthite (e_2 of McConnell,1974) differs somewhat from that present in compositions above 50% anorthite, (e_1 of McConnell, 1974). Recent thoughts on the nature of the e_2 incommensurate structure suggest that, since the modulation relates to the special point $00\frac{1}{2}$, the second modulation component corresponds to a superlattice containing a partly ordered version of the low albite structure based on the distribution of centres and anticentres as in the anorthite representation of the special point. The second component structure in the e_2 incommensurate phase has the Na/Ca type symmetry of the e_1 structure. The albite structure itself is not admissible as a component structure since it does not have a doubled c axis, i.e. its ordering scheme does not exist as a possible representation of the special point $00\frac{1}{2}$ on which the modulation in the intermediate plagioclase feldspars occurs. On the basis of this hypothesis the Boggild loop results from the fact that Al/Si ordering in the anorthite component structure of the e_1 structure occurs at a higher temperature than the partial Al/Si ordering in the pseudoalbite component of the e_2 phase. This results in a closed spinodal loop, McConnell, 1974.

Several different sets of recent data on the intermediate plagioclase feldspars lend weight to this general hypothesis on the nature of the short range order interactions in these phases. In the first place recent experimental data, Carpenter and McConnell, 1984, for the true anorthite-type ordering in the plagioclases indicate that the $C\bar{1}/I\bar{1}$ ordering phase boundary is very steeply dipping in temperature towards the albite end of the diagram and extrapolates to a temperature of approximately 500°C at 40% anorthite. This is well below the temperature at which the e_2 incommensurate structure is likely to have formed. A second line of evidence on the nature of the short range order and ordering interactions in the intermediate plagioclase feldspars comes from thermochemical data. It is apparent from enthalpy data, Carpenter et al.,1985, plotted as a function of composition for the incommensurate structures e_1 and e_2, that there is a marked point of inflexion at 50% anorthite. Above 50% anorthite the enthalpy reduction from the disordered state to the e_1 structure is comparable with that achieved in a mechanical mixture of the ordered end-members albite and anorthite and serves to demonstrate that the e_1 structure is very well ordered. On the other hand, below 50% anorthite and in the e_2 region, the enthalpy lies about half way between disorder and that of a mechanical mixture. This points clearly to the fact that the e_2 incommensurate structures are but partially ordered at best.

A substantial amount of experimental work still remains to be done on the details of the short range order effects in the plagioclases but even at this stage it is apparent that a full understanding of the system will be of immense interest both from a purely theoretical point of view and from the point of view of its importance in igneous and metamorphic petrology.

4.2. The ordering behaviour of mullite

By contrast with the plagioclase feldspars, where there is very little symmetry since they are triclinic, the incommensurate phase mullite has high symmetry and an average structure with space group Pbam. Mullite is modulated on the basis of $00\frac{1}{2}$ and determination of the two component structures is relatively straightforward since there are known oxygen vacancies which must be ordered in one of the components. Space group criteria imply that the only sensible distribution of these vacancies is provided by derivative space group Pnnm with a doubled c axis repeat. By applying the rule that both component structures must have the same symmetry in the group of P_0 it is possible to determine the symmetry of the second component structure uniquely as Pbnm. This is known to be the space group for the ordered structure of sillimanite which also has a doubled c axis, a fact which implies that the second component structure in mullite is likely to possess the Al/Si ordering scheme of sillimanite. A more complete analysis of the nature of the interactions between these two structures leading to the stability of mullite is presented in McConnell and Heine, 1985. The high symmetry of mullite serves to make it a good example of the method of solving incommensurate structures, and the arguments used for the determination of the phase relationships of the two structures are particularly convincing.

4.3. The ordering interactions in yoderite

Having discussed two examples of ordering interactions which lead to incommensurate phases we now turn to the rather special example of short range order as observed in the mineral yoderite. This example has not been analysed previously in group theoretical terms. Since certain of the features of its ordering behaviour are both novel and interesting, the ordering interactions will be discussed in some detail here.

Yoderite is a magnesium substituted alumino-silicate with structural affinities to kyanite and was first described by McKie, 1959. It is a modulated phase and the average structure was determined by Fleet and Megaw, 1962. The mineral is monoclinic with space group $P2_1/m$ and the principal additional non-Bragg reflections occur with vectors $\pm P_0$ centred on the special point $0\frac{1}{2}\frac{1}{2}$ on the Brillouin zone boundary. It is also relevant to note that the structural analysis showed that chains of mainly Al octahedra (sites A_1) exist parallel to the b axis. These chains are crosslinked by ions in two sites (A_2 and A_3) which are in fivefold co-ordination by oxygen atoms. Data on A_2-O bond lengths indicate that the A_2 site, which lies on the mirror plane, is partly substituted by Mg ions. From which we may conclude that the average structural analysis provides good evidence of substitutional disorder.

We now turn to a group theoretical analysis of the data in an attempt to unravel the likely order interactions in this phase. First we note that, since the space group has a screw diad, an essential degeneracy exists at the special point $0\frac{1}{2}\frac{1}{2}$, and only two-dimensional irreducible representations exist at this point since odd and even bands for the screw diad are necessarily degenerate here. A choice exists between two origins for the single two-dimensional representation at the special point $0\frac{1}{2}\frac{1}{2}$ in relation to the origin of the centre of symmetry, and the operations of the mirror plane. Since the site A_2 lies on the mirror plane it is necessary to chose the two-dimensional representation in such a way that the A_2 site goes into itself rather than its negative. If we do otherwise it is not possible to order A_2 sites at all. It will be most convenient, in discussing the nature of the ordering scheme associated with the special point $0\frac{1}{2}\frac{1}{2}$, to begin with the simpler band structure for ordering associated with the special point $00\frac{1}{2}$ which has simple one-dimensional representations as set out in Table I below. It should be noted that the direct product of this group with the group of order two is needed to illustrate the effect of doubling the c repeat. The full group is not essential to our purpose at this point but will become relevant later. The four ordering schemes associated with R_{1-4} are illustrated fully in Figure 2.

Table I

Point $00\frac{1}{2}$		$\{E\|000\}$	$\{C_2\|0\frac{1}{2}0\}$	$\{i\|0\frac{1}{2}0\}$	$\{m\|000\}$
R_1	$P2_1/m$	1	1	1	1
R_2	$P2_1/b$	1	-1	1	-1
R_3	$P2_1/m$	1	-1	-1	1
R_4	$P2_1/b$	1	1	-1	-1

If we now move from the special point at $00\frac{1}{2}$ to the point of interest at $0\frac{1}{2}\frac{1}{2}$ the two-dimensional representations for the latter point which we have already discussed are obtained by combining pairs of $00\frac{1}{2}$ representations each containing one odd and one even representation for the screw diad $\{C_2|0\frac{1}{2}0\}$. By combining R_1 and R_3 we obtain the first two-dimensional representation with even character for the mirror plane. Combining R_2 and R_4 similarly yields the representation with odd character for the original mirror plane. Table II contains the matrices for the two-dimensional representation of the special point $0\frac{1}{2}\frac{1}{2}$. As in case of Table I we have omitted the extra group elements which we need to associate with the doubling of the c repeat.

Table II

Point $0\frac{1}{2}\frac{1}{2}$	$\{E|000\}$	$\{\bar{E}|000\}$	$\{C_2|0\frac{1}{2}0\}$	$\{i|0\frac{1}{2}0\}$	$\{m|000\}$
	$\begin{matrix}1 & 0\\0 & 1\end{matrix}$	$\begin{matrix}-1 & 0\\0 & -1\end{matrix}$	$\begin{matrix}0 & 1\\-1 & 0\end{matrix}$	$\begin{matrix}1 & 0\\0 & -1\end{matrix}$	$\begin{matrix}0 & -1\\-1 & 0\end{matrix}$
			$\begin{matrix}0 & -1\\1 & 0\end{matrix}$	$\begin{matrix}-1 & 0\\0 & 1\end{matrix}$	$\begin{matrix}0 & 1\\1 & 0\end{matrix}$

The nature of the ordering at the special point should now be apparent. Firstly it must involve the effect of doubling the c axis repeat. This does not lead to any novel effects since the point $00\frac{1}{2}$ is a normal special point. What is of most interest is the fact that two ordering schemes related to positive and negative character for the screw diad $\{C_2|0\frac{1}{2}0\}$ are included in the two-dimensional representation. As we have discussed, in dealing with the theoretical aspects of this situation earlier, the positive and negative group characters for the screw diad may be identified with disorder and order on certain structural sites, which in this case is likely to be the site A_2. Apart from this effect the existence of a doubling parallel to the c axis implies that there is a normal ordering or non-equal partition of ordering cations along the c axis. This effect will appear automatically if we take the direct product of the group of Table II with the factor group for translations of order two.

In discussing ordering at an invariant point earlier we noted that the symmetry-imposed degeneracy between order and disorder at this type of special point was unlikely to provide a free energy minimum and that the actual minimum of free energy must lie off the special point. In yoderite the satellite intensity maxima lie at $\pm 0.08b^*$ from the point $0\frac{1}{2}\frac{1}{2}$. At which points the bands for even and odd symmetry for the screw diad need be no longer degenerate as illustrated already in Figure 1. It seems likely that the energy minimum at $0.08b^*$ is associated with the ordering band but this is not a foregone conclusion. One general comment on this type of ordering interaction is relevant at this point in the discussion. Whereas in a normal incommensurate structure, such as mullite or the plagioclase feldspars, a potential accidental degeneracy leads to favourable interactions, the opposite is the case for an invariant point. Here the band concerned prefers to be solitary.

This brief analysis of the likely ordering interactions in the mineral yoderite amply illustrates the power of group theoretical methods as applied to order modulations in minerals. There remains a great deal of further research to be done but it is clear that the field is particularly rich in interesting physical phenomena.

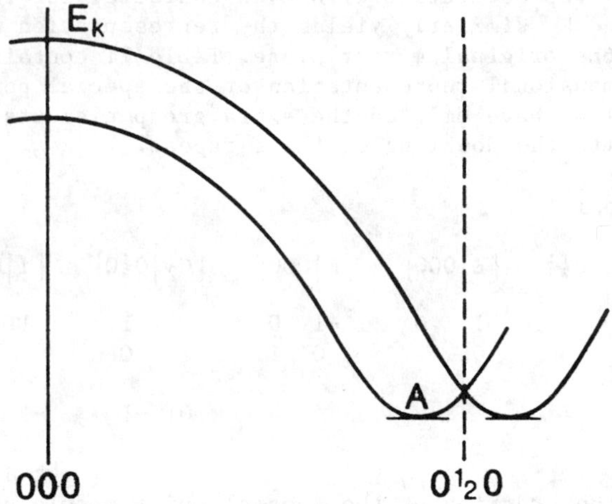

Figure 1. Ordering band structure in the vicinity of an invariant point on the Brillouin zone boundary. It is assumed that the point of essential degeneracy is associated with a screw diad parallel to the b axis of a monoclinic crystal. Note that the energy minimum in the band structure occurs at point A somewhat away from the invariant point.

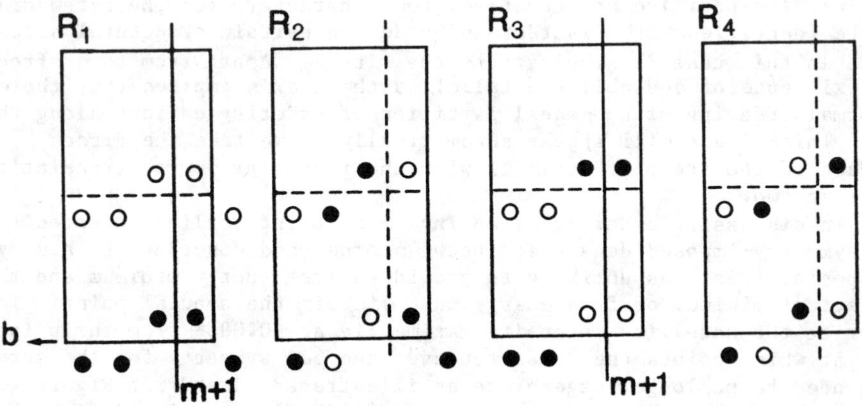

Figure 2. Irreducible representations associated with the special point 00½ in the mineral yoderite with space group $P2_1/m$. R_1 and R_3 relate to even and odd character for the screw diad and both are even in relation to the original mirror plane. These two representations combine at the Brillouin zone boundary point 0½½ to produce a two-dimensional irreducible representation which links order and disorder.

5. ORDER-PHONON (OP) MODULATION STRUCTURES IN MINERALS

Whereas considerable thought has been given in the literature to the origin and occurrence of incommensurate structures (IC) little attention has been paid to the possibility of stable modulated structures based on ordering-phonon interactions. The theory developed earlier in this chapter suggests that such thermodynamically stable modulated structures are possible and this is borne out by an examination of the data existing in the literature on certain pre-martensite transformation structures, McConnell, 1987. In this Section examples of order-phonon modulated structures in minerals will be described and the question of their stability or metastability will be taken up in Section 6. From the point of view of the theory of development of order-phonon modulated structures in general it is not important to distinguish between examples where the transformation behaviour involves substitutional ordering or merely local ordered displacements of the atoms as is observed in a true martensitic transformation. The mineralogical examples of the phenomenon all involve substitutional ordering and appear to be metastable rather than the stable structures frequently observed in certain alloy systems, McConnell,1987.

5.1. The order-phonon modulated structure of potassium feldspar

Potassium feldspar undergoes an ordering transformation from monoclinic (space group C2/m) to triclinic (space group $C\bar{1}$) associated with the ordering of Al and Si atoms on tetrahedral sites. Since this substitutional ordering process is extremely sluggish it is usual to find that potassium feldspar which crystallises at low temperature is monoclinic in symmetry having failed to order on crystallisation. The process of ordering which subsequently takes place on a geological time scale produces the distinctive modulated structure which is the subject of the present discussion. Potassium feldpar which crystallised metastably at low temperature as a monoclinic phase has a distinctive crystal habit and has been described as adularia. The microstructure of adularia was the subject of a very detailed electron-optical study, McConnell, 1965, and this study forms the basis of the present analysis of its transformation behaviour. The relevant details are as follows.

Electron diffraction photographs of thin flakes of adularia in suitable orientation show the presence of strong sharp streaks through the Bragg maxima, both parallel and normal to the diad axis of the monoclinic unit cell. In the diffraction pattern it is also apparent that, for some of the Bragg maxima, certain streaks are missing or at least are extremely weak. Thus streaks parallel to the b axis are missing for spectra 0k0, and streaks normal to the b axis are missing for spectra h0l. The data accord with the existence of two orthogonal transverse distortion waves in which the amplitude vectors also lie in the plane defined by the two wave normals. The direct electron-optical image of a flake of adularia in suitable orientation confirms that the orthogonal distortion waves coexist throughout the crystal and give rise in the image to a very distinctive tweed texture which is somewhat irregular but with a mean wavelength of approximately 100 Angstroms.

It is relevant to note here that a similar tweed texture is observed in pre-martensite microstructures in cubic alloys and that in these materials strong streaks run parallel to $[1\bar{1}0]$ and [110] associated with transverse distortion waves, Tanner et al., 1982.

We now turn to the analysis of the tweed texture observed in adularia using the theory of phonon-order modulations developed in Section 3. This theory implies that the ordering transformation in potassium feldspar, which is associated with a symmetry change from monoclinic to triclinic symmetry, is likely to couple with acoustical modes through the spontaneous shear strain associated with the loss of monoclinic symmetry. The theory also implies that coupling with a single acoustical phonon is unlikely since the coupling must then occur only with the gradient of the displacement wave, a situation which is likely to be uneconomical in energy terms. On the other hand strong coupling is permissible where the product of two phonon irreducible representations gives the symmetry of the ordering, and related shear strain fields.

We begin the analysis by setting out in Table III the group representations for the zone centre vector 000 which define the basis functions Ur for modulation behaviour in space group C2/m.

TABLE III

000		$\{E\|000\}$	$\{C_2\|000\}$	$\{i\|000\}$	$\{m\|000\}$
R_1	$C2/m$	1	1	1	1
R_2	$C\bar{1}$	1	-1	1	-1
R_3	Cm	1	-1	-1	1
R_4	C2	1	1	-1	-1

The special point representation for the ordering transformation in potassium feldspar is R_2 of Table III and from this representation order modulated structures may be derived in the usual way with wave vectors chosen from points throughout the Brillouin zone. We now consider the representations of the transverse acoustical wave vectors which are known to be present in special symmetry positions on the diad axis and in the mirror plane. These may be constructed from the 000 special point representations R_3 (for vectors on the diad) and R_4 (for vectors in the mirror plane) in the usual way. Neither of these displacements can interact directly with order modulations based on the representation R_2 although the associated gradient representations both have symmetry R_2 and may interact with the ordering process to yield second order terms in the Landau free energy expansion, a point to which we may return later. The main interaction between the transverse phonon waves and the ordering process may be demonstrated by taking the direct product of the representations associated with displacements alone. This involves the product of representations R_3 and R_4 and yields the correct ordering representation. It is necessary finally to establish that the sum of k vectors for this proposed triple interaction is equal to zero. This condition is fulfilled where the acoustical wave vectors P_1 and P_2 sum to give the ordering modulation vector P_3. This requires that the ordering wave vectors lie in general positions within the Brillouin zone as indicated in Figure 3. In Figure 4 the effect of interaction between these displacements and the ordering is demonstrated for the vector P_3.

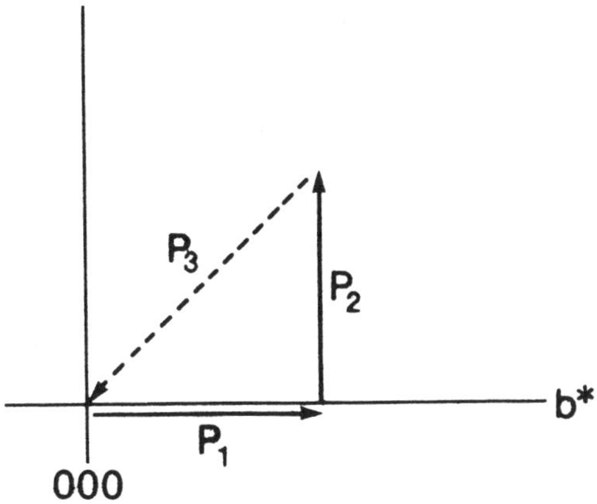

Figure 3. The k-space summation of vectors P_{1-3} in the context of ordering-phonon interactions in adularia. Vectors P_1 and P_2 lie on the diad axis and in the mirror plane respectively. The resultant vector P_3 relates to the ordering component of the modulation.

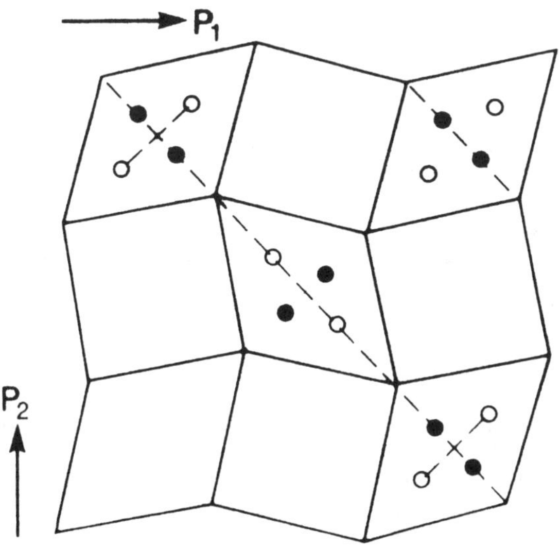

Figure 4. Demonstration of the fact that the combination of two transverse acoustical wave displacements results in a distortion pattern which couples with an ordering mode with wave vector $P_3 = P_1 + P_2$. The ordering effect is illustrated schematically in terms of closed and filled circles for the two species of atoms present.

From Figure 4 it will be apparent that the distortion pattern is derived from the combined displacements associated with the two orthogonal transverse waves. Where this distortion favours the ordering of atoms an additional density modulation appears in the pattern with wave vector P_3 in the diagonal direction, as illustrated in Figure 4. It is important at this stage to comment on the rather obvious fact that the diffracted intensity distribution associated with the transverse displacement waves occurs as streaks rather than at specific points along the high symmetry directions parallel to the diad, and in the mirror plane. Streak intensity has also been noted in the tweed textures observed in premartensitic transformation behaviour in metallic alloys, Tanner et al., 1982. There the streak intensity is practically continuous between the primary Bragg maxima. The reason for this spread is to be found in the fact that the triple interaction which we have already described provides a third order term in the Landau expansion. This means that there is inevitably something equivalent to a nucleation event, involving an associated activation energy, for the formation of strongly localised distortions associated with the order-phonon interactions. Probability factors must therefore act to ensure that a wide spread of component wave vectors participate in the interaction process. At the same time it is likely that there are generally favoured wavelengths for the phenomena, a fact which is apparent on examining electron-optical images of the resulting tweed texture. The role of analogous nucleation phenomena in premartensite behaviour has also been discussed by Yamada, 1987. It will be apparent from this rather brief review of the transformation behaviour in potassium feldspar at low temperature that the observed microstructure can be explained in considerable detail by using the theory of order-phonon interactions developed both here and elsewhere McConnell, 1987. The role of order-phonon interactions in transformation behaviour under both stable or metastable conditions will be taken up again in Section 6. We may anticipate that the phenomena can appear in stable thermodynamic equilibrium as is found in certain alloys, see Figure 5, and also in a purely mechanistic capacity in certain kinetically sluggish transformations, as is the case in adularia.

5.2. Order-phonon interactions in modulated cordierite

The mineral cordierite undergoes a hexagonal to orthorhombic ordering transformation at very high temperatures but crystallises initially from a glass of its own composition in hexagonal form at very much lower temperatures. On annealing, this hexagonal phase develops a modulated structure and shows a characteristic tweed texture in electron-optical images, Putnis, 1980. Group theoretical analysis of the modulation behaviour of cordierite provides strong confirmatory evidence for the order-phonon interaction theory developed here and provides an excellent introduction to the whole subject of pre-martensite transformation behaviour in alloys where order-phonon interactions are of paramount importance, McConnell, 1987. The hexagonal to orthorhombic ordering transformation in cordierite differs essentially from that in potassium feldspar in that the relevant irreducible representation for the zone centre ordering in cordierite involves a two-dimensional irreducible representation since the transformation is associated with the loss of a threefold axis of symmetry. The loss of this axis of symmetry also requires that the transformation is first order in character and therefore involves a discrete nucleation event.

We may now proceed directly to the group theoretical analysis of the origin of the modulation structure. The hexagonal space group of cordierite at high temperature is P6/mcc and it inverts on ordering to an orthorhombic structure with space group Cccm. This is a zone centre transformation and we may use the representation table for the point group 6/mmm alone to establish the functions U_r associated with possible short range order. The representation needed to describe the hexagonal to orthorhombic ordering transformation is E_{2g} which contains component representations which are even, $\{x^2-y^2\}$, and odd $\{xy\}$ in relation to the horizontal diad of the space group P6/mcc. They represent orthorhombic and monoclinic symmetry respectively. It is possible to define both the even and the odd ordering schemes for cordierite, McConnell, 1985, 1987, and they have been illustrated in Figure 6. What is of most importance here, however, is the question of how these two ordering schemes are likely to couple with acoustical phonons in relation to short-range order interactions. The experimental evidence from electron diffraction and optics indicates that there are, as in adularia, two orthogonal transverse waves with wave vectors parallel to a and a*, with amplitude vectors in the same plane as the wave vectors. These distortion waves are necessarily odd under the horizontal diad which, as we have already noted, distinguishes between the two ordering schemes associated with the zone centre representation E_{2g}. From which it will be obvious that the shear strain associated with the dual orthogonal transverse wave system can only couple with the odd ordering variant.

Little remains to be said except to emphasise that, in this case, as in the case of pre-martensite behaviour in cubic alloys, the phase transformation to the order modulated structure utilises what appears at first sight to be the incorrect ordering scheme. That this is possible at temperatures well below the transformation temperature in cordierite is good evidence for the fact that the order-phonon interactions are associated with a substantial free energy reduction. We may also note here that, since the order-phonon interactions in cordierite relate to a third order term in the Landau free energy expansion, the actual transition from the hexagonal state to the order modulated state must be of first order. At the same time we must remember that the symmetry of the order-modulated structure is still hexagonal.

The details of the conversion of the odd order-modulated structure to the final stable orthorhombic structure has already been described in McConnell, 1985 and 1987. It is sufficient to note here that this re-ordering takes place by adapting odd ordering variants. This process is also of first order and results in the appearance of twinning in the final microstructure of the transformation product. It is relevant to note here that recent hard mode Raman studies on the cordierite transformation sequence endorse the characteristics of the transformation from the hexagonal state to the order-modulated state, (personal communication from Dr E.Salje).

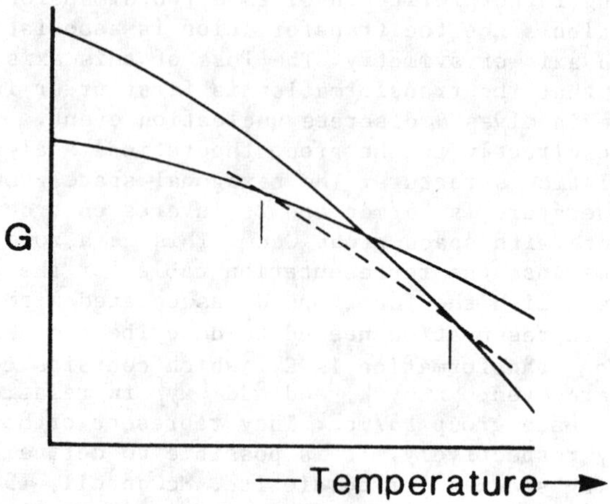

Figure 5. Free energy-temperature diagram illustrating the stable occurrence of the order-phonon modulated structures in the pre-martensite structures found in certain cubic alloys. The field of stability of the modulated structure is indicated with a dashed line. Note that the modulated structure is bounded by first order phase transitions at both the high and low ends of its stability range.

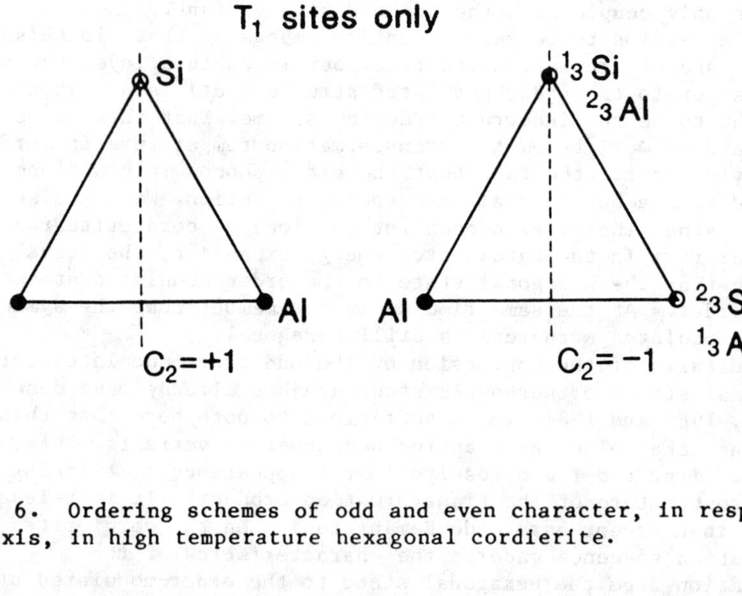

Figure 6. Ordering schemes of odd and even character, in respect of the diad axis, in high temperature hexagonal cordierite.

To complete the analysis of the order-phonon interactions in cordierite it is necessary to note that, unlike the case of adularia, the final symmetry of the spontaneous strain field, which couples with the odd ordering scheme from representation E_{2g}, does not have to have the full symmetry of the hexagonal crystal. Thus if we take wave vectors $\pm P_0$ for one of the transverse wave components it is not necessary to include all the related vectors in the star of P_0. Similarly with the orthogonal wave vectors $\pm P_1$. This implies that the interaction between the full star of P_0, P_1 and the odd ordering mode is strongly anharmonic and that the symmetry degenerates in this interaction leaving the final stable complex as $\pm P_0$ and $\pm P_1$. Notwithstanding, since this choice cannot change the cell dimensions relatively in the plane 0001, at least to first order, the crystal must remain dimensionally hexagonal. Over the crystal as a whole, however, the orientation of the pair of orthogonal wave vectors is likely to range over all the possibilities. With the proviso that the operation of the threefold axis is lost in the anharmonic ordering-strain field interaction we may represent the triple product interaction from the relevant subgroup symmetry as follows.

The augmented group of the two wave vectors P_0 and P_1 is the orthorhombic group mmm but the the group of each of the vectors P_0 and P_1 alone is mm. Using subscripts 1,2 and 3 to represent the directions x,y and z we may write the augmented symmetry for the two displacement waves as follows.

TABLE IV. The augmented symmetry of transverse displacements for orthogonal wave vectors P_0 and P_1.

	E	2_1	2_2	2_3	i	m_1	m_2	m_3
$\pm P_0$	1	-1	(1)	(-1)	(-1)	(1)	-1	1
$\pm P_1$	1	(1)	-1	(-1)	(-1)	-1	(1)	1
Product	1	-1	-1	1	1	-1	-1	1

Note that the elements of symmetry other than those in the groups of P_0 and P_1 have been placed in parenthesis. The final symmetry of the direct product of the two displacement waves is that of the monoclinic function $\{xy\}$ and therefore, as inferred, there is a strong interaction between the strain field due to displacements, and the ordering on the basis of the odd representation of E_{2g}.

The use of subgroup symmetry locally to describe these anharmonic order-phonon interactions is a feature of the problem which becomes more and more important as the symmetry of the parent phase increases. Thus, in the case of the cubic alloys which show pre-martensite behaviour, the diffraction evidence from single crystals indicates that there are three types of domains in which pairs of orthogonal displacements exist in each of the three planes (100), (010) or (001).

This concludes the analysis of order-phonon modulated structures in minerals. It now remains only to discuss the fact that, in certain cases, this type of modulated structure appears metastably in certain transformations while in other examples it is thermodynamically stable.

6. THERMODYNAMIC ASPECTS OF THE OCCURRENCE OF MODULATED STRUCTURES

In this Section we discuss, the relevance of order-modulated structures in the context of thermodynamically stable behaviour, and their role in the mechanism of certain substitutional ordering phase transformations.

6.1. Thermodynamically stable order-modulated structures

In general, in the absence of interband interactions, we expect to find that the energy functional associated with a given ordering band has its minima at the special points, with the exception of cases where the special point itself is an invariant point as discussed in the case of yoderite. The existence of an alternative stable minimum in the band structure over some temperature range, leading to the existence of an order modulated structure, may be associated with an accidental degeneracy between a disordered, and an ordering band in the temperature range close to the temperature of the related normal ordering phase transformation. This type of limited stability for an incommensurate structure is well illustrated by the example of sodium nitrite, as shown in Figure 7. A similar incommensurate structure is observed close to the α/β transition in quartz and results in a modulation structure, visible in the electron microscope close to the transition temperature.

A second and related class of order modulated structure occurs primarily in solid solutions where two different ordering bands are again involved in an accidental degeneracy. This is the case in mullite, where one of the component structures involves the ordering of vacancies, and the other is primarily concerned with the ordering of Al and Si ions. A very similar situation exists in the plagioclase feldspars where two ordering bands are involved in an accidental degeneracy. In this second class of order-modulated structure the stability range of the modulated phase is likely to be much wider in temperature and dependent primarily on the possibilities for alternative behaviour in the binary system such as true exsolution. It is clear from recent thermochemical data, Carpenter et al.,1985, that the incommensurate structure of the intermediate plagioclase feldspars is likely to be stable over a substantial temperature range, giving way eventually to exsolution into the albite and anorthite end-members.

For our third class of stable order-modulated structures we return to the case of yoderite and the zone boundary invariant point. This is a particularly interesting class of behaviour in terms of thermodynamics for the following simple reason. As we noted, the zone boundary degeneracy is an essential degeneracy which must be retained for as long as the system retains its full symmetry. Complete ordering is therefore forbidden at any temperature including absolute zero. It seems likely in this case that the field of stability of the order-modulated structure must be rather broad. Eventually, however, the phase must invert to a completely new compound, or phase assemblage.

One point, of considerable thermodynamic significance, should be stressed here. Co-existing component structures in a stable modulation must be degenerate in free energy. They may, however, differ substantially in relation to their entropy and enthalpy contributions.

6.2. Order-modulation in the mechanism of order-disorder transitions

In this section we turn to the more general problem of discussing the role of order-modulation in the context of the mechanism of order-disorder phase transformations. In particular we will be concerned with the question of distinguishing between order-modulated structures which exist only in the context of irreversible behaviour close to a normal phase transformation, and similar structures which have a true thermodynamic stability range as discussed in section 6.1.

We may start the discussion with a simple observation. If we are dealing with an order-disorder phase transformation, where the process of cation exchange involves a very large free energy of activation, the question of whether its transformation behaviour is ideally of first or second order is largely academic. Thus it is possible to prove quite simply that the ordering transformation in potassium feldspar is likely, on symmetry grounds alone, to be of second order. Ideally, therefore, the transformation should take place in a continuous manner. Nothing is less likely since we know that the transformation only proceeds under highly irreversible conditions in nature. It follows that we must, in the case of the transformation in adularia and similar phases, treat the transformation as though it were definitely of first order. Formally this may be done, in terms of the band theory of ordering, by defining the excess (positive) value of ΔG at some temperature interval ΔT below the transformation temperature, where $\Delta G = \Delta S . \Delta T$, and then constructing a free energy diagram similar to that in Figure 8 to define the temperature dependence of the onset of order modulation for given P_0 values. This treatment is analogous to the corresponding theory of spinodal behaviour in solid solutions. Rate data may also be obtained if one inputs data on the diffusion coefficients for interchange of cations on ordering. There is, however, one marked difference between the theory for order-modulation and conventional spinodal theory. Whereas short wavelength fluctuations are favoured by the diffusion mechanism in conventional spinodal theory, the factor favouring short wavelengths in order-fluctuations is the possibility of shear strain-order coupling away from the zone centre, 000.

A comparison of the relevant factors involved in stable and metastable order-modulation behaviour is made in the set of diagrams presented in Figures 7 and 8. Figure 7 shows the ordering band structure at the onset of order-modulation for the thermodynamically stable case dealt with in section 6.1. The corresponding free energy-temperature diagram shows that a field of true stability for the order-modulated structure appears in a temperature range which includes the normal special point phase transformation. By contrast the appearance of a metastable order-modulated structure, as part of the mechanism of a substitutional phase transformation, is shown in Figure 8. At the same time the free energy-temperature diagram shows that the order-modulation occurs well below the ordering temperature.

It remains only to note that the distinction between truly stable and metastable order-modulation must in the end be a matter for careful analysis and experiment, particularly by attempted reversal of all of the ordering behaviour in well designed thermal experiments.

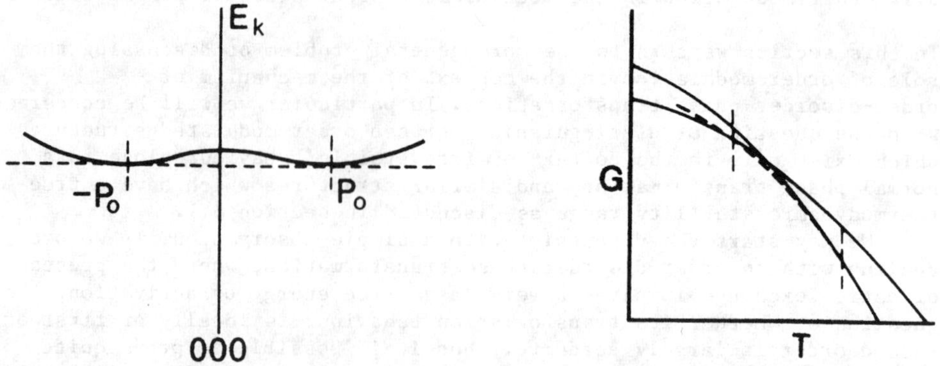

Figure 7. Diagram illustrating the characteristics of the ordering band structure for a stable order-modulated structure at a temperature just above that for the onset of order-modulation. In the free energy-temperature diagram the true stability range of the modulated structure is shown with a dashed line which straddles the temperature for a conventional special point transformation.

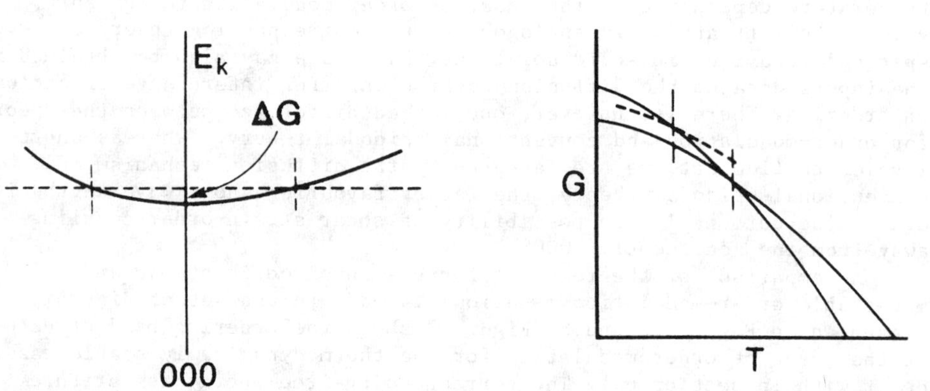

Figure 8. Diagrams illustrating the appearance of an order-modulated structure as part of the mechanism of a normal special point ordering transformation. We assume that the temperature lies below the true ordering temperature and that there is an associated irreversible change in free energy, ΔG, in the process. Note that the temperature at which the order-modulation occurs lies below the ordering temperature.

7. REFERENCES

Bradley, C.J. and Cracknell, A.P., (1972), **The Mathematical Theory of Symmetry in Solids**, Clarendon Press, Oxford.

Birman, J.L., (1962), Space Group Selection Rules: Diamond and Zinc Blende, Physical Review, **127**, 1093-1106.

Birman, J.L., (1974), Theory of Crystal Space Groups and Infra-red and Raman Lattice Processes of Insulating Crystals. In: **Handbuch der Physik, XXV/2b**, S.Flugge, ed., Springer Verlag, Berlin.

Carpenter, M.A., and McConnell, J.D.C., (1984), Experimental Delineation of the Cl = Il Transformation in Intermediate Plagioclase Feldspars, Amer. Mineral. **69**,112-121.

Carpenter, M.A. McConnell, J.D.C. and Navrotsky A., (1985), Enthalpies of Ordering in the Plagioclase Feldspar Solid Solution. Geochim. Cosomochim. Acta **49**, 947-966.

Cotton, F.A., (1971), **Chemical Applications of Group Theory**, John Wiley and Sons, New York.

Cracknell,A.P., (1968), **Applied Group Theory**, Pergamon Press,Oxford.

Fleet, S.G. and Megaw, H.D., (1962), The Crystal Structure of Yoderite, Acta Cryst., **15**, 721-728.

Heine, V., and McConnell, J.D.C.,(1984), The Origin of Incommensurate Structures in Insulators, J.Phy.C: Solid State Physics, **17**, 1199-1220.

Landau, L.D. and Lifshitz, E.M., (1958) **Statistical Physics**, Addison Wesley, Reading, Massachusetts.

McConnell, J.D.C., (1965), Electron Optical Study of Effects associated with Partial Inversion in a Silicate Phase, Phil Mag., **11**, 1289-1301.

McConnell, J.D.C., (1985), Symmetry Aspects of Order-Disorder and the Application of Landau Theory, Reviews in Mineralogy, **14 Microscopic to Macroscopic**, American Mineralogical Society.

McConnell, J.D.C., (1987), Symmetry Aspects of Pretransformation Behaviour in Metallic Alloys, Metall. Trans. AIME., in press.

McConnell, J.D.C., (1974), Analysis of the Time Temperature Transformation Behaviour of the Plagioclase Feldspars, in **The Feldspars** ed. W.S. Mackenzie and J.Zussman, Manchester University Press.

McConnell, J.D.C., (1978), The Intermediate Plagioclase Feldspars: an Example of a Structural Resonance, Zeit. für Krist., **147**, 45-62.

McConnell, J.D.C. and Heine, V., (1982) Origin of Incommensurate Structure in the Cooperative Jahn-Teller system $K_2PbCu(NO_2)_6$, J.Phy.C: Solid State Physics, **15** 2387-2402.

McConnell, J.D.C. and Heine, V., (1984), An Aid to the Structural Analysis of Incommensurate Phases, Acta Cryst. **A40** 473-482.

McConnell, J.D.C. and Heine, V., (1985), Incommensurate Structure and Stability of Mullite, Physical Review B, **31**, 6140-6142.

McKie, D., (1959) Yoderite, a New Hydrous Magnesium Iron Aluminosilicate from Mautia Hill, Tanganyika, Min. Mag., **32**, 282-307.

Putnis, A., (1980) The Distortion Index in Anhydrous Mg-Cordierite, Contrib. to Mineral. and Petrol., **74**, 135-141.

Tanner, L.E., Pelton, A.R. and Gronsky, R, (1982), The Characterization of Pretransformation Morphologies: Periodic Strain Modulations, Jour. de Physique, C4 **43**, C4 169-172.

Yamada, Y., (1987), 'Modulated Lattice Relaxation' in β-based Premartensitic Phase, Metallurgical Transactions, AIME, in press.

INCOMMENSURABILITY IN TWO CLASSES OF SOLIDS : MODULATED INSULATORS AND QUASICRYSTALLINE ALLOYS

F. Dénoyer
Laboratoire de Physique des Solides, Bâtiment 510
Université Paris-Sud
91405 Orsay
France

ABSTRACT. The aim of this paper is i) to present an oversimplified view of the present state of the studies of incommensurate insulator systems and ii) to introduce quasicrystals which appear in a certain manner as a special case of incommensurate solids.

1. INTRODUCTION

The lack of translational periodicity associated with long range order has focussed attention on incommensurate materials during the past decade and more recently on quasicrystals. From the point of view of basic physics, a number of exciting phenomena which are not found in translationally periodic crystals have been theoretically predicted in incommensurate systems and have stimulated this active area of research. However, in "real" incommensurate crystals, the situation is complicated by the presence of defects. This paper, based on a few-selected typical examples, will try to give a schematic present picture of incommensurate insulators (sections 2-5). A parallel between incommensurate crystals and icosahedral quasicrystals will be made in section 6.

2. CHARACTERIZATION OF INCOMMENSURATE CRYSTALS - "STANDARD" BEHAVIOUR

Since von Laue (1912) demonstrated by means of X-ray the earlier hypothesis that crystals were built of identical unit cell, it was generally assumed that space group symmetry is a characteristic feature of solids. This means that is was taken for granted that thermodynamic ground state of a solid is periodic in three dimensions. More recently, it has been discovered a large number of systems where the diffraction peaks are not located on a lattice like for an ordinary crystal; moreover, the sharpness of the peaks shows that these systems are not disordered. The difference with ordinary crystals lies in fact that the diffraction peaks need more than 3 integer numbers for indexing. In the simplest situation, the diffraction wavevectors $\vec{Q}$ can be represented by

E. K. H. Salje (ed.), Physical Properties and Thermodynamic Behaviour of Minerals, 49–74.

$$\vec{Q} = h\, \vec{a}^*_1 + k\, \vec{a}^*_2 + \ell\, \vec{a}^*_3 + n\, \vec{q} \qquad (1)$$

in a certain range of thermodynamic variables like temperature and pressure. The interpretation is that there is a deviation from a perfect periodic structure with 3 reciprocal lattice vectors $\vec{a}^*_1$, $\vec{a}^*_2$, $\vec{a}^*_3$, such that this deviation itself has a periodicity (wavevector $\vec{q}$). When this does not fit with the original three periodicities, the system is called *incommensurate* (sometimes *modulated*) system. Because the incommensurate phase lacks translational symmetry, there is no three-dimensional space group. One way of describing the structure is by introducing a space with more than 3 dimensions (1). Figure 1 gives an illustration of the method, showing the embedding of a one-dimensional modulated crystal (sinusoidal modulation) in a 2-dimensional periodic superspace. The symmetry groups become crystallographic space groups in more than 3 dimensions.

The first examples reported were magnetic systems (e.g. $MnAu_2$,...) where the magnetization has a helix structure such that the pitch of the helix does not have a rational relation to the underlying lattice. Diffraction studies indicating an incommensurate structure were also found in systems with charge density wave (e.g. KCP, TTF-TCNQ,...). In the structure analysis of insulating crystals, the phenomenon was observed in Na_2CO_3, $NaNO_2$, thiourea, later in a whole family of A_2BX_4 compounds for which K_2SeO_4 is a typical representative, and more recently in $BaMnF_4$, $ThBr_4$, $Ba_2NaNb_5O_{15}$, in layer-structure perovskites, in quartz, etc.. : in all these cases, the incommensurability results from a small lattice distorsion incommensurate with the underlying lattice. Since last ten years, it has been discovered that many other physical properties could be incommensurate with the underlying lattice, and the list of incommensurate materials increases day by day.

In this paper, one will take most of examples among *insulating compounds* in which the modulation is of *"displacive" nature*.

Neutron and X-ray diffraction experiments are direct methods to study the characteristic features of the incommensurate (modulation) wave (amplitude, shape, wavevector, symmetry of atomic displacements,.). A diffraction pattern will be constituted i) by a 3-dimensional periodic array of "main" Bragg reflections ((h k ℓ) reflections where h, k, ℓ are integers and correspond to Miller indices), ii) every "main" Bragg reflections will be accompanied by a serie of "satellite" reflections located in $n\vec{q}$ (n = ± 1, ± 2 indicates the order of the satellite reflection, in this additional periodicity

$$\vec{q} = \sum_{i=1}^{3} \delta_i\, \vec{a}^*_i$$

at least one value of δ_i is an irrational number) [cf. figure 2a where $\vec{q}$ is parallel to the $\vec{a}_2$ direction].

In many examples, the incommensurate phase is found to possess a range of stability limited by two transitions T_i and T_L; T_i corresponds to a transition between the "normal" crystalline phase and the incommensurate phase. It marks the onset of the incommensurate modulation

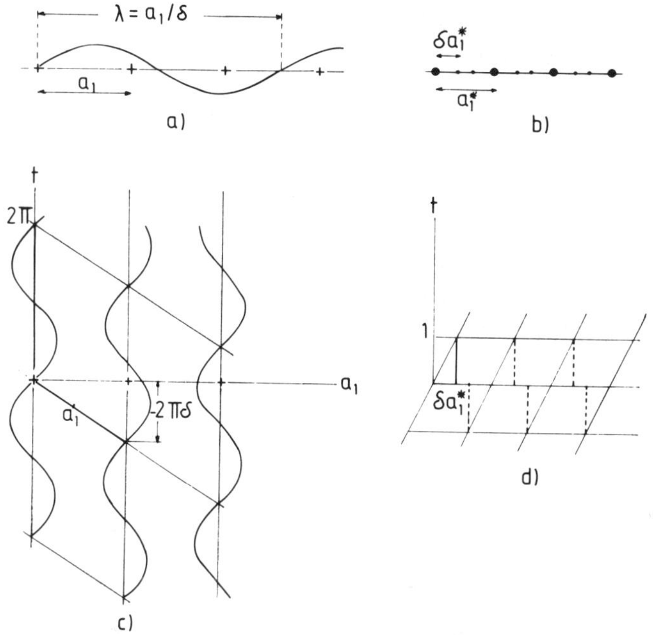

Figure 1. Embedding of a 1-dimensional modulated crystal in 2d-dimensional superspace

a) one dimensional displacive incommensurate structure : a_1 is the periodicity of the underlying lattice parallel to the $\vec{a}_1$ direction and λ is the wavelength of the sinusoidal modulation. For simplification only one chain has been drawn (the chains are located at the points of a lattice defined by a unit cell ($\vec{a}_2$, $\vec{a}_3$).

b) Fourier transform along the $\vec{a}^*_1$ reciprocal direction

c) bidimensional supercrystal : the translation symmetry is recovered in artificially adding one supplementary dimension t. The periodicities of the supercrystal are $a'_1 = (a_1, -2\pi\delta)$, [$a'_2 = (a_2, 0)$, $a'_3 = (a_3, 0)$ not drawn on the figure] and $t = (0, 2\pi)$. The crystal (fig. 1a) corresponds to the t = 0 section of the supercrystal

d) the bidimensional reciprocal superspace associated to figure 1c has for periodicities $a'^*_1 = (a^*_1, 0)$ [$a'^*_2 = (a^*_2, 0)$, $a'^*_3 = (a^*_3, 0)$ not drawn on the figure] and $t^* = (q, 1)$

The diffraction spectrum $\vec{Q} = h\vec{a}^*_1 + [k\,\vec{a}^*_2 + \ell\,\vec{a}^*_3] + n\,\vec{q}$ is viewed as a projection of the bidimensional diffraction pattern and is identical to figure 1b.

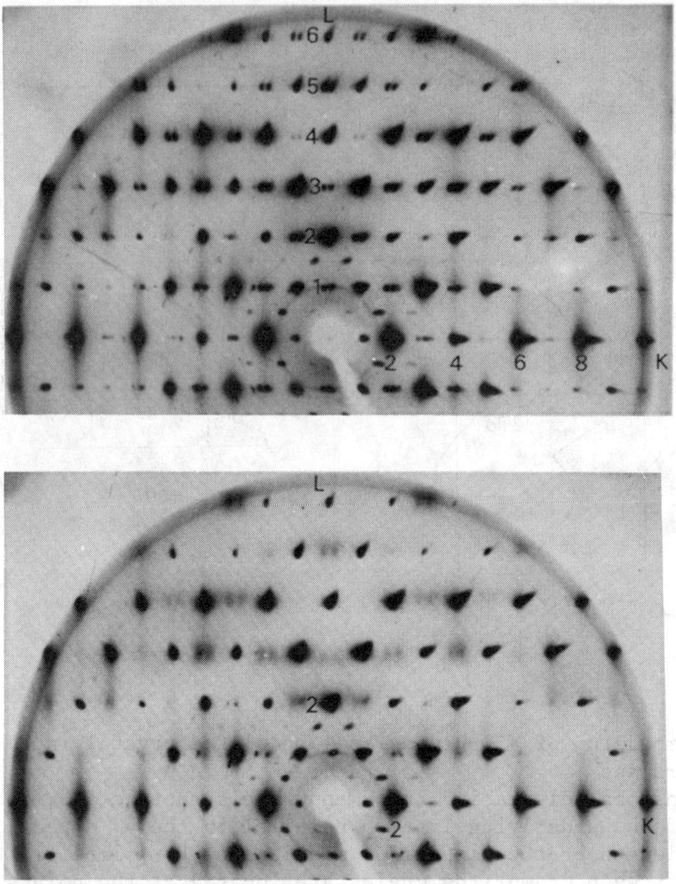

a-

b-

Figure 2. a) Typical diffraction pattern of incommensurate thiourea : X-ray precession photograph obtained near T_i for the zero-level reciprocal plane with the $\vec{a}$ (= $\vec{a}_1$) axis parallel to the precession axis. The basic Pnma structure gives for the main Bragg reflections (o k ℓ) extinction rules obeying $k + \ell = 2p \pm 1$; the satellite reflections (o,k $\pm$ δ,ℓ) appears along the $\vec{b}$ (= $\vec{a}_2$) direction when $k + \ell = 2p + 1$, ($\delta = \delta_2$). In thiourea, near T_i, the incommensurate phase can be viewed as a perturbation of the basic Pnma structure by means of a frozen-in soft-mode normal coordinate Q ($\vec{q}$, τ_4) whose symmetry properties are characterized by its wavevector $\vec{q}$ and its irreducible representation τ_4. The proper choice between the possible irreducible representations is dictated by the observed systematic extinction rules for the satellite diffraction peaks .

b) Precursor effects of the incommensurate phase in the "normal" crystalline phase of thiourea, just above T_i : evidence of critical diffuse scattering.

with a definite wavelength and a vanishingly small amplitude. This onset is announced in the normal phase by precursor effects [i.e. cf. figure 2b]. For the displacive type of modulations considered, these effects consist in an instability of the lattice (i.e. soft mode). The transition at T_L corresponds to the locking of the modulation on a simple multiple of the period of the underlying crystal structure. This corresponds to "standard" behaviour which is found in many incommensurate materials.

3. INCOMMENSURATE PHASE TRANSITION : THE "SOFT-MODE" PRECURSOR EFFECT

K_2SeO_4 is the prototypical example of a soft-mode instability. Figure (3a) illustrates the temperature dependent softening of its unstable phonon branch in the "normal" phase as the incommensurate transition is approached from above (2, 3). The minimum in the phonon dispersion is $q \simeq (0, 0, \xi)$ with $\xi \simeq 0.69$. Below the incommensurate transition temperature T_i, static displacements associated with the condensed phonon give rise to incommensurate satellite reflections. As first pointed out by Moncton et al. (4), there are still interactions in the incommensurate phase which tend to restore periodicity. One of the way these forces manifest themselves, is well-illustrated in figure 3b which shows a smooth decrease of the modulation wavevector as a function of temperature; at T_C (= T_L) the wavevector locks-in onto the rational value (δ = 0), giving rise to a unit cell tripled along the c axis. In every case, the mechanism of incommensurate instabilities is

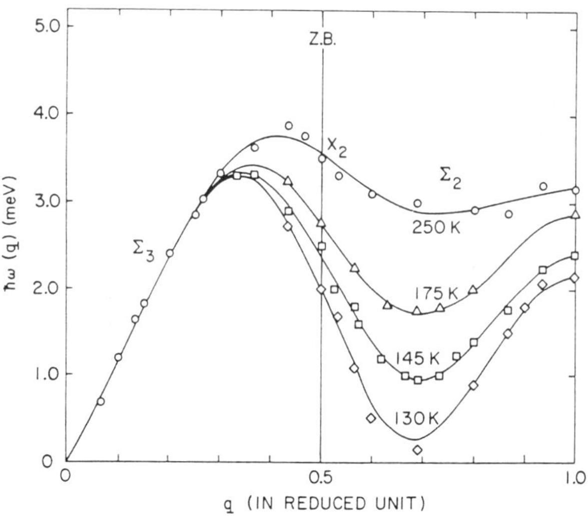

(a)

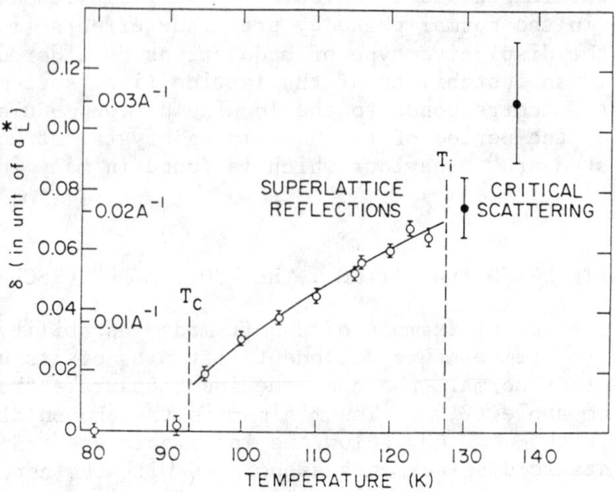

(b)

Figure 3. K_2SeO_4
a) phonon dispersion in the soft branch as a function of temperature in the extended zone scheme (2, 3)
b) temperature dependence of the soft phonon q and incommensurate satellite wavevector $q = (0, 0, 1 - \frac{1}{3}(1 - \delta))$. Below $T_C = 93$ K ($= T_L$) $\delta = 0$ the structure is no longer incommensurate, the unit cell is tripled along the c-axis (2, 3).

attributed to the presence in the system of competing interactions of varying range. For example, from the analysis of the phonon dispersion, Iizumi et al. (2) have shown that force constants between first and second layers of atoms decrease strongly with decreasing temperature above T_i; however, third-neighbours interlayers forces slightly increase in the same temperature interval; it is thought (2) that the competition between these short-range forces is responsible for the observed transition.

Since dynamic studies have been performed on almost incommensurate insulators, however, clear evidence of well defined soft-mode behaviour has been found essentially restricted to biphenyl (5) and $ThBr_4$ (6).

4. SPECIFIC DYNAMIC PROPERTIES OF INCOMMENSURATE PHASES : "PHASONS".

New types of excitations specific to the incommensurate state and corresponding respectively to fluctuations in the phase and amplitude of the static modulated wave have been predicted by Overhauser (7). It is evident that a homogeneous arbitrary shift of the phase of the static incommensurate wave (of wavevector $\vec{q}$) does not modify the global configuration of the incommensurate system; this overall phase shift costs no energy. Now, if we impose an inhomogeneous periodic shift associated to the wavevector $\vec{k}$, we expect that the corresponding energy of this excitation will vanish with $k \to 0$. For a simple sinusoidally modulated solid with displacive dynamics above T_i, Axe (8) has shown that using a simple Landau-type quasi-harmonic approach, the added presence of static displacements below T_i require new harmonic modes to be constructed from linear combinations of phonons with wavevector $(\vec{q} + \vec{k})$and $(-\vec{q} + \vec{k})$ respectively. The result is an upper branch in which the displacements are

$$u_1(\vec{r},t) \approx [e^{i(-\vec{q} + \vec{k})\vec{r}} + e^{-i(\vec{q} + \vec{k})\vec{r}}]e^{i\omega_1 t}$$

$$\approx \cos \vec{q}.\vec{r} \quad e^{i(\vec{k}.\vec{r} + \omega_1 t)}$$

and a lower branch with associated displacements :

$$u_2(\vec{r},t) \approx [e^{i(-\vec{q} + \vec{k})\vec{r}} - e^{-i(\vec{q} + \vec{k})\vec{r}}]\ e^{i\omega_2 t}$$

$$\approx \sin (\vec{q}.\vec{r}) \quad e^{i(\vec{k}.\vec{r} + \omega_2 t)}$$

$u_1(\vec{r},t)$ and $u_2(\vec{r},t)$ represent respectively a modulation of the amplitude A and a modulation of the phase $\emptyset$ of a static distorsion

$\langle u(\vec{r})\rangle = A \cos (\vec{q}.\vec{r})$

(an amplitude fluctuation leads to $\delta u_{ampl} \approx \delta A \cos(\vec{q}.\vec{r})$ and a phase fluctuation gives $\delta u_{ph} = A\delta\emptyset \sin (\vec{q}.\vec{r})$).

For the amplitude mode, the dispersion relation $\omega_1^{\,2}(T)=\Delta(T)+v^2k^2$ is found to have a temperature dependent gap similar to a classic soft-mode. Emanating from the first order satellite reflection, the phason mode has an acoustic-like dispersion relation, which is temperature independent: $\omega_2^{\,2}(T) = v_2^{\,2}k^2$.

The above description has been extended by Bruce and Cowley (9) to allow for the effect of higher-order distorsion harmonics and by Golovko and Levanyuk (10, 11) to include damping effects. Thus, an incommensurate phase is expected to possess in addition to its acoustic branches, one (or several) *acoustic-like excitations (phason branch(es) in the direction (s) of the modulation.* It is important to remark that the system considered above has not more degrees of freedom in the incommensurate phase than in the normal phase. Although phasons exhibit linear acoustic-like dispersion, the velocity v has no relation to the velocity of sound and the phasons are not to be confused with acoustic phonons.

The existence of phason has been searched in vain by the usual light-scattering technique (Brillouin technique) which allows to probe conveniently the similar acoustic modes. The theoretical works of Golovko et al (10, 11) and Zeyher and Finger (1982) (12) have allowed to clarify this situation. They have shown that while the damping Γ_{ac} of the acoustic modes tends to zero for $k \to 0$, the damping $\Gamma_{ph.}$ of the phason modes has no pronounced k dependence and in particular it remains finite when $k \to 0$. Moreover, Γ_{ph} is of the order of the damping of the soft-mode which drives the mechanism of the normal - incommensurate transition (it is more often a heavily damped mode of several hundred of GHz). Thus for Brillouin scattering the frequency $\omega \sim 10$ GHz both for the acoustic and phason branches is much larger than Γ_{ac} ($\sim$ 100 MHz) and much smaller than Γ_{ph} ($\sim$ 500 GHz); the acoustic mode will appear as a thin line in the spectrum while the phason mode will be unobservable. The last possibility to detect the phason mode is to measure at higher k values using inelastic neutron scattering technique, unfortunately due to the decreasing of the phason inelastic structure factor with increasing $\vec{k}$, another difficulty arises. Consequently one will be able to observe the phason branch only in very special cases i) if the studied sample shows an unusually low damping of the soft-mode ii) if the incommensurate phase can be examined at low temperature (Γ_{ph}(T) decreases when $T \to 0$). Up to now, satisfactory conditions for the direct observation of the phason branch by neutron scattering have been only fulfilled in three substances : biphenyl (5, 13) $ThBr_4$ (6, 14) and K_2SeO_4 (15). Figures (4a, 5a, 6a) show the frequency

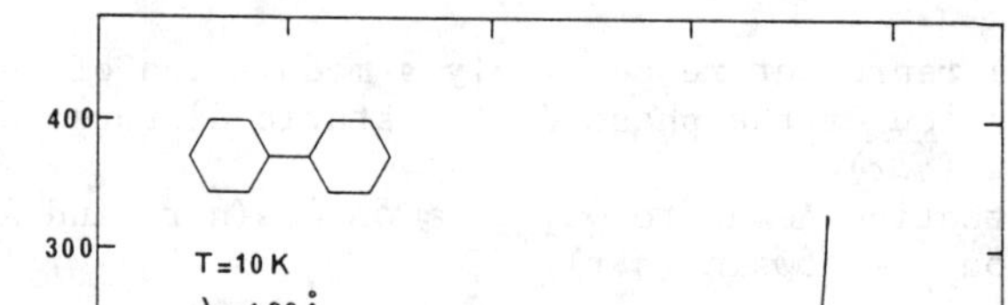

(a)

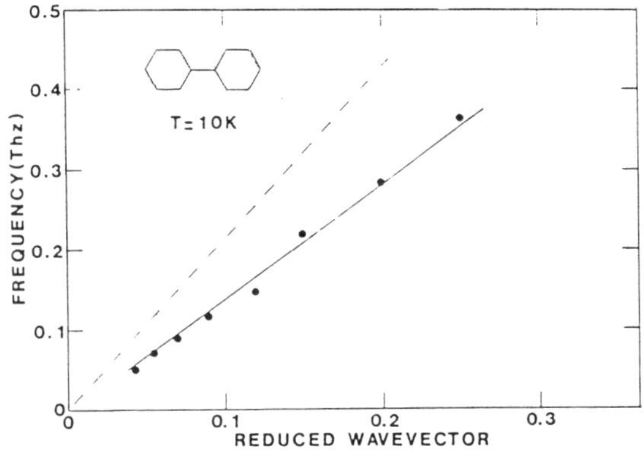

(b)

Figure 4. Incommensurate biphenyl from (5, 13).
a) energy scan close to a satellite reflection (phason mode) compared to energy scan close to a main Bragg reflection (acoustic mode).
b) dispersion law for the phason branch (solid line) compared to the lowest acoustic branch (dotted line) in the $\vec{a}^*$ direction.

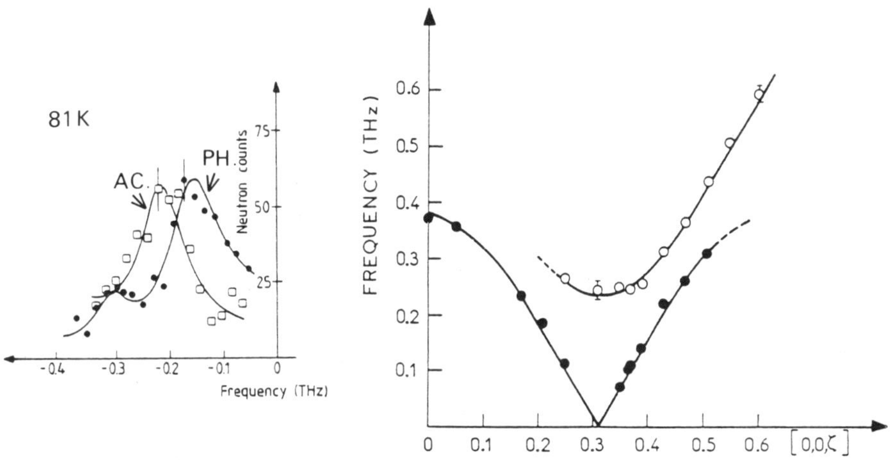

(a)

(b)

Figure 5. Incommensurate $ThBr_4$ from (6, 14).
a) comparison between energy scans close to a satellite reflection (●) and close to a main Bragg reflection (□) at T = 81 K ($T_i \simeq 95$ K)
b) phase (●) and amplitude (○) mode dispersions at 81 K.

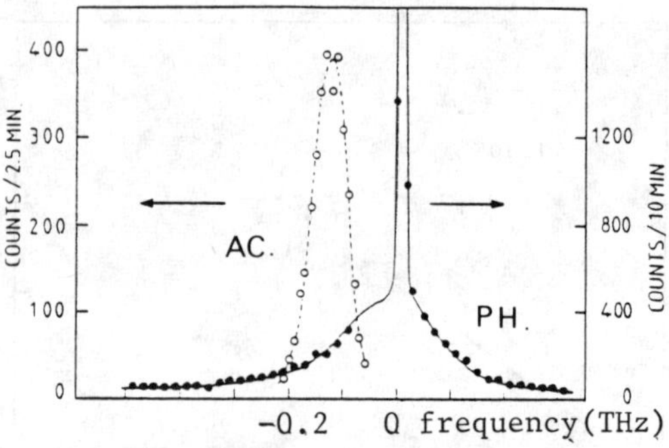

(a)

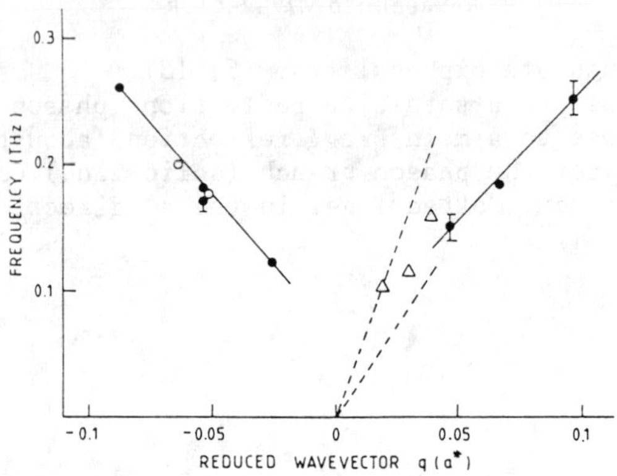

(b)

Figure 6. Incommensurate K_2SeO_4 from (15).
a) comparison of transverse acoustic-phonon (o) and phason (•) line-shapes at equivalent reduced wavevector $\vec{k} = 0.046\ \vec{a}^*$
b) phason dispersion for $\vec{q}$ // $\vec{a}^*$ at 120 K (•) and (o) 100 K, ($T_i \simeq$ 130 K)
acoustic slopes are shown as broken lines.

spectra relative to the acoustic (AC.) and phason (Ph.) branches for these three compounds and figures (4b - 5b - 6b) reproduce the phason branches. In biphenyl and $ThBr_4$ the extrapolated phason branch passes through the first order satellite reflections. In the case of K_2SeO_4, where the soft-mode damping is fairly large ($\Gamma \simeq$ 250 GHZ) (15) direct evidence of overdamped phase mode has been found at small k wavevector.

5. PRESENT SITUATION CONCERNING THE STATIC PROPERTIES OF INCOMMENSURATE PHASES

5.1. Some examples of experimental typical phase diagrams

During past decade, a wide variety of effects which differ from the standard behaviour discussed in § 2 has been discovered and has made the situation rather complicated. In some cases, the lock-in transition at low temperature between the incommensurate and the commensurate phases is absent (for example $BaMnF_4$ (16), $ThBr_4$ (14), biphenyl (5),..). In these systems, the incommensurate modulation is observed down to the lowest temperatures (4.2 K) and the modulation wavevector is nearly independent of temperature. In other cases, accurate diffraction experiments have revealed that the modulation wavevector displays singularities : for example varying pertinent external parameters, *high-order commensurate phases appear inside the incommensurate region*; thiourea is a good illustration of this type of situation [cf. figures (7a, 7b): $\vec{q} = \delta(T, P)\ \vec{b}^*$, (P, T) phase diagram and figures (8a, 8b) : $\vec{q} = \delta(E, T)\ \vec{b}^*$, (E, T) phase diagram]. A succession of steps on commensurate values 3/7, possibly 1/2 (depending of samples), 1/3 is

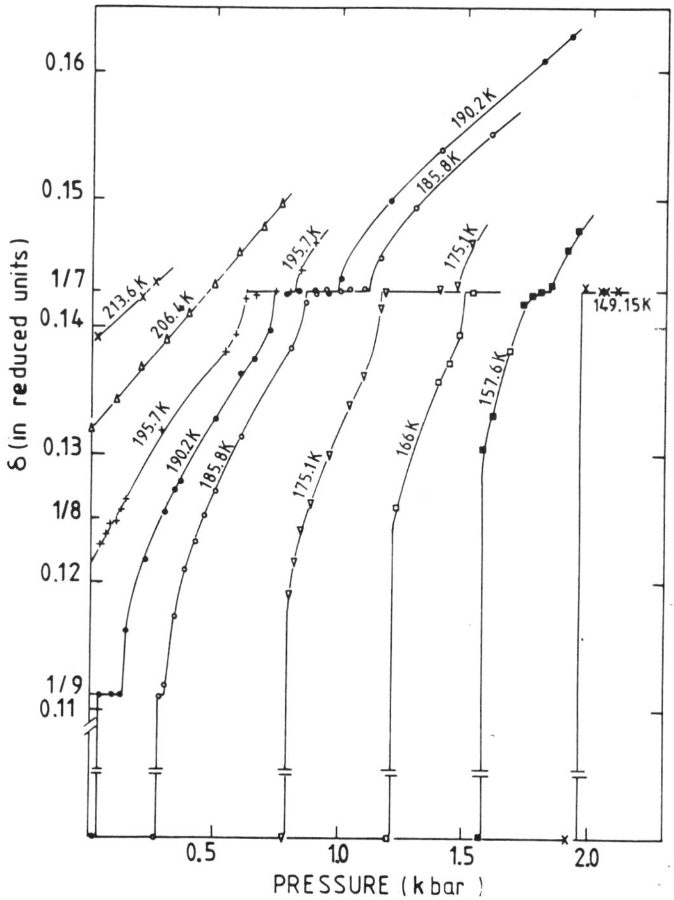

(a)

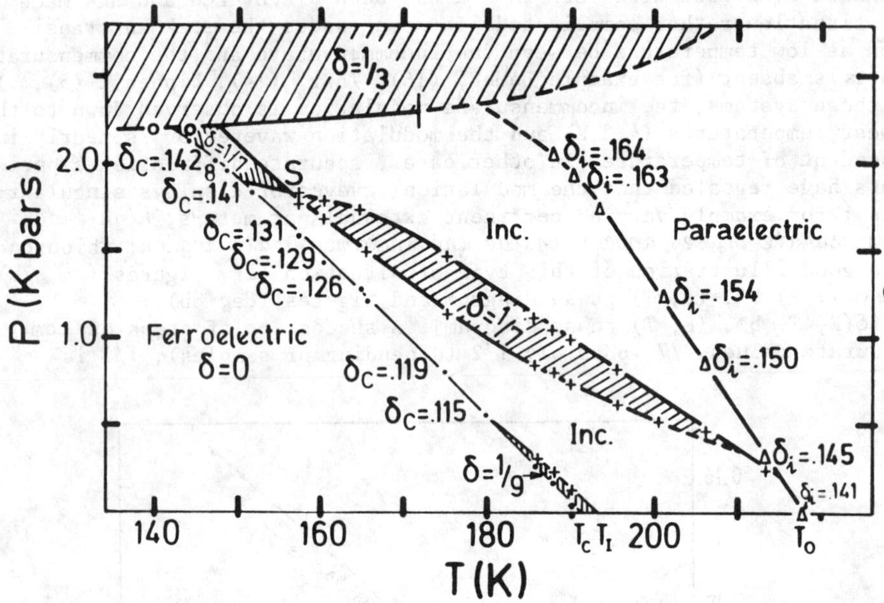

(b)

Figure 7. Modulated $SC(ND_2)_2$ from (17, 18)
a) Isothermal variations of the modulation wave number δ(P, T) as a function of pressure (neutron diffraction results)
b) (P, T) phase diagram. The hatched areas correspond to commensurate phases δ = 1/3, 1/7 and 1/9).

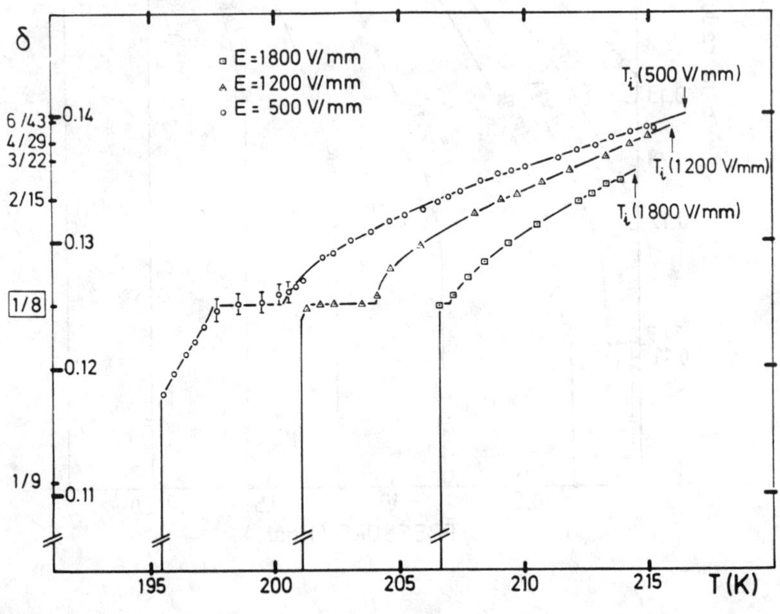

(a)

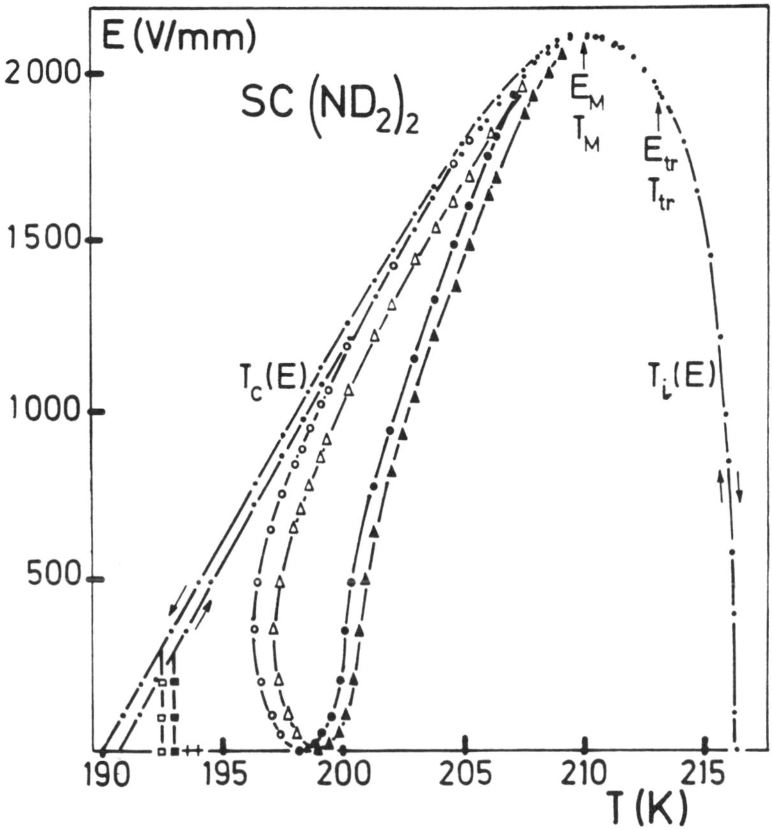

(b)

Figure 8. Modulated thiourea
a) Iso-electric field variations of the modulation wave number $\delta(E,T)$ as a function of temperature from (19) (neutron diffraction results)
b) (E, T) phase diagram from (20). The solid and open squares are related to the (δ = 1/9) - incommensurate transitions for increasing and decreasing temperatures. The solid and open triangles (or circles) delimit the boundaries of a δ = 1/8 commensurate phase inserted into the incommensurate phase for increasing (or decreasing) temperature.

observed in the wavevector temperature dependence of d-TMATC-Zn (21,22), application of pressure (23), and electric-field (24) stabilizing the 2/5 commensurate value. Similar phenomena are found in other compounds of the same family (TMATC-Co, TMATC-Fe,... for a review see for example ref. 13). In every case, the (P, T) phase diagram is complex; several types of regions are found; there are i) incommensurate regions where the wavevector smoothly varies, ii) commensurate regions (1/3, 2/5, 3/7, 1/2) iii) regions where the wavevector is pinned on an incommensurate value. In other compounds (such as Rb_2ZnBr_4) a pinning of the modulation wavevector on a rational value (5/17!) is observed in a large part of the modulated phase (25).

5.2. Other specific properties : global hysteresis, satellite broadening,...

Over the last few years, it has become apparent that most of the incommensurate phases are accompanied by a characteristic thermal hysteresis (qualified by S. Aubry of "*global*" *hysteresis* (26); not only the lock-in transition temperature in a heating run differs from that in a cooling run (normal hysteresis) but also the heating and cooling curves (δ (T ↑) and δ(T ↓)) do not coincide. The most remarkable feature is that the hysteresis appear even when heating and cooling runs are cycled within the incommensurate phase. In many materials, this global hysteresis is observed in a region near T_L but there are systems as for example thiourea [cf. figure 9] where it is observed in all the

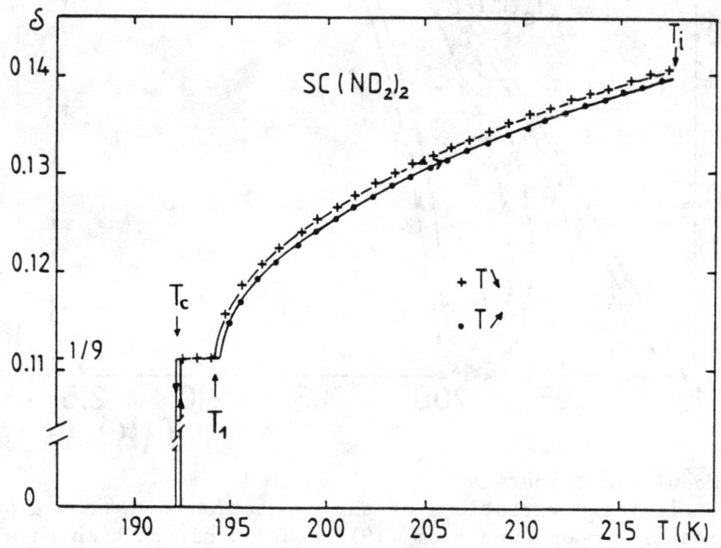

Figure 9. $SC(ND_2)_2$ from (27) : temperature dependence of the modulation wave number
. increasing temperature
+ decreasing temperature
(neutron diffraction results)

region where the modulated phase is stable. Finally, it has been shown that the macroscopic properties which can be related to the incommensurate modulation such as dielectric susceptibility, birefringence are also affected by global hysteresis (see for example (28))

In a few substances, *anomalous linewidth of the satellites* have been discovered. In $BaMnF_4$ the satellites are 4 times wider than the fundamental Bragg reflections in the whole range of the incommensurate phase (29) whereas in Rb_2ZnCl_4, the broadening of satellite is sharply peaked near T_L (30). In these systems, the *long-range phase coherence of the modulated wave is broken.*

5.3. Discussion

5.3.1. Phenomenological theory. In order to describe the different observed phenomena, several theories have been developped. The least ambitious approach for discussing the stability of modulated structures consists in writing down a phenomenological expression for the free energy of the system as a functional of the order parameter and of its spatial derivatives. This Landau-Ginzburg *continuum approach* relies on the assumption that the spatial variations of the order parameter are slow on an atomic scale (the underlying crystal structure can be considered as a continuum). For a complete description see for example references (31-34). Such a formalism accounts for i) an incommensurate phase stable in a temperature interval $[T_i, T_L]$, ii) a continuous transition at T_i corresponding to the appearance of a modulated wave (wavevector $\vec{q}_i$). The modulated wave just below T_i has generally a sinusoidal profile with a small amplitude; when the temperature decreases the modulation wavevector smoothly decreases, the amplitude of the wave increases and often harmonics develop which change the profile of the wave. On approaching T_L, the incommensurate state is expected to be constituted by cristalline regions identical to the commensurate phase below T_L (wavevector q_L) separated by walls (called discommensurations or solitons), these walls form a periodic array, the distance between two discommensurations being equal to $\pi/q-q_L$. The theory predicts that the wavevector q will decrease more or less abruptly on approaching T_L, iii) with such a description, a continuous transition at T_L is found. One possibility to make the incommensurate-commensurate phase transition discontinuous, is to include, in the development of the thermodynamic potential, additional parameters (components of strain tensor or polarisation for example) which couple with the order parameter. In short, Landau-Ginzburg phenomenological theory is well-suited to describe the main characteristics of incommensurate "standard" insulators; in some cases, addition of Umklapp lock-in terms (of order 7, 8 or 9 (!) in thiourea for example) in the free energy expansion (in these terms the microscopic lattice periodicity occurs) allow a theoretical description of high order commensurate phases which appear inside the incommensurate region (17). The complete (E, T) phase diagram of thiourea, including the high-order commensurate - incommensurate boundaries has been mapped out using this phenomenological approach (20).

5.3.2. "Discrete" models : rôle of underlying lattice. In many examexamples, the validity of the "continuous limit" being possibly questionable, various microscopic models which take into account the *discrete atomic structure of the underlying lattice* and which involve competing interactions have been developed. In the Frenkel-Kontorova theory (initially adapted to incommensurate systems by S. Aubry (26), then developped by Pokrovsky (35), Bak (36), etc...), the model consists of a linear array of atoms connected with harmonic springs in a periodic potential of period "b". In the absence of periodic potential, a lattice constant "a_o" is favoured which in general would be incommensurate with "b". If the periodic potential is strong enough, it may be favourable for the lattice to relax into a commensurate structure where the average lattice spacing "a" is a rational fraction of the period "b". The atoms lose elastic energy but gain potential energy. In the "anisotropic next nearest neighbour Ising" models ("ANNNI" models) a predominant antiferromagnetic interaction between second neighbour spins competes with a ferromagnetic interaction between nearest neighbour spins [cf. for example Selke (37)]. A third model has been developped by Janssen and Tjon (38, 39) where competition results from the presence of harmonic forces between first, second and third neighbours and anharmonic forces between first neighbours. Each of these models with short-range interactions in competition lead to a specific phase diagram. However, common features can be drawn i) the main characteristic is the appearance of commensurate phases inside the modulated region. The temperature behaviour of the wavevector displays two regimes : in the vicinity of T_i, the q(T) curve is well-described by an incomplete devil's staircase (the curve corresponds to sequences of commensurate plateaus separated by true incommensurate phases); near T_L the devil's staircase becomes complete and the curve q(T) results in an infinite succession of commensurate plateaus. ii) global hysteresis is predicted iii) as in "continuous" theory, the concept of discommensurations is of central importance; in the region where discommensurations are effective, the stable state corresponds to a periodic array of discommensurations; moreover, the possibility of metastable states consisting in an irregular array of the discommensurations with an average periodicity λ is also predicted by these models; a random pinning of the discommensurations on the lattice sites of the underlying crystal structure occurs giving rise to a broadening of the satellite diffraction peaks. (This effect has been qualified initially by Bak (40) of chaotic state). This broadening can be reduced with time if the metastable state decays towards the stable one.

The models discussed above qualitatively describe the global features of phase diagrams;moreover, the discrete character of these last models introduces a number of interesting new features not present in phenomenological theory such as global hysteresis, satellite broadening. In the next section, we shall see that the defects can lead to similar effects and unfortunately in many incommensurate systems studied, the question remains open to what extent the discrete nature of the underlying lattice plays a rôle.

5.3.3. Rôle of defects. Intensive experimental studies during past decade have led in some modulated materials to *conflicting results*. It has been shown that results could depend in a very pronounced way on a number of conditions : they have a sample-dependence, a time-dependence sometimes an instrumental dependence (!), X-ray or neutron diffraction can lead to results in total disagreement [see for example the comparison between figures 8a and 10]; more often they depend of the thermal history of the sample (see below, for example the memory effect), sometimes combined-actions of several effects can give surprising results.

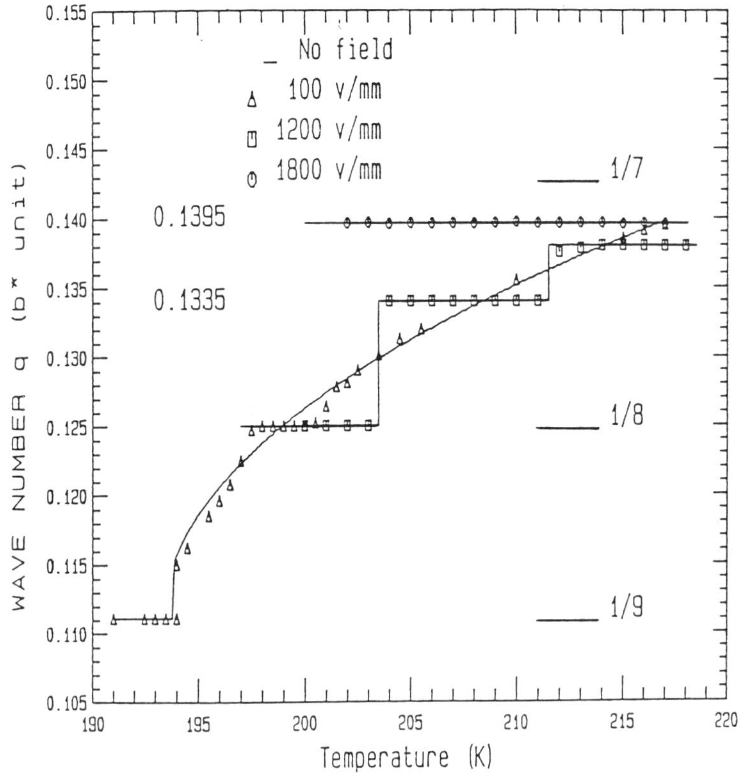

Figure 10. $SC(ND_2)_2$ from (50)
Isoelectric field variations of the modulation wave number δ(E, T) as a function of temperature (X-ray diffraction results)

As a consequence, one has realized that defects could play an important rôle modifying the properties of "pure" systems and one has to be careful in studying many of non-recent available experimental results in which the experimental conditions have not been always well-controlled. Simultaneously theories have been developped. All the models are based on the presence of defects interacting with the modulated wave, they generally assume the existence of *randomly fixed-defects* pinning the phase of the modulated wave (41 - 49). We shall briefly describe their main results. A first consequence is a loss of long-range order for the modulated wave, giving rise to satellite broadening; another consequence is the modification of the temperature decreasing of the modulation wavevector; generally convex in a pure system, it can become concave in presence of defects. In the region where discommensurations are predominant (generally in the vicinity of T_L) the influence of defects is much more important. The pinning of discommensuration by defects generates metastable states, giving rise when temperature varies to pinning of the modulation wavevector; in addition, metastable states related to defects can be at the origin of a global hysteresis. During last few years, it has been shown in a few cases that hysteresis effects, satellite broadening, wave-vector pinning observable on nominally-pure systems become more pronounced with the addition of well-controlled chemical defects (e.g. Rb_2ZnCl_4 (51), $BaMnF_4$ (16)).Study of radiation damage in thiourea (52), TMATC-family (53- 54) has shown that the disagreements between X-ray and neutron diffraction results (cf. figures 8a and 10) could be attributed to X-ray irradiation defects in the X-ray experiments (in incommensurate systems, in order to avoid radiation damage, it will be better, when it is possible to use thermal neutron diffraction rather than X-ray diffraction).In spite of these clarifications, generally it is not easy to determine what are the contributions - due to the discrete nature of the underlying lattice - and/or due to the presence of defects. In many cases, this remains a real challenge.

Finally, one would like to mention another type of effect which was not predicted by theoreticians but which was revealed by experiment : it is "*memory effects*". Discovered in thiourea (55), then theoretically interpreted (55-57) it has been reported since in $Ba_2NaNb_5O_{15}$ (58), Rb_2ZnCl_4 (35) and quartz (59). Observations are often made on macroscopic properties (e.g. birefringence, dielectric susceptibility), but the microscopic origin of the observed effects is a remarkable *periodicity memory* associated with the presence of *mobile defects*. The migration of defects in the periodic potential created by

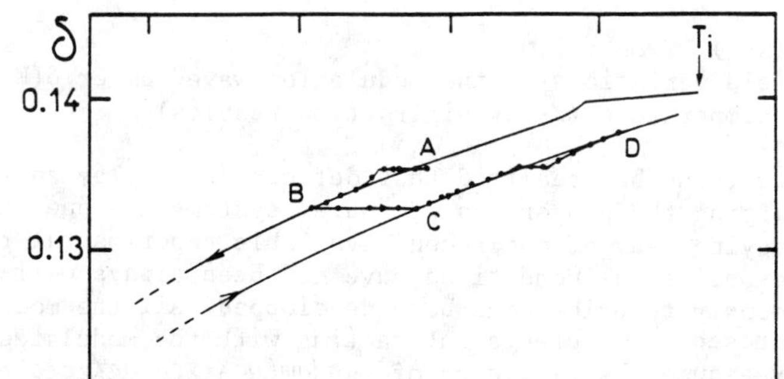

Figure 11. Irradiated $SC(ND_2)_2$ from (52).
Direct evidence for periodicity memory effect.

the modulation gives rise to what has been termed, a defect density wave (DDW) (55-57) of same periodicity and whose amplitude grows with time. Once established, the DDW exerts a pinning potential on the modulation itself, whose effect is to limit the latter's ability to readjust its periodicity in response to a temperature, pressure or field variation. Memory effects are observable on nominally-pure systems, but as in the case of global hysteresis, they become more pronounced with the addition of chemical or irradiation defects (52). Direct evidence for periodicity memory in irradiated thiourea is illustrated in figure 11 : i) starting from room temperature the sample is cooled and kept at a temperature T_A = 209.3 K for 19 h ("writing" of memory effect), during this "writing" period no significant relaxation in satellite position, intensity or width could be noticed, ii) upon cooling run, a plateau of $\sim$ 1.6 K is observed, iii) at point B, the sense of temperature variation is reversed and the wavevector remains constant until the warming curve is reached (point C), iv) the wavevector follows the warming curve until the δ_A value is reached, v) a plateau of $\sim$ 1.1 K is observed at $\delta = \delta_A$ ("reading " of memory effect), vi) after a $\sim$ 2 K crossover region, the normal warming curve is recovered (point D). A similar cycle on the same sample before irradiation, yielded temperature plateaus narrower by a factor 2, on writing, as well as on reading.

6. QUASICRYSTALS

The discovery by Schechtman et al. (60) in 1984 of a metallic phase ($Al_{0.86}Mn_{0.14}$) with long-range icosahedral orientational symmetry has revealed a new class of ordered structures : the quasicrystals. It is well-known that such a point group symmetry is *incompatible with a periodic lattice structure*; nevertheless the sharpness of the diffraction peaks indicates *long-range order*. The difference with incommensurate solids lies in the absence of main Bragg reflections. All these features have led T. Janssen (61) to define quasicrystals as "incommensurable solids without conspicuous lattice of main Bragg reflections".

Up to last year, the new quasicrystalline phases have been mainly observed in micron size grains of quenched alloys. Therefore, the structural studies have been limited to electron diffraction on single domains and X-ray and neutron diffraction on powder samples. Very recently the growth of large quasicrystals of uniform orientation have been reported in the aluminium-lithium-copper system (62). These "single" quasicrystals with a chemical composition near to Al_6Li_3Cu show nice flat faces (as in ordinary crystals) arranged in a triacontahedral shape (63). This supports the idea that the quasicrystalline state may be at thermodynamic equilibrium, at the contrary to what it was expected from the first examples of quenched alloys. Figure (12) shows typical X-ray diffraction patterns obtained with such a "single" quasicrystal of Al_6Li_3Cu (64, 65). The diffraction peaks of the icosahedral phase require *six* integer indices; taking for example a cubic reference frame with basis vectors $\frac{2\pi}{a}$ (1, 0, 0), $\frac{2\pi}{a}$ (0, 1, 0) and

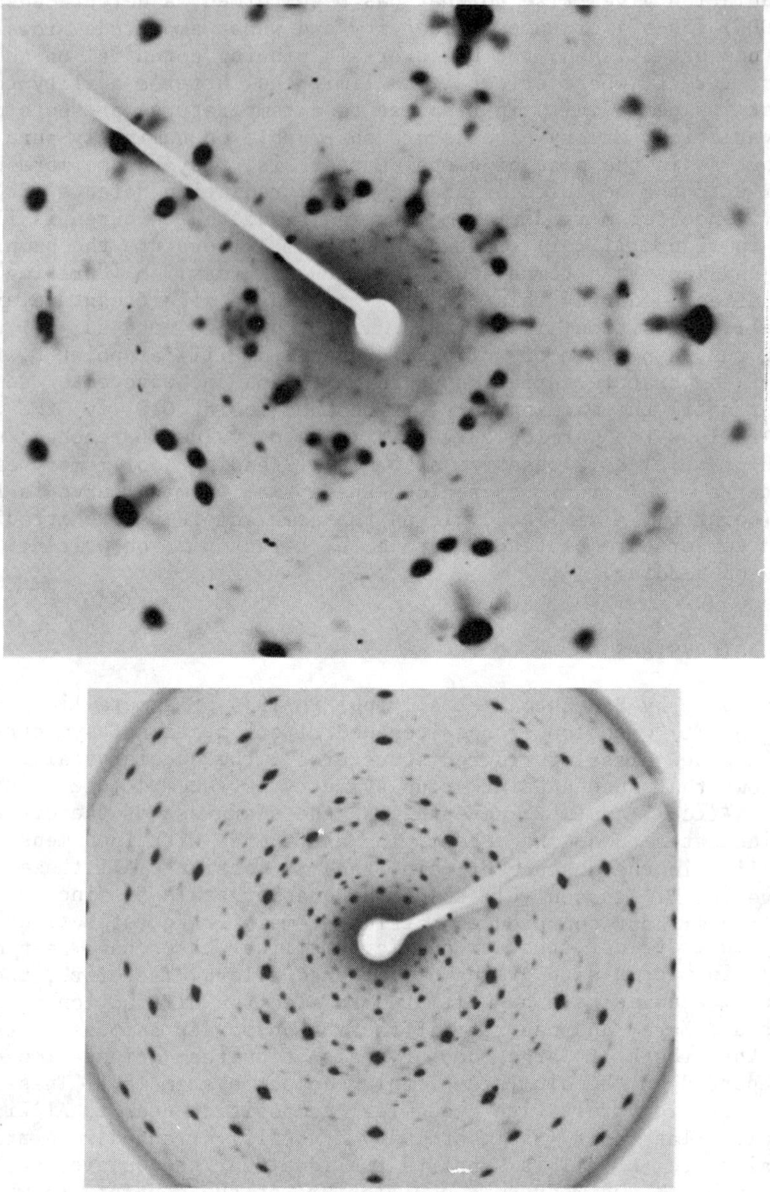

a

b

Figure 12. X-ray diffraction patterns obtained with a "single" quasi-crystal Al_6Li_3Cu (from (64, 65)
a) Monochromatic transmission X-ray Laue diffraction pattern obtained with a 5-*fold axis* of the sample parallel to the incident beam
b) Monochromatic X-ray precession photograph obtained for the zero-level reciprocal plane with a 5-fold axis of the sample parallel to the precession axis.

$\frac{2\pi}{a}(0,0,1)$, all the diffraction peaks can be written : $(h+h'\tau,\ k+k'\tau,\ \ell+\ell'\tau)$ where τ is the golden ratio $\tau=\frac{1}{2}(\sqrt{5}-1)$ and h, h', k, k', ℓ, ℓ' are integers. As introduced in *incommensurate phases (1), periodicity may be restored using higher dimensional space* (D > 3). In the case of icosahedral phase an hypercubic (D = 6) lattice is used (66).

How to construct a model of these quasiperiodic structures. To simplify we consider the one-dimensional model initially introduced by Katz and Duneau (67). It consists in quasiperiodic tiling or patterns obtained by the projection method. One begins by a brief description of this method (illustrated in figure 13a) : one considers in the plane

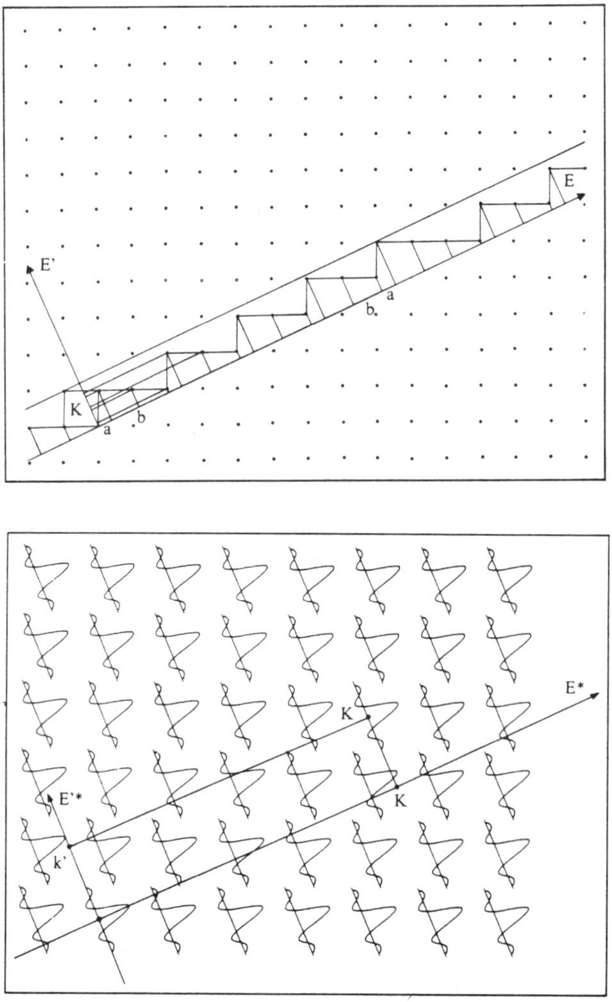

(a)

(b)

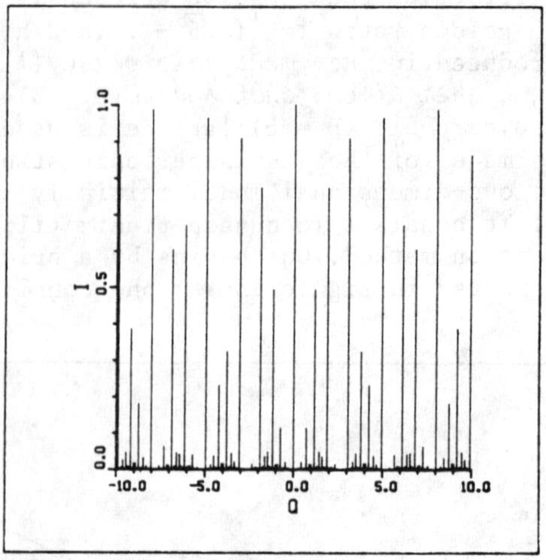

(c)

Figure 13. Illustration of the projection method : one dimensional quasiperiodic tiling
a) in real space from (67)
b) in reciprocal space from (67)
c) diffraction pattern by a Fibonacci chain from (61)

R^2, a square lattice Z^2 and a line E to be tiled; a strip is obtained by shifting the unit square K along E. Therefore, for almost all positions of the line, the strip contains an unique broken line (constituted of edges of the lattice), which exactly joins all the vertices inside the strip. The *orthogonal projection of this line on E* gives a 1D-*tiling*, the two tiles "a and b" being respectively the projections of the vertical and horizontal edges of the unit square K. The choice of an irrational (or rational) slope of the line E with respect to the periodic bidimensional square lattice implies the non-periodicity or (periodicity, possibly long-period) of the tiling. The Fourier transform of such a 1D-quasiperiodic tiling is obtained by putting one Dirac delta at each vertex of the line. This measure is obtained as follows : i) one takes a square lattice of Dirac delta in the plane ii) one multiplies by the characteristic function of the strip and iii) one projects on the line E. The Fourier transform is obtained i) in taking the Fourier transform of the square lattice which is a reciprocal square lattice of Dirac delta ii) convoluted with the Fourier transform of the characteristic function of the strip in the direction E*; the result of these two operations is a Dirac delta $\delta(k)$ in the direction E* and a smooth function $\sim \frac{\sin \pi k'}{\pi k'}$ in the orthogonal E'* direction iii) finally the Fourier image of the projection (1D-quasiperiodic tiling) is merely the *restriction to the line* E*. Such a method is illustrated in figure 13b. The Fibonacci chain is a beautiful and important exemple of a quasicrystal in one dimension; in this special case, the small (a) and long (b) tiles have a length ratio equal to τ. Another way to generate a Fibonacci chain is to start with a short tile (a) and to iterate according the following procedure : a short tile (a) is replaced by a long one (b) and a long one by a pair (b a). Starting from (a) one obtains after some iterations b a b b a b a b. The ordering of the 2 types of tiles is not arbitrary, and in the limit of an infinite number of iterations a Fibonacci chain is found. The diffraction pattern has been calculated (61) and is illustrated in figure 13c. The diffraction peaks are located in $Q = h\,a^*_1 + k\,a^*_2$ with $a^*_1 = \frac{2\pi}{3\tau-1}$, $a^*_2 = \frac{2\pi\tau}{3\tau-1}$ and where h and k are integers; the peaks are Dirac delta (long range order). One notices the absence of a lattice of main Bragg reflections.

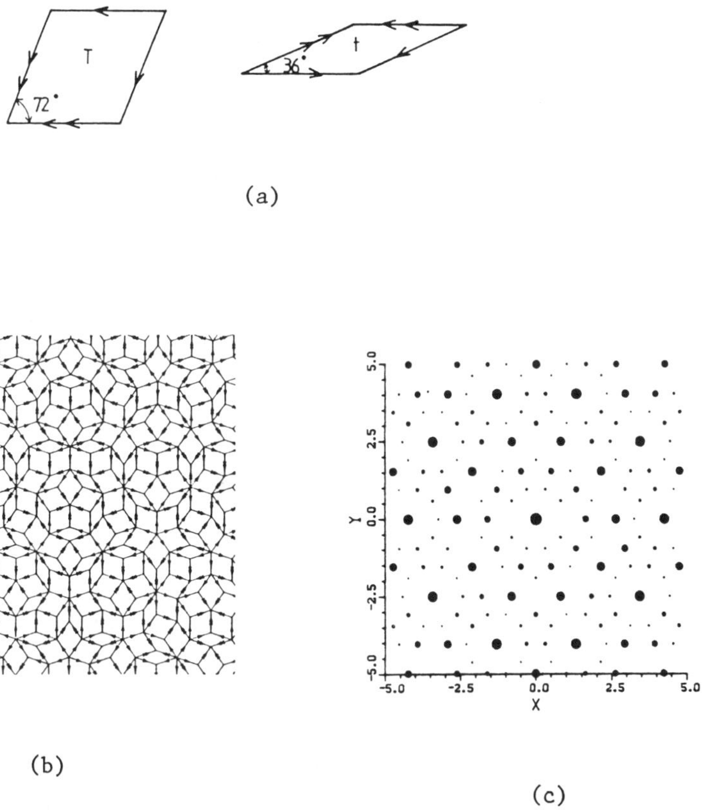

Figure 14. An example of 2D quasiperiodic tiling
a) tiles and their laws of association in usual Penrose tiling (from (70))
b) usual Penrose tiling (from (70))
c) diffraction pattern of the Penrose tiling from (61)

Generalization to 2-dimensional tiling and 3-dimensional tiling have been respectively done by Penrose (68) and Mackay (69). It has been shown (68), that it is possible to cover the plane in a non-periodic way using a set of two tiles : as an example one can use rhombus with equal edges and angles $\pi/5$ and $4\pi/5$ (thin rhombus) or $2\pi/5$ and $4\pi/5$ (thick rhombus); we notice that the golden ratio $\tau = 2\cos 2\pi/5$. Figures 14a, b show these tiles and their laws of association in the usual Penrose tiling and figure 14c shows the diffraction pattern resulting from such a 2-dimensional quasicrystal. The icosahedral phase can be considered as a generalization of the Penrose tiling to 3 dimensions (thick and thin rhombohedra are used in this case) and an elegant way of generating "Penrose filing" in 3 dimensions is to project a 6 periodic lattice (hypercubic lattice) on a 3-dimensional hyperplane. Calculations of the Fourier transform give patterns which look like the experimental diffraction photographs obtained with icosahedral quasicrystalline phase. So far, unfortunately nobody has succeeded in reproducing with precision the experimental diffraction pattern by decorating each tile by an arrangement of atoms. The structure of quasicrystals is not yet known.

REFERENCES

(1) P.M. de Wolff, (1977), Acta Cryst. A 33, 493
A. Janner and T. Janssen, (1977), Phys. Rev. B 15, 643
A. Janner and T. Janssen, (1979), Physica A, 99, 47
P.M. de Wolff, T. Janssen and A. Janner, (1981), Acta Cryst. A 37, 625

(2) M. Iizumi, J.D. Axe, G. Shirane and K. Shimaoka, (1977), Phys. Rev. B 15, 4392

(3) J.D. Axe, M. Iizumi and G. Shirane, (1986), Incommensurate Phases in Dielectrics 2, Edited by R. Blinc and A.P. Levanyuk North Holland, Chapter 10

(4) D.E. Moncton, J.D. Axe and F.J. Disalvo, (1975), Phys. Rev. Lett. 34, 734

(5) H. Cailleau, (1986) Incommensurate Phases in Dielectrics 2, Edited by R. Blinc and A.P. Levanuk, North Holland, chapter 12, and references therein.

(6) L. Bernard, R. Currat, P. Delamoye, C.M.E. Zeyen, S. Hubert and R. de Kouchkovsky, (1983), J. Phys. C 16, 433

(7) A.W. Overhauser, (1971), Phys. Rev. B 3, 3173

(8) J.D. Axe, Proc. of Gatlinburg Neutron Scattering Conference, CONF-760601-P 1 (U.S. Dept. of Comm.) p 353, (1976)

(9) A.D. Bruce and R. Cowley, (1978), J. Phys. C 11, 3609

(10) V.A. Golovko and A.P. Levanuyk, (1981), Ferroelectrics 36, 306

(11) V.A. Golovko and A.P. Levanyuk, (1982), Sov. Phys. JETP 54, 1217

(12) R. Zeyher and W. Finger (1982), Phys. Rev. Lett. 49, 1833

(13) H. Cailleau, F. Moussa, C.M.E. Zeyen and J. Bouillot (1980), Solid State Comm. 33, 407

(14) R. Currat, L. Bernard and P. Delamoye, (1986), Incommensurate Phases in Dielectrics 2, Edited by R. Blinc and A.P. Levanyuk, North Holland, chapter 15

(15) M. Quilichini and R. Currat, (1983), Solid State Comm. 48, 1011

(16) J.F. Scott, (1986), Incommensurate Phases in Dielectrics 2, Edited by R. Blinc and A.P. Levanyuk, North Holland, Chapter 19

(17) F. Dénoyer, A.H. Moudden, R. Currat, C. Vettier and M. Lambert, (1982), Phys. Rev. B 25, 1697

(18) F. Dénoyer and R. Currat, Incommensurate Phases in Dielectrics 2, Edited by R. Blinc and A.P. Levanyuk, North Holland, Chapter 14

(19) D. Durand, F. Dénoyer, R. Currat and C. Vettier, (1984), Phys. Rev. B 30, 1112

(20) M. Barreto, J.P. Jamet and P. Lederer, (1983), Phys. Rev. B 28, 3994

(21) G. Marion, R. Almairac, J. Lefebvre and M. Ribet, (1981), J. Phys. C 14, 3177

(22) M. Iizumi and K. Gesi, (1983), Physica B 120, 291

(23) G. Marion, R. Almairac, M. Ribet, V. Steigenberger and C.Vettier (1984), J. Physique 45, 929

(24) D. Durand, F. Dénoyer, M. More, to be published

(25) M. Iizumi and K. Gesi, (1983), J. Phys. Soc. Jpn 52, 2526

(26) S. Aubry : Solitons and Condensed-Matter Physics, eds. A.R. Bishop and T. Schneider (Springer, New York)

(27) G. André, D. Durand, F. Dénoyer, R. Currat, F. Moussa, (1987), Phys. Rev. B 35, 2909

(28) K. Hamano, (1986), Incommensurate Phases in Dielectrics 1. Fundamentals, Edited by R. Blinc and A.P. Levanyuk, North Holland, chapter 9.

(29) D.E. Cox, S.M. Shapiro, R.J. Nelmes, T.W. Ryan, H. Bleif, M. Eibschutz and H.J. Guggenheim (1983), Phys. Rev. B 28, 1640

(30) H. Mashiyama, S. Tanisaki and K. Hamano (1981), J. Phys. Soc. Jpn. 50, 2139

(31) A.P. Levanyuk, (1986), Incommensurate Phases in Dielectrics 1. Edited by R. Blinc and A.P. Levanyuk, North Holland

(32) R.A. Cowley and A.D. Bruce, (1978)J. Phys. C 11, 3577

(33) A.D. Bruce and R.A. Cowley, (1978), J. Phys. C 11, 3609

(34) A.D. Bruce, R.A. Cowley and A.F. Murray, (1978), J. Phys. C 11 3591

(35) V.L. Pokrovsky, (1981), J. de Physique 42, 761
V.L. Pokrovsky and A.L. Talapov, (1980), Solid State Comm. 33 749

(36) P. Bak and R.Bruinsna, (1982), Phys. Rev. Lett. 49, 249

(37) W. Selke, (1984), Modulated structures in a simple Ising Model Proc. ASI on Modulated Structure Materials ed. Tsakalokos

(38) T. Janssen and J. Tjon, (1983), J. Physics A 16, 673

(39) T. Janssen, (1986), Incommensurate Phases in Dielectrics 1. Fundamentals, chapter 3, Edited by R. Blinc and A.P. Levanyuk, North Holland

(40) P. Bak, (1982) Rep. Prog. Phys. 45, 587

(41) Y. Imry and S.K. Ma, (1974), Phys. Rev. Lett. 35, 1399
(42) L.J. Sham and B.R. Patton, (1976), Phys. Rev. B13, 3151
(43) H. Fukuyama and P.A. Lee, (1978), Phys. Rev. B17, 535
(44) N. Teranishi and R. Kubo, (1979), J. Phys. Soc. Jpn 47 , 720
(45) K. Nakanishi, (1979), J. Phys. Soc. Jpn 46, 1434
(46) T.M. Rice, S. Whitehouse and P. Littlewood, (1981), Phys. Rev. B 24, 2751
(47) P.B. Littlewood and T.M. Rice, (1982), Phys. Rev. Lett. 48, 44
(48) T. Natterman, (1983), J. Phys. C 16, 4113
(49) J. Villain, (1982), J. Physique Lettres 43, L 551
(50) A.H. Moudden, D.E. Moncton and J.D. Axe (1983), Phys. Rev. Lett. 51, 2390
(51) H. Mashiyama, S. Tanisaki and K. Hamano, (1982), J. Phys. Soc. Jpn 51, 2538
(52) G. André, D. Durand, F. Dénoyer, R. Currat and F. Moussa, (1987) Phys. Rev. B 35, 2909
(53) E. Fjaer, R.A. Cowley and T.W. Ryan, (1985) J. Phys. C 18, L 41
(54) M. Bziouet, R. Almairac, P. Saint-Grégoire, to be published in J. Phys. C
(55) J.P. Jamet and P. Lederer, (1983), J. Physique Lettres 44, L 257
(56) P. Lederer, G. Montambaux, J.P. Jamet, M. Chauvin, (1984), J. Physique Lettres 45, L 627
(57) P. Lederer, J.P. Jamet, G. Montambaux, (1986), Ferroel. 66, 25
(58) G. Errandonea, J.C. Toledano, A. Litzler, H. Savary, J. Schneck, J. Aubree, (1984), J. Physique Lettres 45, L 329
(59) G. Dolino, Japanese J. of Applied Physics, (1985), 24, Suppl.24-2, 153
(60) D. Schechtman, I. Blech, D. Gratias and J.W. Cahn, (1984), Phys. Rev. Lett. 53, 195
(61) T. Janssen, (1986), J. de Physique, 47, C3-85
(62) P. Sainfort, B. Dubost, A. Dubus, C.R.A.S. Paris, (1985) 301, série II, 689
B. Dubost, J.M. Lang, M. Tanaka, P. Sainfort and M. Audier, (1986), Nature, 324, 48
P. Sainfort and B. Dubost, (1986), J. de Physique, 47, C3-321
(63) J.M. Lang, M. Audier, B. Dubost and P. Sainfort, Journal of crystal growth, in press
(64) F. Dénoyer, G. Heger, M. Lambert and J.M. Lang C.R.A.S. Paris, (1987), 304, Série II, 625
(65) F. Dénoyer, G. Heger, M. Lambert, J.M. Lang and P. Sainfort, J. de Physique (1987), 48 (in press)
(66) J.W. Cahn, D. Schechtman and D. Gratias, (1986), J. of Material research, 1, 13
(67) A. Katz and M. Duneau, (1986), J. de Physique, 47, C3-103
(68) R. Penrose, (1979), Math. Intelligencer, 2, 32
(69) A.L. Mackay, (1982), Physica A, 114, 609
(70) M. Kleman and A. Pavlovitch, (1986), J. de Physique 47, C3-229

STRUCTURAL PHASE TRANSITIONS AND SPECIFIC HEAT ANOMALIES

Ekhard Salje
Department of Earth Sciences
University of Cambridge
Cambridge CB2 3EQ, England

ABSTRACT. Precise measurements of the excess specific heat associated with structural phase transitions lead to determinations of the temperature evolution of thermodynamic order parameters. Experimental results are discussed within the scope of Landau-Ginzburg theory. The limitations of this theory are described, together with a survey of possible experimental methods for verifying the scaling relations based on specific heat measurements. Emphasis is placed on the discussion of possible influence of defects and secondary order parameters on the phase transition.

1. INTRODUCTION

The consideration of thermodynamic properties as being quantitative rather than qualitative quantities has led in the field of Earth Sciences to the conclusion that phase equilibria between minerals depend sensitively on their structural states. Changes of these structural states are often related to phase transitions which dramatically influence the macroscopic physical properties of the mineral (e.g. its elasticity, enthalpy, ionic conductivity etc.). If we consider common minerals like calcite or feldspar, we find that structural phase transitions significantly effect the enthalpy of the crystal as a function of temperature, pressure or chemical composition. The correlated specific heat of anorthite in Fig. 2 shows a typical anomaly associated with the phase transition at 510.4 K between a phase with space group $I\bar{1}$ and the low temperature form (SG = $P\bar{1}$). A formal description of the excess entropy:

$$\Delta S = \int_{T_1}^{T} \frac{\Delta C_p(\tau)}{\tau}\, d\tau \qquad (1.1)$$

becomes possible if the thermodynamic order parameter of the system is known. Here, ΔCp is the excess specific heat, i.e. that part of the

E. K. H. Salje (ed.), Physical Properties and Thermodynamic Behaviour of Minerals, 75–118.

specific heat which is due to the structural phase transition. Such an order parameter exists for virtually all phase transitions and much theoretical effort has been spent over the years on attempts to understand the dependence of this order parameter on temperature, pressure and chemical poptential (e.g. Landau and Lifshitz, 1979; Bruce and Cowley, 1981). An important conclusion has been that this dependence is mostly independent of the details of the transition mechanism. Structural aspects come into play when we wish to know the geometrical meaning of the order parameter Q, however. The first task is always to determine experimentally the order parameter behaviour, which is in many cases possible from entropy measurements. In the most simple approximation, the relationship between the order parameter Q and the excess entropy ΔS is

$$\Delta S \propto Q^2 \tag{1.2}$$

Furthermore, it is possible to express the excess Gibbs energy of the phase transition within the scope of Landau-Ginzburg theory in terms of the order parameter:

$$\Delta\phi(T,P,\mu,Q) = G(\nabla Q)^2 - HQ + AQ^2 + BQ^4 + CQ^6,\ A = \alpha t \tag{1.3}$$

where $t = (T - T_c)/T_c$ is the reduced temperature, T_c is the Curie temperature and H is the conjugated field of the order parameter. The correlated excess entropy is then in lowest order:

$$\Delta S = -\frac{\partial \Delta\phi}{\partial T} = -\frac{\alpha}{T_c} Q^2 \tag{1.4}$$

in agreement with our basic assumption in equation 1.2. The excess specific heat follows from:

$$\Delta C_p = \left.\frac{\partial \Delta S}{\partial T}\right|_p \tag{1.5}$$

At temperatures close to the transition temperature of a continuous phase transition, fluctuations of the order parameter occur which lead to extra contributions of the specific heat as:

$$\delta\Delta C_p \sim \int \frac{V}{T} \left\langle Q_k Q_{-k} \right\rangle^2 dk \tag{1.6}$$

where V is the volume and Q_k is the k-th component of the space dependent order parameter Q(r). The bracket $\langle\ \rangle$ indicates statistical averaging. The temperature dependence of C_p near T_c allows, in principle, for the direct measurement of the integrated order parameter fluctuations and provides an invaluable tool especially for the investigation of magnetic phase transitions. Moreover, remembering that the introduction of critical indices was largely motivated by the experimental observation of stronger singularities in C_p at T_c than expected from Landau theory, namely

$$\Delta C_p \propto |t|^{-\alpha} \tag{1.7}$$

the central interest in investigating the behaviour of C_p realted to structural phase transitions becomes obvious. Fluctuations also lead to logarithmic correction factors displayed in the asymptolic critical behaviour, e.g. in uniaxial dipolar systems (Meissner et al., 1987)

$$C \propto \left| \ell n |t| \right|^{\frac{1}{3}} \tag{1.8}$$

2. EXPERIMENTAL METHODS

Among numerous experimental methods available for the determination of thermodynamic properties of micro- or macro-samples, three main techniques are used for the investigation of heat capacities in relation to structural phase transitions. Each of these methods has its very specific advantages and limitations as will be briefly reviewed here.

2.1 Scanning Calorimetry

The basic principle of a differential scanning calorimeter (DSC) is based on the determination of the heat flow from a heat bath (normally a resistance heater) into the sample as compared to a reference sample. This is either done dynamically in the "heat flux" designs or using the "null balance" principle where the energy absorbed or evolved by the sample is compensated by adding or subtracting the equivalent heat to the heater which is located as close as possible to the sample holder. The temperature is thereby always kept identical to that of the reference. Heat flow away from the sample and the reference is ensured by a gas stream.

Technically, the basic requirements of an instrument working with high precision are good thermal stability over a long time period, low noise of the heat flux data (0.003mW) and reproducible, reasonably straight baselines of the instrument without sample. Experimentally, a very steady gas flow of a high purity gas and careful calibration of the temperature and energy scale are necessary for the determination of specific heats on an absolute scale. Useful standards for calibration are listed in Table 1. The calibration should be performed using at least three standards in the temperature range under investigation. The specific heats are then usually determined using high purite sapphire as standard (O'Neil, 1966). The C_p values of sapphire are listed by Binnings and Furukawa (1953) and Ditmars and Douglas (1971). In general, several measurements must be made for overlapping temperature ranges in order to reduce the standard deviations. The individual temperature intervals should not be too large if the baseline is not fully linear. The accuracy of such measurements can be 0.1% and the precision of the individual data point is generally 1%.

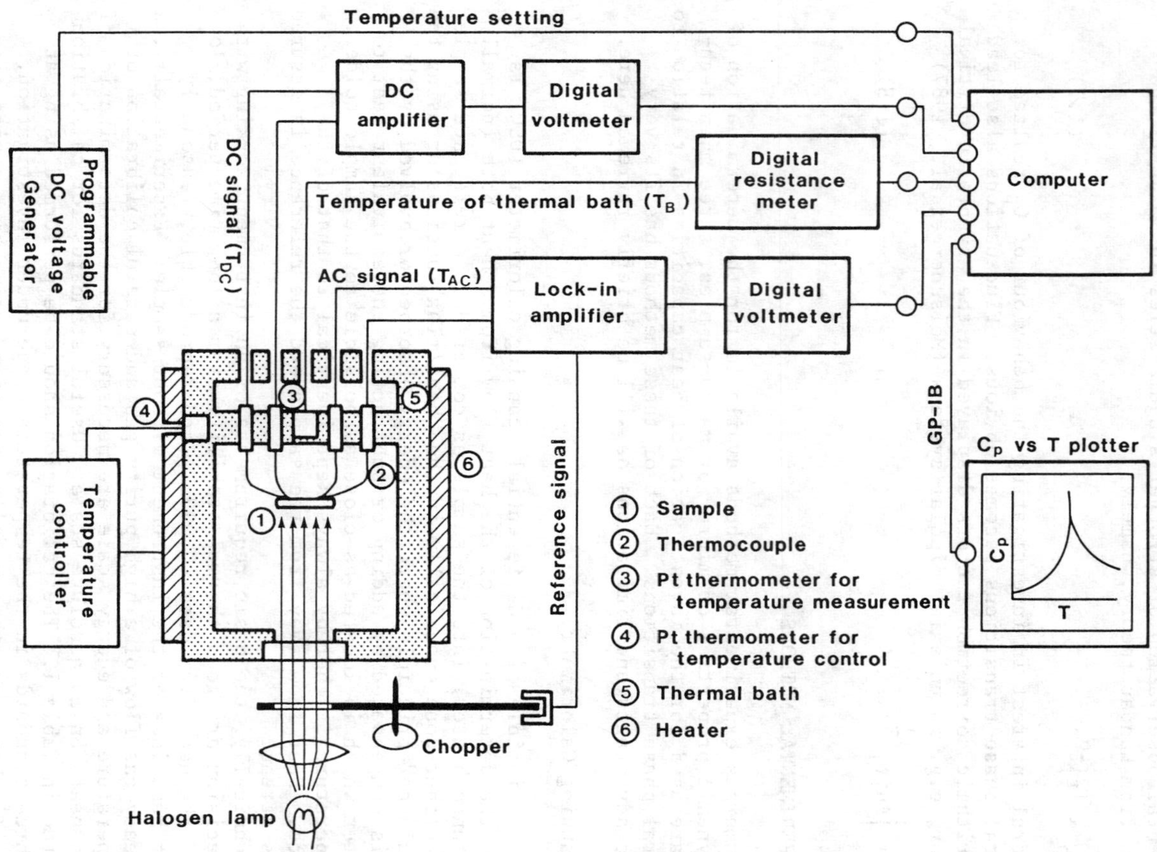

Figure 1. Schematic diagram representing the principle of the AC calorimetric technique (after Hatta and Ikushima, 1981).

The standard deviation after repeated determinations of the specific heat, as calibrated with respect to sapphire, can reach the 0.1% level.

Table 1

Standard	Transition temp. (^{o}C)
Cyclopentane	-151.16
	-135.06
	- 87.06
n-Heptane	- 90.56
Indium	156.60
Tin	231.88
Lead	327.47
Zinc	419.47

The important advantages of the DSC method are, firstly, the possibilities for determining absolute values of the specific heat over a rather large temperature range from ca. -100K to ca. 800K, with standard deviations which are below 0.2%. Secondly, very little sample material is needed (say 10 to 40mg for precise measurements). The disadvantages of this method are that the precision drops at low and high temperatures, and that the precision in temperature is in general not better than 0.1K. Even relative measurements for smaller temperature intervals are difficult to perform. This limits the application of the DSC methods to the investigation of, for example, the critical behaviour of C_p at temperatures very close to T_c.

2.2 AC Calorimetry

This experimental technique was, in the beginning, designed to measure the heat capacities of metals at high temperatures (Kraftmakher, 1962). The basic requirement for the experimental set-up was that, to ensure adiabatic conditions, the thermal response of the system should be of a certain speed in comparison with the thermal relaxation from the sample to the heat bath. This basic idea was extended in modern AC calorimeters to a wider temperature range. The heat source is in general either chopped light or Joule heaters attached directly to the sample. The temperature change in the sample is measured with thermocouples and phase sensitive lock-in amplifiers. A schematic diagram of an AC calorimeter as developed by Hatta and Ikushima (1981) is shown in Fig. 1. At the sample cell, light from a halogen lamp, chopped at a frequency f, irradiates a sample platelet. The sample itself is held at a DC temperature. Following the light induced heat flow, the actual sample temperature varies around the DC temperature by T_{AC} which is inversely proportional to the heat capacity of the sample. The heat exchange between the sample plate and the thermal bath, which is normally a copper block, is achieved by a gas filling the sample holder. Three different sensors are used for temperature measurements of the thermal bath, the temperature difference between the bath and the specimen (DC) and the AC component.

The condition that the experiment is performed under adiabatic conditions requires higher frequencies of the chopped light than the reciprocal dissipation time τ of the heat flow from the sample to the thermal bath.

$$f \gg 1/(2\pi\tau) \tag{2.1}$$

On the other hand, thermal relaxation within the sample must be faster than the oscillation time of the heat source. This means that the wavelength of the temperature wave (Carslaw and Jaeger, 1959), defined as the thermal diffusion length (Rosencwaig and Gersho, 1976), must be larger than the thickness of the sample L. More strictly, the frequency f should be smaller than 90 D/(2πL), where D is the thermal diffusivity of the material under investigation (Sullivan and Seidel, 1968). Experimental procedures to check for these two conditions for the applied frequency of the chopped light are discussed by Hatta and Ikushima (1981). Typical chopping frequencies are in the range from 1 Hz to 10 Hz for thin samples.

The advantage of AC calorimetry is related to the small size of the sample (e.g. 0.1 x 1 x 1mm) and the small variation of the temperature in the sample due to the AC heat flow. This allows experiments to be carried out at temperatures very close to the phase transition point so that the critical behaviour at reduced temperature as small as $t = 10^{-4}$ can be studied. Two disadvantages, on the other hand, limit the application of the AC method for some investigations. Firstly, since only the AC response of the system is detected, the influence of endothermic or exothermic processes at first order phase transitions or chemical reactions, polymorphic transformations etc. can not be measured. Secondly, determination of the absolute specific heat is by no means simple because the absorption cross-section of the sample is in general unknown as is the possible influence of the large addends attached to the sample.

An important extension of AC calorimetry for the investigation of structural phase transitions under pressure was described by Baloga and Garland (1978). Slabs of crystals (1 x 1 x 0.05cm) with a thin epoxy resistance heater on one side and a microbed thermister on the opposite side were mounted in a gas pressure vessel. Experimental results on NH_4Cl and ND_4Cl were obtained with an AC frequency of f = 0.0325 Hz, with an experimental accuracy of 0.5 - 1% (Garland and Baloga, 1977). The maximum gas pressure was 3kbar.

2.3 Adiabatic Calorimetry

Although the adiabatic calorimetry is presumably the best established experimental method for the determination of specific heats, especially at low temperatures, its application for the investigation of excess specific heats due to phase transitions met two major difficulties in the past. Firstly, the temperature resolution close to T_c must be good enough to follow critical phenomena. Secondly, the time needed by a large sample to reach thermal equilibrium can be very long for temperature close to T_c and high thermal stability is required for the calorimeter. In the last decade, various experimental methods were developed and tested and two of them were particularly successful for the investigation of phase transitions. In the case of the "NERST - METHOD" the calorimeter consists essentially of a sandwich sample holder attached to heaters which are mounted in a single or double shielding within a vacuum chamber. A typical instrument is that developed by Gmelin and Roedhammer (1980) which is entirely constructed from copper with a vacuum chamber which is sealed by indium wire. The thermosensors are calibrated germanium and platinum resistors attached to the sample holder. The specimen holder and the specimen are attached by a nylon thread and are suspended from the isothermal shield by cotton threads. This calorimeter works in the temperature range from 0.3 to 320K. During the experiment, the sample is heated stepwise. The specific heat is determined upon division of the applied energy ΔP by the increase in temperature ΔT. The energy supplied by the electric heater is determined by measuring heating time, current and voltage drop across the heater resistance. To obtain the temperature increment ΔT, the temperature drift before and after heating is extrapolated to the midpoint of the heating time.

The principal set-up of an adiabatic "SCANNING-RATIO" calorimeter was developed by several groups (e.g. Buckingham et al. (1973); Thoen et al. ((1978); Lancaster and Barker (1981); Wuerz and Grubic (1980). In this method, the sample is placed in a vacuum insulated thermal enclosure and its temperature is controlled to equal that of the enclosure. The enclosure is then heated at a constant rate and the energy P needed to force the specimen to heat at the same rate T is measured. The specific heat is then given by the ratio

$$C = P/T \qquad (2.2)$$

Alternatively, the heating at the sample can be predetermined and the power needed to equilibrate the enclosure can be measured. The accuracy of the heat capacity measurements in an instrument of this kind depends on the heating rate. At low rates the principal limit is determined by the noise of the regulated temperature difference ΔT. At high scanning rates the accuracy is limited by the impossibility of correctly compensating for thermal losses due to temperature gradients. Water was used by Wuerz and Grubic (1980) as standard for the determination of the accuracy of their set-up. The empirical accuracy of their experiments was about 0.7% for scanning rates between 0.7 K/sec and 70 K/sec. Lanchaster and Baker (1981) report that their instrument gives reproducible results within about 0.1%, but due to the uncertainty in the heat leak correction, the absolute error is likely to be about 0.3%.

The adiabatic calorimetry thereby possesses about the same accuracy as the scanning calorimetry. Both methods differ mainly by the temperature range where they are normally applied, namely adiabatic calorimetry at low temperatures (e.g.<400K) and scanning calorimetry at higher temperatures (e.g. 220K to 900K). The disadvantages of the adiabatic calorimetry are the large amount of sample material required for precise experiments (often more than 10g) and the long time needed by the calorimeter to reach thermal equilibrium. However, there is hope that these problems can be overcome in future by new, computer-controlled instruments with higher temperature resolution.

3. REDUCTION OF BACKGROUND HEAT CAPACITIES

At temperatures well below T_c, the influence of the phase transition on the specific heat becomes very small indeed and even at temperatures close to T_c the excess specific heat for most displacive phase transitions contributes only a few percent to the total. The accuracy of the excess specific heat determination, therefore, depends a great deal on the correct reduction of the total specific heat by the "non-critical" or "background" specific heat, denoted in the following by C_{po}, with $\Delta C_p - C_{po}$. The background specific heat is mainly due to lattice vibrations which do not take part in the phase transition (hard modes). However, the distinction between hard modes and soft modes, is rather difficult. The transformation of the C_p data to C_v for correct calculation of the background specific heat from the phonon density of states, requires exact knowledge of the phonon dispersion. Approximative methods are discussed by Winterfeld and Schaak (1977) and Salje and Werneke (1982).

The direct determination of the background specific heat is generally carried out by extrapolation of the C_p data at temperatures well above the fluctuation regime towards lower temperatures. Three different extrapolation methods are in general use:

a). The background specific heat is compared with an empirical temperature dependence of the form

$$C_{po} = a + b\,T^{-1} + c\,T^{\frac{1}{2}} + d\,T + e\,T^{2} + f\,T^{-2} \qquad (3.1)$$

The parameters a, b, c, d, e and f are determined by a least square fit of the experimental data for $T \gg T_c$. Often, some of the parameters (e.g. c, d, e) can arbitrarily be set to zero depending on the convergency of the least square procedure.

b). The experimental data at $T \gg T_c$ are expressed as Einstein temperatures $\Theta_E(T)$ or Debye temperatures $\Theta_D(T)$. These functions are then extrapolated to lower temperatures using the polynomial fit as in a). This method is, in particular, useful for a check of the baseline obtained using method a).

c). Empirical phonon density of states $g(\omega)$ can be used to give a best fit with the C_p data for $T \gg T_c$ from

$$C_p(T)\,\alpha \int \left(\frac{h\omega}{kT}\right)^2 \frac{e^{\frac{h\omega}{kT}}}{\left(e^{\frac{h\omega}{kT}} - 1\right)^2}\, g(\omega)\, d\omega \qquad (3.2)$$

No further explicit temperature dependence of $g(\omega)$ is needed in the extrapolation and rather precise C_{po} data can be obtained over a wide temperature range. Note, however, that in equation (3.2) the difference between C_p and C_v is ignored and $g(\omega)$ is a purely empirical function which is chosen to fit the C_p values in the high temperature phase. This function is, therefore, close to, but not identical with, the vibrational phonon density of states.

All three extrapolation methods require fit procedures using an extended temperature interval at temperatures well above the fluctuation regime. If the temperature interval is too small or too close to T_c, systematic errors tend to increase C_{po} at lower temperatures and this leads to unrealistic values of the excess entropy in the asymmetric phase. It is very easy to underestimate the excess specific heat at temperatures slightly above T_c and extension of the data collection to higher temperatures is always to be recommended.

In phase transitions with mainly order/disorder character, the distribution functions in the low temperature phase can be highly restricted by symmetry conditions. The related excess entropies are then extremely small and the extrapolation of C_{po} can even lead to negative values of the excess specific heat. This situation was encountered, for example, by Matsuo et al. (1976) in $Cu(HCO_2)_2 \cdot 4H_2O$ where the 8 hydrogen atoms of the water molecules change their distribution spontaneously at T_c but always obey the Bernal-Fowler ice condition in the asymmetric phase - namely that ionic D- and L-faults are absent. There is, therefore, an infinite gap between the ground state and the excited fluctuation state at $T > T_c$ with $\Delta C_p = 0$. At $T > T_c$, fluctuations occur and C_p^+ becomes larger than C_p^-. The extrapolation methods would lead, in this case, to a negative excess specific heat. If the symmetry restriction for fluctuations in the asymmetric phase are removed, as in the structurally closely related compounds $K_2Fe(Cn_6)$ $3H_2O$ or $SnC\ell_2$ $2H_2O$, the excess specific heat increases.

Systematic investigations of the background specific heat become possible if the phase transition process is kinetically so sluggish that both phases coexist metastably over a larger temperature range. Typical examples for such behaviour are provided by silicates, when the ordering of $A\ell$ and Si are involved in phase transitions. In albite, $NaA\ell Si_3O_8$, the continuous phase transition $C2/m$ - $C\bar{1}$ takes place at 1,250K. The transition temperature decreases sharply when Na is replaced by K. The experimentally determined excess specific heat near T_c is 2,739 J/molK (Salje et al. 1985).

Useful constraints can be included in the fitting procedure if the temperature dependence of the order parameter is known for $T \ll T_c$. Since the excess entropy scales as the square of the order parameter, its own relative temperature dependence, and therefore the slope of C_p, serve as boundary conditions for the parameter refinement. In the case of classical mean field behaviour, for example, the linear temperature dependence of C_p, with $\alpha = 0$, is equivalent to the condition that the C_p and C_{po} curves are parallel over a limited temperature range. Further constraints follow from Landau theory and will be discussed later.

Finally, it is noteworthy that the determination of the latent heat

L in the case of discontinuous phase transitions also requires a knowledge of the background specific heat from

$$L = \lim_{\varepsilon\to 0} \int_{T_c-\varepsilon}^{T_c+\varepsilon} (C_p - C_{po})\, dT \qquad (3.3)$$

The common method of determining the area below the C_p curve using arbitrary cut-off temperatures - as provided by most DSC routines - may lead to extensive systematic errors.

4. APPLICATION OF LANDAU THEORY

Expanding the free energy in a Landau polynomial in powers of a one-component, one dimensional order parametner Q as

$$\Delta\phi = \phi - \phi_o = AQ^2 + BQ^4 + CQ^6 \qquad (4.1)$$

with $A = \alpha.t$, $C > 0$ and $B > 0$ or $B < 0$ depending on the order of the phase transition, makes the excess entropy simply

$$\Delta S = -\partial\Delta\phi/\partial T = \frac{\alpha}{T_c} Q^2 \qquad (4.2)$$

In a second order phase transition with C = 0, the excess specific heat is a linear function of temperature with a step at T_c of

$$\Delta C_p(T = T_c) = T_c\, \partial\Delta S/\partial T = \frac{\alpha^2}{2BT_c} \qquad (4.3)$$

In a discontinuous phase transition with $C \neq 0$ the excess entropy and the excess specific heat is

$$\Delta S = \frac{\alpha}{T_c} \frac{1}{3C} \left[-B + \sqrt{B^2 - 3AC}\right] \qquad (4.4)$$

$$\Delta C_p = \frac{\alpha^2 T}{T_c^2\, 2\sqrt{B^2 - 3AC}} \qquad (4.5)$$

If the first order phase transition is close to a tricritical point with small values of B, the difference between the temperature T_o at which $B^2 = 3AC$ and T_c

$$T_c - T_o = -\frac{B^2 T_c}{3\alpha C} \qquad (4.6)$$

tends to zero as the square of B. The singularity of C_p then becomes, in lowest order

$$\Delta C_p = T\left(\frac{\alpha^3}{12CT_c^3}\right)^{\frac{1}{2}} \frac{1}{\sqrt{T_o - T}} \tag{4.7}$$

Thus the specific heat increases near T_c of a tricritical phase transition or a "Gauss model" (Shang-Keng Ma, 1976) as

$$\Delta C_p \sim \frac{1}{\sqrt{|t|}} \quad , \ \alpha = 0.5 \tag{4.8}$$

The experimental data on $NaNO_3$ show $\alpha = 0.56$ for small values of t, the classical value of $\alpha = 0.5$ can be expected for lower temperatures. The phase transition $P\bar{1}$-$I\bar{1}$ in highly ordered anorthite (Val Pasmeda) in Fig. 2 is also tricritical with $\alpha = 0.5$ (Wruck, 1986). Helwig et al. (1977) observed $\alpha = 0.494$ in $AgNa(NO_2)_2$ and other singularities with $\Delta C_p \sim |t|^{-\frac{1}{2}}$ were observed near tricritical points by Sheng and Salomon (1980).

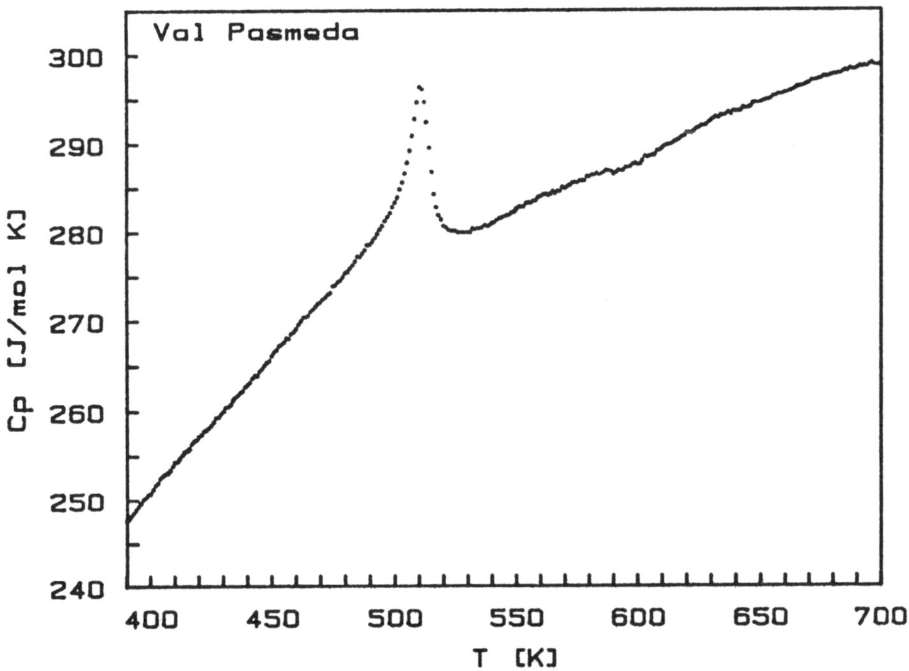

Figure 2. Specific heat anomaly of the $P\bar{1}$ - $I\bar{1}$ phase transition in highly ordered anorthite (Val Pasmeda).

A convenient representation of C_p in slightly first order phase transitions follow from equation (4.7) as

$$\left(\frac{T}{\Delta C_p}\right)^2 = \frac{4T_c^4(B^2 - 3\frac{\alpha C}{T_c}(T_o - T_c))}{\alpha^4} + \frac{12CT_c^3}{\alpha^3}(T_o - T) \qquad (4.9)$$

which is a linear function of temperature. The first term gives the discontinuity of the specific heat at the transition temperature T_o namely

$$\Delta C_p\Big|_{T_o} = \frac{T_o\alpha^2}{2T_c^2}(B^2 - 3\frac{\alpha C}{T_c}(T_o - T_c))^{-\frac{1}{2}} \qquad (4.10)$$

From the equations (4.9) and (4.10), the relative magnitude of the parameters A, B and C in the Landau potential can be determined. In a second order phase transition, the last term in equation (4.9) vanishes and the first reduced to $4B^2T_c^4/\alpha^4$ as in equation (4.3). The closeness of a phase transition to the tricritical point or to a second order phase transition can therefore be estimated from the $(T/\Delta C_p)^2$ versus T plots (e.g. 4.9). Experimental data are discussed by Aleksandrov and Flerov (1979). The development of excess specific heat, without consideration of the additional contributions of fluctuations, is therefore expected to be a smooth function of B when the order of the phase transition is changed. A single step is predicted from Landau theory for a continuous phase transition, a $t^{-\frac{1}{2}}$ singularity at the tricritical point and a discontinuity for first order phase transitions. Although the Landau potential is physically unrealistic near a transition line of the second order, it is noteworthy that the conditions for this approach to be valid are more easily satisfied as the tricritical point is approached and only logarithmic corrections are expected for small values of B.

The most universal way to determine the free parameters in the Landau potential follows from equation (4.4) which can be rewritten as

$$\Delta S = \frac{2}{3}\frac{L}{T_{trans}}\left\{1 + \left[1 - \frac{3}{4}\frac{T - T_c}{T_{trans} - T_c}\right]^{\frac{1}{2}}\right\}$$

or

$$\frac{4}{3}\left[-\left(\frac{\Delta S}{2L/3T_{trans}} - 1\right)^2 + 1\right] = \frac{T - T_c}{T_{trans} - T_c} \qquad (4.11)$$

T_{trans} is the temperature at which both phases are in thermodynamic equilibrium. The latent heat L is determined as described before. If the expression on the left hand side of equation (4.11) is plotted against temperature, the intersection with the temperature scale defines T_c. The slope of the curve is $(T_{trans} - T_c)^{-1}$ which allows the determination of T_{trans}.

An alternative Landau polynomial with second-, third- and fourth-order terms (Bruce and Cowley, 1981) gives the following expression for the excess entropy

$$\Delta S = \frac{9}{8}\left(\frac{L}{T_{trans}}\right)\left\{1 + \left[1 - \frac{8}{9}\left(\frac{T - T_c}{T_{trans} - T_c}\right)\right]^{\frac{1}{2}}\right\} \tag{4.12}$$

which can be fitted to the experimental data points for the determination of T_c and T_{trans}. Differences in the excess entropies expressed by equations (4.11) and (4.12) therefore allow the distinction between first order phase transitions with negative fourth order terms and those involving symmetry breaking third order terms. A typical example is the $P2_13$-$P2_12_12_1$ symmetry reduction in $K_2Cd_2(SO_4)_3$ where third order terms are allowed from symmetry (Dvorak, 1972). Heat capacity measurements by Devarajan and Salje (1984), however, demonstrated that their numerical values are nonetheless small and the first order character of the phase transition results from a large, negative fourth order term.

5. FLUCTUATIONS OF THE ORDER PARAMETER AND CRITICAL EXPONENTS

For an inhomogeneous body with strong time- and space-dependent order parameters, the thermodynamical potential as a macroscopic quantity is defined as

$$\langle\phi\rangle = \frac{1}{3}\iint\phi(t,\vec{r})dt\ dr^3 \tag{5.1}$$

According to Landau-Ginzburg theory, the potential ϕ (P,T,Q) is replaced by $\Omega(\mu,T,Q)$ with the chemical potential μ. The partition function of is then

$$\Omega = \Omega_o - k_BT\ell n\int e^{\frac{-H_{eff}}{k_BT_c}}\ \Pi\ V\ dQ_k'\ dQ_k'' \tag{5.2}$$

where

$$H_{eff} = \int dV\left[\alpha tQ^2 + BQ^4 + CQ^6 + G(\nabla Q)^2 - HQ\right] \tag{5.3}$$

acts as the effective Hamiltonian of the system undergoing the structural phase transition. The parameter Q_k is the k-th component of the spatial Fourier expansion of Q. In the range where the "homogeneous" Landau theory is applicable, the fluctuations are small. This reduces the integral in equation (5.3) to values of $\langle Q\rangle$ which minimize the effective Hamiltonian itself and leads to the classical Landau potential. Close to T_c, however, significant fluctuations with long wavelengths occur and Fourier components with small k-values become important. The temperature interval of the fluctuation regime depends sensitively on the "interaction length" L_c, which characterizes the range of interaction causing the structural phase transition. The Ginzburg criterion relates the interaction length, the temperature interval ΔT of the fluctuation regime and the excess specific heat ΔC_p which should be expected if the Landau theory were valid as follows:

$$\Delta T = \frac{k_B^2 T_c}{32\pi^2(\rho_m \Delta C_p)^2 L_c{}^6} \tag{5.4}$$

where ρ_m is the molar density.

It seems likely that the interaction length ranges from one interatomic distance up to several lattice constants, the corresponding ΔT would then vary between some tens of degrees down to vanishing by small intervals. Equation (5.4) can, on the other hand, be used for the estimation of the interaction length if ΔC_p and the temperature interval in which additional fluctuation induced excess specific heats appear are known from calorimetry.

For vanishing conjugate field H and $t > 0$ the averaged order parameter is zero and the effective Hamiltonian becomes, in lowest order,

$$\langle H_{eff} \rangle = V \sum_{k<k_o} (\alpha t + Gk^2)|Q_k|^2 \tag{5.5}$$

The quadratic dependence of H_{eff} on Q_k implies a Gaussian distribution in the integral (5.3). For such integrals the mean value of the product of several Q_k's is identical to the sum of products of pairs of mean values of the correlation functions of the fluctuations in k-space. As Lenanyuk (1963) has shown, the integral in equation (5.3) using the effective Hamiltonian in (5.5) gives

$$\Omega = \Omega_o - k_B T_c V \sum_{k<k_o} \ell n \frac{\pi T}{(\alpha t + Gk^2)^2} = \Omega_o + k_B T_c V \int_0^{k_o} \ell n \frac{\alpha t + Gk^2}{\pi T} \frac{2\pi k^2}{(2\pi)^3} dk \tag{5.6}$$

Under the assumption that this correction holds equally for ϕ, the resulting correction for the specific heat becomes

$$\partial C_p = C_p - C_p^O - C_p^Q = \frac{V\alpha^2}{4\pi^2} \int_0^\infty \frac{k^2 dk}{(\alpha t + Gk^2)^2} = \frac{V\alpha^{\frac{3}{2}}}{16\pi G^{\frac{3}{2}}} \frac{1}{\sqrt{t}} \tag{5.7}$$

This power law is identical for tricritical behaviour (equation 4.6) and, as will be shown later, is also identical to certain defect induced excess specific heats.

The isotropic Hamiltonian (5.5) is considerably changed if fluctuations are accompanied by long range fields (electric, magnetic, elastic). As a result, not all fluctuation waves increase in sufficiently anisotropic crystals near T_c and this leads to a strong decrease of the phase volume in equation (5.6). The resulting critical exponents are then reduced.

A typical example is the behaviour of C_p near T_c if bilinear coupling between Q and the local elastic strain e is allowed. Starting from the effective Hamiltonian of the form

$$H_{eff} = \alpha Q^2 + BQ^4 + G(\nabla Q)^2 + \sigma Qe + H_{e\ell} \tag{5.8}$$

Levanyuk and Sobyanin (1970) found the resulting pair correlation function

$$<Q_k\ Q_{-k}> = \frac{T}{V\tilde{\alpha}(k)} \tag{5.9}$$

where $\tilde{\alpha}(k)$ is the renormalized coefficient of the quadratic term in the effective Hamiltonian (Levanyuk and Sobyanin, 1970):

$$\tilde{\alpha}(k) = 2\alpha T_c + Gk^2 - \sigma^2\left(\frac{1}{4\mu} - \frac{1}{4\mu}\ (\cos^2\theta + \frac{\lambda + \mu}{\lambda + 2\mu}\ \sin^4\theta \sin^2 2\phi)\right) \tag{5.10}$$

in the case of an elastically isotropic material. The first two terms are identical to the isotropic, uncoupled Hamiltonian, the last term shows from the elastic energy $H_{e\ell}$ and acts as damping. The specific heat correction due to these fluctuations becomes

$$\delta C_p = C_p - C_p^0 - C_p^Q \propto \int \frac{dk}{\tilde{\alpha}^2(k)} \tag{5.11}$$

which obviously converges for $t \to 0$. Only the first derivative with respect to t becomes singular, with

$$\frac{\partial \delta C_p}{\partial T} \propto t^{-\frac{1}{2}} \tag{5.12}$$

The singularity in C_p is accompanied by cusps in C_v according to

$$C_v \propto C_{vo} - X/C_p, \quad t \to 0 \tag{5.13}$$

with constants C_{vo} and X. If C_p increases as the power law $|t|^{-\alpha}$, only $\partial C_v/\partial T$ becomes singular with

$$\partial C_v/\partial T \propto |t|^{\alpha-1} \tag{5.14}$$

Much effort has been expended in the experimental verification of scaling and tests of the universality hypothesis using specific heat data. The most commonly adapted system for such experiments in crystals appears to be Heisenberg ferromagnets, but dipole dominated magnetic systems have also been investigated since the first publication of Ledermann et al. in 1974. In order to extract the parameters which describe ΔC_p in the limit $t \to 0$, the experimental data must have been background reduced and then be fitted to a function of the type

$$\Delta C_p = \frac{A}{\alpha} |t|^{-\alpha} \left[1 + g(t)\right] + B \tag{5.15}$$

The correction g(t) vanishes as t goes to zero. When $\alpha < 0$, C_p is finite at the transition point with a maximum value of B. From the condition that C_p is continuous at T_c it follows that B is identical above (+) and below (-) T_c:

$$B^+ = B^- = B \tag{5.16}$$

Even when $\alpha > 0$, this condition should hold from renormalization theory (Bruce and Aharony, 1974; Wagner, 1972; Swift and Grover, 1974), although for $\alpha = 0$ the result is in strict contradiction to Landau theory where a jump of ΔC_p by B^- is predicted. The inverse conclusion, namely that a violation of the condition (5.16) always implies vanishing (or small) α seems to be unproven but widely accepted.

In addition to the constraint in (5.16) scaling theory predicts the critical exponents to be equal below and above the transition point (Ahlers and Kornblitz, 1975). Experimental data, analysed under these constraints, were obtained on several magnets and some of the resulting parameters are listed in Table 2. The systematic changes of α are

Table 2: Experimental critical exponents α and amplitude ratios A^-/A^+ of some magnetic compounds.

Compound	α	A^-/A^+	Reference
EuO	-0.044	1.22	Ahlers & Kornblit, 1975
Ni	-0.089	1.38	" " " "
Fe	-0.096	1.30	" " " "
	-0.04	-	Aldret & Kouvel, 1977
RbMnFe	-0.137	1.46	Ahlers & Kornblit, 1975
$EuS_{0\cdot5}Se_{0\cdot5}$	-0.15	1.22	Westerholt & Sobotta, 1983
$Eu_{0.9}Sr_{0.1}S_{0.5}Se_{0.5}$	-0.31	1.24	" " " "
$Eu_{0.8}Sr_{0.2}S_{0.5}Se_{0.5}$	-0.72	1.00	" " " "
$Fe_{30}Ni_{45}P_{16}B_{6}A\ell_{3}$	-0.36	-	Beckmann et al., 1981

presumably due to the superposition of the short range behaviour of antiferromagnets like $RbMnF_3$ by long range dipolar forces in the case of isotropic ferromagnetics with low T_c such as EuO (Fisher and Aharony, 1973). In fact, the first four compounds in Table 2 show a systematic decrease of α with decreasing dipolar contribution in the order EuO-Ni-Fe-$RbMnF_3$ as first stated by Ahlers and Kornblit, (1975). Still smaller critical exponents are expected in diluted magnetic systems. Sobotta (1982) showed that within the framework of an ε-expansion renormalization group theory, in the limit of high dilution, the α-value approached the Fisher-renormalized tricritical exponent (Fisher, 1968) namely $\alpha = -1$. This value also coincides with the results of the three-dimensional spherical model of Berlin and Kac (1952). Experimentally, such behaviour was found in the system $Eu_x x Sr_{1-x} S_{0.5} Se_{0.5}$ where Westerholt and Sobotta (1983) fitted the experimental data by the expression (5.14) with g(t) = 0. The results are inclined in Table 2 and show the dramatic decrease of the α toward -1 with increasing dilution (increasing Sr content).

6. THE INFLUENCE OF LATTICE IMPERFECTIONS

Very few systematic investigations of the influence of lattice imperfections on the excess specific heat have been performed. In the development of the theory, most attention has been paid so far to renormalization of critical exponents and to smearing of the critical singularities due to defects (e.g. Lubensky, 1975; Imry and Ma, 1975). These effects take place when the correlation radius r_c of the order parameter exceeds the average distance between the defects which is in general true only for a very limited temperature range around T_c. However, even in a wider temperature range, when defects can be regarded as weakly interacting with Q, corrections to the excess specific heat and the excess entropy can be large (Levenyuk et al. 1979, 1980). Starting from the idea that the order parameter variation close to the lattice imperfection follows the Ornstein-Zernike function

$$Q = Q_c \frac{1}{r} e^{-r/r_c} \tag{6.1}$$

with correlation length r_c of Q, this homogeneous modulation of the crystal can be looked upon as induced by a hypothetical conjugate field $H_d(r)$ with

$$H_{eff} = \alpha t Q^2 + BQ^4 + \ldots\ldots + D(\nabla Q)^2 - H_d(r)Q(r) \tag{6.2}$$

The renormalization of the quadratic term due to the imperfections, e.g. variations of the chemical composition c, becomes

$$\alpha = \alpha_o + \alpha_1\ c(r) \tag{6.3}$$

which is similar to the effect of bilinear strain coupling with the fluctuations in equation (5.9). The influence of the lattice imperfections, namely the kind of imperfections, their concentration and their spatial distribution in the crystal, are described by c(r) in equation (6.3). It must be assumed that c(r) is itself a function of the order parameter Q. The reason for this dependence is due to the observation that the solubility of defects in the high-symmetric and low-symmetric phases is generally different and, moreover, in the low-symmetric phase depends on the actual value of the lattices distortion, degree of order etc.. This dependence may lead to sluggish exsolution processes, especially at temperatures near T_c when defects tend to reequilibrate for any variation of Q. We may expand the defect function into powers of Q as

$$c(r) = C_o(r) + C_1(r)Q + C_2(r)Q^2 \tag{6.4}$$

It is important to note that the recalibration of the order parameter and the lattice imperfection may act on very different timescales. In displacive phase transitions, the variations of Q with temperature are fast in comparison with the time constants of heat flow in a calorimetric experiment. However, such a crystal is not in thermodynamic equilibrium until the defects are allowed to migrate and thus reduce the total free energy, which includes the contributions of the defects. This process is similar to annealing and may require long time constants.

There are very few theoretical predicitons concerning the analytical form of the defect contributions to the excess specific heat after the first pioneering work of Varma and Halprin (1976). Levanyuk et al. (1979) showed that in the case of spatially fixed, randomly distributed defects the C_p-corrections can be calculated within the scope of Landau theory yielding:

$$\delta\Delta C_p^{defect} = \pi N\, r_c^3 \left(\frac{Q_o}{Q^*}\right)^2 \Delta C_p^o \qquad t > 0 \qquad (6.5)$$

$$\delta\Delta C_p^{defect} = \left[1 + 4\pi N \left\{ r_c^3 \left(\frac{Q_o}{Q^*}\right)^2 \left[1 - 2\frac{Q_\infty}{Q_o}\frac{r_c}{d} - 3\frac{Q_\infty}{Q_o}^2\right] + 4d^3\left[\ell n\frac{r_c}{d} - 1.5 + \frac{DC}{2B^3d^2}\right]\right\}\right] \Delta C_p^o \qquad t < 0 \qquad (6.6)$$

ΔC_p^o is the jump of C_p according to Landau theory, r_c is again the correlation radius, Q_o is the defect induced value of Q near the defect position, Q^* is the maximum value of Q including fluctuations $Q^* = \sqrt{(D/Bd^2)}$, and d is the diameter of the defect and N is the defect concentration. Q_∞ is the solution of the undisturbed effective Hamiltonian including the field correction. The strength of the defect can be expressed by the ratio Q_o/Q^*. For strong defects with $Q_o/Q^* \gtrsim 1$ and $Q_o \gg Q$, the expressions (6.5) and (6.6) simplify to

$$\delta\Delta C_p^{defect} \sim r_c^3 \qquad (6.7)$$

If the critical radius depends on temperature as $r_c \sim t^{-\frac{1}{2}}$, the excess specific heat shows a singularity at T_c with

$$\delta\Delta C_p^{defect} \sim |t|^{-3/2} \qquad (6.8)$$

This temperature dependence superimposes the influences of fluctuations and may grossly modify the critical exponent of the "ideal" crystal.

For weak defects, the equations (6.5) and (6.6) describe smeared phase transitions as sketched in Figure 3. The actual critical behaviour in a crystal with lattice imperfections depends critically on the type of imperfections. For non-mobile, frozen-in defects with a statistical distribution, Levanyuk et al. (1979) found

$$\delta\Delta C_p^{defect} \propto |t|^{\nu-1} \qquad Q_o \neq 0 \qquad (6.9)$$

$$\delta\Delta C_p^{defect} \propto |t|^{2\beta-2} \qquad Q_o = 0 \qquad (6.10)$$

For unpolarized point defects, the excess specific heat

$$\delta\Delta C_p^{defect} \propto |t|^{-\frac{1}{2}} \qquad (6.11)$$

shows the same temperature dependence as for fluctuations in the limit of equation (5.6) and the tricritical behaviour in equation (4.6).

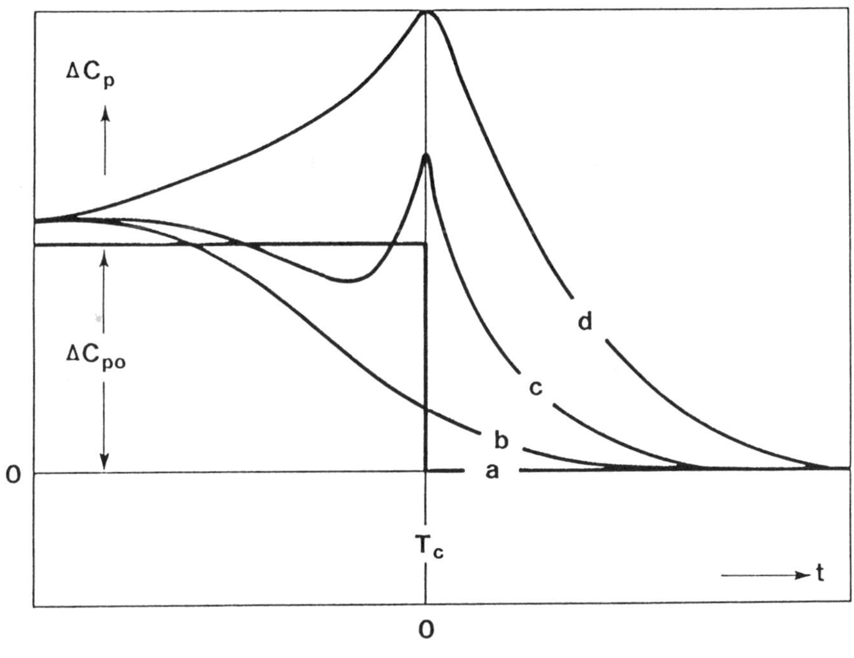

Figure 3. The anticipated influence of defects on the excess specific heat in the region of applicability of the Landau theory of a second order phase transition a) in the absence of defects, and with frozen-in random defects. The shape of ΔC_p depends on the interactions strength; b) weak interactions cause a smearing of the phase transition; c) intermediate interactions with an additional maximum at T_c and d) strong interactions leading to a λ-shaped excess specific heat (after Levanyuk et al., 1979).

In a systematic experimental approach, Strukov et al. (1980) showed that the time hysteresis in c(r) and the direct influence of defects on the excess specific heat of TGS crystals tend to follow the predicted pattern. The form of the C_p anomaly near $T_c = 322.18K$ depends heavily on the crystal history. The "purest" crystals are those grown above T_c but annealed below T_c. Equilibrium distribution of the defects can be assumed in this case and the anomaly in C_p is stepwise and rather sharp. On the other hand, crystals annealed above T_c showed, during their first cooling across T_c, a λ-shaped singularity at slightly reduced maximum temperatures, i.e. in the asymmetric phase if the crystal were in thermodynamical equilibrium. The onset of the increase of C_p in the high symmetric phase is identical for both experiments.

Additional defects were induced by γ-irradiation. For small doses, there is an increase of the specific heat and again a shift of T_c towards lower temperatures. If these curves were interpreted as being due to critical fluctuations, reasonable curve fittings using the expression (5.14) would have been possible with only slightly "exotic" critical exponents α. This example demonstrates clearly the danger of interpreting peaks in the C_p curves at T_c invariably as fluctuation-induced critical phenomena without further consideration of the possible influence of lattice imperfections. Misinterpretations of this kind are even more likely to occur because the power law singularities of C_p that are due to defect structures are expected only for low defect concentrations, whereas high defect concentrations are more likely to smear the phase transition.

A typical example for such behaviour is KH_2PO_4 where the C_p anomaly has been known since 1969 (Reese) and interpreted in terms of a power law singularity with $\alpha = 0.5$ for $t < 0$. At temperatures above T_c, a logarithmic singularity was analysed by Reese (1969). However, Courtens (see Scott et al. 1982) showed that this anomaly can be removes entirely by proper sample preparation and annealing and is therefore dur to lattice imperfections. Another prominant example of the influence of defects on specific heat measurements is the behaviour of $BaMnF_3$ where Scott et al. (1982) found a mean field discontinuity of 0.4 cal/mol K at $T_c = 255.3K$. A power law divergency of 6.4 ± 0.5 cal/mol is superimposed on top of this discontinuity and is entirely related to defects. The power law singularity satisfies the predictions of the defect dominated dynamical theory of Levanyuk et al. (1980) and was further confirmed by comparison of ultrasonic attenuation and soft mode linewidth divergencies with its predictions. The influence of dislocations has been investigated by Dubrovskii and Krivoglaz (1979).

The close similarity between critical scaling and defect induced singularities in C_p, both described by power laws with similar exponents were found previously (e.g. in K_2SeO_4, Lopez Echarri et al. 1980). However, the extreme sensitivity of C_p to defect structure can also be a starting point for systematic investigations of defect structures themselves. The pre-eminence of the heat capacity as a sensitive detector and indicator of dopant level impurities has been demonstrated for the Verwey transition in Fe_3O_4 by Bartel and Westrum (1976). A sharp transition peak appears for pure magnetite with a small high temperature shoulder. Both singularities depend sensitively on dopants, such as Zn, Cd and Mn, and show a large splitting of the two critical temperatures for $Zn_{0.005}Fe_{2.995}O_4$. The attribution of these singularities to structural and electronic phase transitions is not yet fully clarified and strong coupling effects (see next chapter) are likely to occur.

The role of impurities on the metal/insulator transition of V_2O_5 was studied by Keer et al. (1976). Doping V_2O_5 with Aℓ or Cr on a 1% level shifts the temperature of the $\alpha - \beta$ transition near 300K by more than 60K and the low temperature transition near 180K by ca. 10K. The $\alpha - \beta$ transition depends sensitively on thermal history and the dopant concentration. The strong positive entropy change in the "dirty" material seems to indicate that strong electron-lattice coupling over and above the electronic correlation energy dominate this structural phase transition (McWhan et al. 1975).

Similar effects in superconductors (see e.g. Schlottermann, 1975) are normally attributed to Kondo scattering and are not discussed here.

In a different experimental approach, Shimotamai et al. (1985) used the DSC method for the investigation of quenched-in vacancies in NiAℓ. The DSC measures the rate of energy flow into or out of a simple direction as a function of temperature and, with constant scanning rate, as a function of time t. The energy E released by recovery of defects with concentration n is

$$\left.\frac{dE}{dT}\right|_T = \varepsilon \left.\frac{dk}{dt}\right|_T \tag{6.12}$$

with ε being the formation energy. If the recovery process is the m-th order, a rate constant is defined as

$$\left.\frac{dn}{dt}\right|_T = - k_T \, n_T^m \tag{6.13}$$

and can be directly determined from

$$k_T = - \left(\frac{dE}{dt}\right)_T \frac{\varepsilon^{m-1}}{E_T^m} \tag{6.14}$$

If the annealing process is thermally activated, the rate constant becomes

$$k_T = Ae^{-E_A/k_T} \tag{6.15}$$

with the activation energy E_a. Experimentally, the activation energy can be determined directly from the release rate in the DSC by

$$E_A = \frac{Ck_T^2}{\Delta T} \tag{6.16}$$

where ΔT is the line width (FWHM) and C = 2.4 for first order processes and C = 2.9 for second order processes (Granato and Nilan, 1965; Iwata and Nihira, 1976). The experimental results on NiAℓ are shown in Figure 4.

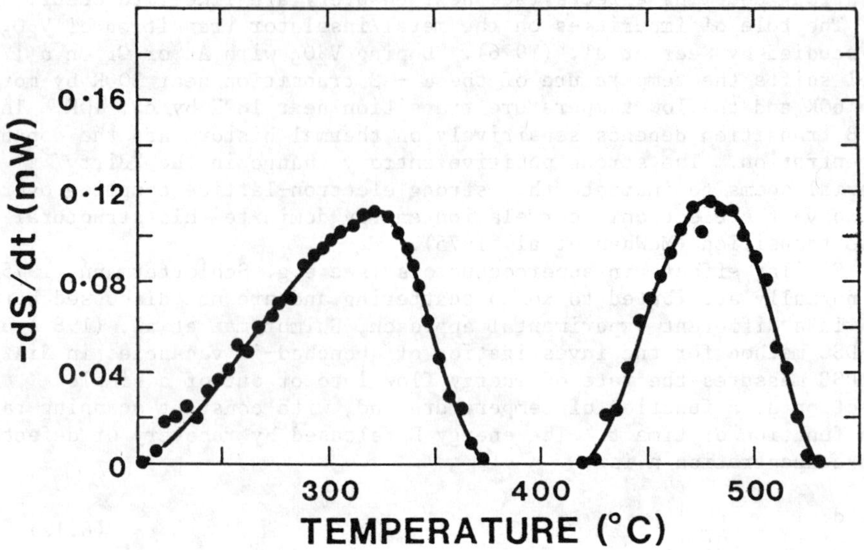

Figure 4. Temperature dependence of the release rate of Ni/Aℓ alloy quenched from 1400°C. The heating rate is 5 K/min (after Shimotamai et al., 1985).

7. ORDER PARAMETER COUPLING AND EXCESS SPECIFIC HEATS

The quantitative behaviour of a crystal undergoing a structural phase transition can only in exceptional cases be described on the basis of the critical behaviour of only a single one-dimensional order parameter. Instead, coupling effects either to other components of the same multi-dimensional order parameter (e.g. Toledano and Toledano, 1977, 1980, 1982) coupling with other order parameters (Imry, 1975; Salje and Devarajan, 1985; Anisimov et al., 1981; Onodera et al., 1985) changes the excess entropy over a large temperature range. The excess specific heat and the excess entropy, being volume quantities, integrate over all temperature dependent contributions to the free energy regardless of a certain order parameter being primary, secondary, or temperature dependently coupled. Most scattering experiments and optical parameters such as birefringence and optical activity, on the other hand, detect specifically the temperature dependence of the interacting order parameter only.

Significantly different behaviour can therefore be expected if the thermal evolution of <Q> determined from specific heat measurements is compared with results from these other experimental techniques.

An example for additional excess entropies due to order parameter coupling is the ferroelastic phase transition in $Pb_3(PO_4)_2$ at T_c = 453.6 K (Salje and Wruck, 1983). The symmetry reduction R$\bar{3}$m-C2/c is described by an effective Hamiltonian (Salje and Devarajan, 1981):

$$H_{eff} = \frac{\alpha}{2}(Q_1^2 + Q_1^2) + \frac{\alpha'}{2} Q_3^2 + (Q_1^2 + Q_2^2)^2 L_2 + Q_3^2(Q_1^2 + Q_2^2)L_3 + $$
$$+ Q_3^4 L_1 + \frac{1}{\sqrt{2}} Q_1 Q_3 (Q_3^2 - \frac{1}{3} Q_2^2) L_4 \qquad (7.1)$$

which can be diagonalized under the assumption of well separated fixed points of the two sets of order parameters $\{Q_1, Q_2\}$ and Q_3. The order parameter Q_3 alone describes a classical second order phase transition with a Landau expression of equation (4.1). The parameters Q_1 and Q_2 follow the Hamiltonian of a Potts-oscillator with

$$\phi'' = (\frac{\alpha}{2} + L_3\pi^2)(Q_1^2 + Q_2^2) + L_2(Q_1^2 + Q_2^2)^2 + \omega Q_1(Q_2^2 - \frac{1}{3} Q_1^2) \qquad (7.2)$$

Structurally, the pseudo spin part of the equation (7.2) describes the orientation of the monoclinic binary axis of the low temperature phase with respect to the trigonal unit cell of the high temperature phase (Bismayer et al. 1982). The related excess entropy follows from the leading terms in equation (7.1):

$$-\Delta S = \frac{\alpha_1}{2}(Q_1^2 + Q_2^2) + \frac{\alpha_2}{2} Q_3^2 \qquad (7.3)$$

with

$$\alpha_1 = \frac{\partial\alpha}{\partial T} \text{ and } \alpha_2 = \frac{\partial\alpha'}{\partial T} \qquad (7.4)$$

The excess entropy can be compared with the optical birefringence induced by the structural phase transition (Bismayer et al. 1981; Salje and Wruck, 1983):

$$\Delta n_{bc} = P_{eff}(\tfrac{1}{2}D^2 + 2G^2)^{\frac{1}{2}}(Q_1^2 + 2\sqrt{2}\,|Q_1Q_3|) \qquad (7.5)$$

In the displacive limit ($Q_3 = 1/\sqrt{2}\, Q_1$), the excess entropy and the morphic birefringence depend quadratically on the same order parameter and they are therefore proportional to each other. At temperatures close to T_c, this proportionality is broken, based upon differences in the temperature dependencies of the order parameters Q_1 and Q_3. In Figure 5 the linear relation holds for $T < 415K$, whereas larger entropy values are found for temperatures closer to T_c. The deviation from proportionality is plotted as a function of temperature in Figure 6 which shows the additional divergency of $|\Delta S|$ at T_c.

This method of rescaling the excess entropy against other macroscopic variables can also be used for the investigation of the order of coupling effects. One illustration is from the cubic to orthorhombic ($P2_13 \rightarrow P2_12_12_1$) phase transition in crystals with Langbeinite structure (e.g. $K_2Cd_2(SO_4)_3$, $K_2Ca_2(SO_4)_3$). In this case multicomponent order parameter behaviour combines with reasonably strong order parameter

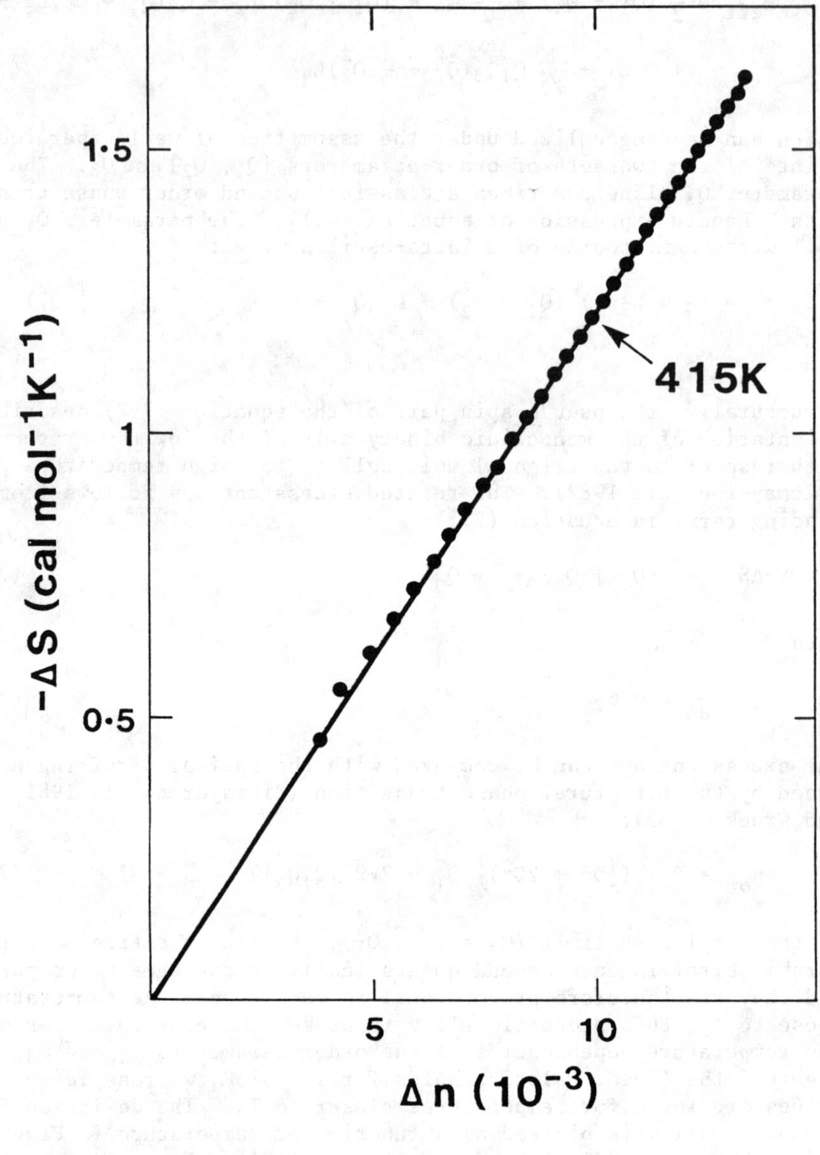

Figure 5. Scaling of the excess entropy with respect to the morphic birefringence in $Pb_3(PO_4)_2$ in the temperature range 377.5 - 453.6 K (T_c). Note the linear correlation at T < 415 K. (after Salje and Wruck, 1983).

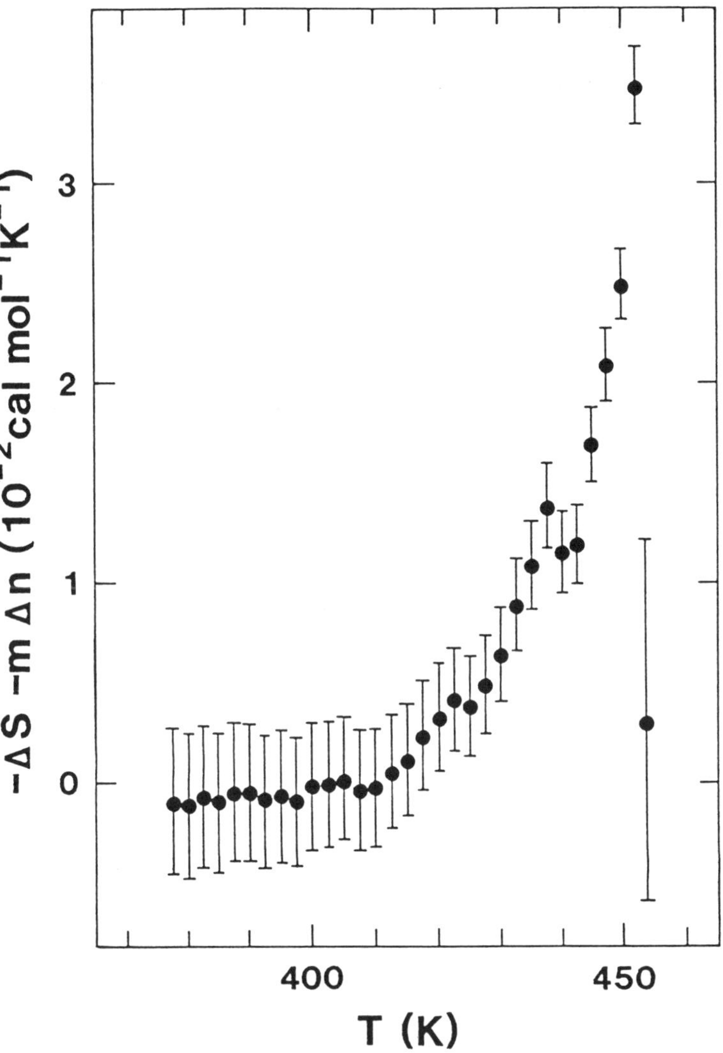

Figure 6. Difference between the experimental excess entropy ΔS and the extrapolated values of the renormalized birefringence of $Pb_3(PO_4)_2$ (deviation from linearity in Fig. 5) (after Salje and Wruck, 1983).

strain coupling (Dvorak, 1972; Maeda, 1979; Speer and Salje, 1985).

The macroscopic strain x_i and the morphic birefringence of Δn follow from minimisation of the Landau theory in a stress free crystal and are higher polynomials of the order parameter components (Devarajan and Salje, 1984). The general assumption, however, that the lowest order terms dominate the temperature dependence of x_i and Δn turned out to be wrong when these two quantities were plotted as functions of the correlated excess entropy (Figure 7a,b). The higher order polynomial terms lead to strong nonlinearities and there is no extrapolated common zero point of simultaneously vanishing excess entropy and lattice strain. The structural interpretation of these findings is based on X-ray observations of Speer and Salje (1985) who found that two different types of structural deformations dominate the transition process. Firstly, the octahedra complexes around the divalent ions distort and expand and this is compensated by a denser packing of oxygen around the potassium positions. Secondly, the SO_4 tetrahedra tilt in order to follow the structural deformations of the larger polyhedra along the trigonal axes. Both movements contribute differently to the excess entropy, the ferroelastic strain and the morphic birefringence resulting in a strong nonproportionality between these parameters.

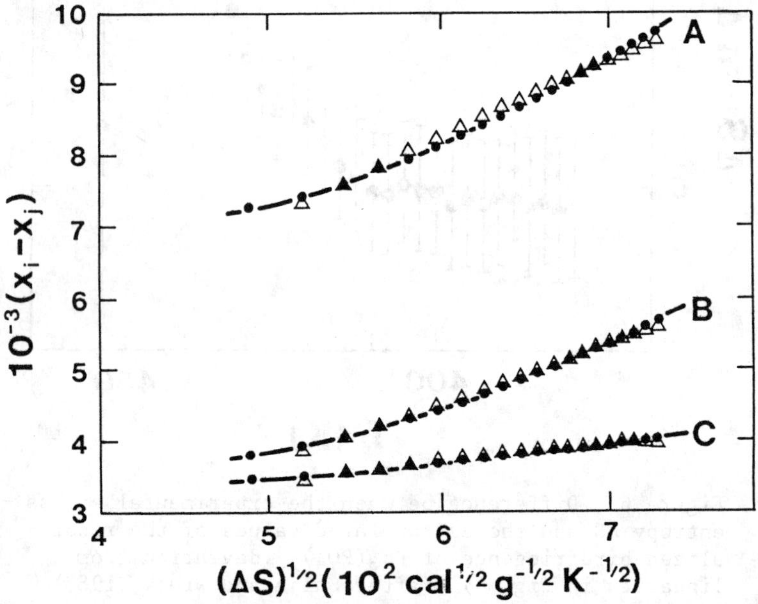

Figure 7a. The non-linear correlation between the spontaneous strains $X_2 - X_3$(A), $X_1 - X_2$(B), $X_1 - X_3$(C) and the square root of the excess entropy of $K_2Cd_2(SO_4)_3$. The full circles represent experimental values and the triangles correspond to results of lattice dynamical calculations (after Devarajan and Salje, 1984).

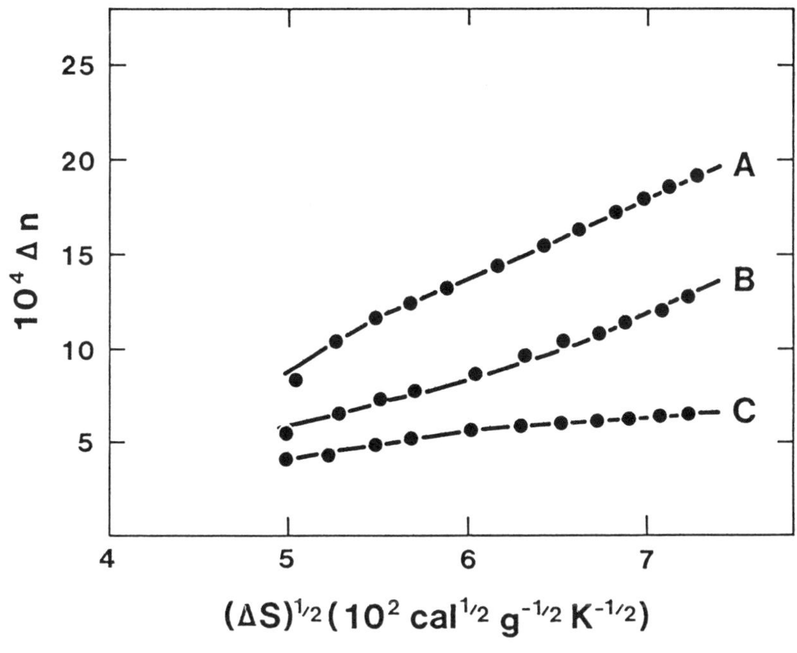

Figure 7b. Plot of optical birefringence of $K_2Cd_2(SO_4)_3$ against $(\Delta S)^{\frac{1}{2}}$ (A: $\Delta n_{\alpha\gamma}$, B: $\Delta n_{\beta\gamma}$, C: $\Delta n_{\alpha\beta}$, after Devarajan and Salje, 1984).

Strain coupling of the order parameter was reported also to cause differences in the critical behaviour of ΔC_p near T_c between single crystals and crystal powders (Hatta et al. 1977). This effect is mainly attributed to boundary conditions by having monodomain crystals similar to uniaxially clamped crystals or polydomain, powder samples respectivley. In powder samples the additional contribution of inhomogeneous strain fields in the individual grains may appear and smear the phase transition. Experimentally, Hatta et al. (1977) observed a peak in the ΔC_p curve at T_c for a single domain crystal of $SrTiO_3$. The critical exponent was found to be 0.14, which is close to the Ising value. On the other hand, a polydomain sample showed only a very small anomaly near T_c which was attributed to the intrinsic strain free C_p value.

According to Landau theory of an uncoupled, one dimensional order parameter, the transition temperature under stress free conditions is shifted under uniaxial stress when the coupling sets in, but the form of the C_p curve remains almost unchanged. Using the generalized Pippard relation (Garland, 1966; Janovec, 1966), the isothermal elastic compliance S_T is related to the thermal expansion coefficient β as follows

$$S_T = S_t + \gamma\beta\Big|_x \qquad (7.6)$$

where S_t is the elastic compliance along the transition line $t = 0$ and γ is a constant. The specific heat under constant stress $C_p\big|_x$ relates to $\beta\big|_x$ as

$$C_p\Big|_x = \frac{T}{\gamma}\left(\beta\Big|_x - \beta_t\right) \qquad (7.7)$$

The thermal expansion along the transiton line is β_t. The specific heat under uniaxial stress $C_p\big|_p$, on the other hand, is

$$C_p\Big|_x - C_p\Big|_p = T\,\frac{\beta^2\Big|_x}{S_t} \qquad (7.8)$$

This equation shows that if neither $\beta\big|_x$ nor S_T posses law singularities at $t = 0$, the critical exponents $C_p\big|_x$ and $C_p\big|_p$ should be identical.

The parallel behaviour of the psecific heat of a stress free crystal and a crystal under uniaxial stress is in obvious contradiction to the experimental results on $SrTiO_3$. However, if the order parameter is allowed to couple with strain and acoustic phonons, larger singularities are expected than in a second order phase transition. (See also the multicritical behaviour of $SrTiO_3$ as discussed by Stokka and Fossheim, 1982). If the coupling is linear in the strain but quadratic in the order parameter, as expected for acoustic phonons, Larkin and Pikin (1969) showed that the phase transition can even become first order although it would have been continuous without coupling.

8. COUPLED ORDER PARAMETERS IN FELDSPAR

The effect of order parameter coupling on the excess specific heat is now demonstrated for the two examples: Na-feldspar and anorthite. In both minerals, excess specific heats have been measured which are associated with the phase transitions $C2/m$ - $C\bar{1}$ and $I\bar{1}$ - $P\bar{1}$, respectively. In the case of the displacive phase transition $C2/m$ - $C\bar{1}$ in fully Aℓ,Si disordered Na-rich alkali feldspar, a step-wise increase of the specific heat is observed during the symmetry reduction (Fig.8). With increasing degree of Aℓ,Si order, this phase transition is supressed and Landau theory has been used in order to predict the temperature evolution of the excess specific heat for Na-feldspar in thermodynamic equilibrium (Salje, 1985).

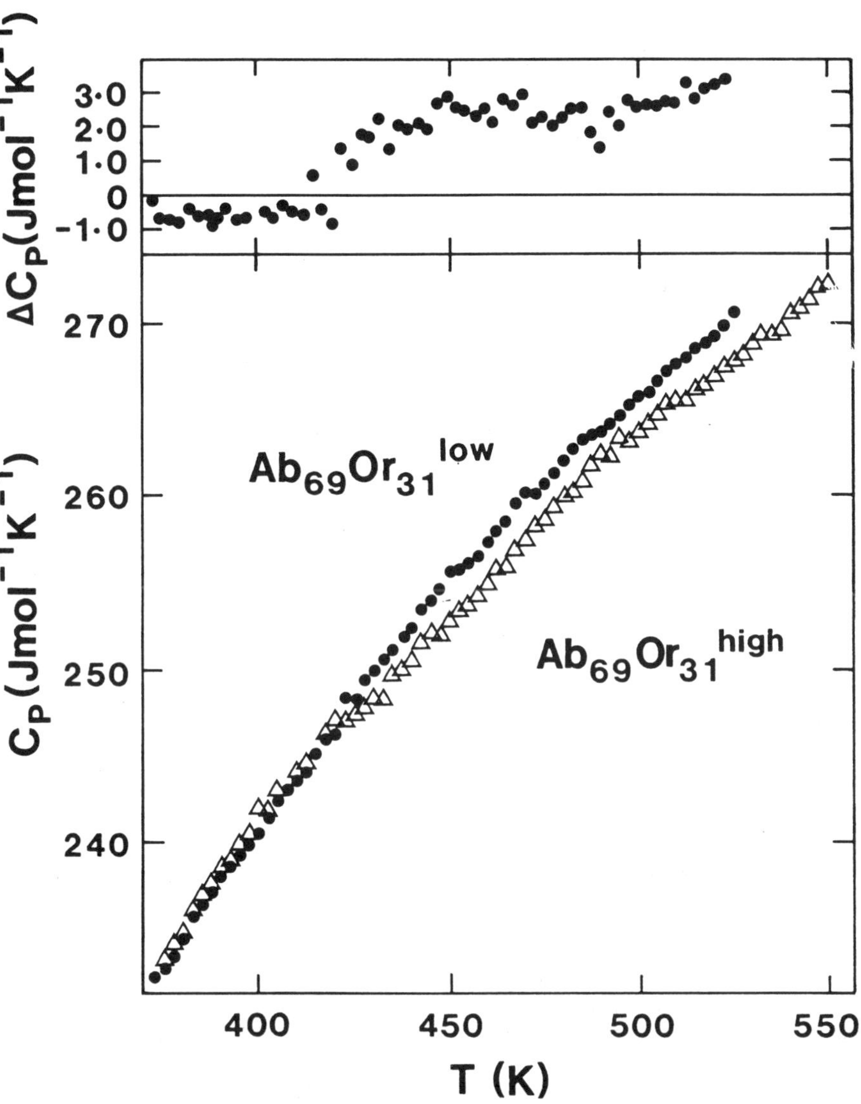

Figure 8a. The specific heats of ordered (low) and disordered (high) $Na_{0.69}K_{0.31}A\ell Si_3O_8$. The structural phase transition only occurs in the disordered sample at $T_c > 416k$.

8.1. Na-feldspar

The free energy of the Na-feldspar can be expressed using the two order parameters Q and Q_{od}:

$$\Delta F_{tot} = \frac{1}{2}aQ^2 + \frac{1}{4}bQ^4 + \frac{1}{6}cQ^6 + \frac{1}{2}a_{od}\,Q_{od}^2 + \frac{1}{4}b_{od}\,Q_{od}^4 +$$

$$+ \frac{1}{6}c_{od}\,Q_{od}^6 + \xi Q((A + C/\chi)Q + (B + D/\chi)Q_{od}) +$$

$$+ \xi^{od}\,Q_{od}((A + C/\chi_{od})Q + (B + D/\chi_{od})Q_{od}) +$$

$$+ \frac{1}{2}c_{44}(AQ + BQ_{od})^2 + \frac{1}{2}c_{66}(CQ + DQ_{od})^2 +$$

$$+ c_{46}(AQ + BQ_{od})(CQ + DQ_{od}) + \lambda QQ_{od} \qquad (8.1)$$

Note that only those terms which depend on both order parameters in the form $Q \cdot Q_{od}$ appear in the free energy expression. This type of coupling is called "bilinear" and occurs only if the irreducible representations of the two order parameters are identical. The thermodynamic equilibrium of Na-feldspar is described by

$$\partial \Delta F_{tot}/\partial Q = 0 \qquad (8.2)$$

$$\partial \Delta F_{tot}/\partial Q_{od} = 0 \qquad (8.3)$$

leading to the two basic equations for the two order parameters

$$a_o(T - T_c^*)\,Q + bQ^3 + cQ^5 + dQ_{od} = 0 \qquad (8.4)$$

$$a_{od,o}(T - T_c^{od*})\,Q_{od} + b_{od}Q_{od}^3 + c_{od}Q_{od}^5 + dQ = 0 \qquad (8.5)$$

The entire thermodynamic behaviour of Na-feldspar depends, therefore, only on 7 independent parameters, and these are all accessible to experimental determinations. Feldspars are often found with nonequilibrium structural states because the two order parameters act on extremely different time scales. The lattice distortions, described by Q, are dynamically induced by phonon processes with typical relaxation times of faster than 10^{-10} secs. The order parameter Q can rapidly follow any change of external parameters, such as temperature or pressure. No metastability is expected for Q. The Aℓ,Si ordering, on the other hand, is described by Q_{od} and changes of Q_{od} require the breaking of tetrahedral (Aℓ-O and Si-O) bonds. For each exchange of Si and Aℓ between sites a threshold energy is needed and, in simple cases, this energy is the activation energy for ordering. The relaxation of Q_{od} is, therefore, sluggish and can easily be influenced by mobile ions, such as H^+ which will act to reduce threshold energy.

Metastable states are described by

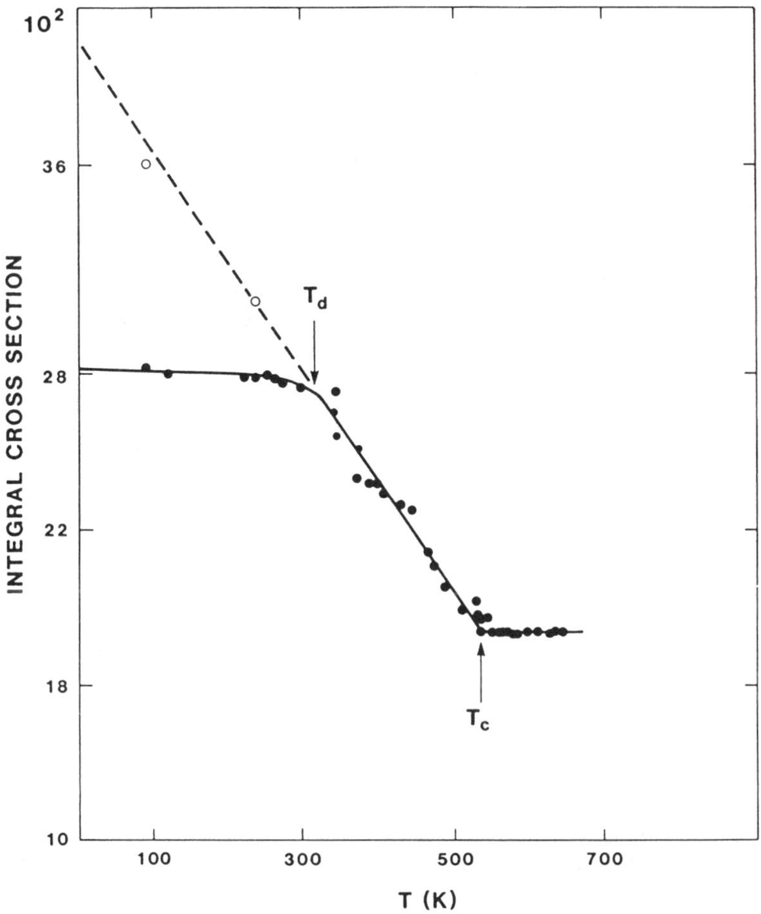

Figure 8b. Integral cross section of the Raman scattering signal at 514 cm^{-1} which is proportional to the excess entropy in the temperature interval between 300 K and T_c.

$$a_o(T - T_c^*)Q + bQ^3 + cQ^5 + dQ_{od} = 0 \quad (8.6)$$

where Q_{od} measures the frozen-in degree of Aℓ,Si order. The parameters a_o, b, c, d, $a_{od,o}$, b_{od} and c_{od} have been determined experimentally by Salje et al. (1985). The total free energy in equation (8.1) can, therefore, be expressed in terms of the order parameter alone and all related thermodynamic functions can be calculated directly: the excess specific heat of the displacive part - in a metastably fully ordered sample - shows a step at T_c and the excess entropy increases at $T \ll T_c$ linearly with decreasing temperature (Figure 8b). In thermodynamic equilibrium, both ordering mechanisms take place in the structural phase transition. The resulting excess specific heat again shows a step at T_c but is now followed by a large peak near the onset temperature of the Aℓ,Si ordering. This peak, although similar to a λ-peak, does not indicate any symmetry breaking and is not related to a structural phase transition but is due to the coupling of the two order parameters. This example illustrates that in the case of coupled order parameters the actual phase transition is not necessarily accompanied by major changes of the excess specific heat. Such changes may well occur at much lower temperatures in the asymmetric phase without being related to a further phase transition process, however.

8.2. Ca-feldspar

In anorthite, the coupling between the displacive order parameter and the Aℓ,Si ordering parameter follows a different symmetry pattern. The hypothetical Aℓ,Si ordering phase transition takes place at high temperatures above the melting point. Its symmetry change is according to our basic assumption of a monoclinic paraphase C2/m → C$\bar{1}$ → I$\bar{1}$ with T_c = 514.5 K. The underlying model for the displacive transition is similar to that of Adlhart et al., (1980), where the driving force of the structural phase transition is essentially identified as the condensation of a soft mode at the boundary of the first Brillouin zone of the I$\bar{1}$ structure. In an extension of this model, clustering and ordering of alkali positions in multi-cell potentials were allowed for by Salje (1987), and this restabilises the soft mode of temperatures close to the transition temperature T_c. The symmetry reduction during the phase transition is restricted to a breaking of the translational symmetry τ = [111] of the I$\bar{1}$ structure which leads to additional superlattice reflections (h + k = even, ℓ = odd and h + k = odd, ℓ = even) to appear in the P$\bar{1}$ phase. The related critical point of the first Brillouin zone is termed "Z" (Fig. 9). The order parameter Q_Z has to transform according to the active representation at the Z-point.

For symmetry reasons, direct coupling between Q_{od} and Q_Z is not allowed. (Their respective active representations are not identical). However, the following three coupling mechanisms are consistent with the symmetry reductions of Q_{od} and Q_Z:

(i) Strain induced coupling. The free energy due to coupling has the form

$$\Delta F = \sum \lambda_i x_i Q_Z^2 + \sum e_i x_i Q_{od}^2 \qquad (8.7)$$

This type of interaction is most common for proper ferroelastics and has been explored theoretically by Salje and Devarajan (1986). Experimentally, strain induced coupling has been found by Bismayer et al., (1985).

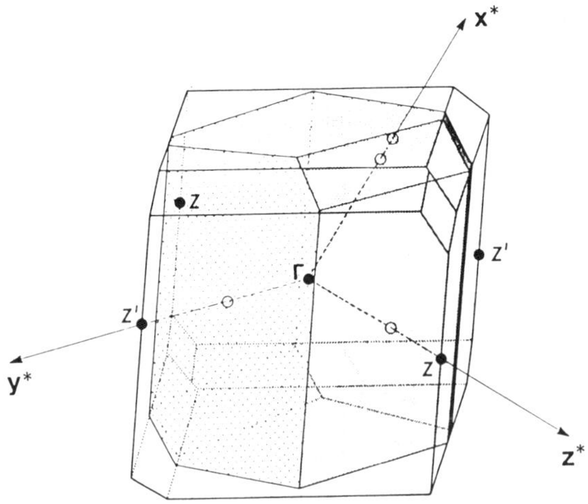

Figure 9. Brillouin zones of anorthite in the $I\bar{1}$ phase (enclosing Brillouin zone) and in the $P\bar{1}$ phase (smaller Brillouin zone with darker faces). The lattice parameters used for the construction of the Brillouin zones (Redfern and Salje 1986) are taken at 250°C ($I\bar{1}$) and room temperature ($P\bar{1}$). The critical points of the phase transition are Z and Z' which become points in the $P\bar{1}$ phase. The points Z and Z' are symmetry equivalent in the ideal $I\bar{1}$ structure.

(ii) Direct biquadratic coupling with $\Delta F = \xi_{12} Q_Z^2 Q_{od}^2$ (8.8) was advocated by Anisomov et al., (1981).

No experimental evidence for this coupling mechanism is yet reported. The resulting thermodynamic behaviour, however, is identical with case (i).

(iii) Inhomogeneous symmetry breaking. The homogeneity of the crystal can be pertubated by fluctuations of Q_{od}. These fluctuations depend on the cooling history of the sample. The degeneracy of the active representation is lifted and the order parameter splits into two components. It can be shown that this splitting can lead to rather periodic antiphase boundaries and modulated structures (Perez-Mato et al. 1984; McConnell, 1985; Salje, 1987).

The free energy of a Ca-rich plagioclase includes the free energies of both order parameters, their coupling and the renormalisation of the critical temperature due to Na-Ca exchange. The expression given by Salje (1987) is:

$$\Delta F_{tot} = \tfrac{1}{2}\alpha(T - (T_c + \gamma n_{Ab}))\, Q_Z^2 + \tfrac{1}{4}B_{eff}\, Q_Z^4 + \tfrac{1}{6}cQ_Z^6 + \sum \lambda_i x_i Q_Z^2 +$$

$$+ \tfrac{1}{2}\sum C_{ik}x_i x_k + \sum e_i x_i Q_{od}^2 + \xi_{12}Q_Z^2 Q_{od}^2 +$$

$$+ \tfrac{1}{2}\alpha_{od}(T - T_c^{od} - n_{Ab}\frac{\partial T_c^{od}}{\partial n_{Ab}})\, Q_{od}^2 + \tfrac{1}{4}b_{od}Q_{od}^4 +$$

$$+ \tfrac{1}{6}c_{od}Q_{od}^6 + \xi_{od}^* n_{Ab}Q_{od} \qquad (8.9)$$

where n_{Ab} is the mole fraction of the Ab in the plagioclase, T_c is the critical temperature of the $I\bar{1}$ - $P\bar{1}$ phase transition and x_i is the i-th compound of the spontaneous strain of the $P\bar{1}$ phase.

The equivalent temperature dependence of the order parameter of the Aℓ,Si ordering ($C\bar{1}$ - $I\bar{1}$ phase transition) follows from:

$$\Delta F_{od} = \tfrac{1}{2}\alpha_{od}(T - T_c^{od} - n_{Ab}\gamma_{od})Q_{od}^2 + \tfrac{1}{4}b_{od}^* Q_{od}^4 + \tfrac{1}{6}c_{od}Q_{od}^6 \; ;$$

$$\partial \Delta F_{od}/\partial Q_{od} = 0 \qquad (8.10)$$

which shows clearly that the phase transition takes place by changes of temperature, chemical composition or both.

The major difference between this free energy expression and the equivalent one for Na-feldspar is that the coupling between the two order parameters is bilinear for Na-feldspar and biquadratic for anorthite. The biquadratic coupling including the role of stress fields changes both the critical temperature T_c of the $I\bar{1}$ - $P\bar{1}$ phase transition and the 4th order term of the Landau expression. This latter renormalisation implies that the conclusion that Aℓ,Si disorder changes the order of the phase transition and the transiton temperature.

The specific heat anomaly of the $I\bar{1}$ - $P\bar{1}$ transition in anorthite has been observed using DSC techniques by Wruck (1986). Highly ordered anorthite from the Val Pasmeda locality shows a large λ-peak with a critical exponent $\alpha = 0.5$. Spontaneous strain measurements of Redfern and Salje (1987) revealed a critical exponent of the order parameter of $\beta = 0.25$. These two results confirm the theoretical prediction that anorthite with large Q_{od} possesses a tricritical point at the $I\bar{1}$ - $P\bar{1}$ phase transition. In samples which have either been heat-treated ($Q_{od} \gg 1$) or contaminated with Na (e.g. the anorthite from the Monte Somma locality), the fourth order term in the potential (8.10) becomes renormalized and the phase transition is second order. The λ-anomaly disappears and a smeared-out stepwise increase of the excess heat capacity was found by Wruck (1987). (See Figures 2 and 10).

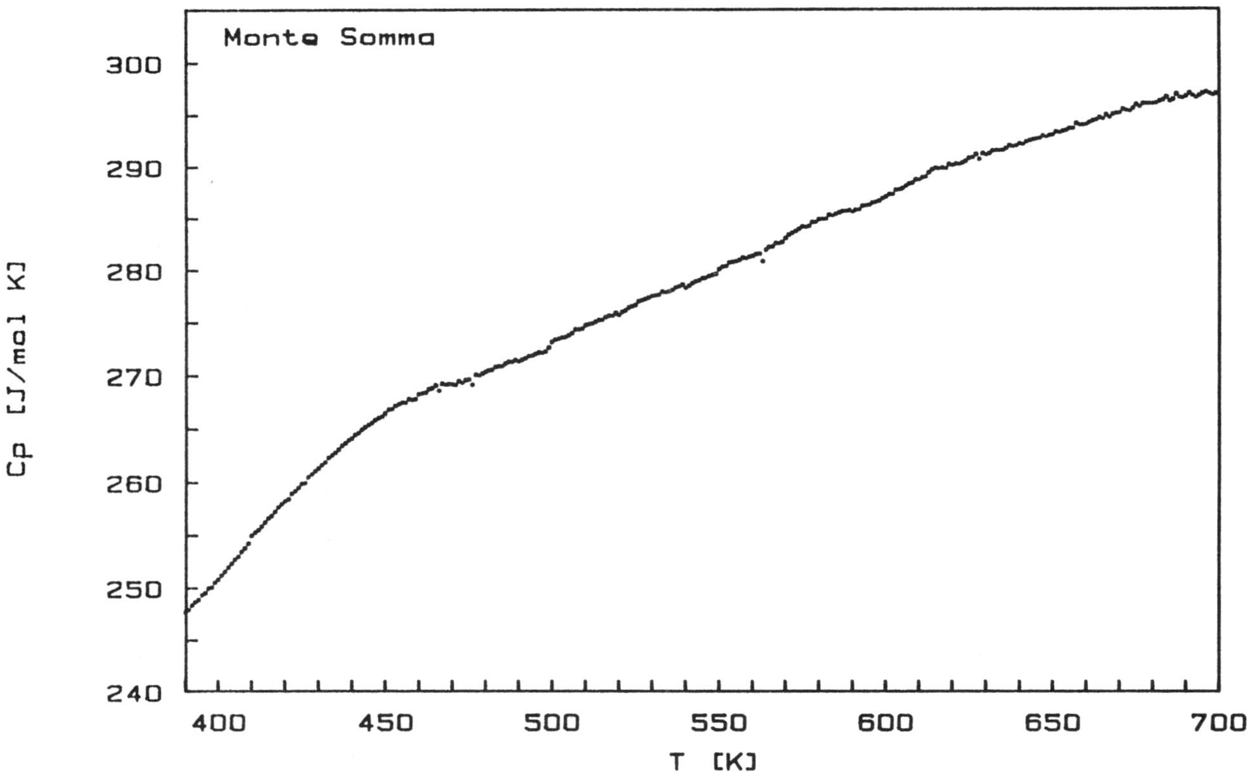

Figure 10. The effect of the $P\bar{1}$ - $I\bar{1}$ phase transition near 480 K in anorthite from the Monte Soma locality.

9. SELECTED EXAMPLES OF STRUCTURAL PHASE TRANSITIONS

Most experimental work on the investigation of structural phase transitions of compounds which do not have direct correlations with order parameters representing ferroic properties are directed towards the determination of the transition temperature, the critical exponents and the temperature dependence of the excess entropy. Some of the recent literature is listed in the references and typical examples are selected for further discussion.

Proton ordering seems to be the driving ordering process of the structural phase transition in crystalline hydronium perchlorate ($C\ell O_4 \cdot H_3O$) at $T_c = 248.4K$ (Janik et al., 1974). The phase transition is strongly first order with a latent heat of 1.30 ± 0.03 Kcal/mol. The high temperature phase (Pnma) is characterized by proton tunneling in a double well potential with typical relaxation times of 3.5×10^{-12} sec whereas the proton positions are fixed in the low temperature phase ($P2_1/n$). This is an extreme case of the symmetry restrictions for the experimental excess entropies described in section 3. It results in much smaller excess specific heats in the low temperature phase than in the disordered phase because only in the latter phase do the additional contributions of the proton jumps to the specific heat appear. The negative step of ΔC_p at T_c is therefore related to the disappearance of the relaxational mode below T_c and can be used for the determination of the (negative) relaxational excess entropy as a function of temperature.

Order/disorder behaviour of protons also dominates the structural phase transition in $SnC_{1 2}2H_2O$ (Kiriyama et al., 1973). The specific heat near $T_c = 218K$ shows a rather symmetrical increase which obeys a power law dependence of equation (5.15) (Matsuo et al., 1973). The critical exponent is close to the magic value of 1/2, namely 0.492 ± 0.02.

The structural phase transition from a rocksalt structure to a rhombohedrally distorted polar phase in the system GeTe - SnTe depends sensitively on the hole concentration n. Variations of T_c by ca. 27K between samples with $n = 8.8 \times 10^{-19}cm^3$ and those with zero free carrier concentration were observed (Iizuma et al., 1975; Kobayashi et al., 1976). Ultrasonic measurements by Rehwald and Lang (1975) on the ternary compounds revealed a jump-like anomaly at T_c, which was interpreted phenomenologically in terms of strong coupling between the strain and the order parameter. A similar step of the heat capacity was observed by Hatta and Rehwald (1977) which was superimposed by weak λ-shape anomalies around T_c. Both discontinuities show the same temperature dependence as the elastic bulk modulus, as predicted from Landau-Ginzburg theory. Deviations from Landau theory without considering fluctuations can be estimated from the Ginzburg criterion that relates the temperature interval with dominant contributions from the fluctuations to the excess heat as expressed by equation (5.4). Hatta and Rehwald used this method of determining L_c in Snx Ge^{1-x} Te and found a characteristic variation of L_c with x from 0.67nm (x = 0.91) to 0.36 (x = 0.62). As the fluctuation interval disappears rapidly (sixth-order power law!) with increasing interaction length L_c, a purely Landau behaviour should be expected for the composition SnTe and a larger fluctuation regime for GeTe.

Other Te containing intermetallic compounds like Cr_3Te_4 possess structural phase transitions, including the ordering of vacancies in every other metal layer in the NiAs type structure. This phase transition shows a maximum in the specific heat at 903K (Gronvold, 1973). The closely related compound Co_3Se_4 has a λ-type phase transition at even higher temperature (T_c = 1005K) with a large tail of the excess specific heat at lower temperatures indicating that the disordering of the defects with increasing temperature starts well below 800K. The phase transition is presumably related to disorder of cobalt atoms and vacancies, affecting 0.25 Co amd 0.25 vacancies per formula unit. The theoretical excess entropy is

$$\Delta S = (R/2)(0.5 \ln 0.5 + 0.5 \ln 0.5) = 2.88 \text{ J/Kmol}$$

These results demonstrate that vacancy ordering is not necessarily so sluggish that calorimetric experiments cannot detect the temperature dependence of the excess entropy. The correlated excess specific heats due to the defect migration are very large indeed and are not restricted to a small temperature interval as in the case of proton tunneling.

Considerable experimental work has been directed towards understanding the excess specific heat of magnetite (Fe_3O_4) at temperatures close to the Vervey transition at ca. 123K. Starting from the self-consistent band calculations of Cullen and Cullen (1971, 1973) multiple ordering, with the interaction of three different order parameters, was advocated by Evans and Westrum (1972). Evidence of the multi-order parameter behaviour was found in the splitting of the excess specific heat at the Vervey transition into two peaks and also in the occurrence of two singularities in the Moessbauer spectra. However, later results of Moessbauer experiments by Dormann et al. (1982) on fairly stoichiometric magnetite were not indicative of multiple ordering. Moreover, Matsui et al. (1977) showed that the bifurcated peaks of the excess specific heat, which were observed by Westrum and Gronwold, (1969); Evans and Westrum, (1972); Kamilov et al., (1975); Bartel. (1976) and Falk et al., (1979), could be removed by annealing the samples. This indicates that stress-coupling and, possibly, defect structures contribute to the bifurcated and broadened specific heat peaks near T_c (see section 6). New experimental data of Gigo et al., (1983) show a small singularity at 111K next to the Vervey transition at 123K even for the purest samples but information about the actual purity of the sample was not given by any of the authors. However, the low temperature peak was also observed in fluorinated magnetite, where no Vervey-transition takes place (Rigo et al., 1978). This could indicate that the second phase transition separates from (but is presumably coupled with) the Vervey transition and is strongly influenced by impurities and strain. Nevertheless, its character may not reflect the long range structural deformation of the Vervey transition but may be related to spin rotation.

The final solution of this problem certainly requires new experiments on better characterized material.

10. CHARGE CARRIER INDUCED STRUCTURAL PHASE TRANSITIONS

The field of investigations of charge carrier induced structural phase transitions, especially of metal to insulator transitions (M/I), has developed rapidly. Experimental studies of the excess specific heats related to the phase transitions represent a link between the electronic properties of the transformations and thermodynamics. If the critical parameter of the M/I transition is the chemical composition, like in most Anderson phase transitions (Mott, 1974; McMillan, 1981; Salje and Guettler, 1984), the excess specific heat has to be determined from comparison of the absolute specific heats of different crystals in the vicinity of zero K. On the other hand, if the M/I transition is temperature induces, the excess specific heat can be measured calorimetrically. In most transition metal oxides, where the M/I transition is correlated with the formation of small polarons in the asymmetric phase, the excess specific heats are large near T_c and the phase transitions are mostly first order (Mott, 1974; Goodenough, 1972; Chankraverty, 1975; Schirmer and Salje, 1980).

Typical examples are the two phase transitions near 154K and 142K in Ti_4O_7 (Lakkis et al., 1976) where an ordering of Ti^{3+} and Ti^{4+} takes place in the low temperature phase such that ordered chains of Ti^{3+} dimers are separated from each other by Ti^{4+}. The intermediate phase is a disordered phase as far as the Ti^{3+} dimers are concerned. The polaronic ground state is, therefore, that of bipolarons (Mott, 1984; Anderson, 1975) and the disorder is considered dynamic by Lakkis et al., (1976). Both phase transitions appear as large peaks in the specific heat curves, the excess entropy of the transition between the fully ordered and the dynamically disordered phase is ca. 70% of the maximum configurational entropy. The reduction of ΔS was attributed to differences in the configurational energies of the possible Ti^{3+} - Ti^{4+} links.

The phase transition at 150K is attributed to the delocalization of the 3d electrons in the high temperature phase. The excess entropy of ca. 3.04 $cal.K^{-1}mol^{-1}$ is very large and Lakkis et al., (1976) assumed that such a large entropy change cannot be due to the lattice deformations alone, but suggested that roughly half of the excess entropy results from an electronic entropy change. This idea is based on the Brinkman-Rice model (1970) of a highly correlated electron gas, where the observed enhancement factor of the susceptability and the electronic excess specific heat are identical. The resulting excess entropy is in fact 1.50 $cal.K^{-1}mol^{-1}$, i.e. half of the total excess entropy of the phase transition.

A charge density wave phase transition related to the formation of an incommensurate phase was observed in the polytypes 2H of $TaSe_2$ and TaS_2. The phase transitions are essentially driven by the coupling between the conduction electrons of a highly anisotropic Fermi surface and lattice phonons (e.g Mattheis, 1973; Wilson et al., 1975; Yoffe, 1976). The stability and structure of these charge density wave states depend upon the shape of the Fermi surface. This, in turn, is related to the symmetry of the polymorph under consideration. The locking wave vector in $TaSe_2$ is the periodon value $k_a = 1/3\ a^*$ and the discommensuration parameter increases with temperature to 0.02 at the high-temperature phase transition point (Moncton et al., 1975). The latter phase transition is assumed to be continuous and the lock-in phase transition is first order. Thermodynamic investigations of these phase transitions are mainly concerned with the open question of how far the excess entropies come from the excitation of electrons across the gap in the band structure, which has been introduced by the charge density wave phase transition itself. Whereas in the Rice-Scott model (1975) a large contribution, similar to the case of Ti_4O_7, can be expected, McMillan (1977) denies such contributions on the assumption that the coherence length of the charge density wave state is too short to produce large excess specific heats. Very careful measurements of the excess specific heats and resistivities around both phase transition by Craven and Meyer (1977) lead them to the conclusion that the excess entropy is mainly due to lattice contributions with short correlation lengths. The excess heat capacity of the normal to incommensurate phase transition shows a power law singularity with a critical exponant α close to the magic value of the 0.5, namely 0.45 ± 0.035.

The structural phase transitions in narrow band IV - VI compounds had been explained by Kristoffel and Konsin (1973) on the basis of an inter-band electron - TO phonon interaction model. In the scope of classical Landau theory, the carrier concentration renormalizes the transition temperatures of the cubic to rhombohedral phase transition. Free energy expressions were derived including order parameter-strain coupling (Rehwald and Lang, 1975; Slonczewski and Thomas, 1970) and Sugimoto et al., (1981) determined the specific heats in the model compounds $Pb_{1-x}\,Ge_x\,Te$ for x ranging from 0.05 to 0.3. A smeared stepwise behaviour was found for low Ge content which becomes sharper and superposed by an additional rounded singularity with increasing x-values. Although it is difficult to attribute the excess specific heat over and above the stepwise behaviour to one of the described mechanisms, the increase of the extrapolated step at T_c with increasing x and increasing transition temperature follows the pattern of a second order phase transition approaching a tricritical point (see equation (4.3 ff)). The extrapolated chemical composition of the tricritical behaviour was found by Sugimoto et al., (1981) to be $x = 0.6 - 0.7$. The continuous behaviour of these phase transitions, similar to the M/I transtions in vanadium bronzes investigated by Kapustkin et al., (1977) is in contrast to the stepwise carrier induced phase transition described before.

ACKNOWLEDGEMENTS

E.S. is indebted to Bernd Wruck for many fruitful and stimulating discussions. He is also thankful to V. Heine, M. A. Car penter and A. Putnis for their encouragemnet and help during the course of his work. He thanks Sheila Tuffnell for her patience and indulgence while editing illegible manuscripts. Work supported by N.E.R.C., E.E.C. and departmental resources.

References

Ahlers, G. and A. Kornblit, 1975, Phys. Rev. B 12, 1938-1946.
Aleksandrov, K. S. and I. N. Flerov, Sov. Phys. Solid State, 21, 195-200.
Baloga, J. D. and C. W. Garland, 1977, Rev. Scientific Instruments, 48, 105.
Bartel, J. J. and E. F. Westrum, 1976, J. Chem. Thermodynamics, 8, 583-600.
Berlin, T. H. and M. Kac, 1952, Phys. Rev., 86, 821.
Bismayer, U., E. Salje, M. Glazer and X. Cosier, 1985, Phase Tran. 6, 126.
Bismayer, U., E. Salje and C. Joffrin, 1982, J. Phys. (Paris) 43, 1379.
Bruce, A. D. and A. Aharony, 1974, Phys. Rev., B 10, 2078.
Bruce, A. D. and R. A. Cowley, 1981, Structural Phase Transitions, Taylor and Francis, London.
Carslaw, H. S. and Jaeger, J. C. 1959, Conduction of Heat in Solids, (The Clarendon Press, Oxford).
Devarajan, V. and E. Salje, 1984, J. Phys. C. 17, 5525-5537.
Ditmers, D. A. and T. B. Douglas, , J. Res. A No. 5, 401-420.
Dvorak, V. 1972, Phys. Stat Solidi, 52, 93-98.
Fisher, M. E. and A. Aharony, Phys. Rev. Lett., 30, 559.
Garland, C. W. 1964, J. Chem. Phys., 41, 1005.
Garland, C. W. and J. D. Baloga, 1977, Phys. Rev. B 16, 331-339.
Ginnings, D. C. and G. T. Furukawa, 1953, J. Am. Chem. Soc., 75, 522.
Granato, A. V. and T. G. Nilan, 1965, Phys. Rev., 137, 1250.
Hatta, I, and A. J. Ikushima, 1981, Japan Journal of Applied Physics, 20, 195-201.
Hatta, I., Y. Shiroishi, K. A. Mueller and W. Berlinger, 1977, Phys. Rev. B 16, 1138-1145.
Imry, Y. 1975, J. Phys. C. 8, 567.
Imry, Y. and S. Ma, 1975, Phys. Rev. Lett., 35, 1399.
Iwata, T. and T. Nihira, 1976, J. Phys. Soc. Japan, 15, 575.
Janovec, V. 1966, J. Chem. Phys., 45, 1874.
Kerr, H. V., D. L. Dickerson, H. Kuwamoto, H. L. C. Barros and J. M. Honig, 1976, J. Solid State Chem., 19, 95-102.
Kraftmakher, Y. A. 1962, Zh. Prikl. Mekh. & Tekh. Fiz., 5, 176.
Landau, L. D. and E. M. Lifshitz, 1980, Statistical Physics, Pergamon.
Larkin, A. I. and S. A. Pikin, 1969, Soc. Phys. JETP, 29, 891-896.
Ledermann, F. L., M. B. Salamon and L. W. Shacklette, 1974, Phys. Rev. B 9, 2981-2988.
Levanyuk, A. P. 1963, see L. D. Landau and E. M. Lifshitz, 1980, Statistical Physics, Vol. 5, Pergamon Press, 3rd. Edition, p.482.
Levanyuk, A. P., V. V. Osipov, A. S. Sigov and A. A. Sogyanin, 1979, Sov. Phys. JETP, 49, 176-188.
Levanyuk, A. P., A. S. Sigov and A. A. Sobyanin, 1980, Ferroelectrics, 24, 61-66.
Levanyuk, A. P. and A. A. Sobyanin, 1970, ZhETF Pis. Red 11, 540-543.
Lopez-Echarri, A., M. J. Tello and P. Gilli, 1980, Solid State Commun., 36, 1021.
Lubensky, T. C. 1975, Phys. Rev., B 11, 3573.
Meissner, G., L. Lasvari and B. Tadic, 1986, Phys. Rev. B 34, 350.

Chaudhuri, B. K., K. Nomoto, T. Ataka and H. Chihara, 1980, Phys. Lett., A 79, 361.
Cheung, K. M. and F. G. Ullmann, 1974, Phys. Rev., B 10, 4760-4764.
Egorov, V. M., S. Kh. Esayan, V. V. Lemanov and T. M. Polkhovskaya, 1981, Sov. Phys. Sol. State, 23, 1455-1457.
Ema, K., K. Hamano, K. Hurihara and I. Hatta, 1977, Journ. Phys. Soc. of Japan, 43, 1954-1961.
Ginnings, D. C. and G. T. Furukawa, 1953, J. Am. Chem. Soc., 75, 522.
Hatta, I. and A. Ikushima, 1976, Journ. Phys. Soc. of Japan, 41, 558-564.
Helwig, J. 1976, Ferroelectrics, 11, 471-477.
Helwig, J., J. Peterson and E. Schneider, 1977, Z. Phys., B 28, 87.
Lopez Echarri, A. and M. J. Tello, 1981, J. Phys., D 14, 71-77.
Matsuo, T., M. Oguni, H. Suga and S. Seki, 1973, Proc. Japan, Acad., 49, 196.
O'Neill, M. J. O. 1966, Anal. Chem., 38, 1331.
Perez Jubindo, M. A., A. Lopez Echarri and M. J. Tello, 1981, Ferroelectrics, 39, 1171.
Reese, M. and L. F. May, 1972, Ferroelectrics, 4, 65.
Stokka, S. and K. Fossheim, 1981, Phys. Rev., B 24, 2807-2811.
Strukov, B. A., S. A. Traskin and V. M. Varikash, 1968, Sov. Phys. Sol. State, 10, 1445.
Taraskin, S. A. 1970, Sov. Phys. Sol. State, 12, 1443.
Taraskin, S. A., B. A. Strukov and V. A. Maleshina, 1970, Sov. Phys. Sol. State, 12, 1089.
Yoshida, H., H. Takemasa, Y. Oshino and Y. Makita, 1984, Journ. Phys. Soc. of Japan, 53, 2600-2605.

References for structural phase transitions -

Barta, Ch., B. S. Zadokhin, Yu. F. Markov and V. V. Tikhonov, 1977, Sov. Phys. Sol. State, 19, 1292-1293.
Bartel, J. J. 1976, J. Chem. Thermodynam., 8, 583.
Chihara, H. and K. Masukane, 1973, Journ. Chem. Phys., 59, 5397-5403.
Cullen, J. R. and E. Callen, 1971, Phys. Rev. Lett., 26, 236.
Davidovich, R. L., P. S. Gordienko, J. Griggs, T. A. Kaudalova, U. Urbonavicius and L. A. Zemnukhova, 1984, Phys. Stat. Sol., A 84, 387-392.
Dormann, J. L., C. Djega-Mariadassou and V. A. M. Brabers, 1982, Proc. ICAME (New Delhi, Indian Nat. Sci. Acad.), p.196.
Evans, B. J. and E. F. Westrum, 1972, Phys. Rev., B 5, 3791.
Falk, B. G., L. Pan, B. J. Evans and E. F. Westrum, 1979, J. Chem. Thermodynamics, 11, 367.
Flerov, I. N. 1976, Sov. Phys. Sol. State, 18, 487-488.
Gorechi, T. 1978, Z. Metallkunde, 69, 782-783.
Gronvold, F. J. 1973, J. Chem. Thermodynamics, 5, 545.
Gronvold, F. J. 1976, J. Chem. Thermodynamics, 8, 757-775.
Hatta, I. and W. Rehwald, 1977, J. Phys. C. 10, 2075-2081.
Heibel, R. and H. Wollenberger, 1976, Scripta Metallurgica, 10, 945-947.
Heibel, R. and H. Wollenberger, 1978, J. of Less Common Metals, 57, 177-183.

Iizusi, M., Y. Hamaguchi, K. F. Komatsubara and Y. Kato, 1975, J. Phys. Soc. Japan, 38, 443-449.
Imai, K., M. Sato and Y. Abe, 1974, J. Electrochem. Soc., Solid State Sci. and Technology, 121, 1674-1676.
Inaba, H., H. Shimizu and K. Naito, 1977, Journ. of Nuclear Materials, 64, 66-70.
Janik, J. M., M. Rachwalsky and J. A. Janik, 1974, Physica, 72, 168-178.
Jewess, M., S. Lee-Bechtold and L. A. K. Stavely, 1980, J. C. S. Faraday, 76, 803-811.
Kamilov, I. K., G. G. Musaev and G. M. Shakhsaev, 1975, Sov. Phys. Dokl., 20, 121.
Kiriyama, H., K. Kiriyama, O. Nakasura and R. Kiriyama, 1973, Bull. Chem. Soc. Japan, 46, 1389.
Knapp, G., S. Knapp, S. D. Bader, H. V. Culbert, F. Y. Fradin and T. E. Klippert, Phys. Rev., B 11, 4331-4338.
Kobayashi, K. L., Y. Kato, Y. Katayama and K. F. Komatsubara, 1976, Phys. Rev. Lett., 37, 772-774.
Leadbetter, A. J. and T. W. Smith, 1976, Phil. Mag., 33, 113-119.
Matsuo, T., M. Tatsumi, H. Suga and S. seki, 1973, Sol. State. Comm., 13, 1829-1831.
Matsuo, T., S. Todo and S. Chikazumi, 1977, J. Phys. Soc. Japan, 42, 15-17.
Nelson, D. R. and E. Domany, 1976, Phys. Rev. B 13, 236-250.
Novotny, V., C. A. Martin, R. L. Armstrong and P. P. Meincke, 1977, Phys. Rev. B 15, 382-389.
Orehotsky, J. and K. Schroeder, 1974, J. Phys. F 4, 196-201.
Rigo, M. O. and J. Kleinclauss, 1980, Phil. Mag., 42, 393.
Rigo, M. O., J. Kleinclauss and A. J. Pointon, 1978, Solid State Comm., 28, 1013.
Rigo, M. O., J. F. Mareche and V. A. M. Brabers, 1983, Phil. Mag. B 48, 421-430.
Sullivan, P. and G. Seidel, 1968, Phys. Rev., 173, 679.
Suzuki, T. and M. Hagiwara, 1975, Trans. JIM, 16, 473-479.
Todo, S. and Chikazumi, 1977, J. Phys. Soc. Japan, 43, 1091.
Umenura, S. and S. Iida, 1976, J. Phys. Soc. Japan, 40, 679.
Urban, S., Z. Tomkovicz, J. Mayer and T. Waluga, Acta Physica Polonica, A 48, 61-67.
Westrum, E. F. and J. F. Gronvold, 1969, J. Chem. Thermodynamics 1, 543.
Westrum, E. F. and S. Henriquez, 1976, Mol. Cryst. Liq. Cryst. 32, 31-35.

References for charge carrier induced structural phase transitions -

Anderson, P. W. 1975, Phys. Rev. Lett., 34, 953.
Brinkman W. F. and T. M. Rice, 1970, Phys. Rev. B 2, 4302.
Chakraverty, B. K. 1975, J. Solid State Chem., 12, 376.
Cohen, M. H., E. N. Economou and C. M. Soukoulis, 1983, Phys. Rev. Lett., 51, 1202.
Cowan, W. B. and M. J. Zuckermann, 1975, Sol. State Comm., 16, 207-210.
Craven, R. A. and S. F. Meyer, 1977, Phys. Rev. B 16, 4583-4593.
Goodenough, J. B. 1972, Prog. Sol. State Chem., 5, 145.

Maeda, M. 1979, J. Phys. Soc. Japan, 47, 1581-1587.
McWhan, D. B., A. Jayaraman, J. P. Remeika and T. M. Rice, 1975, Phys. Rev. Lett., 34, 547.
Matsuo, T., Y. Kume, H. Suga and S. Seki, 1976, J. Phys. Chem. Solids, 37, 499-506.
O'Neill, M. J. 1966, Anal. Chem., 38, 1331.
Reese, W. 1969, Solid State Commun., 7, 969.
Rosencwaig, A. and A. Gersho, 1976, J. Appl. Phys., 47, 64.
Salje, E. 1985, Phys. Chem. Minerals, 12, 93-98.
Salje, E. and V. Devarajan, 1981, J. Phys. C., L1029.
Salje, E. and V. Devarajan, 1985, Phase Trans., 6, 35.
Salje, E., B. Kuscholke, B. Wruck and H. Kroll, 1985, Phys. Chem. Minerals, 12, 99-107.
Salje, E. and Chr. Werneke, 1982, Contrib. Mineral. Petrol., 79, 56-67.
Salje, E. and B. Wruck, 1983, Phys. Rev. B, 28, 6510-6518.
Schlottermann, P. 1975, J. Low Temperature Physics, 20, 123-154.
Scott, J. F., F. Habbal and M. Hidaka, 1982, Phys. Rev. B, 25, 1805-1812.
Shang-Keng, Ma, 1976, Modern Theory of Critical Phenomena, Benjamin, Reading, Mass.
Shimotomai, M., T. Iwata and M. Doyama, 1985, Phil. Mag., 51, L49-53.
Sobatta, G. 1982, J. Magn. Mater., 28, 1.
Speer, D. and E. Salje, 1986, Phys. Chem. Min., 13 17.
Stokka, S. and K. Fossheim 1982, Phys. Rev., B 25, 4896-4898.
Strukov, B. A., S. A. Taraskin, K. A. Minaeva and V. A. Fedorikhin, 1980, Ferroelectrics, 25, 399-402.
Sullivan, P. and G. Seidel, 1968, Phys. Rev., 173, 679.
Swift, J. W. and M. K. Grover, 1974, Phys. Rev., A9, 2579.
Toledano, J. C. and P. Toledano, 1980, Phys. Rev., B 21, 1139-1172.
Toledano, P. and J. C. Toledano, 1977, Phys. Rev., B 16, 386-407.
Toledano, P. and J. C. Toledano, 1982, Phys. Rev., B 25, 1946-1964.
Varma, C. M. amd B. I. Halprin, 1976, Phys. Rev., B 14, 4030.
Wegner, F. J. 1972, Phys. Rev., B 5, 4529.
Westerhold, K. and G. Sobotta, 1983, J. Phys. F 13, 2371-2388.
Winterfield, V. and G. Schaak, 1977, Phys. Stat. Sol., B 80, 559-567.

References for experimental methods -

Buckingham, M. J., C. Edwards and J. Lipa, 1973, Rev. Sci. Instruments, 44, 1167-1172.
Gmelin, E. and P. Roedhammer, 1981, J. Phys. Rev., 15, 778-789.
Lancaster, P. C. amd D. P. Baker, 1981, J. Phys. E 14, 805-809.
Lippa, J. A., C. Edwards and M. J. Buckingham, 1977, Phys. Rev., 15, 778-789.
Wuerz, U. and M. Grubic, 1980, J. Phys., E 13, 525-529.

References for ferroelectrics -

Barth, E., J. Helwig, H. D. Maier, H. E. Muser and J. Peterson, 1979, Z. Phys., B 34, 393-397.
Chaudhuri, B. K., T, Atake and H. Chihara, 1981, Solid State Comm., 38, 927-930.

Kapustkin, V. K., E. D. Pletneva and V. L. Volkov, 1977, Sov. Phys. Sol. State, 19, 429-431.
Kristoffel, N. and P. Konsin, 1973, Ferroelectrics, 6, 3.
Lakkis, S., C. Schlenker, B. K. Chakraverty and R. Buder, 1976, Phys. Rev. B 14, 1429-1440.
Marezio, M., D. Tranqui, S. Lakkis and C. Schlenker, 1977, Phys. Rev. B 16, 2811-2821.
Mattheis, L. F. 1973, Phys. Rev. B 8, 3719.
McMillan, W. L. 1977, Phys. Rev. B 16, 643.
McMillan, W. L. 1981, Phys. Rev. B 24, 2739.
Moncton, D. E., J. D. Axe and F. J. DiSalvo, 1975, Phys. Rev. Lett., 34, 734.
Mott, N. F. 1974, Metal-Insulator Transitions (Taylor and Francis, London, 1974).
Rehwald, W. and G. K. Lang, 1975, J. Phys. Chem., 8, 3287.
Rice, T. M. and G. K. Scott, 1975, Phys. Rev. Lett., 35, 120.
Salje, E. and B. Guettler, 1984, Phil. Mag. B 50, 607-620.
Schirmer, O. F. and E. Salje, 1980, a) Sol. State. Comm., 33, 333: b) J. Phys. Chem., 13, L1067.
Schlenker, C. and M. Marezio, 1980, Phil. Mag., B 42, 453-472.
Slonczewski, J. C. and H. Thomas, 1970, Phys. Rev., B 1, 3599.
Sugimoto, N., T. Matsuda and I. Hatta, 1981, J. Phys. Soc. Japan, 50, 1555-1559.
Surikov, V. I., S. V. Danilov and V. I. Surikov, 1983, Sov. Phys. Sol. State, 25, 909-910.
Surikov, V. I., I. I. Miller, E. M. Yarosh, V. I. Surikov and N. E. Kondrat'eva, 1979, Sov. Phys. Sol. State, 21, 1069-1070.
Wilson, J. A., F. J. DiSalvo and S. Mahajan, 1975, Adv. Phys., 24, 117.
Wilson, J. A. and A. D. Yoffe, 1969, Adv. Phys., 18, 193.
Yoffe, A. D. L976, Chem. Soc. Rev., 5, 51.

WHAT CAN SPIN MODELS TELL US ABOUT THE BEHAVIOUR OF MINERALS?

Julia Yeomans
Department of Theoretical Physics
1, Keble Road
OXFORD OX1 3NP
England

ABSTRACT. I aim to discuss the extent to which model spin Hamiltonians can give insight into the properties of minerals. I argue that although it is difficult to reproduce the exact quantitative phase behaviour the physical principles involved are often clearly elucidated by a simple model.

Examples include the universality of critical exponents at continuous phase transitions, interface properties and modulated phases that result from competing interactions.

1. INTRODUCTION

My aim in this article is to discuss the extent to which rather simple discrete spin models, like the Ising model of magnetism, can be useful in understanding the physics of real materials. The main advantage of using model systems is that they can be chosen to be tractable theoretically and therefore the details of their behaviour can be understood with some confidence. In particular the aim is to extract a clear understanding of the physics leading to this behaviour which, it is hoped, will be mirrored in the real compounds.

Once the basic principles have been established various refinements or perturbations can be included in the models. Examples would be the effect of more complicated interactions, of defects or of more realistic lattice structures. By ascertaining the robustness of the system to these perturbations it should be possible to establish whether they will significantly change the important physics and hence whether they are essential to realistically model a particular experimental system.

The usefulness of models in obtaining a quantitative fit to experimental data, rather than a qualitative understanding of it, depends crucially on the property of the material considered. For example, the value of a critical temperature, T_C, will depend strongly on the details of the interatomic interactions and predictions are the domain of band theorists and quantum chemists who aim to model these interactions accurately. Critical exponents, on the other hand, are Universal, in that they depend only on symmetries and the range of the

E. K. H. Salje (ed.), Physical Properties and Thermodynamic Behaviour of Minerals, 119–142.

interactions. Hence accurate values can be obtained from very simple models which only have to include these features correctly.

I shall, in Section 2, describe some simple models and their experimental counterparts to illustrate these ideas. The first example is the classical case of β-brass and its relation to the Ising model which helped to establish the Universality of critical exponents. A more complicated Ising Hamiltonian, which reproduces the phase diagram of face-centred cubic binary alloys, will then be discussed. The third example, the two-dimensional Potts model, has a different symmetry and has been important in understanding adsorbed layers of krypton on a graphite substrate.

Section 3 is devoted to recent applications of axial Ising models with competing interactions to describe polytypism in minerals and binary alloys. Using silicon carbide as an example it is shown that models can be constructed that reproduce almost all of the salient features of polytypic phases. However, long-period phases can be generated by either entropic effects or long-range interactions and it remains an open question which is most relevant in the various polytypic compounds.

In Section 4 I consider some properties of surfaces and interfaces which have been predicted from the study of model systems. Their relevance to experimental work to date is considered. The aim is to provoke interest in other experimental systems which may show similar features.

2. UNIVERSALITY AND PHASE DIAGRAMS

a. The Ising Model and β-brass

The simplest spin model, which already provides a useful representation of some compounds, is the spin-½ Ising model (Stanley 1971). This comprises a spin variable, S_i, which can take values ±1, positioned on each site, i, of a regular lattice. The spins interact through a Hamiltonian

$$H = -J \sum_{nn} S_i S_j - H \sum_i S_i \tag{2.1}$$

where initially we just consider nearest neighbour pairwise interactions. The model is usefully thought of as the simplest case of an interacting two-state system.

In dimensions, d>1, and zero field there is a second order phase transition at a temperature $k_B T_c/J$ at which the thermodynamic functions diverge with critical exponents, α, β, γ defined by (Stanley 1971, Fisher 1967, 1974)

$$\begin{aligned} &\text{specific heat:} && C \sim |T-T_c|^{-\alpha}, \\ &\text{magnetisation:} && M \sim (T_c-T)^{\beta}, \\ &\text{susceptibility:} && \chi \sim |T-T_c|^{-\gamma}. \end{aligned} \tag{2.2}$$

Table I: Critical exponents, α, β, γ, of the d-dimesional Ising model. Experimental results for β-brass are shown for comparison.

	α	β	γ
d=2	0(log)	$\frac{1}{8}$	$1\frac{3}{4}$
d=3	0.105±0.010	0.328±0.008	1.239±0.002
β=brass		0.305±0.005[a]	1.24±0.015[a]
d⩾4	0	$\frac{1}{2}$	1

[a] Als-Nielsen 1976.

In d=1 ordering only occurs at zero temperature. T_c and the critical exponents are known exactly in d=2 and the critical exponents take their mean-field values for d⩾4. In d=3 numerical approximations have delved the behaviour near the transition to great accuracy. Values of the critical exponents are listed in Table I. Note that for J>0 the ordered state is ferromagnetic whereas J<0 it is antiferromagnetic.

The most obvious physical examples which might be expected to correspond to an Ising model are magnetic phase transitions in systems where the spins have strong uniaxial anisotropy; for example Rb_2CoF_4 is an Ising antiferromagnet. However, any interacting two-state system provides a possible candidate: a famous example is β-brass (Als-Nielsen 1976).

β-brass is a Cu-Zn alloy with equal numbers of Cu and Zn atoms occupying the sites of a body-centred cubic lattice. At high temperatures the distribution of Cu and Zn atoms on the lattice sites is random but, as the temperature is decreased, there is a continuous order-disorder phase transition to a low temperature phase where each Zn atom is surrounded by eight Cu nearest neighbours and vice versa.

As each lattice site, i, can be occupied by either a Cu or a Zn atom the occupation number of each site is a two-state system. Hence it may be mapped onto an Ising model where S_i = +1 or -1 if site i is occupied by a Cu or Zn atom respectively. If the interaction energy of two neighbouring Cu atoms is J_{cucu} and so forth, an interaction Hamiltonian may be written as

$$H = \tfrac{1}{4} \sum_{nn} J_{cucu}(1+S_i)(1+S_j) + \tfrac{1}{4} \sum_{nn} J_{znzn} (1-S_i)(1-S_j)$$

$$+ \tfrac{1}{4} \sum_{nn} J_{cuzn} \{(1+S_i)(1-S_j) + (1-S_i)(1+S_j)\} \qquad (2.3)$$

Collecting terms this maps onto

$$H = -J \sum_{nn} S_i S_j - H \sum_i S_i + C \qquad (2.4)$$

where J, H and C are linear combinations of J_{cucu}, J_{znzn} and J_{cuzn}. For equal numbers of Cu and Zn atoms $\Sigma S_i = 0$ and hence we are led to an Ising model in zero field.

In writing down the Hamiltonians (2.3) and (2.4) we have assumed that interactions between atoms act between nearest neighbour pairs only. Moreover, we have ignored any dependence of the interaction parameters on temperature which may occur because of, for example, thermal expansion. The extent to which these assumptions matter depends on which properties one is tying to extract from the spin Hamiltonian. The important things to calculate are T_c and the critical exponents and the variation of the sublattice magnetisation (or equivalently, the difference in the number of A and B atoms on a given sublattice) below T_c.

Of these the critical exponents fall into a special category because they are Universal and depend, not on the details of the compound considered but only on a small number of parameters, the most important of which are the dimensionality of space, the symmetry of the order parameter and the range of the interactions (Fisher 1974). Hence the Hamiltonian (2.4) is expected to have the same exponents as β-brass solely because the order parameter is a scalar in each case. The addition of multispin or further neighbour interactions, which may be present in the real compound, will not change the critical exponents - they are termed irrelevant variables (unless the missing interactions are sufficiently long-ranged). This has been confirmed by neutron scattering experiments (Als-Nielsen 1976), the results of which are given in Table I for comparison with the three-dimensional Ising values.

Non-universal quantities, such as T_c and the variation of the order parameter with temperature, do depend on the specific details of the compound considered. Hence to obtain quantitatively accurate results the model Hamiltonian must provide a faithful representation of the interatomic interactions. That this is the case can either be confirmed by fitting to the experimental results or by a first principles calculation based on band theory or model atomic potentials. The experimental results for β-brass, shown in Figure 1, indicate that it is well modelled by a nearest neighbour Ising model particularly if the expansion of the lattice with temperature is taken into account (Baker and Essam 1971).

Although in general it is impossible to exactly reproduce the real Hamiltonian of a compound, often this is unimportant as the rôle of the model is to provide a more easily solvable case where the physical principles leading to ordering in the real compound can be elucidated. Hence unimportant terms in the Hamiltonian which change quantitative but not qualitative results are often omitted. An example of this is given in Section 3 where we show how competing interactions can result in modulated phases.

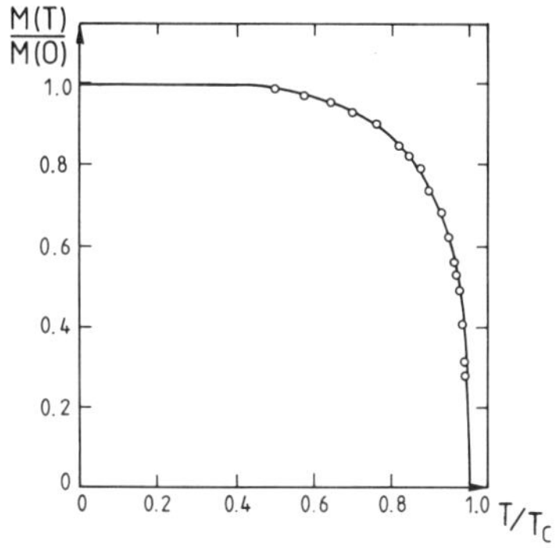

Fig.1: Variation of the reduced order parameter, M(T)/M(0), of β-brass as a function of reduced temperature, T/T_C. The open circles are neutron scattering results; the solid line is the theoretical result for a compressible Ising model (Als-Nielsen 1976).

b. The Ising Model and Face-centred Cubic Binary Alloys

We shall now consider how a more complicated Ising model can be used to study ordering in an AB binary alloy on a face-centred cubic lattice, a standard example of which is CuAu (de Fontaine 1979, Binder 1986). Again we consider only the configurational degrees of freedom and ignore, for example, lattice vibrations and electronic degrees of freedom. A spin variable S_i=+1 or -1 is assigned to lattice site i if it is occupied by an A- or a B-atom respectively. A Hamiltonian like (2.3) can be defined and, collecting terms, we are led as before to an Ising Hamiltonian

$$H = J \sum_{nn} S_i S_j - \alpha J \sum_{nnn} S_i S_j + H \sum_i S_i \qquad (2.5)$$

where we have included both first (nn) and second (nnn) neighbour interactions. The justification that this is necessary and sufficient to give a good approximation to the phase diagram must follow from comparisons to experiment. Many authors have concentrated on the Hamiltonian (2.5) where interesting theoretical problems still remain, although others have included further terms, such as 4-spin couplings.

We are working on a face-centred cubic lattice which can be considered to consist of four interpenetrating sublattices. The nearest neighbour interaction, J>0, is taken to be antiferromagnetic and acts between a site and its twelve neighbours, four on each of the other three sublattices. The ferromagnetic interaction, αJ>0, couples a site and its six second neighbours which lie on its own sublattice. The field, H, acts as a chemical potential which controls the relative concentrations of A and B atoms.

Our aim is to describe the phase diagram of (2.5) pointing out the theoretical techniques available to obtain it and the sort of outstanding problems that concern theoretical physicists. The ground

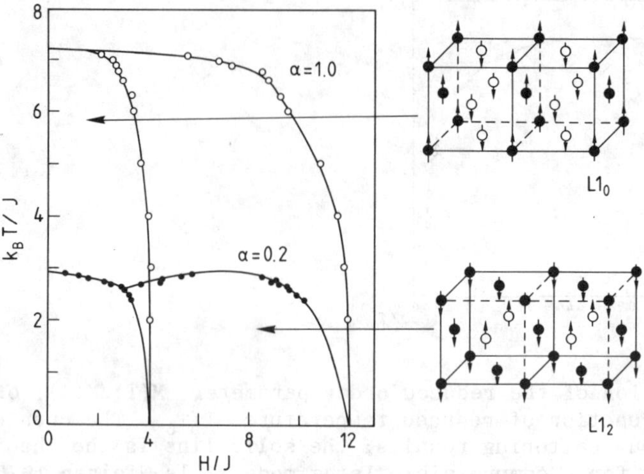

Fig.2: Phase diagram of the Hamiltonian (2.5) from Monte Carlo simulations (Binder 1981, Binder et al. 1983). The $L1_0$ and $L1_2$ structures are illustrated with black circles or spins $S_i = -1$ corresponding to B-atoms and white circles or spins $S_i = +1$ corresponding to A-atoms.

state follows from a comparison of the energies of possible structures. For $\alpha \neq 0$ and H<4J the six-fold degenerate $L1_0$ state, where two sublattices are filled with A atoms and two with B atoms, provides the ground state. For 4J<H<12J the four-fold degenerate $L1_2$ phase, in which one sublattice is populated by A atoms and the other three by B atoms, is stable. These phases are illustrated in Figure 2. For H>12J the ground state consists entirely of B atoms.

For $\alpha=0$ the situation is more complicated as the ground-state is frustrated - it is impossible to satisfy all the bonds at the same time. For H<4J the (½, 0, 0) planes remain antiferromagnetically ordered but there is no ordering between planes. Similarly for 4J<H<12J the (½, 0, 0) planes alternate between ferromagnetic and antiferromagnetic order but there are no long-range correlations between the antiferromagnetically ordered planes. (H=4J, α=0) and (H=12J, α=0) are special points at which the ground-state is so degenerate that the entropy remains finite at zero temperature.

The theoretical methods available to study this and similar models at finite temperatures fall primarily into two categories (Binder 1986). Mean-field theories have the advantage that they are usually tractable but the disadvantage that they are not always reliable in predicting, say, the order of phase transitions and should be compared to other methods before they can be believed. A sophisticated mean-field theory - the cluster variational method - appears to work well for this particular problem but the equations involved present some degree of complexity. Monte Carlo simulations provide an alternative approach and give good results provided finite-size effects and the possibility of getting stuck in metastable states are taken into account. They provide a check on the mean-field theories and often can give greater insight into more subtle aspects of the phase diagram.

The finite temperature phase diagram of the Hamiltonian (2.5) obtained from a Monte Carlo simulation is shown in Figure 2 for two values of α (Binder 1981, Binder et al. 1983). All phase boundaries are first order except at the point H=0, α=1.0. The triple point, where the AB, AB_3 and disordered phases meet, is at finite H for small α, but moves towards H=0 as α increases reaching it at $\alpha_c \approx 0.3$. This point probably coincides with the crossover from first to second order behaviour but this is still controversial. Another outstanding question is what happens for low temperatures near the highly degenerate points (H=4J, α=0) and (H=12J, α=0). It is not known whether the triple point still occurs at a finite temperature or whether the disordered phase persists to zero temperature. No theoretical technique exists which can be trusted in this region and numerical simulations are very difficult (Lebowitz et al. 1985).

Theoretical studies on models like (2.5) demonstrate the type of behaviour which might be observed in experiments on face-centred cubic alloys. Once a model seems, through experimental comparisons, to be a good representation of reality, other properties, such as the effect of competing interactions, discussed in Section 3d, and surface and grain boundary properties, mentioned in Section 4, can be studied and experiments suggested.

c. The Three-state Potts Model and Krypton on Graphite

As a rather different example of a system where models have been very useful in explaining experimental results we consider Kr adsorbed on the basal surface of graphite (Caflisch et al. 1985, Stephens et al. 1984). The graphite provides a stable substrate with hexagonal modulation onto which the inert gas atoms are physisorbed to form an effectively two-dimensional system. Because the Kr atoms are larger than the graphite hexagons nearest neighbour adsorption is unfavourable and the strongest pair adsorption is at second neighbour sites. Accordingly for coverages of around 1/3 the system likes to order in the $\sqrt{3} \times \sqrt{3}$ commensurate structure shown in Figure 3.

The experimental phase diagram is shown in Figure 4 (Stephens et al. 1984). Models of increasing degrees of sophistication have been used to explain various features of this phase diagram and we shall consider various aspects of these in turn. Firstly consider a coverage of one Kr atom per three adsorption sites and note that there are three equivalent positions for the ordered state with the atoms occupying one of the three sublattices, A, B or C, of the triangular adsorption-site lattice. Different possibilities are shown in different regions of Figure 3. Each sublattice is completely equivalent and hence the order parameter has a choice of three values when the symmetry is broken in the ordered phase (compare the Ising model where the magnetisation can be equivalently up or down in zero field). The simplest model with an order parameter of this symmetry is the three-state Potts model (Wu 1982)

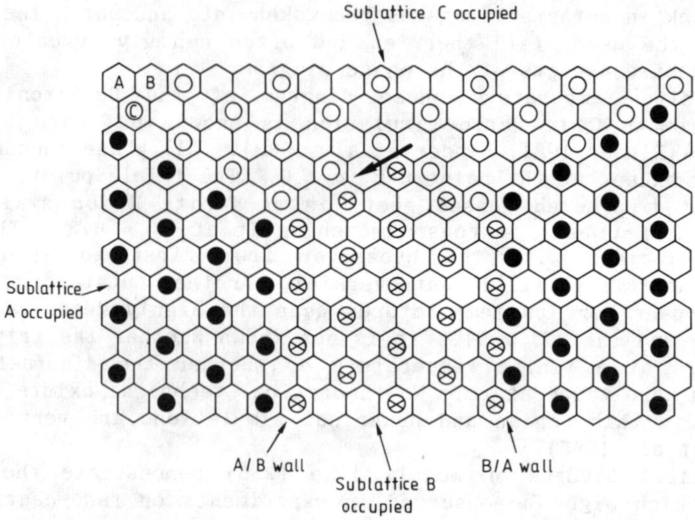

Fig.3: Adsorption of krypton on a graphite substrate showing the three possible domains and walls between them. The black arrow indicates the Burger's vector of a dislocation (Caflisch et al. 1985).

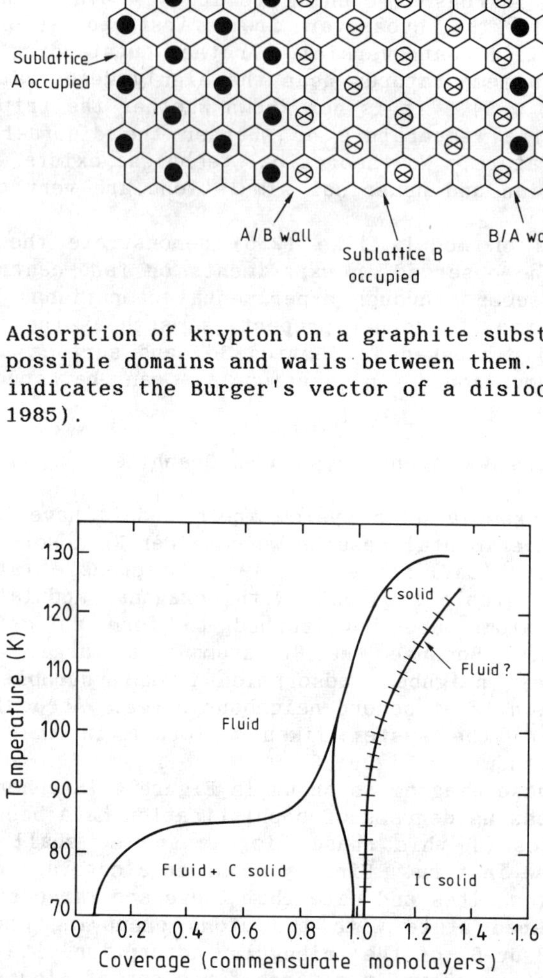

Fig.4: Experimental phase diagram of krypton on graphite (Stephens et al. 1984). C and IC label the commensurate and incommensurate phases respectively.

$$H = -J \sum_{nn} \delta\sigma_i\sigma_j \qquad (2.6)$$

where the variables σ_i can take the three values, A, B or C and the sum is over sites on a triangular lattice, each of which is associated with a triplet, A, B, C, of possible adsorption sites. $\delta\sigma_i\sigma_j$ is the Kronecker delta which takes the value 1 if $\sigma_i=\sigma_j$ and 0 otherwise.

The first steps in obtaining the phase diagram used the Potts model (2.6) but with more realistic potentials and with the additional possibility that a given A, B, C triplet of adsorption sites could be empty (Berker et al. 1978). The coverage could then be controlled by a chemical potential which determined the number of empty sites.

More recently a lot of work has been done to incorporate into the models an interesting feature which is necessary to describe the incommensurate phase correctly (Caflisch et al. 1985). As the coverage increases from $1/3$ extra atoms must be accommodated on the surface. Rather than stick uniformly to form a regular pattern of atoms, most of which would then not lie in the potential minima of the adsorption sites, the extra atoms are accommodated in domain walls which separate commensurate regions of the three different ground states. This is shown in Figure 3. It is typical of an incommensurate phase where the mean atomic spacing, or equivalently the domain wall spacing, vary continuously with the coverage. In the case of Kr on graphite it is favourable for the domain walls to form a hexagonal array.

By noting the incommensurate phase one can see that the Potts model, where all three states are equivalent does not correctly describe the symmetry of this compound for coverages $\neq 1/3$. This is because an A|B domain wall is not entirely equivalent to a B|A wall as the latter contains more densely packed atoms (Caflisch et al. 1985, Huse et al. 1983). Hence the symmetry between the three states is broken by a so-called chiral interaction which differentiates between A|B or B|C and C|B or B|A walls. This may alter the critical exponents if the chiral interaction is relevant for hexagonal walls in two-dimensions. (The same phenomenon leads to an interesting case of domain wall wetting which is described in Section 4.)

Note also that the experimental results in Figure 4 show that a disordered fluid phase persists to zero temperature between the $\sqrt{3} \times \sqrt{3}$ commensurate and the incommensurate phases. This is again a property of the domain walls in the incommensurate phase and occurs because they are unstable to the formation of dislocations, such as that shown in Figure 3 (Coppersmith et al. 1981).

3. MODULATED STRUCTURES

In this Section the aim is to describe in some detail the use of Ising models with competing interactions to model modulated phases in minerals and binary alloys (Yeomans 1987, Yeomans and Price 1986). Two different mechanisms for generating long-period phases will be described. It is still an open question which is most relevant in the real systems.

a. Polytypism in Silicon Carbide

Many compounds can be considered to be composed of one or more structural units. If these units can be stacked in different ways to form stable or metastable states the resulting phases are known as polytypes (Verma and Krishna 1966, Krishna 1984). One of the richest illustrations of polytypism is provided by the classical polytypes, such as SiC and CdI_2, in which the 'ABC' stacking sequence of the close-packed layers can vary to form large numbers of different crystal structures, some with very long periods.

It is conventional to describe the stacking sequence of a given polytypic compound using a notation introduced by Zdhanov and Minervina (1945). Different polytypes can be distinguished by assigning + to represent the stackings A-B, B-C and C-A and - to denote the anticyclic orderings B-A, C-B and A-C. Then, for example, the stacking sequence

A B C B A C A B A C B C A C B A
or + + - - - + + - - - + + - - -   (3.1)

is denoted by $\langle 23\rangle_3$ because two plus stackings are followed by three minus stackings. The subscript 3 outside the brackets appears because the sequence must repeat three times to obtain a complete unit cell of the crystal. More complicated structures, for example $\langle 2223\rangle_3$ are abbreviated as $\langle 2^3 3\rangle_3$. In the following it will be helpful to use the term band to describe each of the figures in the Zdhanov symbol.

Taking SiC as an example, we list the properties of this compound which we shall later attempt to explain using a spin model:

1. The short-period phases, $\langle 3\rangle$, $\langle 2\rangle$, $\langle 23\rangle$, $\langle\infty\rangle$ and $\langle 1\rangle$ are by far the most commonly observed.
2. In the rarer long-period polytypes 3-bands predominate, 2-bands are common and 4-and longer bands also appear. 1-bands are only observed in the structure $\langle 1\rangle$. The sequence of bands is not random and the observed structures can be grouped into well-defined series, as shown in Table II (Pandey and Krishna 1984).
3. Disordered structures, which cannot be assigned a unique wavelength, are also common and may have been erroneously documented as long-period phases in some cases.

It is important to stress, before embarking on the programme of modelling SiC using a spin system, that our formalism assumes that the polytypes are equilibrium or highly metastable states. It is now widely accepted that the short-period phases in SiC can exist as equilibrium structures and indeed phase transitions between them, albeit not reversible ones, have been observed (Jepps and Page 1984). However, there is still considerable controversy about whether the longer-period phases occur just as a result of growth conditions (Pandey and Krishna 1984). One of the arguments against their existence as stable states was the lack of a convincing mechanism and we aim to remedy this deficiency here. It is important to remember, however, that metastability effects are extremely important in polytypic compounds and

Table II:[a] Prominent structural series in SiC

Series	Known Polytypes	Corresponding Values of n
$\langle 23(33)^{n}\rangle$	15R,33R,$51R_1$,$69R_1$ 87R,105R,123R 141R,159R,393R	0,1,2,3,4,5,6,7,8,21
$\langle (33)^{n}34\rangle$	21R,39R,57R,93R 111R,129R	0,1,2,4,5,6
$\langle (22)^{n}23\rangle$	15R,27R,$51R_2$	0,1,3
$\langle (22)^{n}33\rangle$	6H,10H,14H,18H	0,1,2,3
$\langle 22(33)^{n}\rangle$	4H,10H,16H	0,1,2
$\langle 22(23)^{n}\rangle$	4H,19H,27R,267R	0,1,3,17
$\langle (23)^{n}33\rangle$	6H,33R,168R,273R	0,1,10,17
$\langle 2^{3}3(33)^{n}\rangle$	27R,45R,99R	0,1,4
$\langle (23)^{2}(33)^{n}(23)(33)^{n}\rangle$	27H,39H	1,2
$\langle (23)^{n}(32)^{m}\rangle$	10H,45H,75R	
$\langle 23(33)^{n}34(33)^{m}\rangle$	27H,39H	
$\langle (33)^{n}35(33)^{m}34\rangle$	21H,$33H_1$,81H	

[a] after Pandey and Krishna 1984

must have a large bearing on their properties.

b. Axial Ising Models with Competing Interactions

In order to understand how the stacking sequence in SiC can be mapped onto an Ising model it is helpful to notice that Zdhanov notation emphasises that the close-packed sequence can be regarded as an assembly of two-state systems. This occurs because we are working in the limit where the stacking A-A is strictly forbidden; thus two choices, A-B or + and A-C or -, remain at each step and we may regard the Zdhanov symbols as Ising variables.

To guess a sensible Hamiltonian we consider the properties of SiC. Firstly, to preserve the integrity of the close-packed layers, a strong ferromagnetic in-plane interaction, J_0, is needed. Just taking nearest neighbour interactions between the spins in neighbouring planes will allow only ferromagnetic, <∞>, or antiferromagnetic, <1>, ordering and therefore we include pair interactions up to third neighbours along the stacking direction. Thus we are led to the Hamiltonian

$$H=\frac{J_0}{2}\sum_{ijj'} S_{ij}S_{ij'}-J_1\sum_{ij}S_{ij}S_{i+1j}-J_2\sum_{ij}S_{ij}S_{i+2j}-J_3\sum_{ij}S_{ij}S_{i+3j},\quad S_{ij}=\pm 1 \tag{3.2}$$

where i labels the layers perpendicular to the stacking or axial direction and j, j' label nearest neighbour sites within a layer. The ground state of the Hamiltonian (3.2) is shown in Figures 5a and b for $J_1>0$ and $J_1<0$ respectively. Note that all the common short-period structures appear as ground state phases: including axial interactions up to third neighbours ensures that this is the case (Price 1983).

However, only short-period phases are stable as ground states because the interactions are of finite range. A further mechanism is necessary to explain long period structures. Here we describe two possibilities: the effects of entropic terms in the free energy at finite temperature and long-range interactions.

i. <u>Entropy</u>. Six of the zero temperature phase boundaries, shown in bold type in Figure 5, are so-called multiphase lines on which the ground state is infinitely degenerate. For example, we shall be particularly interested in the <2>:<3> boundary for $J_1>0$, where all structures containing only 2-and 3-bands are degenerate, and the <3>:<∞> boundary ($J_1>0$), where all states made up of bands of length ⩾3 have the same energy. At finite temperatures in the vicinity of the multiphase lines very small entropic terms in the free energy can distinguish between different ground state phases and as a result an infinite number of different long-period phases become stable as the interaction parameters and the temperature are varied (Yeomans 1987).

The finite temperature phase diagram as a function of J_2/J_1 ($J_1>0$) for $J_3/J_1=-0.1$ is shown in Figure 6 (Selke et al. 1985). It is only possible to show the dominant phases in the Figure but infinite and regular sequences of phases are stable. An infinite sequence of phases <3^k4>, k=0, 1, 2 ... springs from the multiphase point at J_2/J_1 = 0.35

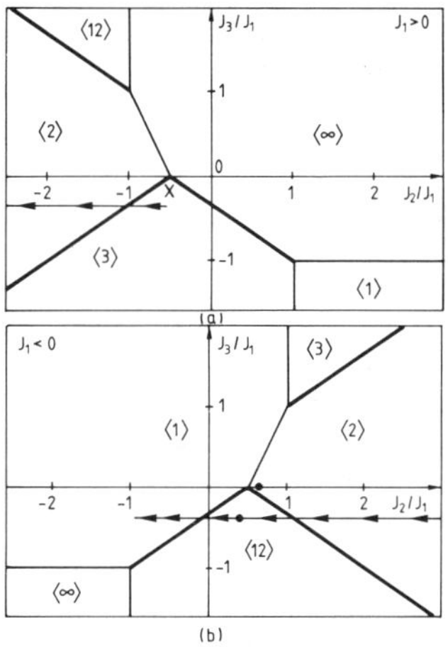

Fig.5: Ground state of the axial Ising model with interactions to third neighbours for (a) $J_1>0$ and (b) $J_1<0$. The cross corresponds to a calculation by Cheng et al. (1987) of the interaction parameters for SiC and the dots to calculations by Price et al. (1985) for the magnesium silicate spinelloids. The arrows show a possible variation of interaction parameters in the spinelloids with pressure.

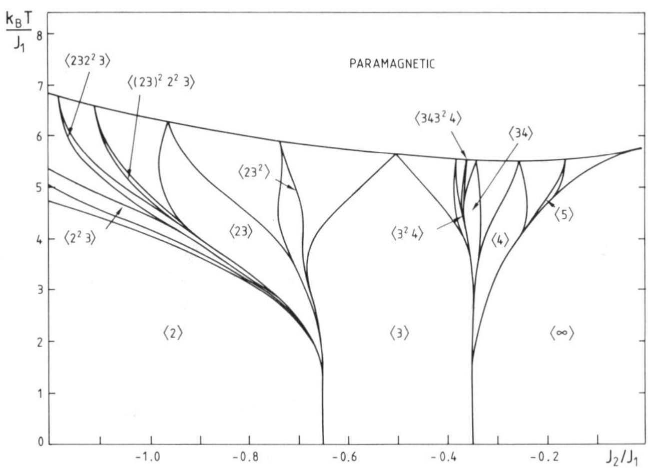

Fig.6: Phase diagram of the axial Ising model with interactions to third neighbours for $J_3/J_1 = -0.1$ (Selke et al. 1985).

between $\langle\infty\rangle$ and <3>. As the temperature is increased new phases appear in a regular manner, with $\langle\nu_1\nu_2\rangle$ always being the first phase to be stabilized between $\langle\nu_1\rangle$ and $\langle\nu_2\rangle$. For example, $\langle 343^2 4\rangle$ appears between <34> and $\langle 3^2 4\rangle$. Similarly the phase sequences $\langle 23^{k+1}\rangle$ and $\langle 23^k 23^{k+1}\rangle$ spring from the multiphase point at $J_2/J_1=0.65$ between <2> and <3>, and, as the temperature is increased new phases appear though the same combination rules. As the paramagnetic phases is approached, incommensurate phases, where the wavevector does not lock to the underlying lattice, are also thought to appear.

A remarkable resemblance to the polytypic phases in SiC is immediately apparent. The most common structures observed in this compound appear as ground state phases and also dominate the phase diagram at finite temperature. 3-bands appear most frequently in the longer-period phases, with 2- and longer bands also being common. No phases containing 1-bands are stable. Moreover the structural sequences listed in Table II closely resemble those appearing in the Ising model.

The free energy differences between the different phases are extremely small and each phase persists as a strongly metastable state even when it is not thermodynamically stable. Hence the achievement of thermodynamic equilibrium in polytypic compounds is, as expected, very difficult and disordered structures are expected to be common.

A useful way of understanding why the entropic terms in the free energy at finite temperatures lead to long period phases is to think of these structures as being comprised of regular arrays of domian walls (Villain and Gordon 1980). The exact definition of a wall can be chosen for convenience: for example in the phase <6> it is most natural to define a wall as a boundary between up- and down- bands, whereas in the $\langle 23^n\rangle$ sequence the 2-bands can be considered as the walls. At finite temperatures the domain walls fluctuate because of local excitations or spin-flips which contribute to the entropic term in the free energy. These fluctuations cause the domain walls to collide and hence interact. The interactions (together with the intrinsic wall energy) determine the spacing of the walls and hence the wavelength of the associated modulated phase.

The main problem in applying the Hamiltonian (3.2) directly to polytypic compounds is that the spin excitations needed to stabilise the long-period phases correspond to large and unphysical rearrangements of atoms. However such large deviations are in fact not necessary to stabilise the phase sequences. One can write down a Hamiltonian in which the spin variables S_{ij} are allowed to deviate from the values ±1 by small deviations u_{ij}. Allowing the same interactions as in (3.2) gives

$$H=-\frac{J_0}{2}\sum_{ijj'}(S_{ij}-u_{ij})(S_{ij'}-u_{ij'})-\sum_{n=1,2,3}J_n\sum_{ij}(S_{ij}-u_{ij})(S_{i+n,j}-u_{i+nj})$$

$$+A\sum_{ij}u_{ij}^2 \qquad (3.3)$$

where the last term, which behaves like an effective temperature, ensures that the u_{ij} remain small. Villain and Gordon (1980) have shown that, at low temperatures within a mean-field approximation (which is thought to be accurate), the Hamiltonian (3.3) maps onto (3.2) and leads to the same phase diagram.

ii. Long Range Interactions. A second and intuitively more obvious mechanism for generating long-period phases is the existence of long-range interactions due to, for example, elastic or electronic terms in the Hamiltonian. The effect of interactions beyond third neighbours, and indeed of multispin interactions which one would also expect to be present, is to lift the degeneracy on the multiphase lines. The maximum period of the resulting structures will be related to the range of the interaction. Although it is obviously not possible to construct the ground-state for all J_n several special cases are understood and enable us to elucidate the behaviour expected.

If the interactions are long-range and oscillatory, for example, elastic interactions between defect planes in crystals, the ground state is a so-called harmless devil's staircase (Villain and Gordon 1980). That is, there are a large number of long-period commensurate phases separated by first order phase transitions. Phases appear in sequences determined by the usual structure combination rules. Indeed the Hamiltonian (3.2) can, at low temperatures, within a mean-field approximation, be mapped onto an Ising model with long-range oscillatory interactions and therefore, for certain ratios of these interactions, one would expect to reproduce the phase diagram shown in Figure 6.

If the interactions are long-range, convex and antiferromagnetic - this includes power law decays and exponentially decaying Coulomb interactions - there is a complete devil's staircase, with all possible commensurate phases being stable (Bak and Bruinsma 1982). It would, however, be impossible experimentally to distinguish between a complete and a harmless staircase unless the latter was very timid indeed, so that, as the instrumental resolution was increased no new phases appeared!

We have discussed two ways in which long-period structures can appear as thermodynamically stable phases: long-range interactions or entropic wandering of the domain walls. Experimentally the most obvious difference between the two mechanisms are that in the former case the walls remain straight, although they might be uniformly displaced by an elastic interaction, and the phases persist as the temperature approaches zero, whereas in the latter the domain walls fluctuate in time and the long-period structures disappear as the temperature is lowered. In SiC it is not understood which mechanism is dominant. The discrete spin-flip model (3.2) seems very unlikely because the corresponding atomic rearrangements are so large as to be energetically virtually impossible. The Hamiltonian (3.3) could provide a plausible explanation with the u_{ij} modelling the effects of phonons on the domain walls. Long-range interactions provide another possibility but it is difficult to propose a convincing physical origin for them in this compound. Other systems are discussed below.

The interaction parameters, J_n, n=0, 1, 2, 3, have been introduced as purely phenomenological variables, expected to depend in general on the temperature, pressure and defect concentration, and their values have been chosen to fit the experimental facts. Recent pseudopotential calculations by Cheng et al. (1987) have lent strong support to the usefulness of representing SiC polytypes by a spin model by relating the J_n to atomic interactions. They find values corresponding to the point marked by a cross in Figure 5, very close to the values of the parameters chosen in calculating the phase diagram in Figure 6.

c. Polytypism in Other Minerals

We have so far, in order to have an example to focus on, considered the properties of SiC. This is one of many polytypic minerals and we now briefly describe two others which fall usefully into the same framework.

i. Cadmium Iodide. Cadmium iodide is a second classical polytype where the different structures are defined by the close-packed stacking sequence of the iodine layers. The phases which have been observed fall into similar sequences as for SiC, but now 1- and 2-bands predominate (Pandey and Krishna 1984). This suggests that it might be useful to interpret CdI_2 as a spin model with $J_2 \sim \frac{1}{2}J_1 < 0$. A first principles calculation to check these assignments would be of great interest.

ii. The Spinelloids. The spinelloid structural family, which has an ideal stoichometry of AB_2O_4, is based upon a structural unit which is defined by the position of the cations within a basic O framework. All spinelloid structures are constructed from this unit, shown in Figure 7, and its inverse. Hence they may be mapped onto an array of Ising spins (Price and Yeomans 1984).

In almost all spinelloids the ordering within two-dimensional layers corresponds to an Ising ground state in which the ordering is ferromagnetic in one direction and antiferromagnetic in the other.

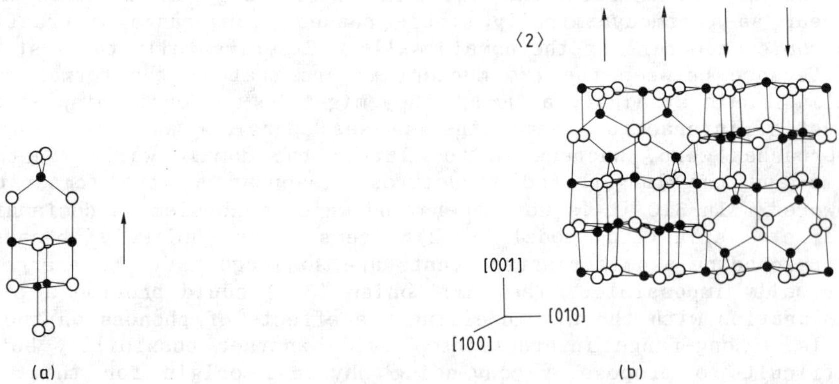

Fig.7: a) Basic spinelloid structural unit. The white circles are oxygen ions, the black, cations. b) Construction of the <2> phase from this unit.

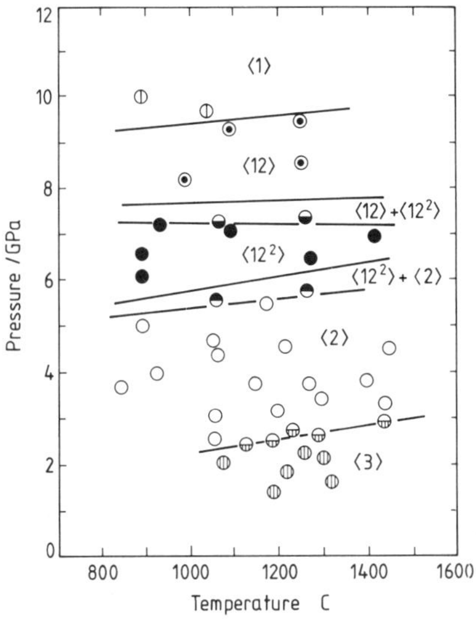

Fig.8: Experimental phase diagram of the Ni_2SiO_4 - $NiAl_2O_4$ system (Akaogi et al. 1982).

Perpendicular to the layers, however, the phases spinel, <1>, the β-phase, <2>, illustrated in Figure 7b, manganostibite, <3>, and three stuctures found only in the Ni_2SiO_4 - $NiAl_2O_4$ system, <12>, <12^2> and <13>, have been observed. The experimental phase diagram of this system is shown in Figure 8. (Akaogi et al. 1982). The phase sequence resembles that along the arrowed line in Figure 5, where <12^2> would be the first phase to be stabilised between <2> and <12> by either finite temperature or longer-range interactions. Which of the two mechanisms is relevant here remains an open question, but the latter is probably more likely as the phase boundaries are essentially temperature independent. It seems that the dependence of the J_i on pressure is chiefly responsible for driving the system between different phases.

Price et al. (1985) have obtained values for J_1, J_2 and J_3 for magnesium silicate spinelloids from atomic modelling calculations. The two sets of values they obtain, which differ because of the difference in the model potentials, are shown by dots in Figure 5b. The results are consistent with the spin picture we are presenting here. We note that the existence of a <13> spinelloid as a stable or metastable state does not fit in to this theory and its existence remains a mystery.

d. Modulated Phases in Binary Alloys

Many binary alloys, which order in the face-centred cubic $L1_2$ structure shown in Figure 2, have been observed to form long-period commensurate or incommensurate phases. Examples are Ag_3Mg, CuAu, Au_4Zn and $TiAl_3$.

The modulation is introduced into the crystal structure by antiphase domain boundaries which correspond to a displacement of the minority sublattice through [½, ½, 0] as shown in Figure 9. Hence each face-centred cube has two positions which can be mapped, as usual, onto an Ising spin variable. Using the usual notation the phase in Figure 9 is $\langle 2 \rangle$.

Many experiments are now in progress in an attempt to establish the extent to which an Ising model with competing interactions provides a valid picture of long-period ordering in binary alloys. Here space restricts us to considering $TiAl_3$ as an example and we summarize results of electron diffraction and high resolution electron microscopy studies by Loiseau et al. (1985). More details and examples are given in Yeomans (1987) and de Fontaine and Kulik (1985).

Long-period phases are stable in $TiAl_3$ for 71-73 at.% Al and temperatures between 700 and 1200K. The phases observed as a function of temperature are listed in Figure 10. The experiments suggest that each structure is stable over a small temperature range, with the shortest period structures having the largest range of stability: about 50K for $\langle 12^2 \rangle$ and $\langle (12^2)^2 12 \rangle$. At each annealing temperature different long-period structures were found in separate areas of the sample which may have been due to variation in concentration across the sample or experimental limitations on the speed of quench, annealing time or temperature control producing metastability effects.

Certainly these results are very reminiscent of the behaviour of an axial Ising model with antiferromagnetic first neighbour interactions and competing second neighbour interactions. The strong temperature dependence of the stable wavevector suggests that the long-period phases are stabilised by entropy as described in Section 3bi. Further evidence of this is the observation of jogs along the antiphase boundaries which becomes more frequent with increasing annealing temperature.

It is important to point out the difference between the construction of the Hamiltonian (2.5), which was used to describe bulk ordering in face-centred cubic binary alloys, and the Ising model used here to represent modulated order. In the former case the spin variables refer to the occupation of a site: here they are used to represent the position of a face-centred cube. It would be informative to try to obtain modulated structures using the former approach perhaps by extending (2.5) to include further neighbour competing interactions. This would lead to an isotropic Hamiltonian which is more realistic than an anisotropic one where one imposes a special direction on the model by hand. Unfortunately, however, it is much harder to treat theoretically.

4. SURFACES AND INTERFACES

In this Section I aim to report some recent results on phase transitions at surfaces and interfaces. This is a very rich field which has received much interest from statistical mechanicians in recent years. Useful reviews are by Sullivan and Telo da Gama (1985) and Dietrich (1987). I shall concentrate on the behaviour of various model systems, many of which have been mentioned earier in the article, but attempt to

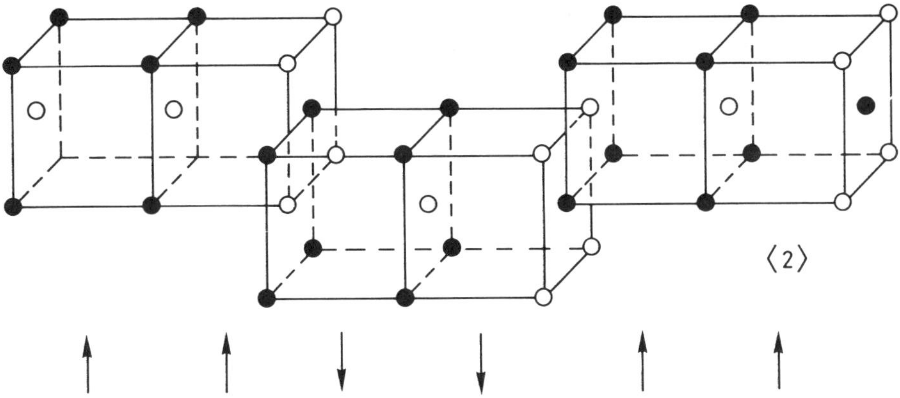

Fig.9: Formation of the <2> phase in a face-centred cubic $L1_2$ structure (intermediate planes which contain only one type of atom are omitted for clarity).

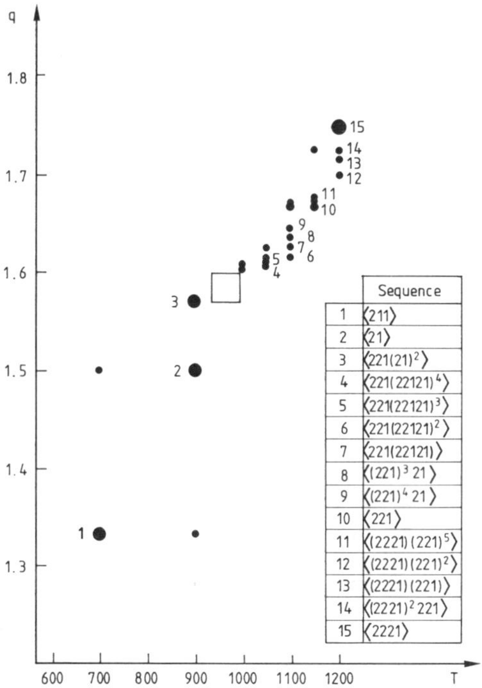

Fig.10: Wavevector of long-period phases observed in $TiAl_3$ as a function of temperature (Loiseau et al. 1985). The square marks a two-phase region where no modulated structures are stable.

give experimental examples where appropriate. One wonders if other relevant experiments exist, well known to metallurgists and mineralogists but not to theoretical physicists.

a. Surface Induced Disorder

Surface induced disorder occurs in a semi-infinite system which disorders through a bulk first-order phase transition at a temperature, T_C (Lipowksy 1982). Because of the missing neighbours at the surface the disordered phase can be favoured there even below T_C. In this case a layer of the disordered phase which has width which diverges as $T \to T_C^-$ appears between the bulk and the surface. The surface order parameter, M_S, goes to zero continuously and hence one may define a critical exponent

$$M_S \sim (T_C - T)^{\beta_1}. \qquad (4.1)$$

Surface induced disorder has been predicted to occur for the Hamiltonian (2.5) on a semi-infinite face-centred cubic lattice, with the effect of the free surface being modelled by allowing the field to take different values, H_1 and H_2, on the two layers nearest the surface. However, the situation is complicated because whether such a transition exists depends sensitively on H_1, H_2 and α, on the direction of the surface considered and on the degree of surface segregation. What is more worrying is that the details of the results depend on the mean-field approximation used and sometimes on the particular authors who implement the technique! A heroic summary is given in Dietrich (1987). The exponent β_1 takes its mean-field value of ½ for surface induced disorder transitions at the (111) and (110) surfaces but is non-universal for the (100) surface.

Experimental evidence for surface induced disorder in a binary alloy has been seen in Cu_3Au where the order parameter at the (100) surface seems to vanish continuously as $T \to T_C^-$ (Jamison et al. 1985, Alvarado et al. 1987). No convincing values of β_1 have been obtained experimentally. The relevance of the model Hamiltonian to experiments is itself debatable because the surface behaviour is expected to depend delicately on the microscopic details of the Hamiltonian, particularly the existence of long-range interactions. Moreover, in comparison to experiments, vacancies may play an important rôle and surface reconstruction is a possibility not easily incorporated into the model systems.

A similar phenomenon occurs at an interface between two equivalent ordered phases as $T \to T_C^-$. The simplest model showing this effect is the three-dimensional Potts model, defined by the Hamiltonian (2.6), where the disorderd phase intrudes at an interface between, say, the A and B phases as the transition temperature is approached from below (Derrida and Schick, 1986). As a more physical example we note that wetting may also occur at antiphase boundaries between different domains of Cu_3Au but the behaviour of the disordered layer is likely to be affected by the existence of long-range forces (Kikuchi and Cahn 1979).

If the bulk transition is not an order-disorder transition but a

melting transition at a temperature T_m a surface or interface may be wet by a liquid layer as $T \to T_m^-$. It has been surmized that the melting of a solid starts at its surface by the wetting of the surface by the melt (van der Veen and Frenken 1986) but there is no unamibiguous theoretical or experimental evidence for this. Similarly the possibility of grain boundary melting below T_m is a fascinating one which has attracted a lot of theoretical, numerical and experimental
studies from which, as yet, no firm conclusions have been reached (Kikuchi and Cahn 1980, Franck et al. 1983, Dietrich 1987).

b. Layering Transitions

At low temperatures lattice effects become important and an iterface may unbind from a surface, or two interfaces may replace one, through a series of layering transitions (de Oliveira and Griffiths 1978, Pandit et al. 1982). As a simple example of this behaviour we consider the Ising Hamiltonian (2.1) in three dimensions (Duxbury and Yeomans 1985). An interface is introduced into the system by imposing boundary conditions which force all the spins in layer i=0 to take values $S_{o,j}$ = -1 and all those in layer i = ∞ to be in a state $S_{\infty,j}$ = +1 as shown in Figure 11a. At zero temperature for H>0 the spins in the bulk point up and the interface lies next to the surface in the position n=1. For H=0, however, any position of the interface parallel to the surface has the same energy.

At finite temperatures the interface can gain entropy by lying further from the surface where it is more able to fluctuate and its position is determined by a balance between the entropic gain and energy loss of moving away from the surface. As a result, as the field tends to zero, the interface unbinds from the surface through an infinite series of layering transitions as shown in Figure 11b. This behaviour

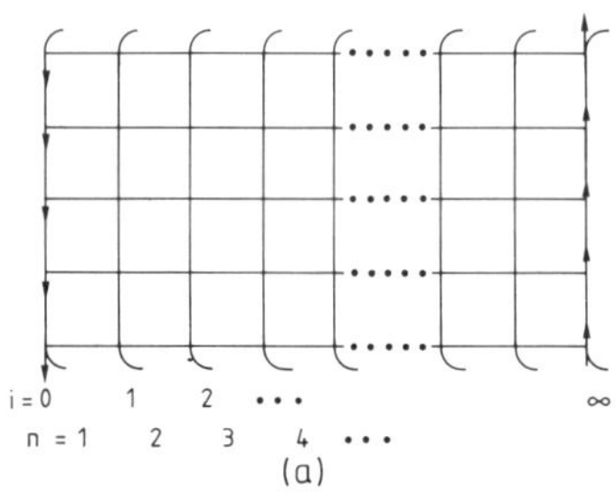

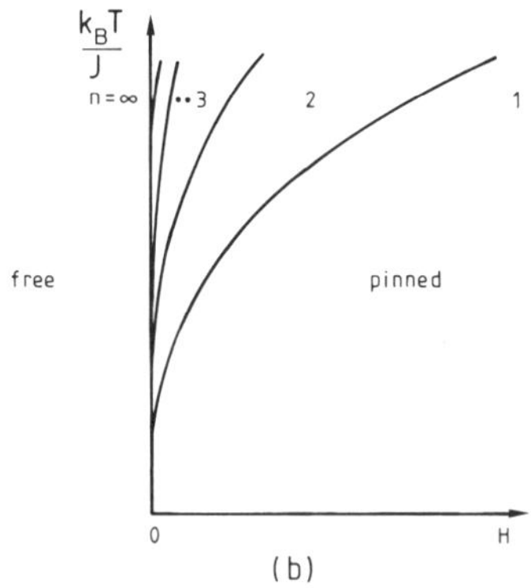

Fig.11: a) Boundary conditions which force an interface to exist in an Ising model. b) Layering transitions which occur when the interface unbinds from the surface as the field, H, tends to zero.

persists up to $T \sim \frac{1}{2}T_c$ where the interface becomes rough, that is it fluctuates so strongly that it is not possible to define its position relative to a particular layer.

An alternative way of binding the interface to the surface is by an attractive surface potential. This is the situation in the case of multilayer adsorption on a graphite substrate. The two coexisting phases are the liquid and gas of which the former is preferentially attracted by the graphite. Hence as the chemical potential tends to its value at coexistence layers of the liquid sequentially appear on the graphite surface and the liquid-gas interface depins from the graphite through a series of layering transitions. The number of layering transitions depends on the relative strengths of the absorbate-absorbate and absorbate-substrate attraction and a wide range of different behaviours is possible in diferent systems (Pandit et al. 1982).

Similar layering transitions are expected at low temperatures in systems where it is favourable for a single interface to split into two or more interfaces (Huse et al. 1983). An example would be a chiral model such as that introduced in Section 2c to model krypton on graphite. If the energy of an A|B (or equivalently B|C or C|A) interface decreases relative to that of a B|A interface it will eventually become favourable for the latter to split up into a ...B|CC ...CC|A ... configuration with two unbound interfaces. There is evidence that this can, in three dimensions, occur through a series of layering transitions (Armitstead et al. 1986). It would be interesting to consider whether the same behaviour would be expected to occur at domain walls in minerals.

In this article I have attempted to present some examples of model systems which have given insight into experimental results. Different possibilities have been considered which answer varying questions. Moreover, I have tried to show which of the features of the models interest theoreticians. In places there have been results which do not yet seem relevant to reality. Hopefully continued interaction between theoretical physicists and applied scientists will suggest experiments to the latter and provide the former with a rational for their pechant for inventing and studying spin models.

REFERENCES

Akaogi M., Akimoto S., Horioka K., Takahashi K. and Horiuchi H. 1982 J. Solid State Chem. 44 257.

Als-Nielsen J. 1976 in "Phase Transitions and Critical Phenomena" Vol. 5A p.87.

Alvarado S.F., Campagna M., Fattah A. and Uelhoff W. 1987 Z. Phys. B66 103.

Armitstead K., Yeomans J.M. and Duxbury P.M. 1986 J. Phys. A19 3165.

Bak P. and Bruinsma R. 1982 Phys. Rev. Lett. 49 249.

Baker G. and Essam J.W. 1971 J. Chem. Phys. 55 861.

Berker A.N., Ostlund S. and Putnam F.A. 1978 Phys. Rev. B17 3650.
Binder K. 1981 Z. Phys. B45 61.
Binder K. 1986 in "Festkörperprobleme - Advances in Solid State Physics" Vol. 26 Ed. P. Grosse p. 133.
Binder K., Kinzel W. and Selke W. 1983 J. Magn. Magn. Mater. 31-34 1445.
Caflisch R.G., Berker A.N. and Kardar M. 1985 Phys. Rev. B31 4527.
Cheng C, Needs R.J., Heine V. and Churcher N. 1987 Europhysics Lett. in press.
Coppersmith S.N., Fisher D.S., Halperin B.I., Lee P.A. and Brinkman W.F. 1981 Phys. Rev. Lett. 46 549.
Derrida B. and Schick M. 1986 J. Phys. A19 1493.
Dietrich S. 1987 in "Phase Transitions and Critical Phenomena" Vol. 12 Ed. C. Domb and J. Lebowitz (Academic Press, London).
Duxbury P.M. and Yeomans J.M. 1985 J. Phys. A18 L983.
Fisher M.E. 1967 Rep. Prog. Phys. 30 615.
Fisher M.E. 1974 Rev. Mod. Phys. 46 597.
de Fontaine D. 1979 in "Solid State Physics" Vol. 34 Eds. H. Ehrenreich, F. Seitz and D. Turnbull (Academic Press, N.Y.) p.73.
de Fontaine D. and Kulik J. 1985 Acta. Metall. 33 145.
de Oliveira M.J. and Griffiths R.B. 1978 Surf. Sci. 71 687.
Franck J.P., Kornelsen K.E. and Manuel J.R. 1983 Phys. Rev. Lett. 50 1463.
Huse D.A., Szpilka A.M. and Fisher M.E. 1983 Physica 121A 363.
Jamison K.D., Lind D.M., Dunning F.B. and Walters G.K. 1985 Surf. Sci. 159 L451.
Jepps N.W. and Page T.F. 1984 J. Cryst. Growth Charact. 7 259.
Kikuchi R. and Cahn J.W. 1979 Acta. Metall. 27 1337.
Kikuchi R. and Cahn J.W. 1980 Phys. Rev. B21 1893.
Krishna P. 1984 (Ed.) "Crystal Growth and Characterisation of Polytypes" J. Cryst. Growth Charact. 7 (Pergamon: special issue).
Lebowitz J.L., Phani M.K. and Styer D.F. 1985 J. Stat Phys. 38 413.
Lipowsky R. 1982 Phys. Rev. Lett. 49 1575.
Loiseau A., van Tendeloo G., Portier R. and Ducastelle F. 1985 J. de Physique 46 595.
Pandey D. and Krishna P. 1984 J. Cryst. Growth Charact. 7 213.
Pandit R., Schick M. and Wortis M. 1982 Phys. Rev. B26 5112.
Price G.D. 1983 Phys. Chem. Miner. 10 77.
Price G.D. and Yeomans J.M. 1984 Acta. Crystal. B40 448.
Price G.D., Parker S.C. and Yeomans J.M. 1985 Acta Cryst. B41 231.
Selke W., Barreto M.N. and Yeomans J.M. 1985 J. Phys. C. 18 L393.
Stanley H.E. 1971 "Introduction to Phase Transitions and Critical Phenomena" Clarendon Press, Oxford.
Stephens P.W., Heiney P.A., Birgeneau R.J., Horn P.M., Moncton D.E. and Brown G.S. 1984 Phys. Rev. B29 3512.
Sullivan D.E. and Telo da Gama M.M. 1985 in "Fluid Interfacial Phenomena" Ed. C.A. Croxton (Wiley).
Van der Veen J.F. and Frenken J.W.H. 1986 Surf. Sci. 178 382.
Verma A.R. and Krishna P. 1966 "Polymorphism and Polytypism in Crystals" (Wiley N.Y.).
Villain J. and Gordon M. 1980 J. Phys. C13 3117.
Wu F.Y. 1982 Rev. Mod. Phys. 54 235.

Yeomans J.M. 1987 in "Solid State Physics" Eds. H. Ehrenreich, F. Seitz and D. Turnbull (Academic Press, N.Y.) in press, and references therein
Yeomans J.M. and Price G.D. 1986 Bull. Min. 109 3.
Zdhanov G.S.and Minervina Z. 1945 J. Phys. (Moscow) 9 151.

NEW DEVELOPMENTS IN RAMAN SPECTROSCOPY ON STRUCTURAL PHASE TRANSITIONS

Ulrich Bismayer
Institut für Kristallographie und Petrographie
Universität Hannover
Welfengarten 1
3000 Hannover 1
Germany

ABSTRACT. The application of Raman spectroscopy for the investigation of phase transitions focused traditionally on the observation of soft modes in displacive systems. Furthergoing studies on displacive and order-disorder systems are based on the observation of systematic changes of the scattering profiles of hard modes during the phase transition. They revealed the temperature evolution of the order parameter, the role of order parameter fluctuations and phonon density of states effects in model systems like alkali feldspars, $Pb_3(P_{1-x}As_xO_4)_2$, As_2O_5 and $NaNO_3$. Allied with supplementary experimental techniques hard-mode Raman spectroscopy (HMRS) is an ideal method for the investigation of order parameter coupling effects.

1. INTRODUCTION

Traditionally the application of Raman spectroscopy for the investigation of structural phase transitions focused on the study of displacive systems. The related cooperative changes of the atomic positions were already described by Cochran (1960) and Anderson (1960) within the scope of the soft-mode concept. A soft mode is a particular mode of vibration of the high symmetry phase whose frequency tends towards zero as the critical temperature is approached. This softening results in the instability of the crystal against the low frequency phonon which requires a critical temperature dependence mainly of the quadratic term in the relevant thermodynamic potential, that is the expansion of the free energy in powers of the order parameter as given by Landau's (1958) phenomenological theory of structural phase transitions. The order parameter Q(T) is nonzero below the transition temperature T_C and increases on cooling according to $Q(T) \sim |T-T_C|^{\beta}$ with its critical exponent β. In terms of thermodynamics, the soft mode theory relates directly to the Landau theory in the quasiharmonic theory in which the order parameter is associated with the mean value of the eigenvector of the soft mode (Blinc and Zeks 1974).

However, most minerals and other materials with a complex chemical composition do not show pure displacive phase transitions. Often the

E. K. H. Salje (ed.), Physical Properties and Thermodynamic Behaviour of Minerals, 143–183.

transition process involves the ordering of atoms or a reorientation of molecules describing fairly large displacements about the equilibrium positions of the high symmetry phase of the crystal. Such an ordering process is not accompanied by critical mode softening and consequently, soft mode Raman spectroscopy cannot be applied. Hard modes, on the other hand do show marked changes of their scattering profiles. These changes are directly related to the mechanisms involved in the structural phase transformation.

Theories of HMRS have first been developed based on results of Raman scattering experiments, investigating the temperature dependence in the frequency and the damping constant (Matsushita 1976) or the change of the mode frequency and the strength (Petzelt and Dvorak 1976 a,b) of nonsoft phonons.

A more detailed theoretical approach for the understanding of Raman scattering near displacive phase transitions involving the profile analysis of nonsoft Raman lines was given by Bruce et al. (1980, see also references therein). They have shown that the integrated intensity of a scattered Raman hard mode is a characteristic quantity which is correlated to the critical behaviour of the macroscopic order parameter. The study of the temperature evolution of the line width, FWHM (full width at half maximum), allows to detect relaxation processes (orientational fluctuations), order-disorder contributions and the formation of micro-domains.

Examples for the successful application of HMRS are the studies on palmierite type compounds (Salje et al. 1983), langbinites (Devarajan and Salje 1986), alkali feldspars (Salje 1986), ammonium fluorosilicate (Jenkins 1986), arsenic pentoxide (Bismayer et al. 1986), sodium nitrate (Poon and Salje 1987a) and magnesium cordierite (Poon et al. 1987).

In this paper we discuss the future possibilities of HMRS starting from a simple introduction of its theoretical background. The objective of this article is not to give a complete monograph on hard-mode spectroscopy but to provide a condensed review for those readers who are interested in this experimental method and who whish to consult the given references for detail. Selected model systems are Na-K feldspar, $Pb_3(P_{1-x}As_xO_4)_2$, As_2O_5 and $NaNO_3$.

2. THEORY

Theoretical treatment of the Raman scattering process was given by Born and Huang (1954). The Raman effect occurs when a beam of monochromatic light with frequency ω_i passes through a crystal leading to inelastic scattering by the phonons of the crystal with frequency shifts ω_{ph} of 10 cm^{-1} to 3000 cm^{-1} according to $\omega_s = \omega_i \pm \omega_{ph}$. ω_s is the scattered beam frequency. The signs - and + represent Stokes and anti-Stokes processes, respectively. In terms of quantum dynamics the Raman scattering is associated with a quantum transition in the crystal. As a consequence of the polarizable nature of the crystal an electric dipol moment is induced by the electric field of the incident light. The correspondence between the dipole moment amplitude m and the quantum mechanical matrix element between the vibrational states $|\nu\rangle$ and $|\nu'\rangle$ is given by

(Born and Huang 1954)

$$m = \langle \nu | P_{\alpha\beta} | \nu' \rangle E \qquad (1)$$

where $P_{\alpha\beta}$ is the polarizability tensor which depends on the nuclear coordinates.

Cowley (1964) presented detailed calculations on Raman spectra using the shell model with anharmonic interactions between the electronic dipole moments. According to Cowley (1964) the Raman scattering cross section is characterized by the correlation tensor which determines the scattering into a polarization $(\beta\delta)$ from an incident beam polarized along $(\alpha\gamma)$

$$I_{\alpha\beta\gamma\delta}(\Omega) = \frac{1}{2\pi N} \int_{-\infty}^{\infty} \langle P^{*}_{\alpha\beta}(0) P_{\gamma\delta}(t) \rangle \exp i\Omega t dt \qquad (2)$$

where Ω is the frequency transfer given by

$$\Omega = - \frac{2\pi n i}{\hbar / k_B T} \qquad (3)$$

$P_{\gamma\delta}(t)$ describes the deviation from the equilibrium polarizability of a crystal of N unit cells. In the harmonic approximation the nuclear motion is discussed on the basis of normal coordinates. The displacements of the atoms from their equilibrium positions are regarded to be small and therefore, the polarizability can be expanded in a Taylor series with respect to the displacements (Bruce and Cowley 1981)

$$P_{\alpha\beta}(K) = P^{0}_{\alpha\beta}(K)\Delta(K) + \left(\frac{1}{N}\right)^{1/2} \sum_{qj} P_{\alpha\beta}\begin{pmatrix} q \\ j \end{pmatrix} Q(qj)\Delta(K+q) + \frac{1}{N}\sum_{q_1 q_2 j_1 j_2} P_{\alpha\beta}\begin{pmatrix} q_1 & q_2 \\ j_1 & j_2 \end{pmatrix} Q(q_1 j_1) Q(q_2 j_2) \Delta(K+q_1+q_2) + \ldots \qquad (4)$$

where K is the light-wave vector transfer and Q(qj) describes the mode coordinates of wave vector q and phonon branch j. In the case of a displacive structural phase transformation the distortions may be described by an n-fold degenerate mode labelled by wave vector q and branch indices $j_s = 1,\ldots n$. The static amplitude of the distortion represents the order parameter. According to the calculations by Bruce et al. (1980) Raman active hard modes which are symmetry equivalent to the soft mode with $q = q_s$ but branch indices $j_h \neq j_s$ contribute by the following term to the first-order Raman scattering using the spectral function G

$$I_{\alpha\beta\gamma\delta}(\Omega) = 4\sum_{j_h j_s} P_{\alpha\beta}\begin{pmatrix} q_s & -q_s \\ j_h & j_s \end{pmatrix} P_{\gamma\delta}\begin{pmatrix} q_s & -q_s \\ j_h & j_s \end{pmatrix} Q^2_{j_s} G^{ph}(q_s j_h \Omega) \qquad (5)$$

Hence, the integrated intensity of the Raman scattering from the hard modes (qj_h) varies with the square of the macroscopic order parameter when T approaches the critical temperature. However, this correlation function represents only that part of scattering which corresponds to the long-range order. In the vicinity of T_C the fluctuations of the coordinates Q_i about their equilibrium value, which are comparable to the order-parameter value itself, dominate and provide the (short-range) precursor order contribution to the scattering. Hence, the scattering at frequencies close to the hard modes arises from terms that involve two coordinates belonging to a hard branch, and two belonging to a soft branch which can be represented after a decoupling approximation (Bruce et al. 1980) by

$$I_{\alpha\beta\gamma\delta}(\Omega) \approx 4\sum_{j_h j_s} P_{\alpha\beta}\begin{pmatrix} q_s & -q_s \\ j_h & j_s \end{pmatrix} P_{\gamma\delta}\begin{pmatrix} q_s & -q_s \\ j_h & j_s \end{pmatrix} \times \sum_q \frac{1}{2\pi N} \int_{-\infty}^{\infty} \langle Q(qj_s,0)Q(-qj_s,t)\rangle$$

$$\times \langle Q(qj_h,0)Q(-qj_h,t)\rangle \exp i\Omega t dt \tag{6}$$

Contributions of this type may also arise from anharmonic interactions between phonons in relaxational systems. The correlation function of the hard-mode coordinates can be expressed as

$$\langle Q(qj_h,0)Q(-qj_h,t)\rangle = \int_{-\infty}^{\infty} G^{ph}(qj_h,\Omega) \exp(-i\Omega t)\, d\Omega. \tag{7}$$

The pair correlation corresponding to the soft coordinates can be separated in static and dynamic contributions

$$\langle Q(qj_s,0)\, Q(-qj_s,t)\rangle = Q^2_{j_s}(q) + \langle\hat{Q}(qj_s,0)\hat{Q}(-qj_s,t)\rangle \tag{8}$$

where $\hat{Q}(qj_s)$ represents the fluctuations in the soft coordinates whose correlation function is given by

$$\langle\hat{Q}(qj_s,0)\hat{Q}(-qj_s,t)\rangle = \int_{-\infty}^{\infty} G(qj_s,\Omega) \exp(-i\Omega t)\, d\Omega. \tag{9}$$

The spectral function leading to a quasi-first-order Raman spectrum close to the critical temperature of a displacive system contains at least three contributions from 1) the soft mode, 2) the central peak and 3) higher order Raman scattering arising from processes involving hard and soft phonons

$$G(qj_s,\Omega) = G^{lr} + G^{cp} + G^{ph'}. \tag{10}$$

G^{lr} is the spectral function of a long range phonon sideband with a quasiharmonic frequency that softens with $T \to T_C$ but which saturates at a finite limiting frequency value at T_C. G^{cp} describes a central component leading to quasi-first-order Raman scattering from the hard modes induced by a slow finit range component related to phonon density of states fluctuations or to cluster dynamics (assuming a defect-free crystal). A measure of the local distortion (quasi static on phonon time scales) within the dynamic clusters is the precursor order parameter. The clusters may be regarded as pre-ordered regions. Accordingly short-range precursor order induces, above T_C, similar effects to those which static long-range order induces below T_C. $G^{ph'}$ represents the conventional higher order Raman contribution to the spectral density involving hard and soft quasiharmonic phonons which gives rise, close to T_C, to quasi-first-order features characterized by overdamped soft phonons.

In the fluctuation regime one may separate the spectral function

$G^{ph'}$ in two terms

$$G^{ph'} \approx \frac{1}{N} \sum_q G^{ph}(qj_s)\overline{G}^{ph}(qj_h,\Omega) \tag{11}$$

where

$$G^{ph}(qj_s) \equiv \int_{-\infty}^{\infty} G^{ph}(qj_s,\Omega)\, d\Omega = \frac{kT}{\omega_\infty^2(qj_s)} . \tag{12}$$

$\overline{G}^{ph}$ is the spectral function of a damped quasiharmonic oscillator with frequency $\omega_\infty(qj_h)$ and damping

$$\tilde{\Gamma}(qj_h) = \Gamma(qj_h) + \frac{\omega_\infty^2(qjs)}{\Gamma(qjs)}. \tag{13}$$

3. APPLICATIONS OF HARD MODE RAMAN SPECTROSCOPY

A detailed experimental description of hard-mode Raman spectroscopy was given by Salje (1986) in the context with studies on the structural phase transition of alkali feldspar. The technique recorded in his paper is essentially transferable to other experimental studies reported in the following section, respectively.

3.1. Displacive phase transition and evidence for Na-K site ordering in alkali feldspar - experimental part

Raman spectroscopical investigations on the silicate structures of alkali feldspars by Salje (1986) were carried out with a Raman spectrometer consisting of a 5 W-argon laser light source (λ = 514.5 nm) with continuously variable polarization direction. The beam was double focused on the sample. The light at 90 degrees scattered from the sample was focused on the entrance slit of a double monochromator and detected with a Peltier cooled photomultiplier coupled to a photon counter. For further line profile analysis the counter was attached to a micro computer.

The Raman spectrum reflects the spectral density of the autocorrelation function of the normal coordinates Q_i. The vibrational part of the profile follows from the quasi-harmonic response function

$$\chi^{-1}(qj\omega) = \omega_\infty(qj)^2 - \Omega^2 - 2i\Omega\Gamma_0 - \frac{\delta^2}{1-i\Omega\tau_0} \tag{14}$$

where

$$\omega_\infty(qj)^2 = \omega_0(qj)^2 + \delta^2. \tag{15}$$

ω_0 is the uncoupled phonon frequency with wave vector q of the phonon branch j. The weakly temperature-dependent damping constant Γ_0 is independent of frequency. Coupled Debye relaxations are expressed by the coupling constant δ and the relaxation time τ_0. For sufficiently small values of the damping constant, the spectral function with the Bose-Einstein factor n(T) and $\omega_Q = \tau^{-1}(\omega_0/\omega_\delta)$ reads

$$G^{ph}(q\Omega) = \frac{n(T)}{\pi} \frac{2\Gamma}{(\omega_\infty^2-\Omega^2)^2+4\Omega^2\Gamma^2} + \frac{\delta^2}{\omega_0^2\omega_\delta^2(\omega_Q^2+\Omega^2)} \tag{16}$$

Ordering processes by alkali ions in the feldspars can originate further line broadening as phonons with wave vectors $q \neq 0$ may contribute to the spectral density. The resulting hard mode line profile G^{hm} is the folding convolution function of two components

$$G^{hm}(q\omega) = \int G^f(\Omega)G^{ph}(q,\Omega'-\Omega)\,d\Omega \tag{17}$$

where G^f is the density of states fluctuation function. With the approximation of a Gaussian fluctuation profile the density of states term becomes

$$G^f = \frac{1}{\sigma\sqrt{2\pi}} e^{-\frac{(\Omega-\omega_\infty)^2}{\sigma^2}} \tag{18}$$

with the half width 2σ of the deconvoluted profile. The line profiles were deconvoluted using the experimental line profile of the Rayleigh line as a standard for the instrumental part of the line broadening. Hence, the experimentally observed convoluted spectrum is

$$G^{obs}(\Omega) = \int_0^\infty G^{hm}(q\Omega)S^{slit}(\Omega-\Omega')\,d\Omega \tag{19}$$

with the slit function $S^{slit}(\Omega)$.

In a first step, the hard-mode scattering profile was determined using the convolution function G^{obs} in equation (19). In a second step, the line profiles were fitted with the theoretical spectral density in equation (16). In case of a small coupling strength δ (which was observed by Salje (1986) in all feldspar samples) Debye relaxations were neglected for further calculations. The theoretically expected temperature dependence of the damping factor $\Gamma(T)$ follows from multi-phonon interaction and can be expressed as (Arora and Umadevi 1982)

$$\Gamma(T) = AT + BT^2 \tag{20}$$

provided that no further broadening effect occurs. In the case of potassium containing alkali feldspars (anorthoclase) Salje (1986) observed additional broadening effects due to density of states fluctuations. For these samples, the convolution using the spectral function G^{hm} (equation (17)) was made and the residual line width 2σ according to equation (18) determined.

3.2. Phase transitions and order parameter treatment in feldspars

The following part contains basic informations on general features of the transition mechanism in alkali-feldspars.

Two different symmetry breaking processes determine the thermodynamic behaviour of sodium-rich alkali-feldspars so that multi-order parameter Landau theory was applied to determine the corresponding free energy expression (Salje 1985). Under thermal equilibrium Na-feldspar possesses monoclinic symmetry at high temperatures with space group C2/c (monalbite). With decreasing temperature a symmetry reduction to a triclinic modification (albite, s.g. $C\bar{1}$) takes place near 1250 K involving a very sluggish Al, Si ordering described by the order parameter Q_{od} and a rapid distortion due to instabilities of the (Si-Al)-O "crankshaft" framework. During the transformation the crankshafts are tilted against each other and the displacive order parameter Q is correlated with the tilt angle. Only this rapid lattice distortion can be observed in situ for Na-rich feldspars. Hence, albite occurs in metastable states. Depending on the degree of order, at lower temperatures thermal regimes with the modifications high albite (with low Al, Si order) and low albite (with high Al,Si order) exist. A thermal crossover betwee high albite and low albite was described to be a smooth function of temperature (Salje et al.1985). Fully disordered (Q_{od} = 0) Na-feldspar is called analbite. In alkali-feldspars containing both, Na and K ions (anothoclase) similar coupling effects can be observed. In these minerals additional Na, K-site ordering and strong local distortions of the Si-Al-O network due to inequivalent alkali positions take place.

The space group $C\bar{1}$ is a subgroup of C2/m and the distortive phase transition is continuous. Consequently, the thermodynamic behaviour can be derived from a Landau-type free energy expression. General solutions of systems with order parameter coupling phenomena were discussed by Imry (1975), Gufan and Larin (1980) and Salje and Devarajan (1986). The symmetry reduction, C2/m -> $C\bar{1}$, in Na feldspar is given by a coset with respect to the subgroup $C\bar{1}$ according to

$$C2/m = C\bar{1} + C_2 \otimes C\bar{1}. \tag{21}$$

C_2 is the symmetry operation of a twofold rotation. The critical point of the Brillouin zone of the monoclinic phase is the origin (Γ). Hence, the translational symmetry remains unchanged during the phase transition. The active representation of the order parameter is that irreducible representation of the supergroup which transforms into the identity representation of the subgroup and which is not the identity representation itself (Janovec et al. 1975). The only representation fulfilling this condition is B_g and, therefore, the order parameter transforms as B_g. Consequently, critical lattice vibrations which are directly related to the order parameter, must be phonons of the B_g species. They are not infrared-active but Raman-active and symmetry equivalent to the components x_4 and x_6 of the spontaneous strain.

Two order parameters, namely Q and Q_{od}, are necessary to describe any structural state of a triclinic sodium feldspar. The only exception is fully disordered analbite with Q_{od} = 0; its thermal behaviour depends only on Q. However, the symmetry breaking process, C2/m -> $C\bar{1}$, which defines the structural phase transformation, can only occur once and the second order parameter Q_{od} cannot break the symmetry again. The only "phases" in the strict sense of the word are therefore the monalbite-phase and the albite-phase. The additional change in the degree of Al,Si order between high albite and low albite does not change the symmetry of the averaged crystal structure and can be described as crossover between the two "regimes" of albite.

The coupling of two order parameters due to common strain components was investigated by Salje and Devarajan (1985). Applying their arguments to the thermodynamics of Na-feldspars the free energy can be written as

$$\Delta F_{tot} = \frac{a}{2}Q^2 + \frac{b}{4}Q^4 + \frac{c}{6}Q^6 + \frac{a_{od}}{2}Q_{od}^2 + \frac{b_{od}}{4}Q_{od}^4 + \frac{c_{od}}{6}Q_{od}^6$$
$$+ \xi Q(x_4 + \frac{x_6}{\chi}) + \xi^{od} Q_{od}(x_4 + \frac{x_6}{\chi_{od}})$$
$$+ \frac{1}{2}\sum_{4,6} C_{ik}x_i x_k + \lambda Q Q_{od} \qquad (22)$$

with the elastic constants C_{ik}, the coupling constants between the order parameter and the components of the spontaneous strain $_i$ and the coefficients $\chi = \xi_4/\xi_6$ and $\chi_{od} = \xi_4^{od} / \xi_6^{od}$. The last term in the equation (22) introduces a bilinear coupling between Q and Q_{od} which possibly exists besides the indirect strain induced coupling. Both order parameters may become the driving order parameter, namely Q in case of analbite and Q_{od} in case of microcline (potassium feldspar). Once the symmetry is broken, both order parameters are nonzero with different temperature dependences.

The thermal equilibrium is defined by the minimum of ΔF_{tot} with respect to both order parameters and the strain components. The leading terms for the temperature dependence of Q and Q_{od} are $a = a_0(T-T_C)$ and $a_{od} = a_{od0}(T-T_C^{od})$. The absolute minimum of the free energy for the equilibrium values of both order parameters is

$$\delta\Delta F_{tot}/\delta Q = \delta\Delta F_{tot}/\delta Q_{od} = 0. \qquad (22)$$

The strain components depend linearly on the order parameters (Salje 1985) and enter in the renormalized critical temperatures T_C^{i*} to give the final equilibrium conditions in renormalized form

$$a_0(T-T_C^*)Q + bQ^3 + cQ^5 + dQ_{od} = 0 \qquad (23)$$

$$a_{od0}(T-T_C^{od*})Q_{od} + b_{od}Q_{od}^3 + c_{od}Q_{od}^5 + dQ = 0. \qquad (24)$$

General features of these solutions are:

1.) At $T > T_C$ the only stable phase is monalbite with $Q = Q_{od} = 0$. T_C^* is not identical with the experimentally determined transition temperature of the monalbite-analbite inversion (T_C) but slightly above this temperature.

2.) At temperatures below T_C^* but above T_C^{od} the order parameter Q dominates over Q_{od} which means that the Al,Si order is small but present. This thermal regime is that of high albite.

3.) With decreasing temperature, the Al,Si order starts to increase more rapidly and a crossover between high albite and low albite occurs as a smooth function of temperature.

Calorimetric studies by Salje et al. (1985) by means of excess specific heat measurements in the compound $Or_{31} = Na_{0.69}K_{0.31}AlSi_3O_8$ revealed the influence of the structural phase transition between the monalbite phase and the albite phase at 416 K. The authors observed a step of 2.74 J/mol K in the c_p-curve corresponding to a continuous phase transition as no latent heat was observed (Fig. 1). The two order parameters Q and Q_{od} act on extremely different time scales and, therefore, metastable phases occur in sodium feldspar. The condition for any metastability is

$$\delta\Delta F_{tot}/\delta Q = 0 \; ; \qquad Q_{od} = \text{const.} \tag{25}$$

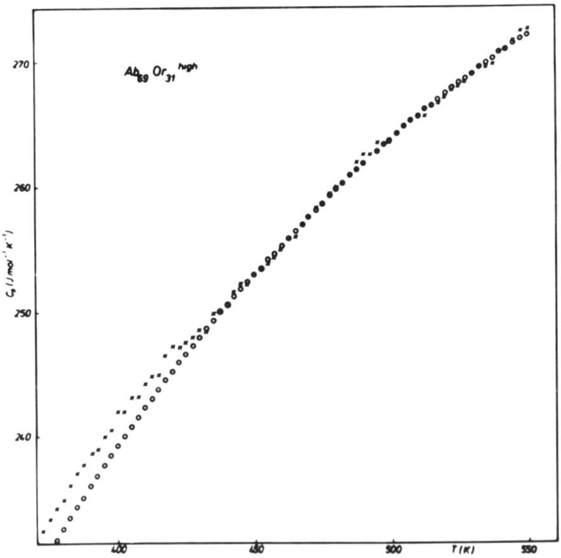

Figure 1. The heat capacities of disordered Or_{31} as compared with the extrapolated values (open circles) of the monoclinic phase at low temperatures (Salje et al. 1985).

Hence, in a metastable state the Al,Si order is described by a constant value of Q_{od} and the solutions for Q follow from its temperature dependence given in equation (23). In figure 2 the solutions for Q are plotted for Q_{od} increasing from 0 to 1 in Q_{od}-steps of 0.1. Figure 2 shows that below the melting point of the material, a structural phase transition can only be achieved for $Q_{od} = 0$ (analbite). Only analbite can transform to monalbite at T_C whereas for any other value of Q_{od} the order parameter Q can never reach zero and therefore it becomes impossible to observe the monoclinic phase even in slightly orderd high albite. In a first approximation Salje et al. (1985) extrapolated the linear temperature dependence of the excess entropy (attributed to the displacive transition) to low temperatures according to

$$|\Delta S| \sim (T_C - T). \tag{26}$$

However, the calculation of phase equilibria in systems with order parameter coupling requires a numerical knowledge of the symmetry breaking quantities at low temperatures where saturation effects must be expected. To avoid inherent uncertainities in the determination of excess specific heats at temperatures lower than 100 K below T_C (Salje and Wruck 1983, Winterfeld and Schaak 1977) HMRS was choosen as an ideal technique which is extremely sensitive to thermal sublattice polarization (Q), ordering processes (Q_{od}) and early stages of exsolution (Salje 1986).

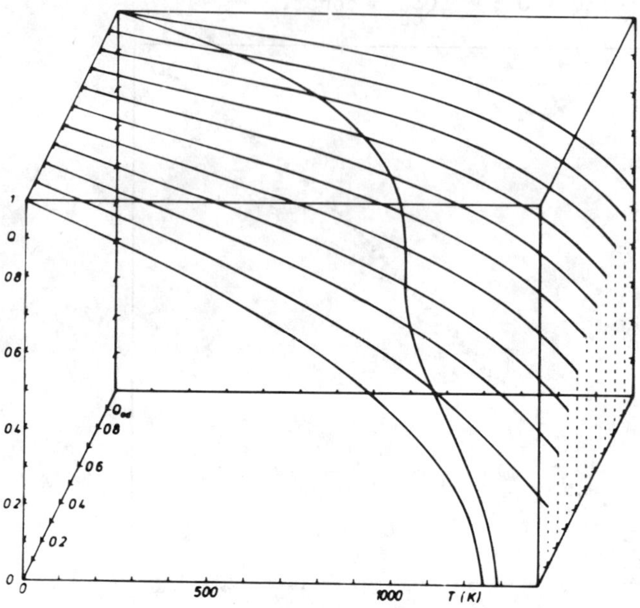

Figure 2. Temperature dependence of the order parameter Q for different Al,Si order (Q_{od}). Each curve represents one metastable solution, e.g. Q_{od} = 0 (analbite) in front and Q_{od} = 1 (fully ordered albite) in the back. The line on the surface of these curves indicates the solutions for thermal equilibrium (Salje et al. 1985).

His Raman spectroscopical measurements at low temperatures revealed distinct differences between Na-feldspar (albite and analbite) on the one side and Na-K feldspar (anorthoclase) on the other side. However, no lattice softening of any optical mode was observed in the minerals and therefore Salje (1986) measured the line profile of intense lines near 293 cm^{-1}, 481.5 cm^{-1}, 510 cm^{-1} and 1104 cm^{-1} selected from the full spectral range (Fig. 3).

During the symmetry reduction the representations A_g and B_g in 2/m transform into A_g in $\bar{1}$. In albite the total cross sections of all four Raman lines are nearly independent of temperature (Fig. 4).

The temperature dependence of the line widths of the bands in the Na-feldspars follow the theoretical predictions of multi-phonon coupling in the quasi-harmonic approximation according to equation (20). All line widths vanish at absolute zero as shown in figure 5 and no fluctuation broadening occurs in Na-feldspar.

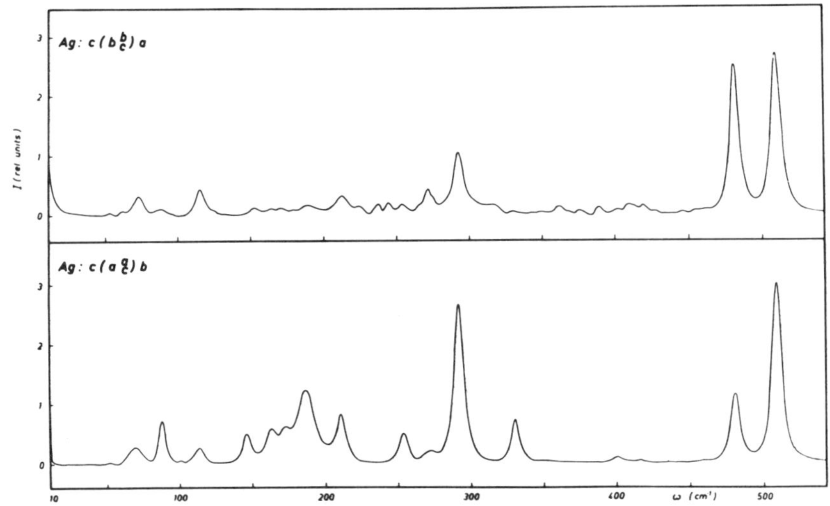

Figure 3. A_g-Raman spectra of albite (scattering geometries and polarization directions in Porto notation),(no analyser was used for the scattered light) (Salje 1986).

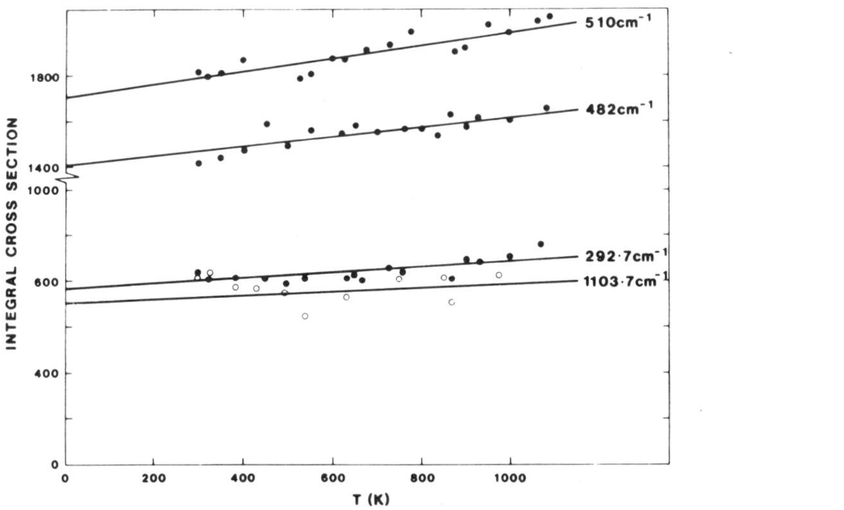

Figure 4. The integral cross-section of Raman hard modes in albite (Salje 1986).

The spectroscopic results show, that the lattice vibrations of Na-feldspar can reasonably well be understood in the harmonic approximation. The Al,Si disorder does not increase the line width of the 510 cm^{-1} band (Fig. 5) perhaps due to very small dispersion of this phonon in k-space.

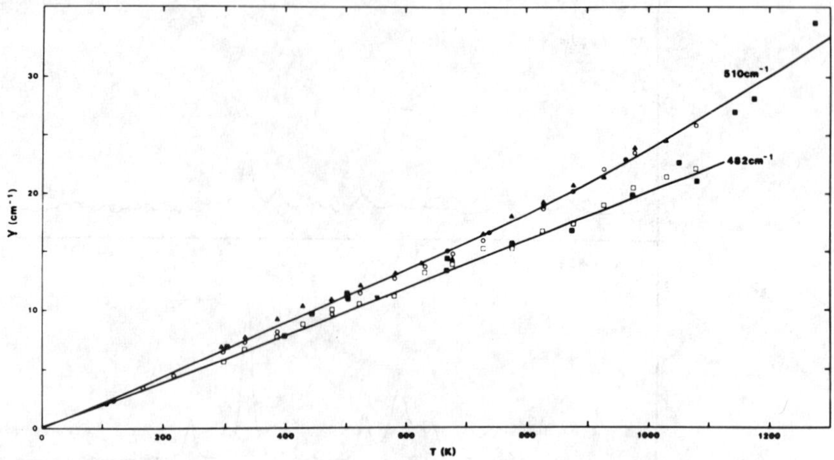

Figure 5. Polynomial temperature dependence of the line width of the 510 cm^{-1} (o), 482 cm^{-1} (□), 292.7 cm^{-1} (Δ) and 1103.7 cm^{-1} (●) modes in low albite. The same behaviour shows the 510 cm^{-1} line (▲) in analbite (Salje 1986).

The Raman spectrum of Na-K-feldspar shows slightly shifted scattering frequencies in comparison to albite at room temperature. Although their temperature dependent frequency shift is very small, the total cross sections of Raman lines in anorthoclase show in striking contrast to all Na-feldspars investigated by Salje (1986) a strong temperature dependence, with a discontinuity at the critical temperature (T_C = 523 K)(Fig. 6).

The structural phase transition, C2/m -> $C\bar{1}$, of anorthoclase leads to a temperature independent intensity I_0 above T_C and a linear intensity behaviour below T_C with saturation effects at $T < T_d$ caused by local lattice distortions. The total cross section in alkali feldspar was defined by Salje (1986)

$$I_{\alpha\beta\gamma\delta}(\Omega) = CG^{ph}(\Gamma,\omega_\infty,\Omega) + DQ^2G^{ph}(\Gamma,\omega_\infty,\Omega) + EQ^2_{pc}G^{ph}(\Gamma,\omega_\infty,\Omega) + F\frac{1}{N}\sum_q G^f(qj)G^{ph}(\tilde{\Gamma},\omega_\infty,\Omega). \quad (27)$$

The first term represents the Raman scattering at $T > T_C$ and can be identified with I_0. The second term explains the temperature dependence of the integrated intensity below T_C as being proportional to the square of the displacive order parameter Q. The last two terms correspond to precursor ordering at $T > T_C$ and dynamical fluctuations near T_C. None of the latter effects was observed near the critical temperature. However, the linear temperature dependence of the total cross section yields the critical behaviour of Q in the form

$$Q^2 \sim I - I_0 = A^2(T_C-T)^{2\beta}, \qquad \beta = 1/2. \quad (28)$$

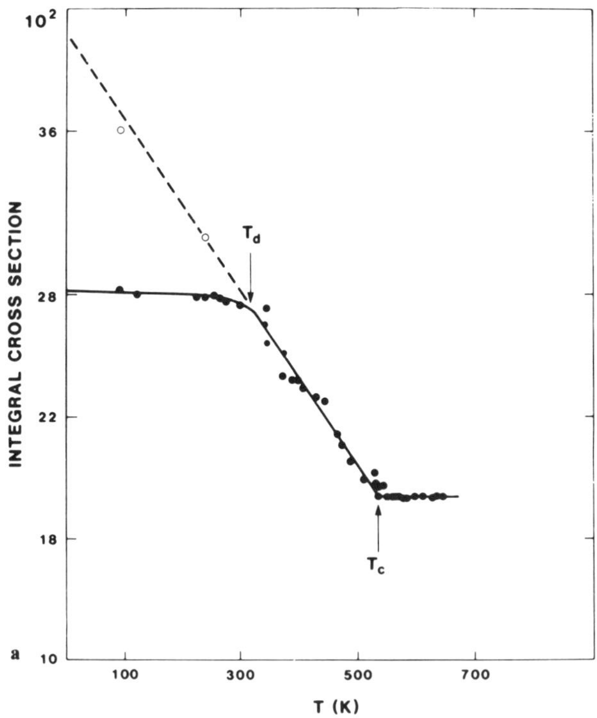

Figure 6. Temperature evolution of the total cross section of the 514 cm^{-1} line in anorthoclase. The dashed line indicates the hypothetical temperature dependence of the Raman intensity of this mode at low temperatures (Salje 1986).

The value of the critical exponent ß is accordant with a second order transition and agrees well with the linear temperature dependence of the excess entropy determined by Salje et al. (1985). The excess entropy of the structural phase transition is proportional to Q^2 (Fig. 7).

Fundamental differences between albite and Na-K-feldspar shows the observed fluctuation broadening of the Raman profiles in anorthoclase. Polynomial behaviour only occurs at high temperatures whereas high saturation values are reached at low temperatures (Fig. 8). The fluctuation broadening, with a Gaussian component of $2\sigma = 10.2$ cm^{-1} for the line at 514 cm^{-1}, is attributable to Na,K site ordering and strong local distortions of the Si-Al-O network due to inequivalent alkali positions (Salje 1986, Fenn and Brown 1977). The fluctuation broadening was observed on Raman lines which primarily involve internal Si-O vibrations. This indicates that local sub-lattice distortions occur not only around the alkali positions, but influence also the nearest tetrahedral SiO_4 and AlO_4 complexes. A coupling of scattering modes with the macroscopic order parameter is symmetry forbidden for modes with non-vanishing wave vectors or localized modes with fracton character.

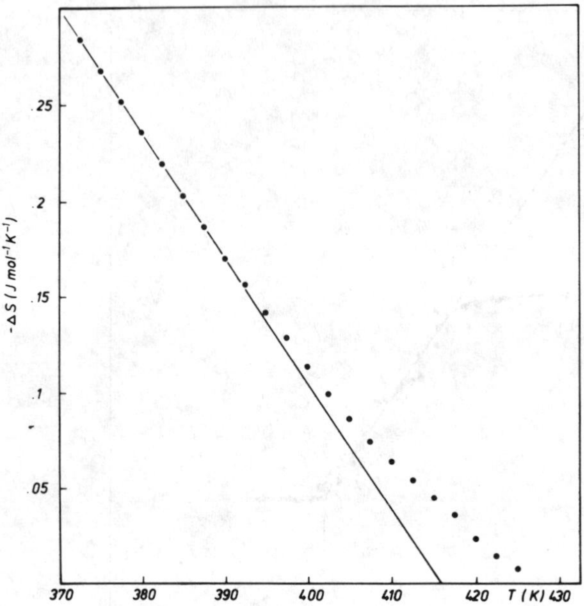

Figure 7. Temperature dependence of the excess entropy $-\Delta S(T) = \int_{T}^{T_C} \frac{\Delta c_p}{T}\, dT$ in disordered Or_{31}. c_p is the heat capacity (Salje et al. 1985).

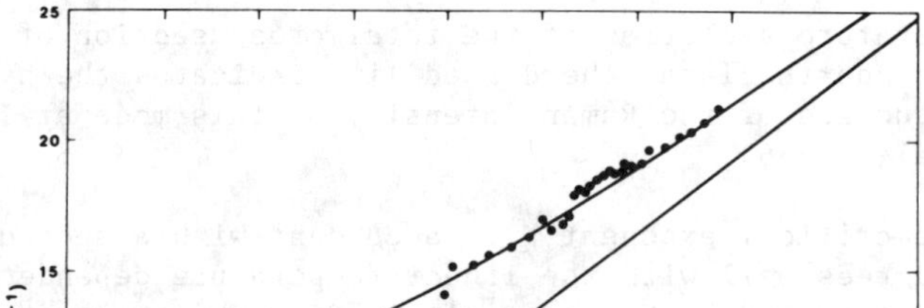

Figure 8. Temperature dependence of the half width of the 514 cm^{-1} line in anorthoclase. The fitted fluctuation broadening (horizontal line), the multi-phonon part (proportional to T) and the resulting total line width are indicated (Salje 1986).

Therefore, the increase of the line width and the saturation effect of the integrated intensity may be attributed to an inhomogeneous crystal (with respect to the macroscopic order parameter) associated with broken long-range translational symmetry resulting in a reduced phonon coherence length of a few Si-O distances.

3.3. Phase transition in ferroelastic As_2O_5

A simple, but significant, example of a pure, proper ferroelastic phase transition is that of arsenic pentoxide, As_2O_5. The tetragonal high temperature modification ($P4_12_12$) transforms in a second order transition at T_C = 575,7 K into ferroelastic orthorhombic phase ($P2_12_12_1$) (Bismayer et al. 1986). The structural changes during the phase transformation are dominated by tilts of edge-sharing AsO6 octahedra. These octahedra form chains along the cristallographic c-axis and their tilt axes are along [100] (Jansen 1978, Jansen 1979). Pure polyhedra tilts would suggest a pure displacive transformation character but recent investigations of the phase transition mechanism revealed, beside the mainly displacive behaviour, aspects of an order-disorder contribution.

HMRS was applied to study the correlation between the ferroelastic strain and local sublattice distortions which correspond to coupling effects between the strain, the elastic constants and the order parameter (Bismayer et al. 1986). Both, the ferroelastic and paraelastic phase are based on the same principle arrangement of corner-sharing AsO_6 octahedra in the network of polyhedra and both have the same primitive unit cell. Maximum displacements during the phase transition are ca. 5 pm for arsenic atoms and ca. 20 pm for oxygen atoms. The supergroup ($P4_12_12$)-subgroup($P2_12_12_1$) relation conserves the translational symmetry and the critical point of the Brillouin zone is the origin (Γ). The active representation of the order parameter is B_1. Consequently, critical vibrations, which are directly related to this order parameter, are phonons of the B_1 species. They are not infrared-active but Raman-active and belong to the same irreducible representation as the spontaneous strain $e_S = e_2 - e_1$.

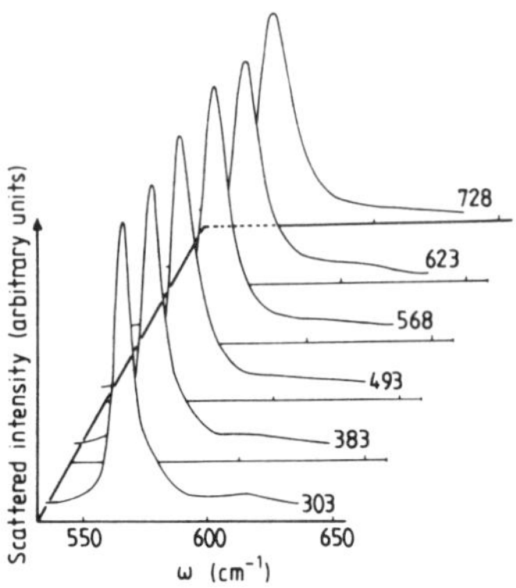

Figure 9. 579 and 567.7 cm^{-1} Raman bands (unpolarized) for As_2O_5 at different temperatures (given by curves in K) (Bismayer et al. 1986).

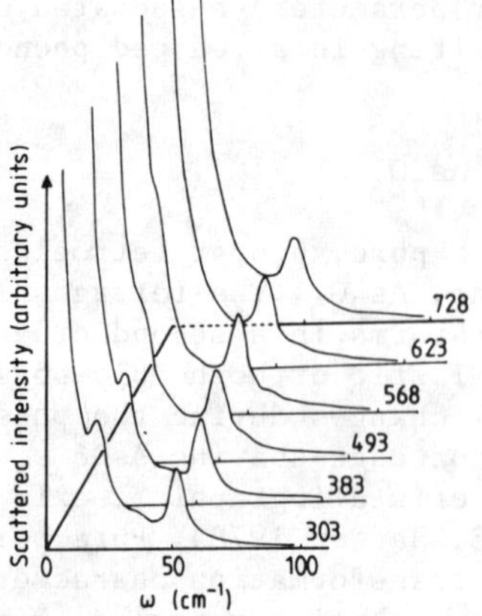

Figure 10. 66.2 and 35.5 cm^{-1} Raman bands (unpol.) for As_2O_5 at some selected temperatures (given by curves in K) (Bismayer et al. 1986).

The critical behaviour of the order parameter was identified through Raman spectroscopy. Raman-active hard modes at 579,0 and 567.7 cm^{-1} (Fig. 9) as well as the low frequency bands (Fig. 10) show considerable shifts and profile changes as a function of temperature. Both, the frequencies ω^2 and the line intensity I_0 of all bands observed by Bismayer et al. (1986) decrease linearly with increasing temperature from 303 K to T_C = 575.7 K. However, the authors did not observe a vibration that shows vanishing frequency with $T \to T_C$. The lowest frequency saturates at 27 cm^{-1} (Fig. 12).

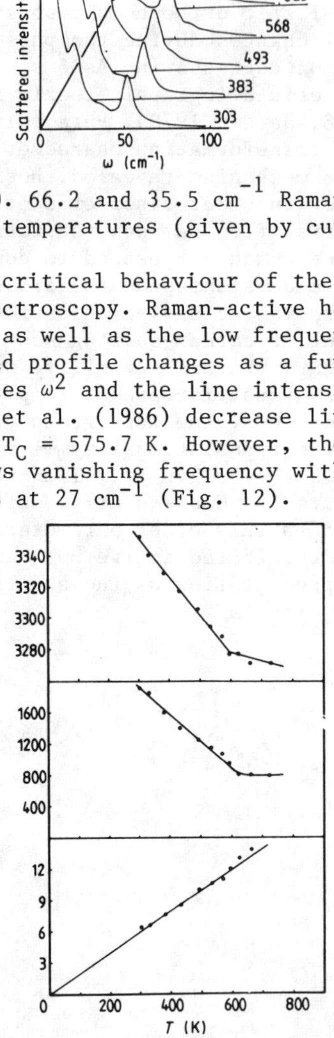

Figure 11. Frequency ω^2, integrated intensity I_0 and spectral linewidth Γ^* of the 579.0 cm^{-1} mode for As_2O_5 versus T (Bismayer et al. 1986).

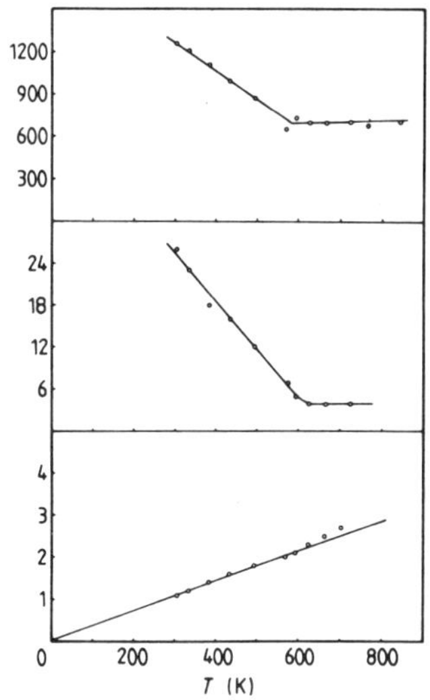

Figure 12. Frequency, integrated intensity and linewidth of the 35.5 cm^{-1} mode for As_2O_5 versus temperature (Bismayer et al. 1986).

The phase transition temperature in figures 11 and 12 is clearly indicated by a discontinuity in the frequency curves. The variation of ω shows for hard modes at higher frequencies a characteristic temperature dependence resulting from anharmonic interactions described by the mode Grüneisen relation (Munn 1975, Salje and Viswanathan 1976)

$$\Delta\omega/\omega = \gamma_i \Delta V/V \tag{29}$$

with the molar volume V. The mode Grüneisen parameter of the phonon at 579.0 cm^{-1} (Fig. 11) was found to be $\gamma = 1.1$ below T_C.

The frequency shift of the lowest lattice mode in As_2O_5 at 35.5 cm^{-1} agrees well with the scalar factor given by Landau theory according to

$$\omega = \omega_0 |T-T_C/T_C|^{\beta} \qquad \text{with } \beta = 1/2. \tag{30}$$

However, no progressive softening of the 35.5 cm^{-1} phonon below 27 cm^{-1} was observed.

The resulting value of the critical exponent obtained by Raman spectroscopy was strongly supported by optical birefringence measurements (Salje et al. 1987) which were chosen because of their outstanding sensitivity to small changes of the order parameter. To avoid uncertainities arising from superposition effects of the spontaneous birefringence with the optical activity a propagation direction for the light beam transmitting the crystal along [110] was selected. The optical birefringence due to the structural phase transition is related to the order parameter via the elastooptic effect and was obtained from the correlated birefringence $\Delta n = n_3 - n_{\bar{1}10}$ (Fig. 13).

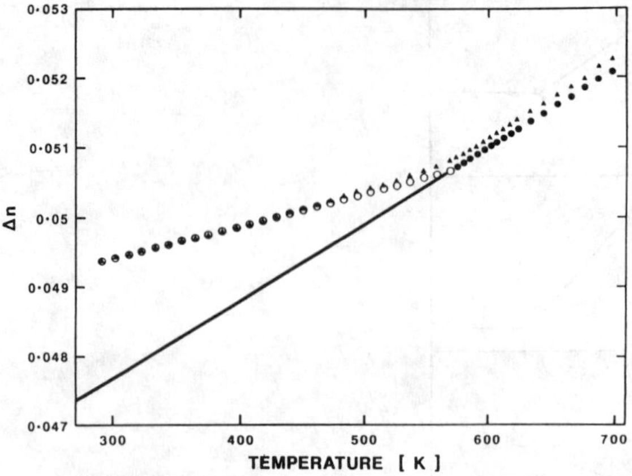

Figure 13. Temperature dependence of the optical birefringence for light propagating along [110]. Triangles: observed birefringence at $T < T_C$; open circles: data corrected for thermal expansion; full circles: observed birefringence at $T > T_C$; full line: extrapolated birefringence of the tetragonal phase at temperatures below T_C (Salje et al. 1987).

For light, traveling along [110] the total excess birefringence depends only on the square of the order parameter according to

$$\delta\Delta n = \chi' Q^2 \sim \chi'(T-T_C)^{2\beta} \qquad (31)$$

with nonvanishing coefficient of the elastooptic tensor χ' in the high temperature phase. The experimental value of 2β determined from log/log plot of the excess birefringence versus the reduced temperature is $\beta = 0.509 \pm 0.01$ (Fig. 14).

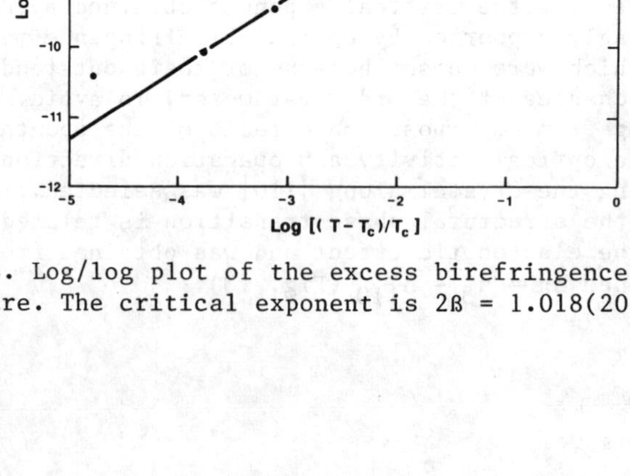

Figure 14. Log/log plot of the excess birefringence versus reduced temperature. The critical exponent is $2\beta = 1.018(20)$ (Salje et al. 1987).

Deviations from the ß-value occur within a temperature interval of 5 K below T_C which is accordant with the behaviour of the temperature dependent frequencies ω^2 of the Raman modes investigated.

The temperatur behaviour of the linewidth Γ* of all measured vibrations in As_2O_5 shows a linear evolution between room temperature and the phase transition point. However, the lines in figures 11 and 12 broaden over and above their anharmonic thermal background at T_C by $\Delta\Gamma^* \cong 1.5\ cm^{-1}$ which is at variance with a pure displacive type of phase-transition mechanism.

Hence, from the temperature dependences of the scattering intensities and linewidths the conclusion was drawn that appreciable order parameter coupling occurs in As_2O_5. A simple model of linear order-parameter-strain coupling was proposed by Bismayer et al. (1986). The difference between the relevant parts of the free energy for the orthorhombic and tetragonal phases leads to the following polynomial

$$\Delta F = aQ^2 + bQ^4 + \ldots + \lambda Q(e_2 - e_1) + \frac{1}{2}\sum_{1,2} C_{ik} e_i e_k . \qquad (32)$$

The order parameter is Q, the elastic constants are C_{ik} and the coupling constant between the order parameter and the components of the spontaneous strain is λ. Parameters in the thermodynamic potential which possibly lead to a structural phase transition are (i) a critical behaviour of the strain components e_i (this is the case in proper ferroelastics), (ii) a local sublattice distortion (e.e. a tilt or ordering of polyhedra in the As_2O_5 structure) represented by the order parameter Q and (iii) the critical coupling between spontaneous strain and Q.

Provided that Q represents the driving order parameter (or is coupled linearly to the driving order parameter) the total scattering cross section is expected to depend quadratically on Q. In fact, a linear dependence of I_0 on temperature occurs in As_2O_5 and the resulting critical exponent of Q is ß = 1/2 as predicted by classical mean field theory. However, the increase of the damping constant Γ* (Figures 11 and 12) at $T \geqq T_C$ can possibly be correlated to density of states effects or fluctuations with relaxation frequencies below 1.5 cm^{-1} (Bruce et al. 1980).

In conclusion, the critical exponent of the order parameter resulting from precise optical birefringence measurements agrees well with the value which corresponds to a pure, proper ferroelastic phase transition. The Raman spectroscopic results indicate further that the transition mechanism in As_2O_5 contains aspects of a displacive behaviour (mode softening, linear temperature dependence of the scattering intensities) and influences of order-disorder effects (finite scattering frequencies at T_C, increase of Γ* above T_C).

3.4 Raman scattering of hard modes and stepwise phase transitions in $Pb_3(P_{1-x}As_xO_4)_2$

Lead phosphate-arsenate crystals, $Pb_3(P_{1-x}As_xO_4)_2$, possess palmierit type structure and undergo a ferroelastic transition between a paraelastic, trigonal phase a (R3m) and a monoclinic phase b (C2/c) (Bismayer and

Salje 1981, Bismayer et al. 1982). Depending on the x-value in As-rich mixed crystals, a further ferroelectric low temperature phase (c) has been observed by Viswanathan and Miehe (1978) with space group $P2_1/c$. The transition $C2/c - P2_1/c$ is first order in pure lead arsenate and becomes smooth for all compounds with $x < 0.85$. This behaviour is a consequence of the strain-induced biquadratic order-parameter coupling which leads at low temperatures to an intermediate phase b_c (Bismayer et al. 1986). However, the transition behaviour of lead phosphate-arsenate at low temperatures shall not be discussed further in this contribution.

The transformation $R\bar{3}m - C2/c$ approaches second-order behaviour with increasing As content (Bismayer and Salje 1981) and fulfills the conditions of an improper ferroelastic phase transition (Torres 1975, Toledano and Toledano 1980). Reinvestigations of the character of this transformation reported in the literature revealed that no direct transformation a ($R\bar{3}m$) - b ($C2/c$) was ever observed. Experimental studies involving x-ray (Joffrin et al. 1977), neutron diffraction (Joffrin et al. 1979), Raman and infrared spectroscopy (Benoit and Chapelle 1974, Benoit 1976, Luspin et al. 1979, Salje and Bismayer 1981, Salje et al. 1983) and electron microscopy (Ayroles et al. 1979, Roucau et al. 1979) as well as calorimetric studies (Salje and Wruck 1983) indicate the existence of an intermediate pseudo-phase a_b. Bismayer et al. (1982) reinvestigated the intermediate pseudo-phase a_b through high-resolution neutron scattering, diffuse x-ray scattering and Raman spectroscopy. They concluded that whereas the deformations in the intermediate phase of $Pb_3(PO_4)_2$ are purely dynamic in character, the deformations in the mixed crystals $Pb_3(P_{1-x}As_xO_4)_2$ are static just above the ferroelastic phase transition and are superposed by reorientational flipping motion at higher temperatures.

According to Torres (1975) critical L-points of the surface of the rhombohedral Brillouin zone transform to the origin (Γ-point) during the symmetry reduction. He expressed the thermodynamic potential in terms of three order parameters stemming from the three inequivalent L-points. Salje and Devarajan (1981) developed a multicomponent order-parameter model (three states Potts model) which explained very well the monoclinic domains in the paraelastic phase with trigonal lattice constants. It starts from the effective Hamoltonian of the n-component Landau-Ginzburg-Wilson Hamiltonian

$$\Phi = \frac{\alpha}{2}(Q_1^2 + Q_2^2) + \frac{\alpha'}{2}Q_3^2 + (Q_1^2 + Q_2^2)^2 L_2 + Q_3^2(Q_1^2 + Q_2^2)L_3 + Q_3^4 L_1 + \frac{1}{\sqrt{2}}Q_1 Q_3 (Q_3^2 - \frac{1}{3}Q_2^2)L_4 \qquad (33)$$

with coefficients defined by Salje and Devarajan (1981). A diagonalization is possible assuming well-separated fixed points of $\{Q_1,Q_2\}$ and $\{Q_3\}$. The order parameters Q_1 and Q_2 then describe the effective Hamiltonian of a three states Potts oscillator (Aharony et al 1977)

$$\Phi'' = (\frac{\alpha}{2} + L_3 M^2)(Q_1^2 + Q_2^2) + L_2(Q_1^2 + Q_2^2)^2 + wQ_1(Q_2^2 - \frac{1}{3}Q_1^2) \qquad (34)$$

with corresponding critical exponents belonging to the three states Potts model (Potts 1952). In fact, the experimental critical exponent value for $Pb_3(PO_4)_2$ as determined from measurements of the birefringence (Torres et al. 1974, Bismayer and Salje 1981, Wood et al. 1980) and spontaneous strain (Guimaraes 1979) is near 1/4 for temperatures not too close to the transition point.

Structurally, the three states described by $\{Q_1,Q_2\}$ are represented by three possible orientations of the monoclinic crystallographic axes of the low-temperature phase (C2/c) with respect to the trigonal unit cell of the high-temperature phase ($R\bar{3}m$) (Bismayer et al. 1982). Salje and Devarajan (1981) suggested that in the case of pure lead phosphate there is far above the ferroelastic transition point T_C a first phase transformation around 563 K involving the order parameter Q_3. This is equivalent to the simultaneous condensation of all three zone-boundary L-modes giving rise to static shifts of Pb positions away from the ternary axis along the three possible directions of the binary axis of the monoclinic symmetry which are still symmetry equivalent. At T_C = 453.6 K the ferroelastic phase transition described by equation (34) takes place corresponding to the critical behaviour of Q_1 and Q_2 which describe the orientation of the binary axis in C2/c. During this first-order process one binary axis is chosen and consequently the symmetry is broken from trigonal to monoclinic. In figure 15 the geometrical relation between the unit cells of the trigonal ($R\bar{3}m$) and monoclinic (C2/c) phases as well as the three possible orientations of the binary axis perpendicular to the ternary axis corresponding to the domains in the ferroelastic modification are depicted.

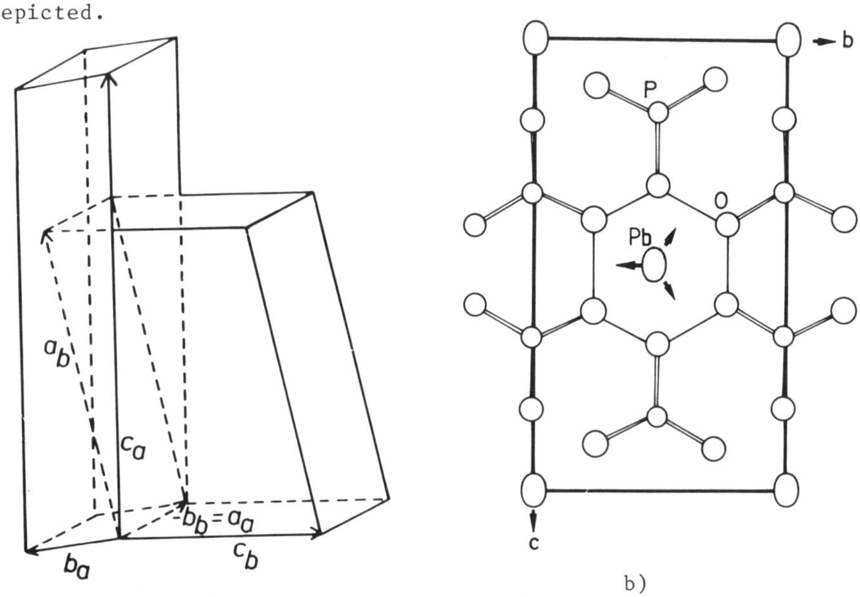

b)

Figure 15. a) Geometrical relation between the unit cells of the phases a and b (Bismayer et al. 1986). b) Pb-displacement corresponding to the three possible orientations of the binary axis perpendicular to the ternary axis of phase a projected onto the b_a-c_a-plane.

The critical mode in the intermediate phase a_b is similar to the tunnelling in KDP (Wilson and Cummins 1971) and represents the spontaneous thermal switching between the three binary axes. As the corresponding dominant atomic motion is that of Pb, no tunnelling in the sense of quantum mechanics is possible and therefore, this reorientation mode was called 'flip mode' (Salje and Devarajan 1981). This mode couples with the L phonons, giving rise to the reduced frequencies of the zone boundary modes in the phase a_b observed by Joffrin et al. (1979). In the ferroelastic phase b, their neutron scattering experiments show the recovery of two non-symmetry-breaking L modes at the zone boundary. The relevant order parameters of the flip mode are Q_1 and Q_2, with non-zero values in this low temperature phase corresponding to the lock in of the monoclinic axis. Salje and Devarajan (1981) correlated this ordering process with a flip relaxational mode similar to the mode suggested by Raich and Hüller (1979) in the case of the NaN_3 phase transition. Consequently the pure relaxational mode was described as being the driving soft mode in $Pb_3(PO_4)_2$ which is likely to couple anharmonically with the lowest A_g mode and prevents the A_g mode in the ternary oxides $Pb_3(P_{1-x}As_xO_4)_2$ from a critical slowing down (figure 16). In $Pb_3(P_{0.97}As_{0.03}O_4)_2$ the A_g and B_g modes near 40 cm^{-1} merge into a single E_g mode when the temperature increases. The critical temperature for this line splitting is that of the a_b-a transformation.

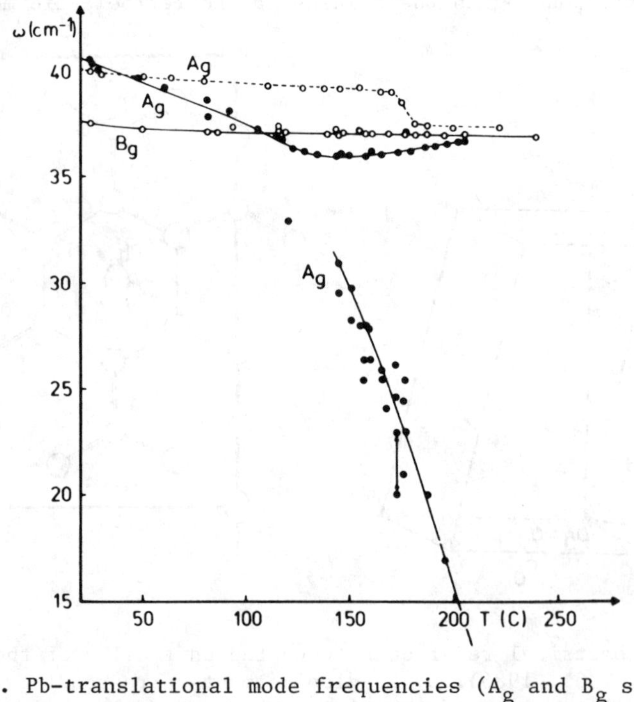

Figure 16. Pb-translational mode frequencies (A_g and B_g species) versus temperature for $Pb_3(PO_4)_2$ (full circles) and $Pb_3(P_{0.97}As_{0.03}O_4)_2$ (open circles) (Bismayer et al. 1982).

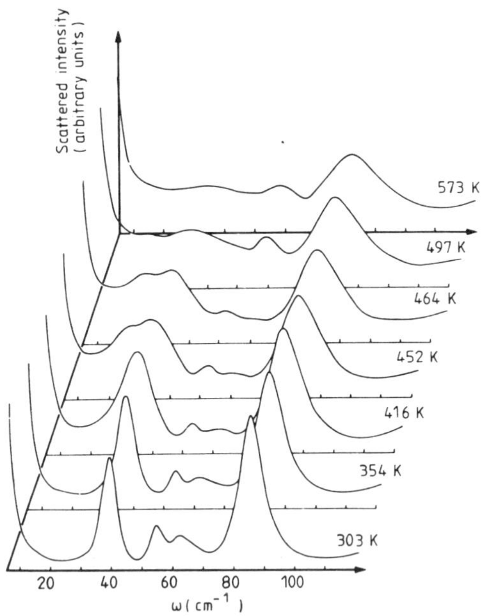

Figure 17. Low frequency A_g Raman band (bb polarisation) for $Pb_3(PO_4)_2$ at different temperatures (Salje et al. 1983).

The low frequency Raman spectrum with A_g scattering symmetry is more complex in the case of pure lead phosphate (Fig. 17). At temperatures above 416 K at low energies a scattering signal was observed with A_g symmetry which is strongly temperature dependent. This mode appeares in addition to the A_g mode at 40 cm^{-1}. The additional mode below 37 cm^{-1} remains underdamped even at temperatures close to the ferroelastic transition (T_C = 453.6) and was also observed in the intermediate phase a_b (Fig. 17).

Hence, even in $Pb_3(PO_4)_2$ the A_g mode is not the driving soft mode of the ferroelastic transformation and therefore hard-mode spectroscopy was chosen by Salje et al. (1983) for the investigation of the structural phase transition of lead phosphate-arsenate.

The authors recorded both, A_g and B_g species Raman spectra for $Pb_3(PO_4)_2$ and the A_g mode for $Pb_3(P_{0.77}As_{0.23}O_4)_2$ and the strong 80 cm^{-1} Raman band which appears in both polarisations was followed as a function of temperature. Whereas the 40 cm^{-1} band vanishes at about 70 K above T_C the 80 cm^{-1} phonon is also strongly present in the high-temperature phase far above T_C. For the mixed crystals, the profile analysis described in section 3. was made. Salje et al. (1983) compared the integrated intensities of the B_g and A_g-Raman modes at 80 cm^{-1} with the order parameter dependence of the morphic birefringence Δn_{bc} on temperature. Both curves, which are displayed in figure 18, are parallel at low temperatures and show in agreement with equation (5) the same quadratic dependence on the order parameter for all A_g modes as for the birefringence Δn_{bc} at $T << T_C$.

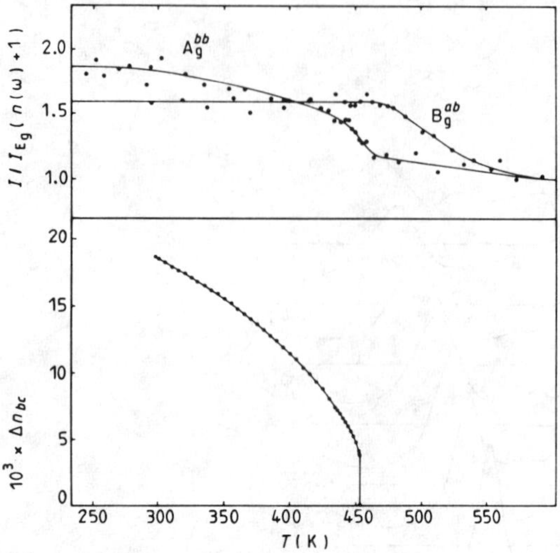

Figure 18. Normalised integrated intensities of the $B_g^{(a*b)}$ and $A_g^{(bb)}$ Raman modes at 80 cm^{-1} for $Pb_3(PO_4)_2$ versus temperature. For comparison the temperature dependence of the spontaneous birefringence Δn_{bc} is shown in the lower part (Salje et al. 1983).

The intensities of the B_g modes do not show any dependence on the order parameter at temperatures below T_C, which can be understood from the fact that the order parameter transforms as A_g thus forbidding the coupling with the B_g hard mode.

Near T_C no discontinuity in the Raman cross section similar to the temperature dependence of the morphic birefringence was observed but a continuous decrease of the $A_g^{(bb)}$ and $B_g^{(a*b)}$ intensities which merge to give the same E_g intensity near 570 K in $Pb_3(PO_4)_2$. This reveals strong local monoclinic symmetry in the intermediate pseudo-phase a_b.

In the mixed crystals an additional small peak occurs in the temperature dependence of the $A_g^{(bb)}$ intensity directly above T_C which does not exist in pure $Pb_3(PO_4)_2$ (Fig. 19). The temperature range for this peak is the same as found by Bismayer et al. (1982) for the existence of a quasi-static central peak and by Salje and Wruck (1983) for increased specific heat in the intermediate pseudo-phase as shown in figure 20 and figure 21 for $Pb_3(P_{0.2}As_{0.8}O_4)_2$.

Both, the results of Raman spectroscopy and specific heat measurements were interpreted in terms of static or quasi-static long-range monoclinic deformations.

A measure for the remaining Raman intensity in the pseudo-phase a_b due to monoclinic lattice distortions directly above T_C is the parameter $C_I = Q(T_C)/Q(T_{1/2})$ first defined by Bruce et al. (1980). The parameters Q(T) are the order parameters calculated from the Raman intensities at T_C and $T_{1/2} = 2T_C/3$. It is $C_I = 0.77$ for $Pb_3(P_{0.77}As_{0.23}O_4)_2$ which shows the large order-disorder character of the phase transition near T_C.

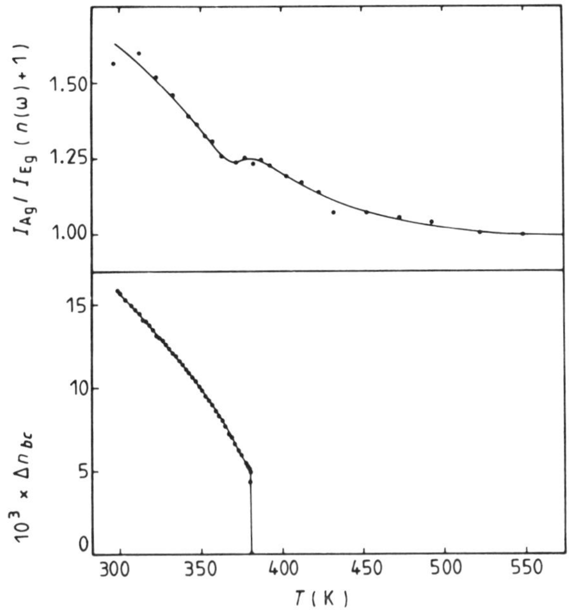

Figure 19. Normalised integrated intensity of the A_g Raman mode at 80 cm^{-1} and Δn_{bc} for $Pb_3(P_{0.77}As_{0.23}O_4)_2$ versus temperature (Salje et al. 1983).

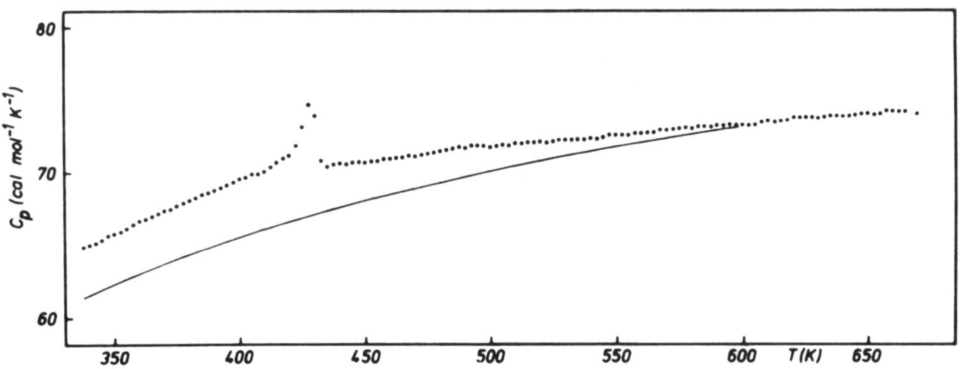

Figure 20. Temperature dependence of the molar heat capacity of $Pb_3(P_{0.77}As_{0.23}O_4)_2$. The solid line shows the background heat capacity obtained from the analysis by a least-square fit using $C_p{}^0 = a + bT^{-1}$. For higher temperatures this curve coincides with the dots representing the measured values (Salje and Wruck 1983).

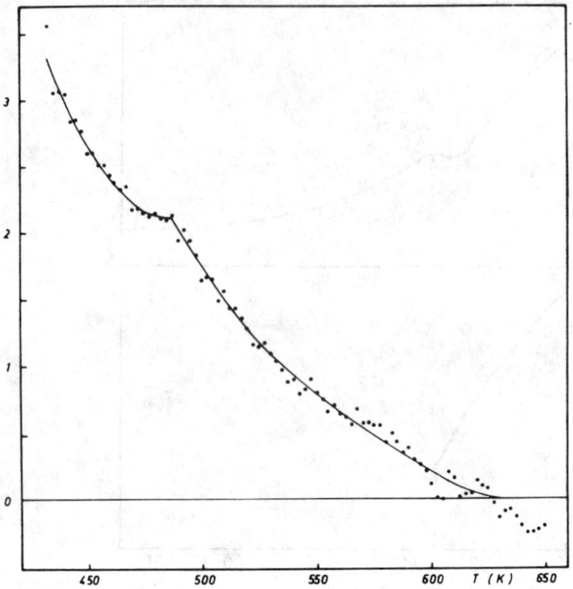

Figure 21. Excess specific heat in the temperature range $T > T_C$ for $Pb_3(P_{0.2}As_{0.8}O_4)_2$ (Salje and Wruck 1983).

The pair correlation function corresponding to equation (8) contains the soft-mode coordinate $Q(qj_s)$ which represents the displacive order parameter of the low temperature phase. This parameter is associated with the order parameter Q_3 defined by Salje and Devarajan (1981) in their Potts model interpretation of the phase transition in lead phosphate. Neglecting fluctuations in Q_3 at temperatures well below T_C the Raman scattering intensity in $Pb_3(P_{1-x}As_xO_4)_2$ is given for low temperatures by

$$I^{low}_{\alpha\beta\gamma\delta}(\Omega) = \sum_{j_h} P_{\alpha\beta}\begin{pmatrix}0\\ j_h\end{pmatrix} P_{\gamma\delta}\begin{pmatrix}0\\ j_h\end{pmatrix} G^{ph}(0j_h,\Omega) + 4\sum_{j_h j_s} P_{\alpha\beta}\begin{pmatrix}0 & 0\\ j_h & j_s\end{pmatrix} P_{\gamma\delta}\begin{pmatrix}0 & 0\\ j_h & j_s\end{pmatrix} Q^2\, G^{ph}(0j_h,\Omega). \qquad (35)$$

In the limit of the high temperature the hard mode and the soft mode in $R\bar{3}m$ belong to different points of the Brillouin zone, namely the Γ point and the L point, respectively. Therefore the second-order polarisability derivatives will vanish at temperatures far above T_C and no hard-mode-soft-mode coupling is possible. Close to T_C, however, this argument does not hold under the condition that large monoclinic domains exist like in $Pb_3(P_{0.77}As_{0.23}O_4)_2$. Nevertheless, it is known for this compound that no measurable softening of Raman modes occurs (Salje and Bismayer 1981, Bismayer et al. 1982) (see also figure 16) and the coupling of the hard mode with the soft mode can be neglected altogether.

The main contributions to the fluctuation correlation are therefore left to the central peak, the flip relaxation and the flip mode. The Raman scattering intensity just below T_C and in the intermediate pseudo-phase was hence expressed by Salje et al. (1983) as

$$I^{int}_{\alpha\beta\gamma\delta}(\Omega) \approx \sum_{j_h} P_{\alpha\beta}\begin{pmatrix}0\\ j_h\end{pmatrix} P_{\gamma\delta}\begin{pmatrix}0\\ j_h\end{pmatrix} G^{ph}(0j_h,\Omega)$$
$$+ A \sum_{j_h} P_{\alpha\beta}\begin{pmatrix}0 & 0\\ j_h & j_{Q3}\end{pmatrix} P_{\gamma\delta}\begin{pmatrix}0 & 0\\ j_h & j_{Q3}\end{pmatrix}$$
$$\times \frac{1}{N}\int_{-\infty}^{\infty} G^{ph}(qj_h,\Omega)(G^{flip}(qj_{Q3}) + G^{cp}(qj_{Q3}). \qquad (36)$$

Q_3 corresponds to the static displacement which persists above T_C and the term

$$P_{\alpha\beta}\begin{pmatrix}0 & 0\\ j_h & j_{Q3}\end{pmatrix}$$

takes the role of

$$P_{\gamma\delta}\begin{pmatrix}0 & 0\\ j_h & j_s\end{pmatrix}$$

of the low temperature phase. The equations (35) and (36) can be written together in analogy to Bruce et al. (1980) as

$$I_{\alpha\beta\gamma\delta}(\Omega) \cong CG^{ph}(\Gamma(0j_h)\omega_\infty(0j_h)\Omega)$$
$$+ DQ^2_{js}G^{ph}(\Gamma(0j_h)\omega_\infty(0j_h)\Omega)$$
$$+ EQ^2_{jQ3}G^{ph}(\Gamma(0j_h)\omega_\infty(0j_h)\Omega)$$
$$+ F\,\frac{1}{N}\sum_q G^{flip}(qj_{Q3})G^{ph}(\tilde{\Gamma}(qj_h)\omega_\infty(qj_h)\Omega) \qquad (37)$$

with parameters C, D, E and F involving components of polarisability devirative tensors. Q_{js} is the displacive order parameter below T_C and Q_{jQ3} corresponds to Q_3 which is present in the intermediate pseudo-phase. The damping constant $\tilde{\Gamma}(qj_h)$ includes the effect of coupling between the hard mode and the flip mode or flip relaxation in equation (36).

The experimental results were compared by Salje et al. (1983) with the theoretical approach in equation (37) as follows. At $T \ll T_C$ the A_g modes, both in $Pb_3(PO_4)_2$ and $Pb_3(P_{0.77}As_{0.23}O_4)_2$, show the usual quadratic dependence on the order parameter as given in the second term of equation (37). Above T_C the observed persistence of the scattering intensity in figures 18 and 19 is attributed to the third and fourth terms of equation (37). A separation of the central peak effect and the influence of the flip mode appears from the comparison of the spectra of $Pb_3(PO_4)_2$ and $Pb_3(P_{0.77}As_{0.23}O_4)_2$. A quasi-static central peak exists in the case of the mixed phosphate-arsenates, but not in pure lead phosphate (Benoit et al. 1981, Bismayer et al. 1982). It is noteworth that the structural transformation a_b - b in $Pb_3(P_{1-x}As_xO_4)_2$ is always accompanied by the appearance of diffuse superreflexions above the ferroelastic transition. Benoit et al. (1981) have shown that the origin of the superreflexions is purely dynamic in case of $Pb_3(PO_4)_2$. In the phosphate-arsenate mixed crystals diffuse x-ray scattering occurs at temperatures close to the ferroelastic transition point with a superposition of two scattering contributions (Bismayer et al. 1982). A comparison of the profiles of the superreflexion $02\bar{1}$ of $Pb_3(PO_4)_2$ and $Pb_3(P_{0.48}As_{0.52}O_4)_2$ is given in figure 22.

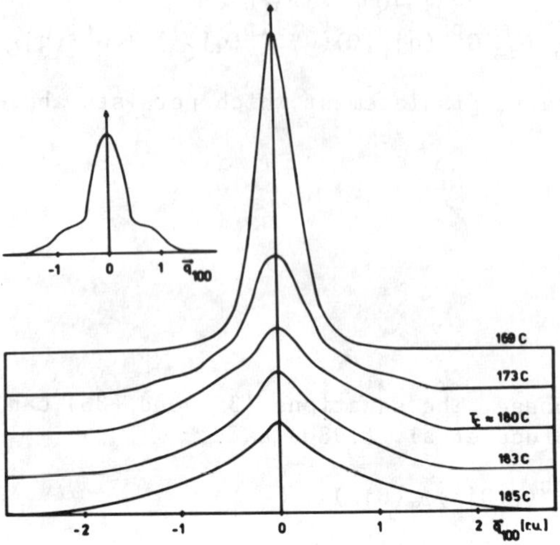

Figure 22. Profiles of the diffuse scattering near the reflex $0\bar{2}1$ of $Pb_3(PO_4)_2$ at different temperatures near T_C. The inset shows the same profile for $Pb_3(P_{0.48}As_{0.52}O_4)_2$ at $T_C + 5$ K (Bismayer et al. 1982)

In contrast to the approximately Lorentzian peaks in the phase a_b of $Pb_3(PO_4)_2$, the profiles in $Pb_3(P_{0.48}As_{0.52}O_4)_2$ are always a superposition of two peaks: a sharp peak with the same profile as the main reflexions and a diffuse peak similar to lead phosphate which is identified as dynamical scattering from the lowest-energy phonon branches (Joffrin et al. 1979) and the effect of the flip mode. Bismayer et al. (1982) correlated the the sharp peak with a static structural distortion (static monoclinic deformations in a trigonal matrix) in the intermediate pseudo-phase which possibly leads to the central peak. Exactly in the same temperature range where the central peak was observed an increased Raman scattering signal can be observed in figure 19 and Salje et al. (1983) associated this peak with the coupling of the hard mode with the central peak.

The coupling of the hard mode with the flip mode or the flip relaxation in the fourth term of equation (37) results in both an increase of scattering intensity and an increase of the linewidth represented by $\tilde{\Gamma}$. In fact, both effects were observed in $Pb_3(PO_4)_2$. They exist in $Pb_3(P_{0.77}As_{0.23}O_4)_2$ but the intensity values are superposed by the above-mentioned influence of the central peak. The linewidths in figures 23 and 24 show deviation from the anharmonic background values just below T_C and in the intermediate pseudo-phase. In $Pb_3(PO_4)_2$ above T_C, where the flip mode is likely to have a resonant character, the flip-mode frequency can be estimated directly from $\Delta\Gamma^* = \Gamma^* - \Gamma^*_{anh}$. This gives a value of $\Delta\Gamma^* \approx 1.1$ cm^{-1} and a flip frequency of about 3×10^{-11} s. This frequency agrees well with the value evaluated by Benoit et al. (1981) from their neutron scattering experiments.

In the case of mixed crystals the deviations from the anharmonic background are as high as 3.5 cm^{-1}. From the results of Salje and Wruck (1983) a quasi static regime remains valid up to about 60 K above the ferrroelastic transition point. Salje et al. (1983) assumed that above this temperature the flip mode presumably has resonant character whereas below it can be characterized as flip relaxation with a relaxation time of the order of $\tau \approx 1 \times 10^{-11}$ s.

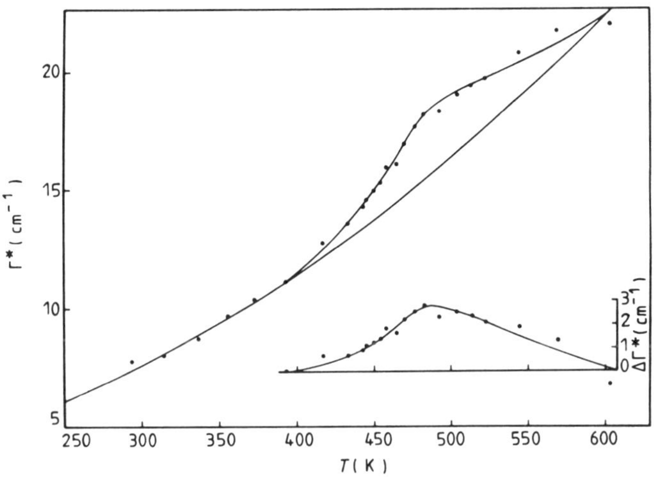

Figure 23. Spectral linewidth Γ* of the 80 cm-1 A_g^{a*a*}mode for $Pb_3(PO_4)_2$ versus temperature (Salje et al. 1983).

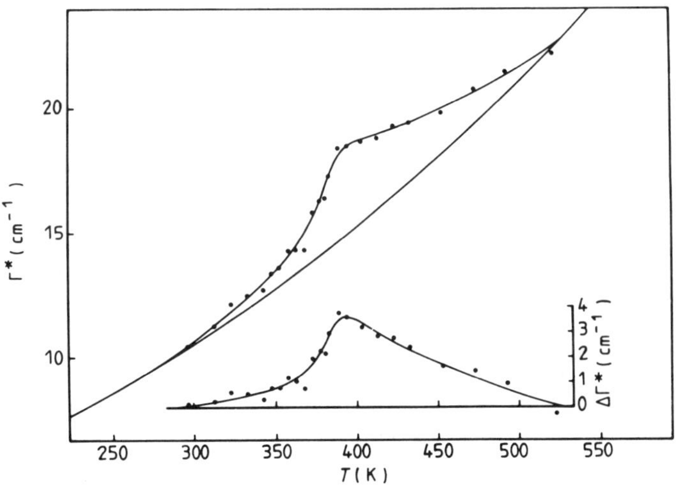

Figure 24. Spectral linewidth Γ* of the 80 cm^{-1} A_g mode for $Pb_3(P_{0.77}As_{0.23}O_4)_2$ versus temperature (Salje et al. 1983).

3.5. Sodium nitrate

Sodium nitrate, $NaNO_3$, undergoes a phase transition at $T_C \sim 552$ K first observed by Kracek (1931). The low temperature form of $NaNO_3$ is isostructural with calcite (s.g. $R\bar{3}c$), in which there are two formula units per primitive unit cell with the two nitrate groups pointing in opposite directions (Fig. 25). The phase transition at 552 K is widely believed to be associated with orientational disorder of these triangular molecular ions, and is regarded as a typical order-disorder transition. Previous investigations of this transition have been extensive (see e.g. references in Poon and Salje 1987a), including work using Raman spectroscopy (Chisler 1969, Prasad Rao et al. 1973, Shen et al. 1975, Neumann and Vogt 1978). However, the precise nature of the transition remains unclear, and the structure of the high temperature phase is still a matter of debate.

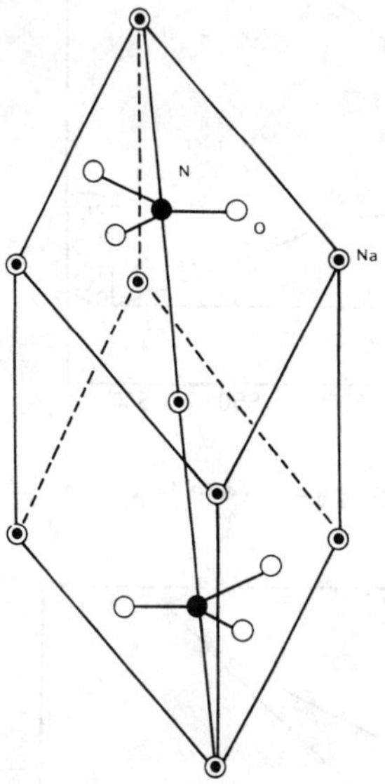

Figure 25. The primitive (rhombohedral) unit cell of $R\bar{3}c$ sodium nitrate. Note the antiparallel orientation of the two nitrate groups (Poon and Salje 1987a).

Recently Poon and Salje (1987b) have reinvestigated this 552 K transition using HMRS. In their work, these authors observed strong temperature dependence for the frequency and linewidth of the E_g libration mode at ca. 184 cm^{-1} (Figures 26 and 27), as well as significant residual intensity above T_C. The space group symmetry of the high temperature form ($R\bar{3}m$) forbids the activity of this mode; the observed intensity is due to 'precursor ordering' - the presence of clusters of $R\bar{3}c$ material above T_C. The step increase in the steep increase in the linewidth (Fig. 27) can be satisfactorily accounted for by the folding of the Brillouin zone as long range orientational correlation is lost.

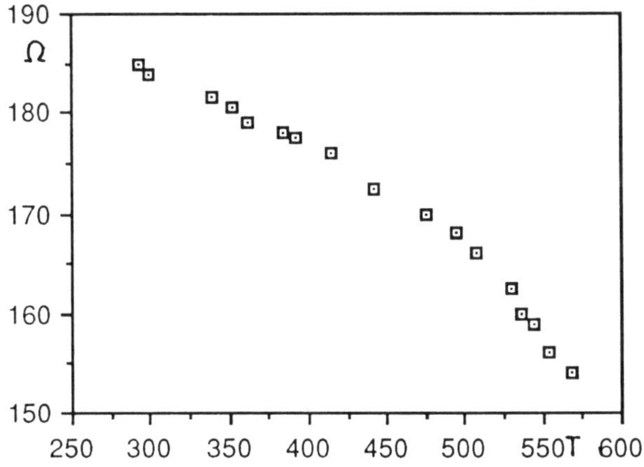

Figure 26a. The temperature evolution of the E_g libration frequency Ω (T given in K) (Poon and Salje 1987a).

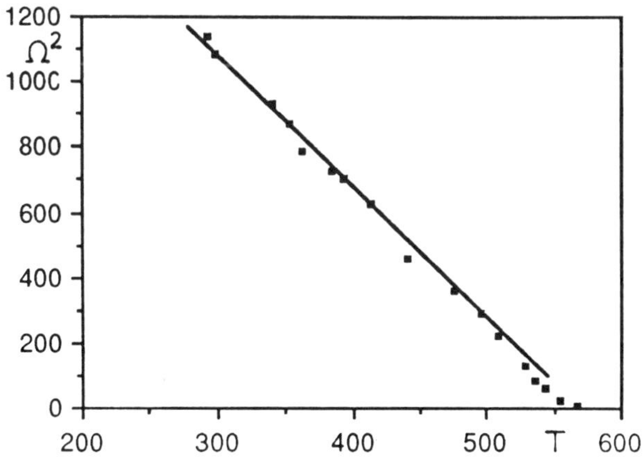

Figure 26b. $\Delta\Omega^2$ vs. T (in K) where $\Delta\Omega = \Omega - \Omega_c$ with $\Omega_c = 151\ cm^{-1}$ (Poon and Salje 1987a).

The large decrease in frequency of this mode has been observed before (e.g. Prasad Rao et al. 1973). Previous authors, however, did not comment on the fact that this behaviour is not typical of a pure order-disorder phase transition. Instead it shows that the E_g libration mode is coupled to the order parameter, and provides a significant contribution to the free energy of the phase transition (the total decrease in frequency from 0 K to T_c is 47 cm^{-1}). The coupling of this mode to the order parameter can be formulated as follows (Ginzburg et al. 1983).

Let ζ be the appropriate normal coordinate and η the order parameter. The ζ-dependence part of the free energy density may then be written as

$$F(\zeta) = a\zeta^2 + b\zeta^2\eta^2 + \ldots \tag{38}$$

where the coefficients a and b are taken as temperature independent. Above T_C, $\eta = 0$, so that

$$\Omega^2_{0\zeta} = F_{\zeta\zeta}/m_\zeta \sim a \tag{39}$$

whereas in the low temperature phase ($\eta \neq 0$) we have

$$\Omega^2_{0\zeta} \sim a + b\eta^2. \tag{40}$$

Therefore the change in Ω^2 is given by

$$\Delta\Omega^2 \sim \eta^2 \tag{41}$$

which can be rewritten as

$$(\Omega - \Omega_c)(\Omega + \Omega_c) \sim \eta^2 \tag{42}$$

where Ω_c is the frequency at T_C. For a hard mode we can take the second bracket as approximately constant, and obtain the result

$$\Delta\Omega = \Omega - \Omega_c \sim \eta^2. \tag{43}$$

The same result was obtained earlier by Petzelt and Dvorak (1976) using group theoretical arguments.

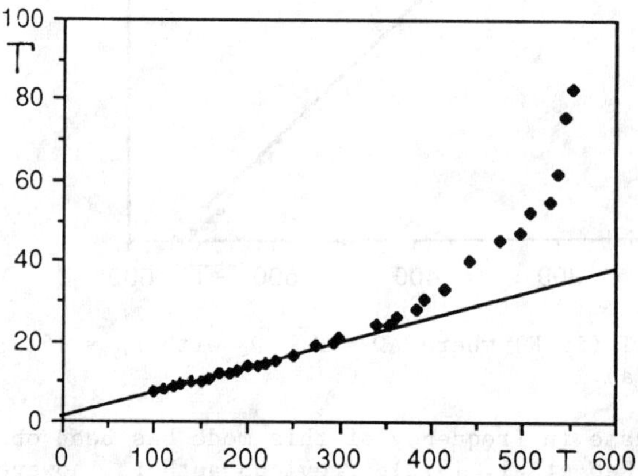

Figure 27. The linewidth of the E_g libration mode versus temperature (in K). The straight line is a guide to the eye to highlight the linewidth step associated with the phase transition (Poon and Salje 1987a).

$\Delta\Omega \sim \eta^2$ implies that $\Delta\Omega$ should be directly proportional to the spontaneous strain $e_3 = \Delta c/c$ (Reeder 1987) as well as the morphic birefringence (Poon and Salje 1987a) due to the phase transition. The proportionality of $\Delta\Omega$ and e_3 is shown in figure 28. Note that the linear relationship holds to above T_C. The Grüneisen relation (equation (29)) is therefore valid over the whole temperature range shown where in this case $\Delta V/V = \Delta c/c$. The mode Grüneisen parameter given by the slope of the line in figure 28 is $\gamma \approx 3.9$.

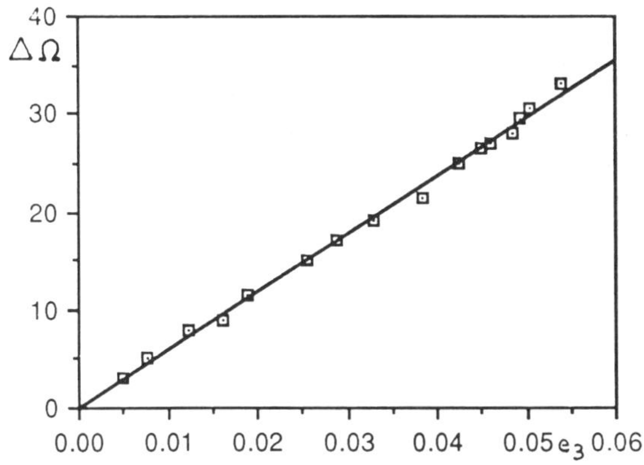

Figure 28. $\Delta\Omega$ versus the spontaneous strain e_3. The two points closest to the origin are data from $T > T_C$. The linearity demonstrates the validity of the Grüneisen relation (Poon and Salje 1987a).

$\ln(\Delta\Omega)$ against $\ln[(T_C-T)/T_C]$ (T_C = 552 K) is shown in figure 29. The slope of this plot is twice the critical exponent ß

$$\eta^2 \sim (T - T_C)^{2\beta} \qquad (44)$$

$$\Delta\Omega \sim \eta^2 \qquad (45)$$

so that 2ß = 0.43, and ß = 0.22 $\pm$ 0.01, which is the critical exponent expected for the three-dimensional three state Potts model.

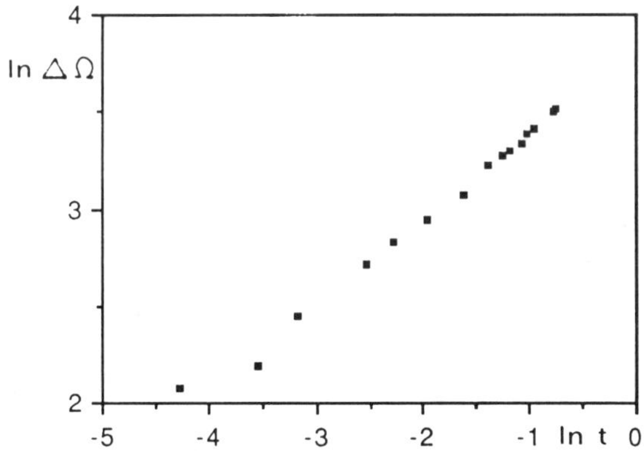

Figure 29. $\ln\Delta\Omega$ versus $\ln t$ where $t=(T_C-T)/T_C$. The slope, which is twice the critical exponent ß, is 0.43, so that ß=0.22 (Poon and Salje 1987a).

The work of Poon and Salje (1987a) clearly contradicts the claim by Shen et al. (1975) that the Bragg-Williams model, for which $\beta = 0.5$, provides a good description for this phase transition. Another very common way of describing this transition in $NaNO_3$ is the 'flip' model in which the 60° flipping of nitrate ions between two equivalent positions leads to the loss of long range correlation between the two positions in the $R\overline{3}c$ unit cell (see Fig. 25). This model would lead one to a pseudospin Hamiltonian of the Ising type. Furthermore, since the nitrate ions occur in sheets in the structure, one may rather naively think that the 2-dimensional Ising model is appropriate. However, the value of β for this model (1/8) clearly does not correspond to that obtained for $NaNO_3$.

Further away from T_C it is expected from theoretical grounds that there will be a crossover to Gaussian behaviour with a different extrapolated transition temperature. A plot of $\Delta\Omega^2$ versus temperature is shown in figure 26. Tricritical behaviour ($\beta = 0.25$) should manifest itself as a linear region

$$\Delta\Omega^2 \sim \eta^4 \sim (T - T_C)^{4\beta} \tag{46}$$

It can be seen from this plot that indeed $\Delta\Omega^2$ is a linear function of temperature below $\approx$ 500 K indicating tricriticality.

In their work Poon and Salje (1987a) probed the order parameter behaviour by observing the temperature evolution of the hard mode frequency. Such an experiment is intrinsically easier than the measurement of absolute intensities and is capable of high accuracy.

3.6. Magnesium cordierite

Synthetic magnesium cordierite ($Mg_2Al_4Si_5O_{18}$) has been the subject of many recent investigations (see e.g. references in Putnis et al. 1985). Central to all these investigation is the characterisation and study of Al-Si order in various structural states obtained from progressively longer periods of annealing at high temperatures (≧1100 °C). HMRS has made interesting contributions to this field (Poon et al. 1987).

The annealing of Mg-cordierite glass produces the following sequence of structural states: hexagonal → modulated → orthorhombic. The Raman band centered at ca. 560 cm^{-1} for one such sequence of samples (annealing temperature 1290 °C) is shown in figure 30.

At annealing times inaccessible experimentally (<< 1 minute) one obtains a homogeneous, disordered hexagonal phase in which the distribution of Al an Si within each of the two sets of tetrahedral sites in the unit cell is random. Longer annealing times bring about increasing degrees of Al-Si ordering, which is accommodated into the hexagonal phase as spatial inhomogeneities until a specific degree of local ordering is reached, at which point a modulated phase appears. This hexagonal → modulated transition has been widely believed to be a continuous transformation. A study of the linewidth of the 567 cm^{-1} band, however, suggests otherwise.

The degree of local order is related to the number of Al-O-Al bonds N, which had been determined by magic angle NMR by Putnis et al. (1985). These authors defined an order parameter

$$Q_{od} = \sqrt{1 - \frac{N}{3.3}} \qquad (47)$$

The value of Q_{od} of all the samples used in the study of Poon et al. (1987) have already been determined by this method. These authors studied the linewidth of the 567 cm^{-1} band (Γ_{567}) as a function of the short range order parameter

$$\sigma = Q_{od}^2. \qquad (48)$$

Assuming the contribution to Γ_{567} from normal multiphonon processes at room temperature with $\Gamma_{anh} \approx 8\ cm^{-1}$, the behaviour of the remaining extra linewidth $\Delta\Gamma = \Gamma_{567} - \Gamma_{anh}$ versus σ is shown in figure 31. It can be seen that $\Delta\Gamma(\sigma)$ extrapolates linearly to zero at $\sigma = 0$, and shows a discontinuity at $\sigma \approx 0.78$ (corresponding to $Q_{od} \approx 0.88$).

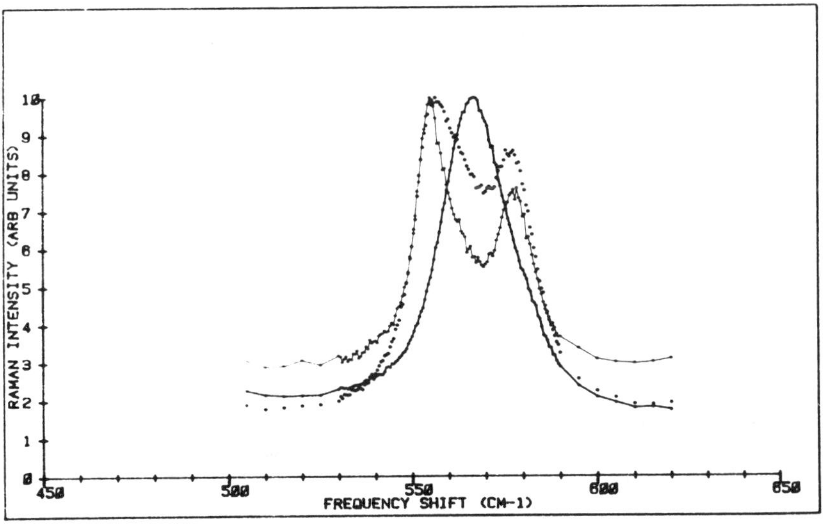

Figure 30. Raman scattering near 560 cm^{-1} from Mg-cordierites annealed at 1290 °C for progressively longer periods; (—) 20 minutes, (···) 1 hour, (-•-•-) 168 hours (Poon et al. 1987).

The linewidth results are significant for cordierite investigations in two important ways:

(1) the linear behaviour immediates suggests a simple means for determining Q_{od}, which hitherto has been measured by magic angle NMR;

(2) the discontinuity at $Q_{od} \approx 0.88$ shows that the hexagonal → modulated transition is in fact discontinuous. Furthermore, since $\Delta\Gamma$ is a measure of the degree of inhomogeneities due to local ordering (extrapolating to zero at complete order), these results indicate that the transformation is an attempt of the system to lower its energy by re-arranging a catastrophic accumulation of inhomogeneities into a spatially modulated but homogeneous structure.

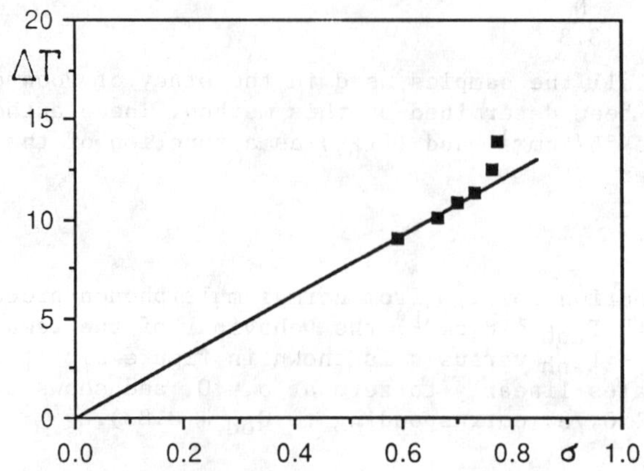

Figure 31. $\Delta\Gamma(=\Gamma_{567} - \Gamma_{anh}$ with $\Gamma_{anh} = 7$ cm^{-1}) versus short range order σ. Note that $\Delta\Gamma$ extrapolates to zero at $\sigma = 0$. The discontinuity at $\sigma \approx 0.78$ indicates that the hexagonal → modulated transition is not continuous as previously been supposed (Poon et al. 1987).

Poon et al. (1987) also investigated in detail the Raman scattering from the modulated phase. Assuming that the relevant modulation wavelengths are longer than the ≈ 20 Å Raman correlation radius suggested by Yacoby (1978), these authors explained the observed band profile as the convolution of the spectra contributed by different regions of the modulated structure

$$I(\omega) = f(\sigma) \otimes P(\sigma) \otimes G_{ph}(\omega,\Gamma) \quad (49)$$

where ⊗ denotes convolution and $G_{ph}(\omega,\Gamma)$ is the scattering profile.

$f(\sigma)$ is the contribution from regions in the modulated structure with short range order σ. The modulations in local order throughout the structure gives rise to modulations of the strain field, which have the same symmetry as the Raman tensor and can therefore be coupled directly. These authors assumed that

$$f(\sigma) = a\delta(\omega-(\omega_0^2+\frac{\alpha}{2}\sigma)^{0.5}) + (1-a)\delta(\omega-(\omega_0^2-\frac{\alpha}{2}\sigma)^{0.5})$$

$$\approx a\delta(\omega-(\omega_0+\frac{\alpha}{4\omega_0}\sigma)) + (1-a)\delta(\omega-(\omega_0-\frac{\alpha}{4\omega_0}\sigma)) \quad (50)$$

where the values of the constants a and α were determined from the spectrum of the orthorhombic structure (infinite annealing time). In other words, the contribution from a region with local order σ is taken to be two delta functions of intensity ratio a : 1-a displaced equally to either side of ω_0 by an amount proportional to σ, so that the splitting reaches a maximum for the fully ordered, orthorhombic structure.

Regions of different σ occur with varying probabilities in the modulated structure. The probability distribution function $P(\sigma)$ is therefore directly a consequence of the form of the spatial modulations

$\sigma(r)$. Poon et al. (1987) obtained $P(\sigma)$ by fitting observed spectra, and deduced the forms of $\sigma(r)$ for different samples. Their procedure is illustrated schematically in figure 32.

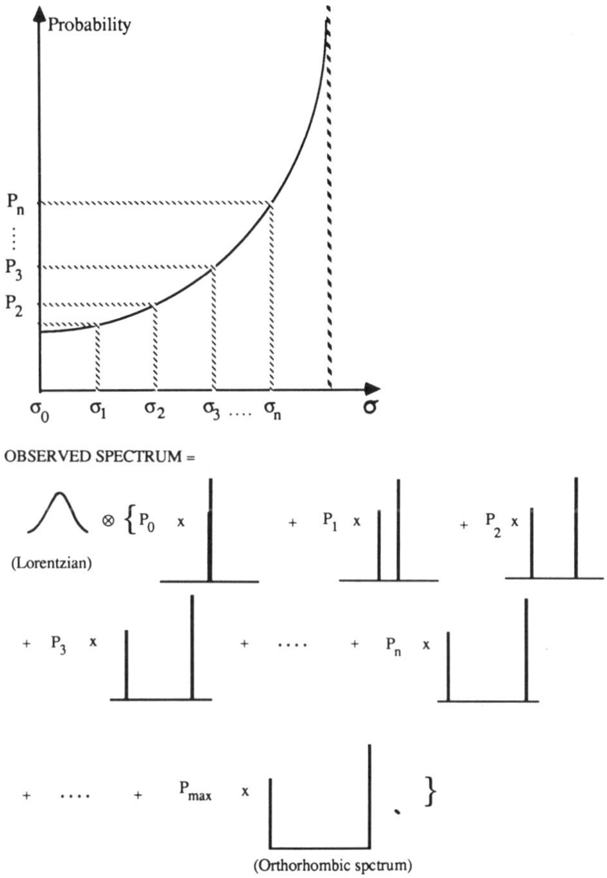

Figure 32. Schematic diagram indicating the procedure used in Poon et al. (1987) for analysing Raman spectra of the modulated phase in Mg-cordierite.

Their results show that starting from the emergence of the modulated phase at T_{anneal} = 1 hour, progressively longer annealing times have the effect of sharpening $P(\sigma)$ at $\sigma = 1$, leading finally to the orthorhombic structure for which

$$P_{orth}(\sigma) = \delta(1). \tag{51}$$

Besides being a novel example of the application of HMRS, this work also raises importent questions concerning Raman scattering from modulated structures in general, and much work is to be done in this area.

4. CONCLUSIONS

The different examples presented in this review show that even in the case when soft-mode Raman studies are inadequate in revealing the transition process, the observation of the behaviour of the hard modes can be of great help in eliciting an understanding of the phase transition mechanisms involved. Therefore hard mode Raman spectroscopy is a very powerfull experimental method when allied with supplementary techniques.

5. REFERENCES

Achiam Y and Imry Y 1975 Phys. Rev. B 12 2768

Anderson P W 1960 Fizika dielektrikov ed. by Skanavi G I (Acad. Nauk. SSR Moscow) 290

Arora A K and Umadevi V 1982 Appl. Spectrosc. 36 424

Ayroles R, Torres J, Aubree J, Roucau C and Tanaka M 1979 Appl. Phys. Lett. 34 4

Benoit J P and Chapelle J P 1974 Solid State Commun. 15 531

Benoit J P 1976 Ferroelectrics 13 331

Benoit J P, Hennion B and Lambert M 1981 Phase Transitions 2 102

Bismayer U and Salje E 1981 Acta Cryst. A 37 145

Bismayer U, Salje E and Joffrin C 1982 J. Physique 43 1379

Bismayer U, Salje E, Jansen M and Dreher S 1986 J. Phys. C: Solid State Phys. 19 4537

Bismayer U, Salje E, Glazer A M and Cosier J 1986 Phase Transitions 6 129

Born M and Huang K 1954 Dynamical Theory of Crystal Lattices (London; Oxford University Press)

Bruce A D, Taylor W and Murray A F 1980 J. Phys. C: Solid State Phys. 13 483

Bruce A D and Cowley R A 1981 Structural Phase Transitions (London; Taylor and Francis)

Chisler E V 1969 Sov. Phys.: Solid State Phys. 11 1032

Cochran W 1960 Adv. Phys. 9 387

Cowley R A 1964 Proc. Phys. Soc. 84 281

Devarajan V and Salje E 1986 Phys. Chem Minerals 13 25

Fenn P M and Brown G E 1977 Z. Kristallogr. 145 124

Ginzburg V L, Sobyanin A A and Levanyuk A P 1983 Chapter 1 in Light Scattering near Phase Transitions ed H Z Cummins and A P Levanyuk (Amsterdam: North Holland Publishing C.)

Gufan Yu M and Larin E S 1980 Sov. Phys. Solid State 22 270

Guimaraes D M C 1979 Acta Crystallogr. A 35 108

Imry Y 1975 J. Phys. C: Solid State Phys. 8 567

Janovec V, Dvorak V and Petzelt J 1975 Czech. J. Phys. B 25 1362

Jansen M 1978 Z. Anorg. Allg. Chem. 441 5

Jansen M 1979 Z. Naturf. B 34 10

Jenkins T E 1986 J. Phys. C: Solid State Phys. 19 1065

Joffrin C, Benoit J P, Deschamps L and Lambert M 1977 J. Physique 38 205

Joffrin C, Benoit J P, Currat R and Lambert M 1979 J. Physique 40 1185

Kracek F C 1931 J. Am. Chem. Soc. 53 2609

Landau L D and Lifshitz E M 1958 Statistical Physics (Reading Mass: Addison-Wesley)

Luspin Y, Servoin J L and Gervais F 1979 J. Phys. Chem. Solids 40 661

Matsushita M 1976 J. Chem. Phys. 65 23

Munn R W 1975 Phys. Rev. B 12 3491

Neumann G and Vogt H 1978 Phys. Stat. Solidi B 85 179

Petzelt J and Dvorak V 1976 J. Phys. C: Solid State Phys. 9 1571

Petzelt J and Dvorak V 1976 J. Phys. C: Solid State Phys. 9 1587

Poon W C-K and Salje E 1987 a submitted to J. Phys. C: Solid State Phys.

Poon W C-K and Salje E 1987 b to be submitted to J. Phys. C: Solid State Phys.

Poon W C-K, Putnis A and Salje E 1987 to be submitted to Phys. Chem. Minerals

Prasad Rao A D, Katiyar R S and Porto SPS 1973 Adv. Raman Spectrosc. 1 174

Putnis A, Fyfe C A and Gobbi G C 1985 Phys. Chem Minerals 12 211

Raich J C and Hüller A 1979 J. Chem. Phys. 70 3669

Reeder R J 1987 results presented at the Fourth Meeting of the European Union of Geosciences, Strasbourg, Abstract in Terra Cognita 7 264

Roucau C, Tanaka M, Torres J and Ayroles R 1979 J. Micros. Spectrosc. Electron. 4 603

Scagliotti M, Jouanne M, Balkanski M, Ouvrard G and Benedek G 1987 Phys. Rev. B 35 7097

Salje E and Viswanathan K 1976 Contrib. Miner. Petrol. 55 55

Salje E and Devarajan V 1981 J. Phys. C: Solid State Phys. 14 L1029

Salje E, Devarajan V, Bismayer U and Guimaraes D M C 1983 J. Phys. C: Solid State Phys. 16 5233

Salje E and Wruck B 1983 Phys. Rev. B 28 6510

Salje E 1985 Phys. Chem. Minerals 12 93

Salje E, Kuscholke, Wruck B and Kroll H 1985 Phys. Chem. Minerals 12 99

Salje E and Devarajan V 1986 Phase Transitions 6 235

Salje E 1986 Phys. Chem. Minerals 13 340

Salje E, Bismayer U and Jansen M 1987 J. Phys. C: Solid State Phys. in press

Shen T Y, Mitra S S, Casella R C and Trevion S F 1975 Phys. Rev. B 12 4530

Toledano J C and Toledano P 1980 Phys. Rev. B 21 1139

Torres J, Aubree J and Brandon J 1974 J. Opt. Commun. 12 416

Torres J 1975 Phys. Status Solidi 71 141

Viswanathan K and Miehe G 1978 Z. Kristallogr. 148 275

Wilson C M and Cummins H Z 1971 Light Scattering in Solids ed M Balkanski (Paris: Flammarion) 420

Winterfeld V and Schaak G 1977 Phys. Status Solidi B 80 559

Wood I, Wadhavan V K and Glazer A M 1980 J. Phys. C: Solid State Phys. 13 5155

Yacoby Y 1978 Z. Physik B 31 275

LINEAR AND CIRCULAR BIREFRINGENCE AND CRYSTAL STRUCTURES

A. M. GLAZER
Clarendon Laboratory
University of Oxford
Parks Road
Oxford OX1 3PU, U.K.

ABSTRACT. The measurement of linear birefringence of crystalline materials is described, together with some examples, showing how it is linked to crystal structures, particularly through the study of phase transitions. Circular birefringence (more commonly known as optical activity) has now also been related to structures of inorganic materials.

1. INTRODUCTION

When light passes through a medium it is slowed down to a velocity v with respect to its velocity in a vacuum c and it is usual to denote this by the refractive index

$$n = c/v \tag{1}$$

In general, however, materials are often anisotropic in their properties and so the refractive index can vary, depending on the direction of propagation of the light and the orientation of its electric vectors or polarization state (we ignore here the influence of magnetic structures). In such situations, it is convenient to measure the linear birefringence defined by

$$\Delta n = n_X - n_Y \tag{2}$$

where n_X and n_Y refer to refractive indices measured for light polarized with electric vectors parallel to the X and Y -directions of the crystal, respectively. Being a difference in refractive indices, its measurement is a far more sensitive indicator of changes in the crystal structure than is the measurement of refractive index alone. It is usual for the linear birefringence to be called simply the 'birefringence' in much of the literature, and its measurement has been a standard analytical tool of mineralogists for a considerable time. Indeed, before the discovery of x-ray diffraction, linear birefringence measurements were the normal means of obtaining structural information

E. K. H. Salje (ed.), Physical Properties and Thermodynamic Behaviour of Minerals, 185–212.

about crystals. Although x-ray diffraction has changed the scene somewhat, birefringence studies have in recent years enjoyed a renaissance, especially with the development of new, automatic and highly precise methods of measurement.

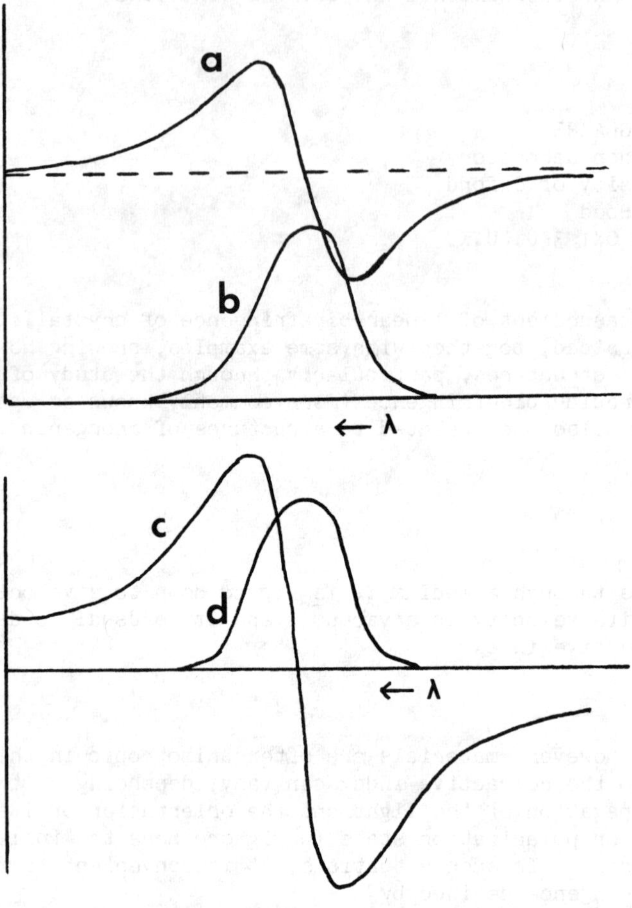

Figure 1. Real and imaginary birefringence as a function of wavelength a) real linear b) imaginary linear c) real circular d) imaginary circular.

The other topic of this article is circular birefringence, more usually called optical activity or optical rotation. This is the rotation of the plane of polarised light when a ray of light passes through a suitable medium (a crystal in our case). The reason why I shall refer to the phenomenon here as circular birefringence is to emphasise its analogy with linear birefringence. In the former case, as we have seen, we deal with the difference between two linear refractive indices. In the latter, as first pointed out by Fresnel

(1824), linearly polarized light can be decomposed into the sum of two oppositely rotating (left and right) circular components (see for example, Nye, 1957), and that the slowing down of one circular component relative to the other provides a net rotation of the plane of polarization: thus the optical rotatory power is measured by the difference in circular refractive indices

$$\rho = (\pi/\lambda)(n_L - n_R) \tag{3}$$

where n_L and n_R are the refractive indices for left and right circular polarized light, respectively.

A particularly convenient way of putting into perspective the two types of birefringence is shown in Figure 1. Now, refractive index can in general be expressed as a sum of real and imaginary parts

$$n = n' + in'' \tag{4}$$

Figure 1(a) shows the difference in the real parts of the linear refractive index, i.e. the real part of the linear birefringence, plotted as a function of wavelength in order to show its dispersion. This displays the characteristic resonance typical of any forced oscillator. Corresponding to this resonance, a plot of the difference in imaginary refractive indices (Figure 1(b)) gives rise to an absorption peak: this phenomenon is known as linear dichroism, i.e. the selective absorption of one linear polarization state with respect to another. Similar curves can be plotted for the circular polarizations. Thus Figure 1(c) shows a plot against wavelength of the difference between real circular refractive indices, and again we see the typical resonance curve. Such a plot can be measured in terms of the angle of optical rotation (equation 3) as a function of wavelength, the so-called optical rotatory dispersion (ORD). Correspondingly, the difference in imaginary circular refractive indices shows a peak (Figure 1(d)) at a characteristic wavelength and is a measure of the selective absorption of one circular component relative to the other. This is known as circular dichroism (CD).

2. LINEAR BIREFRINGENCE

2.1 Methods of Measurement

The measurement of linear birefringence has long been a standard technique used by mineralogists, and a great number of methods have evolved over the last 100 years or so. We can roughly divide the methods into two classes, absolute and relative.

In absolute methods, the linear birefringence is measured through the insertion of some kind of compensation device into the light path above the specimen viewed between crossed polars in a microscope. The simplest form of this is a wedge cut from a birefringent section of quartz. With a crystal sample oriented in the so-called 45° position with respect to the polarizer of the microscope, two waves with perpendicular polarizations propagate through the crystal. If the crystal is optically anisotropic i.e. linearly birefringent, the two waves are shifted with respect to each other to give a phase difference (one is slowed with respect to the other) and this is given by

$$\delta = (2\pi/\lambda)\Delta n \ell \tag{5}$$

where ℓ is the thickness of the crystal. The term $\Delta n\ell$ is known as the

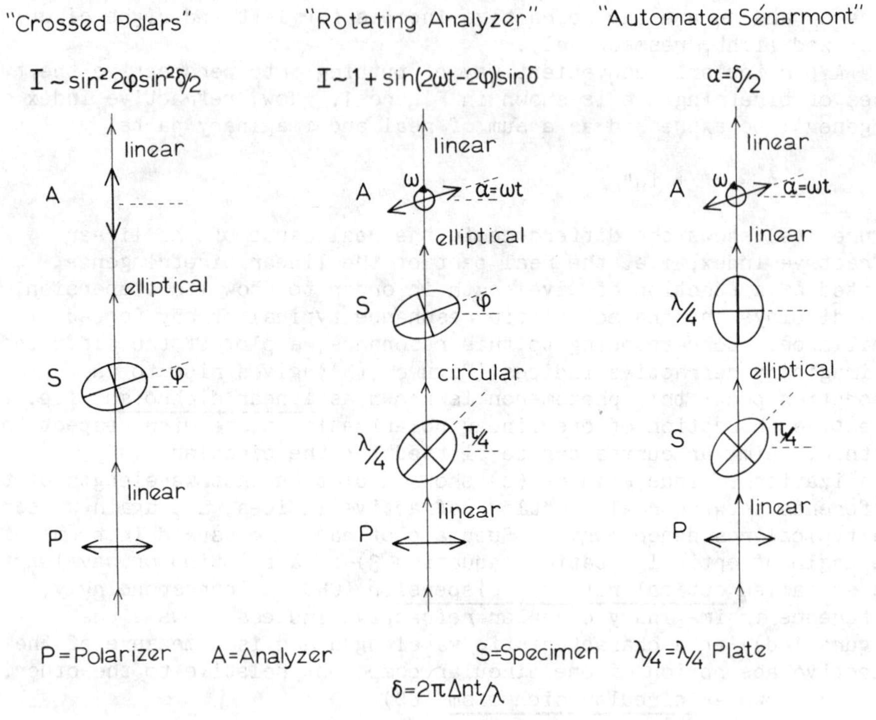

Figure 2. Three modes of operation for relative measurements of linear birefringence.

optical retardation and is the quantity that is normally found in any experiment. Converting to birefringence requires an accurate knowledge of the thickness of the specimen, and this is often a limiting factor in determining the birefringence. By insertion of the quartz wedge in the 45° position, the retardation of the quartz will either add to or subtract from that of the specimen. On passing the resultant light through an analyser crossed with respect to the polarizer the specimen appears black when the quartz wedge is inserted to cancel exactly the retardation due to the specimen. Since the retardation of the quartz wedge is known, this compensation position directly gives the retardation of the specimen. There are many variants of this basic method: some of the more precise employ birefringent plates that are tilted relative to the light path (as in the Berek and Ehringhaus compensators), the angle of tilt being related to the effective retardation of the section seen in projection. A thorough account of these and other traditional techniques can be found in standard books on optical crystallography, such as those by Hartshorne and Stuart (1970), Gay (1982) and Wahlstrohm (1965), and we shall not dwell further on them, except to point out that they can measure retardations typically to about 1 part in 10^4.

Relative methods, while more complicated, are capable of much greater precision, reaching in some of the best systems to precisions of the order of 1 part in 10^7, thus making these measurements among the most sensitive to changes in the crystal that are currently known (a good review applied to magnetic crystals is given by Ferré and Gehring, 1984). Other advantages are their experimental convenience and their ability to make continuous, rather than point-by-point measurements. These methods lend themselves particularly well to obtaining information about the changes in retardation, when for instance a crystal is heated or in some other way altered, and are especially useful for studying phase transitions. Figure 2 summarises three basic forms that are in current use (there are in existence several variations of these forms, but these will serve to illustrate the idea).

The first of these, the crossed-polars method (as used for example by Harley and Macfarlane, 1975), is the simplest development over the compensator plate. A photodetector, usually a photomultiplier, is used to measure the intensity of light passing through crossed polars with the specimen inserted between. This can be set up on an optical bench, but is more conveniently done simply by attaching the photomultiplier to the eyepiece of a polarizing microscope. The linear polarized light from the polarizer P, is converted to elliptical polarization by the birefringent specimen S. The elliptical state is then reconverted to linear polarization by the analyzer A. It can then be shown for such an optical system (Born and Wolf, 1975, p.696) that the resultant intensity is given by

$$I = I_0 \sin^2 2\phi \sin^2 \delta/2 \qquad (6)$$

where I_0 is the incident intensity and ϕ is the angle between the polarizer and an allowed vibration direction in the crystal (usually set to 45° in order to obtain maximum signal). If now the specimen is heated, say, any change in retardation will provide an intensity signal proportional to $\sin^2\delta/2$, so that when plotted on a chart recorder a series of fringes are seen. By counting the maxima and minima, exact measurements of the retardation relative to that before heating can be made. The only error is in the measurement of the temperature. The crossed-polars method is capable of great precision, especially when combined with fringe-counting in this way. It does have the disadvantages that

(i) it is sensitive to fluctuations in the incident light intensity and to noise from stray light entering the photomultiplier,
(ii) for small δ the dependence of the signal is nonlinear, and
(iii) the specimen must be oriented to give maximum signal.

A more sensitive method still is to make use of modulation techniques. A simple scheme, currently used in the author's laboratory, is shown in Figure 2. We call it the rotating-analyser technique (Wood and Glazer, 1980). Various versions of this have been described by a number of workers (Suits, 1971; Chudakov et al., 1972), although these have been used usually for studying magneto-optical effects. The optical components are

(a) an initial polarizer P,
(b) a $\lambda/4$ plate set at 45° to the polarizer, thus producing circularly-polarized light incident on the sample S,
(c) the specimen, with orientation ϕ and phase lag δ, thus converting the incident circularly-polarized state into an elliptically-polarized state,
(d) an analyzer A rotating at a frequency ω.

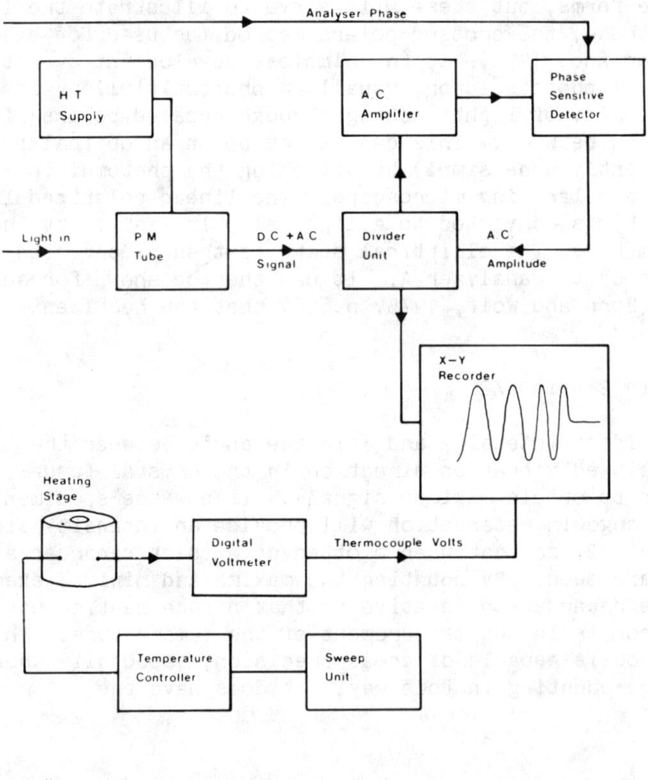

Figure 3. Block diagram for rotating-analyser technique.

Under these conditions the light intensity passing the analyzer is given by

$$I = (I_0/2)[1 + \sin 2(\omega t - \phi)\sin \delta] \qquad (7)$$

where t is time.

In our setup, the rotating analyzer is a simple piece of Polaroid set in a rotating bearing (in other forms of this basic technique, the rotating analyzer is replaced by some electrooptic crystal whose polarization properties can be modulated by an oscillating electric field). The experimental arrangement used is shown schematically in Figure 3. The apparatus is assembled around a conventional polarizing microscope, with the rotating-analyzer attached to the camera-port of the microscope (it could equally be placed over the eyepiece tube). The light source is usually a small He/Ne laser, although a filtered white source can also be used in most cases, depending on the degree of optical dispersion showed by the specimen. The rotating polaroid is driven by an elastic belt at about 80-100Hz. The AC amplitude of the light-intensity signal is measured by a phase-sensitive detector locked to the analyzer frequency by an optical chopper. In order to eliminate the effects of incident light variation and other non-birefringent intensity, the AC signal is finally divided by the DC component to give an output proportional to $\sin\delta$. The principal advantages of this system are that

(i) since circular-polarized light is used below the microscope stage, the orientation of the specimen is immaterial (instead, the signal is maximised by shifting its phase electronically in the phase-sensitive detector), and

(ii) the signal is normalised against non-birefringent intensity variations, such as those caused by changes in laser light intensity or in the normal absorption of the specimen.

The resulting signal as a function of some external variable, such as temperature, electric field applied to the specimen, or displacement of the specimen across an aperture, etc. consists of sinusoidal fringes which can then be counted to obtain the retardation change. Another advantage, useful in some cases, is that if the retardation is small, we can then set $\sin\delta$ equal to δ itself. In this case, the intensity output is linear in retardation and so the retardation change is plotted out directly on the chart-recorder. The rotating analyser-technique is therefore best suited to specimens where the retardation is very small or very large (where a large number of fringes can be counted). On the other hand, for specimens with an intermediate retardation, only a small number of fringes can be obtained and thus only a limited amount of information on the retardation can easily be extracted.

A convenient method of dealing with the intermediate case is illustrated in Figure 2, the automated Sénarmont method. Here, the specimen is set in the 45° position above a polarizer. Above this is a $\lambda/4$ plate followed by a rotating-analyzer as before. With this arrangement, the phase of the AC signal is proportional directly to the phase lag δ and so it is possible to obtain a continuous measurement of the optical retardation by monitoring the phase. In our laboratory, this is done automatically by a phase-meter, constructed to produce a voltage on a chart-recorder proportional to the phase.

2.2. Applications to Crystals

It is well known that the optical anisotropy provides information about crystal symmetry and we shall not discuss this matter here as it is well covered by the standard books on crystal optics. Instead, we shall discuss a few examples of modern work on linear birefringence of crystals, most of which are taken from the author's own laboratory.

One particularly complex case is that of the perovskite $NaNbO_3$ (Ishida and Glazer, 1974). Measurements of birefringence using traditional compensator techniques as a function of temperature revealed the existence of six different phases above room temperature (Figure 4(a)). These are labelled P, R, S, T_1, T_2 and cubic. In P and R the structure is characterised by displacements of the Nb atoms along [110] and [100] respectively, with at the same time corner-linked oxygen octahedra tilted about the three crystallographic axes to form very complicated structures (Glazer, 1972). In phases S, T_1 and T_2 only tilted octahedra occur, with tilts about all three axes, about two axes and about only one axis respectively. Finally, in the cubic phase, there are no displacements or tilts.

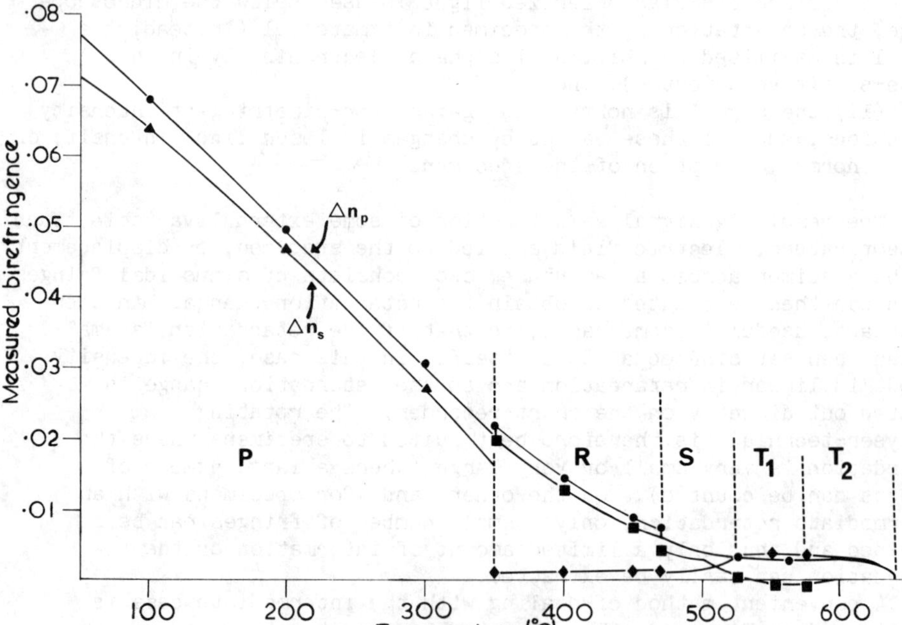

Figure 4. a) Plot of experimental measurements of linear birefringence along all three axes of $NaNbO_3$ as a function of temperature.

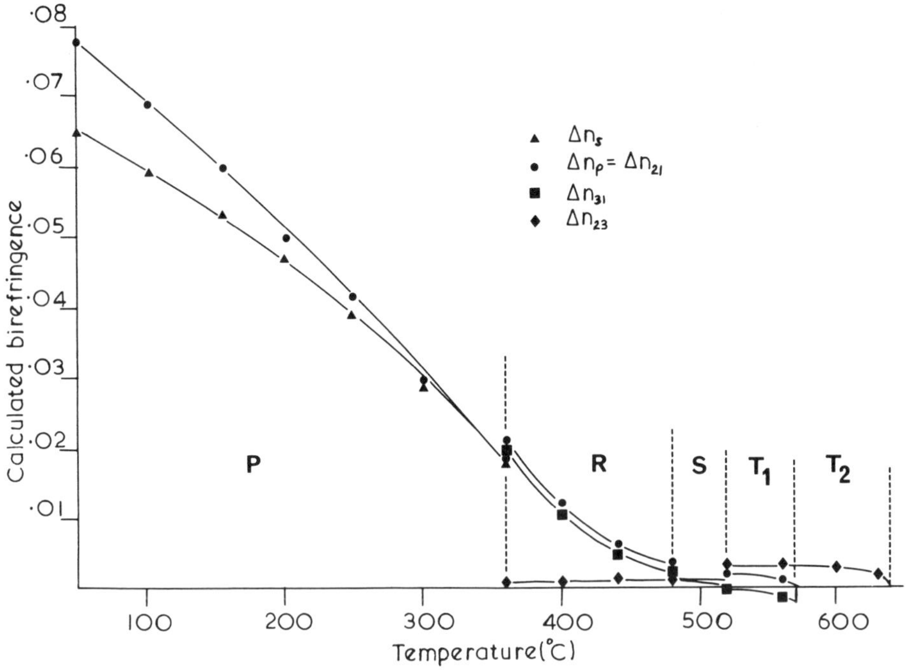

Figure 4. b) Calculated linear birefringence by Ishida and Glazer (1974).

The measured birefringence (Figure 4(a)) shows a very complicated behaviour with temperature, generally falling towards higher temperature. Ishida and Glazer showed that the birefringence could be traced to the sum of a sublattice polarization (caused by antiparallel displacements of the Nb ions in phases P and R) and strain (caused by the unit cell distortion from the cubic phase). The former effect could be measured from x-ray reflections that gave a direct measure of the displacements and hence the sublattice polarization. The strain was obtained from unit cell measurements as a function of temperature. It was found that with only four fitting parameters, the birefringence could be calculated very satisfactorily from these effects (Figure 4(b)) and that estimates could be made of elastooptic coefficients (from the strain part) and quadratic electrooptic coefficients (from the sublattice polarization part). In phase P and R, the birefringence is dominated by the sublattice terms and hence it is much larger than in the higher-temperature phases. Thus, a study of the linear birefringence of this material shows, on raising the temperature, the successive loss of cation displacement components and octahedron tilts.

Perhaps one of the best known applications of birefringence measurements is in the study of the phase transition at 105K in the perovskite material $SrTiO_3$. This is a crystal which undergoes a tetragonal to cubic transition, characterised by a structural distortion in which corner-linked oxygen octahedra tilt in the tetragonal phase to form the $a^0a^0c^-$ structure (Glazer, 1972, 1975).

The tilting of the octahedra causes a reduction of the a and b axes relative to the c axis, thus introducing a 'spontaneous strain' into the crystal (Aizu, 1970). In rough terms this means that the unit cell is distorted relative to the cubic unit cell of the high-temperature phase. In this case, a measurement of the birefringence (along a direction perpendicular to the tetragonal c-axis), which is sensitive to the strain, gives information about the angle of tilt of the

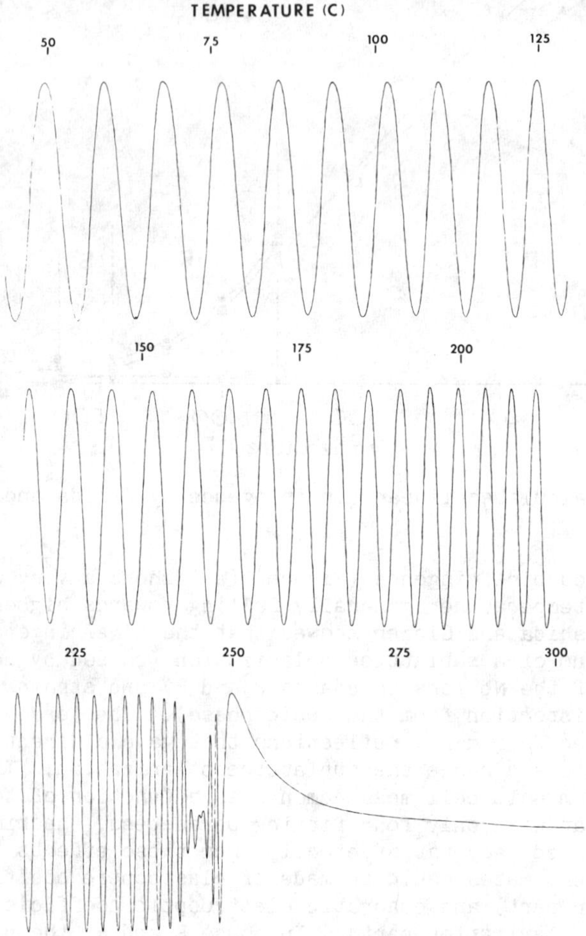

Figure 5. Signal output from the rotating-analyzer apparatus for a crystal of $BiVO_4$ viewed down the pseudo-tetragonal axis. The loss of signal just below the transition at 250°C is caused by the spontaneous onset of fine-scale domains. This does not affect the calculation of retardation from these fringes, as it is the position of the fringes that matters, not their heights. A small amount of birefringence can be seen to persist into the high-temperature phase, probably caused by residual strain.

octahedra, and so provides a useful indicator of the way in which the octahedra tilt as a function of temperature. The measurements were made by Courtens (1972, 1973) using a Berek compensator and a photomultiplier to measure the intensity as a function of compensator angle. By fitting the intensity curves to a simple analytic function, Courtens was able to extract an order of magnitude increase in precision out of the Berek compensator, although the precision was still less than is now possible with relative methods.

This transition had been studied before by electron paramagnetic resonance. Courtens showed that the birefringence was related to the square of the order parameter, with a reasonable agreement with the EPR results, and so birefringence could be used as an alternative method of studying the microscopic behaviour of the order parameter. Small differences between the birefringence curve and order-parameter squared curve were interpreted in terms of fluctuations of the order parameter.

Turning to examples of the use of relative measurements, a particularly nice case is that of the study of the ferroelastic monoclinic to tetragonal phase transition of $BiVO_4$ (Wood and Glazer, 1980). Figure 5 shows the signal output from the rotating analyzer

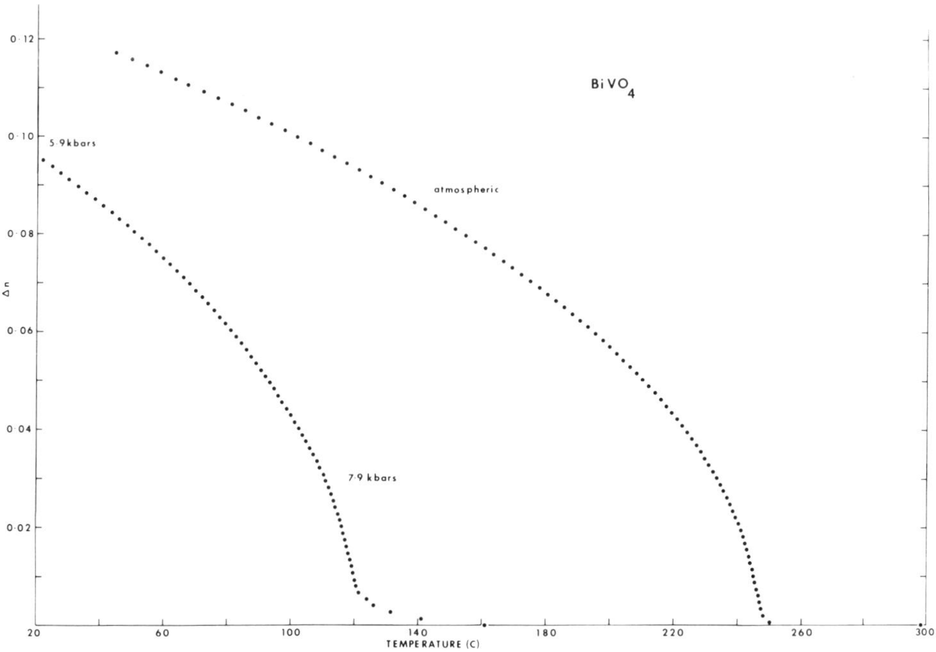

Figure 6. Birefringence of $BiVO_4$ as a function of temperature. Two curves are shown, one at atmospheric pressure and another measured in a diamond-anvil cell.

method for a (001) slice of $BiVO_4$ as a function of temperature. A beautiful set of fringes below about 517K shows this section of the crystal to be birefringent, with the birefringence vanishing above. Note how the fringes are all normalised to the same heights, except for a small region below the transition. This behaviour below the transition is characteristic of this material and is caused by the onset of fine-scale domains that spontaneously appear in this temperature region. Fortunately, this does not affect the measurements of the retardation, since here one is interested in the positions of the maxima and minima and not their heights. By counting fringes and using the known thickness, the birefringence was plotted (Figure 6) as a function of temperature. This shows a clear example of a second-order transition. We were able to show that, in this case, the order parameter for the transition was represented by the spontaneous strain and hence by the birefringence. According to the Landau theory

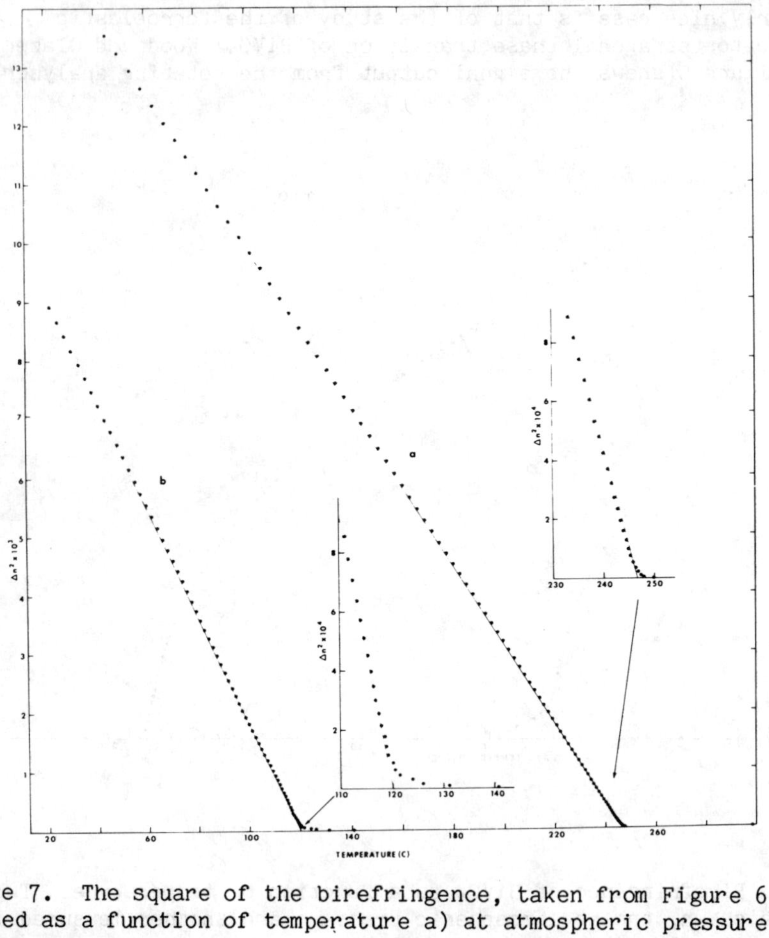

Figure 7. The square of the birefringence, taken from Figure 6, plotted as a function of temperature a) at atmospheric pressure, and b) at high pressure.

parameter would be linear with respect to temperature, and Figure 7 shows what was obtained for a plot of Δn^2 against temperature. An excellent straight line is obtained. The birefringence then clearly demonstrates something about the way in which the crystal structure is changing. Here, the monoclinic distortion from the tetragonal phase is derived from distortions of the oxygen tetrahedra coupled with displacements of the Bi atoms along the c-axis (David, Glazer and Hewat, 1979), and so birefringence measurements like this are able to follow the progress of such distortions to high precision. Experiments conducted at high pressure with the temperature changing gave similar results and enabled us to construct a complete Δn-P-T phase diagram for $BiVO_4$.

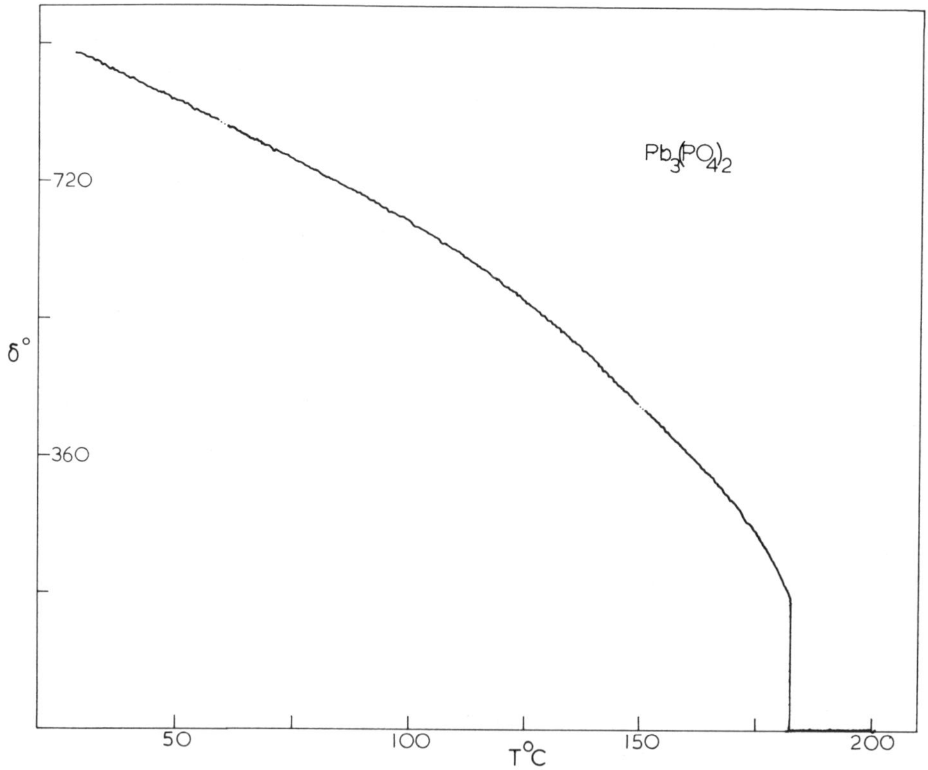

Figure 8. The output from the automated Sénarmont technique for a crystal of $Pb_3(PO_4)_2$.

As an example of the automated Sénarmont method, Figure 8 shows a direct plot of phase lag δ for $Pb_3(PO_4)_2$. This undergoes a monoclinic to rhombohedral phase transition at about 180°C, involving a complicated structural change. The essence of the structural change is that the 3-fold axis in the rhombohedral phase becomes a pseudo-3-fold axis in the monoclinic phase. On looking along this pseudo-3-fold axis, one finds Pb atoms slightly displaced in directions perpendicular axis, one finds Pb atoms slightly displaced in directions perpendicular to the axis. Measurements of birefringence then give a sensitive indication of these displacements. The birefringence plot shown in Figure 8 agrees closely with similar plots made with a much thicker crystal using the rotating-analyzer technique and counting fringes (Wood et al., 1980),and shows an abrupt drop at the transition, indicating a first-order change.

3. CIRCULAR BIREFRINGENCE

3.1 Measurement and observation

Circular birefringence was first observed in 1811 by Arago, and since that time an enormous amount of observation and measurement of the effect has been made. During the last century, this was a fashionable topic, occupying no lesser scientists than Biot, Fresnel, Pasteur, Herschel (John), and many other well known figures. For an excellent review of the history of the subject and early methods of measurement, see the classic book by Lowry (1935). It is an interesting fact too that in the first half of the last century much interest was in the variation of circular birefringence (or optical rotatory power) with wavelength (the optical rotatory dispersion ORD). However, when Bunsen produced his famous burner, it became such an easy matter to obtain monochromatic sodium light that measurements with wavelength almost ceased!

In general, the standard way of measuring optical rotation is through the use of a polarimeter. Many of these have been described in the literature (Lowry, 1935) and although the details of design vary, they all depend on the same principle. That is, the incident light is monochromatized and passed through a polarizer. The resulting polarized light passes through the specimen and then through another polarizing medium (the analyzer) which is rotated to a position where no light is observed through the instrument. The angle made by this 'extinction' position with respect to the crossed-polars position is a measure of the angle of rotation. The optical rotatory power is then given by this angle divided by the thickness of the crystal (usually in $^{\circ}$/mm).

The same basic technique is sometimes used with a polarizing microscope, but the precision with which the angle of measurement can be made is limited. The polarizing microscope is best used for observing the sign of the rotation rather than its magnitude. Provided that the crystal is optically isotropic (cubic) or that one is looking at a non-birefringent section (such as along the optic axis of a uniaxial crystal i.e. tetragonal, trigonal or hexagonal, or along one of the optic axes of a biaxial crystal, i.e. triclinic, monoclinic or orthorhombic) and that the crystal is optically active, it is often possible to observe the rotation of the plane of polarization simply by rotating the analyzer (or the polarizer in the opposite direction) to the extinction position. However, how does one know that this angle corresponds to a rotation of θ° to the left, say, and not to a rotation of 180 - θ^{o} to the right? This can be sorted out by repeating the experiment with a crystal of a different thickness.

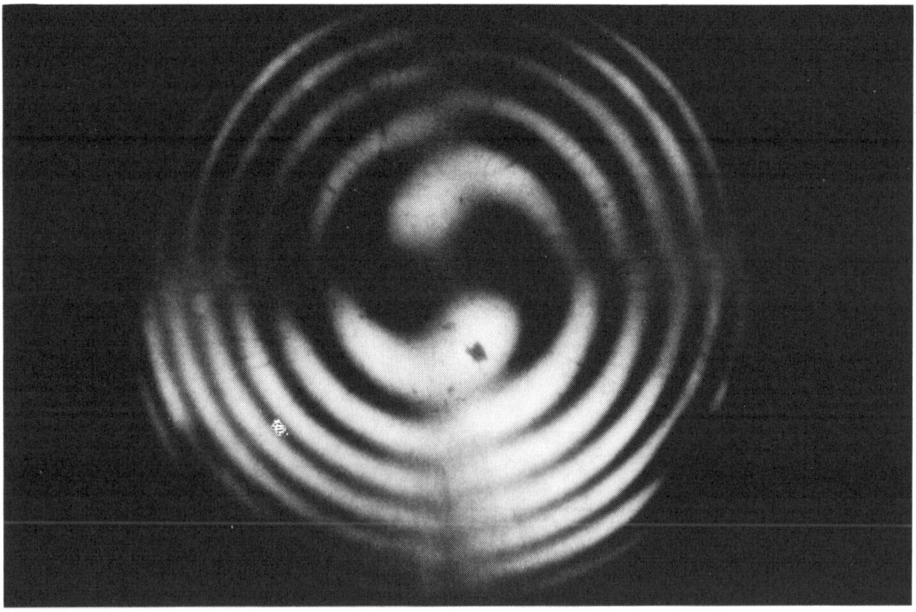

Figure 9. Airy's spiral for a dextrorotatory crystal of $LiIO_3$.

Another easy way to determine the sign, which has been used with much success in the author's laboratory, is to observe Airy's spiral. In this technique, the microscope is set up to form a conoscopic image (see Hartshorne and Stuart, 1970) and a quarter-wave plate is inserted above the specimen. The result is an interference pattern in the form of a spiral (Figure 9) and immediately the optical chirality is determined. A variant of this (and so far I have been unable to find any publication of this method) is to grind the crystal into a sphere. On observing the sphere between crossed polars and looking down the unique axis - a binocular microscope is particularly good - a conoscopic image can usually be seen on the top of the crystal sphere. Insertion of a quarter-wave plate above the sphere again gives Airy's spiral. This is a good method when trying to relate optical chirality with structural chirality using x-ray anomalous scattering, as often the x-ray crystallographer grinds his crystal to a small sphere in order to simplify x-ray absorption corrections, and so the optical chirality can then be determined on the same crystal used in the x-ray work. Almost all the observations of circular birefringence have been carried out on optically isotropic sections. However, suppose one wishes to measure the effect in sections that display linear birefringence. This is much harder to do, since the linear effect is usually much larger than the circular effect, thus masking it (although note that in one or two unusual cases, where there is a strong dispersion of the linear birefringence, it has been found that at a particular wavelength the linear birefringence changes sign on passing through zero, and so at this wavelength only the circular birefringence is left. A good example of this was given for $AgGaS_2$ by Hobden, 1968). Until recently, the only crystal for which reliable measurements of the circular birefringence in non-uniaxial directions had been made was in quartz (Szivessy and Münster, 1934).

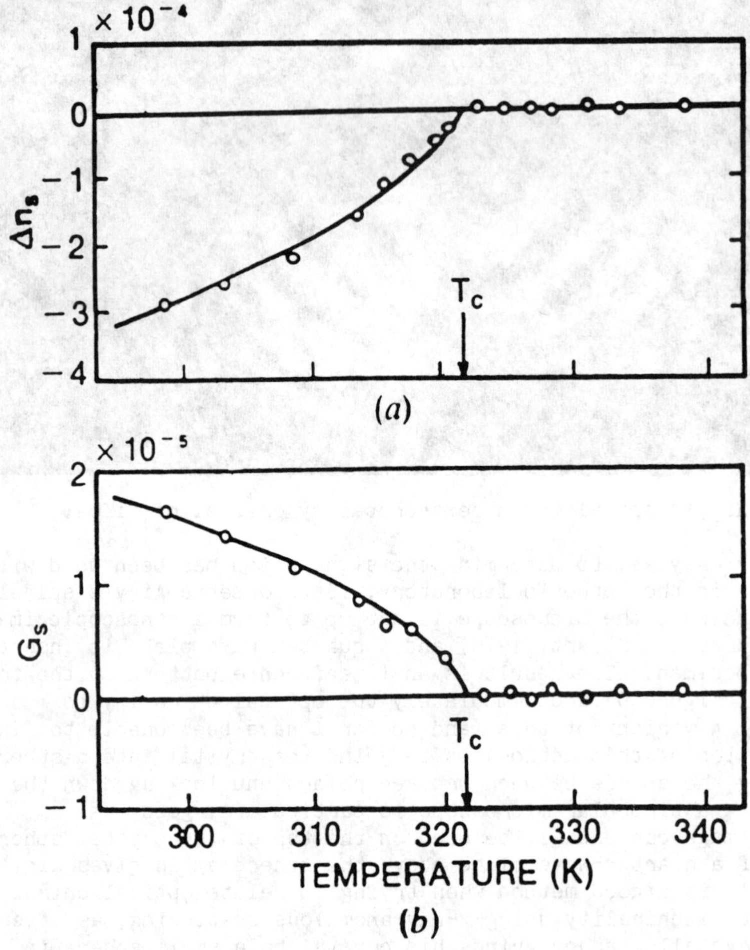

Figure 10. Temperature dependence of a) spontaneous birefringence Δn_s and b) spontaneous gyration G_s of TGS.

Recently, however, Kobayashi and Uesu (1983) have devised a method for separating the two effects in any crystal direction. The method is still based around the polarimeter and is called HAUP (High Accuracy Universal Polarimeter). The idea of the experiment is to measure by a photon counting system the intensity of light transmitted through polarizer, specimen and analyzer as a function of two angles, say the analyzer and that of the specimen. The polarizer and analyzer must be made from very high-quality polarizing prisms, typically of the Glan-Thompson type. Kobayashi and Uesu showed that provided only small angles about the extinction position were used, the optical equations governing the intensity as a function of the two angles could be expressed in a simple biquadratic form, thus lending themselves to linear least-squares fitting procedures. From these, the two components, retardation (hence linear birefringence) and gyration (hence circular birefringence) can be separated. Kobayashi has now used this method in several crystals and has shown how the gyration components behave in certain crystals undergoing phase transitions (e.g. Figure 10). The HAUP technique can be regarded as something of a breakthrough in this field of study and it should now be possible to determine the gyration tensors of many crystals for the first time in order to compare with theories of optical rotation.

3.2 Relationship to Crystal Structure

As mentioned earlier, circular birefringence is the phenomenon known more commonly as optical activity (OA). Although this effect was discovered as long ago as 1811, it has until now been relatively less studied than the linear effect, and the establishment of its relationship to the underlying crystal structure has been elusive. There have been many theories to explain OA, some of them of great complexity, but it is true to say that until recently, these have rarely worked well. Probably the main obstacle to applying theory to this phenomenon has been the lack of reliable information about the crystal structures of optically active materials. The problem is that OA is associated with the chirality or hand of a structure, and this is not an easy quantity to determine in a crystal. Thus, it is well known that crystals of quartz are found in either left or right-hand forms. It is seen that crystals of the two forms are mirror images of each other, and this is reflected in their internal arrangements of atoms (the crystal structure). At the same time, it has been known for about 150 years, that the two morphological or enantiomorphic forms of quartz rotate polarized light passing along the c-axis in opposite ways. And yet, there remains the question: which hand does the structure have for a particular sense of rotation of light?

The usual way of determining crystal structures is through x-ray diffraction. However, under normal circumstances, x-ray diffraction patterns are centrosymmetric (Friedel's Law) and so are incapable of distinguishing between up and down or left and right. Thus any information about polarity or chirality is lost. However, if some atoms in the crystal strongly absorb x-rays, it is found that Friedel's Law is no longer quite true, and the pattern becomes slightly non-centrosymmetric: it is then possible to distinguish between up, down, left and right. This 'anomalous scattering' effect was first used to determine the absolute configuration (i.e. the true relationship between structural and optical chirality) in a series of tartrates (Bijvoet, 1949). Since then, the incorporation of anomalous scattering has become standard practice in most x-ray crystal structure determinations. In theory then there should by now have been a large accumulation of experimentally determined crystal structures to compare with the OA in order to test the available theories. Nevertheless, most attempts at applying the theories seemed to fail.

Glazer and Stadnicka (1986) made a systematic study of all optically-active inorganic crystals for which the absolute configurations had been reported, and found to their surprise that the literature was full of errors. Typical errors included confusion over the meaning of right and left-handed, incompatible space group symbols and failures to perform optical rotation measurements on the same crystals used for the x-ray anomalous scattering measurements. In many cases, the x-ray measurements taking into account anomalous scattering merely served to indicate the way in which the particular crystal was mounted on the diffractometer used to collect the data, rather than to provide any absolute information about the structural chirality! On removing all dubious cases, Glazer and Stadnicka found only about 10 optically-active inorganic crystals whose absolute configuration determinations could be trusted. From this they were able for the

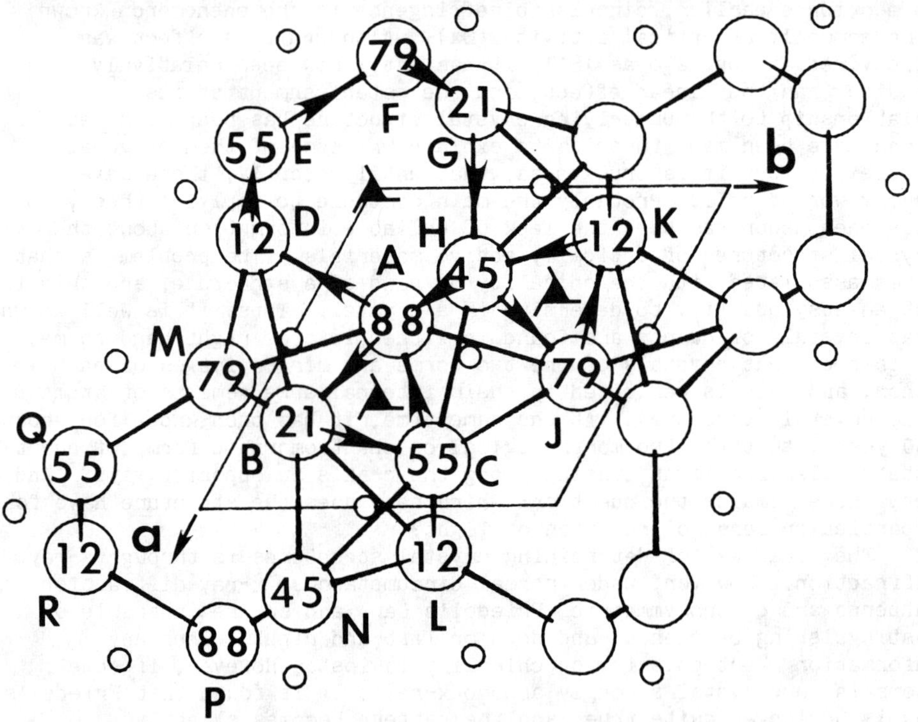

Figure 11. The (0001) projection of the structure of α-quartz. a) Large circles are oxygens, small circles silicon. The numbers within the large circles represent the fractional heights of the oxygens in 1/100'ths of the c axis.

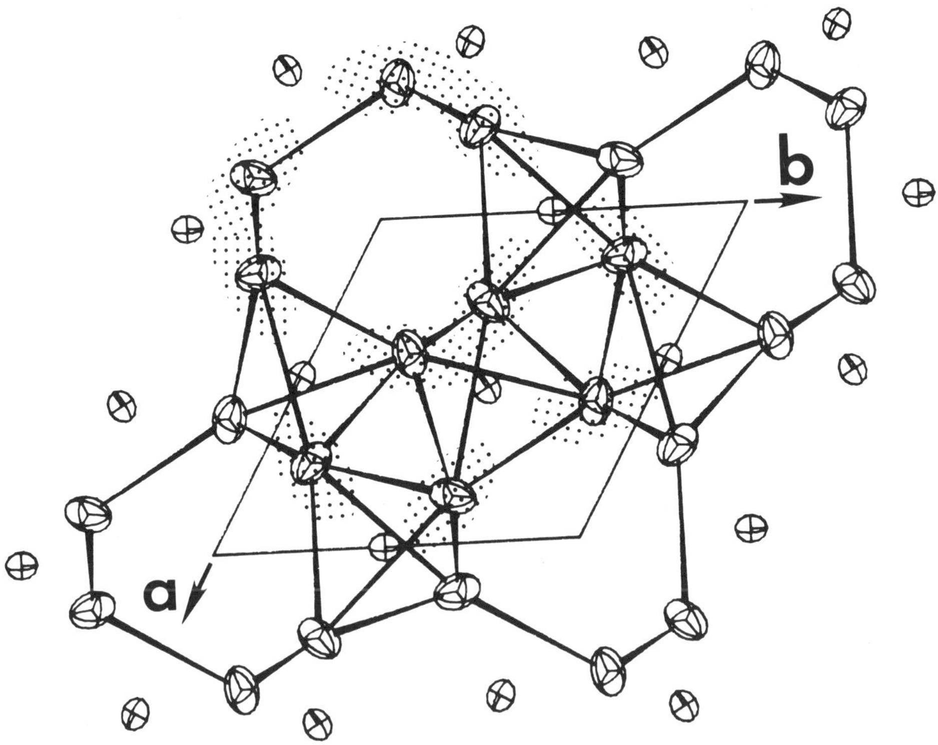

Figure 11 b) Ellipsoids drawn in continuous lines are thermal ellipsoids representing atomic vibration. Shaded ellipses represent anisotropic polarizability ellipsoids as determined by Glazer and Stadnicka (1986).

first time to associate the sign of the optical rotation with the hand of certain helical arrangements in the crystal structures, and they gave a set of simple principles which allowed the sign to be explained in every known example. Their theory is a visual one, concentrating on finding structural helices to correlate with the direction of rotation of the light. It is based principally on the importance played by electronic polarizabilities of individual atoms. Full details can be found in the original paper, but here we shall summarize the findings. The principles are:

(I) Highly polarizable atoms will be the main contributors to the OA.

(II) The directions of the shortest distances between polarizable atoms will tend to correspond to the directions of highest polarizability, since a dipole on one will induce a dipole on the other, and vice-versa.

(IV) The sign of the OA will be determined by the directions of the anisotropic polarizabilities in the plane perpendicular to the helix axis. When the component tangential to the helix is larger, the incident polarized light will rotate in the same sense as that of the helix. When the radial component is larger, the light will rotate in the opposite sense.

To illustrate these simple ideas, consider the crystal structures of quartz and cinnabar.

Figure 11(a) shows the crystal structure of laevorotatory (left-rotating) quartz projected down the c-axis, with the fractional heights of the oxygens marked. Applying principle (I) we consider only the oxygen atoms as being significant in this problem. Principle (II) allows us to determine in which directions the oxygens will be most easily polarized by light propagating along the c-axis i.e. out of the plane of the Figure. Taking oxygen A as an example, we find three close oxygens, C, J and H to the right and three D, M and B to the left. Since these will influence the polarizability of A most strongly, we can imagine then that the most polarizable direction for oxygen A will point towards these two groups of atoms. If we denote this on the drawing by an elongated ellipsoid and repeat the exercise for all oxygens, we obtain the drawing in Figure 11(b) where the shaded regions are intended to indicate these anisotropic polarizability ellipsoids (in this Figure the atoms are also represented by anisotropic thermal ellipsoids, derived from the crystal structure determination, and we find that these are often roughly perpendicular to the anisotropic polarizability ellipsoids). Examining the structure, we find that there are three helices of oxygens in the unit cell: a large left-handed helix ADEFGH and two smaller right-handed helices ABC and KHJ. The question now is, with which helix does the rotation of light correspond? We note that for all three helices the polarizability ellipsoids are roughly tangential, and so principle (IV) says that the rotation should be in the same sense as all three helices. In this case, there are two right-handed helices to one left-handed one, and so we might expect that the right-handed helices will dominate. Thus we expect the laevorotation to correlate with the right-handed helix, i.e. to turn in the same sense as the right-handed helix (note a possible confusion here. A right-handed helix turns to the right on going away from the observer but turns to the left towards the observer. Rotation of light is normally described in terms of the direction of rotation for light coming towards the observer, so that if the rotation is to the left and in the same sense as a structural helix, that helix will be right-handed, as here in quartz). Left-rotating light in this case correlates with a right-handed structural helix.

An even better example is given by cinnabar (HgS). The laevorotatory structure along c is shown in Figure 12(a). Using principle (I) we consider only the sulphur atoms, which are the most polarizable atoms. As in quartz, we again find three helices, this time a 6-fold right-handed helix BCDEFG and two 3-fold left-handed helices BHG and ABC. Appealing to principle (II) we mark

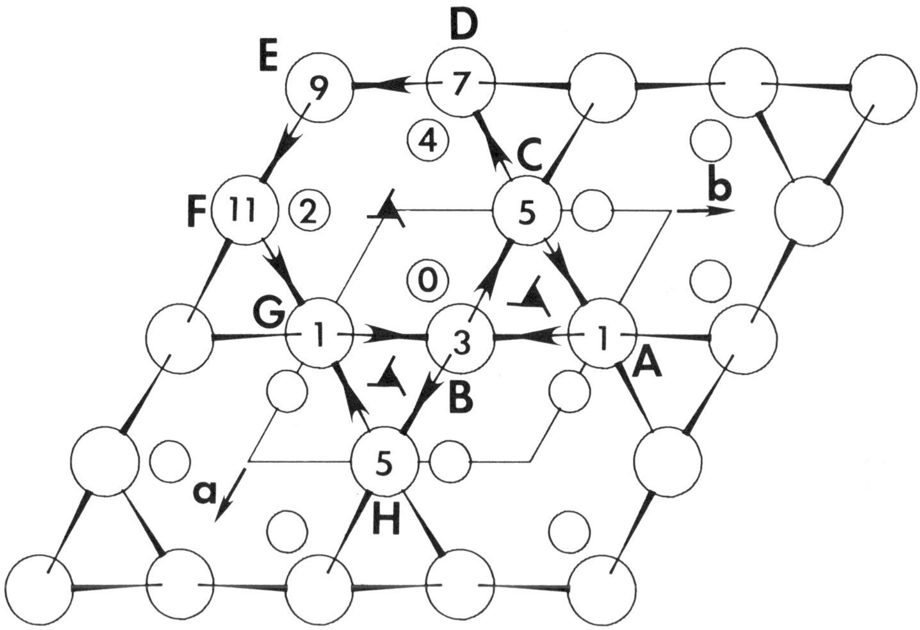

Figure 12. The (0001) projection of the structure of cinnabar. a) Large circles are sulphurs, small circles mercury. The numbers within the large circles represent the fractional heights of the sulphurs in 1/6'ths of the c axis.

polarizability ellipsoids on the diagram: note that starting from sulphur B there are two close sulphurs G and H on one side and A and C on the other. Thus the direction of highest polarizability should point along the average GH to AC line. Repeating this for all sulphur atoms gives the picture in Figure 12(b) (again note that thermal ellipsoids are perpendicular to the polarizability ellipsoids). This time we find that the polarizability ellipsoids are tangential to the right-handed helix but radial to the two left-handed helices. The former will lead to light rotating in the same sense and the latter to light rotating in the opposite sense (principle IV). In this case therefore all helices contribute to light rotating to the left, and indeed cinnabar has a massive optical rotation.

The same ideas were used by Glazer and Stadnicka for all the other known cases and they worked well for all except $LiIO_3$. As a result, a new measurement of the anomalous scattering and optical rotation of $LiIO_3$ was carried out (Stadnicka et al., 1985) and this showed that the original determination of absolute configuration must be in error, thus providing further support for the visual theory.

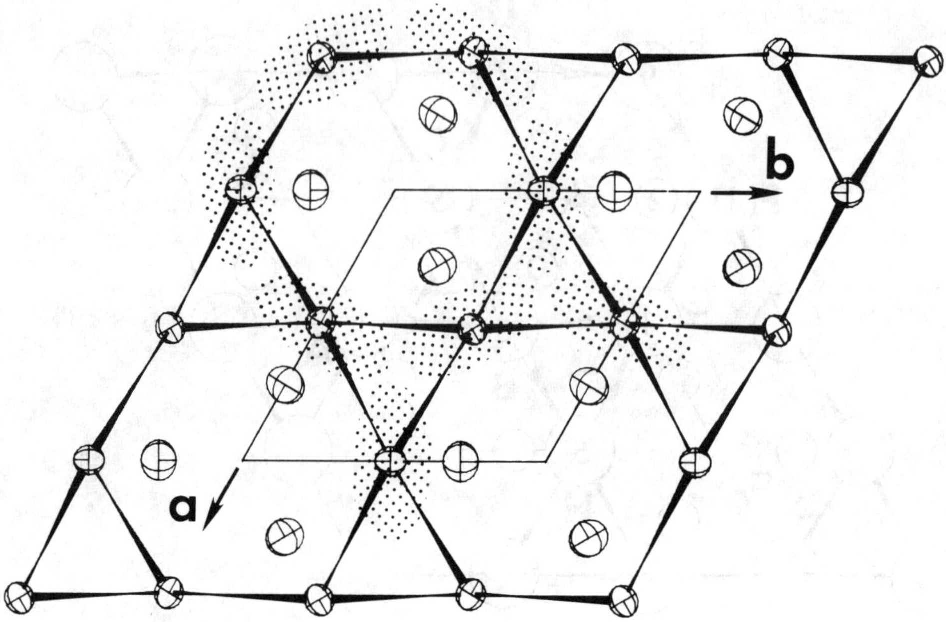

Figure 12. b) Ellipsoids drawn in continuous lines are thermal ellipsoids. Shaded ellipses represent anisotropic polarizability ellipsoids as determined by Glazer and Stadnicka (1986).

3.3 Calculation of Optical Rotation

There have been many attempts at producing a theory that enables the optical rotatory power to be calculated, but, in general, for the reasons given earlier, these have not been too successful. However, given that the fault has been mainly in the experimental work rather than in the theory, Devarajan and Glazer (1986) went back to the early polarizability theory due to Born and Goeppert-Mayer (1933) and applied by van Laar et al. (1968) and by Reijnhart (1970). Rather than give a complete formulation of the theory here, the reader is referred to the paper by Devarajan and Glazer for full details.

The basic idea is to assign polarizability values to each individual atom in the structure and calculate the dipole field on each atom due to every other one and its influence on an electromagnetic disturbance passing through the crystal. The theory is a classical one, and while a more up-to-date quantum-mechanical treatment could be equally well applied, we find this simple theory appealing as it lends itself more easily to interpretation. The starting point of the theory is to consider the dielectric displacement in an optically-active, and non-absorbing crystal:

$$\underline{D} = \varepsilon_0 \varepsilon_r \underline{E} + i\underline{G} \times \underline{E} \quad (8)$$

where ε_r is the relative dielectric permittivity tensor, $\underline{E}$ the electric field and $\underline{G}$ the so-called gyration vector parallel to the wave normal of the light. An effective dielectric tensor can then be defined

$$\varepsilon_{eff} = \varepsilon_0 \varepsilon_r + i[G] \quad (9)$$

where [G] is an antisymmetric tensor. Thus the problem becomes one of calculating the dielectric permittivity whose imaginary part

Table I. Results of calculations for different crystals

Crystal	λ(Å)	$\alpha' = \alpha/4\pi\varepsilon_0$ (Å^3)	n_o^{obs}	n_o^{cal}	n_e^{obs}	n_e^{cal}	ρ_o^{obs} (mm^{-1})	ρ_o^{cal} (mm^{-1})
α-SiO_2	5080	Si=0.207 O=1.213	1.548	1.543	1.557	1.551	-29.8	-29.2
α-HgS	6328	Hg=0.000 S=8.000	2.89	2.74	3.24	3.42	-320	-312
α-$AlPO_4$	6328	Al=0.05 P=0.05 O=1.37	1.521	1.541	1.530	1.545	-15	-11.2
$Bi_{12}SiO_{20}$	6500	Bi=0.15 Si=0.001 O=3.54	2.550	2.503			-20.5	-19.5
$Bi_{12}GeO_{20}$	6500	Bi=0.025 Ge=0.001 O=3.725	2.540	2.545			-19.5	-33.5
α-$LiIO_3$	6328	Li=0.03 I=1.32 O=1.55	1.88	1.93	1.74	1.60	-86.7	-87.4
$NaClO_3$	6328	Na=0.29 Cl=0.01 O=1.60	1.531	1.526			-2.44	-3.70
$NaBrO_3$	6328	Na=0.29 Br=0.10 O=1.60	1.643	1.498			-1.65	+1.50

contributes to the optical rotation. Assuming that the atoms behave as if they were free, so that the response of an atom s to an electromagnetic field E will be described by a point-dipole oscillator at its centre, the induced dipole moment vector is given by

$$\underline{p}_s = \alpha_s \underline{E}(\underline{r}_s) \quad (10)$$

where α_s is the polarizability tensor for atom s and $E(\underline{r}_s)$ is the electric field at the atom due to the dipole waves emanating from all other atoms. It is now necessary to evaluate this field and this is done by the standard Ewald summation (Ewald, 1921). As shown by Devarajan and Glazer, this eventually leads to results that can be used to evaluate both the refractive indices and the gyration tensor. The latter then gives the optical rotatory power.

A computer program was written to do this. The procedure is then to define the crystal structure in terms of the positions of all the atoms in the unit cell. To each atom a polarizability tensor is assigned, usually taken to be isotropic in the first instance and using values lying within normally accepted ranges from the literature. The program then calculates the refractive indices and the polarizabilities are adjusted to give the best fit. Then the calculated optical rotatory power is compared with experiment and further adjustment of the polarizabilities is carried out in an attempt to match both refractive indices and optical rotation. The results of this treatment

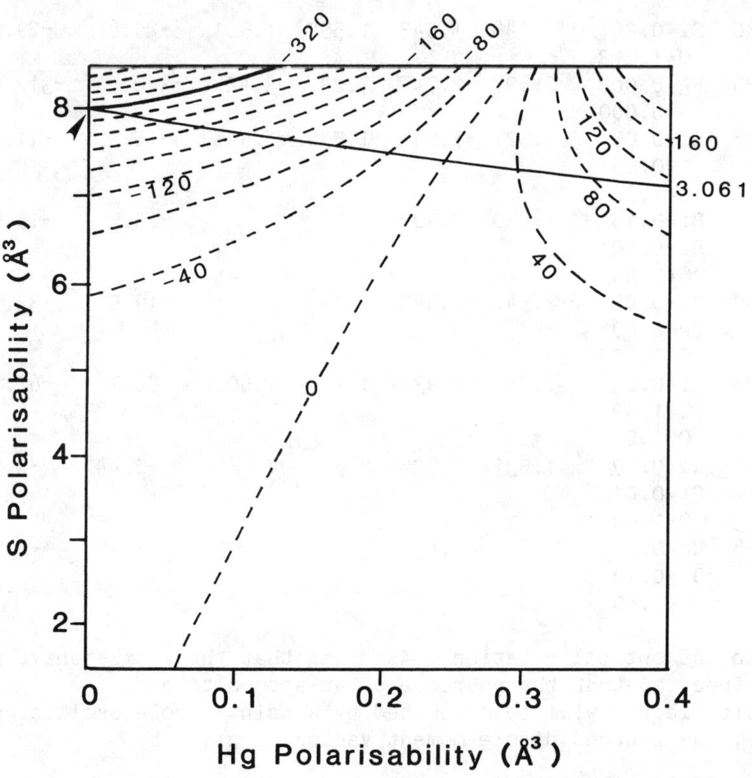

Figure 13. Contour diagram showing how the calculated rotatory power varies with assumed polarizabilities for Hg and S in cinnabar. The heavy full line indicates the contour corresponding to the observed rotation and the thin line indicates the path for the average refractive index.

are very encouraging. It is more or less automatic to obtain the correct sign of rotation, and in most cases excellent agreement can be obtained for the magnitudes with small changes in the input polarizabilities. Table I gives the results for several crystals.

Figure 13 illustrates graphically for cinnabar how the optical properties depend on polarizabilities. In the diagram, contours of calculated optical rotation are plotted as a function of the Hg and S polarizabilities. We note that it changes sign in the middle. However, if we plot a curve representing those polarizabilities which give the correct average refractive index (thin line), we find that it meets the contour corresponding to the observed optical rotation (-320°/mm) on the axis where the Hg polarizability is zero. This had hitherto been predicted by Glazer and Stadnicka who stated that the optical rotation in cinnabar was due entirely to the S atoms. As a by-product of the theory, the dipole field has the effect of

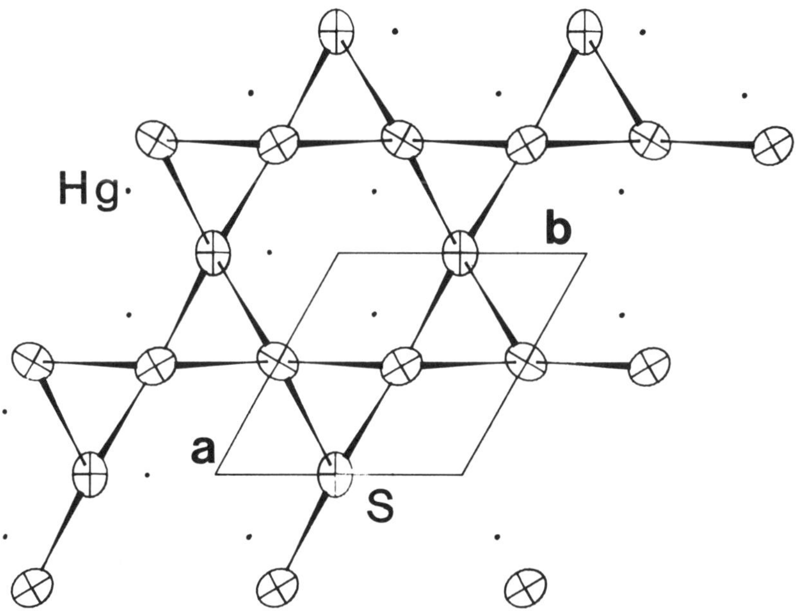

Figure 14. Anisotropic polarizability ellipsoids calculated by Devarajan and Glazer for cinnabar.

renormalizing the original isotropic polarizabilities on each atom to make them anisotropic. From the theory we can extract these anisotropic polarizabilities and plot them on each atom as ellipsoids. Figure 14 shows this for cinnabar and should be compared with the "guesses" made by Glazer and Stadnicka shown in Figure 12(b).

Finally in Figure 15, the particularly complicated example of α-nickel sulphate hexahydrate is shown. It can be seen that there are several ordinary and CD absorption peaks in the visible range and that the observed ORD changes sign twice. Recently, Stadnicka et al. (1987) determined the absolute configuration for the first time from optical

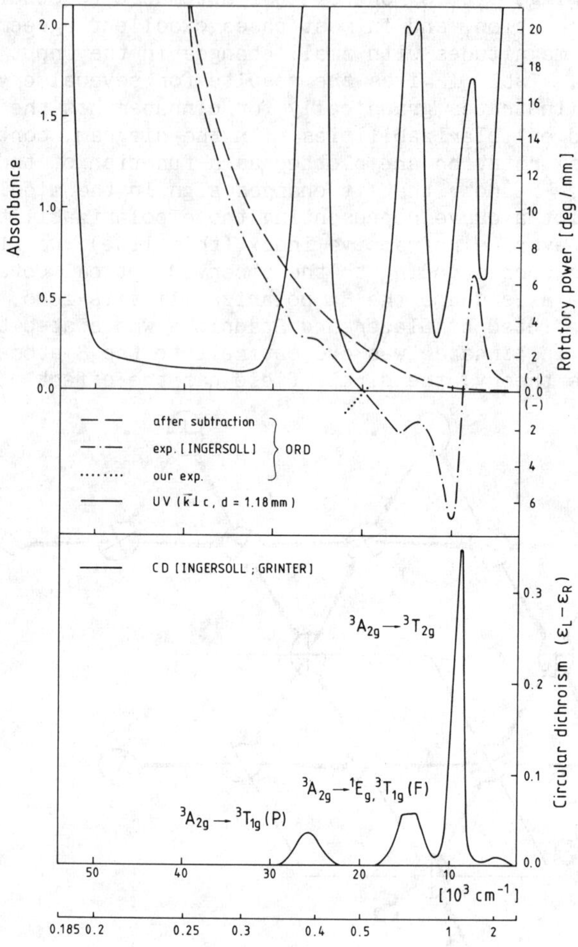

Figure 15. ORD, CD and absorption spectra of α-$NiSO_4.6H_2O$. The measured ORD is marked in dash-dot lines, and the ORD calculated after removal of the contribution from the Ni^{2+} transitions is given by the curve marked in dashes. The small line of dots around 0.5μm are the measurements of Stadnicka et al. (1987), showing that the crystal used there was of opposite sign.

measurements (the ORD is indicated by the line of dots over a small wavelength range in the upper diagram of Figure 15: note that the actual crystal used was of opposite sign and hence the dotted curve is shown opposite to the full one) and anomalous x-ray scattering. It had already been established by others that the anomalies in the ORD were essentially due to d-d transitions within the $Ni(O_2H)_6$ clusters in the structure. Stadnicka et al. demonstrated that if the contributions from these were calculated by applying a Kramers-Kronig calculation to the CD peaks (imaginary part, as in Figure 1(d)) in order to obtain the ORD (real part, as in Figure 1(c)) and subtracted, the resulting curve (thin continuous curve) showed a normal ORD. This curve corresponds then only to the oxygens. Calculations with the Devarajan and Glazer program revealed that it was possible to evaluate this residual optical rotatory power and that the oxygens belonging to the water molecules played a more important part in the ORD than the oxygens in the sulphate groups.

4. SUMMARY

In this brief review, we have seen that in recent years considerable progress has been made in the techniques of measuring linear and circular birefringence and in the theoretical link with the underlying crystal structures. We are now at the stage where we are beginning to discover how to use the vast wealth of crystal structure information that has been amassed since the discovery of x-ray diffraction in order to explain the various optical properties of the crystals themselves. Whereas optical studies of crystals, to a great extent, went out of fashion with the introduction of x-ray diffraction, it is gratifying to see that this older, more traditional tool of the crystallographer, is returning to favour. The new and exciting field of optical communications, using electrooptic, acoustooptic or elastooptic materials, is now of prime importance throughout most of the developed world and it is going to be essential, if new optical materials are to be found, that we should understand how the properties arise.

5. REFERENCES

Aizu K. (1970) J. Phys. Soc. Jpn., 28, 706

Bijvoet J.M. (1949) Proc. K. Ned. Akad. Wet., B52, 313

Born M. and Goeppert-Mayer M. (1933) Dynamische Gittertheorie der Kristalle. In Handbuch der Physik, Vol. 24, pp 623-794

Born M. and Wolf E. (1975) Principles of Optics, 5th ed. New York: Pergamon

Chudakov V.S., Grechushnikov B.N. and Prave G.G. (1972) Sov. Phys. Crystallogr., 16 813

David W.I.F., Glazer A.M. and Hewat A. (1979) Phase Trans., 1, 155

Ewald P.P. (1921) Ann. Phys. (Leipzig), 64 253

Ferré J. and Gehring G.A. (1984) Rep. Prog. Phys., 47, 513

Fresnel A. (1824) Bull. Soc. Philomath. Paris., 147

Gay P. (1982) An Introduction to Crystal Optics. London: Longmans

Glazer A.M. (1972) Acta Crystallogr., B28, 3384

Glazer A.M. (1975) Acta Crystallogr., A31, 756

Glazer A.M. and Ishida K. (1974) Ferroelectrics, 6, 293
Glazer A.M. and Stadnicka K. (1986) J. Appl.Cryst., 19 108
Harley R.T. and Macfarlane R.M. (1975) J. Phys. C., 8 L451
Hartshorne N.H. and Stuart A. (1970) Crystals and the Polarizing Microscope, 4th ed. London: Arnold
Hobden M.V. (1968) Acta Crystallogr., A24, 676
Ishida K. and Glazer A.M. (1974) Ferroelectrics, 6, 293
Kobayashi J. and Uesu Y. (1983) J. Appl. Cryst., 16 204
Laar J. van, Endeman H.J. and Bijvoet J.M. (1968) Acta Cryst., A24, 52
Lowry T.M. (1935) Optical Rotatory Power. London: Longmans. Reprinted in 1964 by Dover, New York.
Nye J.F. (1969) Physical Properties of Crystals. Oxford University Press
Reijnhart R. (1970) Classical Calculations Concerning the Double Refraction, Optical Rotation and Absolute Configuration of Te, Se, Cinnabar, α- and β-Quartz, β-Cristobalite, $NaNO_2$, $NaClO_3$ and $NaBrO_3$. Dissertation, Delft
Stadnicka K., Glazer A.M. and Moxon J.R. (1985) J. Appl. Cryst., 18, 237
Stadnicka K., Koralewski M. and Glazer A.M. (1987) Acta Cryst., in press
Suits J.C. (1971) Rev. Sci. Instrum., 42, 19
Szivessy G.and Münster C. (1934) Ann. Phys. (Leipzig), 20, 703
Wahlstrohm E.E. (1965) Optical Crystallography. New York: Wiley
Wood I.G. and Glazer A.M. (1980) J. Appl.Cryst., 13, 217
Wood I.G., Wadhawan V.K. and Glazer A.M. (1980) J. Phys. C., 13, 5155

EXSOLUTION, ORDERING AND STRUCTURAL TRANSFORMATIONS: SYSTEMATICS AND SYNERGISTICS

D. E. Laughlin
Dept. of Metallurgical Engineering & Materials Science
Carnegie Mellon University
Pittsburgh, PA 15213 USA

W. A. Soffa
Dept. of Materials Science and Engineering
University of Pittsburgh
Pittsburgh, PA 15261 USA

ABSTRACT. Various paths of transformation can occur in metallic and non-metallic solid solutions depending on the thermodynamic stability of the initial state with respect to the relevant order parameters describing composition, atomic ordering, magnetic ordering, strain, *etc.* The particular reaction path followed by the system can have a profound influence on the resultant microstructure and properties of the material. Of particular importance in real systems is often the influence of the different possible ordering tendencies on the phase diagram and their complex interplay during transformation. In this paper reactions such as exsolution or phase separation, atomic ordering, magnetic ordering, *etc.* will be discussed with emphasis on the interaction of higher order transitions with first order phase boundaries in the phase diagram and on the synergistics of different ordering tendencies during transformation. Extensive use of free energy diagrams is employed as a tool to systematize the phase equilibria and the various multistage reaction paths accessible to the system.

1. INTRODUCTION

A single phase solid solution, α, which is stable when it has a particular set of intensive thermodynamic parameters, T,P,... may become metastable or unstable with respect to the formation of another phase, β, when one or more of these parameters are altered. The reaction which ensues may be termed exsolution, precipitation or decomposition.* A description of the paths that

---

*In the metallurgical literature decomposition is used in a synonymous fashion to "exsolution". Many metallic alloys when they undergo exsolution form solid solutions that are based on their elemental constituents. Hence the term "decomposition" is appropriate.

E. K. H. Salje (ed.), Physical Properties and Thermodynamic Behaviour of Minerals, 213–264.

such a solid solution passes through in order to attain thermodynamic equilibrium involves a knowledge of the phase diagram, the symmetry of the phases α and β, interphase interfacial free energies, as well as kinetic parameters such as diffusivity, *etc.* In this article we discuss several types of exsolution reactions that commonly occur in metallic or nonmetallic solid solutions.

We begin with the simplest exsolution reaction, namely one in which both the α and β phases have the same crystal structure, differing only in their respective compositions (**2.1**). This *isostructural* reaction can be modeled by only one "order parameter", namely the difference in composition between the two ensuing phases. The equilibrium value of this order parameter changes discontinuously at all points along the phase boundary (except at the critical point) and hence these transformations are thermodynamically classified as *first order*. We consider such exsolution reactions at both small supersaturations (**2.2**) and large supersaturations (**2.3**), namely under conditions in which the reaction may be modeled as nucleation or spinodal decomposition, respectively. We also discuss the case in which the supersaturation is of intermediate value, *viz.* the reaction occurs near the spinodal. In the discussion of spinodal decomposition, we differentiate between the microstructure of the reaction occurring in isotropic elastic media and that occurring in an anisotropic (cubic) elastic medium.

Another simple solid state reaction occurs when the parent α phase transforms entirely into an atomically ordered phase β, which is a crystallographic derivative (Buerger, 1947) of the α phase (**3.1**). In this case the reaction can be described by the long range (atomic) order parameter η. Such ordering reactions may be thermodynamically of either *first order* (**3.2.2**) or *higher order*** (**3.2.1**).

When a line of higher order transitions ends on a first order phase boundary, a *critical end* point exists (Allen and Cahn, 1982). If such a line terminates uniquely at the critical point of a two phase region, a *tricritical* point is produced at which three phases become indistinguishable (Griffiths, 1970 and 1974). Phase diagrams exhibiting *tricritical points* are herein presented, and the various possible reaction paths that exist for solid solutions exhibiting *tricritical* behavior are discussed (**3.2.1**).

The final sections (**4** and **5**) introduce another ordering reaction, namely "magnetic ordering". This entails an additional order parameter (**M**,

---

**We prefer to use the term "higher order" because of the difficulty in distinguishing between "second", "third", etc. order transitions. All "higher order" transitions are continuous at the transition temperature.

magnetization, for the ferromagnetic case). This ordering process interacts with first order phase boundaries in a similar fashion as does the higher order atomic order/disorder reactions. Examples of such phase diagrams are discussed (**4**). Finally, phase diagrams exhibiting atomic and magnetic ordering are discussed. Here, the *bicritical* and tetracritical points are introduced, (**5**) and the salient features of such reactions discussed.

The emphasis of the paper is to discuss how various ordering reactions in solid solutions interact with each other. Of particular concern is how their interaction affects phase equilibria, phase diagrams, free energy curves, as well as how their interaction affects the paths through which the solid solution passes in attaining thermodynamic equilibrium (metastable or stable). Although the diagrams used in the paper are schematic, many are idealizations of actual phase diagrams that describe minerals and/or metallic alloys. Of course many equilibrium phase diagrams are much more complicated than those discussed herein. In these diagrams, the conjugate phases may not be crystallographic derivatives of the high temperature disordered phase. Nevertheless, *metastable* phase equilibria may obtain in these systems because of kinetic considerations. Only after extended times will the equilibrium phases appear. In such cases, the *metastable equilibria* and corresponding *metastable* transition sequence may be modeled as done herein. Indeed, in metallurgical systems, many of the phase transitions that give rise to important changes in mechanical and physical properties (mechanical strengthening, magnetic hardening, *etc.*) are metastable ones, the equilibrium phases or structures often being undesirable.

2. ISOSTRUCTURAL DECOMPOSITION

2.1. Introduction

We begin our discussion of isostructural decomposition by considering the phase diagram shown in Figure 1. An alloy of composition C_0 is in a single phase region of the diagram at T_0 and within a two phase region at temperature T_A. From the point of view of *configurational entropy*, the high temperature state is *disordered*, while the equilibrium low temperature state consisting of two phases is *ordered*. The two phases differ from one another only in their composition; they have identical crystal structures (*isostructural decomposition*). Thus, in following the decomposition of the high temperature phase the relevant order parameter is the difference in composition, ΔC, between the various regions within the material.

Two distinct decomposition behaviours can be observed for a

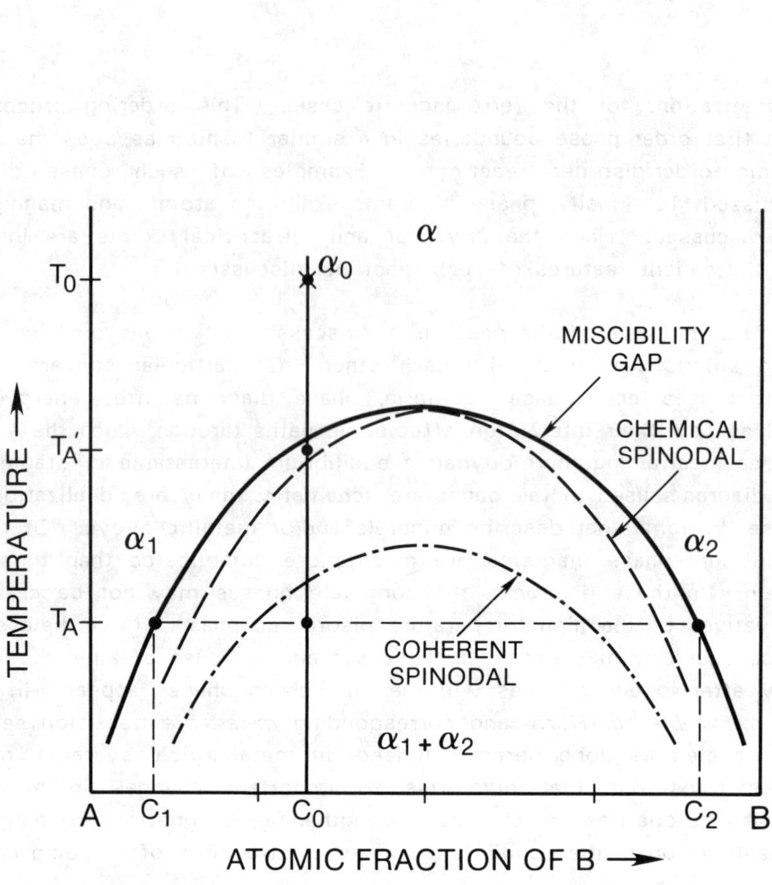

Figure 1. Schematic phase diagram of a binary solution exhibiting a miscibility gap in the solid state. Both the chemical and coherent spinodal lines are depicted.

supersaturated solid solution of an alloy that is quenched from the single phase region into the two phase region. For shallow quenches (small supersaturations) the initiation of decomposition derives from fluctuations in which the order parameter, ΔC, undergoes relatively large changes within regions of small spatial extent. In such a process the embryos or "nuclei" of the new phase are formed randomly within the matrix. This is the decomposition path that has been described by the classical nucleation and growth models (Christian, 1975). At large undercoolings, the alloy decomposes along paths in which the change of the order parameter, ΔC, is initially small, but the extent over which it changes is large. Such a process involving the progressive development of initially low amplitude, highly extended fluctuations has been termed *spinodal decomposition* (Cahn, 1961, 1962a). (For reviews on spinodal decomposition see Cahn, 1968; Hilliard, 1970; and de Fontaine, 1970.)

Classical nucleation and spinodal decomposition represent extremes in a decomposition spectrum characterized by a gradual transition from metastability to instability. The Cahn-Hilliard nonclassical homogeneous nucleation theory (Cahn and Hilliard, 1958; 1959) essentially provides the critical linkage for understanding the progressive change in the nature of the behavior of the decomposition process as the supersaturation or undercooling is increased. As the supersaturation is increased the critical nucleus rapidly becomes diffuse and progressively exhibits characteristics which are very different from those of the "classical nucleus". The amplitude of the critical fluctuation deviates from the equilibrium composition and the interface between the incipient precipitate and parent phase becomes markedly extended. Furthermore, as the spinodal is approached the "nucleation barrier" approaches kT, and a range of finite-amplitude fluctuations against which the system is unstable becomes part of the spectrum of frequently occurring fluctuations. Hence the concept of a well-defined critical nucleus loses its meaning. A variety of fluctuations of varying spatial extent and amplitude readily leads to decomposition of the supersaturated state. This transition regime extends both above and below the spinodal. Thus, the transition from metastability to instability is characterized by a process in which the reaction path does not necessarily minimize the free energy of formation of the critical fluctuation but rather maximizes the rate of decomposition (Cahn, 1962b; Langer, 1975; Aaronson and Russell, 1982). Although it has been repeatedly pointed out (Hillert, 1956; de Fontaine, 1969; Martin, 1979) that the boundary between the extreme behavior is not sharp, it has been the practice of experimentalists to attempt to denote the temperature at which the behaviour changes.

2.2 Nucleation and Growth

Let us first look at the nucleation behaviour for coherent *isostructural* exsolution in a *cubic* system. The free energy change accompanying the formation of an embryo of the new phase may be written as:

$$\Delta G = (g_v + E_s)V + \sigma A$$

where g_v is the bulk chemical free energy per unit volume of the new phase

E_s is the strain energy per unit volume produced by the new phase

σ is the surface energy of the interphase interphase between the new phase and the matrix

V is the volume and A the surface area of the new phase

The equation shows that thermodynamically the optimum shape of the new phase generally will represent a compromise between the surface and strain energies. For the case of small atomic misfit or mismatch the particle of the new phase is expected to be a sphere if σ is isotropic. As the particle grows the strain energy becomes more important, and may cause the sphere to change its shape. The elastic anisotropy of the cubic matrix may result in facets normal to the elastically "soft" directions, (e.g. ⟨100⟩) producing cuboids. However, if σ is anisotropic, initially the new phase would take a shape dictated by the surface energy anisotropy, *viz.* facets would be present on the "critical nuclei". The symmetry of the shape of the particle of the new phase would include the symmetry of the point group of its crystal structure, *viz.*, it would be at least $m\bar{3}m$. If, for example, the {111} planes minimize σ, an octahedral particle would be favored. If the {100} planes, on the other hand, represent a minimum for σ, a cuboid would form. Finally, if the {100} and {111} planes are of similar magnitude, a tetrakaidecahedral (14 sided) particle would form. As the particle grows, the strain energy which scales with the volume begins to impose shape effects which eventually may dominate the morphology.

For large atomic misfit, the shape of the particle of the new phase would be dominated by strain energy considerations from the start of its existence. This often gives rise to plate like precipitates of the new phase. The normals to the plates lie along the elastically soft directions of the matrix. Examples of isostructural decomposition with large misfit in metallic alloys include the so-called Guinier-Preston zones in Al-Cu and Cu-Be. As the volume fraction of the second phase increases (either with time at temperature or with increasing supersaturation) the strain fields of the precipitates begin to interact. This may lead to particle alignment along the elastically soft directions and even to the development of spatially correlated precipitates (Tiapkin, 1977).

2.3 Spinodal Decomposition

At large supersaturations, the decomposition process proceeds in a manner which has been termed *spinodal* decomposition. As discussed above, the spinodal reaction is a spontaneous unmixing or diffusional clustering whose behaviour is distinct from classical nucleation and growth in metastable solutions. This different kinetic behavior, which does not require a nucleation step, was first described by J.W. Gibbs in his treatment of the thermodynamic stability of undercooled or supersaturated phases. The *spinodal* line in Figure 1 indicates an instability with respect to the response of the system to compositional fluctuations. The locus, called the "chemical spinodal", is defined by the inflexion points of the isothermal free energy (G) composition curves ($\partial^2G/\partial C^2 = G'' = 0$). Within the spinodes where $G'' < 0$, the supersaturated solution is unstable with respect to unmixing and spinodal decomposition can occur. It should be remembered, as discussed above, that this stability limit is by no means sharp; rather a region exists in which the metastable solution gradually begins to display nonclassical decomposition behavior. Spinodal decomposition or continuous phase separation essentially involves the selective amplification of long wavelength concentration waves within the supersaturated state resulting from uphill diffusion as a consequence of the inherent thermodynamic instability of the system. The transformation occurs homogeneously throughout the alloy *via* the gradual buildup of regions alternatively enriched or depleted in solute, resulting in a two-phase modulated structure.

The essential features of the spinodal process can be understood by considering this diffusional clustering as the inverse of the homogenization of a nonuniform solid solution exhibiting a sinusoidal variation of composition with distance. (A sinusoidal composition wave is rather general, because it may be viewed as a Fourier component of an arbitrary composition variation.) In metastable solutions the small deviations from the average concentration, C_0, will decay with time according to the equation

$$\Delta C = \Delta C_0 \; exp(-t/\tau)$$

where ΔC_0 is the amplitude, the relaxation time $\tau \sim \lambda^2/\tilde{D}$; λ is the wavelength of the fluctuation or Fourier component and $\tilde{D}$ is the appropriate diffusion coefficient. In a binary system $\tilde{D} \propto G''$, and within the spinodes $G'' < 0$; that is, the curvature of the free energy *versus* composition curve is negative. Therefore, in an unstable solid solution $\tilde{D}$ is negative, and "uphill" diffusion occurs, resulting in spontaneous unmixing. The amplitude of the concentration wave grows with time, that is, $\Delta C = \Delta C_0 \exp(+R(\beta)t)$, where the amplification factor $R(\beta)$ is a function of the wave number $\beta = 2\pi/\lambda$. The factor $R(\beta)$ is a maximum for intermediate wavelengths. Long wavelength fluctuations grow sluggishly because of the large diffusion distances; short wavelength fluctuations are suppressed by the so-called gradient or surface energy of the diffuse or incipient interfaces that evolve during phase separation. Therefore, the microstructure that develops during spinodal decomposition has a characteristic periodicity that is typically 2.5 to 10 nm (25 to 100 Å) in metallic systems.

The factors controlling the spinodal reaction and resultant structures are clarified by examining the energetics of small-amplitude composition fluctuations in solid solutions. The free energy of an inhomogeneous solution expressed as an integral over the volume, V, of the crystal can be written as:

$$G = \int \left\{ f(C) + K\nabla C^2 + E_s \right\} dV$$

where $f(C)$ is the free energy per unit volume of a uniform solution of composition C, K is the gradient energy parameter, and E_s is a strain energy term that depends on the elastic constants and misfit (difference in lattice parameter) between the solute-enriched and solute-depleted regions. For a sinusoidal composition fluctuation

$$C - C_0 = A \sin \beta x$$

(where A is the amplitude of the sine wave), the gradient or surface energy term varies as $K\beta^2$ and prohibits decomposition on too fine a scale. The wavelength of the dominant concentration wave that essentially determines the scale of decomposition varies as $K^{1/2}(\Delta T)^{-1/2}$, where $\Delta T = T_s - T_A$, in which T_s is the *spinodal temperature* and T_A is the reaction temperature. The coherency strain energy term is independent of wavelength, but can vary markedly with crystallographic direction in elastically anisotropic crystals. Therefore, the dominant concentration waves will develop along elastically "soft" directions in anisotropic systems. For most cubic materials, the <100> directions are preferred, although <111> waves are predicted in certain alloys, depending on the so-called anisotropy factor. The strain energy can also stabilize the system against exsolution and effectively displace the spinodal curve (and the solvus), thus defining a "coherent spinodal" (Cahn, 1962a; See Figure 1).

Periodic composition fluctuations in the decomposing solid solution cause diffraction effects known as "satellites" or "sidebands". The fundamental reflections in reciprocal space are flanked by satellites or secondary maxima, and the distance of the satellites from the fundamental varies inversely with the wavelength of the growing concentration wave. This diffuse scattering arises from the periodic variation of the lattice parameter and/or scattering factor. The strain effects are negligible around the origin of reciprocal space. Thus, small-angle x-ray and neutron scattering can be used to study quantitatively the kinetics of the reaction by monitoring the changes in the intensity distribution around the direct beam due to changes in the structure factor modulations, assuming that double diffraction is negligible. The electron diffraction pattern of a spinodally decomposed copper-titanium alloy shown in Figure 2 (Datta and Soffa, 1976) reveals the dominant ⟨100⟩ concentration waves that develop during the early stages of phase separation. The extinctions of the satellites at certain reciprocal lattice nodes (where $\vec{g} \cdot \vec{R} = 0$) show that the satellites arise primarily from *longitudinal displacive* waves induced by the concentration waves.

If the strain energy term in the free energy expression is negligible (small misfit) or if the elastic modulus is isotropic, the resultant microstructure will be isotropic, similar to the morphologies evolving in phase-separated glasses (Cahn and Charles, 1965; Cahn, 1965). In Figure 3, an isotropic spinodal structure developed in a phase-separated iron-chromium-cobalt permanent magnet alloy is clearly revealed by transmission electron microscopy. The two-phase mixture is interconnected in three dimensions and exhibits little or no directionality. The microstructure is comparable to the computer simulation of an isotropically decomposed alloy shown in Figure 4. On the other hand, in the microstructure shown in Figure 5, the dominant composition waves have developed preferentially along the ⟨100⟩ matrix directions to produce an aligned modulated structure in a copper-nickel-chromium alloy (Chou *et al.*, 1978). Its electron diffraction pattern would be similar to that shown in Figure 2. Because the homogeneous phase separation process is relatively structure-insensitive, the spinodal product is generally uniform within the grains up to the grain boundaries, and independent of such defects as dislocations, stacking faults, twins, *etc.*

Finally we should consider decomposition in the region near the *mean field spinodal*. In this region, the classical models for understanding both nucleation and spinodal decomposition break down quantitatively. The classical nucleation theory rests on the major assumption that the *only* nuclei to form are those whose barriers to formation, ΔG^*, are minima. At

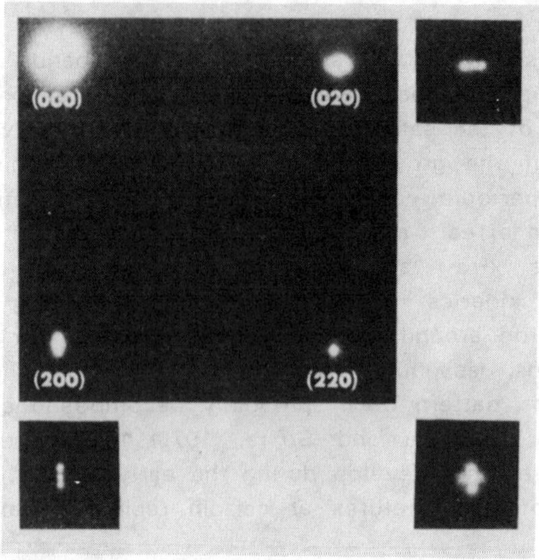

Figure 2. A [001] electron diffraction pattern taken from a Cu-4 wt/% Ti alloy aged 100 min at 400 C, showing <100> satellites flanking the fundamental reflections. Note the extinction of the [010] satellites at the (200) position and the [100] satellites at the (020) position, consistant with the presence of <100> longitudinal strain waves.

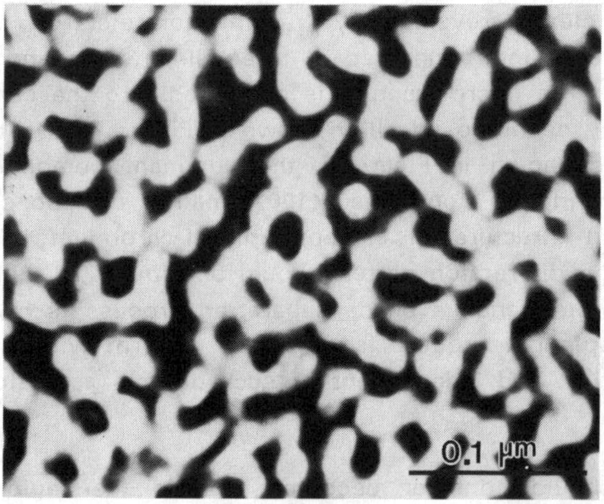

Figure 3. A transmission electron micrograph of Fe-28.5 wt/% Cr-10.6 wt/% Co alloy aged 4 hrs at 600 C. Note that the contrast appears to be isotropic and derives from structure factor variations.

Figure 4. A computer simulation of an spinodally decomposed solution, exhibiting isotropic elasticity (after Cahn, 1965).

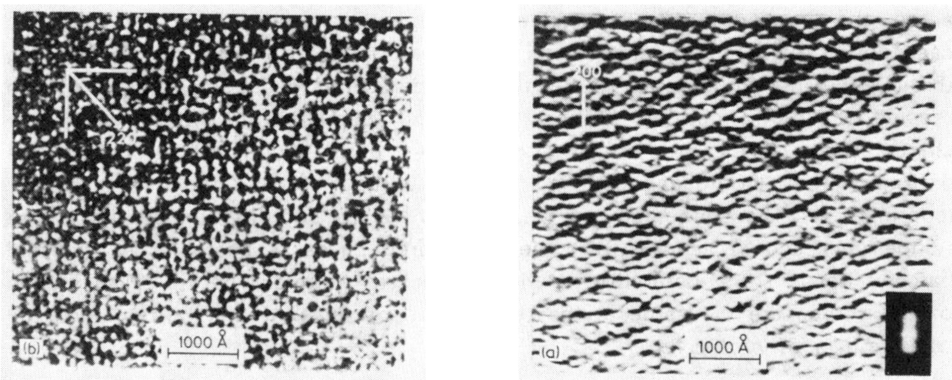

Figure 5. The microstucture of a Cu- 31.6 wt/% Ni-1.7 wt/% Cr alloy, aged at 650 C for 1 hr. The figure on the left is taken under diffraction conditions such that two sets of ⟨100⟩ modulations are visible. In the figure on the right, only one set of modulations is visible. This shows that the contrast derives from longitudinal displacive waves, induced by compositional modulations.

undercoolings near the mean field spinodal, such an assumption is unreasonable. Many decomposition paths exist with barriers of the order of 1-10 kT. These paths appear to be barrierless, because they can be readily surmounted by composition fluctuations induced by thermal agitation. Thus the decomposition process appears to be continuous in many aspects. However, this decomposition cannot be modeled quantitatively by classical spinodal theories, because the assumptions involved in these theories also break down as the spinodal is approached. Of course, the decomposition process occurs, whether we can model it or not! Recent advances in the theory of decomposition have been made along the lines of extending either the nucleation type approaches (cluster kinetics) to larger supersaturations or extending the spinodal decompositional approaches to smaller supersaturations. The common finding of all the theories is that there is not an abrupt transition between the two extremes in decomposition behavior as described above. Rather, the process changes progressively from nucleation-like behavior to spinodal like behavior through a region in which decomposition exhibits characteristics of both processes (Cahn and Hilliard, 1958; 1959; Binder *et al.*, 1978 and Aaronson and Russell, 1982).

3. ATOMIC ORDERING

3.1 Single Order Parameter

A simple phase diagram for atomic ordering of a bcc phase (A2) is shown in Figure 6. The high temperature disordered phase is stable at temperatures above the hatched line and *unstable* at temperatures below the line. In this case only first neighbor interactions have been taken into account. On cooling through the instability temperature, the disordered phase becomes unstable with respect to the formation of short wavelength composition waves, which are characteristic of the ordered structure. The ordering builds up throughout the entire system (homogeneous) and occurs without a thermodynamic barrier (continuous). The accompanying free energy *versus* composition curves at the temperature shown denotes that the disordered phase is unstable by inverted hatched lines. This instability is with respect to atomic ordering. The slopes of the curves are the same at the termination of lines of instability, since the transition is of *higher order*.

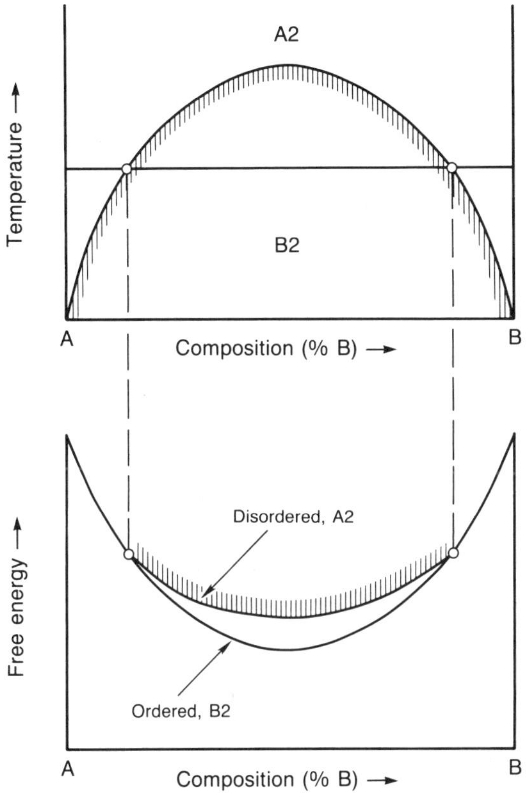

Figure 6. A schematic phase diagram and accompanying free energy *versus* composition diagram of a solution exhibiting a higher order atomic ordering reaction. The hatched lines denote where the solution is unstable with respect to (continuous) ordering.

3.2 Two Order Parameters

3.2.1 Higher Order Order/Disorder Transition

If second neighbor interactions are taken into account, the low temperature equilibrium state of off stoichiometric compositions must be two phase, since the ordered structure has opposite second neighbors (Ino, 1978; Kubo and Wayman, 1980). To minimize the energy in such cases, the solute atoms must segregate and then atomically order. The clustering of solute allows for the maximum number of similar *second* neighbors within the ordered phase. This shows that clustering and ordering are not opposite processes; in fact, they very often occur together in decomposition reactions (Soffa and Laughlin, 1982).

Figure 7 shows the phase diagram of the order/disorder system when the second neighbors interaction energies are taken into account. At low temperatures, the equilibrium state is two phase, one of which is the ordered B2 phase (CsCl, $Pm\overline{3}m$); the other is the disordered A2 phase (W,$Im\overline{3}m$). The line of higher order transitions intersects the miscibility gap at a *critical end point*. The free energy curves at temperatures T_3, T_2 and T_1 are shown in Figures 8, 9 and 10 respectively. At T_3 (Figure 8), the ordered phase breaks away at the composition denoted C_λ, the slopes of both free energy curves are identical at this point, while their curvatures are different. At the lower temperature T_2 (Figure 9), the ordered phase develops a miscibility gap. Spinodal decomposition of the ordered phase is now possible. At lower temperatures yet (T_1, Figure 10), the equilibrium consist of an ordered phase (B2) and a disordered phase, A2. Figure 10 can be used to delineate various decomposition paths. For solute compositions up to but less than C_λ, the disordered phase upon entering the two phase region is metastable with respect to the formation of the ordered phase; *viz.* the ordered phase must nucleate. For compositions between C_λ and C_s', the disordered phase first becomes unstable with respect to homogeneous ordering and then the ordered phase becomes metastable with respect to the formation of the equilibrium disordered phase. Finally, for compositions between C_s' and C_s'' the disordered phase at T_1 first becomes unstable with respect to the continuous formation of B2, which in turn is unstable with respect to phase separation into two ordered phases. Eventually one of the ordered phases disorders, as its composition becomes less than C_λ. The latter spinodal decomposition has been termed "conditional" by Allen and Cahn (1976), since it is contingent on the prior ordering of the disordered phase.

A limiting case of the *critical end point* diagram is shown in Figure 11. Here the critical *end point* has become a *tricritical* point, since the line of

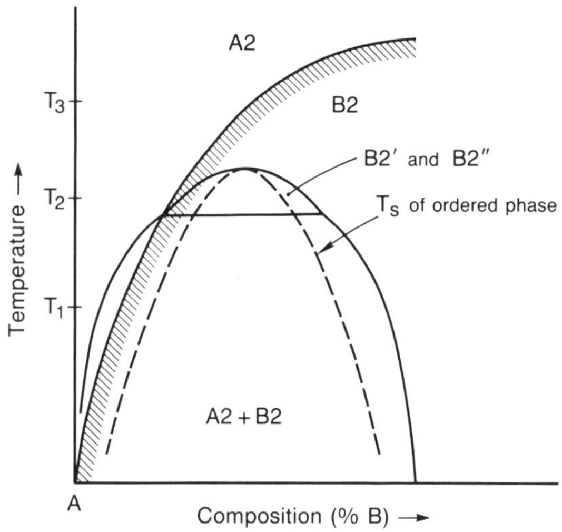

Figure 7. A schematic phase diagram of a solution exhibiting a *critical end point.* Within the two phase region, a quenched A2 solution could be either metastable or unstable with respect to phase separation. The spinodal region delineates instability with respect to phase separation contingent on prior atomic ordering, *i.e.* a conditional spinodal.

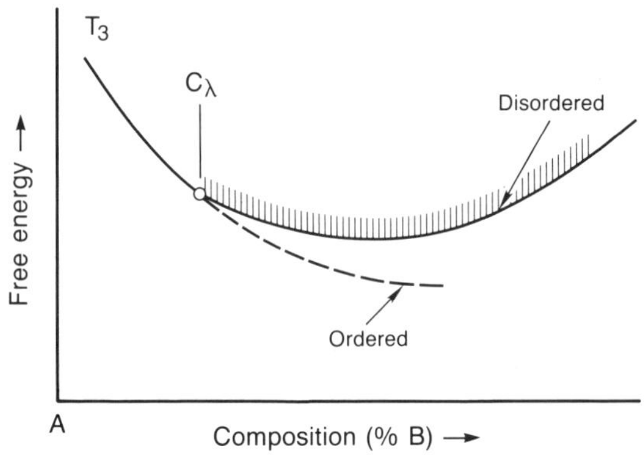

Figure 8. A schematic free energy *versus* composition diagram of the solution shown in Figure 7, at $T=T_3$. Note that the ordered solution is everywhere stable with respect to phase separation.

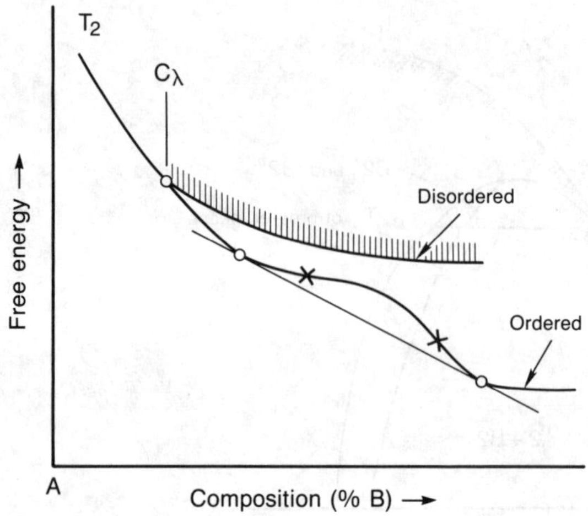

Figure 9. A schematic free energy *versus* composition diagram of the solution shown in Figure 7, at $T=T_2$. Note that the ordered solution exhibits an unstable region between the x's.

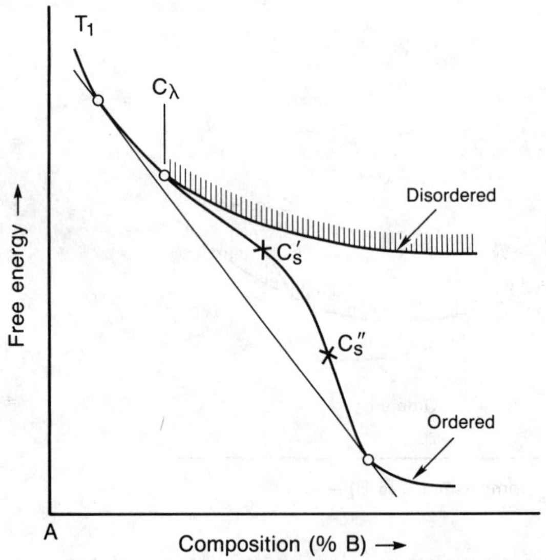

Figure 10. A schematic free energy *versus* composition diagram of the solution shown in Figure 7, at $T=T_1$. Note that the ordered solution exhibits an unstable region, and also that the equilibrium conjugate phases are ordered and disordered respectively.

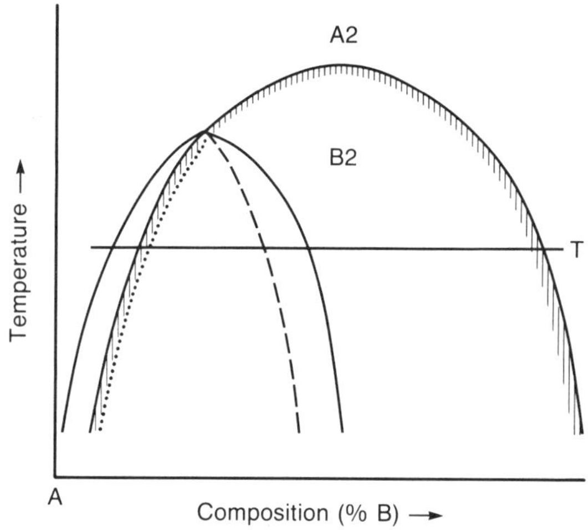

Figure 11. A schematic phase diagram of a binary system exhibiting a *tricritical point.* Here the line of higher order transitions terminates uniquely at the top of the two phase region. The dashed curve represents a spinodal line (G"=0) for the ordered phase, whereas the dotted line (coincident with the A2→B2 transition line) represents a limit of metastability (G" changes sign discontinuously from positive to negative).

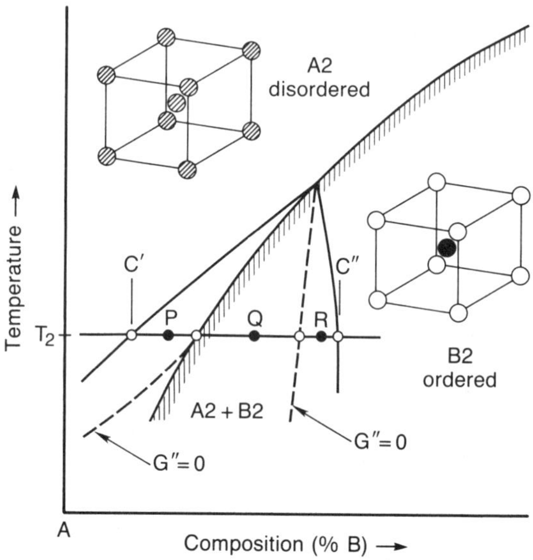

Figure 12. Region near the *tricritical point,* showing the emergence of a spinodal in the disordered solid solution at low temperature. Points P,Q and R represent initial states of a quenched solid solution, exhibiting different thermodynamic stabilities with respect to clustering and ordering.

higher order transitions is drawn so as to terminate uniquely at the critical point of the miscibility gap. As shown in the diagram, the instability of order now coincides with the instability for phase separation on the left hand side of the diagram. Importantly, the limit of metastability on the left represents a discontinuity of G", across which it changes sign abruptly, from positive to negative without going through zero.

We now look at the tricritical case in more detail. Figure 12 shows the region near the *tricritical point* and the free energy *versus* composition curve at T_2 is shown in Figure 13. T_2 is the temperature at which the spinodal of the disordered phase coincides with C_λ. As shown in Figure 12, below T_2 a region exists in which a quenched disordered phase is unstable with respect to phase separation into two disordered phases. This spinodal reaction will be followed by the continuous ordering of the solute-enriched phase.

Let us examine the decomposition paths of disordered solid solutions quenched to points P, Q and R, respectively. A disordered solution quenched to P is metastable with respect to the formation of the ordered B2 phase. A disordered alloy quenched to Q is unstable with respect to homogeneous, continuous ordering into a single order phase, Q'. This ordered phase is unstable with respect to spinodal decomposition into two ordered phases. Eventually, the composition of the solute depleted phase passes through the line of critical points, and it disorders. Finally, a disordered phase quenched to R is unstable with respect to homogeneous ordering to R'. The ordered phase is then metastable with respect to the formation of the disordered A2 phase. These three different decomposition paths all take the initially disordered phase into a state where disordered and ordered phases are in equilibrium.

3.2.2 First Order Order/Disorder Transformations

Spinodal processes occurring in solid solutions in conjunction with an apparent metastable or stable order/disorder transition thermodynamically of first order are not generally well understood. However, these continuous processes involving concomitant ordering and phase separation appear to be common in metallic and non-metallic systems. The following discussion is essentially after Kuikarni *et al.*, 1986 and Khachaturyan *et al.*, 1987.

Consider a metastable or stable phase diagram configuration in a binary system based on an fcc crystal structure as shown in Figure 14. The γ phase is an ordered phase which is a crystallographic derivative of the high temperature α solid solution; that is, the superstructure can be produced by a simple atomic rearrangement on the sites of the fcc lattice, thus reducing the

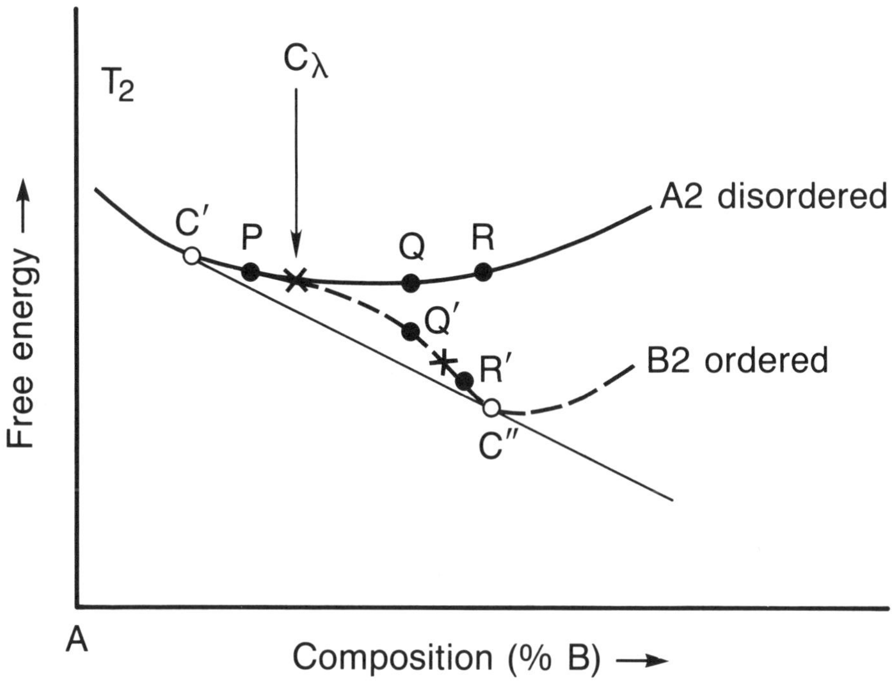

Figure 13. Free energy *versus* composition diagram at T_2, showing the disordered and ordered solutions corresponding to the states P, Q and R of Figure 12.

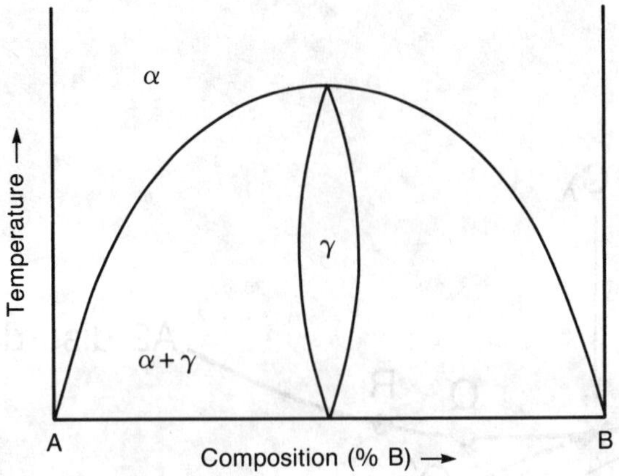

Figure 14. Binary system exhibiting a first order order/disorder reaction. Here, a stable two phase region exists at all temperatures below the critical temperature for the γ phase.

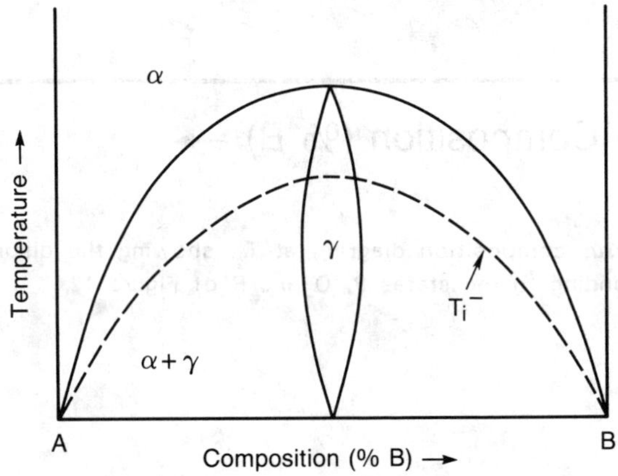

Figure 15. First order reaction showing the *locus* of ordering instability (T_i^-). Below T_i^- a quenched disordered phase will homogeneously or continuously order.

symmetry of the disordered phase. The ordered phase will have a stoichiometry A_xB_y, for example, A_3B or AB. It is well-known that at small departures from equilibrium the ordered phase can only emerge in an undercooled α phase phase by a heterogeneous reaction involving the nucleation and growth of small, highly ordered regions within a disordered matrix. This is termed "heterogeneous ordering". However, at large degrees of undercooling or large departures from equilibrium, the disordered phase can become thermodynamically unstable and continuously order by the progressive amplification of "ordering waves" uniformly throughout the system, *i.e.*, "homogeneous or continuous ordering". The homogeneous ordering reaction is the counterpart of spinodal decomposition in the scheme of reaction paths involving order/disorder processes. The continuous ordering process can occur for $T \leq T_i^-$ where T_i^- is the "ordering instability temperature" on cooling. In this regime away from equilibrium, the phase transformation has many of the thermodynamic and kinetic properties of a second or higher order transition. See Figure 15.

The instability locus can be calculated thermodynamically in a straightforward manner using the method of concentration waves (Khachaturyan, 1978 and 1983) and is given by:

$$T_i^- = \frac{-2V(\vec{k})c(1-c)}{k_B}$$

where $V(\vec{k})$ for the wave vector $\vec{k}$ is related to the Fourier transform of the pair interaction parameter

$$V_i = E^i_{AB} - \frac{1}{2}\left(E^i_{AA} + E^i_{BB}\right)$$

where E^i_{AB} , E^i_{AA} and E^i_{BB} are the AB, AA, and BB bond interaction energies in the ith nearest-neighbor shell, respectively. Ordering instabilities are associated with "special points" in $\vec{k}$-space where the function $V(\vec{k})$ is an *extremum* (de Fontaine, 1975). Clustering instabilities are associated with the term V(0) in this formalism (Khachaturyan, 1978 and 1983).

Consider the free-energy *versus* composition schemes with decreasing temperature and associated phase diagram shown in Figures 16-19. The common tangent construction establishes the equilibrium compositions of the conjugate phases in the two-phase region in the usual manner. A solution of composition C_o rapidly quenched from T_o to T_1 would form the equilibrium $\alpha+\gamma$ two-phase mixture by nucleation and growth of the ordered γ phase within the supersaturated disordered phase (Figure 17). However, if an alloy of

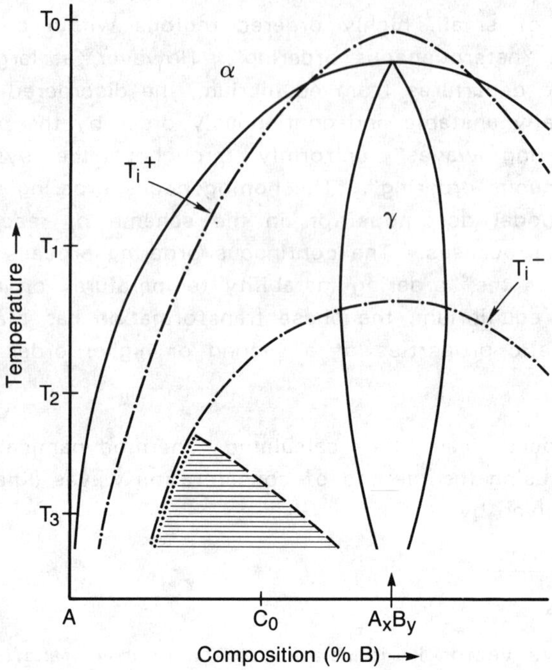

Figure 16. Schematic of a first order order/disorder phase diagram with various instability lines included. In the crosshatched region the homogeneously ordered phase is unstable with respect to phase separation. The *locus* T_i^+ is the instability line of the ordered phase with respect to disordering upon heating.

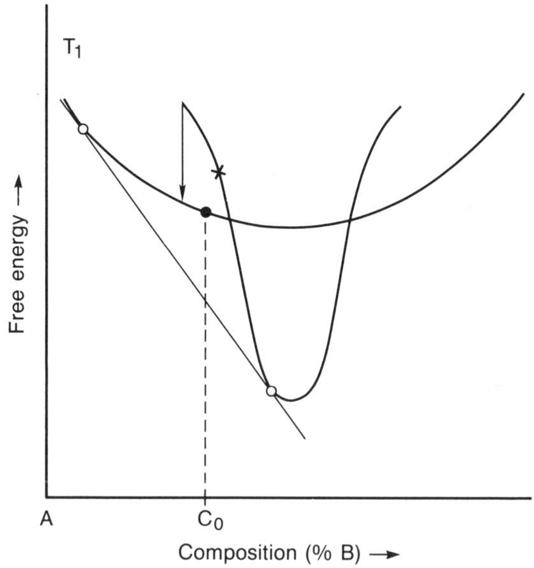

Figure 17. Schematic free energy *versus* composition diagram for the ordered and disordered phases at T_1. The arrow denotes the limit of stability of the ordered phase.

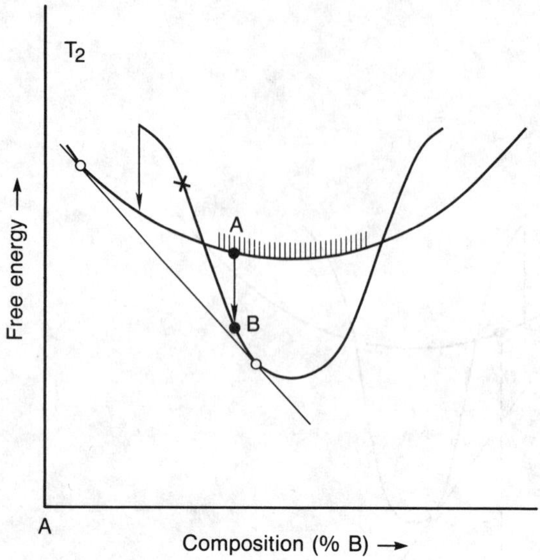

Figure 18. Schematic free energy *versus* composition diagram for the ordered and disordered phases at T_2. The hatched segment of the free energy curve for the disordered phase corresponds to solutions which are unstable with respect to continuous ordering.

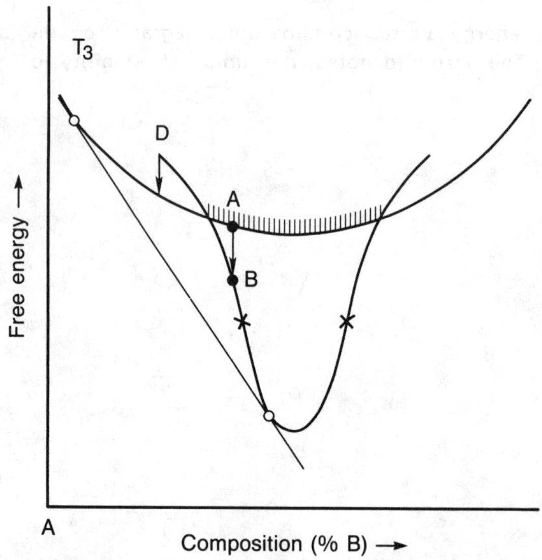

Figure 19. Schematic free energy *versus* composition diagram for the ordered and disordered phases at T_3. Note the appearance of a spinodal (G″=0) in the ordered phase and a region of thermodynamic instability with respect to phase separation contingent on prior atomic ordering.

composition C_o is quenched to the temperature T_2 from the single phase region, the free energy *versus* composition scheme shown in Figure 18 obtains. Here, another reaction path will occur. The hatched region of the α curve now delineates compositions which are unstable with respect to continuous ordering. The α phase will order homogeneously or continuously, bringing the free energy of the solution from point A to point B. The equilibrium two-phase mixture will now be produced by classical nucleation of disordered α within the ordered solution. See Figure 18.

Now consider the temperature T_3 and the free energy *versus* composition scheme shown in Figure 19. At T_3 the solution of composition Co is unstable with respect to ordering and the continuous ordering process will take the free energy from point A to point B. But the ordered state of the solution at B is unstable with respect to phase separation and will undergo spinodal decomposition or continuous phase separation. The compositions of the emerging phases move along the free energy curve for the ordered phase and at point D the solute depleted regions spontaneously disorder. This essentially defines a locus T_i^+ or an instability with respect to disordering on heating in the phase diagram (Khachaturyan *et al.*, 1987).

The different regimes of thermodynamic stability with respect to <u>ordering</u>, <u>clustering</u> and <u>disordering</u> can be summarized as follows: (See Figure 16) In the cross-hatched region a <u>conditional</u> <u>spinodal</u> reaction can occur in the ordered phase contingent on the prior ordering of the solution. The supersaturated states in the cross-hatched regions are actually thermodynamically unstable with respect to ordering and then clustering.

Interesting variations in conjunction with <u>two</u> fundamentally first order transformations are shown in Figures 20a and 20b. All supersaturated or non-equilibrium undercooled states which fall in the cross-hatched region in Figure 20a are unstable with respect to continuous ordering and phase separation within the ordered phase. The monotectoid configuration derives from a miscibility gap in the ordered phase. At $T=T_1$ the free energy *versus* composition diagram of the ordered phase exhibits two regimes of negative curvature as shown in Figure 21. At lower temperatures the regimes converge and the free energy *versus* composition diagram at T_2 is as shown in Figure 22. A metastable miscibility gap can also occur in the ordered phase, and produce the metastable diagram shown in Figure 20b.

It is also possible that the disordered phase could be unstable with respect to phase separation at higher temperatures, with ordering occurring at low temperatures as shown in Figure 23a. This produces a syntectoid isotherm

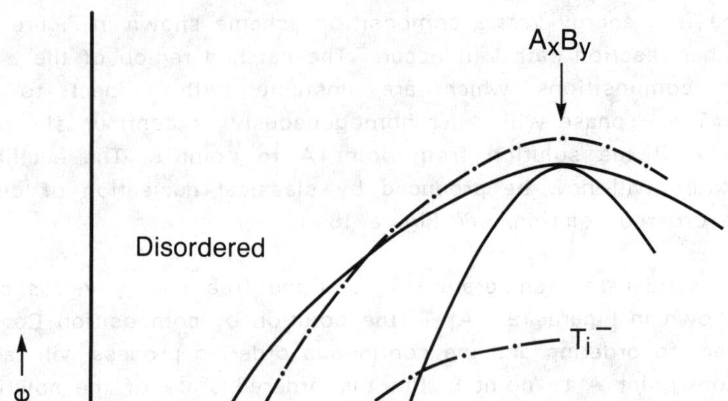

Figure 20a. Phase diagram configuration showing a monotectoid reaction involving the ordered and disordered phases. The crosshatched region denotes the region of thermodynamic instability with respect to phase separation.

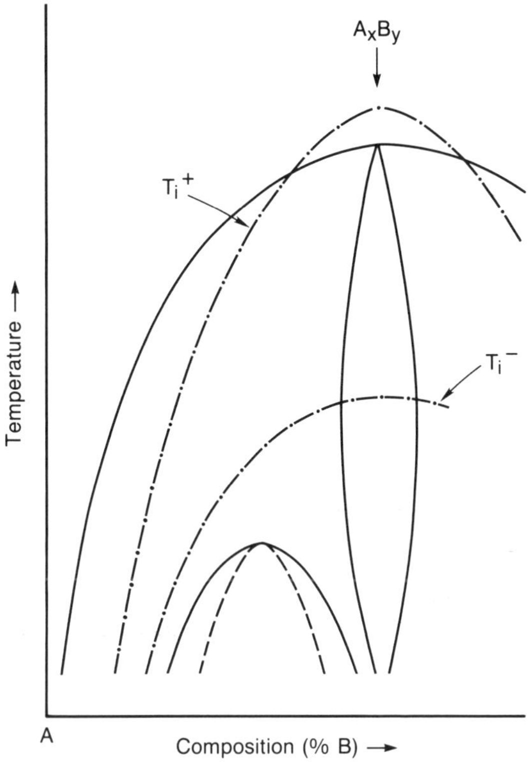

Figure 20b. A phase diagram for a first order order/disorder reaction showing a metastable miscibility gap in the homogeneously ordered phase.

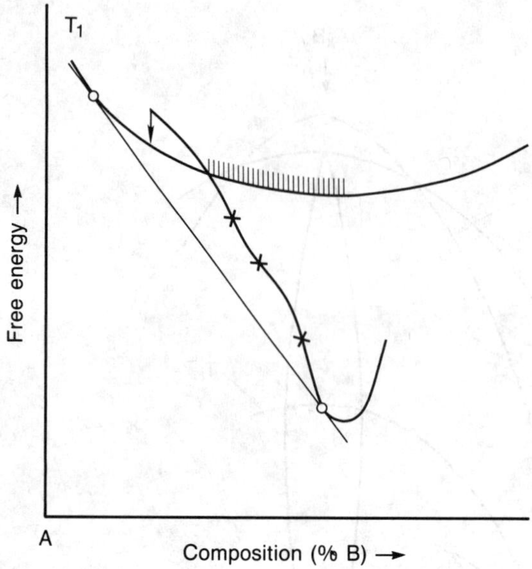

Figure 21. Free energy *versus* composition diagram at T_1 for the phase diagram shown in Figure 20a. Note the two regions of negative curvature appearing in the free energy curve of the ordered phase.

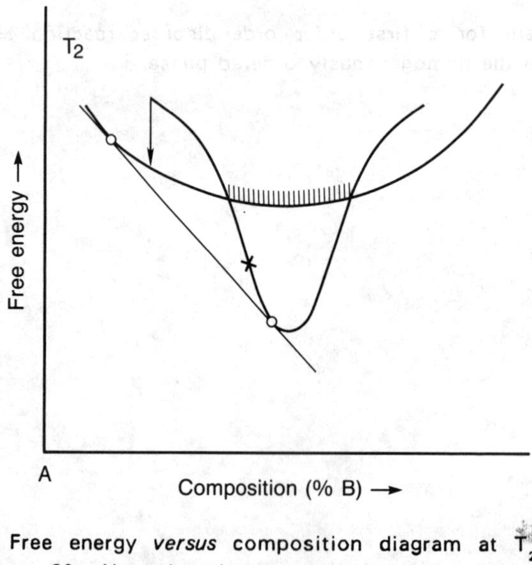

Figure 22. Free energy *versus* composition diagram at T_2 for the phase diagram shown in Figure 20a. Note that the two spinodal regions have merged.

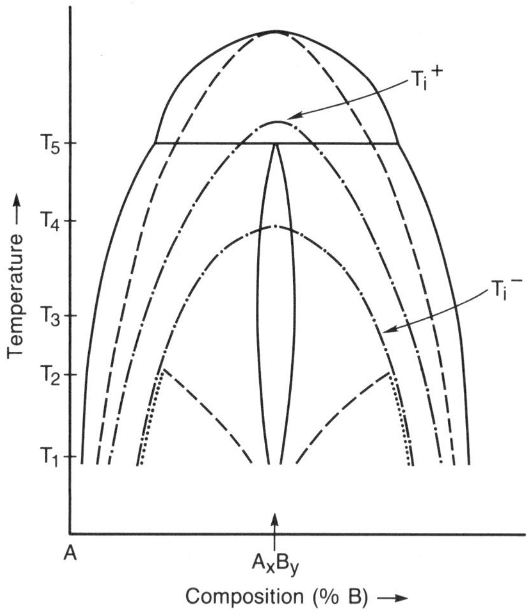

Figure 23a. Schematic phase diagram showing a miscibility gap in the disordered phase at higher temperatures and a first order order/disorder reaction at lower temperatures, combining to produce a *syntectoid* reaction isotherm. The various clustering and ordering instabilities have been included.

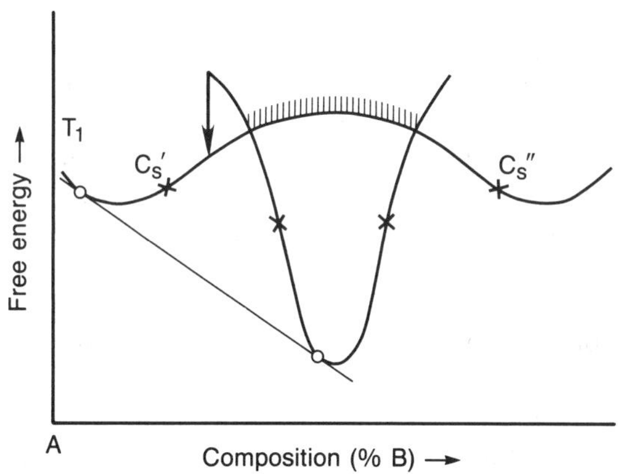

Figure 23b. Free energy *versus* composition diagram at T_1 for the phase diagram shown in Figure 23a. This free energy scheme allows for various reaction paths involving clustering and ordering.

at T_5. The free energy *versus* composition diagram at temperature $T=T_1$ is shown schematically in Figure 23b. Various reaction paths involving clustering and ordering can occur. If the high temperature, disordered phase α of composition just to the right of C_s' is quenched from the single phase region to produce a supersaturated state, it is unstable with respect to phase separation into two disordered phases. The solute-enriched regions will continuously order to γ beyond the point where the cross-hatching begins, and a two-phase mixture consisting of a disordered α phase and an ordered phase A_xB_y will obtain at equilibrium. The supersaturated state inside the hatched region, on the other hand, is unstable with respect to continuous ordering and phase separation; however, short-range atomic exchange will rapidly bring the solution to the ordered state. The ordered solution will spinodally decompose into two ordered phases if it is within the range of negative curvature with the solute-depleted regions spontaneously disordering at the composition where the arrow points downward. Again a two-phase mixture consisting of a disordered and an ordered phase is established at equilibrium. If the ordered phase is outside of the range of negative curvature it is metastable with respect to formation of a second phase and the disordered phase will nucleate within the supersaturated ordered phase.

In Figures 23c and 23d the free energy *versus* composition curves are displayed for the higher temperatures T_2 and T_3. At temperatures below the special temperature T_2, the ordered phase exhibits a region of instability with respect to phase separation. It is interesting to note that at the higher temperature T_3, solutions with compositions on the free energy curve for the disordered phase which are initially thermodynamically unstable with respect to both clustering and ordering can become metastable with respect to clustering *after* undergoing continuous ordering. The continuous ordering reaction will occur faster because of the shorter diffusional distances involved in the process. Thus, a supersaturated solid solution can initially be unstable with respect to both ordering and phase separation, but be rendered metastable with respect to phase separation by a prior ordering process. The metastable ordered state will precipitate a disordered phase by a classical nucleation process.

In Figure 20, the ordered phase was shown to exhibit a miscibility gap which produced a monotectoid configuration. A different monotectoid configuration can arise if the disordered phase exhibits a miscibility gap, as shown in Figure 24. It has been suggested that this configuration occurs in the commercially important Al-Li alloy (Gayle and Vander Sande, 1984).

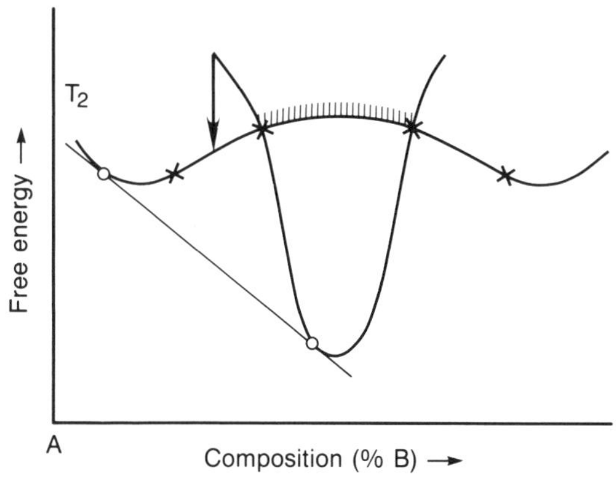

Figure 23c. Free energy *versus* composition diagram at T_2 for the phase diagram shown in Figure 23a. At this special temperature a region of instability with respect to phase separation begins to appear.

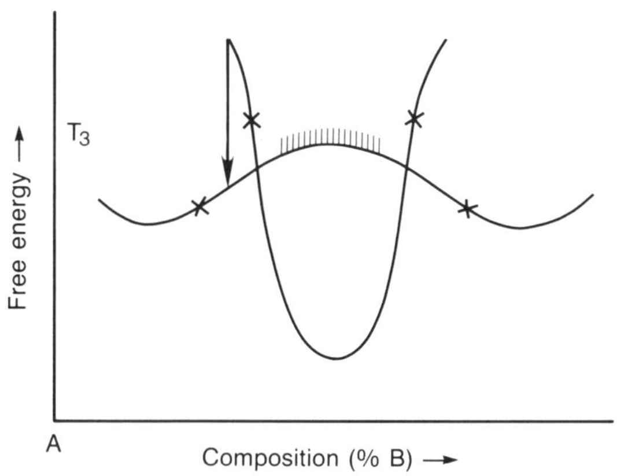

Figure 23d. Free energy *versus* composition diagram at T_3 for the phase diagram shown in Figure 23a. Note the limited range of compositions on the free energy curve of the disordered phase which exhibit ordering instabilities.

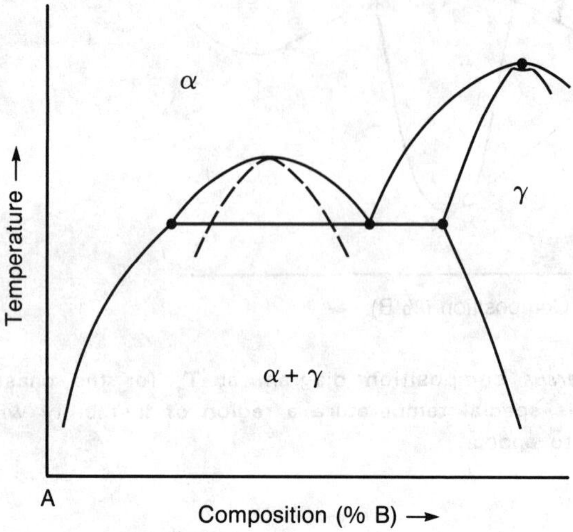

Figure 24. Monotectoid configuration associated with the ordered phase and a miscibility gap in the disordered phase.

4. MAGNETIC ORDERING

4.1 Introduction

The interaction of lines of higher order transformations associated with magnetic and/or atomic ordering with the usual first order boundaries in phase diagrams can produce unusual configurations of heterogeneous equilibria (Meijering, 1963, Miodownik, 1982 and Inden, 1982). Also, these effects can exert a significant influence on the variation of the free energy curve of a given phase with composition and temperature and thus markedly affect the thermodynamic stability of single phase states and the reaction paths leading to multi-phase states accessible to the system (Nishizawa *et al.*, 1979; Lin and Chang, 1987). The magnetic transitions (e.g. ferromagnetic, antiferromagnetic) involving ordering of elementary moments/spins require no mass transport and cannot be suppressed by rapid quenching. Also, magnetic contributions to the free energy can be quite large and thus these higher order transitions are expected to have an unavoidable primary influence on the phase diagram configurations and transformation mechanisms. Higher order/continuous atomic ordering processes which in principle, of course, can produce similar thermodynamic changes (see above) in the phase equilibria require at least short range atomic exchanges and thus can be suppressed by rapid cooling producing unstable or metastable disordered states. Many interesting effects are expected resulting from the simultaneous and synergistic effect of magnetic and atomic order. An important result of the interplay of these various influences on free energies of different phases is the appearance of multi-critical points and unusual critical point and miscibility gap configurations (Allen and Cahn, 1982).

4.2. Miscibility Gaps in Ferromagnetic Binary Systems

A simple but instructive example to use to begin our discussion is the *magnetically induced miscibility gap and tricritical point* appearing in a binary system as a result of the magnetic contributions to the free energy of the solid solution. Consider a binary fcc system which essentially forms an ideal solution with respect to atomic or chemical interactions, but one of the components, A, is ferromagnetic and exhibits a variation of the Curie temperature with concentration of the non-ferromagnetic component as shown in Figure 25. In the absence of magnetic contributions to the binary solution the face-centered cubic components form a complete series of solid solutions over the temperature range of interest and the free energy *versus* composition curve at a given temperature is everywhere concave upward as shown in Figure 26. However, if the influence of the magnetic transition on the solution thermodynamics is included, the free energy curves will show a markedly different behavior with decreasing temperature illustrated schematically in

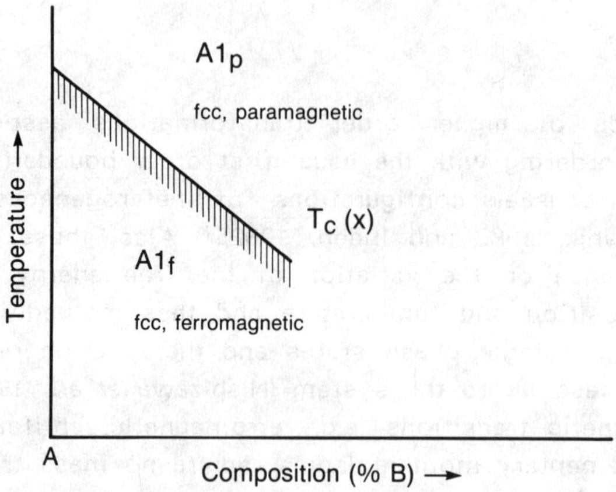

Figure 25. Schematic diagram showing the variation of the Curie temperature with composition in a binary system.

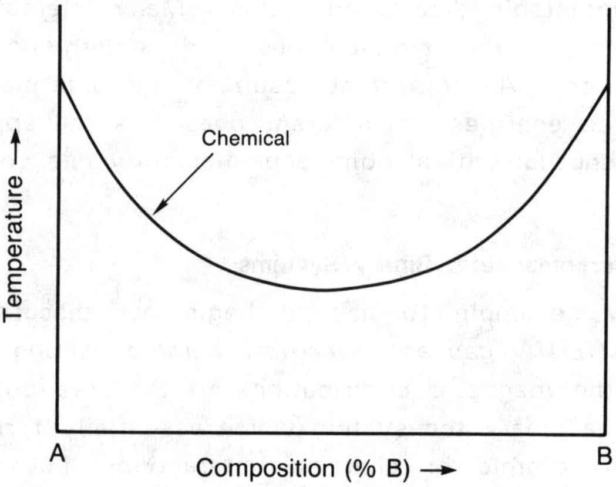

Figure 26. Free energy *versus* composition curve for a binary solution for which the chemical free energy exhibits nearly ideal behavior.

Figure 27. This variation of the free energy of the solutions rich in the ferromagnetic species results in a two-phase region at some critical temperature as shown in Figures 27b and 28. The Curie line ends at a sharp critical point T_c', a so-called *tricritical point.* Below the tricritical point a two phase region appears in which the conjugate phases are a ferromagnetic phase $A1_f$ enriched in the ferromagnetic component and a paramagnetic phase $A1_p$, enriched in the non-magnetic component. It should also be noted that the magnetically induced miscibility gap gives rise to a *spinodal curve* (G"=0) and a limit of metastability separating metastable and unstable single phase states. The extension of the Curie line produces a locus along which the second derivative of the free energy with respect to composition changes sign abruptly from negative to positive with increasing concentration of B - a limit of metastability. However at lower temperatures, a spinodal curve (G"=0) for the paramagnetic/disordered phase appears associated with the free energy *versus* composition scheme shown in Figure 29. (Compare with Figure 12.)

Above the "break away" temperature where the spinodal appears in the disordered phase only magnetically long range ordered single phase states (*ferromagnetic*) can be unstable with respect to phase separation and exhibit spinodal decomposition. However, below the "break away" temperature, single phase short range order paramagnetic states above the Curie temperature, $T_c(x)$, can spinodally decompose.

Figure 30 shows several alloys, and two aging temperatures within the two phase region. Note that if one were monitoring the decomposition behavior of solutions over a composition range within the two phase region of the phase diagram, using a technique which is sensitive to compositional changes as well as magnetic ordering (e.g. Mössbauer spectroscopy, neutron diffraction, *etc.*) the following reaction paths must be recognized and distinguished.

1. A solution of composition X_1 when quenched rapidly to temperature T_A results in the production of a single phase state at point Q (see Figure 30). This phase is magnetic, $A1_f$, and is unstable with respect to the formation of the paramagnetic phase $A1_p$ *via* continuous phase separation; *i.e.*, spinodal decomposition into two *ferromagnetic* phases occurs, the one which is enriched in the non-magnetic component eventually becoming *paramagnetic.*

2. A solution of composition X_2, on the other hand, when quenched to T_A- (i.e. point R) is *metastable* with respect to the formation of the *paramagnetic* phase; the paramagnetic phase will form by a nucleation and growth process, either homogeneously or heterogeneously, depending on the specific energy considerations.

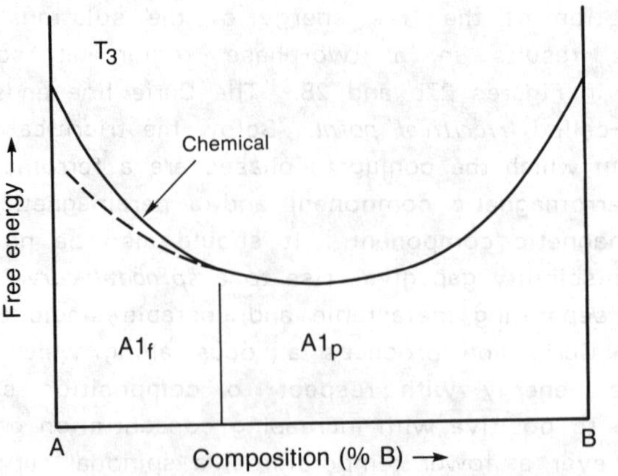

Figure 27a. Free energy *versus* composition curve for a binary solution showing the lowering of the free energy of the ferromagnetic phase over a range of compositions.

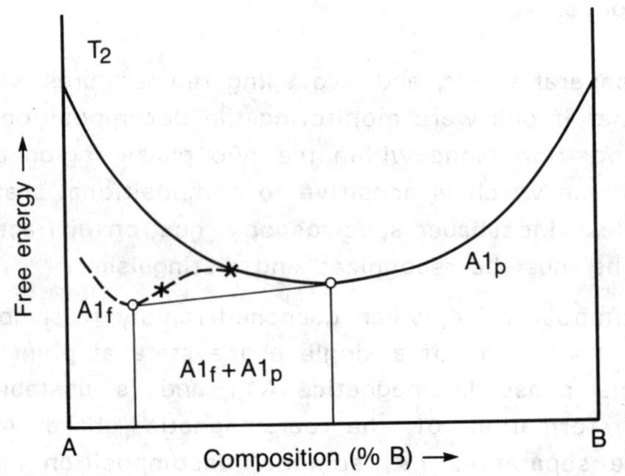

Figure 27b. Free energy *versus* composition curve for a binary solution showing the formation of a two phase region resulting from the magnetic contributions to the solution thermodynamics.

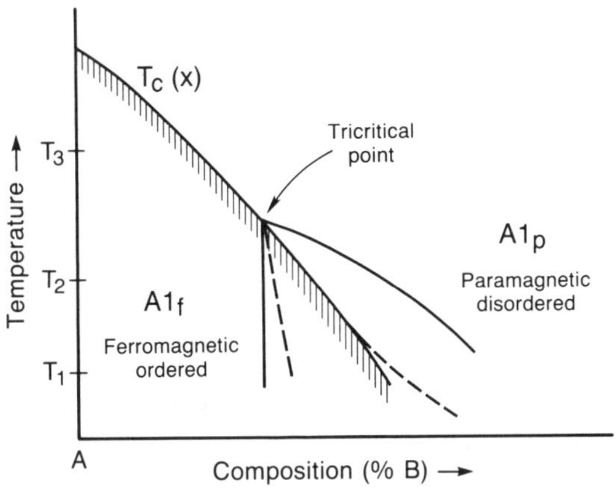

Figure 28. Magnetically induced two phase region and a *tricritical* point. Free energy *versus* composition curves at the temperatures T_2 and T_3 are shown in Figures 27a and 27b.

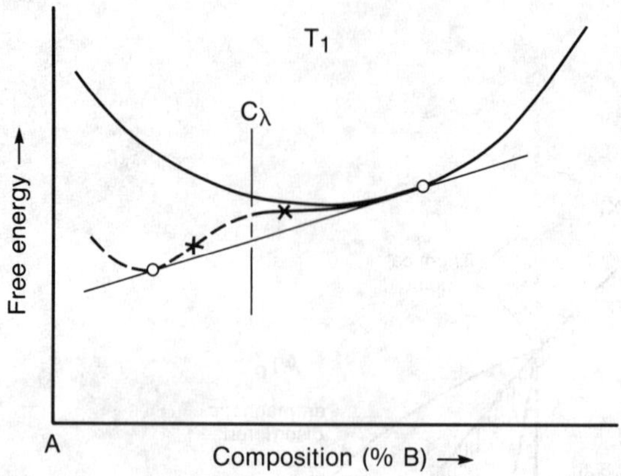

Figure 29. Free energy *versus* composition curve for the phase diagram shown in Figure 28 showing the emergence of a spinodal in the paramagnetic phase.

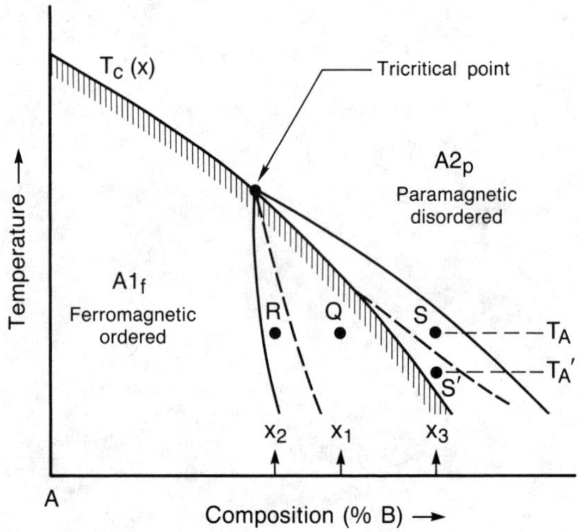

Figure 30. A phase diagram exhibiting a *tricritical* configuration and points Q, R and S representing different initial non-equilibrium states.

It should be noted that in both of these cases, the initial single phase state is *ferromagnetic*, because the spin ordering occurs immediately as the temperature falls below the Curie line ($T_c(x)$).

3. A solution of composition X_3, when quenched to T_A (point S) initially remains paramagnetic, being metastable with respect to the formation of the *ferromagnetic* phase, $A1_f$. At this temperature, the *ferromagnetic* phase can form only by means of nucleation and growth. If, however, the temperature is lowered to T_A' (S'), the single phase $A1_p$ is unstable with respect to continuous phase separation, *viz*, spinodal decomposition. Two *paramagnetic* phases initially form; one is enriched in the nonmagnetic component, B, and the other is enriched in the magnetic component, A. The regions that are enriched with the magnetic component become *ferromagnetic* as their composition crosses the Curie line.

It should be noted that the final state of each of the alloys discussed above is the same: namely a two phase mixture of *paramagnetic* and *ferromagnetic* phases. Only the paths to the equilibrium state varied. Of course, such differences would give rise to different microstructures, and possibly to different physical and mechanical properties of the transformed alloy.

4.3 Further Examples Including Magnetic Transitions

An example of a eutectoid*** diagram exhibiting a *tricritical* point is shown in Figure 31. This configuration seems to be particularly important in the Fe-Ni system and plays a central role in the behavior of the so called INVAR alloys as well as in the microstructural development in meteorites (Lin and Chang, 1987, Reuter, 1987).

Two further examples of the intersection of magnetic transitions with first order phase boundaries are shown in Figures 32 and 33. In Figure 32, the interaction of the magnetic transition with a chemical gap produces a *critical end point* and only a slight distortion of the phase boundary. The metastable extension of the Curie line is shown to produce a *metastable tricritical point* and a significant convolution of the spinodal curve. In Figure 33, the magnetic contribution to the free energy produces a *stable tricritical* point and attendant magnetically induced two-phase region which terminates on a monotectoid reaction isotherm. Again we see that the magnetic transition pulls the chemical spinodal upward producing the contorted regions of instability.

***This diagram would be denoted as a monotectoid if the lowering of the symmetry of the γ_1 phase on magnetic ordering were ignored.

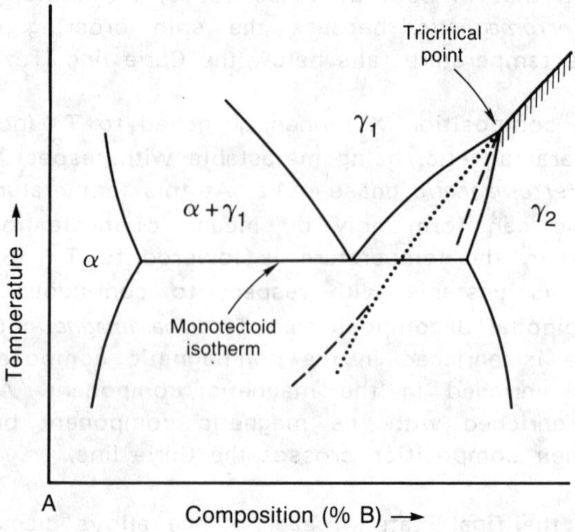

Figure 31. Eutectoid configuration associated with a *tricritical* point in a binary system.

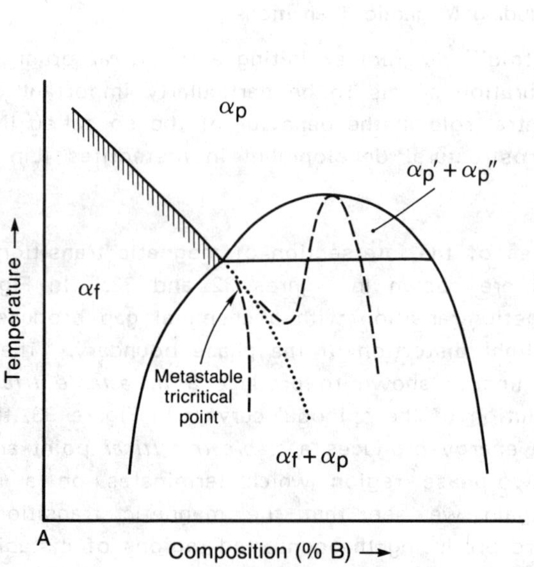

Figure 32. Interaction of a magnetic transition with a chemical miscibility gap producing a critical end point and a metastable *tricritical* point.

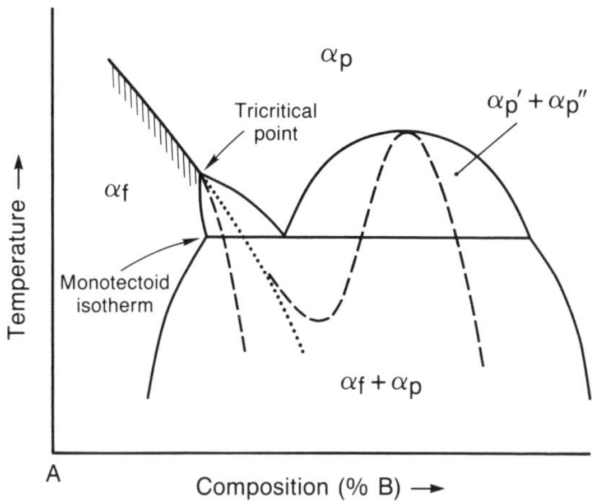

Figure 33. Interaction of a magnetic transition with a chemical miscibility gap producing a stable *tricritical* point and associated monotectoid isotherm.

5. BICRITICAL AND TETRACRITICAL POINTS

5.1 Introduction

The importance of multicritical phenomena in understanding phase equilibria and the transformation behavior of real systems with various ordering tendencies has been recognized over the past decade. (Griffiths, 1970; 1974; Allen and Cahn, 1982; de Fontaine, 1979). Allen and Cahn (1982) have discussed multicritical points in alloys based on a simple Landau-type free energy expansion and have characterized the essential features of bicritical, tricritical, and tetracritical points.

5.2 Bicritical Points

If we consider systems in which both magnetic ordering and atomic ordering ("higher order") can occur, the intersection of the two lines of criticality in a temperature *versus* composition diagram gives rise to a special multicritical point. If the phase which is produced by the combination of *both* ordering processes is *not* stable at low temperatures, the point of intersection is denoted a *bicritical* point. Such a diagram is shown in Figure 34. Here the *bicritical* point can be seen to exist at the top of the two phase region consisting of the different ordered phases. In this example, the paramagnetic bcc phase ($A2_p$) can order magnetically to the ferromagnetic phase $A2_f$ or atomically to the paramagnetic phase $B2_p$. These two phases form the two phase region below the *bicritical* point. The extensions of the ordering instability lines are also denoted in the diagram. A free energy *versus* composition curve which obtains for temperatures below the bicritical point is shown in Figure 35. Here it can be seen that the doubly ordered phase (ferromagnetic $B2_f$) is always unstable with respect to phase separation into two "singly" ordered phases ($A2_f$ and $B2_p$).

An interesting variant of a phase diagram containing a *bicritical* point along with the corresponding free energy *versus* composition curve is shown in Figure 36. Here, the doubly ordered phase becomes stable at low temperatures and a two-phase region between $B2_p$ and $B2_f$ is formed. The extension of the atomic ordering instability emerges from the two phase region producing the isotherm separating the $A2_f$ + $B2_f$ regions. Also, it should be noted that above this isotherm the doubly ordered phase is metastable over a limited temperature or composition range and a spinodal line appears.

An important metallurgical system which apparently exhibits a *bicritical*

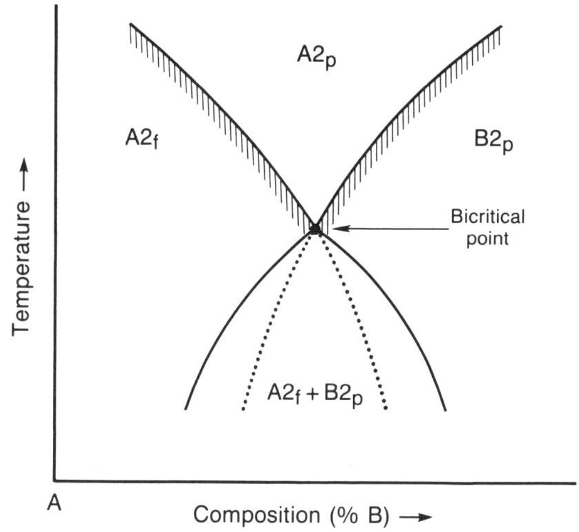

Figure 34. Bicritical point in a binary phase diagram deriving from two ordering tendencies, namely magnetic ordering and atomic ordering. Note that the extensions of the higher order transition lines project into the two phase region.

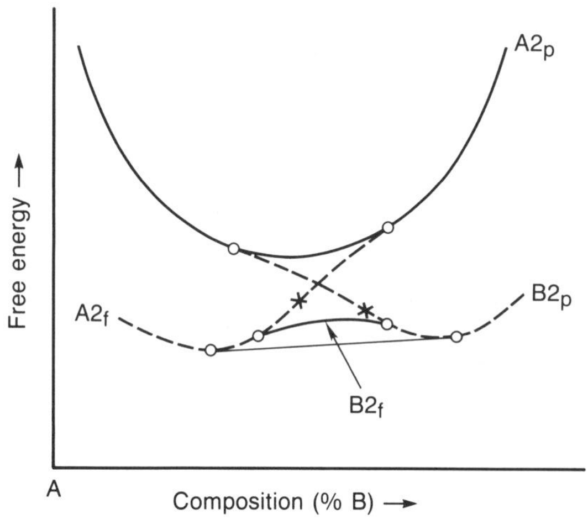

Figure 35. Free energy *versus* composition diagram for a temperature below the *bicritical* point. The common tangent denotes the equilibrium between conjugate phases $A2_f$ and $B2_p$. Note that the doubly ordered phase, $B2_f$, is thermodynamically unstable with respect to phase separation.

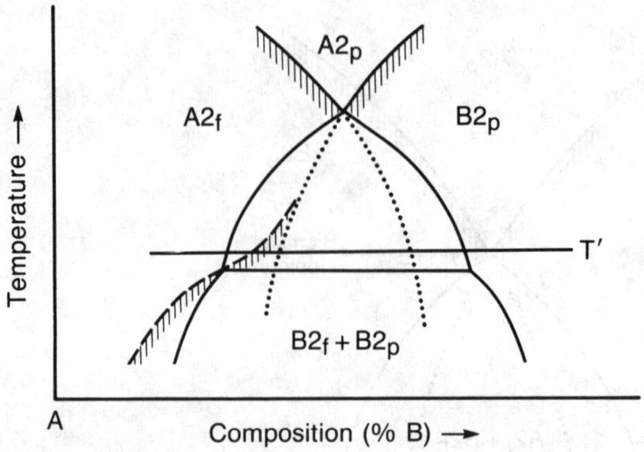

Figure 36a. Bicritical configuration and emergence of the stable doubly ordered phase at low temperature.

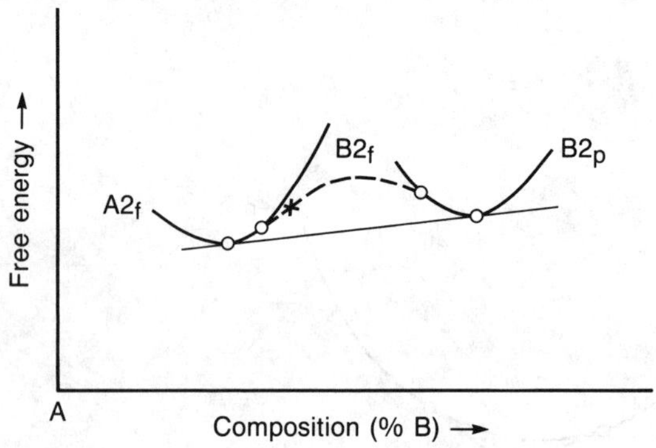

Figure 36b. Free energy *versus* composition diagram corresponding to T = T′. The doubly ordered phase is metastable over a small composition range and exhibits a spinodal point (x) and a limit of metastability.

point and several intersections of higher order transitions and first order boundaries is the Fe-Al binary shown schematically in Figure 37 (Semenovskaya, 1974). The system exhibits several atomically and magnetically ordered phases with corresponding critical end points and appropriate isotherms delineating the boundaries between the various two-phase fields. In such a system continuous phase separation or spinodal decomposition in different regions of temperature and composition is intimately related to the ordering tendencies which prevail. Furthermore, concomitant ordering and clustering phenomena have been revealed experimentally (Allen and Cahn, 1976; Allen, 1977).

5.3 Tetracritical Point

A *tetracritical point* is a special configuration where a "doubly ordered" phase becomes stable and four phases become indistinguishable, having the same free energy. A simple example of such a phase diagram in a binary system is shown in Figure 38a.

The four phases to be distinguished are:

$A2_p$ paramagnetic disordered bcc solid solution

$A2_f$ ferromagnetic, spin-ordered bcc solid solution

$B2_p$ atomically ordered CsCl-type paramagnetic phase

$B2_f$ ferromagnetic, atomically ordered solid solution

Below the *tetracritical* temperature the free energy *versus* composition diagram looks essentially as shown in Figure 38b. At the *tetracritical* temperature the free energy *versus* composition scheme is expected to look like Figure 39, where the free energies of the ordered and disordered phases are identical at a fixed composition. The extensions of the ordering transitions $A2_p \rightarrow A2_f$ and $A2_p \rightarrow B2_p$ may project into the doubly ordered single phase region (Figure 38a) or outside this region; see Figure 40. These two possibilities have been treated quantitatively in Landau formulation by Allen and Cahn (1982).

The Fe-Si system depicted in Figure 41 shows a composite of several of the critical points discussed above, including a *tetracritical* point. The complex ordering and clustering behavior of such a system can be interpreted simply along the lines developed in this overview. Without a systematic understanding of the individual multicritical behavior and convergence of first

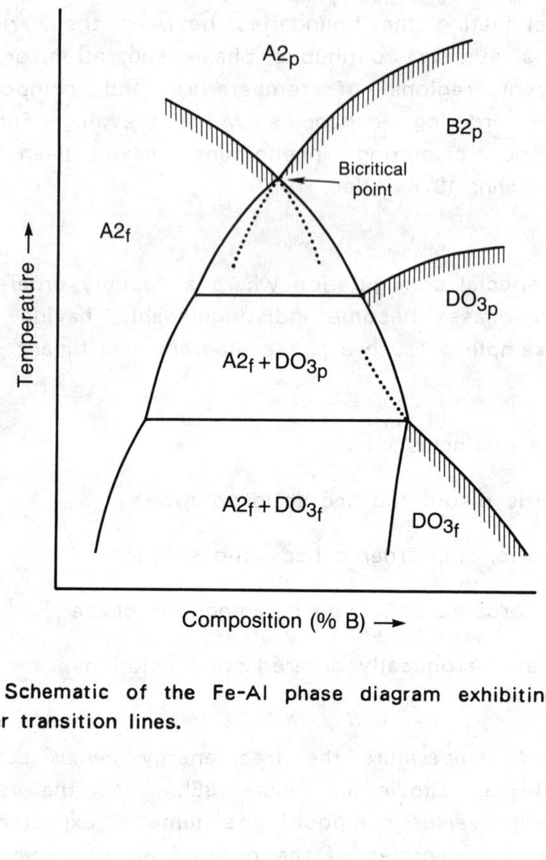

Figure 37. Schematic of the Fe-Al phase diagram exhibiting a bicritical point and higher order transition lines.

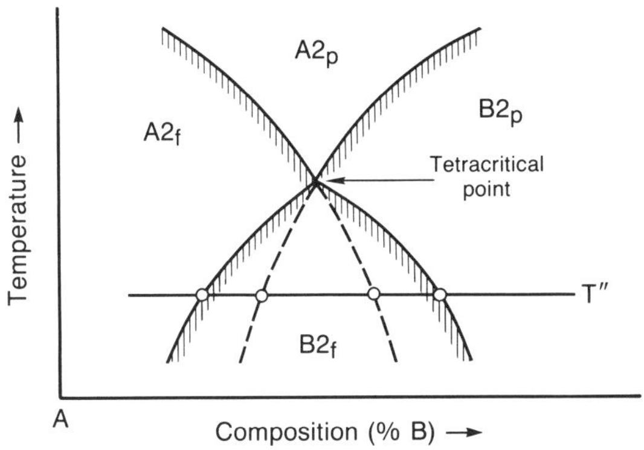

Figure 38a. A schematic of a phase diagram exhibiting a tetracritical point.

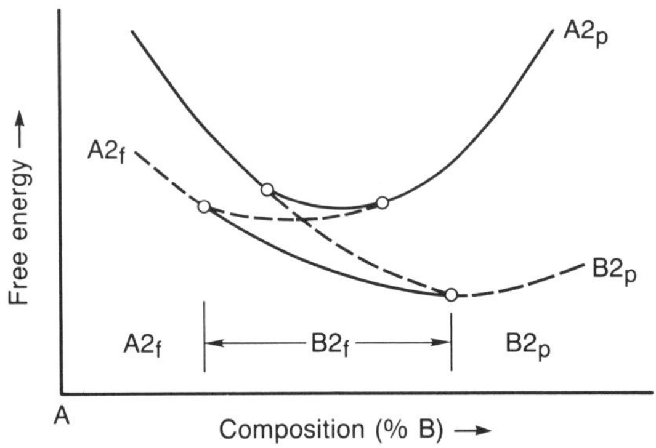

Figure 38b. Free energy *versus* composition curves at T = T″.

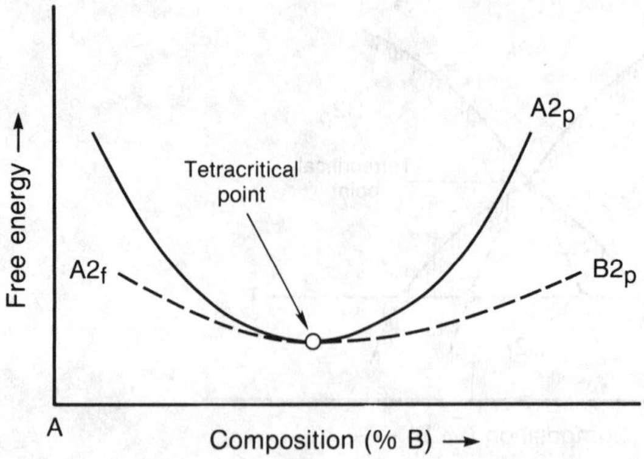

Figure 39. Free energy *versus* composition curves at the tetracritical temperature.

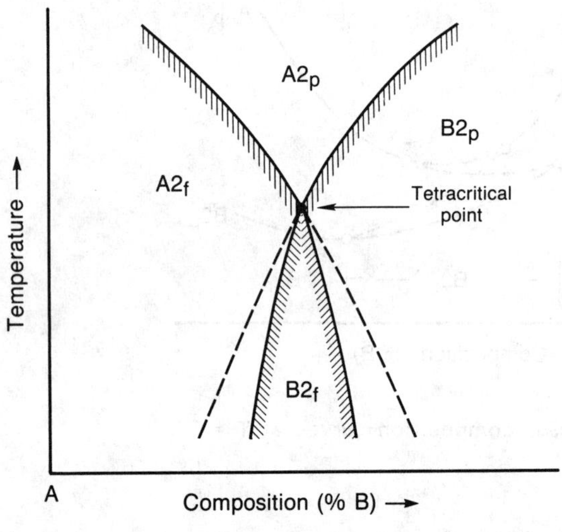

Figure 40. Schematic of a phase diagram of a tetracritical point showing an alternative projection of the lines of higher order transition.

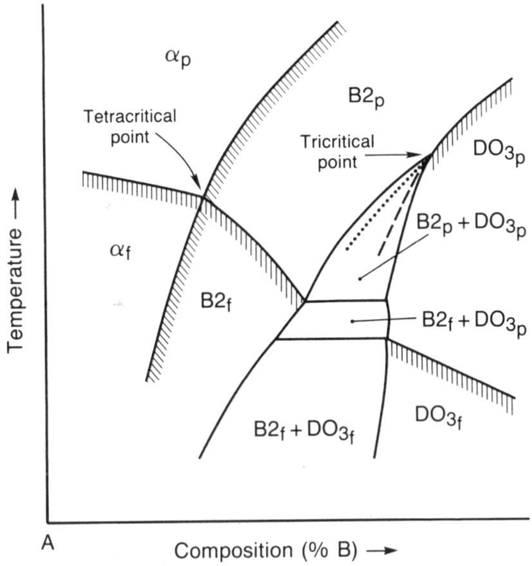

Figure 41. A schematic of the Fe-Si binary diagram exhibiting multicritical points.

order and higher order transitions, the sequences of ordering and phase separation in such a system would appear hopelessly complex.

6. CLOSURE

We have shown that the complex multistage reaction paths exhibited by solid solutions during their approach to equilibrium can be understood largely in terms of the thermodynamic stability of their initial states. The stability of a single phase state is determined by the behavior of the free energy as a function of the relevant order parameters. Continuous transformation of the initial state to a state of lower free energy derives from various thermodynamic instabilities which can be deduced from the phase diagram and its corresponding free energy curves. The simple geometric or graphical thermodynamic approach discussed herein is independent of particular solution models but reveals the essential features of the possible transformation behavior and phase equilibria. In this paper the interplay of atomic ordering, magnetic ordering and exsolution/clustering has been used to illustrate a range of phase transformation behavior that can be understood and systematized using this simple and essentially qualitative approach. Finally, an understanding of the full range of behavior exhibited by real systems with different ordering tendencies is only possible if lines of higher order transition and instability are appropriately included in the phase diagram, along with the usual first order phase boundaries.

ACKNOWLEDGEMENTS

The financial support from the National Science Foundation, Grants DMR-8413115 at Carnegie Mellon University and DMR-8611807 at the University of Pittsburgh is gratefully acknowledged.

REFERENCES

Aaronson, H.I. and K.C. Russell, in *Proceedings of International Conference on Solid→Solid Phase Transformations*, ed. by H.I. Aaronson, D.E. Laughlin, R.F. Sekerka and C.M. Wayman, Warrendale, PA, 371 (1982).

Allen, S.M. and J.W. Cahn, *Acta Met.* **24**, 425 (1976).

Allen, S.M., *Phil. Mag.* **36**, 181 (1977).

Allen, S.M. and J.W. Cahn, *Bulletin of Alloy Phase Diagrams*, **3**(3), 287 (1982).

Binder, K., C. Billotet and P. Mirold, *Z. Physik*, **B 30**, 183 (1978).

Buerger, M.J., *J. Chem. Phys.*, **15**, 1 (1947).

Cahn, J.W. and J.E. Hilliard, *J. Chem. Phys.* **28**, 258 (1958).

Cahn, J.W. and J.E. Hilliard, *J. Chem. Phys.* **31**, 688 (1959).

Cahn, J.W., *Acta Met.* **9**, 795 (1961).

Cahn, J.W., *Acta Met.* **10**, 179 (1962a).

Cahn, J.W., *Acta Met.* **10**, 907 (1962b).

Cahn, J.W. and R.J. Charles, *Phys. Chem. Glasses*, **6**, 181 (1965).

Cahn, J.W., *J. Chem. Phys.*, **42**, 93 (1965).

Cahn, J.W., *Trans. TMS-AIME*, **242**, 166 (1968).

Chou, A., A. Datta, G.H. Meier and W.A. Soffa, *J. Mat. Sci.*, **13**, 541 (1978).

Christian, J.W., *The Theory of Transformations in Metals and Alloys*, Pergamon Press, Oxford (1975).

Datta, A. and W.A. Soffa, *Acta Met.*, **24**, 987 (1976).

de Fontaine, D., *Solid State Physics*, **34**, 73 (1979).

de Fontaine, D., *Trans. of Met. Soc. AIME*, **245**, 1703 (1969).

de Fontaine, D. in *Ultrafine-Grain Metals*, ed. J.J. Burke and V. Weiss, Syracuse U. Press, Syracuse, N.Y., 93 (1970).

de Fontaine, D., *Acta Met.*, **23**, 553 (1975).

Gayle, F.W. and J.B. Vander Sande, *Bulletin of Alloy Phase Diagrams*, **5**(1), 19 (1984).

Griffiths, R.B., *Phys. Rev. Lett.*, **24**, 715 (1970).

Griffiths, R.B., *J. Chem. Phys.*, **60**, 195 (1974).

Hillert, M., Sc.D. Thesis, Massachusetts Institute of Technology (1956).

Hilliard, J., in *Phase Transformations*, ASM, 497 (1970).

Inden, G., *Bulletin of Alloy Phase Diagrams*, **2**(4), 412 (1982).

Ino, H., *Acta Met*. **26**, 827 (1978).

Khachaturyan, A.G., *Progress in Mat. Sci*., **22**, 1 (1978).

Khachaturyan, A.G., *Theory of Structural Transformations In Solids*, John Wiley and Sons (1983).

Khachaturyan, A.G., T.F. Lindsey and J.W. Morris, *Met. Trans.* in press (1987).

Kubo, H. and C.M. Wayman, *Acta Met*., **80**, 395 (1980).

Kulkarni, U.D., S. Banerjee and R. Krishnan, *Materials Science Forum* **3**, 111 (1985).

Langer, J.S., in *Fluctuations, Instabilities and Phase Transitions*, ed. by T. Riste, p. 19, Plenum Press, N.Y. (1975).

Lin, J-C. and Y.A. Chang, *Met. Trans*., in press (1988).

Martin, G., in *Solid State Phase Transformations in Metals and Alloys, Les Editions de Physique*, Orsay, France, 337 (1979).

Meijering, J.L., *Philips Res. Reports*, **18**, 138 (1963).

Miodownik, A.P., *Bulletin of Alloy Phase Diagrams* **2**(4), 406 (1982).

Nishizawa, T., M. Hasebe and M. Ko, *Acta Met*. **27**, 817 (1979).

Reuter, K.B., Ph.D. Thesis, Lehigh University (1987).

Semenovskaya, S.V., *Phys. Status Solidi*, **64**(b), 627 (1974).

Soffa, W.A. and D.E. Laughlin, in *Proceedings of an International Conference on Solid→Solid Phase Transformations*, ed. by H.I. Aaronson, D.E. Laughlin, R.F. Sekerka and C.M. Wayman, Warrendale, PA, 159 (1982).

Tiapkin, Y.D., *Ann. Rev. Mater. Sci*. **7**, 209 (1977).

THERMOCHEMISTRY OF ALUMINIUM/SILICON ORDERING IN FELDSPAR MINERALS

M. A. Carpenter
Department of Earth Sciences
University of Cambridge
Downing Street
Cambridge CB2 3EQ
England

ABSTRACT. Landau theory provides the basis for a unified thermodynamic treatment of structural phase transitions in minerals. It has recently been used, with effect, to describe the excess thermodynamic properties due to displacive and Aℓ/Si ordering transitions in feldspars by Salje and coworkers. Semi-flexible T-O-T bonds (T = Aℓ, Si) in the framework of the feldspar structure allow the influence of local site configurations to extend over large correlation lengths. As a consequence, the transition behaviour of natural feldspars can be described for a wide temperature range using simple Landau expansions. Two order parameters are frequently required to describe the individual contributions of atomic displacements and Aℓ/Si ordering effects; coupling between them occurs via the spontaneous strain and may be effectively bilinear or biquadratic. These concepts, which are relatively new to earth sciences, are introduced, and their application to the thermodynamic analysis of alkali feldspar and plagioclase feldspar solid solutions is outlined. Emphasis is placed on data obtained by calorimetric methods.

1. INTRODUCTION

In laboratory experiments it is possible to subject minerals to carefully controlled pressures, temperatures, stresses, etc., and to determine the time scales and physical conditions required to induce structural changes or chemical reactions. The same minerals may be found in nature but, almost invariably, they will have experienced long and relatively complex thermal histories. One of the primary objectives of recent research within the earth sciences has been to use formal theories of kinetics and equilibrium thermodynamics to bridge the divide between experience from such experiments and the actual behaviour of minerals and rocks in nature.

The response of any assemblage of minerals to changes in pressure and temperature will be to attempt to re-equilibrate both chemically and structurally. Textural evidence for the nature and direction of the re-equilibration may sometimes be preserved in a rock, either as reaction zones between crystals (Fig. 1a) or as characteristic microstructures within the crystals (Fig. 1b). The two categories of response, inter-

E. K. H. Salje (ed.), Physical Properties and Thermodynamic Behaviour of Minerals, 265–323.

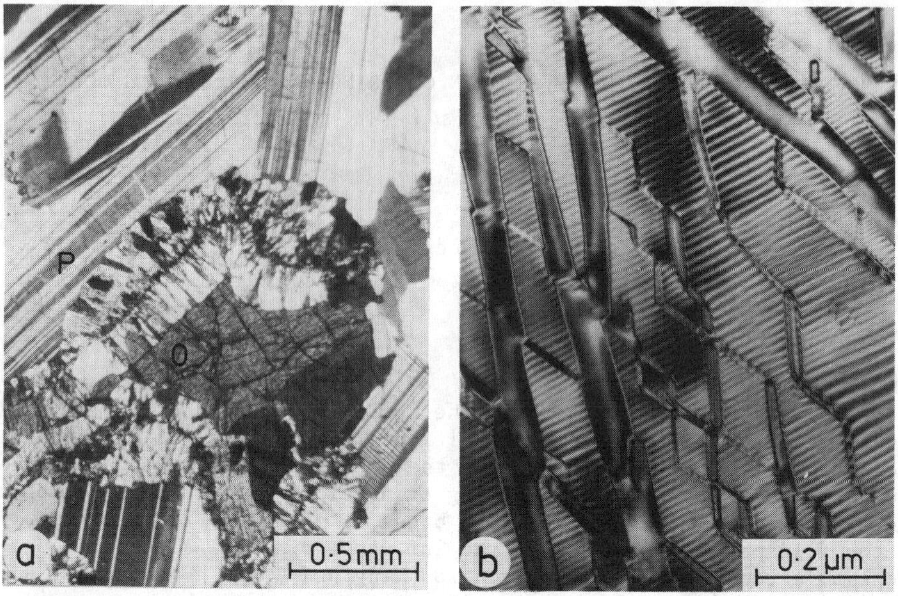

Figure 1. (a) Transmitted light optical micrograph of a thin section from a rock showing reaction rims of fibrous hypersthene [$(Mg, Fe)_2Si_2O_6$] and diopside + spinel [$Ca(Mg, Fe)Si_2O_6$ + $(Mg, Fe)Al_2O_4$] between crystals of olivine [$(Mg, Fe)_2SiO_4$], labelled O, and twinned plagioclase feldspar, labelled P. The reactions occurred in response to changes in pressure and temperature during the evolution of the rock. (Sample from Risor, Norway). (b) Bright field transmission electron micrograph of a two phase intergrowth within alkali feldspar from Klokken, Greenland. The $NaAlSi_3O_8$ rich component has fine lamellar twinning and the $KAlSi_3O_8$ rich component has twin planes at the intersections between oblique lamellae. The microstructure is the result of exsolution combined with displacive and order/disorder transitions during cooling from high temperatures in nature. (From Parsons and Brown, 1984; reproduced by kind permission of I. Parsons).

mineral (chemical reactions) and intra-mineral (exsolution, cation ordering, polymorphic changes, etc.), are associated with energy effects of comparable magnitudes. For example, the reaction:

$$\underset{\text{(jadeite)}}{NaAlSi_2O_6} + \underset{\text{(quartz)}}{SiO_2} \rightleftarrows \underset{\text{(albite)}}{NaAlSi_3O_8}$$

has a standard enthalpy and entropy change (at 298 K, 1 bar) of ~12 kJ mole^{-1} and ~45 J mole^{-1} K^{-1} respectively (Holland and Powell, 1985), while the excess enthalpy and entropy associated with a phase transition in albite can be as much as ~14 kJ mole^{-1} and ~17 J mole^{-1} K^{-1} (in the temperature range 298 - 1290 K; Salje et al., 1985a). It follows that the thermodynamic contributions of intramineral structural changes can be significant in determining the equilibrium state of a rock and the manner in which it evolves. Aside from this essentially geological motivation for studying processes within minerals, however, naturally occurring crystalline materials provide model systems for testing general theories of structural phase transitions.

There are many types of phase transitions which can occur in minerals. One of the more important of these, in terms of the energy involved and the frequency of its occurrence, involves Aℓ/Si order/disorder. The major elements of the earth's crust are essentially Aℓ,Si and O, with lesser proportions of Na, Ca, Fe, Mg, K and H, and they are mainly held in silicates. Typically, alkalis and Fe or Mg occupy sites between $A\ell O_4$ and SiO_4 tetrahedra. Ordering of Aℓ and Si among the tetrahedral sites is known to be responsible for substantially stabilising some of the structures. In the case of cordierite ($Mg_2A\ell_4Si_5O_{18}$), more than 60% of the enthalpy of formation from the component oxides is due to Aℓ/Si ordering (Carpenter et al., 1983). In the past, however, study of this group of phase transitions has been hampered by the problem of distinguishing Aℓ from Si using conventional diffraction techniques. The X-ray scattering factors of Aℓ and Si are similar, and neutron scattering experiments require crystals of a size and perfection that is not always available. Recent developments in the theory of structural phase transitions, based on the approach of Landau, have largely circumvented these difficulties and allow much improved quantitative descriptions of phase transitions in silicates. Detailed structure models can also now be tested by the use of magic angle sample spinning NMR spectroscopy (MASNMR) as a direct method of determining local Aℓ/Si distributions (Smith, K. A., et al., 1983; Fyfe et al., 1983; Oldfield and Kirkpatrick, 1985; Kirkpatrick et al., 1985; Carpenter, 1986a; Putnis, this volume).

Landau theory applies on a macroscopic level and relates many of the physical properties of crystals through the order parameter, Q (see, for example, Landau and Lifshitz, 1958; Rao and Rao, 1978; Parsonage and Stavely, 1978; Bruce and Cowley, 1981; Salje, this volume). Salje and coworkers (Salje, 1985, 1987a, b; Salje et al., 1985a, b; Redfern and Salje, 1987a; Putnis et al., 1987) have shown how these relationships and their temperature dependences can be expressed formally using parameters which are relatively simple to measure for silicates. Which properties are important in each case depends on the specific character of the phase transition under investigation but, for Aℓ/Si ordering, the most useful properties appear to be spontaneous strain, determined from lattice parameters, heat capacity, determined by differential scanning calorimetry, and enthalpy, determined by solution calorimetry. The purpose of this chapter is to provide an introduction to the overall approach. In the first section, some of the general properties of Aℓ/Si ordering transitions are summarised. Appropriate Landau formalism is then introduced and the relationships involving enthalpy, in particular, are discussed. The final section contains the results of recent thermochemical studies on feldspar minerals. Alkali feldspars ($NaAlSi_3O_8$ - $KAlSi_3O_8$) and plagioclase feldspars ($NaAlSi_3O_8$ - $CaAl_2Si_2O_8$) show a remarkable diversity of structural phase transitions. Discrete Al/Si ordering transitions, displacive transitions and phase segregation all occur, but, in addition, there are mixed transitions which involve aspects of all three of these processes.

2. SOME GENERAL FEATURES OF Al/Si ORDERING TRANSITIONS IN MINERALS

Mineralogists are brought up with the idea that, wherever possible, Al^{3+} ions will avoid occupying neighbouring tetrahedral sites in silicates. This empirical observation is usually referred to as "Loewenstein's rule" (after Loewenstein, 1954) or as the "Al-avoidance rule". Ordered distributions of Al and Si that are consistent with Al-avoidance can be strongly favoured energetically, as reflected by the high order/disorder transition temperatures of minerals with Al:Si close to 1:1. For anorthite ($CaAl_2Si_2O_8$), the equilibrium ordering transition temperature is estimated to be $\sim 2040^\circ C$, well above the melting point (see below). For cordierite, the transition temperature is $\sim 1450^\circ C$ (Smart and Glasser, 1977; Putnis et al., 1987), and for sillimanite (Al_2SiO_5) it is above $\sim 1700^\circ C$ (Greenwood, 1972; Holland and Carpenter, 1986). However, neither gehlenite ($Ca_2Al_2SiO_7$) (Louisnathan, 1971) nor calcium tschermak's pyroxene ($CaAl_2SiO_6$) (Okamura et al., 1974; Grove and Burnham, 1974; Cohen and Burnham, 1985) appear to develop ordered superstructures so that Al-avoidance should not be taken as a hard and fast rule (see Smith, 1974, p.79).

A second general feature of Al/Si ordering familiar to earth scientists is the slowness of Al^{3+} and Si^{4+} diffusion in silicates. To achieve substantial changes of the degree of Al/Si order in feldspars by heating them in air at one atmosphere, annealing times on the order of hours or days are required at $\sim 1300^\circ C$, days or weeks at $\sim 1200^\circ C$ and months at $\sim 1000^\circ C$ (McKie and McConnell, 1963; Kroll and Müller, 1980; Carpenter and McConnell, 1984). Typical activation energies are in the range 250 - 500 kJ $mole^{-1}$ (Freer, 1981). The presence of water can have a dramatic catalytic effect, however, and, under hydrothermal conditions, ordering can be achieved on a sensible laboratory time scale down to temperatures of $\sim 700^\circ C$ (Yund and Tullis, 1980; Goldsmith and Jenkins, 1985; Carpenter, 1986b; and references therein). High pressures ($P > 10$ kb) further enhance reaction rates when water is present (Goldsmith and Newton, 1974; Yund and Tullis, 1980; Yund, 1986) and when a carbonate flux is present (Goldsmith and Jenkins, 1985). Goldsmith (1986, 1987) has suggested that the catalytic agent at high pressures is actually hydrogen. These kinetic constraints often preclude the synthesis of crystals with equilibrium structural states at low temperatures, but they do allow an almost infinite variety of metastable states to be produced. The metastable structures may persist indefinitely during subsequent experiments, thereby opening up a whole range of non-equilibrium phenomena to investigation.

Many framework silicates also undergo both displacive and Aℓ/Si ordering transitions (see reviews by Ribbe, 1984; Merlino, 1984; Henderson, 1984). Albite ($NaA\ell Si_3O_8$), for example, has a high temperature displacive transition and undergoes Aℓ/Si ordering at lower temperatures, while anorthite is ordered at high temperatures but has a displacive transition at a lower temperature. These order/disorder and displacive transitions are typically coupled and may exert a profound influence on each other (Salje, 1985, 1987a, b; Salje et al., 1985a; Salje and Devarajan, 1986; Redfern and Salje, 1987a). Significant coupling may occur even when the critical temperatures of two transitions in a single crystal are hundreds, if not thousands of degrees apart. There is an extreme contrast between the rate of response to changes in temperature of these processes. Aℓ/Si ordering occurs at a slow to negligible rate while atomic displacements are effectively instantaneous and the two effects are therefore easy to distinguish in experimental studies.

As will be discussed at greater length below, the thermodynamics of phase transitions in aluminosilicate systems appear to conform to the predictions of Landau theory over a remarkably wide range of temperatures, including the region within a few degrees of the equilibrium transition temperatures. The temperature interval over which deviations from mean field behaviour are expected to occur as $T \to T_c$ may be very small if the correlation length is large (Als-Nielson and Birgeneau, 1977). One mechanism by which one tetrahedral cation in a silicate could influence the occupancy of a tetrahedral site some distance away might be due to large electrostatic interactions, since both Aℓ and Si have large nominal charges. A more effective means of transmitting information over large correlation lengths, however, would be via strain fields associated with the local cation configurations. It is known from observations of twin boundary morphologies in ferroelastic materials that such strain fields can be extensive (Salje et al., 1985b). In the case of alkali feldspars, comparison of idealised morphologies (e.g. Salje et al., 1985b) with observed morphologies (e.g. Fitzgerald and McLaren, 1982; McLaren, 1984) suggests that the strain fields extend to at least ∿0.1μm. Cowley (1976), Folk et al. (1976), Als-Nielsen and Birgeneau (1977) and Salje et al. (1987) have argued that mean field behaviour should also predominate in systems with strong coupling between the order parameter and the elastic strain. If these arguments hold for transitions in framework silicates, relatively simple Landau expansions should provide adequate descriptions of their excess thermodynamic properties.

Underlying these general features of Aℓ/Si ordering processes, of course, is the nature of the Aℓ-O and Si-O chemical bonds. Recent ab initio molecular orbital calculations on molecules such as $H_6Si_2O_7$, $H_6A\ell_2O_7^{2-}$ and $H_6A\ell SiO_7^{1-}$, selected to mimic fragments of crystal structures, show that Aℓ-O-Si linkages are substantially favoured energetically, relative to $\frac{1}{2}$(Aℓ-O-Aℓ) + $\frac{1}{2}$(Si-O-Si) (de Jong and Brown, 1980; Gibbs et al., 1981; Geisinger et al., 1985; Navrotsky et al., 1985; Lasaga and Gibbs, 1987). Navrotsky et al. (1985) have suggested that this stabilisation might be understood in terms of the polarisation of the bridging oxygen in the asymmetric linkage, which results in less competition for oxygen between the tetrahedral cations. The calculations also show that variations in T-O bond lengths give much larger changes in energy than variations in T-O-T bond angles (Fig. 2a) (Geisinger et al., 1985, and references therein). The T-O-T bond angles are constrained by only rather shallow potential wells (Fig. 2b) and, as a result, framework silicate structures are likely to behave as if made up of rigid SiO_4 and $A\ell O_4$ tetrahedra with more or less flexible linkages between the tetrahedra. Strain effects would then be transmitted by flexing of the T-O-T

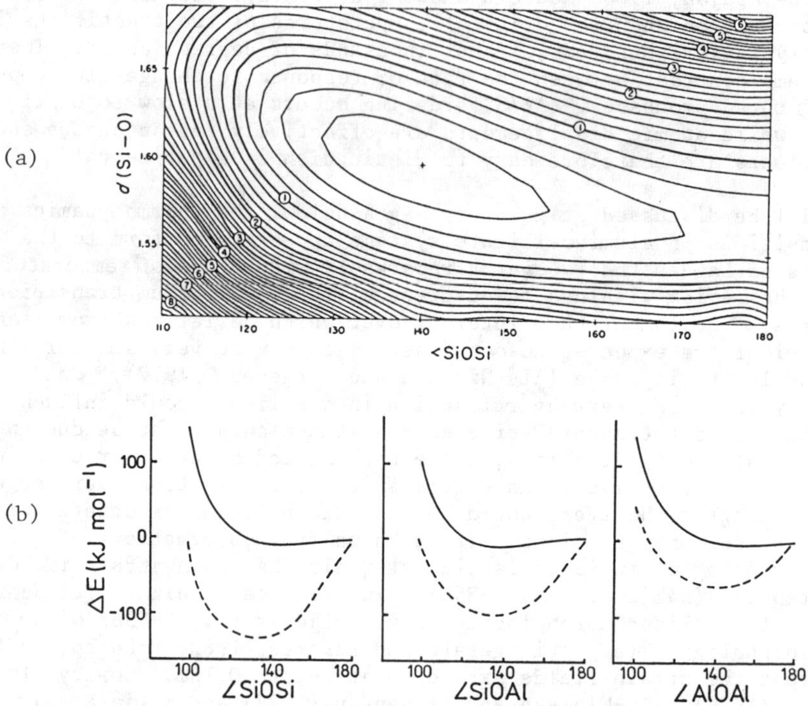

Figure 2. (a) Calculated potential energy surface for $H_6Si_2O_7$ plotted as a function of the bridging oxygen bond length, d(Si-O), in Å, and the SiOSi bond angle, ∠SiOSi, in degrees. Contours represent energy increments of 2.6 kJ $mole^{-1}$ relative to a minimum energy point (-2.9×10^6 kJ $mole^{-1}$) denoted by the cross. 13 kJ $mole^{-1}$ increments are indicated by heavier lines labelled 1 to 8 in order of increasing energy. (From Gibbs et al., 1981; reproduced by kind permission of G. V. Gibbs). (b) Calculated potential energy for $H_6Si_2O_7$, $H_6A\ell SiO_7^{1-}$ and $H_6A\ell_2O_7^{2-}$ molecules as a function of, respectively, SiOSi, SiOAℓ and AℓOAℓ bond angles (solid lines). Dashed lines indicate the same variation, but with an additional proton coordinated to the bridging oxygen of the molecules (after Geisinger et al., 1985).

bonds. The effect of increasing the coordination of bridging oxygens, by Na^+, K^+ or Ca^{2+} in feldspars, for example, is to reduce the range of energetically feasible T-O-T angles (Geisinger et al., 1985), thereby reducing the flexibility of the structure and increasing the likely extent of strain fields.

3. MACROSCOPIC THERMODYNAMIC PROPERTIES

Excess entropies and enthalpies due to displacive transitions in framework silicates can be determined from the excess heat capacity as measured by adiabatic calorimetry (between ∿5 K and 300 K), or by differential scanning calorimetry (up to ∿1,000 K). Aℓ/Si ordering, however, occurs too slowly to be measurable in situ in a calorimeter and the properties of crystals equilibrated at different temperatures have to be measured using static techniques. The configurational entropy can be particularly awkward to determine because, aside from the experimental problem of distinguishing Aℓ from Si, there is the difficulty of assigning entropies to structural states which possess local variations or fluctuations in the degree of order. Enthalpy variations, on the other hand, can be followed by solution calorimetry, though not to the same precision as the excess enthalpies of displacive transitions from scanning calorimetry.

It is convenient to define excess thermodynamic properties associated with displacive and order/disorder processes* in terms of macroscopic order parameters, Q and Q_{od} respectively, which vary between one for the low symmetry phase at 0 K, and zero for the high symmetry phase. Formal relationships between the order parameters and free energy, entropy and enthalpy are given by Landau theory, and the order parameters themselves can be determined through their dependence on related properties, such as spontaneous strain (ε). Lattice parameters of crystals with different states of order may be measured by conventional methods and used to determine the spontaneous strain, for example. In the case of a zone centre phase transition, the order parameter varies linearly with ε, and, since the excess entropy varies with Q^2, the problem of determining configurational entropies directly is avoided. Similarly, the intensity of superlattice reflections is proportional to Q^2, providing an alternative method of following entropy changes.

The basis of this macroscopic approach has only relatively recently been introduced into the mineralogical literature (McConnell, 1978, 1985; Salje, 1985, 1987a, b; Salje et al., 1985a; Carpenter, 1985; Redfern and Salje, 1987a; Putnis et al., 1987) and it is therefore outlined in this section.

---

* The terms displacive and order/disorder are used in the mineralogical literature to distinguish between transitions which involve only small displacements of atoms and transitions which involve ordering of cations between discrete crystallographic sites. This is a somewhat looser distinction than is commonly used in the physics literature.

3.1 Landau theory

As already mentioned, Landau theory seems to describe phase transitions in framework silicates rather well. In the feldspars, the transitions tend to be second order, tricritical or just first order. Because of the low symmetries involved (monoclinic or triclinic), the relevant order parameters usually contain only one component. The required Landau potentials therefore have the form:

$$G = \frac{1}{2}a(T - T_c)Q^2 + \frac{1}{4}bQ^4 + \frac{1}{6}cQ^6 \quad (1)$$

where G is the excess free energy due to the transition, a, b, and c are coefficients which do not depend explicitly on temperature, T, and T_c is a critical temperature. (Transitions which, by symmetry, have odd order terms in the expansion are not considered in this chapter).

The a coefficient is always positive. For a second order transition, b is positive and the Q^6 term is dropped. For a tricritical transition, b = 0 and the Q^6 term is needed for stability, so that c must be positive. In both cases, the equilibrium transition temperature is given by T_c. When b becomes negative, the transition becomes first order and the Q^6 term is again needed for stability (c positive). The equilibrium transition temperature then occurs at a temperature T_{tr} which is higher than T_c. Solutions for these three alternatives are obtained by requiring the equilibrium condition, dG/dQ = 0, and setting Q = 1 at 0 K. For the second order and tricritical cases, Q goes continuously to zero at T_c while, in the first order case, there is a jump from Q = 0 to Q = Q_o at T_{tr}. The appropriate solutions are as follows:

Second order:

$$G = \frac{1}{2}a(T - T_c)Q^2 + \frac{1}{4}bQ^4 \quad (2)$$

$$T_c = \frac{b}{a} \quad (3)$$

$$Q = \left[\frac{a}{b}(T_c - T)\right]^{\frac{1}{2}} = \left[1 - \frac{T}{T_c}\right]^{\frac{1}{2}} \quad \text{for } T < T_c \quad (4)$$

$$S = -\frac{dG}{dT} = -\frac{1}{2}aQ^2 \quad (5)$$

$$H = G + TS = -\frac{1}{2}aT_cQ^2 + \frac{1}{4}bQ^4 \quad (6)$$

$$C_p = \frac{a^2}{2b} \cdot T \quad \text{for } T \leqslant T_c \quad (7)$$

Note that G, H, S and C_p are defined here as the excess free energy, excess enthalpy, excess entropy and excess heat capacity (at constant pressure) due to the transition.

Tricritical:

$$G = \frac{1}{2}a(T - T_c)Q^2 + \frac{1}{6}cQ^6 \quad (8)$$

$$T_c = \frac{c}{a} \quad (9)$$

$$Q = \left[\frac{a}{c}(T_c - T)\right]^{\frac{1}{4}} = \left[1 - \frac{T}{T_c}\right]^{\frac{1}{4}} \quad \text{for } T < T_c \quad (10)$$

$$S = -\frac{1}{2}aQ^2 \quad (11)$$

$$H = -\frac{1}{2}aT_cQ^2 + \frac{1}{6}cQ^6 \quad (12)$$

$$C_p = \frac{aT}{4\sqrt{T_c}}(T_c - T)^{-\frac{1}{2}} \quad \text{for } T < T_c \quad (13)$$

First order: (for transitions which, by symmetry, cannot have odd order terms in the free energy expansion in terms of Q).

$$G = \frac{1}{2}a(T - T_c)Q^2 + \frac{1}{4}bQ^4 + \frac{1}{6}cQ^6 \quad (14)$$

The equilibrium transition temperature, T_{tr}, is given by:

$$T_{tr} = T_c + \frac{3}{16} \cdot \frac{b^2}{ac} \quad (15)$$

A jump in the order parameter occurs at T_{tr}, from $Q = 0$ to $Q = Q_o$, where:

$$Q_o = \pm\left[-\frac{4a(T_{tr} - T_c)}{b}\right]^{\frac{1}{2}} \quad (16)$$

Below T_{tr}:

$$Q^2 = \frac{2}{3}Q_o^2\left\{1 + \left[1 - \frac{3}{4}\left(\frac{T - T_c}{T_{tr} - T_c}\right)\right]^{\frac{1}{2}}\right\} \quad (17)$$

$$S = -\frac{2}{3}\frac{L}{T_{tr}}\left\{1 + \left[1 - \frac{3}{4}\left(\frac{T - T_c}{T_{tr} - T_c}\right)\right]^{\frac{1}{2}}\right\} \quad (18)$$

$$H = -\frac{1}{2}aT_cQ^2 + \frac{1}{4}bQ^4 + \frac{1}{6}cQ^6 \quad (19)$$

The latent heat of the transformation, L, is given by:

$$L = \frac{1}{2}aQ_o^2\, T_{tr} \quad (20)$$

Solutions for Q as a function of T/T_c are shown in Figure 3. The first order example has been calculated for a relatively small jump in Q at T_{tr} ($Q_o = 0.4$) and a small temperature difference between T_{tr} and T_c ($T_c = 0.99\ T_{tr}$). In the second order and tricritical cases, because Q goes continuously to zero, G, H and S must also go continuously to zero. For the first order case, there is a latent heat and latent entropy, and there is a discontinuity in the enthalpy and entropy at $T = T_{tr}$.

The experimental problem is to follow the variation of Q with temperature and to determine values of the coefficients by whichever technique is most appropriate for the material under investigation.

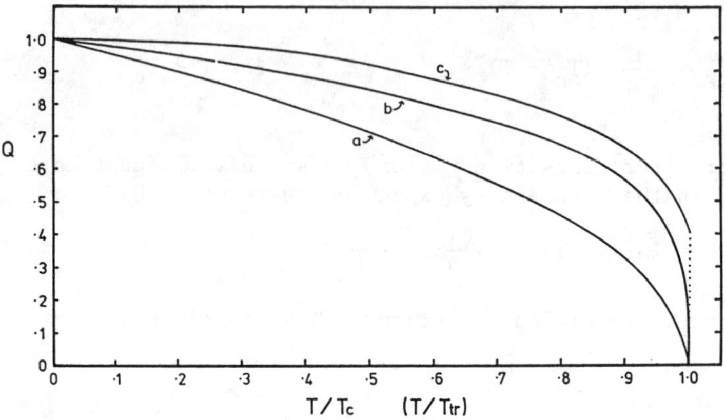

Figure 3. Variation of Q as a function of T/T_c for second order (a) and tricritical (b) Landau expansions, and as a function of T/T_{tr} for a first order transition (c) with $Q_o = 0.4$ and $T_c = 0.99\ T_{tr}$ (from equations 4, 10, 17).

3.2 Enthalpy measurements

For displacive transitions, direct measurements of C_p and T_c may be sufficient to characterise the thermodynamics. Further measurements of intensities of superlattice reflections, lattice parameters, birefringence, etc., as functions of temperature, may further define the character of the phase transitions. In the case of Aℓ/Si ordering transitions, enthalpies measured by solution calorimetry can provide useful constraints on the Landau coefficients.

Many silicates, including feldspars, dissolve readily in lead borate flux at 700°C and the heats of solution (ΔH_{soln}) can be measured using a Calvet type solution calorimeter. Alternatively, the heats of solution may be measured in hydrofluoric acid at ∿50°C. The experimental techniques are reviewed by Robie and Hemingway (1972), Navrotsky (1977, 1979, this volume) and Kleppa (1982). Absolute heats of solution are usually of little consequence, but differences between the values obtained for samples with differing degrees of Aℓ/Si order provide a measure of the enthalpy changes due to the ordering. Enthalpy changes observed in this way can be thought of as having several origins. Firstly, there are ΔC_p effects associated with the difference between the calorimeter temperature and the equilibration temperatures of the samples. These arise because the lattice vibrations and, hence, heat capacities of Aℓ/Si ordered and disordered crystals are not identical. Using C_p data from Robie et al. (1978), the enthalpy difference due to phonon contributions between sanidine ($KAℓSi_3O_8$ with Aℓ/Si disorder) and microcline ($KAℓSi_3O_8$ with Aℓ/Si order) is $\sim$1.3 kJ $mole^{-1}$ at 1000 K. This value indicates the order of magnitude expected for what might loosely be termed "thermal energies".

Secondly, it is possible to define an elastic energy associated with distortions of the framework structures. In general, this contribution may be expressed as:

$$\frac{1}{2} \sum_{i,k} C_{ik} \, x_i \, x_k \qquad (21)$$

where C_{ik} are elastic constants, x_i, x_k are components of the strain and i, k = 1 to 6 (Nye, 1957). Crystals with different degrees of order will have different degrees of distortion and, therefore, different elastic energies.

Remaining enthalpy differences between ordered and disordered structures might be referred to as "chemical bond energies" since they arise from the changes in bonding. While the distinctions are clearly artificial, in the sense that all the sources of enthalpy are interdependent, they can in fact be measured separately. As an example, the chemical bond energy and elastic energy contributions to the enthalpy of Aℓ/Si ordering in cordierite may be compared.

Putnis and coworkers (Putnis, 1980a, b; Putnis and Bish, 1983; Putnis et al., 1985, 1987; Putnis and Angel, 1985; McConnell, 1985) have investigated the Aℓ/Si ordering transition, from hexagonal to orthorhombic, in pure magnesian cordierite. Crystals produced by annealing glass of cordierite composition for short times at temperatures of 1100-1400°C are initially hexagonal. With further annealing they develop strain modulations on a scale of hundreds of Å while remaining hexagonal. At a critical point in the process, the crystals become orthorhombic. Solution calorimetry on a sequence of samples prepared at 1200°C showed that there is a total enthalpy change of $\gtrsim$ 40.8 ± 6.5 kJ $mole^{-1}$ due to the ordering (Carpenter et al., 1983). The enthalpy appeared to change linearly with the logarithm of annealing time (Fig. 4a). MASNMR spectroscopic studies have subsequently shown that the number of Aℓ-O-Aℓ linkages changes (decreases) linearly with log t (Putnis et al., 1985; and see Putnis, this volume) (Fig. 4b) and correlation of the slopes gives an enthalpy effect of $\sim$ -34 kJ per mole of Aℓ/Si exchanges (Putnis and Angel, 1985). Each such exchange results in a reaction of the form:

$$(-Aℓ\text{-}O\text{-}Aℓ-) + (-Si\text{-}O\text{-}Si-) \rightarrow 2(-Si\text{-}O\text{-}Aℓ-)$$

(a)

(b)

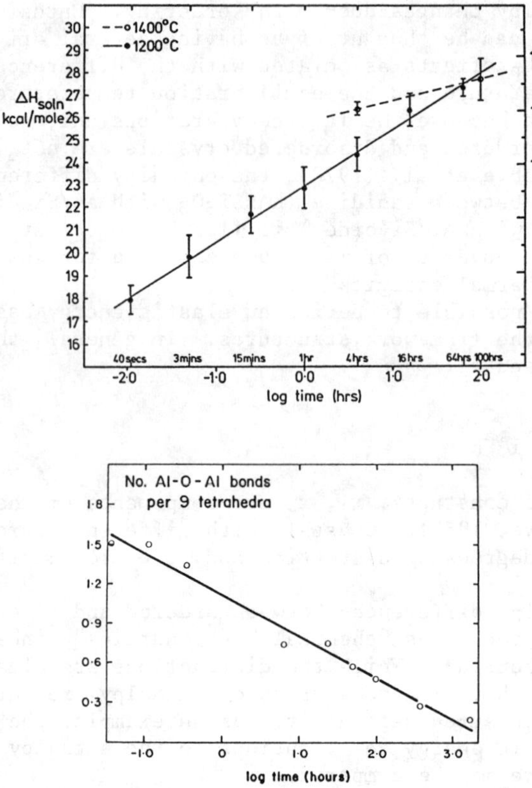

Figure 4. (a) Heat of solution (ΔH_{soln}) in lead borate at $\sim 700^{\circ}C$ as a function of log (annealing time) for cordierite crystals prepared from glass at 1200°C and 1400°C. After short annealing times the crystals are hexagonal (disordered) and after long annealing times they are orthorhombic (ordered). (From Carpenter et al., 1983). The enthalpy due to Aℓ/Si ordering appears to change continuously through the hexagonal → orthorhombic transition. (b) The number of Aℓ-O-Aℓ bonds per unit cell of cordierite (from NMR data), as a function of log (annealing time) for crystals prepared from glass and annealed at 1185°C. (From Putnis and Angel, 1985: reproduced by kind permission of A. Putnis).

While the cordierite crystals remain hexagonal, their macroscopic spontaneous strain is zero, and the calorimeter results give some idea of the total "chemical bond energy" involved in the ordering transition. On transforming from hexagonal to orthorhombic symmetry, the crystals should release elastic energy equal to $\frac{1}{2}.C_{11}.x_1^2$, where C_{11} is the relevant elastic constant and x_1 is the spontaneous strain (Salje, 1987b; Putnis et al., 1987). In terms of the orthorhombic unit cell parameters:

$$x_1 = \frac{a - \frac{b}{\sqrt{3}}}{a + \frac{b}{\sqrt{3}}} \qquad (22)$$

(Putnis et al., 1987) and, at room temperatures for well ordered cordierite, $x_1 = 0.007$. If it is assumed that C_{11} has a value somewhere between 1 and 100 GPa, the elastic energy released during the hexagonal → orthorhombic transition is between 6 and 600 J mole^{-1}. The lattice parameters, and hence the value of x_1, may of course be different at the calorimeter temperature (700°C) but this elastic energy is negligibly small in comparison with the "chemical bond energy". That such a small effect is not detected by solution calorimetry at the point in the ordering sequence where the symmetry change occurs (Fig. 4a) is not surprising given that one standard deviation of the calorimetric data is ∿3 kJ mole^{-1}.

In contrast with cordierite, the elastic energy associated with the C2/m ⇄ C$\bar{1}$ transition in albite is ∿6 kJ mole^{-1} at room temperature (Salje et al., 1985b). This is ∿40% of the total excess enthalpy at room temperature, showing that the macroscopic elastic energy can, in some circumstances, be as important as the "chemical bond energy" in magnitude. As discussed below, however, this transition involves both a displacive transition and Aℓ/Si ordering, which are coupled, and gives rise to a spontaneous strain which is an order of magnitude greater than seems to be typical. For the C$\bar{1}$ ⇄ I$\bar{1}$ transition in anorthite, which is associated with Aℓ/Si ordering, the elastic energy is << 1 kJ mole^{-1} out of a total enthalpy change on the order of ∿16 kJ mole^{-1}. The displacive I$\bar{1}$ ⇄ P$\bar{1}$ transition in anorthite involves elastic energies which are similarly small and, in general, the elastic energies are expected to be small even though the elastic constants are usually not well known.

3.3 Spontaneous strain

In the present context, the importance of spontaneous strain, ε, lies much less in its absolute contribution to the energetics of phase transitions than in its role as providing both a mechanism for long range interactions within a crystal and a mechanism for coupling between different order parameters. In addition, it provides an indirect measure of the behaviour of Q as a function of temperature, pressure, chemical composition, etc. ε is a relatively straightforward parameter to determine since it is defined as that part of the distortion of a crystal which results from a phase transition, and can be obtained from measurements of lattice parameters (Aizu, 1970; Guimaraes, 1979; Tolédano and Tolédano, 1980; Bismayer and Salje, 1981; Salje, 1986a; Salje et al., 1985a; Redfern and Salje, 1987a; Putnis et al., 1987). For Aℓ/Si ordering problems it is especially useful because of the difficulty in obtaining excess entropies by means other than through the behaviour of Q^2.

For phase transitions in which more than one order parameter operates, coupling of individual order parameters to the common strain provides a mechanism for coupling between the order parameters (Salje, 1985, 1986a, 1987a; Salje and Devarajan, 1986; Bismayer et al., 1986). Under these circumstances, the Landau potential may be expanded to include the contributions of ε explicitly (after Salje and Devarajan, 1986, and references therein) as:

$$G = \frac{1}{2}aQ^2 + \frac{1}{4}bQ^4 + \frac{1}{6}cQ^6 + dQ\varepsilon + eQ^2\varepsilon + f\varepsilon^2 \quad (23)$$

The coefficients d and e are coupling constants which express the strength of coupling of Q with ε (linear coupling) and Q^2 with ε (quadratic coupling); $f\varepsilon^2$ gives the elastic energy, which could be written more formally as in equation 21. It can be shown that, for a transition with a critical point at the boundary of the Brillouin zone, linear coupling between Q and ε is disallowed by symmetry (see for example, Tolédano and Tolédano, 1980). For a zone centre transition, linear coupling is allowed; it is then usually assumed that the dQε term is large relative to $eQ^2\varepsilon$, and the quadratic term is dropped from the expansion. The two criteria for equilibrium are dG/dQ = 0 and dG/dε = 0.

Applying the condition that dG/dε = 0 gives, for a zone boundary transition:

$$\varepsilon = -\frac{eQ^2}{2f} \quad (24)$$

and, for a zone centre transition:

$$\varepsilon = -\frac{dQ}{2f} \quad (25)$$

The simple relationships, $\varepsilon \propto Q^2$ and $\varepsilon \propto Q$ for zone boundary and zone centre transitions, respectively, are thus obtained.

Any strain can be expressed as a second rank tensor based on a Cartesian coordinate system (see Redfern and Salje, 1987a). For triclinic feldspars, a suitable coordinate system has:

Y parallel to the crystallographic b axis
Z parallel to the crystallographic c* axis
X perpendicular to both

For the general case of this orientation, the relationships between the strain tensor elements (x_1 to x_6) and the lattice parameters are given by Redfern and Salje (1987a) as:

$$x_1 = \frac{a \sin\gamma}{a_o \sin\gamma_o} - 1 \quad (26a)$$

$$x_2 = \frac{b}{b_o} - 1 \tag{26b}$$

$$x_3 = \frac{c \sin\alpha \sin\beta^*}{c_o \sin\alpha_o \sin\beta_o^*} - 1 \tag{26c}$$

$$x_4 = \left[\frac{c \cos\alpha}{c_o \sin\alpha_o \sin\beta_o^*} - \frac{b \cos\alpha_o}{b_o \sin\alpha_o \sin\beta_o^*} + \right.$$

$$\left. + \frac{\cos\beta_o^*}{\sin\beta_o^* \sin\gamma_o} \cdot \left(\frac{a \cos\gamma}{a_o} - \frac{b \cos\gamma_o}{b_o}\right)\right] \tag{26d}$$

$$x_5 = \left(\frac{a \sin\gamma \cos\beta_o^*}{a_o \sin\gamma_o \sin\beta_o^*} - \frac{c \sin\alpha \cos\beta^*}{c_o \sin\alpha_o \sin\beta_o^*}\right) \tag{26e}$$

$$x_6 = \left(\frac{a \cos\gamma}{a_o \sin\gamma_o} - \frac{b \cos\gamma_o}{b_o \sin\gamma_o}\right) \tag{26f}$$

a, b, c, α, β and γ are lattice parameters of the low symmetry phase at some pressure and temperature while a_o, b_o, c_o, α_o, β_o and γ_o are the lattice parameters that the high symmetry phase would have under the same conditions of P and T. Since strain is a symmetric tensor it may be referred to its principal axes, giving the principal strains. Three principal strains, ε_1, ε_2, ε_3, along the three mutually perpendicular axes are obtained by diagonalising the strain matrix. Finally, a scalar spontaneous strain may be defined as (Redfern and Salje, 1987a):

$$\varepsilon_s = \sqrt{\sum_i \varepsilon_i^2} \qquad (i = 1 - 3) \tag{27}$$

For transitions involving phases with high symmetry, the definition of ε in terms of the spontaneous strain elements (x_1 to x_6) depends on the symmetries of both the high and low symmetry forms. For the monoclinic (C2/m) $\rightleftharpoons$ triclinic ($C\bar{1}$) transition in feldspars, the relevant strain elements are x_4 and x_6 and the amplitude of the spontaneous strain is given by Salje et al., (1985a) as:

$$\varepsilon = \frac{1}{\sqrt{2}} (x_4^2 + x_6^2)^{\frac{1}{2}} \tag{28}$$

From these definitions of ε it should be clear that the spontaneous strain is zero in the high symmetry phase (above the transition temperature) but varies explicitly with Q, and hence with pressure, temperature, composition, etc., in the low symmetry phase (below the transition point).

4. TRANSFORMATION BEHAVIOUR IN FELDSPARS

Feldspars are amongst the most abundant minerals in the earth's crust and, although their basic structure was solved more than fifty years ago by Taylor (1933) our understanding of their thermochemical properties has remained rather limited. Both theory and available experimental data are more advanced for alkali feldspars than for plagioclases. For the latter, even qualitative descriptions of the stability relations are in dispute (Grove et al., 1983, 1986; Smith, 1983, 1984; Carpenter, 1985, 1986a, b).

The feldspar structure consists of a framework of SiO_4 and $A\ell O_4$ tetrahedra linked in three dimensions, with alkali or alkaline earth ions filling large cavity sites between the tetrahedra. The framework can also be described as consisting of chains of tetrahedra, shaped like double crankshafts, which are approximately parallel to the crystallographic a axis. Many different compositions have been prepared (e.g. with Rb^+, Na^+, K^+, Ca^{2+}, Sr^{2+} or Ba^{2+} as the large cations and $A\ell^{3+}$, Si^{4+}, Ga^{3+} or Ge^{4+} as tetrahedral cations), but most feldspars in nature fall within the ternary system, $CaA\ell_2Si_2O_8$-$NaA\ell Si_3O_8$-$KA\ell Si_3O_8$. Most natural crystals are further restricted in composition, and can be effectively described in terms of two binary systems: $NaA\ell Si_3O_8$-$KA\ell Si_3O_8$ and $NaA\ell Si_3O_8$-$CaA\ell_2Si_2O_8$. Along these two binary joins there are two distinct displacive transitions, three distinct commensurate ordered phases, an incommensurate structure and four separate miscibility gaps. There is also evidence for significant coupling, not only between the different ordering transitions, but also between the ordering transitions and the miscibility gaps. Schematic phase diagrams are shown in figure 5, but, for a complete description of the structure and chemistry of feldspars, readers are referred to reviews by Smith (1974, 1983, 1984), Ribbe (1983a, b, c, 1984).

4.1 Alkali feldspars

4.1.1 Albite

At high temperatures ($T \gtrsim 978°C$), the equilibrium state of albite is monoclinic, space group C2/m. On cooling, it transforms to a triclinic ($C\bar{1}$) structure by a transition which involves both a displacive component and an $A\ell/Si$ ordering component. Each component part of the transition would separately give rise to the same symmetry reduction and, if the displacive transition failed to occur, $A\ell/Si$ ordering would still take place, as in potassium feldspar (see reviews by Thompson and Hovis, 1979a; Kroll et al., 1980, 1986; Kroll, 1984; Brown et al., 1984). In reality the two processes are strongly coupled, with the displacive transition having a significant influence on the ordering and vice versa. Solution calorimetry has consistently indicated enthalpy differences of the order of 12 kJ $mole^{-1}$ between ordered and disordered albites (Holm and Kleppa, 1968; Newton et al., 1980; Blinova and Kiseleva, 1982; Carpenter et al., 1985), and the maximum expected configurational entropy associated with $A\ell/Si$ ordering is -18.7 J $mole^{-1}$ K^{-1} (the entropy associated with spreading 3$A\ell$ atoms and 1Si atom randomly between 4 tetrahedral sites).

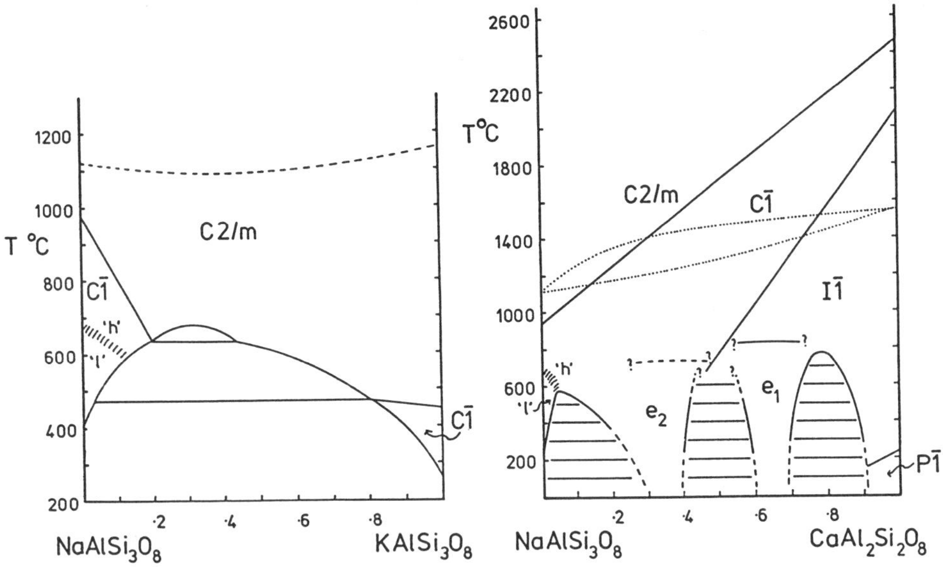

(a) (b)

Figure 5. (a) Schematic phase diagram for the alkali feldspar solid solution. The crossover region for high ('h') ⇌ low ('l') albite is indicated by shading and the C2/m ⇌ C$\bar{1}$ transition in $KA\ell Si_3O_8$ is shown as being thermodynamically continuous. The onset of melting is shown approximately by a dashed line.
(b) Schematic phase diagram for plagioclase feldspars. Position of the C$\bar{1}$ ⇌ I$\bar{1}$ line from Carpenter and McConnell (1984); position of the C2/m ⇌ C$\bar{1}$ transition line estimated from data of Kroll and Bambauer (1981); position of the C$\bar{1}$ ⇌ e_2 and I$\bar{1}$ ⇌ e_1 transition lines from Carpenter (1986b). The C$\bar{1}$ ⇌ e_2 transition was not reversed by Carpenter and could be at lower temperatures than shown. The crossover transition from high ('h') to low ('l') albite is shown by shading. Question marks indicate uncertainties in topology. Miscibility gaps are indicated by horizontal lines but their limits with respect to composition and temperature are not known precisely. Dotted lines indicate the solidus and liquidus; note that the C2/m ⇌ C$\bar{1}$ and C$\bar{1}$ ⇌ I$\bar{1}$ transition lines have been extrapolated into the stability field of liquid. (After Carpenter, 1985).

Salje (1985) assigned an order parameter, Q, to describe the displacive transition and a separate order parameter, Q_{od}, to describe the Aℓ/Si ordering. It follows from the relative transition temperatures of the displacive and order/disorder processes that the order parameter which drives the symmetry breaking, from C2/m to $C\bar{1}$, must have essentially the character of Q. Coupling between Q and Q_{od} can be easily understood in physical terms since Aℓ/Si ordering will obviously occur to a different extent in a framework which is already distorted, relative to one which is undistorted. Conversely, the degree of Aℓ/Si order must influence the manner in which the framework distorts. The coupling is described formally by the coupling of each of Q and Q_{od} with the common strain, ε. Given that the critical point of the transition is at the centre of the Brillouin zone, the appropriate Landau expansion has the form:

$$G = \frac{1}{2}a(T - T_c)Q^2 + \frac{1}{4}bQ^4 + \frac{1}{6}cQ^6 +$$
$$+ \frac{1}{2}a_{od}(T - T_c^{od})Q_{od}^2 + \frac{1}{4}b_{od}Q_{od}^4 + \frac{1}{6}c_{od}Q_{od}^6 +$$
$$+ dQ\varepsilon + d_{od}Q_{od}\varepsilon + e\varepsilon^2 \qquad (29)$$

where T_c is the critical temperature associated with the displacive transition, T_c^{od} is the critical temperature associated with the ordering, d and d_{od} are coupling coefficients, and $e\varepsilon^2$ is the elastic energy (Salje, 1985). A further term $fQ.Q_{od}$ may be required if there is some direct coupling between Q and Q_{od} in addition to the indirect coupling via the strain.

For a C2/m ⇄ $C\bar{1}$ transition, the relevant components of the spontaneous strain tensor are x_4 and x_6, and the elastic energy is $\frac{1}{2}\sum_{4,6} C_{ik}x_i x_k$. Salje (1985) showed that, under the conditions $dG/dx_4 = dG/dx_6 = 0$, the strain components are linearly dependent on Q and Q_{od} as:

$$x_4 = AQ + BQ_{od} \qquad (30)$$

$$x_6 = CQ + DQ_{od} \qquad (31)$$

These new coefficients, A, B, C and D, depend upon the elastic constants, C_{44}, C_{46}, C_{66}, and the coupling coefficients which describe the strength of coupling between x_4, x_6 and Q, Q_{od} (see equation 26 of Salje, 1985). Replacing the generalised strain, ε, with the correct strain components, x_4 and x_6, and substituting equations 30, 31 into equation 29, then gives the excess free energy in terms of Q, Q_{od}, Landau coefficients, coupling coefficients and the coefficients A, B, C, D. Gathering together all terms in Q^2, Q_{od}^2 and $Q.Q_{od}$, and renormalising, gives:

$$G = \frac{1}{2}a(T - T_c^*)Q^2 + \frac{1}{4}bQ^4 + \frac{1}{6}cQ^6 +$$
$$+ \frac{1}{2}a_{od}(T - T_c^{od*})Q_{od}^2 + \frac{1}{4}b_{od}Q_{od}^4 + \frac{1}{6}c_{od}Q_{od}^6 +$$
$$+ \lambda Q.Q_{od} \qquad (32)$$

In this expression, only T_c and T_c^{od} have been renormalised, to T_c^* and T_c^{od*}, as a consequence of coupling between the order parameters and the spontaneous strain. T_c^* is the critical temperature for the displacive transition in a crystal with $Q_{od} = 0$ which is allowed to deform and T_c^{od*} is the critical temperature for the order/disorder transition in crystals which are allowed to deform but have $Q = 0$. T_c and T_c^{od}, as defined in equation 29, are in fact hypothetical critical temperatures for transitions in crystals which are clamped in such a way as to prevent any spontaneous deformation ($\varepsilon = 0$). It should be noted also that the initial model of linear coupling between ε and Q and between ε and Q_{od} gives an effective bilinear coupling directly between Q and Q_{od}, as described by the term $\lambda Q.Q_{od}$.

Under equilibrium conditions, $dG/dQ = dG/dQ_{od} = 0$ and hence:

$$a(T - T_c^*)Q + bQ^3 + cQ^5 + \lambda Q_{od} = 0 \tag{33}$$

$$a_{od}(T - T_c^{od*})Q_{od} + b_{od}Q_{od}^3 + c_{od}Q_{od}^5 + \lambda Q = 0 \tag{34}$$

It is now only a matter of obtaining values for T_c^*, T_c^{od*} and the various coefficients in order to describe the thermodynamic behaviour of albite quantitatively. Salje et al. (1985a) measured the excess heat capacity for the displacive transition in crystals with $Q_{od} = 0$ and composition $Or_{31}Ab_{69}$. These crystals had a second order transition at 416 K and a step in C_p of 2.739 J mole^{-1} K^{-1} at the transition temperature. Salje et al. (1985a) then assumed that the size in the step in C_p at the transition does not vary with composition, giving, for pure albite, a = 5.479 J mole^{-1} K^{-1} and b = 6854 J mole^{-1} (using $T_c^* = \frac{b}{a} =$ 1251 K and $\Delta C_p = \frac{a^2}{2b} T_c^*$ at $T = T_c^*$).

Values of the coefficients a_{od}, b_{od}, c_{od} and λ were extracted from observations of lattice parameter variations with temperature in crystals with different values of Q_{od}. These crystals had been previously annealed at high temperatures to induce partial disorder. By definition (equations 26d and 26f):

$$x_4 = -\frac{c}{c_o}\cos\alpha^* \tag{35}$$

$$x_6 = \frac{b}{b_o}\cos\gamma \tag{36}$$

and the values of A, B, C and D in equations 30, 31 turn out to be such that x_4 is more strongly dependent on Q than Q_{od}, and x_6 is more strongly dependent on Q_{od} than Q. The lattice constants b and c (of the $C\bar{1}$ structure) vary only a little relative to b_o and c_o (of the C2/m structure extrapolated to the same temperature) and two useful approximate proportionalities are therefore obtained:

$$Q \;\alpha\; x_4 \approx -\cos\alpha^* \tag{37}$$

$$Q_{od} \;\alpha\; x_6 \approx \cos\gamma \tag{38}$$

From the variation of x_4 and x_6 with Q_{od} and temperature, Salje et al. (1985a) fit: $a_{od} = 41.62$ J mole^{-1} K^{-1}, $b_{od} = -9301$ J mole^{-1}, $c_{od} = 43600$ J mole^{-1}, $T_c^{od*} = 824.1$ K. The coupling coefficient, λ, obtained by Salje et al. is temperature dependent and this dependence can be expressed as:

$$\lambda = -2.171 - 3.043T - 1.569 . 10^{-3}T^2 + + 2.109 . 10^{-6}T^3 \quad (\text{J mole}^{-1}) \quad (39)$$

The final Landau expansion given by Salje et al. (1985a) to describe the excess free energy for the $C2/m \rightleftarrows C\bar{1}$ transition in albite is:

$$G = \tfrac{1}{2}a(T - T_c^*)Q^2 + \tfrac{1}{4}bQ^4 + \tfrac{1}{2}a_{od}(T - T_c^{od*})Q_{od}^2 + + \tfrac{1}{4}b_{od}Q_{od}^4 + \tfrac{1}{6}c_{od}Q_{od}^6 + \lambda Q.Q_{od} \quad (40)$$

from which the entropy and enthalpy may be derived in the usual way. Equilibrium values of Q and Q_{od}, as a function of temperature, are given by simultaneous solutions to equations 33 and 34 and the predicted behaviour is remarkably consistent with the experimental results of Goldsmith and Jenkins (1985), who independently bracketed the ordering reaction down to temperatures of ∿933 K.

Several important points emerge:

(a) In effect, Salje et al. (1985a) used the displacive transition as a probe to determine the order/disorder behaviour. The two processes occur on such different time scales that the variation of Q with temperature for each albite sample (from lattice parameters) refers to a fixed value of Q_{od}. Equations 33 and 34 define two surfaces in Q-Q_{od}-T space representing all possible stable and metastable states of albite, as shown in figure 6. The intersection of these surfaces represents the equilibrium states as a function of temperature.

(b) The coefficients a_{od}, b_{od}, c_{od} and T_c^{od*} define the order/disorder behaviour of albite when Q = 0, that is independently of the displacive transition. The values obtained describe a first order transition with a jump from Q = 0 to $Q_o = 0.4$ at $T_{tr}^{od*} = 833$ K. This transition, however, is only just first order (as in the first order example shown in figure 3) and is very close to being tricritical (b_{od} is small in comparison with c_{od} and T_{tr}^{od*} is within 1% of T_c^{od*}). For tricritical behaviour b_{od} would be zero.

The entropy at absolute zero is given by $-a_{od}/2 = -20.81$ J mole^{-1}K^{-1} and this is close to the maximum configurational entropy expected for Aℓ/Si ordering (-18.7 J mole^{-1} K^{-1}). The difference could easily be due to non-configurational (phonon) contributions.

(c) Coupling of Q_{od} with Q causes the first order transition in crystals with Q = 0 to be smeared out in crystals with Q > 0, to the extent that the step in Q_{od} becomes a continuous variation with temperature (Fig. 6a). Instead of a discrete order/disorder transition, there is a "crossover" at ∿950 K between a regime with Q dominant and a regime in which Q_{od} increases more rapidly than Q (Fig. 7). The symmetry of the

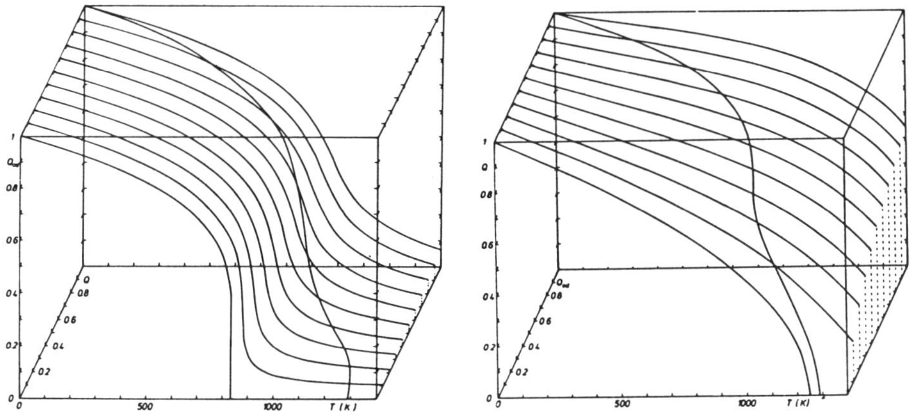

(a) (b)

Figure 6. Temperature dependence of (a) Aℓ/Si order parameter (Q_{od}) for different values of Q, and (b) Q for different values of Q_{od}, in pure albite. Each curve represents one metastable solution and the line across these curves indicates the solutions for thermal equilibrium. (From Salje et al., 1985a; reproduced by kind permission of E. Salje).

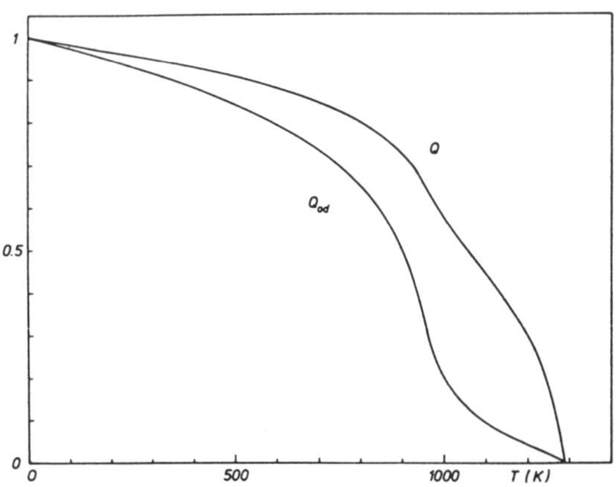

Figure 7. Temperature dependence of the displacive order parameter (Q) and Aℓ/Si order parameter (Q_{od}) at equilibrium for albite. A single, second order transition occurs at 1290 K, and a crossover occurs near 950 K between a regime of rapidly changing Q and a regime of rapidly changing Q_{od}. (From Salje et al., 1985a; reproduced by kind permission of E. Salje).

crystals remains the same ($C\bar{1}$) in both regimes, and, physically, the smoothing effect is due to the superposition of the distortions caused by the displacive and Aℓ/Si ordering components of the transition.
(d) The equilibrium transition temperature is at 1290 K, according to Salje et al. The transition is second order, as for fully disordered crystals, and occurs ∿40 K above T_c^*, the (metastable) displacive transition temperature in pure albite crystals with $Q_{od} = 0$. In other words, coupling of Q_{od} with Q raises the equilibrium transition temperature slightly. (The value of T_c^* actually used by Salje et al., 1251 K, is probably the equilibrium transition temperature (Kroll et al., 1980, 1986). A more appropriate value is 1238 K, from the data of Grundy and Brown (1969) and Kroll et al. (1980), for disordered albite synthesised at 1273 K (see figure 8a), though the precision in this value is only ± ∿10 K).
(e) Bilinear coupling, as described by Salje (1985) and Salje et al. (1985a), causes a renormalisation of the critical temperatures of the individual order/disorder and displacive transitions, but does not affect the higher order terms. The change in thermodynamic character of the Aℓ/Si ordering from a first order transition to a continuous crossover is not due to renormalisation of the high order Landau co-efficients. Rather, it arises from modifications of the free energy surface with respect to Q and Q_{od}, caused by the bilinear coupling term, $\lambda Q.Q_{od}$.
(f) The relationship between the spontaneous strain elements and lattice parameters, and the empirical proportionality observed between the strain elements and order parameters provide a useful means of distinguishing between the effects of ordering and atomic displacements in the feldspar structure (equations 37, 38). x_4 is particularly sensitive to displacive effects and x_6 to Aℓ/Si ordering. Cos α^* varies between 0 in the monoclinic structrue and ∿ -0.06 in low albite at room temperature, and cos γ varies between 0 and 0.04.
(g) The elastic energy, E, due to the strain is given by:

$$E = \tfrac{1}{2}C_{44}x_4^2 + \tfrac{1}{2}C_{66}x_6^2 + C_{46}x_4x_6 \qquad (41)$$

where the elastic constants are estimated as C_{44} = 17.3 GPa, C_{66} = 32.0 GPa, C_{46} = -2.5 GPa (Landolt-Börnstein, 1979, in Salje et al., 1985b). The elastic energy of low albite at room temperature is therefore ∿6 kJ mole^{-1} out of a total measured enthalpy of ∿12 kJ mole^{-1} (Salje et al., 1985b). This large energy is due to the large spontaneous strain. Ferroelastic materials typically show strains of up to 10^{-2}.
(h) The driving mechanism for the displacive transition has commonly been ascribed to vibrations of Na^+ ions in the large cavity sites (see Kroll et al., 1980; Ribbe, 1983a, 1984; Yund and Tullis, 1983). With increasing temperature, the amplitude of such vibrations would increase until the framework becomes monoclinic. No soft mode or anomalous softening of related hard modes has been observed in Raman spectroscopic studies, however, (Salje, 1986b) and it is therefore likely that the transition is due to some instability of the tetrahedral framework (Salje and Kuscholke, 1984). This alternative mechanism depends on the elastic constants of albite varying with temperature, but no data are yet available to test the predicted dependence (Salje, 1985; Salje et al., 1985b).

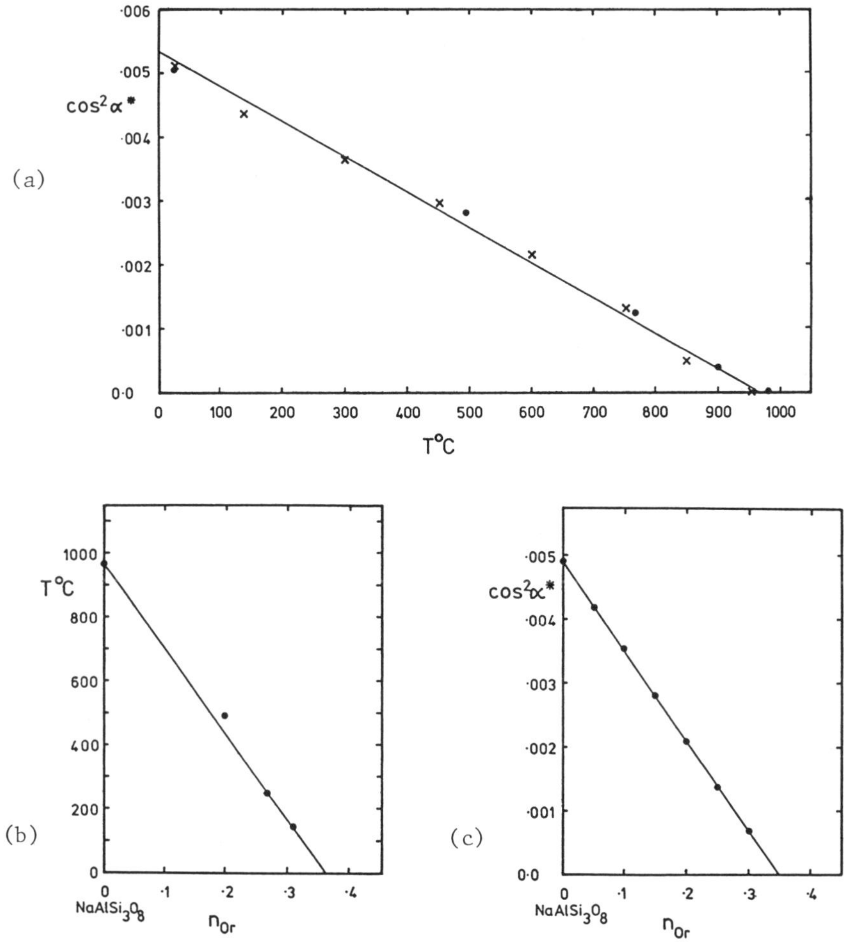

Figure 8. (a) Variation of $\cos^2\alpha^*$ ($\propto \sim Q^2$) as a function of temperature for crystals of pure albite synthesised at 1000°C (crosses = data from Grundy and Brown, 1969; filled circles = data from Kroll et al., 1980). $\mathrm{Cos}^2\alpha^*$ and, hence, Q^2 go to zero at $T_c = 965^\circ$C. (b) Variation of T_c (in $^\circ$C) with composition, for crystals with $Q_{od} = 0$. Data from figure 8a, Kroll et al. (1980), Salje (1986b), Salje et al. (1985a). n_{Or} = mole fraction of $KA\ell Si_3O_8$. (c) Variation of $\cos^2\alpha^*$ as a function of composition, at room temperature, for crystals with $Q_{od} = 0$. Data from Kroll et al. (1986). (see also: Kroll et al., 1980, 1986; Harlow, 1982; Kroll, 1984).

4.1.2 Potassium feldspar

Substitution of K^+ for Na^+ in the feldspar structure suppresses the $C2/m \rightleftharpoons C\bar{1}$ displacive transition, as shown by the steep decrease in T_c^* with increasing orthoclase content (Kroll et al., 1980), (Fig. 8b). For example, T_c^* is 416 K at $Or_{31}Ab_{69}$ but 1238 K in pure albite (Fig. 8a, b). On the other hand, Aℓ/Si ordering in crystals with Q = 0 is much less affected by $Na^+ \rightleftharpoons K^+$ substitutions. An estimate of the equilibrium order/disorder temperature for $KA\ell Si_3O_8$ has been given as ~723 K by Bernotat and Bambauer (1982) and Bambauer and Bernotat (1982), based on observations of the structural states of K-feldspar crystals in a well characterised metamorphic terrain. This may be compared with T_{tr}^{od*} = 833 K predicted by Salje et al. (1985a) for albite. The sequence of transitions with falling temperature is reversed in the potassium end-member, relative to the sodium end-member, though the thermodynamic relationship between the displacive and Aℓ/Si ordering transitions should not be affected in any fundamental way.

Figure 9 shows, in schematic form, the possible variations across the whole solid solution of T_c^* (Q_{od} = 0), T_c^{od*} (Q = 0), the equilibrium transition temperature for the $C2/m \rightleftharpoons C\bar{1}$ transition and the temperature of the "crossover". By analogy with albite, the equilibrium transition temperature in K-feldspar should be slightly higher than the pure ordering transition temperature, due to coupling with the displacive transition. T_c^* for K-feldspar would be well below 0 K but, again, coupling between Q and Q_{od} will change the second order character of the displacive transition into a continuous crossover at some much higher temperature. If it is assumed that T_c^*, T_c^{od*} and the crossover temperature vary linearly with composition, the crossover from Q_{od} dominant (high T) to Q dominant (low T) in pure $KA\ell Si_3O_8$ should occur between room temperature and 0 K. In this context, it is interesting to note that Openshaw et al. (1979a, b) and Wyncke et al. (1981) have described evidence for a displacive transition at 245 ± 5 K. In a powdered sample of natural microcline, the transition appeared to be reversible, whereas, in microcline produced by ion exchanging K^+ for Na^+ in albite, there was a hysteresis effect. The hysteresis, which Wyncke et al. (1981) and Brown et al. (1984) suggested might be due to steric hindrances, allowed Openshaw et al. (1979b) to observe lattice parameter changes as the low temperature form relaxed back to the high temperature form at room temperature. The spontaneous strain associated with this relaxation can be calculated using equations 26a-f, since both forms are triclinic. The data from Openshaw et al. (1979b) suggest that the dominant strain components are x_4 and x_6, with x_4 showing the largest variation. This behaviour is exactly as would be predicted for a crossover transition driven principally by Q, and it may not, therefore, be necessary to propose a totally different type of transition (e.g. see Brown et al., 1984). The transition behaviour for intermediate compositions shown in figure 9 will be obscured to some extent by the influence of Na, K segregation within the solvus.

An alternative approach to the thermochemistry of Aℓ/Si ordering in K-feldspar, as compared with that of Hovis (1974, 1986), Thompson et al. (1974) and Thompson and Hovis (1978, 1979a), is prompted by Salje's

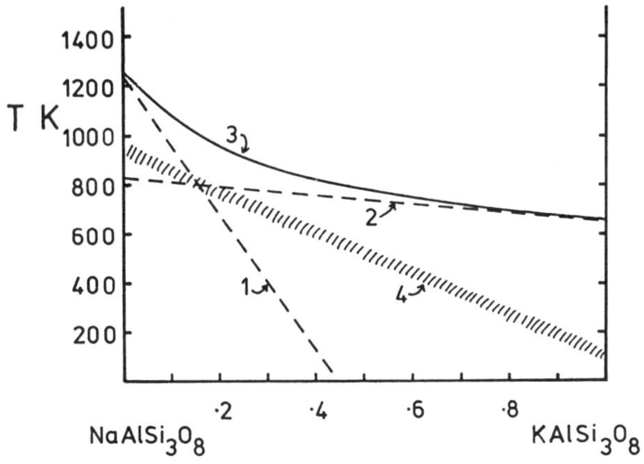

Figure 9. Schematic relations for $C2/m \rightleftharpoons C\bar{1}$ transitions in the alkali feldspar solid solution: 1. Displacive transition in crystals with $Q_{od} = 0$. 2. Order/disorder transition in crystals with $Q = 0$. 3. Equilibrium transition (Q and Q_{od} unconstrained). 4. Crossover, under equilibrium condititions, between regimes of Q varying rapidly with T and Q_{od} varying rapidly with T. 2 and 4 have been assumed to be linear to show that the displacive transition observed in microcline below room temperature could be related to the crossover.

analysis of albite. Since the pure order/disorder transition in albite appears to be close to tricritical, the simplest possible assumption is to treat pure ordering in $KA\ell Si_3O_8$ as also following a tricritical pattern. Using $a_{od} = -2S^o$, where S^o is taken to be the ideal configurational entropy change due to ordering 1Aℓ and 3Si atoms between four tetrahedral sites (-18.7 J mole^{-1} K^{-1}), and $T_c^{od*} = 723$ K (i.e. ignoring the effect of coupling with Q) gives $c_{od} = 27040$ J mole^{-1}. From equation 12 the excess enthalpy for a crystal with $Q_{od} = 0.92$ would be -8711 J mole^{-1} and this compares favourably with an enthalpy difference between microcline ($C\bar{1}$) and sanidine ($C2/m$) of 8160 ± 1340 J mole^{-1}, as measured by solution calorimetry at 50°C (Waldbaum and Robie, 1971). The microcline sample used by Waldbaum and Robie was prepared by ion exchanging K^+ for Na^+ in Amelia albite for which $Q_{od} \approx 0.92 \pm 0.04$, using site occupancies refined from neutron diffraction data by Harlow and Brown (1980). (Q_{od} may be estimated from tetrahedral site occupancies as $(A\ell_{T1O}-A\ell_{T1M})/(A\ell_{T1O}+A\ell_{T1M})$, where $A\ell_{T1O}$ is the Aℓ content of the T1O site and $A\ell_{T1M}$ is the Aℓ content of TIM. These two sites are related by symmetry in the C2/m structure but not in the $C\bar{1}$ structure. According to Harlow and Brown, both the T2O and T2M sites in Amelia albite are fully occupied by Si). The sanidine used by Waldbaum and Robie was prepared by disordering the ion exchanged crystals at 1333 K. The tricritical model predicts $Q_{od} = 0.89$ for crystals equilibrated at 273 K and Q_{od} = 0.77 for crystals equilibrated at 473 K. True equilibrium structural states at these temperatures should have higher degrees of order, however, because of coupling between Q_{od} and Q, as in albite, and natural ordered microclines have values of Q_{od} of $\geqslant$ 0.95 based on site occupancy estimates (Blasi et al., 1984a; Kroll, 1984).

Finally, the important contribution of elastic energy to the energetics of the phase transition in albite prompts a brief analysis of spontaneous strain in K-feldspar. In albite, symmetry breaking occurs by a predominantly displacive mechanism and the Aℓ/Si ordering develops within crystals which are already triclinic. In K-feldspar, the symmetry breaking is due primarily to Aℓ/Si ordering and is, therefore, subject to quite different mechanistic constraints. Being slow, the Aℓ/Si ordering does not necessarily follow an equilibrium pathway and there is a tendency for triclinic ordered domains to develop via a strain modulated structure (McConnell, 1965, 1971; Eggleton and Buseck, 1980; and see reviews by Ribbe, 1983b, and McLaren, 1984). Some natural crystals are thought to have grown at temperatures below ∿723 K with metastably disordered states, and to have then ordered by the development of two orthogonal distortion waves superimposed on a monoclinic lattice (see Smith, 1974; McConnell, 1965, 1971; Černý and Chapman, 1984; Carpenter and Putnis, 1985; and references therein).

The spontaneous strain of crystals with strain modulated structures is zero, of course, while their macroscopic symmetry remains monoclinic. Triclinic crystals have spontaneous strains and elastic energies given by equations 35,36 and 41. Taking the lattice parameters for a microcline sample described by Waldbaum and Robie (1971) and elastic constants from Ryzhova and Aleksandrov (1965) (in Landolt-Börnstein, 1979; C_{44} = 14.3 GPa, C_{66} = 37.4 GPa, C_{46} = -2.8 GPa) an elastic energy of ∿3400 J $mole^{-1}$ is obtained. From solution calorimetry at 50°C, the enthalpy of ordering from sanidine (C2/m) to microcline ($C\bar{1}$) is 8160 ± 1340 J $mole^{-1}$ (Waldbaum and Robie, 1971) and from sanidine to adularia (C2/m, strain modulated) it is 5150 ± 540 J $mole^{-1}$ (Hovis, 1974). The difference between these two excess enthalpies, 3010 ± 1880 J $mole^{-1}$, almost exactly matches the estimated elastic energy of microcline and is therefore consistent with the view that the strain modulated structures have a high degree of order (and low configurational entropy), comparable with ordered microcline itself. The only significant energy difference between triclinic and modulated monoclinic crystals appears to be the strain energy stored in the modulations (McConnell, 1971; Eggleton and Buseck, 1980). Recently, a high degree of Aℓ/Si order has been measured directly from the modulated region of a K-feldspar crystal by analytical electron microscopy (McLaren and Fitzgerald, 1987). As in the case of albite, the elastic energy associated with the C2/m ⇌ $C\bar{1}$ transition forms a high proportion of the total excess enthalpy.

From this outline, it should be apparent that the Landau treatment of Salje (1985) for albite is equally applicable to the transformation behaviour of K-feldspar.

4.1.3 C2/m ⇌ $C\bar{1}$ displacive transition as a function of composition

Salje et al. (1985a) showed that the variation of x_4 provides a reasonable approximation to the variation of Q and, since $x_4 \approx -\cos\alpha^*$, $\cos^2\alpha^*$ should provide a measure of Q^2. For crystals disordered at high temperatures ($Q_{od} = 0$), $\cos^2\alpha^*$ is linear both with temperature (Fig. 8a) and, at room temperature, with composition (Fig. 8c) below the $C2/m \rightleftharpoons C\bar{1}$ displacive transition temperature, T_c^* (see also: Hovis 1980; Kroll et al., 1980, 1986; Harlow, 1982; Kroll, 1984). T_c^* is itself also linear with composition (Fig. 8b) and, given these relationships, a single Landau expansion is adequate to describe the transition at all compositions. The effect of orthoclase content (n_{Or}, the mole fraction of orthoclase) on the transition is included by renormalising the transition temperature of pure albite (T_c^{*Ab}):

$$G = \frac{1}{2}a\left[T - \left(T_c^{*Ab} + \frac{dT_c^*}{dn_{Or}} \cdot n_{Or}\right)\right]Q^2 + \frac{1}{4}bQ^4 \qquad (42)$$

It has been assumed here that the coefficients a and b do not vary with composition. Salje et al. (1985a) assumed, rather, that the step in C_p at the transition temperature is constant, in which case a and b must vary. Further experimental observations of C_p in crystals with different compositions are required to resolve this issue but either result can be readily incorporated into an expression with the general form of equation 42. For the present, the values of a and b given by Salje et al. (1985a) are used because they give a good fit with independent calorimetric results for order/disorder in pure albite (see figure 12 of Salje et al., 1985a). From figure 8b, $dT_c^*/dn_{Or} = -2690$ K per 0.01 orthoclase component.

The solution of equation 42 for Q as a function of temperature and composition is:

$$Q = \frac{a}{b}\left[\left(T_c^{*Ab} + \frac{dT_c^*}{dn_{Or}} \cdot n_{Or}\right) - T\right]^{\frac{1}{2}} \qquad (43)$$

and the excess entropy and excess enthalpy can be obtained by making the same substitution for T_c in equations 5 and 6. The excess enthalpy can also be written as a function of T and n_{Or} as:

$$H = \frac{1}{4}\frac{a^2}{b}\left[T^2 - \left(T_c^{*Ab} + \frac{dT_c^*}{dn_{Or}} \cdot n_{Or}\right)^2\right] \qquad (44)$$

As pointed out by Thompson and Waldbaum (1968), Waldbaum and Thompson (1968), Merkel and Blencoe (1982) and Haselton et al. (1983), the mixing properties for disordered ($Q_{od} = 0$) alkali feldspars will be modified by the effects of the displacive transition. The effects will be asymmetric because only sodium-rich crystals have the lower symmetry (see also Carpenter and McConnell, 1984; Carpenter and Ferry, 1984; Carpenter, 1985). For example, two separate contributions to the enthalpy of mixing can be distinguished, as illustrated in figure 10a with data from Hovis and Waldbaum (1977). A monoclinic solid solution

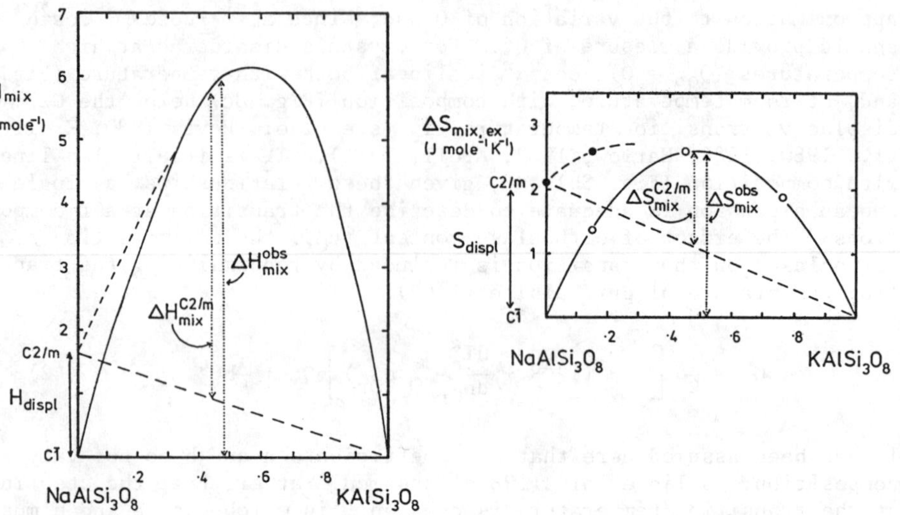

(a) (b)

Figure 10. (a) Enthalpy of mixing (ΔH_{mix}) for an alkali feldspar solid solution with $Q_{od} = 0$. The observed enthalpy of mixing (ΔH^{obs}_{mix}, Hovis and Waldbaum, 1977) consists of the effects of the $C2/m \rightleftharpoons C\bar{1}$ displacive transition at albite-rich compositions (H_{displ}) superimposed on the effect of Na $\rightleftharpoons$ K substitutions in a C2/m structure ($\Delta H^{C2/m}_{mix}$). The effect of a displacive transition in one end-member is to increase the apparent excess enthalpy of mixing. (b) Excess entropy of mixing ($\Delta S_{mix,ex}$) for crystals with $Q_{od} = 0$ at 298 K. The observed excess entropy (ΔS^{obs}_{mix}) consists of the effects of the $C2/m \rightleftharpoons C\bar{1}$ displacive transition (S_{displ}) superimposed on the excess entropy of mixing in a C2/m solid solution ($\Delta S^{C2/m}_{mix}$). Open circles represent data from Haselton et al. (1983) obtained from measurements of heat capacity for crystals with different compositions across the solid solution. Filled circles are the same data plus the excess entropy due to the displacive transition.

will have some positive excess enthalpy of mixing but this is effectively increased by the contribution of the displacive transition. The net excess enthalpy is necessarily temperature dependent because the displacive contribution varies with temperature (from zero at $T > T^*_c$ to ~ -1600 J mole^{-1} in Ab_{100} at 50°C, using the data of Salje et al., 1985a). The transition in albite-rich crystals will, similarly, give rise to an apparent, and temperature dependent, positive excess entropy of mixing (Fig. 10b). At least part of the excess entropy of Na-K mixing reported by Thompson and Hovis (1979b) and Haselton et al. (1983) may be accounted for in this way, since the maximum excess observed for intermediate compositions is ∿2.6 J mole^{-1} K^{-1} (Haselton et al., 1983) and the entropy change due to the purely displacive transition in end-member albite is ∿2.1 J mole^{-1} K^{-1} at 298 K (Salje et al., 1985a).

For ordered solid solutions the transition in end-member albite still gives rise to an apparent asymmetric contribution to the enthalpy of mixing, in spite of both low albite and microcline having the same symmetry. This is because the excess enthalpy for the $C2/m \rightleftharpoons C\bar{1}$ transition in the end-members is greater for $NaAlSi_3O_8$ than for $KAlSi_3O_8$, due to the coupling of Q_{od} with Q in albite. The effect can, in principle, be accounted for by renormalising T_c^{od*} and the coupling constant, λ , in equation 40, to include the effects of composition on the pure order/disorder transition and on the strength of the coupling.

This unified Landau treatment allows quantitative distinctions to be made between the contributions to the mixing properties of the displacive and Al/Si ordering transitions, as opposed to the effects of small local structural relaxations associated purely with substituting Na^+ for K^+. For second order transitions, the mixing properties will, of course, be continuous through the critical composition at which the transition occurs (Merkel and Blencoe, 1982; Carpenter, 1985).

4.2 Plagioclase feldspars

Anorthite forms the end member phase of the plagioclase solid solution (Fig. 5b). It undergoes a displacive transition near 510 K giving a symmetry change of $I\bar{1} \rightarrow P\bar{1}$ with falling temperature (Laves and Goldsmith, 1954; Brown et al., 1963; Bruno and Gazzoni, 1967; Smith and Ribbe, 1969; Laves et al., 1970; Czank et al., 1973; Czank, 1973; Staehli and Brinkmann, 1974; Smith, 1974, 1984; Frey et al., 1977; Adlhart et al., 1980a; Wruck, 1986; Redfern and Salje, 1987a). In the absence of melting, an Al/Si order/disorder transition would occur at some high temperature ($C\bar{1} \rightarrow I\bar{1}$ with falling temperature) and, at a higher temperature still, the crystals would become monoclinic (C2/m), (Fig. 5b). Equilibrium transition temperatures are estimated for the order/disorder and monoclinic/triclinic transitions by extrapolating the $C\bar{1} \rightleftharpoons I\bar{1}$ and $C2/m \rightleftharpoons C\bar{1}$ transition lines to pure anorthite from albite-rich compositions within the plagioclase solid solution (Fig. 5b).

4.2.1 $I\bar{1} \rightleftharpoons P\bar{1}$ transition

The underlying driving force for the $I\bar{1} \rightleftharpoons P\bar{1}$ transition is thought to be related to the condensation of a soft mode at the Brillouin zone boundary of the $I\bar{1}$ structure (Adlhart et al., 1980a, b; Salje, 1987a; Redfern and Salje, 1987a). In spite of uncertainties over some structural details, particularly in relation to the role of Ca^{2+} movements between split positions on the large cation sites (Czank, 1973; Foit and Peacor, 1973; Smith, 1974; Staehli and Brinkmann, 1974; Bruno et al., 1976; Adlhart et al., 1980a, b; Salje, 1987a), it is possible to treat the macroscopic processes involved in the transition using Landau theory, as shown by Adlhart et al. (1980a, b) and Salje (1987a).

Landau theory

The transition near 510 K in pure $CaA\ell_2Si_2O_8$ crystals derived from a metamorphic rock (i.e. with a high degree of Aℓ/Si order) is tricritical in character (Wruck, 1986; Salje, 1987a; Redfern and Salje, 1987a). If a small amount of albite component is present in solid solution, or if a degree of Aℓ/Si disorder is introduced by equilibrating the crystals at high temperatures, the transition becomes second order and the transition temperature is changed (Adlhart et al., 1980a, b; Wruck, 1986; Salje, 1987a; Redfern and Salje, 1987b). This behaviour can be explained if there is effective coupling between the order parameter for the $I\bar{1} \rightleftharpoons P\bar{1}$ transition (Q) and the order parameter for the $C\bar{1} \rightleftharpoons I\bar{1}$ transition (Q_{od}). As in the $C2/m \rightleftharpoons C\bar{1}$ transition described above, Q and Q_{od} operate on dramatically different time scales; Q responds almost instantaneously to changes in temperature, but the value of Q_{od} is fixed in the crystals during some prior, high temperature heat treatment and remains unchanged during subsequent observations of the variation of Q. By way of contrast with the $C2/m \rightleftharpoons C\bar{1}$ transition, however, bilinear coupling should not occur. The $C\bar{1} \rightleftharpoons I\bar{1}$ and $I\bar{1} \rightleftharpoons P\bar{1}$ transitions each have a critical point on the Brillouin zone boundary, though in different places, and linear coupling between Q and ε and between Q_{od} and ε is forbidden by symmetry (Salje, 1987a). Biquadratic coupling is not forbidden by symmetry and coupling between Q^2 and Q_{od}^2 can occur through the coupling of each with the common strain, ε , (Salje and Devarajan, 1986; Salje, 1987a).

In order to illustrate the most general consequences of the coupling effects, we may take the simplest case considered by Salje (1987a), that for sodium-free crystals with homogeneous Aℓ/Si distributions. The excess free energy expression is expanded to include coupling of Q^2 with ε :

$$G = \frac{1}{2}a(T - T_c)Q^2 + \frac{1}{4}bQ^4 + \frac{1}{6}cQ^6 + dQ^2\varepsilon + e\varepsilon^2 \qquad (45)$$

where d is a coupling coefficient and $e\varepsilon^2$ is the elastic energy. The physical effect of changing Q_{od} can be likened to a "deformation cloud" within the crystal and can be treated as an independent stress field, since changing Aℓ/Si order changes ε (Salje, 1987a). In other words, Q_{od} is fixed at some high temperature and then acts to modify the evolution of Q with T through this effective stress.

At equilibrium the crystals are stress free and the condition dG/dε = 0 applies. Thus:

$$0 = dQ^2 + 2e\varepsilon \quad \Rightarrow \quad \varepsilon = -\frac{d}{2e}Q^2 \qquad (46)$$

Substituting equation 46 into equation 45 gives:

$$G = \frac{1}{2}a(T - T_c)Q^2 + \frac{1}{4}(b - \frac{d^2}{e})Q^4 + \frac{1}{6}cQ^6 \qquad (47)$$

Both the $P\bar{1}$ and $I\bar{1}$ phases are triclinic and all six of the spontaneous strain elements are needed to describe the macroscopic spontaneous strain. The elastic energy term ($e\varepsilon^2$) should, strictly speaking, be replaced by

$\frac{1}{2}\Sigma_{i,k} C_{ik} x_i x_k$, the spontaneous strain, ε, by $\Sigma_i x_i$, and the coupling coefficient, d, by d_i, where i,k = 1 to 6. This complicates only the algebraic manipulations, however, and the form of the resulting expression remains the same (see Salje, 1987a).

A direct consequence of the coupling between Q^2 and ε is, therefore, that the fourth order term in the free energy expansion is renormalised. The new coefficient $[b - (d^2/e)]$ will be dependent on the degree of Aℓ/Si order because the elastic constants of anorthite, which make up the coefficient e, and the strength of coupling, d, between Q^2 and the spontaneous strain, will themselves vary with the degree of order. This result predicts that the thermodynamic character of the $I\bar{1} \rightleftharpoons P\bar{1}$ transition may be different in crystals with different states of Aℓ/Si order (Salje, 1987a). The prediction is borne out by the experimental studies which show that, for highly ordered crystals ($Q_{od} \to 1$), the transition is tricritical, i.e. $[b - (d^2/e)] = 0$, and, for crystals with $Q_{od} < 1$, the transition becomes second order, i.e. $[b - (d^2/e)] > 0$ (Wruck, 1986; Redfern and Salje, 1987a, b).

A second prediction from equation 47 is that the equilibrium transition temperature, T_c in the case of second order and tricritical transitions, should be independent of Q_{od}, since the second order term is not renormalised by the quadratic coupling. In reality, however, there are other factors to consider and Salje (1987a) showed that fluctuations in the local degree of order within a crystal, as opposed to having a totally homogeneous Aℓ/Si distribution, would lead to a renormalisation of the second order term, and, hence, to a change of T_c.

Most natural anorthites have some albite in solid solution and the $I\bar{1} \rightleftharpoons P\bar{1}$ transition is further modified by the $Na^+ + Si^{4+} \rightleftharpoons Ca^{2+} + A\ell^{3+}$ substitutions (Smith, 1974; Adlhart et al., 1980b; Wruck, 1986; Redfern and Salje, 1987b). In particular, the transition temperature is lowered as the proportion of albite component increases, and the transition itself appears to become smeared out over a temperature range (Adlhart et al., 1980b). If the transition is driven by a soft mode, the effect of straight substitutions of Na for Ca would be to diminish the driving force to an extent which should depend both on Q and on the occupancy (σ) by Na and Ca of the possible split positions of the large cation sites (Adlhart et al., 1980a, b; Salje, 1987a). Salje added a term $f\sigma Q$ to the initial free energy expansion (equation 45) to allow for this contribution. In addition, however, the solid solution mechanism involves substitution of Si for Aℓ and this results in a concomitant decrease in the degree of Aℓ/Si order, as reflected in the steep decrease in the $C\bar{1} \rightleftharpoons I\bar{1}$ transition temperature (Fig. 5b). This effect can be accounted for by expressing Q_{od} explicitly in terms of temperature and composition with respect to the $C\bar{1} \rightleftharpoons I\bar{1}$ transition (see below). The final free energy expansion is necessarily more complex than that given in equation 47, therefore, but the underlying principles of quadratic coupling remain the same, and the general conclusions may be summarised:
(a) Quadratic coupling of Q^2 with ε, and hence with Q_{od}^2 in the more complete formulation of Salje (1987a), leads to a change in thermodynamic character of the displacive transition, from tricritical to second order, depending on the magnitude of Q_{od}.
(b) This coupling exists and has significant consequences even though

the equilibrium transition temperatures for the $C\bar{1} \rightleftharpoons I\bar{1}$ and $I\bar{1} \rightleftharpoons P\bar{1}$ transitions are $\sim 1800^{\circ}C$ apart.
(c) The transition temperature for the $I\bar{1} \rightleftharpoons P\bar{1}$ transition varies due to the renormalisation of the second order term through the contributions of inhomogeneous Aℓ/Si distributions and $Na^{+} \rightleftharpoons Ca^{2+}$ substitutions.
(d) The displacive transition may, in principle, be used as a rather sensitive probe to characterise the structural states of natural and synthetic anorthite crystals. In other words, a relatively easy property to determine, such as T_c for the displacive transition, might be used to determine a much less accessible property, such as the value of Q_{od}, in any crystal of anorthite.

Spontaneous strain data

The $I\bar{1} \rightleftharpoons P\bar{1}$ transition is an example of an improper ferroelastic transition, and the temperature evolution of the spontaneous strain is as revealing of the thermodynamic behaviour as the variation of C_p. Redfern and Salje (1987a) have determined the spontaneous strain of metamorphic anorthite crystals from measurements of lattice parameters at high temperatures. Given that both the $I\bar{1}$ and $P\bar{1}$ phases are triclinic, there are no a priori reasons for predicting which individual lattice parameters will show the greatest variation as a consequence of the transition. The six spontaneous strain elements are obtained by comparing the lattice parameters below the transition temperature with those of untransformed crystals at the same temperature (equations 26a - f). Lattice parameters for untransformed crystals are estimated by extrapolating data from above the transition temperatures. The three principal strains, ε_1, ε_2, and ε_3, are then given by diagonalising the strain tensor, and are shown in figure 11a (from Redfern and Salje, 1987a). Equation 27 gives the definition of the scalar spontaneous strain, ε_s. Redfern and Salje (1987a) found that the relationship $\varepsilon_s^2 \propto T$ describes the temperature evolution of the strain quite accurately (Fig. 11b). Since $\varepsilon_s \propto Q^2$ for a zone boundary transition, their result implies:

$$Q \propto (T_c - T)^{\frac{1}{4}} \qquad \text{for } T < T_c \tag{48}$$

This tricritical behaviour is consistent both with the heat capacity data of Wruck (1986) and the intensity measurements of superlattice reflections from Adlhart et al. (1980a), (Fig. 11c).

Spontaneous strain values are an order of magnitude smaller for the $I\bar{1} \rightleftharpoons P\bar{1}$ transition in anorthite than they are for the $C2/m \rightleftharpoons C\bar{1}$ transition in albite, as is frequently the case for zone boundary, as opposed to zone centre, transitions. The associated elastic energies are correspondingly two orders of magnitude smaller. Using the spontaneous strain parameters of Redfern and Salje (1987a), and the elastic constants for plagioclase crystals with composition An_{56} from Ryzhova (1964), the elastic energy due to the transition in metamorphic anorthite crystals may be estimated as < 100 J $mole^{-1}$ at room temperature. This value may be compared with the total excess enthalpy and entropy at 390 K, due to the transition, of -1040 J $mole^{-1}$ and -2.17 J $mole^{-1}$ K^{-1}, respectively (Wruck, 1986).

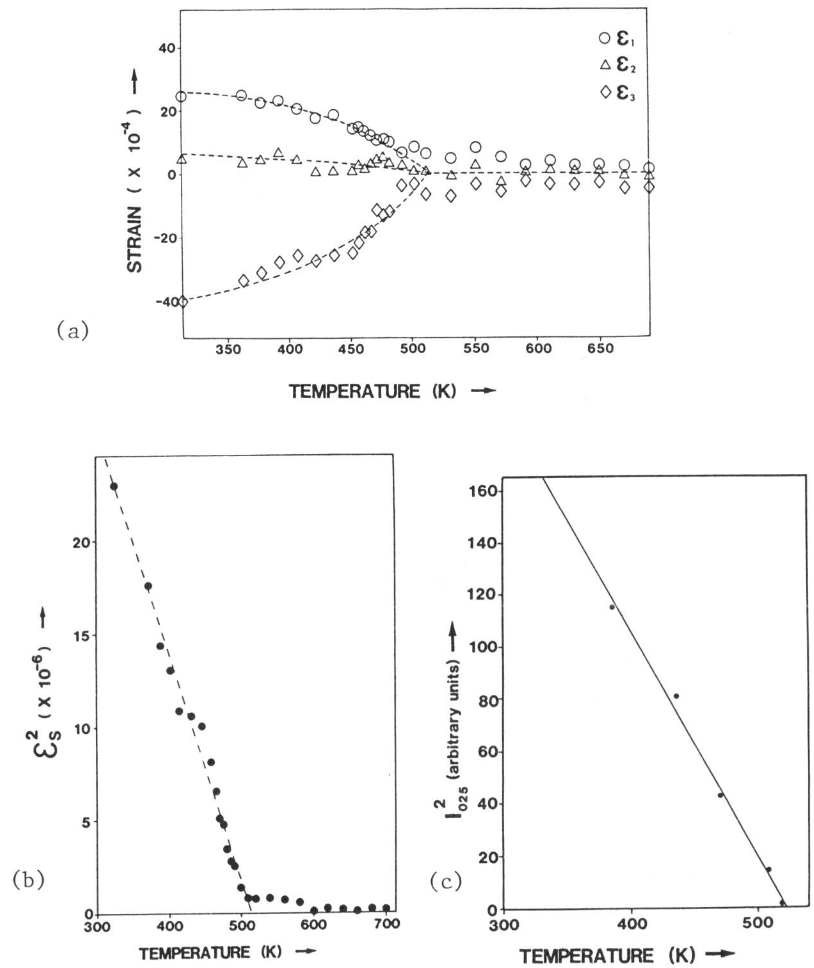

Figure 11. Spontaneous strain measurements for the $I\bar{1} \rightleftharpoons P\bar{1}$ transition in anorthite with a high degree of Aℓ/Si order; T_c = 510 K. (From Redfern and Salje, 1987a; reproduced by kind permission of S.A.T. Redfern). (a) Temperature dependence of the diagonalised strain tensor components. (b) Variation with temperature of the square of the spontaneous strain (ε_s^2), which is proportional to Q^4. The observed linear dependence indicates $Q \propto (T_c - T)^{\frac{1}{4}}$ and reflects the tricritical nature of the phase transition. (c) Temperature variation of the intensity of elastic scattering of the 025 reflection (data of Adlhart et al., 1980a), again showing tricritical behaviour: $I_{025}^2 \propto Q^4 \propto (T_c - T)$.

4.2.2 $C\bar{1} \rightleftharpoons I\bar{1}$ transition

On heating, anorthite melts before it lowers its symmetry by disordering Aℓ and Si among the tetrahedral sites. Equilibrium properties close to the order/disorder transition itself can, therefore, only be investigated in crystals with substantial amounts of albite component in solid solution, for which the transition temperatures are much lower (Fig. 5b). Crystals with intermediate compositions annealed at temperatures just below the $C\bar{1} \rightleftharpoons I\bar{1}$ transition line, and then quenched, give weak but sharp superlattice reflections in electron diffraction patterns (Carpenter and McConnell, 1984). Qualitatively, there is a decrease in intensity of the reflections as the transition is approached from below and the transition is then marked by a change from sharp superlattice reflections to highly diffuse reflections (Carpenter and McConnell, 1984). While there are fewer quantitative data for this transition than for either the $C2/m \rightleftharpoons C\bar{1}$ transition in albite or the $I\bar{1} \rightleftharpoons P\bar{1}$ transition in anorthite, available lattice parameter and solution calorimetric data are sufficient to suggest that the transition is close to being tricritical.

Enthalpy and lattice parameter measurements have been made on a suite of natural plagioclase samples separated from igneous and metamorphic rocks, and on the same samples after they had been heat treated at temperatures of between 1070°C and 1375°C for times of 20 to 42 days (Carpenter et al., 1985). The natural, or "low", series had relatively high degrees of Aℓ/Si order; the heat treated, or "high", series were relatively disordered, with $C\bar{1}$ structures in the composition range An_0 - An_{75} and $I\bar{1}$ structures in the range An_{75} - An_{100} (Fig. 12a).

(a)

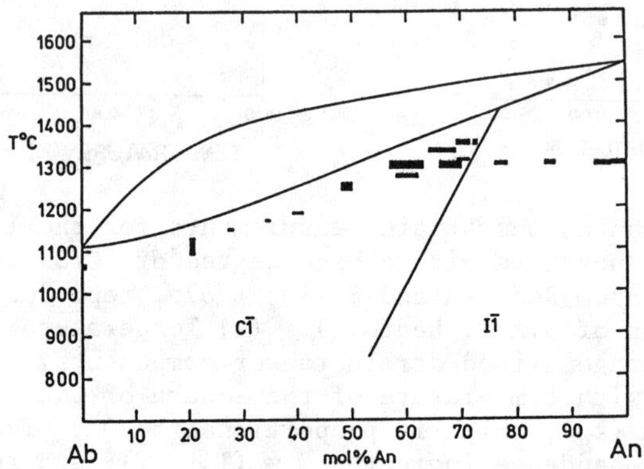

(b)

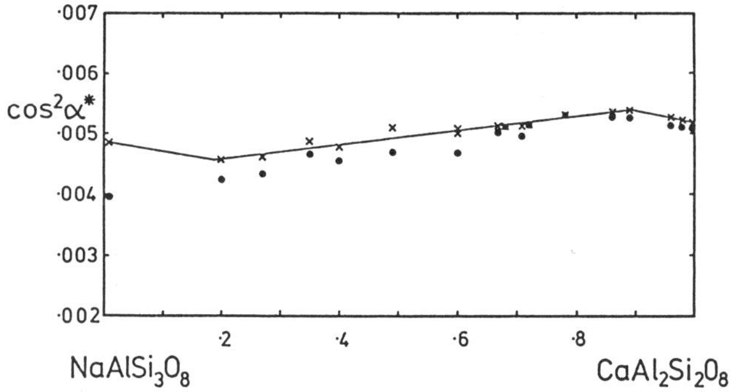

(c)

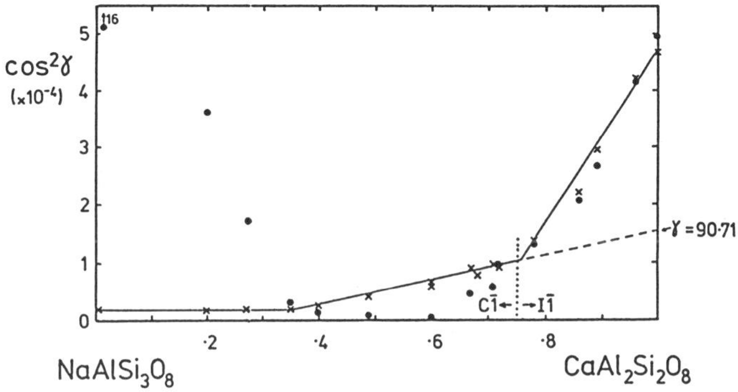

Figure 12. (a) Compositions and temperatures of heat treated "high" plagioclases (filled rectangles) used for lattice parameter measurements, in relation to the $C\bar{1} \rightleftharpoons I\bar{1}$ transition line and solidus/liquidus (from Carpenter et al., 1985). (b) Variation of $\cos^2\alpha^*$ (at room temperature) as a function of composition for high plagioclases (crosses) and natural "low" plagioclases (filled circles). $\mathrm{Cos}^2\alpha^*$ is linear for the high series over almost the entire composition range, with only small deviations at albite-rich and anorthite-rich compositions (solid line). (c) Variation of $\cos^2\gamma$ (at room temperature) as a function of composition. $\mathrm{Cos}^2\gamma$ has three linear segments for the high series (crosses and solid lines). The break in slope at $\sim An_{75}$ coincides with a change in structural state from $C\bar{1}$ to $I\bar{1}$ and extrapolation of the $C\bar{1}$ line (dashed line) to An_{100} suggests that a $C\bar{1}$ anorthite would have $\gamma = 90.71^\circ$. Filled circles = low series; low albite plots off the scale at $\cos^2\gamma = 1.6 \times 10^{-3}$.

Room temperature lattice parameters for "high" plagioclases show systematic trends (Bambauer et al., 1967; Smith, 1974; Kroll and Müller,

1980; Kroll and Ribbe, 1980; Kroll, 1983) but do not individually reflect the behaviour of the order parameter for Aℓ/Si ordering. On the other hand, the spontaneous strain is a thermodynamic quantity which should reveal the underlying variations of Q_{od}. In this context, it is convenient initially to take the monoclinic (C2/m) structure as the reference state for strain calculations. (The ideal monoclinic structure would almost certainly exist at high temperatures over the entire plagioclase solid solution if melting did not occur; Kroll and Bambauer, 1981; and see Fig. 5b). The quantitative results of Salje et al. (1985a) for albite showed that $\cos\alpha^*$ ($\propto$ - x_4) is most sensitive to displacive effects and $\cos\gamma$ ($\propto x_6$) to Aℓ/Si ordering. The same also appears to be true for other feldspars, and these parameters may therefore be used to distinguish between the two structural processes across the plagioclase solid solution. $\cos^2\alpha^*$ is linear over almost the entire range of the high series (Fig. 12b); there is a small deviation at pure albite, perhaps due to the proximity of the $C2/m \rightleftharpoons C\bar{1}$ displacive transition, and another at anorthite-rich compositions, possibly due to the influence of the $I\bar{1} \rightleftharpoons P\bar{1}$ transition. On the other hand, $\cos^2\gamma$ has three distinct linear ranges, and an obvious break in slope which coincides with the change from $C\bar{1}$ to $I\bar{1}$ symmetry at $\sim An_{75}$ (Fig. 12c). The $C\bar{1} \rightleftharpoons I\bar{1}$ transition is clearly marked by an increase in the spontaneous strain and this increase appears to be continuous through the transition point. The second break in slope of $\cos^2\gamma$ occurs at $\sim An_{35}$ in figure 12c and is perhaps related to changes in short range ordering of Aℓ and Si within the $C\bar{1}$ structure. Crystals annealed close to the $I\bar{1}$ field give diffuse superlattice reflections in electron diffraction patterns, indicative of local $I\bar{1}$ ordering, and the intensity of these diffuse reflections drops off with increasing albite content (Carpenter et al., 1985). The most albite-rich crystals are likely to have local ordering patterns more similar to low albite than to $I\bar{1}$ anorthite. Finally, extrapolation of the $C\bar{1}$ trend to An_{100} suggests that Aℓ/Si disordered anorthite crystals with $C\bar{1}$ symmetry would have $\gamma = 90.71^\circ$ at room temperature.

It is tempting to infer more from the linearity of the trends in figure 12, as for the case of $\cos^2\alpha^*$ against composition through the $C2/m \rightleftharpoons C\bar{1}$ transition in the alkali feldspar solid solution (Fig. 8). The analogy with alkali feldspars is incomplete, however, since they were at equilibrium with respect to the displacive transition when their lattice parameters were measured at room temperature. The equilibration temperatures of the high plagioclase series were 1100 - 1400°C and the measured lattice parameters refer to room temperature. In addition, a formal strain analysis for the $C\bar{1} \rightleftharpoons I\bar{1}$ transition must include the use of $C\bar{1}$ (triclinic) structures for reference states.

The variation of enthalpy through the $C\bar{1} \rightleftharpoons I\bar{1}$ transition may also be followed as a function of composition. Figure 13a shows heats of solution, in lead borate at 700°C, for both the ordered "low" plagioclases and the heat treated "high" plagioclases (Carpenter et al., 1985). Absolute heat of solution values are to some extent dependent on impurities and inclusions in the crystals being investigated, but the values of ΔH_{soln} for "high" $C\bar{1}$ crystals nevertheless define an approximately linear trend which closely matches the heats of solution of synthetic high plagioclase crystals (Newton et al., 1980). The implication of a linear trend in this case is that the $C\bar{1}$ solid solution shows ideal mixing (zero enthalpy of mixing). Other mixing data are consistent with this interpretation (Orville, 1972; Windom and Boettcher, 1976; Cermignani and Anderson, 1983; Carpenter and Ferry, 1984). Extrapolation of the $C\bar{1}$ trend to pure anorthite then gives predicted heats of solution for anorthite-rich crystals with $C\bar{1}$ structural states. Differences between these extrapolated values and the measured heats of solution of $I\bar{1}$ crystals are the excess enthalpies (as measured at 700°C) due to Aℓ/Si ordering, at 1300°C if the crystals were annealed at 1300°C or at some lower temperature if the crystals equilibrated in nature. As with the spontaneous strain (Fig. 12c) the excess enthalpy appears to show a continuous variation through the $C\bar{1} \rightleftharpoons I\bar{1}$ transition when followed as a function of composition (Fig. 13a, b).

Enthalpy and spontaneous strain data as a function of composition are thus consistent with the $C\bar{1} \rightleftharpoons I\bar{1}$ transition being continuous. Second order and tricritical models may be tested against the observed enthalpies using T_c^{An} = 2312 K for the transition in pure anorthite (see Fig. 5b), and an excess enthalpy of 13000 ± 2500 J mole^{-1} at 1573 K (see Fig. 13a). For the tricritical case:

$$Q_{od} = \left[1 - \frac{1573}{2312}\right]^{\frac{1}{4}} = 0\cdot75 \quad \text{at } 1573\text{ K} \tag{49}$$

$$H_{1573K} = -13000 = -\frac{1}{2}a_{od} \cdot 2312 \cdot 0\cdot75^2 + \frac{1}{6}c_{od} \cdot 0\cdot75^6 \tag{50}$$

and
$$2312 = \frac{c_{od}}{a_{od}} \tag{51}$$

from which a_{od} = 22.3 J mole^{-1} K^{-1} and c_{od} = 51600 J mole^{-1}. For the

(a)

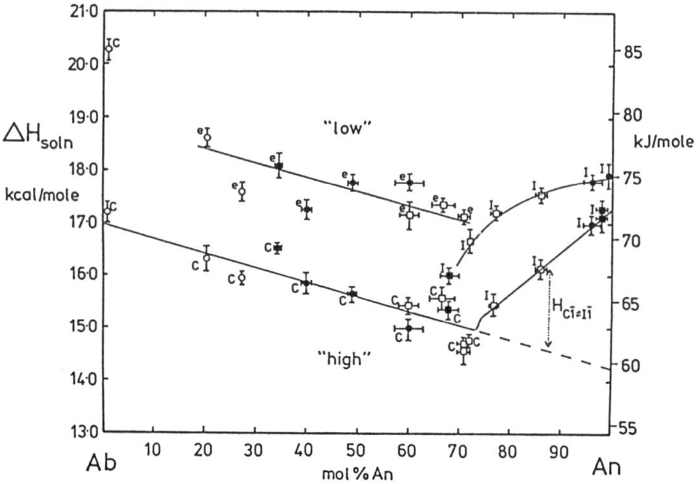

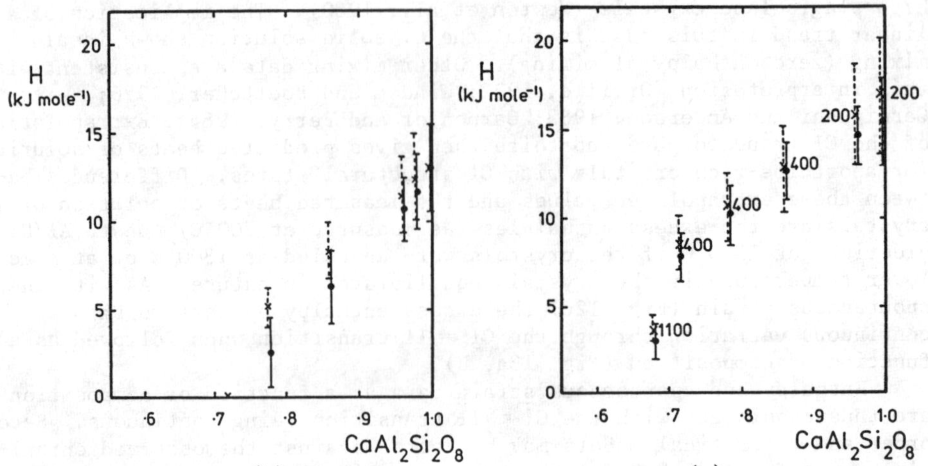

(b) (c)

Figure 13. (a) Heat of solution (ΔH_{soln}) of low and high plagioclases in lead borate at 700°C (from Carpenter et al., 1985). C = C$\bar{1}$ structure; I = I$\bar{1}$ structure. Open circles = samples from pegmatites and their heat treated equivalents; filled circles = metamorphic samples and their heat treated equivalents; open squares = igneous (plutonic) and heat treated equivalents; filled squares = igneous (volcanic) samples and heat treated equivalents. The dashed line represents a linear extrapolation of the ΔH_{soln} trend for "high" structures. Trends for I$\bar{1}$ "low" and I$\bar{1}$ "high" structures are also drawn in; the line for e samples is rather arbitrarily included to show their most general trend. The excess enthalpy due to the C$\bar{1}$ ⇌ I$\bar{1}$ transition ($H_{C\bar{1}\rightleftharpoons I\bar{1}}$) is given by the difference between the dashed line and the measured heat of solution for that sample. (b) Excess enthalpy due to the C$\bar{1}$ ⇌ I$\bar{1}$ transition as a function of composition (mole fraction of $CaA\ell_2Si_2O_8$ component) for high plagioclases equilibrated at 1300°C. Filled circles = observed values; crosses = values calculated using a tricritical model with constant coefficients. (c) Excess enthalpies due to C$\bar{1}$ ⇌ I$\bar{1}$ transition in natural I$\bar{1}$ crystals. Filled circles = observed values; crosses = values calculated using the same tricritical model as in (b), assuming equilibration temperature (in °C) given beside each point.

second order model, equations 3, 4, 6 give the analogous result that a_{od} = 41.3 J mole^{-1} K^{-1} and b_{od} = 95500 J mole^{-1} (using Q_{od} = 0.57 at 1573 K).

If Q_{od} is taken as being equal to 1 in pure anorthite at 0 K, and the coefficients are assumed to be independent of composition (as discussed above for the C2/m ⇌ C$\bar{1}$ transition in alkali feldspar crystals with Q_{od} = 0), a solution for the tricritical case is:

$$G = \frac{1}{2}a_{od}\left[T - \left(T_c^{An} + \frac{dT_c}{dn_{Ab}} \cdot n_{Ab}\right)\right]Q_{od}^2 + \frac{1}{6}c_{od}Q_{od}^6 \tag{52}$$

$$Q_{od} = \left[\left(1 + \frac{dT_c}{dn_{Ab}} \cdot \frac{n_{Ab}}{T_c^{An}}\right) - \frac{T}{T_c^{An}}\right]^{\frac{1}{4}} \tag{53}$$

$$H = -\frac{1}{2}a_{od}\left(T_c^{An} + \frac{dT_c}{dn_{Ab}} \cdot n_{Ab}\right)Q_{od}^2 + \frac{1}{6}c_{od}Q_{od}^6 \tag{54}$$

and, for the second order case:

$$G = \frac{1}{2}a_{od}\left[T - \left(T_c^{An} + \frac{dT_c}{dn_{Ab}} \cdot n_{Ab}\right)\right]Q_{od}^2 + \frac{1}{4}b_{od}Q_{od}^4 \tag{55}$$

$$Q_{od} = \left[\left(1 + \frac{dT_c}{dn_{Ab}} \cdot \frac{n_{Ab}}{T_c^{An}}\right) - \frac{T}{T_c^{An}}\right]^{\frac{1}{2}} \tag{56}$$

$$H = -\frac{1}{2}a_{od}\left(T_c^{An} + \frac{dT_c}{dn_{Ab}} \cdot n_{Ab}\right)Q_{od}^2 + \frac{1}{4}b_{od}Q_{od}^4 \tag{57}$$

where n_{Ab} is the mole fraction of albite in solid solution, and dT_c/dn_{Ab} = -2580 K n_{Ab}^{-1}, the slope of the $C\bar{1} \rightleftharpoons I\bar{1}$ transition line as a function of composition, from figure 5b.

Observed and calculated values for the tricritical model are shown in figure 13b. There is reasonable overlap for $I\bar{1}$ samples equilibrated at 1300°C. Rather better agreement is obtained for the low plagioclases after assuming that their equilibration temperatures (with respect to Aℓ/Si ordering) were ∿200°C for metamorphic rocks and ∿400°C for slowly cooled igneous rocks (Fig. 13b) (∿1100°C for the single volcanic sample). The greatest deviation occurs for the excess enthalpy of crystals with composition An_{78} which had been annealed at 1300°C, and this may point to limitations of the tricritical description at temperatures which are relatively close to T_c. Calculated excess enthalpies from the second order model provide a good fit with data from samples equilibrated at high temperatures but are significantly too high (by ∿7000 J mole^{-1}) for natural low anorthite. The transition thus appears to be close to tricritical but there is some evidence for a change in thermodynamic character with composition and temperature.

Further evidence favouring the tricritical description at $T \ll T_c$ is provided by estimates of Q_{od} from structure refinements of anorthite-rich crystals. The Aℓ and Si occupancies of tetrahedral sites cannot be refined directly from X-ray diffraction intensities but they can be estimated from mean bond lengths of the tetrahedral cations with oxygen (Ribbe and Gibbs, 1969; Smith, 1974; Kroll and Ribbe, 1980, 1983; Kroll, 1983;

Ribbe, 1984; Wenk and Kroll, 1984). Q_{od} may be defined in terms of the site occupancies as:

$$Q_{od} = \frac{(Q_{T1O} + Q_{T1M} + Q_{T2O} + Q_{T2M})}{4} \tag{58}$$

where

$$Q_{T1O} = \frac{\left(A\ell_{T1Oz0} + A\ell_{T1Ozi}\right) - \left(A\ell_{T1O00} + A\ell_{T1O0i}\right)}{\left(A\ell_{T1Oz0} + A\ell_{T1Ozi} + A\ell_{T1O00} + A\ell_{T1O0i}\right)} \tag{59}$$

and similarly for Q_{T1M}, Q_{T2O}, Q_{T2M}. $A\ell_{T1Oz0}$ is the proportion of Aℓ on the T1Oz0 site, $A\ell_{T1Ozi}$ is the proportion of Aℓ on the T1Ozi site etc. In almost every case where crystallographic data are available, the value of Q_{od} predicted by the tricritical model produces the closest agreement with observation (Table 1). The second order model gives values of Q_{od} which are too low by a significant margin.

An independent estimate of the total enthalpy of ordering in pure anorthite is provided by solution calorimetric data for synthetic crystals. If glass of anorthite composition is annealed for short times at temperatures of 1100 - 1400°C, crystals with highly diffuse ordering reflections and small values of cosγ are produced. Subsequent annealing of these substantially disordered crystals causes a progressive increase in the degree of Aℓ/Si order (Laves and Goldsmith, 1955; Goldsmith and Laves, 1956; Kroll and Müller, 1980; Carpenter, unpublished data). The most disordered crystals so far produced have $\gamma = 90.71 \pm 0.03°$ at room temperature, corresponding closely to the value of γ predicted above for $C\bar{1}$ anorthite (Fig. 12c). The enthalpy difference between crystals with $\gamma = 90.77$ and crystals equilibrated at 1673 K, with sharp superlattice reflections, has been measured as 13330 ± 1690 J $mole^{-1}$ (Carpenter, unpublished data). This compares with 12470 J $mole^{-1}$ for the excess enthalpy at 1673 K, due to a tricritical $C\bar{1} \rightleftharpoons I\bar{1}$ transition (equation 12, with the tricritical coefficients given above).

Although this thermodynamic treatment of the $C\bar{1} \rightleftharpoons I\bar{1}$ transition is clearly incomplete, a number of general conclusions may be drawn relating to the overall approach:

(a) The original method of determining coefficients in a Landau expansion using observations of T_c, C_p, spontaneous strain, etc., has to be adapted to the particular circumstances of solid solution behaviour and slow Aℓ/Si equilibration rates. The resulting equations can be used to describe the consequences of the order/disorder both for individual plagioclase crystals and for the solid solution as a whole.

(b) A more refined model must account explicitly for the spontaneous strain. As with the $I\bar{1} \rightleftharpoons P\bar{1}$ transition, the spontaneous strain energy is likely to be small in comparison with the total energy of the transition. The importance of the strain is, once again, that it provides a coupling mechanism, in this case for coupling between the $C2/m \rightleftharpoons C\bar{1}$ order parameter and the $C\bar{1} \rightleftharpoons I\bar{1}$ order parameter. Such coupling will be analogous to the coupling between Q and Q_{od} for the $I\bar{1} \rightleftharpoons P\bar{1}$ transition and should cause a renormalisation of the fourth order term in the Landau free energy expansion. A variation in the apparent thermodynamic

Table 1. Order parameter for Aℓ/Si ordering (Q_{od}) estimated from structural data (equation 58) and calculated using simple Landau expansions (equations 53 and 56).

Composition and locality	Reference	Q_{od} from estimated Aℓ occupancies	Calculated Values of Q_{od}		
			Tricritical Model	(T)	2nd Order Model
An_{100} Val Pasmeda (metamorphic)	Wainwright and Starkey (1971) in Smith 1974	0.96	0.94	(473 K)	0.89
∿An_{98} Vesuvius (volcanic)	Czank (1973), Megaw (1962), Ribbe (1963 } in Smith 1974	0.92, 0.92, 0.89	0.79 0.83	(1373 K) (1173 K)	0.62 0.69
∿An_{98} heat treated at 1803 K	Bruno et al. (1976)	0.73	0.67	(1803 K)	0.45
An_{85} Traversella	Chiari et al. (1984)	0.94	0.89	(473 K)	0.79
An_{85} heat treated at 1723 K	Chiari et al. (1984), Facchinelli et al. (1979)	0.69, 0.56	0.86 0.54	(673 K) (1723 K)	0.74 0.30
An_{80} St. Louis Co. Minnesota	Fleet et al. (1966) in Smith 1974	0.89	0.87 0.83	(473 K) (673 K)	0.76 0.70
An_{76} Crystal Bay Minnesota	Wainwright (1969) in Smith 1974	0.78	0.81 0.85	(673 K) (473 K)	0.66 0.73
An_{66} Lake Co.	Tagai et al. (1980)	0.78	0.40 0.58	(1373 K) (1173 K)	0.16 0.34

character of the $C\bar{1} \rightleftharpoons I\bar{1}$ transition with temperature and composition, as suggested by the enthalpy data, is therefore to be expected. Since the spontaneous strain is defined in terms of lattice parameters, it should be possible to calibrate the coupling from lattice constants, measured at high temperatures, for crystals with various compositions and structural states.

(c) Many of the plagioclase crystals quenched from high temperatures within the $C\bar{1}$ stability field give diffuse superlattice reflections in electron diffraction patterns (Carpenter and McConnell, 1984; Christofferson and Yund, 1984; Carpenter et al., 1985; Carpenter, 1986b). These are indicative of short range Aℓ/Si ordering or of quenched-in fluctuations in the degree of order. Structure refinements of heat treated crystals also suggest high degrees of residual order (Kroll, 1978; Kroll and Ribbe, 1980; Ribbe, 1983b, 1984; Benna et al., 1985). Ordering within crystals belonging to the $C\bar{1}$ solid solution will be influenced by the $C2/m \rightleftharpoons C\bar{1}$ transition, as in albite, and by precursor effects relating to the $C\bar{1} \rightleftharpoons I\bar{1}$ and $C\bar{1} \rightleftharpoons$ incommensurate transitions. It is to be expected that these local ordering variations will produce measurable lattice distortions, and, therefore, that they may also be characterised through spontaneous strain parameters.

4.3 Incommensurate ordering

Incommensurate superstructures have been found in many stoichiometric compounds, and they typically exist over a small temperature interval between the stability field of a high symmetry "para" phase and a low symmetry commensurate phase (Shiba and Ishibashi, 1978; Bruce et al., 1978; Cowley and Bruce, 1978; Heine and McConnell, 1984). The incommensurate superstructure of plagioclases, however, develops in crystals with a range of intermediate compositions in a solid solution (see Ribbe, 1984; Smith, 1984; Kitamura and Morimoto, 1984; and references therein). In the past, this incommensurate structure has been treated as some kind of metastable, compromise ordered state which develops in crystals that have Aℓ:Si ratios inappropriate for ordering either on the same basis as low albite (ideally with Aℓ:Si = 1:3) or $I\bar{1}$ anorthite (ideally with Aℓ:Si = 1:1), (Smith, 1974, 1983, 1984; Wenk et al., 1980; Wenk and Nakajima, 1980; Horst, 1984; Jagodzinski, 1984; and references therein). Recent calorimetric and experimental studies suggest that the incommensurate structure is substantially stabilised by cation ordering and that it may exist over a considerable range of temperature and composition in the equilibrium phase diagram (Carpenter et al., 1985; Carpenter, 1986b).

Thermodynamic relationships between the incommensurate structure and the $C\bar{1}$ and $I\bar{1}$ structures have yet to be fully resolved but, as in the other transitions in feldspars, spontaneous strain probably represents an important factor in controlling the transformation behaviour. With a monoclinic ($C2/m$) structure as the reference state, $\cos\alpha^*$ and $\cos\gamma$ for ordered ("low") plagioclases should reveal the contributions of displacive and Aℓ/Si order/disorder effects as a function of composition. Figure 12b shows that $\cos^2\alpha^*$ is approximately linear over the entire low plagioclase range and is barely different from values for the high plagioclase structures; there is, in particular, no significant difference between commensurate and incommensurate ordered structures. On the other hand, cosγ is of opposite sign for anorthite and low albite and is almost zero for the incommensurate phase. Cos$^2\gamma$ therefore has a pronounced minimum at intermediate composition in figure 12c and the incommensurate phase is characterised by having effectively zero strain associated with the ordering.

Solution calorimetric data show that the most ordered natural crystals with incommensurate structures have larger excess enthalpies of ordering if they are anorthite-rich than if they are albite-rich (Carpenter et al., 1985). This difference is also reflected in the peak temperature at which ordering occurs (Fig. 14), and a clear distinction can be made between "e_1" and "e_2" incommensurate fields on either side of the $C\bar{1} \rightleftharpoons I\bar{1}$ line. The distinction perhaps arises from the fact that to the anorthite side of the $C\bar{1} \rightleftharpoons I\bar{1}$ line there exists a driving force for Aℓ/Si ordering before the incommensurate phase appears, but that this driving force is absent on the albite side, i.e. in the $C\bar{1}$ field.

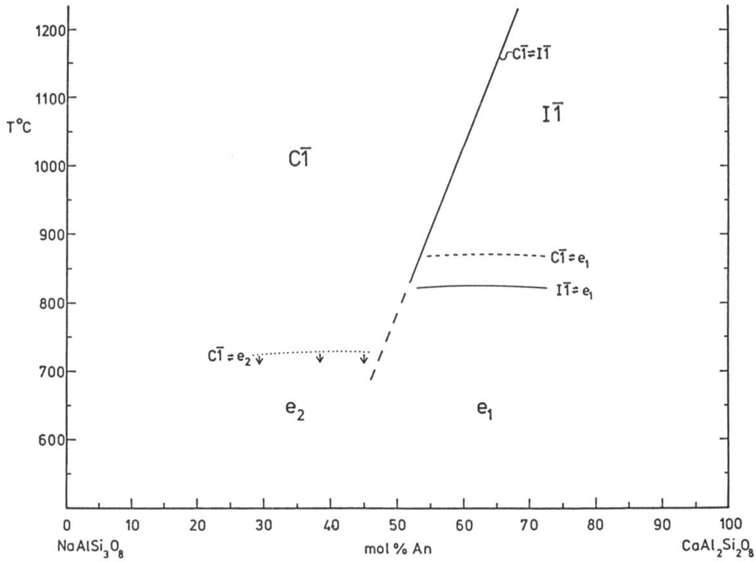

Figure 14. Summary of results from hydrothermal annealing experiments to delineate order/disorder transitions in the plagioclase solid solution (from Carpenter, 1986b). The dotted line for $C\bar{1} \rightleftharpoons e_2$ represents only an upper temperature limit; the dashed line represents a metastable $C\bar{1} \rightleftharpoons e_1$ transition. No attempt has been made to display a possible thermodynamic relationship between the different transitions or to include the miscibility gaps.

Structural differences between crystals from the e_1 and e_2 fields are also implied by a break in the variation with composition at $\sim An_{50}$ of various lattice parameters (Smith, 1974, 1984; Kroll, 1983; and references therein), and of the vector separating individual incommensurate reflections (Slimming, 1976). There is also a change in the character of ^{29}Si MASNMR spectra of incommensurate crystals at about the same composition (Kirkpatrick et al., 1987). These spectra have a number of well resolved and sharp peaks (Fig. 15) showing that the crystals possess a rather high degree of Aℓ/Si order.

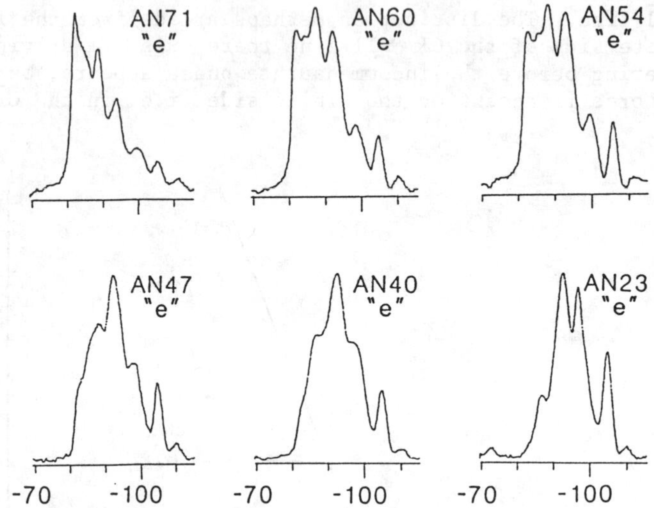

Figure 15. ^{29}Si MASNMR spectra for natural plagioclases with incommensurate structures. Note that the An_{60} and An_{54} ("e_1") spectra are almost identical with each other. The An_{47} and An_{40} ("e_2") spectra are also almost identical with each other but there is a marked difference between the two pairs. Chemical shift, relative to TMS, in p.p.m. (From Kirkpatrick et al., 1987).

By analogy with incommensurate phases in stoichiometric insulators, one might expect the incommensurate plagioclase structure to develop in a small range of temperatures just above the $C\bar{1} \rightleftharpoons I\bar{1}$ transition line. This would give a sequence, on cooling, of $C\bar{1}$ (disordered) → incommensurate (ordered) → $I\bar{1}$ (commensurate, ordered) phases. For compositions between $\sim An_{50}$ and $\sim An_{75}$, the actual sequence is $C\bar{1} \rightarrow I\bar{1} \rightarrow$ incommensurate, however, implying that the incommensurate phase gives the most stable ordering scheme possible in this range (Carpenter, 1986b). Incommensuate ordering may be a particularly appropriate mechanism for stabilising solid solutions since changes in temperature and composition can be accommodated by varying the repeat distance of the ordering periodicity.

Recent theories of incommensurate phases suggest that they are stabilised by interactions between two coexisting component structures (McConnell, 1978, 1981a, b, 1983, 1985; McConnell and Heine, 1985; Heine and McConnell, 1981, 1984). A quantitative thermodynamic model to explain the particular stability of the plagioclase incommensurate structure is not yet available, however.

Breaks in the ordering behaviour as a function of composition are almost certainly responsible for the miscibility gaps which exist in the solid solution at low temperatures. These occur between the fields of low albite ordering and e_2 ordering (peristerite gap), between the e_2 and e_1 fields (Bøggild gap) and between the e_1 and $I\bar{1}$ fields (Huttenlocher gap). The exact relationships between order/disorder and unmixing reactions remain unclear, though qualitative descriptions have been attempted (e.g. McConnell, 1974; Nord et al., 1974; Smith, 1974, 1983, 1984; Wenk and Nakajima, 1980; Wenk et al., 1980; Carpenter, 1981, 1985; Grove et al., 1983, 1986).

5. KINETICS

Owing to the sluggishness of many cation order/disorder processes in minerals, it is possible to quench and investigate a large range of intermediate structural states formed during isothermal annealing experiments. These are typically metastable but may or may not be related to equilibrium states, depending on the mechanisms and thermodynamics of the processes involved. An appropriate method of examining the relationship between equilibrium and non-equilibrium behaviour is to follow the macroscopic order parameters as functions of both time and temperature. The approach can be illustrated briefly for the kinetics of Aℓ/Si order/disorder in potassium feldspar and anorthite.

5.1 Disordering in potassium feldspar

Isothermal annealing of microcline ($C\bar{1}$, triclinic) at high temperatures causes progressive disordering until the equilibrium monoclinic state is achieved. Lattice parameter data obtained at room temperature for a sequence of samples annealed at 1050°C (Blasi et al., 1984b) show that $\cos^2\alpha^*$ ($\propto x_4^2$) and $\cos^2\gamma$ ($\propto x_6^2$) vary linearly with ln (annealing time), (Fig. 16). At short annealing times and long annealing times the trend tails off, since it must approach constant values asymptotically at $\pm\infty$ on a log time scale. Since $Q \propto x_4$ and $Q_{od} \propto x_6$, the result implies $Q^2 \propto \ln t$, and hence that the excess entropy varies linearly with lnt. An exactly analogous result has been obtained for disordering of cations in the mineral omphacite, where the variation of Q^2 was followed through the decreasing intensities of superlattice reflections with annealing time (Carpenter et al., 1987).

5.2 Ordering in anorthite

Figure 17a shows the variation of ΔH_{soln} with annealing time for a sequence of anorthite crystals synthesised and annealed at 1200°C.

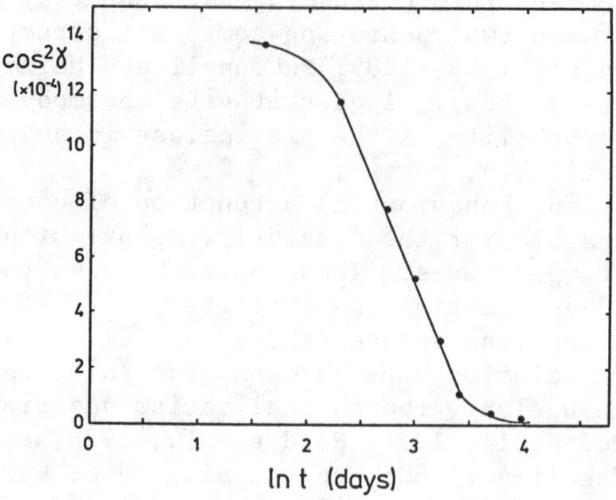

Figure 16. Variation of $\cos^2\gamma$ ($\propto \sim Q_{od}^2$) for Aℓ/Si disordering ($C\bar{1} \rightarrow C2/m$) as a function of annealing time (t) in $KA\ell Si_3O_8$ annealed at 1050°C. (Data from Blasi et al., 1984b).

Within experimental error, ΔH_{soln} is linear with lnt. The crystals, which were synthesised from glass, all had fine scale growth twins. At short annealing times their diffraction patterns had highly diffuse superlattice reflections (h + k = odd, ℓ = odd). Longer annealing causes these reflections to sharpen up, corresponding to steadily increasing Q_{od}, until antiphase domains can be imaged in dark field electron micrographs. The antiphase domains then increase in size with annealing time (Kroll and Müller, 1980; Carpenter, unpublished data). Since it has been shown that the excess enthalpy of ordering in anorthite can be reasonably well described by:

$$H = -\frac{1}{2} . 22{\cdot}3 . 2312 . Q_{od}^2 + \frac{1}{6} . 51600 . Q_{od}^6 \tag{60}$$

it is evident that H is approximately proportional to Q_{od}^2 because the term in Q_{od}^2 is dominant, especially for small values of Q_{od}. The observation (Fig. 17a) is once again that Q_{od}^2 is linear with lnt.

An independent measure of the excess entropy is provided by the variation of the spontaneous strain with time. Both the starting and product crystals are triclinic. The spontaneous strain depends on all six strain elements (equation 26a - f), and the scalar spontaneous strain, ε_s, is given by equation 27. For a reference state, the most disordered crystals ($\gamma = 90.71°$) produced in the experiments were used (Carpenter, unpublished data). If the spontaneous strain is associated entirely with the zone boundary ordering transition, the relationship between strain and order should be given by $\varepsilon_s \propto Q_{od}^2$. The only significant strain elements observed in this case were x_4 and x_6, as if the strain is being dominated by coupling to the monoclinic ⇌ triclinic (zone centre) transition, however, for which the relationship would be $\varepsilon_s \propto Q_{od}$. Figure 17b shows an approximately linear relationship between ε_s^2 and lnt for crystals annealed at 1200°C. To be consistent with the enthalpy data, the excess entropy must be proportional to ε_s^2.

(a)

(b)

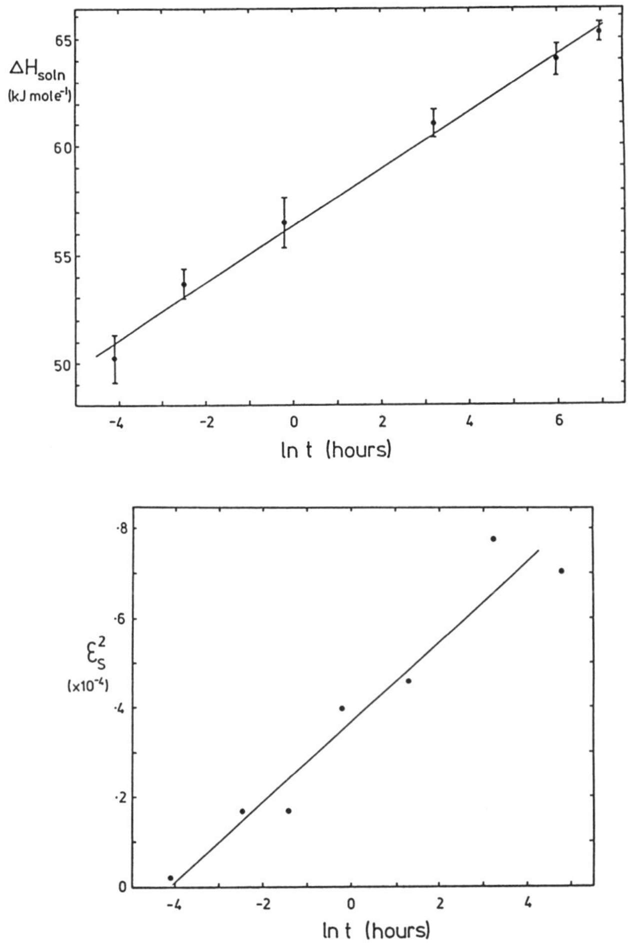

Figure 17. Variation, as a function of annealing time (t), of (a) ΔH_{soln} measured at $700^{o}C$, and (b) the square of spontaneous strain (ε_s^2), measured at room temperature, for Aℓ/Si ordering in anorthite at $1200^{o}C$. Starting material = glass of anorthite composition. (Carpenter, unpublished data).

These empirical relationships remain to be explored in detail, but it is notable that synthetic anorthites show a range of excess enthalpies of ordering which overlap with the equilibrium values. The most disordered synthetic crystals also have the same value of γ as predicted for $C\bar{1}$ anorthite using the extrapolation from $C\bar{1}$ crystals with intermediate compositions (Fig. 12c). This raises the intriguing question as to whether, during isothermal annealing, metastable anorthite structures evolve down the valley of a free energy surface which also defines equilibrium states as a function of temperature.

6. CONCLUSIONS

For geological problems, the importance of Landau theory lies in its description of thermodynamic properties in terms of other, more readily accessible macroscopic properties. The precision of conventional thermodynamic models for minerals has been limited by the difficulties, particularly for feldspars, of obtaining entropies for disordered and partly ordered crystals. Definitions of S in terms of an order parameter which, in turn, is determined from lattice parameters, obviate the necessity of extracting entropies directly from mean site occupancy information. Moreover, the use of order parameter relationships provides an alternative approach to the kinetics of solid state processes which is similarly independent of local structural details.

In a wider context, framework silicates provide models for general studies of structural phase transitions. As shown by Salje and coworkers, feldspars conform remarkably closely to the predictions of the theory. The semi-flexible three dimensional network of T-O-T bonds in these structures allows the influence of individual site configurations to extend over long distances and hence generates large correlation lengths. These strain effects are also responsible for the coupling between different order parameters. The associated elastic energies may represent a substantial proportion of the excess energy due to a phase transition, but, even in cases where the elastic energies are small, the role of spontaneous strain may be to determine the nature of energy minima with respect to lattice geometry, and hence to dominate the structural response to changes in pressure and temperature. It is significant that the influence of phase transitions on thermodynamic properties can extend to many hundreds of degrees away from the actual transition temperatures.

ACKNOWLEDGEMENTS

This review could not have been attempted without the advice and encouragement of E. Salje. I would like also to thank many colleagues for discussions and criticisms of the manuscript including, in particular, C. Capobianco, M. T. Dove, T. J. B. Holland, A. Putnis and B. Wruck. G. V. Gibbs, I. Parsons, A. Putnis, S. A. T. Redfern and E. Salje kindly gave permission for figures from their published work to be reproduced, and Sheila Tuffnell typed the manuscript, a daunting task for which many thanks are due.

Much of the solution calorimetry work was supported through grants from the Natural Environment Research Council of Great Britain (GR3/4404 and GR3/5547) which are gratefully acknowledged.

Cambridge Earth Sciences contribution number ES1012.

REFERENCES

Adlhart, W., Frey, F., Jagodzinski, H. (1980a) X-ray and neutron investigations of the P$\bar{1}$ - I$\bar{1}$ transition in pure anorthite. Acta Cryst. A36, 450-460.

Adlhart, W., Frey, F., Jagodzinski, H. (1980b) X-ray and neutron investigations of the P$\bar{1}$ - I$\bar{1}$ transition in anorthite with low albite content. Acta Cryst. A36, 461-470.

Aizu, K. (1970) Determination of the state parameters and formulation of spontaneous strain for ferroelastics. J. Phys. Soc. Japan 28, 706-716.

Als-Nielsen, J., Birgeneau, R. J. (1977) Mean field theory, the Ginzburg criterion, and marginal dimensionality of phase transitions. Am. J. Phys. 45, 554-560.

Bambauer, H. U., Bernotat, W. H. (1982) The microcline/sanidine transformation isograd in metamorphic regions. I. Composition and structural state of alkali feldspars from granitoid rocks of two N - S traverses across the Aar Massif and Gotthard «Massif», Swiss Alps. Schweiz. Min. Pet. 62, 185-230.

Bambauer, H. U., Eberhard, E., Viswanathan, K. (1967) The lattice constants and related parameters of "plagioclase (low)". Schweiz. Min. Pet. 47, 351-364.

Benna, P., Zanini, G., Bruno, E. (1985) Cell parameters of thermally treated anorthite. Aℓ, Si configurations in the average structures of the high temperature calcic plagioclases. Contrib. Mineral. Petrol. 90, 381-385.

Bernotat, W. H., Bambauer, H. U. (1982) The microcline/sanidine transformation isograd im metamorphic regions. II. The region of Lepontine metamorphism, Central Swiss Alps. Schweiz. Min. Pet. 62, 231-244.

Bismayer, U., Salje, E. (1981) Ferroelastic phases in $Pb_3(PO_4)_2$-$Pb_3(AsO_4)_2$; X-ray and optical experiments. Acta Cryst. A37, 145-153.

Bismayer, U., Salje, E., Glazer, A. M., Cosier, J. (1986) Effect of strain-induced order-parameter coupling on the ferroelastic behaviour of lead phosphate-arsenate. Phase Transitions 6, 129-151.

Blasi, A., Brajkovic, A., de Pol Blasi, C., Foord, E. E., Martin, R. F., Zanazzi, P. F. (1984a) Structure refinement and genetic aspects of a microcline overgrowth on amazonite from Pikes Peak batholith, Colorado, U.S.A. Bull. Minéral. 107, 411-422.

Blasi, A., Brajkovic, A., de Pol Blasi, C. (1984b) Dry-heating conversion of low microcline to high sanidine via a one-step disordering process. Bull Minéral. 107, 423-435.

Blinova, G. K., Kiseleva, I. A. (1982) A calorimetric study of the structural transitions in plagioclase. Geochem. Internat. 19, 74-80.

Brown, W. L., Hoffman, W., Laves, F. (1963) Über kontinuierliche und reversible Transformation des Anorthits ($CaA\ell_2Si_2O_8$) zwischen 25 und 350°C. Naturwiss. 50, 221.

Brown, W. L., Openshaw, R. E., McMillan, P. F., Henderson, C. M. B. (1984) A review of the expansion behavior of alkali feldspars: coupled variations in cell parameters and possible phase transi-

tions. Am. Mineral. 69, 1058-1071.
Brown, W. L., Parsons, I. (1984) Exsolution and coarsening mechanisms and kinetics in an ordered cryptoperthite series. Contrib. Mineral. Petrol. 86, 3-18.
Bruce, A. D., Cowley, R. A. (1981) Structural phase transitions. Taylor and Francis, London.
Bruce, A. D., Cowley, R. A., Murray, A. F. (1978) The theory of structurally incommensurate systems: II. Commensurate-incommensurate phase transitions. J. Phys. C11, 3591-3608.
Bruno, E., Chiari, G., Facchinelli, A. (1976) Anorthite quenched from 1530°C. I. Structure refinement. Acta Cryst. B32, 3270-3280.
Bruno, E., Gazzoni, G. (1967) Ricerche roentgenografiche su plagioclasi bytownico-anoritici tra 15°C e 1300°C. Per. di Min. 36, 683-698.
Carpenter, M. A. (1981) A "conditional spinodal" within the peristerite miscibility gap of plagioclase feldspars. Am. Mineral. 66, 553-560.
Carpenter, M. A. (1985) Order-disorder transformations in mineral solid solutions. In: Kieffer, S. W., Navrotsky, A. (eds.) Microscopic to macroscopic: atomic environments to mineral thermodynamics. Reviews in Mineralogy 14, 187-223, Min. Soc. Am.
Carpenter, M. A. (1986a) Some recent advances in the characterisation of silicates. In: Freer, R., Dennis, P. F. (eds.) Kinetics and mass transport in silicate and oxide systems. Mat. Sci. Forum 7, 19-34. Trans. Tech. Publications: Switzerland, Germany, U.K., U.S.A..
Carpenter, M. A. (1986b) Experimental delineation of the "e" ⇌ $I\bar{1}$ and "e" ⇌ $C\bar{1}$ transformations in intermediate plagioclase feldspars. Phys. Chem. Min. 13, 119-139.
Carpenter, M. A., Domeneghetti, C., Tazzoli, V. (1987) Kinetics of cation disordering in omphacite (abstract). Terra Cognita 7, 263.
Carpenter, M. A., Ferry, J. M. (1984) Constraints on the thermodynamic mixing properties of plagioclase feldspars. Contrib. Mineral. Petrol. 87, 138-148.
Carpenter, M. A., McConnell, J. D. C. (1984) Experimental delineation of the $C\bar{1}$ ⇌ $I\bar{1}$ transformation in intermediate plagioclase feldspars. Am. Mineral. 69, 112-121.
Carpenter, M. A., McConnell, J. D. C., Navrotsky, A. (1985) Enthalpies of ordering in the plagioclase feldspar solid solution. Geochim. Cosmochim. Acta 49, 947-966.
Carpenter, M. A., Putnis, A., Navrotsky, A., McConnell, J. D. C. (1983) Enthalpy effects associated with Aℓ/Si ordering in anhydrous Mg-cordierite. Geochim. Cosmochim. Acta 47, 899-906.
Carpenter, M. A., Putnis, A. (1985) Cation order and disorder during crystal growth: some implications for natural mineral assemblages. In: Thompson, A. B., Rubie, D. C. (eds.) Metamorphic reactions. Advances in physical geochemistry 4, 1-26. Springer-Verlag: New York, Berlin, Heidelberg, Tokyo.
Cermignani, C., Anderson, G. M. (1983) The plagioclase exchange reaction in carbonate solutions and applications to nephelinization. Am. J. Sci. 283A, 314-327.

Černý, P., Chapman, R. (1984) Paragenesis, chemistry, and structural state of adularia from granitic pegmatites. Bull. Minéral. 107, 369-384.
Chiari, G., Benna, P., Bruno, E. (1984) The structure of bytownite (An_{85}). A new refinement. Zeit. Krist. 169, 35-49.
Cohen, R. E., Burnham, C. W. (1985) Energetics of ordering in aluminous pyroxenes. Am. Mineral. 70, 559-567.
Cowley, R. A. (1976) Acoustic phonon instabilities and structural phase transitions. Phys. Rev. B13, 4877-4885.
Cowley, R. A., Bruce, A. D. (1978) The theory of structurally incommensurate systems: I. Disordered-incommensurate phase transitions. J. Phys. C11, 3577-3590.
Christofferson, R. G., Yund, R. A. (1984) An experimental study of the Aℓ,Si distribution in the average structure of andesine. Ns. Jahrb. Miner. Abh. 14, 91-104.
Czank, M. (1973) Strukturuntersuchungen von Anorthit im Temperatur bereich von 20°C bis 1430°C. Doctoral thesis, E.T.H., Zürich.
Czank, M., van Landuyt, J., Schulz, H., Laves, F., Amelinckx, S. (1973) Electron microscopic study of the structural changes as a function of temperature in anorthite. Zeit. Krist. 138, 403-418.
Eggleton, R. A., Buseck, P. R. (1980) The orthoclase-microcline inversion: a high-resolution transmission electron microscope study and strain analysis. Contrib. Mineral. Petrol. 74, 123-133.
Facchinelli, A., Bruno, E., Chiari, G. (1979) The structure of bytownite quenched from 1723 K. Acta Cryst. B35, 34-42.
Fitz Gerald, J. D., McLaren, A. C. (1982) The microstructures of microcline from some granitic rocks and pegmatites. Contrib. Mineral. Petrol. 80, 219-229.
Fleet, S. G., Chandrasekhar, S., Megaw, H. D. (1966) The structure of bytownite ('body-centred anorthite'). Acta Cryst. 21, 782-801.
Foit, F. F., Jr., Peacor, D. R. (1973) The anorthite crystal structure at 410 and 830°C. Am. Mineral. 58, 665-675.
Folk, R., Iro, H., Schwabl, F. (1976) Critical statics of elastic phase transitions. Zeit. Physik. B25, 69-81.
Freer, R. (1981) Diffusion in silicate minerals and glasses: a data digest and guide to the literature. Contrib. Mineral. Petrol. 76, 440-454.
Frey, F., Jagodzinski, H., Prandl, W., Yelon, W. B. (1977) Dynamic character of the primitive to body-centred phase transition in anorthite. Phys. Chem. Min. 1, 227-231.
Fyfe, C. A., Thomas, J. M., Klinowski, J., Gobbi, G. C. (1983) Magic-angle-spinning NMR (MAS-NMR) spectroscopy and the structure of zeolites. Angew. Chem. 22, 259-275.
Geisinger, K. L., Gibbs, G. V., Navrotsky, A. (1985) A molecular orbital study of bond length and angle variations in framework structures. Phys. Chem. Min. 11, 266-283.
Gibbs, G. V., Meagher, E. P., Newton, M. D., Swanson, D. K. (1981) A comparison of experimental and theoretical bond length and angle variations for minerals, inorganic solids, and molecules. In: O'Keefe, M., Navrotsky, A. (eds.) Structure and bonding in crystals, 1, 195-225. Academic Press: New York, London, Toronto,

Sydney, San Francisco.
Goldsmith, J. R. (1986) The role of hydrogen in promoting Aℓ-Si interdiffusion in albite ($NaA\ell Si_3O_8$) at high pressures. Earth Planet. Sci. Lett. 80, 135-138.
Goldsmith, J. R. (1987) Aℓ/Si interdiffusion in albite: effect of pressure and the role of hydrogen. Contrib. Mineral. Petrol. 95, 311-321.
Goldsmith, J. R., Jenkins, D. M. (1985) The high-low albite relations revealed by reversal of degree of order at high pressures. Am. Mineral. 70, 911-923.
Goldsmith, J. R., Laves, F. (1956) Crystallization of metastable disordered anorthite at "low temperatures". Zeit. Krist. 107, 396-405.
Goldsmith, J. R., Newton, R. C. (1974) An experimental determination of the alkali feldspar solvus. In: MacKenzie, W. S., Zussman, J. (eds.) The feldspars, 337-359. Manchester University Press.
Greenwood, H. J. (1972) $A\ell^{IV}$-Si^{IV} disorder in sillimanite and its effect on phase relations of the aluminium silicate minerals. Geol. Soc. Am. Memoir 132, 553-571.
Grove, T. L., Burnham, C. W. (1974) Aℓ-Si disorder in calcium Tschermak's pyroxene, $CaA\ell_2SiO_6$ (abstract). Trans. Am. Geophys. Un. 55, 1202.
Grove, T. L., Ferry, J. M., Spear, F. S. (1983) Phase transitions and decomposition relations in calcic plagioclase. Am. Mineral. 68, 41-59.
Grove, T. L., Ferry, J. M., Spear, F. S. (1986) Phase transitions in calcic plagioclase: a correction and further discussion. Am. Mineral. 71, 1049-1050.
Grundy, H. D., Brown, W. L. (1969) A high-temperature X-ray study of the equilibrium forms of albite. Mineral. Mag. 37, 156-172.
Guimaraes, D. M. C. (1979) Temperature dependence of lattice parameters and spontaneous strain in $Pb_3(PO_4)_2$. Phase Transitions 1, 143-154.
Harlow, G. E. (1982) The anorthoclase structures: the effects of temperature and composition. Am. Mineral. 67, 975-996.
Harlow, G. E., Brown, G. E., Jr. (1980) Low albite: an X-ray and neutron diffraction study. Am. Mineral. 65, 986-995.
Haselton, H. T., Jr., Hovis, G. L., Hemingway, B. S., Robie, R. A. (1983) Calorimetric investigation of the excess entropy of mixing in analbite-sanidine solid solutions: lack of evidence for Na, K short-range order and implications for two-feldspar thermometry. Am. Mineral. 68, 398-413.
Heine, V., McConnell, J. D. C. (1981) Origin of modulated incommensurate phases in insulators. Phys. Rev. Lett. 46, 1092-1095.
Heine, V., McConnell, J. D. C. (1984) The origin of incommensurate structures in insulators. J. Phys. C17, 1199-1220.
Henderson, C. M. B. (1984) Feldspathoid stabilities and phase inversions - a review. In: Brown, W. L. (ed.) Feldspars and feldspathoids. NATO Adv. Stud. Inst. C137, 471-499.
Holland, T. J. B., Carpenter, M. A. (1986) Aluminium/silicon disordering and melting in sillimanite at high pressures. Nature 320, 151-153.

Holland, T. J. B., Powell, R. (1985) An internally consistent thermodynamic dataset with uncertainties and correlations: 2. Data and results. J. Met. Geol. 3, 343-370.

Holm, J. L., Kleppa, O. J. (1968) Thermodynamics of the disordering process in albite. Am. Mineral. 53, 123-133.

Horst, W. (1984) Modulated structures in intermediate plagioclases - history, ideas and perspectives. Fortschr, Miner. 62, 107-154.

Hovis, G. L. (1974) A solution calorimetric and X-ray investigation of Aℓ-Si distribution in monoclinic potassium feldspars. In: MacKenzie, W. S., Zussman, J. (eds.) The feldspars, 114-144. Manchester University Press: Manchester.

Hovis, G. L. (1980) Angular relations of alkali feldspar series and the triclinic-monoclinic displacive transformation. Am. Mineral. 65, 770-778.

Hovis, G. L. (1986) Behavior of alkali feldspars: Crystallographic properties and characterization of composition and Aℓ-Si distribution. Am. Mineral. 71, 869-890.

Hovis, G. L., Waldbaum, D. R. (1977) A solution calorimetric investigation of K-Na mixing in a sanidine-analbite ion-exchange series. Am. Mineral. 62, 680-686.

Jagodzinski, H. (1984) Determination of modulated feldspar structures. Bull. Minéral, 107, 455-466.

de Jong, B. H. W. S., Brown, G. E., Jr. (1980) Polymerization of silicate and aluminate tetrahedra in glasses, melts, and aqueous solutions - I. Electronic structure of $H_6Si_2O_7$, $H_6A\ell SiO_7^{1-}$, and $H_6A\ell_2O_7^{2-}$. Geochim. Cosmochim. Acta 44, 491-511.

Kirkpatrick, R. J., Carpenter, M. A., Yang, W.-H., Montez, B. (1987) ^{29}Si magic-angle NMR spectroscopy of low-temperature ordered plagioclase feldspars. Nature 325, 236-238.

Kirkpatrick, R. J., Smith, K. A., Schramm, S., Turner, G., Yang, W. -H. (1985) Solid-state nuclear magnetic resonance spectroscopy of minerals. Ann. Rev. Earth Planet. Sci. 13, 29-47.

Kitamura, M., Morimoto, N. (1984) The modulated structure of the intermediate plagioclases and its change with composition. In: Brown, W. L. (ed.) Feldspars and feldspathoids. NATO Adv. Stud. Inst. C137, 95-119. Reidel: Dordrecht, Boston, Lancaster.

Kleppa, O. J. (1982) Thermochemistry in mineralogy. In: Schreyer, W. E. (ed.) High pressure researches in geoscience, 461-473. E. Schweizerbart'sche Verlagsbuchhandlung (Nägele u. Obermiller): Stuttgart.

Kroll, H. (1978) The structure of heat-treated plagioclases An_{28}, An_{52}, An_{69} and the estimation of Aℓ, Si order from lattice parameters. Phys. Chem. Min. 3,76-77.

Kroll, H. (1983) Lattice parameters and determinative methods for plagioclase and ternary feldspars. In: Ribbe, P. H. (ed.) Feldspar mineralogy. Reviews in mineralogy 2, (2nd. $ed^{n.}$), 101-119. Min. Soc. Am.

Kroll, H. (1984) Thermal expansion of alkali feldspars. In: Brown, W. L. (ed.) Feldspars and feldspathoids. NATO Adv. Stud. Inst. C137, 163-205. Reidel: Dordrecht, Boston, Lancaster.

Kroll, H., Bambauer, H. -U., Schirmer, U. (1980) The high albite-monalbite and analbite-monalbite transitions. Am. Mineral. 65, 1192-1211.

Kroll, H., Bambauer, H. -U. (1981) Diffusive and displacive transformation in plagioclase and ternary feldspar series. Am. Mineral. 66, 763-769.

Kroll, H., Müller, W. F. (1980) X-ray and electron-optical investigation of synthetic high-temperature plagioclases. Phys. Chem. Min. 5, 255-277.

Kroll, H., Ribbe, P. H. (1980) Determinative diagrams for $A\ell$,Si order in plagioclases. Am. Mineral. 65, 449-457.

Kroll, H., Ribbe, P. H. (1983) Lattice parameters, composition and $A\ell$,Si order in alkali feldspars. In: Ribbe, P. H. (ed.) Feldspar mineralogy. Reviews in mineralogy 2, (2nd edn.), 57-99. Min. Soc. Am.

Kroll, H., Schmiemann, I., von Cölln, G. (1986) Feldspar solid solutions. Am. Mineral. 71, 1-16.

Landau, L. D., Lifschitz, E. M. (1958) Statistical physics. Addison Wesley: Reading, Massachusetts.

Landolt-Börnstein (1979) Numerical data and functional relationships in science and technology. New Series Group 3, 11:72.

Lasaga, A. C., Gibbs, G. V. (1987) Aplications of quantum mechanical potential surfaces to mineral physics calculations. Phys. Chem. Min. 14, 107-117.

Laves, F., Goldsmith, J. R. (1954) Long-range-short-range order in calcic plagioclases as a continuous and reversible function of temperature. Acta Cryst. 7, 465-472.

Laves, F., Goldsmith, J. R. (1955) The effect of temperature and composition on the $A\ell$-Si distribution in anorthite. Zeit. Krist. 106, 227-235.

Laves, F., Czank, M., Schulz, H. (1970) The temperature dependence of the reflection intensities of anorthite ($CaA\ell_2Si_2O_8$) and the corresponding formation of domains. Schweiz. Min. Pet. 50, 519-525.

Loewenstein, W. (1954) The distribution of aluminium in the tetrahedra of silicates and aluminates. Am. Mineral. 39, 92-96.

Louisnathan, S. J. (1971) Refinement of the crystal structure of a natural gehlenite, $Ca_2A\ell(A\ell,Si)_2O_7$. Can. Mineral. 10, 822-837.

McConnell, J. D. C. (1965) Electron optical study of effects associated with partial inversion in a silicate phase. Phil. Mag. 11, 1289-1301.

McConnell, J. D. C. (1971) Electron-optical study of phase transformations. Mineral. Mag. 38, 1-20.

McConnell, J. D. C. (1974) Analysis of the time-temperature-transformation behaviour of the plagioclase feldspars. In: MacKenzie, W. S., Zussman, J. (eds.) The feldspars, 460-477. Manchester University Press: Manchester.

McConnell, J. D. C. (1978) The intermediate plagioclase feldspars: an example of a structural resonance. Zeit. Krist. 147, 45-62.

McConnell, J. D. C. (1981a) Time-temperature study of the intensity of satellite reflections in nepheline. Am. Mineral. 66, 990-996.

McConnell, J. D. C. (1981b) Electron-optical study of modulated

mineral solid solutions. Bull. Minéral. 104, 231-235.
McConnell, J. D. C. (1983) A review of structural resonance and the nature of long-range interactions in modulated mineral structures. Am. Mineral. 68, 1-10.
McConnell, J. D. C. (1985) Symmetry aspects of order-disorder and the application of Landau theory. In: Kieffer, S. W., Navrotsky, A. (eds.) Macroscopic to microscopic: atomic environments to mineral thermodynamics. Reviews in mineralogy 14, 165-186. Min. Soc. Am.
McConnell, J. D. C., Heine, V. (1985) Incommensurate structure and stability of mullite. Phys. Rev. B31, 6140-6142.
McKie, D., McConnell, J. D. C. (1963) The kinetics of the low → high transformation in albite. I: Amelia albite under dry conditions. Mineral. Mag. 33, 581-588.
McLaren, A. C. (1984) Transmission electron microscope investigations of the microstructures of microclines. In: Brown, W. L. (ed.) Feldspars and feldspathoids. NATO Adv. Stud. Inst. C137, 373-409. Reidel: Dordrecht, Boston, Lancaster.
McLaren, A. C., Fitz Gerald, J. D. (1987) CBED and ALCHEMI investigation of local symmetry and Aℓ,Si ordering in K-feldspars. Phys. Chem. Min. 14, 281-292.
Megaw, H. D. (1962) Order and disorder in feldspars. Norsk. Geol. Tidsskr. 42, 2, 104-137.
Merkel, G. A., Blencoe, J. G. (1982) Thermodynamic procedures for treating the monoclinic/triclinic inversion as a high-order phase transition in equations of state for binary analbite-sanidine feldspars. In: Saxena, S. K. (ed.) Advances in Physical Geochemistry 2, 243-284. Springer Verlag: New York, Heidelberg, Berlin.
Merlino, S. (1984) Feldspathoids: their average and real structures. In: Brown, W. L. (ed.) Feldspars and Feldspathoids. NATO Adv. Stud. Inst. C137, 435-470. Reidel: Dordrecht, Boston, Lancaster.
Navrotsky, A. (1977) Progress and new directions in high temperature calorimetry. Phys. Chem. Min. 2, 89-104.
Navrotsky, A. (1979) Calorimetry: its application to petrology. Ann. Rev. Earth Planet. Sci. 7, 93-115.
Navrotsky, A. Experimental studies of mineral energetics (this volume).
Navrotsky, A., Geisinger, K. L., McMillan, P., Gibbs, G. V. (1985) The tetrahedral framework in glasses and melts - inferences from molecular orbital calculations and implications for structure, thermodynamics, and physical properties. Phys. Chem. Min. 11, 284-298.
Newton, R. C., Charlu, T. V., Kleppa, O. J. (1980) Thermochemistry of the high structural state plagioclases. Geochim. Cosmochim. Acta 44, 933-941.
Nord, G. L. Jr., Heuer, A. H., Lally, J. S. (1974) Transmission electron microscopy of substructures in Stillwater bytownites. In: MacKenzie, W. S., Zussman, J. (eds.) The feldspars, 522-535. Manchester University Press: Manchester.
Nye, J. F. (1957) Physical properties of crystals. Oxford University Press: Oxford.
Okamura, F. P., Ghose, S., Ohashi, H. (1974) Structure and crystal chemistry of calcium Tschermak's pyroxene, $CaAℓAℓSiO_6$. Am. Mineral.

59, 549-557.
Oldfield, E., Kirkpatrick, R. J. (1985) High-resolution nuclear magnetic resonance of inorganic solids. Science 227, 1537-1544.
Openshaw, R. E., Hemingway, B. S., Robie, R. A., Krupka, K. M. (1979a) A room-temperature phase transition in maximum microcline. Heat capacity measurements. Phys. Chem. Min. 5, 83-93.
Openshaw, R. E., Hemingway, B. S., Robie, R. A., Krupka, K. M. (1979b) A room-temperature phase transition in maximum microcline. Unit-cell parameters and thermal expansion. Phys. Chem. Min. 5, 95-104.
Orville, P. M. (1972) Plagioclase cation exchange equilibria with aqueous chloride solution: results at 700°C and 2000 bars in the presence of quartz. Am. J. Sci. 272, 234-272.
Parsonage, N. G., Staveley, L. A. K. (1978) Disorder in crystals. Oxford University Press: Oxford.
Parsons, I., Brown, W. L. (1984) Feldspars and the thermal history of igneous rocks. In: Brown, W. L. (ed.) Feldspars and feldspathoids. NATO Adv. Stud. Inst. C137, 317-371.
Putnis, A. (1980a) The distortion index in anhydrous Mg-cordierite. Contrib. Mineral. Petrol. 74, 135-141.
Putnis, A. (1980b) Order-modulated structures and the thermodynamics of cordierite reactions. Nature 287, 128-131.
Putnis, A. Solid state NMR spectroscopy and phase transitions in minerals. (this volume).
Putnis, A., Angel, R. J. (1985) Aℓ,Si ordering in cordierite using "magic angle spinning" NMR. II: Models of Aℓ,Si order from NMR Data. Phys. Chem. Min. 12, 217-222.
Putnis, A., Bish, D. L. (1983) The mechanism and kinetics of Aℓ,Si ordering in Mg-cordierite. Am. Mineral. 68, 60-65.
Putnis, A., Fyfe, C. A., Gobbi, G. C. (1985) Aℓ,Si ordering in cordierite using "magic angle spinning" NMR. I: Si^{29} spectra of synthetic cordierites. Phys. Chem. Min. 12, 211-216.
Putnis, A., Salje, E., Redfern, S. A. T., Fyfe, C. A. (1987) Structural states of Mg-cordierite. I: Order parameters from synchrotron X-ray and NMR data. Phys. Chem. Min. (in press).
Rao, C. N. R., Rao, K. J. (1978) Phase transitions in solids. McGraw-Hill: New York.
Redfern, S. A. T., Salje, E. (1987a) Thermodynamics of plagioclase II: temperature evolution of the spontaneous strain at the $I\bar{1} \leftrightarrow P\bar{1}$ phase transition in anorthite. Phys. Chem. Min. 14, 189-195.
Redfern, S. A. T., Salje, E. (1987b) Spontaneous strain at the $I\bar{1} \rightleftharpoons P\bar{1}$ transition in anorthite (abstract). Terra Cognita 7, 264.
Ribbe, P. H. (1963) Structural studies of plagioclase feldspars. Ph.D. thesis. University of Cambridge.
Ribbe, P. H. (1983a) Chemistry, structure and nomenclature of feldspars. In: Ribbe, P. H. (ed.) Feldspar mineralogy. Reviews in mineralogy 2, (2nd edn.), 1-19. Min. Soc. Am.
Ribbe, P. H. (1983b) Aluminum - silicon order in feldspars; domain textures and diffraction patterns. In: Ribbe, P. H. (ed.) Feldspar mineralogy. Reviews in mineralogy 2, (2nd edn.), 21-55. Min. Soc. Am.

Ribbe, P. H. (1983c) Exsolution textures in ternary and plagioclase feldspars; interference colours. In: Ribbe, P. H. (ed.), Feldspar mineralogy. Reviews in mineralogy 2, (2nd edn.), 241-270. Min. Soc. Am.

Ribbe, P. H. (1984) Average structures of alkali and plagioclase feldspars: systematics and applications. In: Brown, W. L. (ed.) Feldspars and feldspathoids. NATO Adv. Stud. Inst. C137, 1-54. Reidel: Dordrecht, Boston, Lancaster.

Ribbe, P. H., Gibbs, G. V. (1969) Statistical analysis and discussion of mean Aℓ/Si - O bond distances and the aluminium content of tetrahedra in feldspars. Am. Mineral. 54, 85-94.

Robie, R. A., Hemingway, B. S. (1972) Calorimeters for heat of solution and low temperature heat capacity measurements. U.S. Geol. Surv. Prof. Papers 755 , 32pp.

Robie, R. A., Hemingway, B. S., Fisher, J. R. (1978) Thermodynamic properties of minerals and related substances at 298.15° K and 1 bar (10^5 pascals) pressure and at higher temperatures. U.S. Geol. Surv. Bull. 1452, 456pp.

Ryzhova, T. V. (1964) Elastic properties of plagioclase. Akad. Nauk SSSR, Bull. Geophys. Ser. 7, 633-635. (pp. 1049-1051 in Russian).

Ryzhova, T. V., Aleksandrov, K. S. (1965) The elastic properties of potassium-sodium feldspars. Akad. Nauk SSSR, Bull. Geophys. Ser. 1, 53-56. (pp. 98-102 in Russian).

Salje, E. (1985) Thermodynamics of sodium feldspar I: order parameter treatment and strain induced coupling effects. Phys. Chem. Min. 12, 93-98.

Salje, E. (1986a) Application of high resolution powder diffractometry to the investigation of structural phase transitions. In: Catlow, C. R. A. (ed.) High resolution powder diffraction. Mat. Sci. Forum 9, 57-68. Trans. Tech. Publications: Switzerland, Germany, U.K., U.S.A.

Salje, E. (1986b) Raman spectroscopic investigation of the order parameter behaviour in hypersolvus alkali feldspar: displacive phase transition and evidence for Na-K site ordering. Phys. Chem. Min. 13, 340-346.

Salje, E. Structural phase transitions and specific heat anomalies. (this volume).

Salje, E. (1987a) Thermodynamics of plagioclases I: Theory of the $I\bar{1}$ - $P\bar{1}$ phase transition in anorthite and Ca-rich plagioclases. Phys. Chem. Min. 14, 181-188.

Salje, E. (1987b) Structural states of Mg-cordierite II: Landau theory. Phys. Chem. Min. (in press).

Salje, E., Bismayer, U., Jansen, M. (1987) Temperature evolution of the ferroelastic order parameter of As_2O_5 as determined from optical birefringences. J. Phys. C. (in press).

Salje, E., Devarajan, V. (1986) Phase transitions in systems with strain-induced coupling between two order parameters. Phase Transitions 6, 235-248.

Salje, E., Kuscholke, B. (1984) On the structural phase transitions in alkali feldspars. Bull. Minéral. 107, 539.

Salje, E., Kuscholke, B., Wruck, B., Kroll, H. (1985a) Thermodynamics of sodium feldspar II: experimental results and numerical calculations. Phys. Chem. Min. 12, 99-107.

Salje, E., Kuscholke, B., Wruck, B. (1985b) Domain wall formation in minerals: I. Theory of twin boundary shapes in Na-feldspar. Phys. Chem. Min. 12, 132-140.

Shiba, H., Ishibashi, Y. (1978) Incommensurate-commensurate phase transitions in ferroelectric substances. J. Phys. Soc. Japan 44, 1592-1599.

Slimming, E. H. (1976) An electron diffraction study of some intermediate plagioclases. Am. Mineral. 61, 54-59.

Smart, R. M., Glasser, F. P. (1977) Stable cordierite solid solutions in the $MgO-A\ell_2O_3-SiO_2$ system: composition, polymorphism and thermal expansion. Science of Ceramics 9, 256-263.

Smith, J. V. (1974) Feldspar minerals, Vol. 1, Crystal structure and physical properties. Springer-Verlag: Berlin, Heidelberg, New York.

Smith, J. V. (1983) Phase equilibria of plagioclase. In: Ribbe, P. H. (ed.) Feldspar minerals. Reviews in mineralogy 2, (2nd ed^{n}.), 223-239. Min. Soc. Am.

Smith, J. V. (1984) Phase relations of plagioclase feldspars. In: Brown, W. L. (ed.) Feldspars and feldspathoids. NATO Adv. Stud. Inst. C137, 55-94. Reidel: Dordrecht, Boston, Lancaster.

Smith, J. V., Ribbe, P. H. (1969) Atomic movements in plagioclase feldspars: kinetic interpretation. Contrib. Mineral. Petrol. 21, 157-202.

Smith, K. A., Kirkpatrick, R. J., Oldfield, E., Henderson, D. M. (1983) High-resolution silicon-29 nuclear magnetic resonance spectroscopic study of rock forming silicates. Am. Mineral. 68, 1206-1215.

Staehli, J. L., Brinkmann, D. (1974) A nuclear magnetic resonance study of the phase transition in anorthite. Zeit. Krist. 140, 360-373.

Tagai, T., Joswig, W., Korekawa, M., Wenk, H. R. (1980) Die Bestimming der Aℓ/Si-Verteilung Mittels Neutronenbeugung in einem plagioklas An_{66}. Zeit. Krist. 151, 77-89.

Taylor, W. H. (1933) The structure of sanidine and other felspars. Zeit. Krist. 85, 425-442.

Thompson, J. B., Jr., Hovis, G. L. (1978) Triclinic feldspars: Angular relations and the representation of feldspar series. Am. Mineral. 63, 981-990.

Thompson, J. B., Jr., Hovis, G. L. (1979a) Structural-thermodynamic relations of the alkali feldspars. Trans. Am. Cryst. Assoc. 15, 1-25.

Thompson, J. B., Jr., Hovis, G. L. (1979b) Entropy of mixing in sanidine. Am. Mineral. 64, 57-65.

Thompson, J. B., Jr., Waldbaum, D. R. (1968) Mixing properties of sanidine crystalline solutions: I. Calculations based on ion-exchange data. Am. Mineral. 53, 1965-1999.

Thompson, J. B., Jr., Waldbaum, D. R., Hovis, G. L. (1974) Thermodynamic properties related to ordering in end-member alkali feldspars. In: MacKenzie, W. S., Zussman, J. (eds.) The feldspars, 218-248. Manchester University Press: Manchester.

Tolédano, J. C., Tolédano, P. (1980) Order parameter symmetries and free-energy expansions for purely ferroelastic transitions. Phys. Rev. B21, 1139-1172.
Wainwright, J. E. (1969) A refined structure for bytownite. Progr. Eighth Int. Congr. Crystallogr. (abstr.) XII - 48.
Wainwright, J. E., Starkey, J. (1971) A refinement of the structure of anorthite. Zeit. Krist. 133, 75-84.
Waldbaum, D. R., Robie, R. A. (1971) Calorimetric investigation of Na-K mixing and polymorphism in the alkali feldspars. Zeit. Krist. 134, 381-420.
Waldbaum, D. R., Thompson, J. B., Jr. (1968) Mixing properties of sanidine crystalline solutions: II. Calculations based on volume data. Am. Mineral. 53, 2000-2017.
Wenk, H. -R., Joswig, W., Tagai, T., Korekawa, M., Smith, B. K. (1980) The average structure of An62-66 labradorite. Am. Mineral. 65, 81-95.
Wenk, H. -R., Kroll, H. (1984) Analysis of $P\bar{1}$, $I\bar{1}$ and $C\bar{1}$ plagioclase structures. Bull. Minéral. 107, 467-487.
Wenk, H. -R., Nakajima, Y. (1980) Structure, formation, and decomposition of APB's in calcic plagioclase. Phys. Chem. Min. 6, 169-186.
Windom, K. E., Boettcher, A. L. (1976) The effects of reduced activity of anorthite on the reaction grossular + quartz = anorthite + wollastonite: a model for plagioclase in the earth's crust and upper mantle. Am. Mineral. 61, 889-896.
Wruck, B. (1986) Einfluss des Na-Gehaltes und der Aℓ,Si Fehlordnung auf das thermodynamische Verhalten der Phasenumwandlung $P\bar{1}$ -$I\bar{1}$ in Anorthit. Ph.D. Dissertation, University of Hannover.
Wyncke, B., McMillan, P. F., Brown, W. L., Openshaw, R. E., Bréhat, F. (1981) A room-temperature phase transition in maximum microcline. Absorption in the far infrared (10 - 200cm^{-1}) in the temperature range 110 - 300 K. Phys. Chem. Min. 7, 31-34.
Yund, R. A. (1986) Interdiffusion of $NaS\bar{i}$-CaAℓ in peristerite. Phys. Chem. Min. 13, 11-16.
Yund, R. A., Tullis, J. (1980) The effect of water, pressure and strain on Aℓ/Si order-disorder kinetics in feldspar. Contrib. Mineral. Petrol. 72, 297-302.
Yund, R. A., Tullis, J. (1983) Subsolidus phase relations in the alkali feldspars with emphasis on coherent phases. In: Ribbe, P. H. (ed.) Feldspar mineralogy. Reviews in Mineralogy 2, (2nd. $ed^{n.}$), 141-176. Min. Soc. Am.

SOLID STATE NMR SPECTROSCOPY AND PHASE TRANSITIONS IN MINERALS

Andrew Putnis
Department of Earth Sciences
Downing Street
Cambridge CB2 3EQ
England

ABSTRACT. The application of solid state NMR spectroscopy in minerals is reviewed, specifically in relation to the study of phase transitions. After an introduction to the background theory and terminology, displacive transitions in perovskites and anorthite and the orientational order-disorder transition in sodium nitrite are used to illustrate some studies of the temperature dependence of magnetic resonance phenomena. ^{29}Si magic-angle spinning NMR (MAS NMR) is discussed in terms of the correlation between chemical shift and Si site environment, particularly during Al,Si order-disorder, with examples from a number of aluminosilicates. A study of Al,Si ordering in cordierite, in which ^{29}Si spectra are particularly informative is briefly reviewed.

1. INTRODUCTION.

Although NMR spectroscopy has been applied to the study of local structural properties of minerals for many years, recent theoretical and experimental developments in high resolution NMR, and the commercial availability of high field Fourier Transform NMR spectrometers has led to a dramatic increase in research activity in this area. The development of magic-angle sample spinning (MAS NMR) techniques has enabled the measurement of narrow line spectra of solids, particularly for dipolar nuclei such as ^{29}Si , and the general protocol for the study of local structural and chemical environments has been established. A number of reviews of the applications of MAS NMR have been recently published (Kirkpatrick et al.,1985a; Oldfield and Kirkpatrick,1985) and progress in this field is rapid. Single crystal studies of minerals have not received comparable attention recently and one of the aims of this paper is to outline the type of information that can be obtained from both techniques, particularly in relation to the study of structural phase transitions.

The theory of NMR spectroscopy is introduced in an elementary way by Davis (1965); more advanced and complete treatments are given by Poole and Farach (1972), Akitt (1983) and Fyfe (1983). Only the

E. K. H. Salje (ed.), Physical Properties and Thermodynamic Behaviour of Minerals, 325–358.

background theory essential to an understanding of this paper will be summarized here.

2. BACKGROUND THEORY AND TERMINOLOGY.

2.1. Magnetic resonance.

The general field of magnetic resonance involves the interaction of magnetic moments with an applied magnetic field. Most nuclei have magnetic moments due to spin, denoted I. Nuclei with odd mass numbers have I = n/2 where n is an integer, e.g. ^{1}H (I = 1/2), ^{17}O (I = 5/2), ^{23}Na (I = 3/2), ^{27}Al (I = 5/2), ^{29}Si (I = 1/2). Nuclei with even mass numbers but odd atomic numbers have integral values of I, e.g ^{14}N (I = 1). If both mass and atomic numbers are even, I = 0 and the nucleus is is non-magnetic e.g ^{16}O.

When a nuclear spin with a magnitude I is placed in a strong magnetic field H_o, 2I + 1 energy levels exist, from -I to +I. When I = 1/2 the nucleus has two energy levels -1/2 and +1/2 and behaves as a magnetic dipole; when I $\geqslant$ 1 the nucleus has more than two energy levels and behaves as a magnetic multipole.

The differences in nuclear energy levels are in the radio frequency range, and transitions between spin states may be induced by applying a radio frequency field to the sample in a large static magnetic field. A magnetic resonance experiment employs a frequency ω_o equal to the difference in the energy levels divided by $\hbar$ (Figure 1).

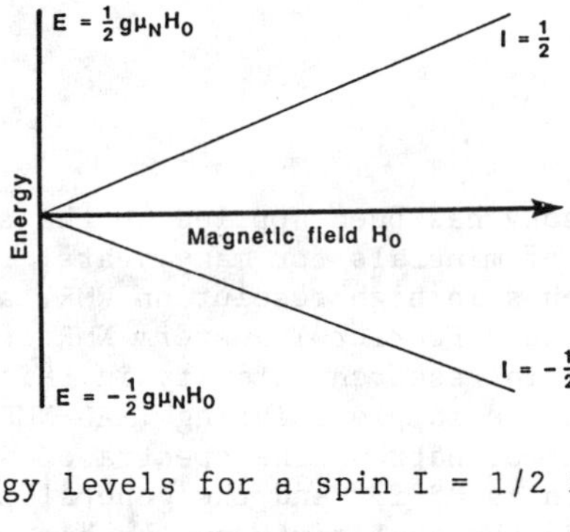

Figure 1. Zeeman energy levels for a spin I = 1/2 in a magnetic field H_o.

2.2. Chemical shift.

All nuclei of the same isotope in the same static magnetic field H_o should have the same resonance frequency. However the exact resonance frequency depends on the local electronic environment of the particular nucleus, because electrons in the vicinity of the nucleus shield it to varying degrees from the applied magnetic field. Thus the local chemical and structural environment can be probed by a nuclear method.

The resonance frequencies are reported as *chemical shifts*, which are differences in parts per million (ppm) relative to a suitable standard. For ^{29}Si the standard used is tetramethyl silane, $(CH_3)_4Si$ (TMS) and the overall range of chemical shifts for ^{29}Si in various mineral structures is from -60 to -120 ppm. Less negative chemical shifts indicate less shielding, in other words the nuclei experience a larger value of the external field and thus resonate at a higher frequency. Each peak in the ^{29}Si NMR spectrum represents Si in a specific chemical environment, and the intensity is proportional to the number of atoms in that environment.

2.3. NMR of solids.

In crystalline solids at rest there are a number of factors which result in the broadening of resonance peaks obscuring the structural details. If the shielding of the applied field H_o for a particular nucleus in a site is markedly anisotropic the orientational dependence of the chemical shift for a certain site produces chemical shift anisotropy (CSA). For spin $I = 1/2$ magnetic dipole - dipole interactions between nuclei, and for $I \geqslant 1$, quadrupole field-gradient interactions, both greatly broaden the NMR spectrum. In liquids, rapid molecular tumbling averages these interactions to zero and sharp spectral lines are readily obtainable, making NMR spectroscopy a principal technique for the study of species in solution.

There are two principal strategies in NMR spectroscopy of solids. Currently receiving most attention are experimental methods, principally magic angle spinning, which attempt to average these line broadening effects and in the simple case of $I = 1/2$, reduce them to zero, resulting in sharp resonance peaks and hence *high resolution NMR*. In the course of this averaging, symmetry properties of the nucleus are lost, although the positions, widths and relative intensities of the lines in the spectrum may provide very detailed information on the local chemical and structural environment around the nucleus. High resolution NMR requires the use of sophisticated, high field spectrometers only recently commercially available.

The second, and older, strategy is *wide-line NMR* in which single crystals are used, specifically to determine the local symmetry properties of the nucleus. In this context the quadrupolar interactions have been of most use in the study of phase transitions.

2.4. Nuclear quadrupole resonance (NQR).

A nucleus with spin $I > 1/2$ has a nuclear quadrupole moment eQ. This quadrupole moment interacts with the local electric field gradients in a solid, and the result is a set of quadrupole energy levels. The measurement of transitions between these energy levels is often designated separately as NQR spectroscopy.

The splitting of quadrupole energy levels is very sensitive to the electric field gradient around the site in the crystal. No quadrupole effects are observed at a cubic site. The electric field gradient at lower symmetry is described in terms of a tensor with components V_{xx},

Vyy and *Vzz* , and an asymmetry parameter $\eta = (V_{xx} - V_{yy})/V_{zz}$ where the principal axes are chosen such that $V_{xx} < V_{yy} < V_{zz}$. For axial symmetry ($\eta = 0$) there are I + 1 energy levels with spacings proportional to the axial field gradient *Vzz*. For lower symmetry ($0 < \eta < 1$) there are 2I + 1 energy levels with splitting dependent on the deviation from axial symmetry measured by η . The interaction of the nuclear quadrupole moment *eQ* with this field gradient is characterized by the quadrupole coupling constant $C = e^2qQ/h$ (where *eq* is the largest component of the electric field gradient tensor) and the asymmetry parameter η .

This coupling of the nuclei to the lattice by magnetic and electric interactions makes NMR and NQR very useful tools to investigate phase transitions and critical phenomena. For magnetic (dipolar) interactions the temperature dependent shift in the NMR frequency , and in the case of electric interactions the temperature dependence of the quadrupole coupling constant and the relaxation time provide valuable parameters to describe the thermodynamics of phase transitions. Several examples will be described below.

2.5. Relaxation times.

The relaxation time is a measure of the rate at which spins in excited states relax back to the ground state, and is sensitive to the immediate surroundings of the nucleus. Hence structural phase transitions can produce significant changes in their values. In solids, the spin - lattice relaxation time T_1, which is associated with the time required for the spin system to come to thermal equilibrium with the lattice is the significant relaxation process. The spin - spin relaxation time T_2 (the time constant for the spin system to establish thermal equilibrium with itself) is very much shorter than T_1 in solids, but is comparable to T_1 in viscous liquids.

Spin lattice relaxation times are generally measured using pulse methods, in which the spin system is saturated by a high power pulse and returns exponentially to its equilibrium Boltzman condition with a time constant T_1. The amplitude of the signal is then recorded as a function of time. Nuclear relaxation time measurements are particularly suitable for investigating the local fluctuations occurring around critical points in structural phase transitions.

2.6. Fourier Transform techniques.

The decay signal as the spin system relaxes to equilibrium between radio frequency pulses is measured in the *NMR* experiment as a function of time. This *time domain* information is Fourier transformed via an interfaced computer into the *frequency domain* to produce the NMR spectrum. The time domain measurement uses a pulse sufficiently wide to span the entire spectrum and hence the decay of all spectral lines is recorded simultaneously. Earlier, less sensitive, techniques particularly those used on single crystal measurements of minerals in the 1970's, entailed sweeping the external magnetic field at fixed frequencies to observe the resonances.

2.7. High resolution solid state NMR.

As mentioned above, NMR spectra in solids are broadened by dipolar and quadrupolar interactions which may be so strong as to obscure all structural information, and a variety of means have been sought to obtain high resolution spectra by removing these broadening effects. One approach has been to use elaborate sequences of radio frequency pulses designed to reduce the interaction times by averaging the spins. The advantage of these methods is that full interaction tensor components in the crystal are preserved, while local structural detail is resolved. Currently the most popular approach however is magic angle sample spinning (MAS NMR) which involves physically spinning the sample at kilohertz frequencies about an axis oriented at 54.7° (the magic angle) to the applied static field H_o. Since dipole - dipole interactions, which are a major cause of line broadening, are described by equations which contain $(3\cos^2\Theta - 1)$ terms where Θ is the angle between the laboratory field and the internuclear vector, they may be reduced to zero when $\Theta = 54.70$. Although different internuclear vectors in the sample have different values of Θ and hence are rotated on different cones, by fast rotation their time-averaged direction lies on the spinning axis (Figure 2). The spinning frequency must be at least comparable to the frequency differences which are present in the sample, which are in turn proportional to H_o. The result of MAS is that resonance peaks almost as narrow as in liquids may be obtained.

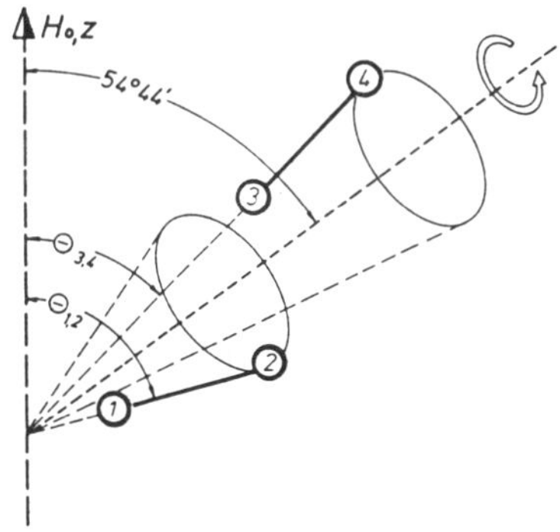

Figure 2. Effect of magic angle spinning. By rotation about the magic angle the time averaged value of all internuclear vectors becomes 54.70 e.g. $\langle\Theta_{1,2}\rangle = \langle\Theta_{3,4}\rangle = 54.70$ although $\Theta_{1,2} \neq \Theta_{3,4}$.

First order quadrupole interactions associated with the central (1/2, -1/2) transition are also averaged, although if quadrupole coupling constants are large, line broadening is only partially averaged by sample spinning. The interpretation of spectra is considerably more straightforward for dipolar nuclei such as ^{29}Si. Most work on MAS NMR in minerals has concentrated on ^{29}Si which can also be observed at natural abundance (4.7%).

MAS NMR also averages the chemical shift anisotropy, although spinning side-bands which are additional peaks on either side of the central peak appear at multiples of the spinning frequency. This makes it possible to identify them and to determine various parameters related to the electronic shielding of the nucleus by the local environment (Herzfeld and Berger, 1980).

3. STRUCTURAL PHASE TRANSITIONS.

There are many different ways in which NMR spectroscopy can be applied to the study of phase transitions, depending on the type of structural information we require. At the present time most mineral studies involve MAS NMR on powdered samples, and in the context of phase transitions, it is possible to determine local Si environments and hence study Al,Si ordering. The determination of Al,Si distributions in aluminosilicate minerals is of fundamental importance in that the state of Al,Si order has a very significant effect on their enthalpy and entropy. Furthermore, the mechanism and kinetics of the Al,Si ordering process and its relationship to structural phase transformations lies at the root of many unresolved problems in mineralogy. Traditionally X-ray diffraction techniques have provided most data on Al,Si distributions in ordered, single crystals but the similarity in scattering factors of Al and Si, the fact that often experimental samples are fine-grained powders, and most importantly the limitations of diffraction techniques when long-range order is poorly developed or non-existent, has emphasised the potential importance of NMR methods. Progress in this field will be reviewed below.

A second important area in mineralogy is the study of displacive phase transitions, particularly since the realization (e.g Salje and Devarajan, 1986) that a detailed understanding of rapid and hence experimentally accessible displacive transitions can provide considerable information on states of cation order, which in the case of Al,Si are subject to very sluggish kinetics and are difficult to study directly. This is due to the coupling of order parameters associated with the displacive transition with the order parameter describing the state of cation order (see Carpenter, this volume).

Displacive transitions have been profitably studied by single crystal methods since the parameters such as quadrupole coupling constant and spin - lattice relaxation times are very sensitive to local static and dynamic critical phenomena.

In the rest of this paper these two main areas will be discussed by giving examples of case studies which illustrate the scope of NMR spectroscopy in problems of current interest.

4. DISPLACIVE PHASE TRANSITIONS.

Fundamental to all studies of displacive phase transitions is the determination of the way in which various thermodynamic parameters vary with temperature. The temperature dependence of these response functions such as the susceptibility , the specific heat C_H and the order parameter Q completely specify the thermodynamic properties of the transition. In displacive transitions the order parameter measures the extent to which the atomic configuration in the less symmetrical phase differs from that in the more symmetrical phase, and is normalized to be zero above T_c and non-zero below T_c.

The temperature dependence of the order parameter in a continuous phase transition is described by a critical exponent β such that

$$Q \propto (T_c - T)^{\beta}$$

The experimental value of β is used to test the validity of theories describing the thermodynamics of phase transitions (Bruce and Cowley, 1981). Systems undergoing structural phase transitions usually approximate to a mean-field model in which all interactions experienced by the local order parameter average out to give an effective field. For such a model $\beta = 1/2$. Critical phenomena around T_c where large fluctuations in order parameter occur, result in deviations from mean-field approximations.

4.1 Displacive transitions in perovskites, ABX_3.

Displacive transitions in perovskites have been the subject of intensive study using various techniques for many years. They are characterized by a variety of transitions associated with small distortions from the ideal cubic structure. The ideal structure consists of corner linked octahedra of X ions with B cations in their centres and A cations at the corners of a simple cubic lattice enclosing the octahedra (Fig.3). The observed distortions usually involve tilts of the octahedra around their symmetry axes, sometimes accompanied by small distortions of the octahedra themselves.

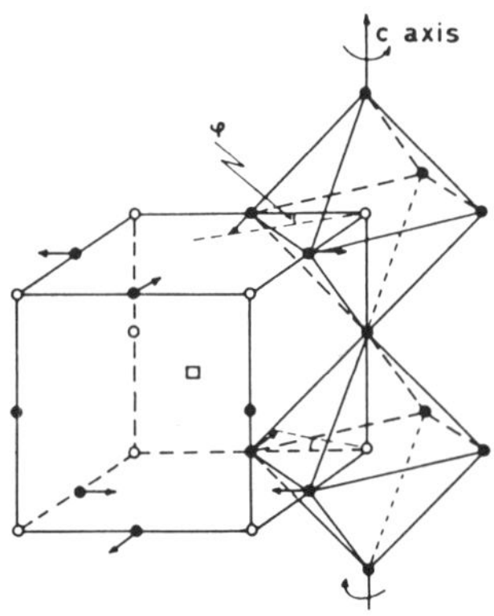

Figure 3. Basic perovskite structure of ABX_3. The atomic displacements correspond to rotations of BX_6 octahedra in opposite senses as shown. The angle φ defines the order parameter. Open square - A; open circle - B; closed circle - X.

In $KMnF_3$ a transition from a cubic to a tetragonal phase occurs at 186°K. The transition involves rotations of the MnF_6 octahedra in the opposite sense in adjacent (001) planes perpendicular to the c axis. The transition may be described by an order parameter φ which coincides with the angular rotation of the octahedra around the tetragonal axis.

Borsa (1973) has carried out an NMR study of the quadrupole effects on the ^{39}K nucleus and determined the temperature dependence of the quadrupole coupling constant C. In the tetragonal phase the asymmetry parameter $\eta = 0$ and the quadrupole coupling constant $C = eQVzz/h$. From symmetry considerations it may be shown that Vzz and hence C are proportional to φ^2.

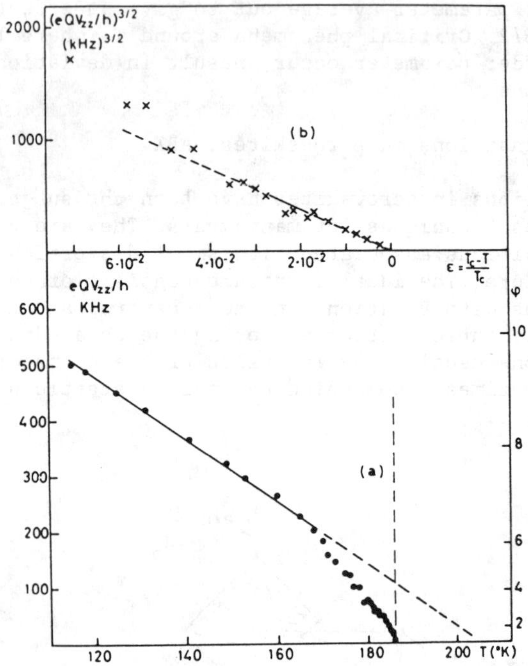

Figure 4. Temperature dependence of the ^{39}K quadrupole coupling constant C in $KMnF_3$. In the lower figure $C \alpha \varphi^2$ Vs. T ; in the upper figure $C^{3/2} \alpha \varphi^3$ Vs.the reduced temperature $(T_c-T)/T_c$. (from Borsa, 1973).

Figure 4 shows the temperature dependence of the ^{39}K quadrupole coupling constant. At low temperatures φ^2 varies linearly with temperature i.e. $\beta = 1/2$. Close to the transition temperature φ^3 varies linearly with temperature i.e. $\beta = 1/3$. This departure from mean-field behaviour is attributed to critical effects in which fluctuations of the order parameter within a coherence length are greater or equal to the order parameter itself. The critical temperature range $(T_c - T)/T_c$ is about 10^{-1} which is a related to the coherence length via the Ginzburg relation (see Salje, this volume). This in turn provides information on the range of the forces driving the phase transition in the material in that the critical temperature range will be narrow if long-range forces are important.

4.2 The $P\bar{1}$ - $I\bar{1}$ phase transition in anorthite, $CaAl_2Si_2O_8$.

The nature of this phase transition had been the subject of considerable controversy, finally clarified by Adlhart et al. (1980). They measured the temperature dependence of sharp and diffuse *c* and *d* reflections using X-ray and neutron diffraction and found a complete, reversible and continuous phase transition from $P\bar{1}$ - $I\bar{1}$ at 240 ± 4°C.

In the context of the present review it is worth noting that a very similar conclusion was reached in an NMR study by Staehli and Brinkmann (1974a,b), but owing to apparently contradictory X-ray diffraction data the issues remained unresolved for several years. Staehli and Brinkmann observed ^{27}Al resonances for the central line (1/2,-1/2) of the quadrupole splitting at temperatures between 25°C and 305°C in a single crystal of anorthite by sweeping the magnetic field at fixed frequency. Although such experimental procedures have now been superseded, the eight inequivalent Al sites of the low temperature phase could be resolved. The experimentally determined electric field gradient tensors at the ^{27}Al sites were compared with theoretical values calculated from a simple point charge model. The measured tensors could be assigned to the corresponding crystallographic sites.

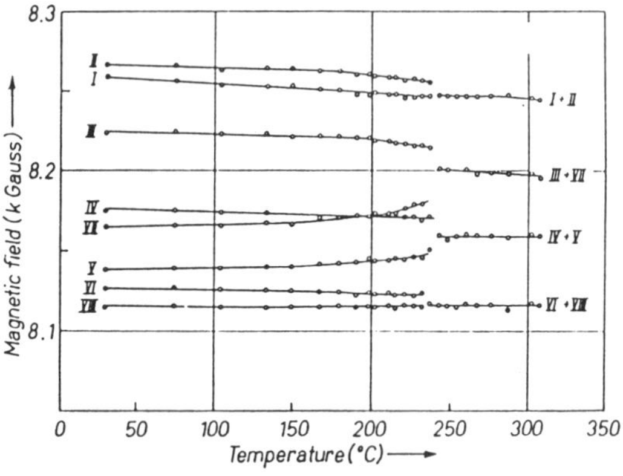

Figure 5. Temperature dependence of the positions of the Al central signals at a fixed orientation of anorthite crystal. Eight assigned peaks in the low temperature form change to four peaks in the high temperature form.

At the critical temperature of 241 ± 4°C a completely reversible structural transition takes place. Two sites each out of the eight low temperature sites become equal, resulting in only four inequivalent sites in the high temperature phase, confirming the body-centred symmetry. Figure 5 shows the temperature dependence of the positions of the Al resonances at a fixed crystal orientation. The discontinuity is clearly defined although the order of the transition could not be determined as the signal intensity was too low to be detected within a temperature range of about 8°C around T_c.

For a review of our current understanding of this transition see Carpenter, this volume.

5. ORIENTATIONAL ORDER - DISORDER TRANSITIONS.

5.1 NO_2 ordering in sodium nitrite, $NaNO_2$.

Sodium nitrite undergoes a classical ferroelectric order - disorder transition which has been very extensively studied and may serve as a type example for mineral systems with orientational disorder (e.g the ordering of CO_3 groups in carbonates). In the high temperature paraelectric phase the NO_2 dipoles undergo reorientational fluctuations along the b axis (Fig 6). The flipping between the two equilibrium sites involves a sudden rotation of NO_2 groups around the c axis (the O - O direction). Below 164° the material is ferroelectric with NO_2 groups pointing in the same direction . Alternate *bc* layers are shown light and dark in Figure 6 stacked in the crystallographic *a* direction.

In a narrow temperature range around T_c a modulated antiferroelectric structure exists with the NO_2 dipoles apparently sinusoidally distributed along the a axis with a period of a few unit cells.

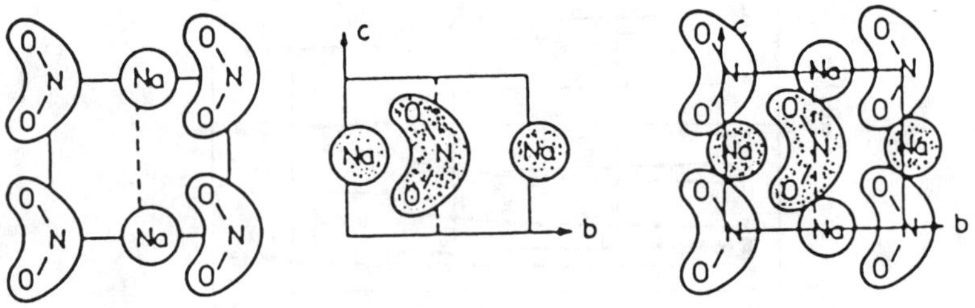

Figure 6. The ferroelectric phase of $NaNO_2$. The shaded atoms indicate layers at height a/2.

An excellent tool for investigating the critical phenomena in $NaNO_2$ is the quadrupole resonance and relaxation of the ^{14}N nucleus. Ambrosetti et al. (1977) carried out such experiments over a temperature range around T_c using high resolution pulse and Fourier transform methods.

Figure 7 shows the quadrupole frequencies ν^+ and ν^- as a function of temperature. The transition from the paraelectric to the modulated antiferroelectric phase appears to be second order, while that to the ferroelectric has a small discontinuity in the order parameter of about 5%. In the antiferroelectric phase the presence of two distinct components of the resonance lines, rather than broadening, supports a model of this structure as regions of ferroelectric short-range order separated by paraelectric regions, rather than due to a continuous modulation in the amplitude of the local polarization.

For a recent discussion on the origin of the modulated phase see Durand et al., (1986) and references therein.

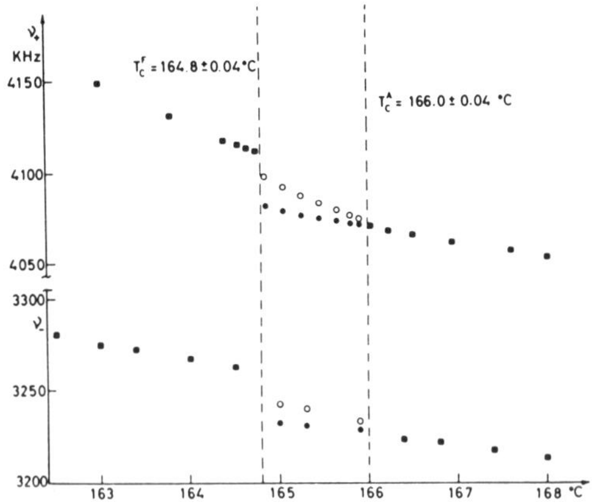

Figure 7. ^{14}N NQR frequencies ν^+ and ν^- in $NaNO_2$ around the ferro- and antiferroelectric transitions (from Ambrosetti et al.,1977).

6. MAGIC - ANGLE SPINNING NMR

The rest of this review will be concerned with recent work on MAS NMR in mineral systems, emphasising those aspects relating to the determination of the degree of cation order, particularly Al,Si order. As ^{29}Si spectra are better understood in this context at the present time, and have been more widely studied, these will be discussed in most detail. The use of other quadrupolar nuclides to determine local chemical and structural states will be considered briefly.

6.1. ^{29}Si NMR spectra

The first investigation of ^{29}Si MASNMR spectroscopy of minerals was by Lippmaa et al. (1980) who investigated a range of silicate structures with different polymerization states. A correlation was found between more negative chemical shifts and increasing degree of polymerization (Q^0 - Q^4) where the Q^n notation represents the number of bridging oxygens per Si tetrahedron. In this early study distinct ranges of chemical shifts were found for Si in Q^0 (isolated tetrahedra), Q^1 (dimers), Q^2 (chains), Q^3 (sheets) and Q^4 (frameworks) for those minerals with no Al in tetrahedral sites. As more silicate phases were studied (K.A.Smith et al. 1983) it has become clear that there is considerable overlap in the range of isotropic ^{29}Si chemical shifts for

Si sites with different polymerization states.

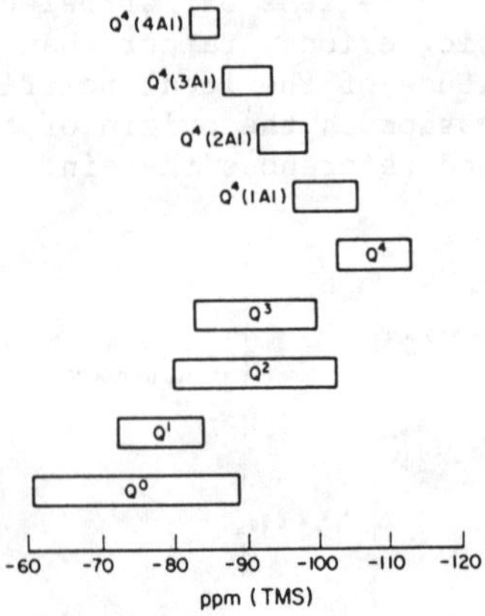

Figure 8. Ranges of isotropic chemical shifts for Si sites with different polymerizations for phases with Si in tetrahedral coordination (from K.A. Smith et al.,1983).

In framework aluminosilicates ^{29}Si chemical shifts have been found to fall into five ranges depending on whether a given SiO_4 tetrahedron is linked to four, three, two, one or no AlO_4 tetrahedra (Lippmaa et al. 1980; Engelhardt et al. 1981; Fyfe et al. 1983a; K.A. Smith et al. 1983). These local environments or structural units are usually denoted Si(nAl) where n is the number of AlO_4 tetrahedra linked to the SiO_4 tetrahedron. There is a constant and additive effect on the shift of about 5ppm for each aluminium atom added to the local environment. The range of isotopic chemical shifts for ^{29}Si in tetrahedral coordination is shown in Figure 8.

6.1.1 Peak areas and Si site populations. As long as sufficient time is allowed between r.f. pulses for the ^{29}Si nuclear spin system to fully relax, peak areas under such spectra are directly proportional to the total number of Si atoms in each distinguishable site. This is the key for determining Al,Si ordering schemes from ^{29}Si spectra.

For framework aluminosilicates Engelhardt et al. (1981) and Klinowski et al. (1982) have described a simple method for determining Si,Al ratios from ^{29}Si spectra, if the Loewenstein rule (or Al-avoidance principle) is obeyed, and if two or more resolvable Si(nAl) peaks occur in the spectrum. Peak intensities are normalized to 100, i.e.

$$\Sigma I_{Si(nAl)} = 100$$

Given that the nearest neighbourhood of every Al atom is Al(4Si), each Si-O-Al linkage in a Si(nAl) structural unit is equivalent to Al/4 atoms. In a tetrahedrally bonded framework structure therefore

$$(Si/Al)_{nmr} = \frac{\Sigma I_{Si(nAl)}}{\Sigma\, 0.25\, n\, I_{Si(nAl)}}$$

If the Si/Al ratio derived from the ^{29}Si spectra is greater than the ratio determined from chemical analysis it follows that the aluminium avoidance rule does not apply (assuming of course that all Si and Al are tetrahedrally coordinated in a single phase), i.e. it implies the existence of Al-O-Al linkages in the framework. In Al-rich aluminosilicates e.g. cordierite, $Mg_2Al_4Si_5O_{18}$; anorthite, $CaAl_2Si_2O_8$, in which Al,Si order is not perfect, the presence of such linkages is inevitable. A method for calculating the number of Al-O-Al bonds in cordierite was devised by Putnis and Angel (1985) and a more general method, applicable to any aluminosilicate framework structure by Fyfe et al.(1986). The expression for the number of Al-O-Al bonds per Al atom is simply

$$N_{Al-Al} \text{ per Al} = 2 - \frac{2(Si/Al)_{actual}}{(Si/Al)_{nmr}}$$

This forms a valuable basis for describing the degree of Al,Si order in Al-rich aluminosilicates and has been applied to cordierite (see 6.4)

6.2 Empirical correlations of ^{29}Si chemical shifts.

Clearly, to be able to use NMR spectra to solve problems of Al,Si order we need to be able to unambiguously assign each peak to a particular Si site in the structure. Ideally we would wish to be able to calculate the magnitude of the chemical shift for any particular local structural environment. This is not yet possible, although progress in the theoretical prediction of NMR chemical shifts is being made (Tossell 1984; Engelhardt and Radeglia, 1984; Radeglia and Engelhardt, 1985; Tossell and Lazeretti, 1986). In the meantime various empirical relationships have been drawn from the databases of NMR spectra of materials with known composition and structure. The variations in chemical shifts for a particular nuclide appear to be due primarily to bond angle, bond length differences in the nearest and next-nearest neighbour co-ordination.

Smith and Blackwell (1983) have presented a linear relationship between the ^{29}Si chemical shift and the mean secant of the Si-O-Si bond angle for the SiO2 polymorphs and also applied this to Si-O-T angles in alkali feldspars (Smith et al., 1984). This approach has been extended to a wider range of aluminosilicates by Oldfield and Kikpatrick (1985) and as shown in Figure 9 excellent correlations have been obtained.

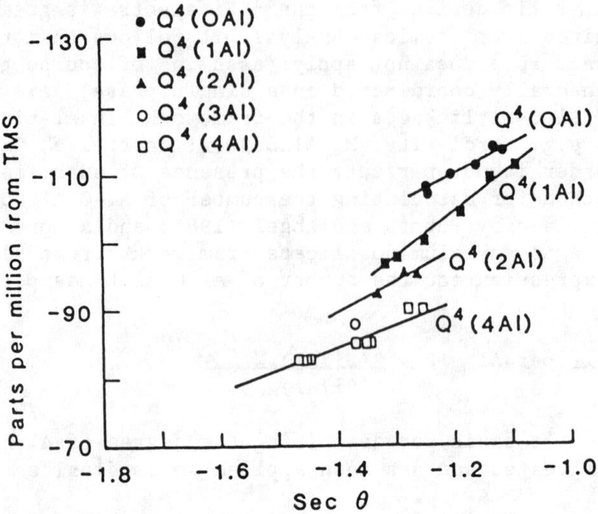

Figure 9. ^{29}Si chemical shifts plotted as a function of Si-O-T bond angle for aluminosilicate frameworks (from Oldfield and Kirkpatrick, 1985).

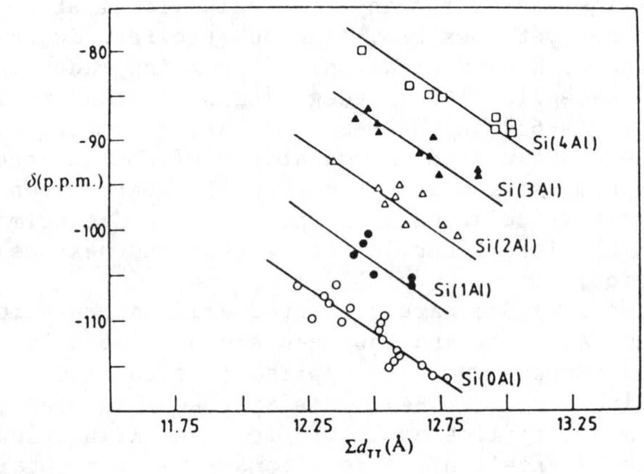

Figure 10. ^{29}Si chemical shifts plotted against the total Si..T non-bonded distance for Si in different Si(nAl) environments (from Ramdas and Klinowski (1984).

Ramdas and Klinowski (1984) introduced a simple semi-empirical relationship for framework aluminosilicates between the chemical shift and the sum total of the four Si..T bond lengths around the central Si atom in a Si(nAl) unit. (Figure 10). K.A.Smith et al.1983 have studied a wide range of natural and synthetic silicates and found a correlation between the ^{29}Si chemical shift and the total cation - oxygen bond strengths for the four oxygens of each Si tetrahedron (using Brown and Shannon bond strengths) (Fig 11).

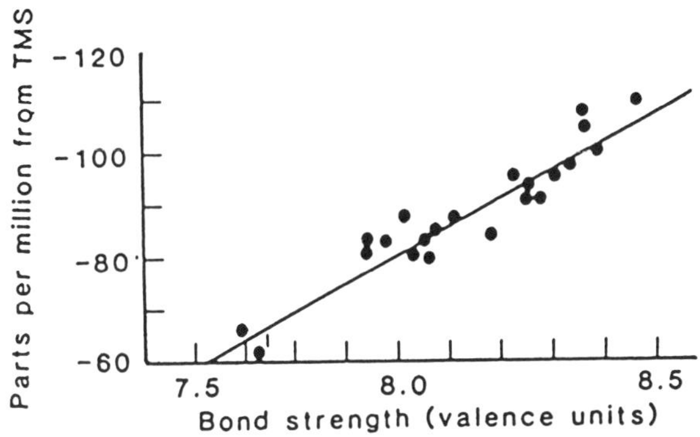

Figure 11. ^{29}Si chemical shift plotted as a function of Brown-Shannon bond strength (from K.A.Smith et al.,1983).

Although these correlations are an important first step in the process of ^{29}Si site assignments, their uncritical application may lead to errors, as will be mentioned below.

6.3 ^{29}Si site assignments.

To illustrate the general approach currently used to assign sites in ^{29}Si MAS NMR spectra a number of examples will be discussed below, ranging from cases where unambiguous assignments can be made to minerals whose low symmetry and structural complexity make assignments difficult. Well ordered structures have the simplest spectra. Increasing the degree of Al,Si disorder increases the number of chemically non-equivalent sites producing distinguishable NMR chemical shifts. Note that crystallographically equivalent sites (in space group terms) may be chemically non-equivalent when local NMR environments are considered. For further elaboration of this point see Yang et al.(1986).

6.3.1. Al,Si order-disorder in albite, $NaAlSi_3O_8$. Figure 12 (from Yang et al., 1986) shows a sequence of ^{29}Si MAS NMR spectra of well ordered Amelia low albite and progressively disordered Amelia albite after annealing at 1073°C for various lengths of time up to 100 days.

The spectrum of low albite has three well-resolved peaks at chemical shifts of -93ppm, -97ppm and -105ppm (relative to TMS). From the known structure of low albite there are three distinct Si sites, labelled T_2m with Si(2Al), and T_2o and T_1m with Si(1Al). Accurate neutron diffraction data have been used to calculate the mean secant of the Si-O-T bond angles (Smith et al., 1984). The positive correlation between mean secant Si-O-T and the chemical shift leads to peak assignments of -93ppm to T_2m, -97ppm to T_2o and -105 to T_1m. The same peak assignments could be made using the cation - oxygen bond strength correlation of K.A.Smith et al.,1983.

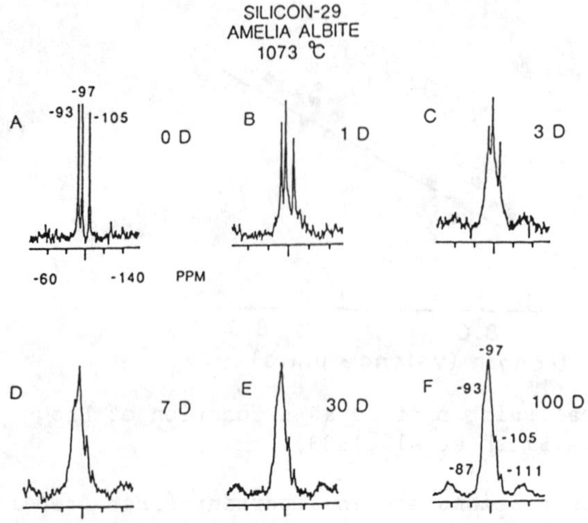

Figure 12. ^{29}Si MAS NMR spectra of unheated Amelia albite (A) and of Amelia albite annealed at 1073°C for various times from 1 to 100 days (from Yang et al.,1986).

Increasing the degree of Al,Si disorder complicates the spectrum by producing Si environments with a greater number of Al nearest neighbours, and individual peaks are not readily resolved. An idea of the complexity of the problem may be seen in Figure 13 in which the ^{29}Si spectrum of albite annealed for 100 days is compared with a simulation using a peak-fitting program, and the corresponding deconvolution. The spectrum is fitted by nine relatively broad peaks. In triclinic albite there are 4 crystallographically distinct sites and during disordering each may be occupied by Si(nAl) where $0 < n < 4$, giving 20 possible crystallochemically non-equivalent sites. At high degrees of disorder the departure from monoclinic symmetry (in which there are 10 such possible sites) has been assumed to be slight and assignments have been made on this basis. Nevertheless the large spontaneous strain associated with small differences in the degree of Al,Si order in albite (Salje et al., 1985, Carpenter, this volume) makes this assumption somewhat questionable. However, the assignment proceeds by using the known empirical relationship between chemical shift and each additional Al atom added to the local environment, as well as the observation that in ordered albite T_1m (1Al) is 8ppm more shielded than T_2m (1Al). This data may be used to derive a site assignment for each of the nine fitted peaks in the spectrum, with the ordered structure as a starting point.

As the area under each deconvoluted peak represents the relative population of Si in each type of site, the sequence of disordering can potentially be followed at a level not previously possible. This opens the possibility for measuring activation energies for AlSi interchange directly, and for correlating local order parameters for Al,Si order with macroscopic data from diffraction and calorimetric results.

However, the extra complexity of the spectra of Al,Si disordered albite demonstrates the problems of dealing with disordered phases of many minerals.

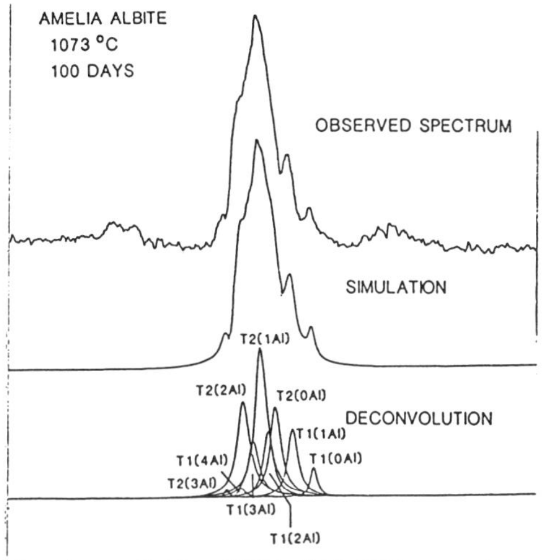

Figure 13. Simulation of the ^{29}Si MAS NMR spectrum of Amelia albite annealed at 1073°C for 100 days.

6.3.2 Si site assignments in microcline, $KAlSi_3O_8$. When Amelia albite is ion-exchanged to give ordered maximum microcline the changes in the NMR spectra can be monitored as a function of K content, and changing Si-O-T angle. The continuity of each peak across the series is used to unambiguously assign the three ^{29}Si peaks in the microcline (Smith et al.,1984; Phillips and Kirkpatrick, in prep). The average rate of change of ^{29}Si chemical shift with mean secant Si-O-T is different for each site, and as Figure 14 shows, is twice as large for T_1m as for T_20 sites. This suggests that factors other than Si-O-T bond angle contribute to the variations in chemical shifts.

The total cation - oxygen bond strength variation with ^{29}Si chemical shift (K.A.Smith et al.,1983) does not correctly predict the microcline peak assignment, since for the T_1m site the correlation is negative, rather than the general positive trend shown in Figure 11 , and followed by the T_2m and T_20 Si sites. Even allowing for ambiguities regarding the co-ordination of K in microcline, this negative correlation is not readily explained and suggests the need for caution.

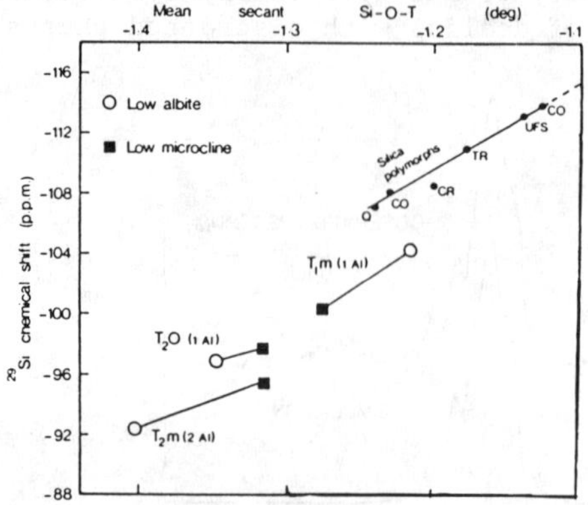

Figure 14. Relationship between ^{29}Si isotropic chemical shift and mean secant Si-O-T angle for low albite, low microcline and the silica polymorphs (from Smith et al.,1984).

6.3.3 <u>Si site assignments in anorthite, $CaAl_2Si_2O_8$.</u> The ^{29}Si spectrum of anorthite is inevitably more complex than that for albite or microcline, as there are eight non-equivalent Si bearing sites in the low temperature $P\bar{1}$ structure. A fully ordered anorthite should give a spectrum with eight peaks of equal intensity, and is the obvious starting point for any NMR study of Al,Si order in anorthite and the plagioclase feldspars. Kirkpatrick et al. (1987) have examined a series of plagioclases which had been characterized by transmission electron microscopy, electron microprobe, X-ray diffraction and high-temperature solution calorimetry (Carpenter et al., 1985; Carpenter, 1986). Such detailed characterization is vital if the complexities of the NMR spectra are to be correctly interpreted.

Figure 15a shows the ^{29}Si spectrum of a well ordered, pure anorthite from Val Pasmeda. Six peaks can be clearly seen, in three pairs, closely spaced over a 10ppm range. The intensity ratio of the three groups is 3:3:2, and the spectrum can be simulated by eight peaks with approximately equal intensities if two of the peaks have the same chemical shift. In the fully ordered structure each Si site has 4Al nearest neighbours, and presumably the differences in chemical shift are due to variations in the local framework distortions. Uncertainties in published refinements as well as the caution which must be exercised in applying Si-O-T bond angle correlations at this level, preclude the assignment of each peak to a specific Si site.

Introducing a rather small amount of Al,Si disorder in anorthite has a considerable effect on the ^{29}Si spectrum suggesting that framework distortions as well as different Si environments affect the chemical shifts. Spectra of volcanic anorthites and well annealed synthetic anorthites show significant broadening and overlap effects relative to metamorphic anorthite even though they only have a small amount of Al,Si disorder. Thus ^{29}Si spectra potentially provide information on the nature of framework distortions, as well as different site occupancies.

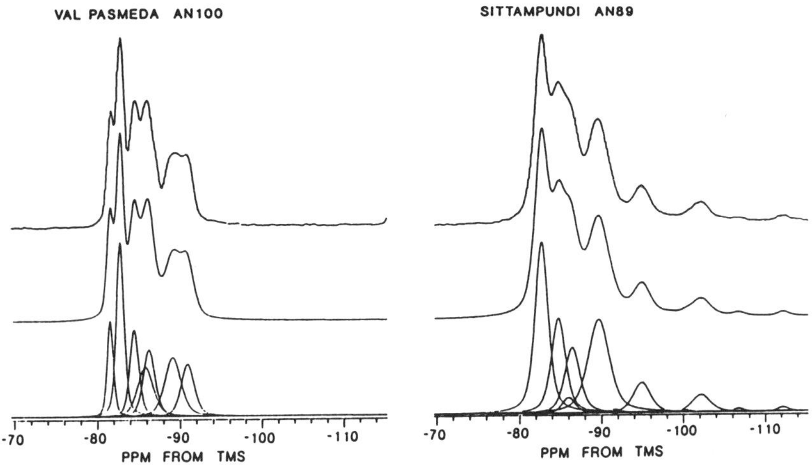

Figure 15. Observed and simulated spectra and spectral components for (a) the ^{29}Si MAS NMR spectrum of well ordered An_{100} and (b) of metamorphic bytownite (from Kirkpatrick et al.,1987).

6.3.4 <u>^{29}Si spectra of plagioclase feldspars.</u> The effect of substitution of some albite component into anorthite is to increase the Si/Al ratio as well as substitute Na for Ca. Figure 15b is the ^{29}Si spectrum of a well-ordered metamorphic bytownite (An_{89}) from Sittampundi, and it can be used to illustrate the strategy adopted by Kirkpatrick et al.(1987) on venturing into the uncertain territory of plagioclase NMR. The spectrum is simulated assuming local $I\bar{1}$ structure, and the four strongest peaks have virtually the same chemical shift as the Val Pasmeda anorthite. Other peaks in the spectrum occur at similar positions to peaks in high albite and have been assigned in the same way; the rest have been assigned on the basis that each Al atom added to the nearest neighbour environment of the Si atom adds 5ppm to the chemical shift.

Despite the assumptions made in this assignment it does stand up to closer scrutiny. Because of the alternation of Si and Al sites in fully ordered anorthite, each additional Si atom replacing an Al atom forms a Si(0Al) site, and causes the neighbouring four Si sites to become Si(3Al). Extending this substitution mechanism to An_{89} should result in 5.2% of the Si sites being Si(0Al) which compares very well with the total intensity of the two peaks assigned to this type of site. Furthermore the apparent preponderance of T1(3Al) sites relative to T2(3Al) sites and of T2(0Al) relative to T1(0Al) suggests that Si substitutes preferentially for Al on T2 sites rather than on T1 sites. This is confirmed by recent structure refinements.

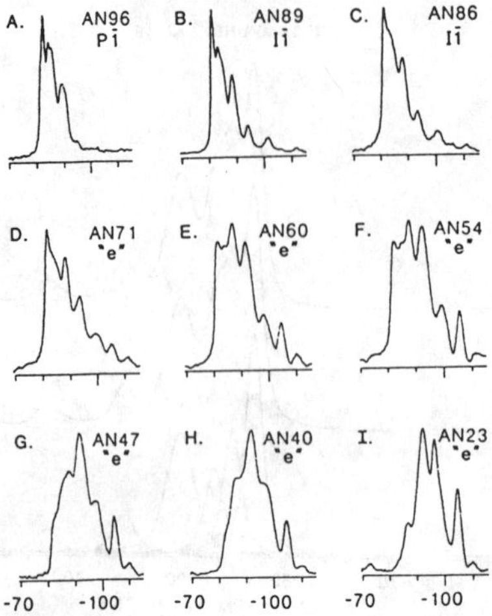

Figure 16. ^{29}Si MAS NMR spectra of a suite of low-temperature, relatively well-ordered P$\bar{1}$, I$\bar{1}$ and "e" plagioclase feldspars (from Kirkpatrick et al.,1987).

Extending the study across the intermediate plagioclase field, a sequence of ^{29}Si NMR spectra for a range of compositions between albite and anorthite is shown in Figure 16. Although the spectra are complex due to the number of overlapping peaks, the peaks are relatively intense and narrow, suggesting a high degree of Al,Si order in the incommensurate structures of the intermediate plagioclases. Furthermore, significant structural differences between e_1 and e_2 plagioclases are revealed, supporting the suggestions from solution calorimetry and lattice parameter data (see Carpenter, this volume).

While accepting the uncertainties in this overall approach, the consistency of the assignments must be acknowledged, and given the complexity of the mineral, the work of Kirkpatrick et al.(1987) has been described as "state of the art" at the present time (Fyfe,1987).

6.3.5 Si site assignments in ordered cordierite $MgAl_4Si_5O_{18}$. The structure of cordierite is somewhat unusual for a framework aluminosilicate, being made up of 6-fold rings of tetrahedra which form vertical channels. The ring tetrahedra (T_2) are connected laterally and vertically by non-ring tetrahedra (T_1) (Figure 17) so that the ratio of T_1 to T_2 sites is 2:1. In the fully ordered structure the 6 T_2 sites per unit cell contain 4 Si and 2 Al atoms; the 3 T_1 sites contain 1 Si and 2 Al atoms. This ordering requires orthorhombic symmetry with a splitting of the T_1 and T_2 sites as shown in Table 1.

TABLE 1

Tetrahedral sites in cordierite

Hexagonal form	Orthorhombic form	Site content	Si environment
3 T_1 sites	2 $T_1$1 sites	2 Al	
	1 $T_1$6 site	1 Si	Si(4Al)
6 T_2 sites	2 $T_2$1 sites	2 Si	Si(3Al)
	2 $T_2$3 sites	2 Si	Si(3Al)
	2 $T_2$6 sites	2 Al	

TABLE 2

Observed isotropic 29Si chemical shifts compared with structural parameters for ordered Mg-cordierite (data from Gibbs,1966)

Si site	Si(nAl)	Chemical shift	Mean sec T-O-T	Σ Bond strength	Σ Si..T distance
$T_1$6	Si(4Al)	-79 ppm	-1.46	8.244	12.44
$T_2$1	Si(3Al)	-100ppm	-1.314	8.212	12.60
$T_2$3	Si(3Al)	-100ppm	-1.314	8.211	12.57

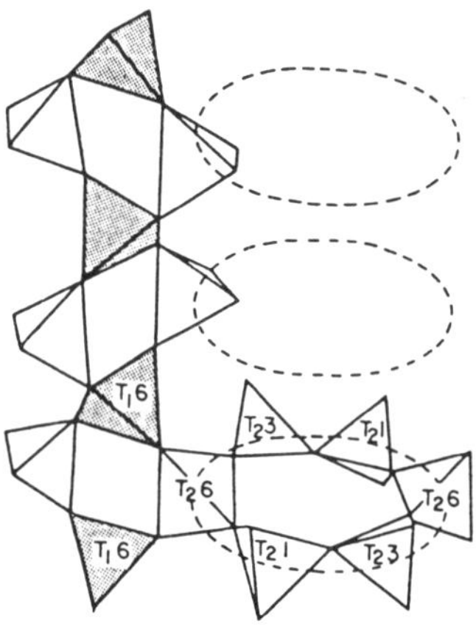

Figure 17. The arrangement of tetrahedra in the cordierite structure. Each T_2 tetrahedron is joined via a T_1 tetrahedron (shaded) to the rings above, below and adjacent rings (not shown) such that all tetrahedra are corner sharing.

The ordering scheme strictly obeys the Al-avoidance principle with each Si atom in the T_1 site having an environment Si(3Al) while the Si atoms in the T_2 sites have Si(4Al). The relevant structural parameters used in ^{29}Si chemical shift correlations discussed in section 6.2. are given in Table 2. The chemical and structural similarity of the T_21 and T_23 Si sites suggest that these would not be readily resolved , and hence only 2 types of Si sites, T_1 and T_2 might be expected in the NMR spectrum.

Figure 18 shows the ^{29}Si MAS NMR spectrum of a synthetic well-ordered Mg-cordierite (Putnis et al. 1985; Fyfe et al. 1986), which consists of two major peaks at -79.3 and -100.2 ppm (from TMS) with intensities 0.144 and 0.698, respectively. The assignment of these peaks may be made with reference to the structure and the tetrahedral site occupancies discussed above. The two main peaks are assigned to Si(4Al) for the T_1 sites and Si(3Al) for the T_2 ring tetrahedra as indicated in the figure. The ratio of the number of T_2 Si atoms to T_1 Si atoms for this sample is 4.41 compared with 4.00 for a fully ordered cordierite. This suggests that the specimen is not fully ordered, and a closer inspection of the figure reveals that there are small contributions to the spectrum from other minor peaks. The origin of these will be discussed in the following section.

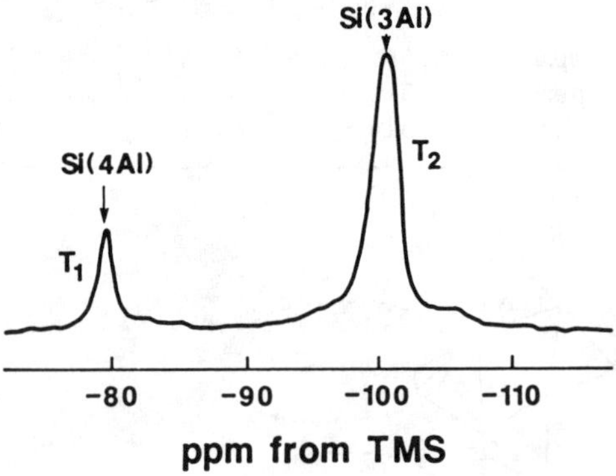

Figure 18. ^{29}Si MAS NMR spectrum of well-ordered synthetic cordierite (from Putnis et al.,1985).

The line-width of the T_2 resonance at -100.2ppm (189 Hz) is larger than that of the T_1 peak at -79.3ppm (97 Hz). This is due to the overlap of T_21 and T_23 Si sites which could possibly be resolved in a perfectly ordered sample.

A remarkable feature of the ^{29}Si spectrum of well ordered cordierite is the large chemical shift difference between the two peaks. Although as shown in Table 2 the T_1 and T_2 Si sites are structurally and chemically different the correlations of chemical shift with structural parameters such as mean secant T-O-T (Smith et al.,1984) and Brown-Shannon bond strength (K.A.Smith et al., 1983) are poor. There is a negative correlation with both mean secant T-O-T and bond strength, compared with the positive correlations shown in Figures 9 and 11.

Although no explanation can be proposed for this at present, there is no doubt that the assignments are correct. Synthetic cordierite samples prepared in different laboratories give the same ^{29}Si spectrum (R.J.Kirkpatrick pers. comm; P.Mirwald pers. comm.), the crystal structure is well known and the spectrum of the ordered structure is consistent with the general trend from disordered to ordered as described in the following section.

This spectrum of ordered cordierite is used as the starting point for an NMR study of Al,Si ordering in cordierite. Fortuitously, the large chemical shift difference between the two Si sites in the ordered structure of cordierite makes a study of disordered structures possible, compared with the situation in anorthite for example where there are difficulties in resolution and assignment even in the ordered structure.

6.4 Al,Si ORDERING IN Mg-CORDIERITE FROM ^{29}Si NMR SPECTRA.

Cordierite, $Mg_2Al_4Si_5O_{18}$ is particularly suitable for a ^{29}Si NMR study of Al,Si ordering because the spectrum of the fully ordered structure is simple and has two very different resonance peaks, as discussed above. Furthermore, it is experimentally straightforward to prepare samples with increasing degrees of Al,Si order by crystallizing a glass within the stability field of the fully ordered phase. The first-formed crystalline phase is always hexagonal with no long-range Al,Si order; subsequent annealing at temperatures below 1450°C (the equilibrium hexagonal - orthorhombic transformation temperature) ultimately results in a fully ordered orthorhombic structure.

Experimental details regarding sample preparation, characterization of the crystalline products by transmission electron microscopy, and the recording of ^{29}Si MAS NMR spectra are given in Putnis (1980), Fyfe et al. (1983b), Putnis et al. (1985) and Fyfe et al. (1986).

On heating cordierite glass at temperatures above 1100°C , crystallization is complete within less than one minute, and X-ray diffraction reveals the formation of well crystalline hexagonal cordierite. Figure 19 shows the sequence of ^{29}Si NMR spectra for crystallization at 1185°C and subsequent annealing for up to 2000 hours. The 2000 hour spectrum is the same as that described in section 6.3.5 above and is the most highly ordered in the sequence. The assigned T_1 Si(4Al) and T_2 Si(3Al) peaks are arrowed. Spectra of samples with shorter annealing times show that while these two T_1 and T_2 groupings are preserved, the number of peaks in each group has increased. This is entirely to be expected as the number of Si chemical environments increases with more disordered Al,Si. For each of the two types of sites a maximum of five ^{29}Si peaks are theoretically possible, namely, Si(4Al), Si(3Al), Si(2Al), Si(1Al) and Si(0Al). within a few minutes of crystallization from the glass 8 of these 10 possible peaks are observed.

The local Si environments of these peaks can be self-consistently assigned using the well-established observation that the addition of each extra Al atom to the Si(nAl) unit shifts the ^{29}Si resonance by approximately 5ppm. These assignments are labelled in Figure 19a. No

Si(OAl) groupings are observed, which is not unreasonable in a material with Si/Al ratio of 1.25.

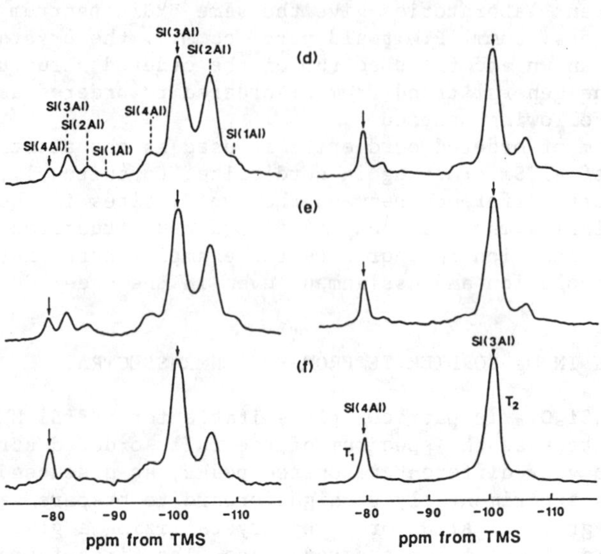

Figure 19. A sequence of ^{29}Si MAS NMR spectra of synthetic cordierites isothermally annealed at 1185°C for increasing time from (a) 2 minutes to (f) 2000 hours (from Putnis et al.,1987).

Measurement of the peak intensities (Putnis et al. 1985) allows the quantitative determination of the population and depopulation of all the individual sites during the Al,Si ordering process in terms of individual local Si environments, information clearly not available by other techniques.

6.4.1 <u>Al-O-Al linkages in cordierite.</u> The NMR data can also be used to demonstrate the existence of Al-O-Al bonds in cordierites without complete Al,Si order, using the method described in 6.1.1. Figure 20 shows the plots of the average number of Al-O-Al bonds per Al_4Si_5 formula unit, N(Al-Al), evaluated from ^{29}Si spectra of synthetic cordierites crystallized from glass at three different temperatures for increasing times (Putnis et al. 1987). The N(Al-Al) bonds are progressively eliminated as the degree of order increases, and hence we can use this as an order parameter to define the temperature - time evolution of the cordierite structure, as well as to constrain models of short-range Al,Si order (Putnis and Angel, 1985).

We can immediately comment on the state of Al,Si order in the first-formed cordierite after crystallization times of 2 minutes. If the cordierite was totally disordered but with partitioning of Si and Al between T_1 and T_2 as in the orthorhombic form, the calculated number of Al-O-Al bonds would be 3.3 per formula unit. The NMR data show that for each of the temperatures N(Al-Al) is less than 1.5 indicating that the first-formed cordierite has considerable local Al,Si order. Given the short crystallization time, the implication is that considerable Al,Si order exists in the glass starting material.

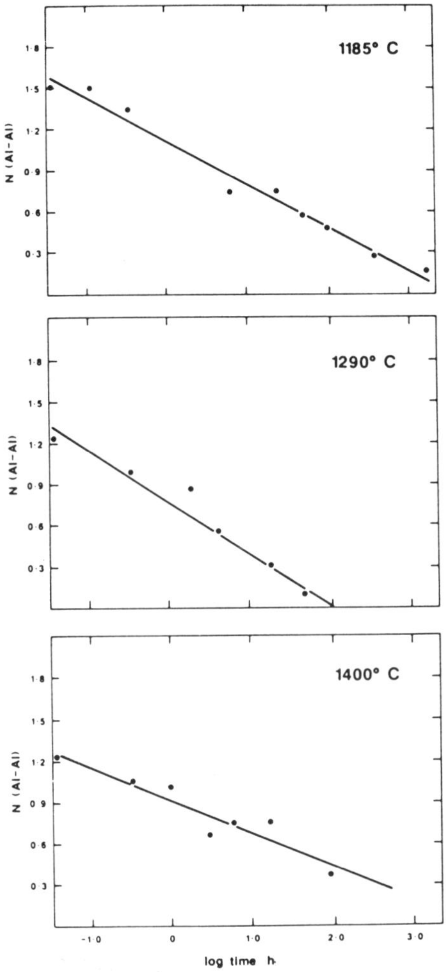

Figure 20. The number of Al-O-Al bonds, N(Al-Al), per formula unit plotted as a function of annealing time at three different temperatures (from Putnis et al., 1987).

Given the known high energy of Al-O-Al bonds in aluminosilicates, and the inevitability of the existence of such bonds in cordierite whenever Al,Si order is not perfect, the ordering process can be considered as being driven by the reduction in enthalpy produced by eliminating these linkages.

The number of Al-O-Al bonds per formula unit, N (normalized to 1 for complete order and 0 for complete disorder) defines the order parameter as

$$Q_{od} = [\ 1 - N/3.3]^{1/2}$$

6.4.2 Correlation of N(Al-Al) with $H_{soln.}$ data in cordierite.
Carpenter et al.(1983) have measured the heat of solution in lead borate of synthetic cordierites prepared at 1185°C in the same way as the samples for NMR spectroscopy. The plot of $\Delta H_{soln.}$ against logarithm of annealing time at 1185°C is shown in Figure 21. From a comparison of the slopes of Figure 20a and Figure 21 , and assuming that the enthalpy differences are due to the elimination of Al-O-Al bonds, we can derive a value of the enthalpy change associated with a single Al-Si interchange, h, i.e. for the reaction

(Al-O-Al) + (Si-O-Si) → 2 (Al-O-Si).

For cordierite h is 34 kjoules $mole^{-1}$.

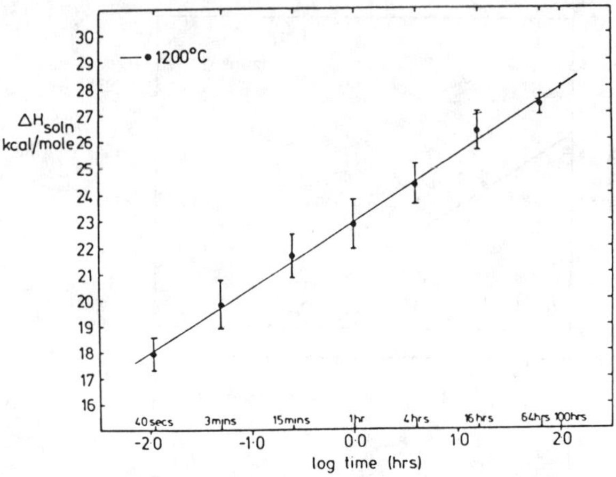

Figure 21. Plot of $\Delta H_{soln.}$ against log annealing time (from Carpenter et al.,1983).

In the case of cordierite the contribution of the elastic energy to the enthalpy differences of samples in different structural states is negligibly small (see Carpenter, this volume).

6.4.3 Correlation of N(Al-Al) with X-ray synchrotron data in cordierite.
As mentioned above, one of the problems in studying an Al,Si ordering process of this type, in which ordering results in a symmetry change, is to define the point at which the structure transforms on a macroscopic scale to the low symmetry form and to relate this to the degree of Al,Si order. The same samples as those described above were studied using X-ray synchrotron radiation to monitor the profile of a single peak (211_{hex}) in the spectrum (Putnis et al.1987). The lattice distortion is defined by an order parameter Q which is determined by the spontaneous strain *e* (Salje et al. 1985). In cordierite *e* is proportional to the splitting of the (211_{hex}) powder line, traditionally described by the distortion index Δ (Miyashiro,1957). Q is defined as 0 in the hexagonal form (Δ = 0) and normalized to 1 for Δ = 0.25 (the maximum distortion).

The relationship between Q and Q_{od} for all the structural states which could be observed may be seen by plotting an order parameter vector diagram (Figure 22). The main conclusion is that cordierite achieves a high degree of Al,Si order while still macroscopically hexagonal. The intervention of a strain-modulated hexagonal phase, observed by transmission electron microscopy (Putnis, 1980; Putnis et al. 1987) is significant in this context, accommodating local order within an overall hexagonal structure.

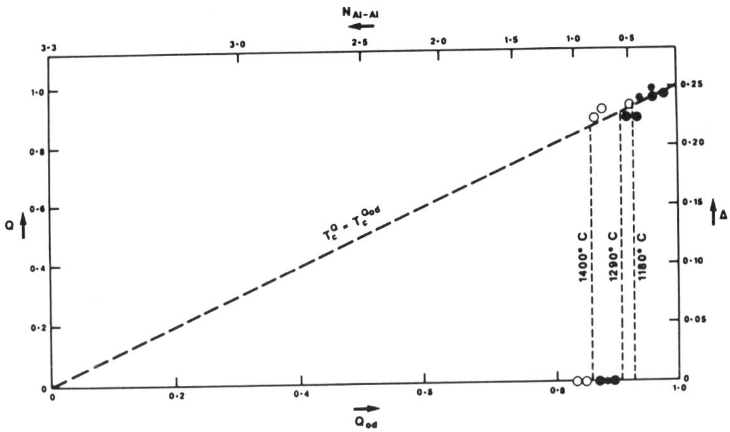

Figure 22. The experimentally determined variation of the order parameter Q, defined by the Δ index plotted against the order parameter Q_{od}, defined as a function of N(Al-Al) during the annealing sequence at different temperatures. The short-dashed lines show the experimentally determined discontinuities in Q. (from Putnis et al., 1987).

The structural phase transition between modulated hexagonal and orthorhombic cordierite is characterized by a spontaneous increase in Q. The order parameter Q_{od}, on the other hand, does not reveal any singularity at the phase transition point. Both parameters Q and Q_{od} show a strong correlation in the orthorhombic phase which follows exactly the theoretical predictions of a bilinear coupling process (Salje, 1987). The phase transition can be looked upon as being "triggered" by a certain threshold value of Q_{od}, which itself depends on the annealing temperature.

6.4.4 Conclusions. The ^{29}Si NMR spectra provide vital information on the progress of local Al,Si order in a system where diffraction data define only one aspect of the structural behaviour. The combination of NMR and diffraction data is a powerful technique for unravelling the structural states in such complex materials. Of particular importance is an understanding of the relationship between the spontaneous strain and the local Al,Si order, without which the structural states cannot be described.

7. NMR SPECTRA OF QUADRUPOLAR NUCLEI.

This brief review of MAS NMR has emphasised ^{29}Si, since quadrupole effects are not present and the MAS technique produces relatively simple, interpretable spectra. Consequently it has been the most widely studied nucleus in minerals. However a multinuclear approach has obvious advantages, and as most nuclei are quadrupolar, there is wide interest in developing techniques of obtaining high resolution spectra for I > 1/2 nuclei (e.g. Kirkpatrick et al.1985a; Oldfield and Kirkpatrick,1985)

The main problem with NMR spectra of quadrupole nuclei is the broadening effect on the resonances due to the interaction of the nucleus with surrounding electric field gradients. Fortunately most of these quadrupolar nuclei have non-integral spin and the central transition (1/2,-1/2) is unaffected to first order by the quadrupole broadening effects. However, second order effects become very significant in structures where the quadrupole coupling constants are large and this substantially broadens and shifts the central transition. This second order effect is diminished and modified by MAS NMR but it is not possible to average it to zero. Thus the prominent feature of the MAS NMR spectrum of a non-integral spin nucleus is a generally complicated central transition line shifted from the true isotropic chemical shift value. The magnitude of this quadrupolar shift is dependent on the quadrupole coupling constants for the site, and inversely on the strength H0 of the magnetic field. Hence the largest available field strengths are used.

The non-central transitions e.g. (± 3/2, ± 1/2) are broadened to first-order by quadrupole effects and lead to extra intensity in spinning sidebands, another prominent feature of quadrupole spectra.

7.1 ^{27}Al MAS NMR spectra.

^{27}Al is the most thoroughly investigated quadrupolar nuclide in minerals, although the difficulty in determining accurate isotropic chemical shifts, free from quadrupolar effects has limited the quantitative interpretation of such spectra. The range of chemical shift for tetrahedral Al is only about 30 ppm (cf. Fig. 8 for ^{29}Si) and hence corrected chemical shifts are particularly important for ^{27}Al.

Recent advances in instrumentation and techniques have made a precise determination of isotropic ^{27}Al chemical shifts possible (Samosan,1985; Lippmaa et al., 1986). Lippmaa et al. have outlined the strategies for high resolution ^{27}Al MAS NMR and other non-integral quadrupolar nuclei in powder samples. For ^{27}Al a major advance is the method of Samosan (1985) in which it is shown that the satellite spinning sidebands corresponding to (± 3/2, ± 1/2) transitions are less affected by second order quadrupole effects than the central transition. The linewidth of this satellite transition is hence smaller by a factor of 24/7 and is shifted from its isotropic value by a factor of 8 less than thre central transition. By measuring the position of the sidebands relative to the central transition the attainable precision of ^{27}Al chemical shifts in aluminosilicates is comparable to that of 29Si chemical shifts in solids.

^{27}Al chemical shifts in framework aluminosilicates (i.e. all 4-coordinated) closely parallel those of ^{29}Si and display a similar dependence on the mean bond angles Θ = <Al-O-Si>. This is shown in Figure 23, and sets the basis for future ^{27}Al MAS NMR studies.

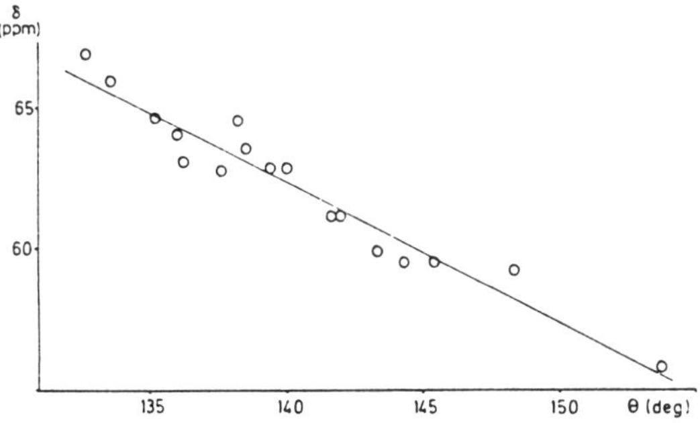

Figure 23. The dependence of ^{27}Al chemical shifts on the mean [Al-O-Si] bond angles in framework aluminosilicates (from Lippmaa et al., 1986).

7.1.1 Al(4) / Al(6) coordination. Al often occurs in octahedral Al(6) as well as tetrahedral Al(4) coordination in many minerals, and one of the most important uses of ^{27}Al MAS NMR is the detection of Al in these two environments and the determination of Al(4)/Al(6) ratios.

Muller et al.(1981) showed that chemical shifts for Al(4) and Al(6) fall into very distinct ranges, making their detection relatively simple. At a field strength of H_o = 11.7 Tesla, Al(4) is in the range +50 to +80 ppm, while Al(6) is in the range -10 to +20 ppm. Quantification of Al(4)/Al(6) ratios using spectral simulations and integrations of ^{27}Al MAS NMR spectra leads to a number of important potential applications, since this ratio could previously only be determined by chemical analysis and allocation of components into sites by calculating idealized structural formulae.

Kinsey et al. (1985) discuss a number of these analytical applications with reference to layer silicates. Determination of Al site ratios and their changes during burial diagenesis have always been problematic in these minerals and yet are critical to an understanding of their behaviour.

Another application, nearer to the subject of this review, is the study of ordering of Al between octahedral and tetrahedral sites as occurs in spinels.

7.1.2 Al(4) - Al(6) order-disorder in $MgAl_2O_4$ spinel. In the spinel structure metal cations occupy one-eighth of the available tetrahedral sites and one half of the available octahedral sites. In a "normal" 2-3 spinel the divalent cation occupies the tetrahedral sites and the trivalent cation the octahedral sites. In an "inverse" spinel one half of the trivalent cations occupy tetrahedral sites, while the divalent cations and the rest of the trivalent cations are in octahedral sites.

$MgAl_2O_4$ spinel is known to exhibit at least partial inverse character and it is of considerable geological interest to determine the nature of the cation distribution and its temperature dependence. The configurational contribution to the entropy of spinel has significant consequences to the P,T slopes of a number of important metamorphic reactions. As Mg^{2+} and Al^{3+} have similar X-ray scattering factors it is very difficult to distinguish between them using X-ray methods, and ^{27}Al MAS NMR spectra provide a convenient technique for the direct investigation of Al coordination.

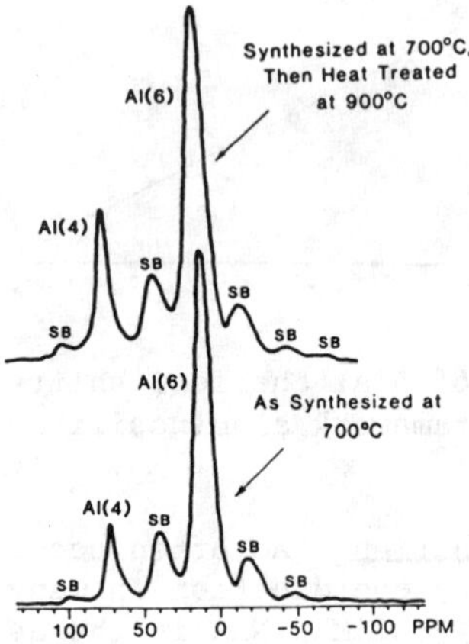

Figure 24. High-resolution 11.7-T ^{27}Al MAS NMR spectrum of two heat-treated spinels. There is a marked increase in intensity of the tetrahedral peak at 900°C. SB denotes spinning sidebands (from Wood et al., 1986).

Wood et al.(1986) have studied the Al(4) - Al(6) distribution in quenched samples of synthetic $MgAl_2O_4$ spinel as a function of annealing temperature and time. Figure 24 shows a typical ^{27}Al spectrum of two heat treated spinels. Estimation of octahedral - tetrahedral distribution was determined from the areas under the tetrahedral and octahedral peaks and is reproducible to better than ±10 %. The temperature dependence of Al(4)/Al(6) ratio is clearly detectable using this method.

8. CONCLUSIONS.

There is no doubt that NMR spectroscopy is a powerful tool for determining the local structural environment of a wide variety of elements in minerals. This opens new possibilities in the study of the way in which structures respond to changes in composition and temperature, and in the relationship between local structure and macroscopic properties. Developments in this field are rapid, and the current efforts in several laboratories to combine MAS NMR with sample heating are likely to herald a very considerable advance in our capability of directly studying temperature dependent phenomena on this scale.

In complex materials such as aluminosilicate minerals in which cation ordering, exsolution effects and displacive transitions all play a role, it is essential that the samples studied be properly characterized by other techniques such as X-ray/neutron diffraction and electron microscopy and diffraction if the NMR spectra are to be correctly interpreted and assigned.

Finally, the main limitations of MAS NMR spectroscopy of natural minerals is that samples must be available in relatively large amounts for the more dilute nuclei e.g. ≈ 500mg for ^{29}Si (compared with ≈ 10 mg for ^{27}Al), and they must be relatively free of paramagnetic impurities. Paramagnetic ions such as Fe^{2+} and Fe^{3+} can have pronounced effects on NMR spectra due to the effect of the large magnetic moment of the unpaired electrons on nuclear spin transitions. The presence of some Fe enhances ^{29}Si NMR spectra by decreasing the spin-lattice relaxation time T_1 and hence allowing more rapid repetition of radio frequency pulsing (Watanabe et al. 1983). However, too much Fe causes broadening of resonance peaks (Oldfield et al.1983) and loss of information. Sherriff and Hartman (1985) have discussed the effects of paramagnetic centres on ^{29}Si MAS NMR spectra of feldspars.

Despite these limitations, the ability to obtain high resolution NMR spectra from solid samples will considerably influence future developments in mineralogy.

9. ACKNOWLEDGEMENTS.

I am grateful to Colin Fyfe for the collaboration which introduced me to the possibilities of studying Al,Si ordering by ^{29}Si MAS NMR, and to Jim Kirkpatrick for many discussions during his sabbatical visit to Cambridge.

10. REFERENCES.

Adlhart W, Frey F, Jagodzinski H (1980) X-ray and neutron investigations of the P1-I1 transition in pure anorthite. Acta Crystallogr. **A36**: 450-460.

Akitt JW (1983) *NMR and Chemistry : an introduction to the Fourier Transform - multinuclear era*. Chapman and Hall, London.

Ambrosetti R, Angelone R, Colligiani A, Rigamonti A (1977) ^{14}N NQR and relaxation at the ferroelectric transition in $NaNO_2$. Phys. Rev. B **15**: 4318-4327.

Bruce AD, Cowley R (1981) *Structural Phase Transitions*. Taylor and Francis, London.

Borsa F (1973) ^{39}K NMR in $KMnF_3$ and the temperature dependence of the generalized order parameter at the structural phase transition. Phys. Rev. B **7**: 913-917.

Carpenter MA, Putnis A, Navrotsky A, McConnell JDC (1983) Enthalpy effects associated with Al,Si ordering in anhydrous Mg cordierite. Geochim. Cosmochim. Acta **47**: 899-906.

Carpenter MA, McConnell JDC, Navrotsky A (1985) Enthalpies of ordering in the plagioclase feldspar solid solution. Geochim. Cosmochim. Acta. **49**: 947-966,

Carpenter MA (1986) Experimental delineation of the "e" I1 and "e" C1 transformations in intermediate plagioclase feldspars. Phys. Chem. Minerals **13**: 119-139.

Davis JC (1985) *Advanced Physical Chemistry* New York: Ronald Press.

Durand D, Denoyer F, Currat F, Lambert M (1986) Incommensurate phase in $NaNO_2$. In: *Incommensurate phases in Dielectrics 2.* Materials Ed. Blinc R, Levanjuk AP. Elsevier-North-Holland : Amsterdam.

Engelhardt G, Lohse U, Lippmaa E, Tarmak M, Magi M (1981) ^{29}Si-Untersuchungen zur Verteilung der Silicium und Aluminiomatome in Aluminosilicatgitter von Zeolithen mit Faujasit Struktur. Z. Anorg. Allg. Chem. **482**: 49-64.

Engelhardt G, Radeglia R (1984) A semi-empirical quantum-chemical rationalization of the correlation between SiOSi bond angles and ^{29}Si NMR chemical shifts of silica polymorphs and framework aluminosilicates (zeolites). Chem. Phys. Lett. **108**: 271-274.

Fyfe CA (1983) *Solid state NMR for Chemists*. CRC Press, Guelph, Ontario, Canada.

Fyfe CA (1987) Solid state NMR comes of age. Nature **325**: 194.

Fyfe CA, Thomas JM, Klinowski J, Gobbi GC (1983a) Magic angle spinning NMR spectroscopy and the structure of zeolites. Angew Chem **22**: 259- 336.

Fyfe CA, Gobbi GC, Klinowski J, Putnis A, Thomas JM (1983b) Characterization of local atomic environments and quantitative determination of changes in site occupancies during the formation of ordered synthetic cordierite by Si and Al magic angle spinning NMR spectroscopy. Chem. Commum. **1983**: 556.

Fyfe CA, Gobbi GC, Putnis A (1986) Elucidation and mechanism and kinetics of the Si,Al ordering process in synthetic Mg-cordierite

by ^{29}Si MAS NMR spectroscopy. Journ. Amer. Chem. Soc. **108**: 3218-3223.
Gibbs GV (1966) The polymorphism of cordierite I: The crystal structure of low cordierite. Amer. Mineral. 51: 1068-1087.
Herzfeld J, Berger AE (1980) Sideband intensities in NMR spectra of samples spinning at the magic angle. Journ. Chem. Physics 73: 6021- 6030.
Kinsey RA, Kirkpatrick RJ, Hower J, Smith KA, Oldfield E (1985) High resolution ^{27}Al and ^{29}Si NMR spectroscopic study of layer silicates, including clay minerals. Amer. Mineral. **70**: 537-548.
Kirkpatrick RJ, Smith KA, Schramm S, Turner G, Yang W-H (1985a) Solid state NMR spectroscopy of minerals. Ann. Rev. Earth Planet. Sci. **13**: 29-47.
Kirkpatrick RJ, Kinsey RA, Smith KA, Henderson DM, Oldfield E (1985b) High resolution solid state ^{23}Na, ^{27}Al,and ^{29}Si NMR spectroscopic reconnaissance of alkali and plagioclase feldspars. Amer. Mineral. **70**: 106-123.
Kirkpatrick RJ, Carpenter MA, Yang W-H, Montez B (1987) ^{29}Si magic angle NMR spectroscopy of low temperature ordered plagioclase feldspars. Nature **325**: 236-238.
Klinowski J, Ramdas S, Thomas JM, Fyfe CA, Hartman CA (1983) A reexamination of Si,Al ordering in Zeolites NaX and NaY. J. Chem Soc. Faraday Trans. 2, **78**: 1025-1050.
Lippmaa E, Magi M, Samosan A, Engelhardt G, Grimmer A-R (1980) Structural studies of silicates by solid state high resolution ^{29}Si NMR. Journ. Amer. Chem. Soc. **102**: 4889-4893.
Lippmaa E, Samosan A, Magi M (1986) High resolution ^{27}Al NMR of aluminosilicates. Journ. Amer. Chem. Soc. **108**: 1730-1735
Miyashiro A (1957) Cordierite - indialite relations. Am. Journ. Science **255**: 43-62.
Muller D, Gessner W, Behrens HJ, Scheler G (1981) Determination of aluminium coordination in aluminium-oxygen compounds by solid state high resolution ^{27}Al NMR. Chem. Phys. Lett. **79**: 59-62.
Oldfield E, Kirkpatrick RJ (1985) High resolution NMR of inorganic solids. Science **227**: 1537-1544.
Oldfield E, Kinsey RA, Smith KA, Nichols JA, Kirkpatrick RJ (1983) High resolution NMR of inorganic solids. Influence of magnetic centres on magic angle sample spinning lineshapes in some natural aluminosilicates. Journ. Magn. Reson. **51**: 325-329.
Poole CP Jnr, Farach HA (1972) *Theory of Magnetic Resonance* Wiley (Interscience) New York.
Putnis A (1980) The distortion index in anhydrous Mg-cordierite. Contrib. Mineral. Petrol. **74**: 135-141
Putnis A, Fyfe CA, Gobbi GC (1985) Al,Si ordering in cordierite using magic angle spinning NMR I: ^{29}Si spectra of synthetic cordierites. Phys. Chem. Minerals **12**: 211-216.
Putnis A, Angel RA (1985) Al,Si ordering in cordierite using magic angle spinning NMR II: Models of Al,Si order from NMR data. Phys. Chem. Minerals **12**: 217-222.
Putnis A, Salje E, Redfern SAT, Fyfe CA, Strobl H (1987) Structural states of Mg-cordierite I: Order parameters from synchrotron X-ray

and NMR data. Phys. Chem. Minerals **14**:446-454.

Radeglia R, Engelhardt G (1985) Correlation of Si-O-T angles and ^{29}Si NMR chemical shifts in silicates and aluminosilicates. Interpretation by semi-empirical quantum chemical considerations. Chem. Phys. Lett. **114**: 28-30.

Ramdas S, Klinowski J (1984) A simple correlation between isotropic ^{29}Si-NMR chemical shifts and T-O-T angles in zeolite frameworks. Nature **308**: 521-523.

Salje E, Devarajan V (1986) Phase transitions in systems with strain induced coupling between two order parameters. Phase Transitions **6**: 235-248.

Salje E, Kuscholke B, Wruck B (1985) Domain wall formation in minerals: theory of twin boundary shapes in Na-feldspar. Phys Chem Minerals **12**: 132-140.

Salje E (1987) Structural states of cordierite II: Landau theory. Phys. Chem. Minerals. **14**:455-460.

Samosan A (1985) Satellite transition high resolution NMR of quadrupolar nuclei in powders. Chem. Phys. Lett. **119**: 29-32.

Sherriff BL, Hartman JS (1985) Solid state high resolution ^{29}Si NMR of feldspars: Al,Si disorder and the effects of paramagnetic centres. Canad. Mineral. **23**: 205-212.

Smith KA, Kirkpatrick RJ, Oldfield E, Henderson DM (1983) High resolution ^{29}Si NMR spectroscopic study of rock forming silicates. Amer. Mineral. **68**: 1206-1215.

Smith JV, Blackwell CA (1983) Nuclear magnetic resonance of silica polymorphs. Nature **303**: 223-225.

Smith JV, Blackwell CA, Hovis GL (1984) NMR of albite-microcline series. Nature **309**: 140-142.

Staehli JL, Brinkmann D (1974a) A nuclear magnetic resonance study of the phase transition in anorthite, $CaAl_2Si_2O_8$. Zeits. fur Kristallogr. **140**: 360-373.

Staehli Jl, Brinkmann D (1974b) Assignment and structural dependence of electric field gradients in anorthite and simple field gradient calculations in some aluminosilicates. Zeits. fur Kristallogr. **140**: 374-392.

Tossell JA (1984) Correlation of ^{29}Si NMR chemical shifts in silicates with orbital energy differences obtained from X-ray spectra. Phys. Chem. Minerals **10**: 137-141.

Tossell JA, Lazeretti P (1986) Ab initio calculations of ^{29}Si chemical shifts for some gas phase and solid state silica fluorides and oxides. Journ. Chem. Phys. **84**: 369-374.

Watanabe T, Shimizu H, Masuda A, Saito H (1983) Studies of ^{29}Si spin-lattice relaxation times and paramagnetic impurities in clay minerals by magic angle spinning ^{29}Si-NMR and EPR. Chem. Lett. 1293-1296.

Wood BJ, Kirkpatrick RJ, Montez B (1986) Order-disorder phenomena in $MgAl_2O_4$ spinel. Amer. Mineral. **71**: 999-1006.

Yang W-H, Kirkpatrick RJ, Henderson DM (1986) High resolution ^{29}Si, ^{27}Al and ^{23}Na NMR spectroscopic study of Al,Si disordering in annealed albite and oligoclase. Amer. Mineral. **71**: 712-726.

NONLINEAR DYNAMICS OF LATTICE MODELS FOR ELASTIC MEDIA

J. POUGET
Laboratoire de Mecanique Theorique associe au C.N.R.S.,
Universite Pierre et Marie Curie, Tour 66, 4 Place Jussieu,
75252 Paris Cedex 05, FRANCE.

Abstract. Nonlinear excitations in crystals exhibiting refined structures are important to material sciences. The study of two models based on lattice dynamics allows us to examine nonlinear excitations of the soliton type. The models present two degrees of sophistication corresponding to two particular descriptions of the microstructure properties. The presence of different length scales and then competitive interactions permitting nonlinear dynamics are particularly crutial to allow for the propagation of solitons in such media. The first model consisting in a shear deformable atomic chain describes martinsitic transformations occuring in alloys. The problem of the structure of interfaces which separate martensitic or austenitic domains is reported in terms of solitary waves and the migration of these are investigated. The second model aims at studying the coupled rotation-displacement motions of an atomic chain equipped with microscopic dipoles, the latter undergoing rotations of large amplitudes. This model enables the description of the reversal motion of molecules passing through a domain wall in molecular crystals where additional degrees-of freedom associated with the libration motion of the molecules are introduced. Based on this model nonlinear excitations lead to the propagation of coupled rotation-displacement solitons along the chain according to two peculiar configurations. The influence of an external field on the coupled solitons is presented as well. Furthermore, all the significante results, for both models, are greatly illustrated by numerical simulations.

1.- Introduction.

Many materials present *critical phenomena* often relating to phase transitions in crystals since below a certain transition temperature there exists an ordered phase and above this temperature a disorded structure may occur. That means, at the microscopic scale, the arrangement of atoms bound by interaction energies forms nonhomogeneous structures consisting in domains separated by transition layers: walls. In order to maintain such structures, *competitive* interactions between particles are necessary, in particular, if spatial changes in a parameter occur over a small distance (some lattice spacings) from one state to the other one. However, the dynamics of such structure becomes more complex if the transition zone can propagate almost without changing its shape. In fact, this transition layer can appear as a sort of defect localized in the well ordered crystalline structure. On the other hand, critical phenomena induces *nonlinear* behaviour in the neighbourhood of the critical region and nonvanishing order parameters can coexist accordingly. Because of the importance of the scale lengths in materials, important competitive or cooperative interactions can be

E. K. H. Salje (ed.), Physical Properties and Thermodynamic Behaviour of Minerals, 359–401.

suggested which introduce the notion of *dispersion* . Starting from these two notions, nonlinearity and dispersion, *nonlinear excitations of the soliton type* can be expected in materials involving critical phenomena.

Soliton emerge from a paradox, since, at first glance, it seems strange that a *nonlinear dispersive* equation could have soliton solutions. In fact, if either of these effects appears alone, there are no soliton solutions. It is only when the effects of nonlinearity are balanced by dispersion that solitons result. When only nonlinear terms are present, the wave steepens because of the continual supply of higher frequency components. The profile of the wave then steepens until the function representing the wave profile is no longer single-valued and discontinuities are therefore formed (those are shock waves). If only dispersion is present, different harmonic components (of the Fourier expansion of the initial signal) of the wave propagate at different velocities with higher frequencies travelling more slowly so that the shape of the original signal spreads while propagating and then it is no longer recognizable. Now, if both are put together in the system the *competition* between the steepening due to nonlinearities and the spreading due to dispersion favours a travelling wave of constant profile and velocity [1,2]. However, it would be better to refer to a *solitary wave* than to a soliton since to have solitons additional conditions are required. Soliton is a solitary wave which preserves its shape and velocity in a collision with another such solitary waves.

The soliton concept occupies a key position in physics since solitons are elementary nonlinear excitations which allow one to model the nonlinear dynamics of the real world. Solitons are found in a wide variety of areas such as hydrodynamics, nonlinear plasma dynamics, dynamics of anharmonic lattices, structural phase transition as well as in biological materials and neurophysics models, to quote only a few examples [1-4]. A very large class of nonlinear dispersive partial differential systems possess solitons as exact solutions, most of these are the Korteweg-de-Vries equation, the Boussinesq equation, the nonlinear Schrodinger equation, the sine-Gordon equation, the Toda lattice equation, etc.... to quote only the most famous [1-3,5,6]. Most of these equations are integrable by means of the inverse scattering transform (IST) or the Backlund transformation or also Hirota's method [1-3,5-7].

The present study aims at illustrating the concept of solitary wave by means of two models of different complexities based on lattice dynamics. Futhermore, an attempt to relate the models to real physical media such as crystal presenting *martensitic transformations* or *molecular crystals* is made. We in fact confine our attention on the contruction of one-dimensional lattice models including all the ingredients which give rise to soliton excitations. Although simple these models can exhibit a rather fine description of materials endowed with *internal degrees-of-freedom* (molecular crystal) or *weak nonlocal behaviour* yielding the competitive interactions (strain gradient elasticity). In fact, such materials can be characterized as microstructured media since they present both crystallographic structures at microscopic and macroscopic scales and some rafinements in their description are necessary. The corresponding continuum version of these microscopic models, of course to the extent that the discreteness effects due to the microscopic scale are neglected, leads to the notion of *media with microstructure* [8,9] where the additional degrees-of-freedom and weak nonlocality play predominant roles in the existence of nonlinear excitations. A finer kinematical description of microstructured media requires then the introduction of additional degrees-of-freedom (for instance, the microgyration of molecular groups in molecular crystals) which, in fact, seems to be important to nonlinear dynamics, since the behaviour of the material, at the micro-scale, is mostly subject to the critical phenomena. On the other hand, competitive interactions induce rapid changes in crystalline structures over a rather small length, so that a finer kinematical description must introduce how the change of the lattice distorsion is strong over the characteristic length, in other word, that means that the usual notion of deformation is not efficient and the introduction of the change in lattice deformation over the characteristic length is needed. This leads to the notion of nonlocal behaviour (here, weak nonlocality) [10]. Obviously, the microstructure description depends strongly on the underlying micro-physics of the specific material. These ideas are developed with the help of two lattice models and their continuum version which involve interactions of special interest permitting particular nonlinear excitations of the soliton type.

This work comprises two parts devoted to two models corresponding to two particular types of microstructures and nonlinear excitations. In part 1, a model of deformable one-dimensional lattice involving only the *shear deformation of atomic layers* is presented. This model allows one to understand the martensitic transformation occuring in alloys associated with the transformation from b.c.c. lattice structure to the f.c.c. lattice structure by means of the shear displacement of the close-packed atomic layers. During the transformation the crystal displays rather large shear deformations which ascertains the importance of anharmonicities in the process. The main characteristic properties of such martinsitic transformations are (i) possible softening of the phonon dispersion branch, (ii) the transformation is due to a first-order transition, (iii) according to temperature the material exhibits elastic, pseudo-elastic and hysteresis (ferroelasticity) behaviours, (iv) shape memory effect is therefore induced and (v) the phase transition is characterized by *mechanical twinnings.* In the present part we are interested in examining the possible structures of the mechanical twinning combining domains where martensitic and austenitic phases coexist and phase transition proceeds by domain nucleation. More precisely, we focus on moving interfaces which can be described by solitary waves. Passing to the continuum approximation, three types of solitary waves are found and correspond to particular domain structures relating to the martensitic transformation. However, the equations of the microscopic model allow us to perform numerical illustrations of these solitary waves.

In part II, a further step in the sophisticated description of the microstructure is done. We propose a one-dimensional lattice model equipped with rotatory microstructures. The latter are modelled simply by a microscopic elastic dipole undergoing rotations of large amplitudes.The interatomic interactions are such they permit the deformations of the atomic chain in both longitudinal and transverse directions as well as the rotation of the dipoles about two axes. In its continuum version, the model provides a so-called *micropolar elasticity* but, here the micropolar variable, that is the microgyration of the dipoles, behaves nonlinearly. Then, the *coupled rotation-displacement motions* in such a medium are governed by a set of coupled equations for the displacements and microgyrations. Along with this model we study the soliton propagation for two types of configuration. For the first one, referred to as A, the dipoles rotate in the plane of displacements whereas in the second configuration, referred to as B, the dipoles rotate about the chain axis. In both configurations, we investigate nonlinear excitations consisting in coupled solitons in rotation and elastic deformations. Furthermore, numerical illustrations can be performed by means of the microscopic model which ascertains the validity of the continuum approximation. The important problem of the *influence of an external field on a moving soliton* is examined and numerically illustrated. This leads to the study of the transient motion of the coupled rotation-displacement soliton from rest.

2.- PART I- Lattice model for martensitic transformations.

2.1- The model.

This section aims at constructing a model based on anharmonic lattice dynamics with the view of describing the interface migration connected with *martinsitic structures* . There are evidently crystals which, when temperature is reduced to a finite value T_C, undergo a phase transition into a state characterized by a spontaneous deformation. Since ferromagnetic or ferroelectric crystals show the same behaviour with respect to the magnetization or the polarization, these alloys are called *ferroelastic crystals* . At high temperature such crystals show a *linear elastic* behaviour. At intermediate temperature the crystal is so-called *pseudoelastic* (or superelastic) whereas at low temperature the material behaves ferroelastically with a residual strain. We therefore have a *hysteresis phenomenon* for which the stress-strain paths on loading and unloading differ [11-14]. This remarkable behaviour is caused by a first-order phase transition [14]. Many martensitic transformations are characterized by the precursor *softening* of the relating lattice, this leads to the *instability* of the acoustic branch of phonons.

Anharmonicity -the higher order elastic terms- in nucleation process involved in martensitic transformations is of great importance. Although the perturbation method seems to be a powerful technique, the latter is not adequate to tackle the problem of nucleation, because the deformation involved in the process are extremely large. In martensitic transformations which are accompagnied by appreciable deformations of the crystalline cell, the process is known to show the propagation of distinct interfaces between the deformed phase (martinsitic or product phase) and the undeformed phase (austenitic or parent phase). Crystals as NiTi, CuSn, CuZn, AgCd, AuCd, $AuCuZn_2$ or NiAl provide good prototypes for martensitic transformations, these possess, at hight temperature an ordered b.c.c. (body centered cubic) structure, called austenitic phase [15,16]. On cooling the b.c.c. lattice structure a large elastic anisotropy appears near the first-order transition temperature. Especially, shear constant $C'=(C_{11}-C_{12})/2$ decreases drastically giving rise to an instability in the transverse elastic dispersion branch. This instability, in turn, at lower temperature induces a large elastic anisotropy due to the small shear constant C' which yields an anomalous thermal vibrations along the direction <Il0>. Then, the crystal lattice can be seen as *stacking of close-packed atomic planes* , i.e. the (Il0) plane in the b.c.c. structure. Thus, a particle in the one-dimensional atomic chain denotes a close-packed atomic layer. Accordingly, the study focus on the deformation of the crystal with respect to the the shear displacement in the plane of the close-packed atomic layers. This has been experimentally confirmed since in metallic alloys the microscopic slips of the atomic planes take place, under the influence of an external shear stress, mostly along the close-packed atomic planes [17]. This means that the barrier for the shear strain in the close-packed atomic plane is much lower compared to the barriers for the shear strain in the other planes. On the other hand, it can be pointed out that a b.c.c. structure can be obtained from a f.c.c. structure by applying a relative shear strain in the close-packed atomic plane of a f.c.c. structure (Fig.1.1a). Therefore, in the present study, only the transformation process in terms of the *relative shear displacement* of the close-packed atomic planes is considered. Fig.1.1c shows a two-dimensional model consisting in stacking of parallel crystal layers ((Il0) plane in b.c.c. structure) deduced from the undeformed crystal (Fig.1.1b) by applying a shear strain in the <Il0> direction in the close-packed atomic plane. The undeformed state corresponding to a vanishing shear strain is the austenitic phase. The shear strain is suppose to be homogeneous in the close-packed layers and the deformation can occur in either directions leading to two equivalent martensitic phases.

According to the above-mentioned hypotheses we can extract from the two-dimensional lattice of stacked crystal planes a one-dimensional chain in the stacking direction (Fig.1.1d). Nevertheless, the structure obtained by shearing the close-packed atomic planes of a f.c.c. structure does not coincide exactly with the b.c.c. structure, this slight disagreement can be solved by considering a weak inhomogeneous structure in the close-packed atomic planes. The particles of the close-packed atomic layers are supposed to have two kinds of interactions with their neighbours. The first kind of interaction potential φ_1 is considered to be *function of particle pairs* between the nearest atomic planes, this reads as

$$\varphi_1 = \varphi_1 (\mathbf{x}_{n+1} , \mathbf{x}_n) \tag{2.1}$$

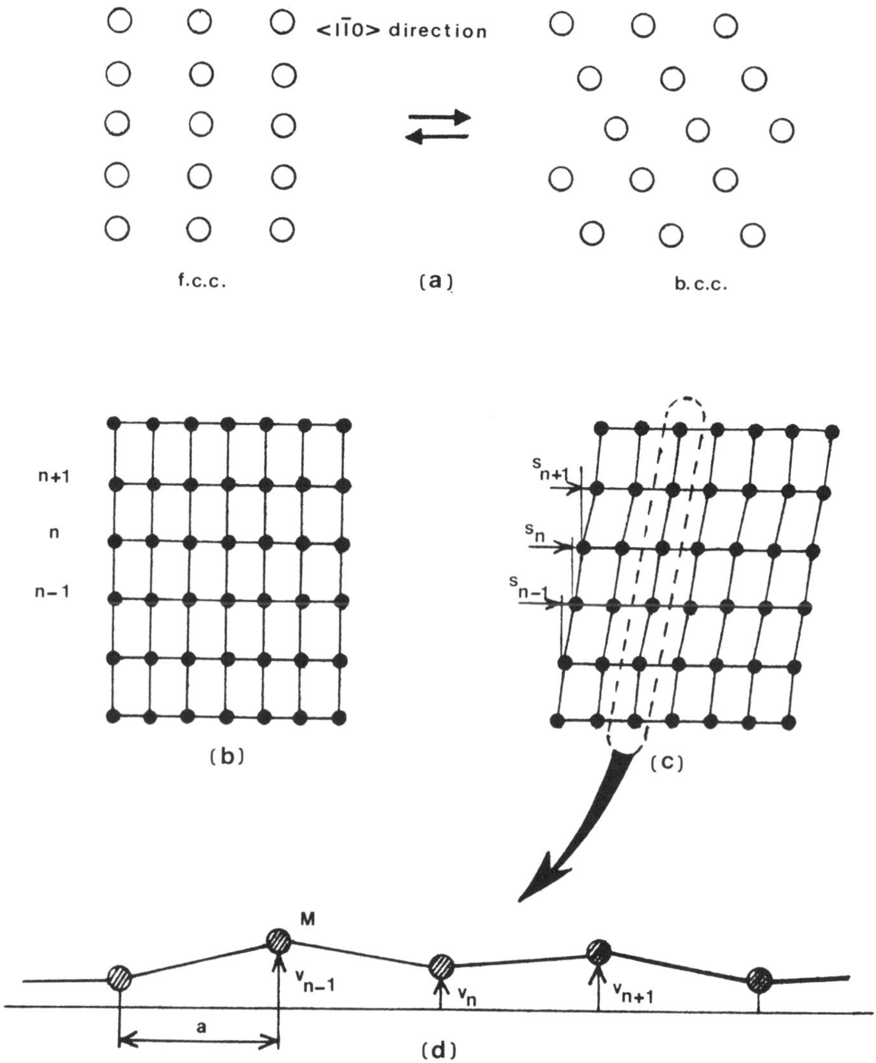

Fig. 1.1 - *One-dimensional lattice model. The martensitic transformation is described in terms of the relative shear displacement of the close-packed atomic layers -(a) transformation of a b.c.c. structure to a f.c.c. structure -(b) two-dimensional undeformed lattice -(c) shear deformed lattice by applying a shear strain on the (110) plane in <110> direction -(d) one-dimensional lattice extracted from the 2D-lattice.*

where $\mathbf{x}_n$ and $\mathbf{x}_{n+1}$ are the absolute positions of the nth and (n+1)th particles in the deformed configuration, with

$$\mathbf{x}_n = na\mathbf{i} + v_n\mathbf{j} \, , \tag{2.2}$$

where *a* is the lattice spacing and v_n is the transverse displacement of the nth particle in the close-packed atomic plane. The second kind of interaction potential φ_2 describes the interactions of the next nearest neighbouring atomic planes and it is considered as *three-boby interaction* . This potential is given by

$$\varphi_2 = \varphi_2\,(\mathbf{x}_{n+1}, \mathbf{x}_n, \mathbf{x}_{n-1}) \, , \tag{2.3}$$

The role of this interaction potential in the structural transformation process is of particular importance. In fact, this type of interaction leads to the notion of *competitive interactions* needed for the instability of the transverse acoustic phonon mode. Moreover, we have a so-called *non-central interaction* type from which we can introduce the notion of micro-stress used in the strain gradient theory of elasticity [18-20]. The presence of such terms has been pointed out ealier in the problem of couple stress in molecular crystals [21].

The energy potential for the whole system is then the sum of the potentials (2.1) and (2.3) over all the particles or close-packed atomic planes. Along with the classical axioms of the lattice invariance, we deduce that the interaction potentials φ_1 and φ_2 depend then on the relative positions of the particles. So that the energy potential of the crystal reads as

$$V = \sum_n \left[\varphi_1(\, |\mathbf{x}_{n+1}-\mathbf{x}_n| \,) + \varphi_2(\, |\mathbf{x}_{n+1}+\mathbf{x}_{n-1}-2\mathbf{x}_n| \,)\right] , \tag{2.4}$$

We assume that the relative shear displacements are small enough to allow for the expansion of the interaction potentials φ_1 and φ_2 with respect to the shear strain. Then, we obtain up to the sixth order in s_n and second order in r_n

$$\mathcal{V} = V - V_0 = \sum_n \left[\frac{1}{2}As_n^2 - \frac{1}{4}Bs_n^4 + \frac{1}{6}Cs_n^6 + \frac{1}{2}Dr_n^2 \right] , \tag{2.5}$$

where it has been set the the discrete shear strain defined by the first order finite difference of the particle displacement of the close-packed layers

$$s_n = \frac{v_n - v_{n-1}}{a} \, , \tag{2.6a}$$

and the second-order finite difference

$$r_n = \frac{v_{n+1} + v_{n-1} - 2v_n}{a^2} \, , \tag{2.6b}$$

and

$$A = a\varphi_1'(a) \, , \qquad B = \frac{1}{2}(\, \varphi_1'(a) - a\varphi_1''(a)) \, ,$$

$$C = \frac{3}{8}a(\, \varphi_1'(a) - a\varphi_1''(a) + \frac{a^2}{3}\varphi_1'''(a)) \, , \qquad D = \varphi_2''(a) \, , \tag{2.6c}$$

and where V_0 is a reference energy. We notice that the microscopic description leads, in a natural way, to the expansion of the energy potential in even terms. The simplest

representation for φ_2 is the harmonic approximation which gives the last term in eq.(2.5). In addition, we introduce the kinetic energy
where M is the mass of the particles.

$$K = \sum_n \frac{1}{2} M \dot{v}_n^2 \quad , \tag{2.7}$$

The Lagrangian of the whole system can be written as

$$L = \sum_n \left[\frac{1}{2} \dot{V}_n^2 - \frac{1}{2} C_T^2 S_n^2 + \frac{1}{4} S_n^4 - \frac{1}{6} S_n^6 - \frac{1}{2} \delta R_n^2 \right] \quad , \tag{2.8}$$

where the following changes of notation have been considered

$$\tau = |B| \sqrt{MC} \; t \, , \tag{2.9a}$$

$$(v_n \, , s_n) = (V_n \, , S_n) \sqrt{B/C} \; , \tag{2.9b}$$

$$C_T^2 = AC/B^2 \qquad \text{and} \qquad \delta = DC/B^2 \, . \tag{2.9c}$$

We assume that A, B and C are all positive constants, nevertheless A may depend on temperature according to the Curie-Weiss law $A=\alpha(T-T_c)$ where T_c is a some temperature of transition. C_T denotes the transverse elastic velocity.

Some comments on the form of the energies are in order. The energy (2.5) which represents the interactions between the nearest neighbours (bond stretching interaction) must have *three minima* as a function of shear deformation in order to describe the transformation from the b.c.c. structure [15]. One of the minima corresponds to the austenitic phase whereas the other minima are related to the martensitic phase since the martensitic phase can be obtained by slipping the close-packed atomic planes in the b.c.c. structure in either direction. Moreover, this interaction energy possesses the following properties: (i) it is invariant by changing S_n into $-S_n$ then the minima corresponding to the martensitic phases are symmetrically located with respect to the minimum of the austenitic phase, (ii) at hight temperature only the minimum $S_n=0$ (austenitic phase) is stable, (iii) at low temperature the nonvanishing minima (martensitic phase) are stable and (iv) at intermediate temperature austenitic and martensitic phases are stable or metastable. Such a potential is sketched in Fig.1.2. Finally, the three-body interaction potential (bond bending interaction) between nearest neighbours denotes the non-central interaction in metal and it is assumed to be *harmonic*.

2.2- Equations of motion.

The equations of motion for the transverse displacements V_n are easely obtained from the Lagrangian (2.8) and the definition of the discrete strain (2.6a). However, it is more convenient to work with the equations governing only the shear deformation, these equations are

$$\ddot{S}_n = C_T^2 (S_{n+1} + S_{n-1} - 2S_n) - (S_{n+1}^3 + S_{n-1}^3 - 2 S_n^3) + (S_{n+1}^5 + S_{n-1}^5 - 2 S_n^5)$$

$$+ \delta (S_{n+2} - 4 S_{n+1} + 6 S_n - 4 S_{n-1} + S_{n-2}) \, . \tag{2.10}$$

This is a system of coupled nonlinear ordinary differential equations which describe the shear

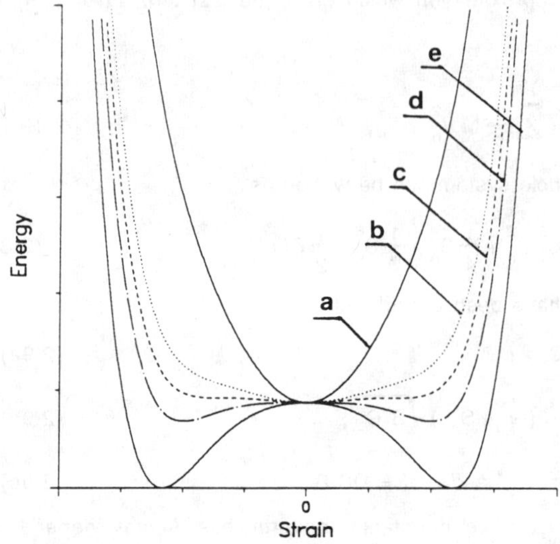

Fig. 1.2 *Elastic energy as function of strain -(a) and (b) high temperature: austenitic phase is stable, (c) and (d) intermediate temperature: austenitic and martensitic phases are stable or metastable and (e) low temperature: martensitic phase is stable.*

deformation excitations travelling perpendicularly to the close-packed atomic planes. In the discrete case, these equations are complicated to solve except for the linear case which will be examined in a further section. The strongly nonlinear feature of eq.(2.10) can be however investigated by means of a numerical simulation. An alternative situation happens in the case of the continuum approximation.

2.3- Continuum model.

a- Equation. Since the present study deals with dynamic processes, we assume the existence of a characteristic length D (for instance, the width of any dynamic disturbance) associated with a dynamic process and D is large as compared to the lattice spacing a. This means that the discrete shear deformation S_n is slowly varying in space and time (if T denotes a characteristic time, we have T=D/C where C is a characteristic velocity connected with the dynamic process). Then, the quantity S_n can be expanded about the point x=na . After some classical algebric manipulations we obtain a partial differential equation for the continuous shear deformation S(x,t) given by

$$S_{,tt} = C_T^2 S_{,xx} - (S^3)_{,xx} + (S^5)_{,xx} - \gamma S_{,4x} , \qquad (2.11)$$

where we have set

$$\gamma = -(\delta + C_T^2/12) , \qquad (2.12)$$

which represents the coefficient of the dispersive term. Going back to the transverse displacement V, we have

$$S = V_{,x} .$$

b. Remarks -(i) *Strain gradient theory .* We can compare the present lattice approach to that of the direct phenomenological continuum theory of nonlinear elasticity in the framework of the strain gradient theory [19, 20]. Then, we introduce an *elastic potential* of the form

$$\psi = \psi(S, S_{,x} , \theta) , \tag{2.13}$$

which is, in fact, function of the shear strain S as in the classical elasticity, but also the strain gradient and obviously the temperature θ. The Lagrangian density of the continuum model is then given by

$$L = K - \psi , \tag{2.14}$$

where K is the density of kinetic energy. By minimizing the functional L depending on $V_{,t}$, $V_{,x}$ and $V_{,xx}$ we arrive at the equation of motion for the transverse displacement

$$\ddot{V} = \Sigma_{,x} , \tag{2.15}$$

in this notation the mass density is set equal to one. Σ is then a shear stress or the response of the system to a change in the shear deformation. The stress can be decomposed into

$$\Sigma = \sigma - \chi_{,x} , \tag{2.16a}$$

with

$$\sigma = \frac{\partial \psi}{\partial S} , \tag{2.16b}$$

$$\chi = \frac{\partial \psi}{\partial S_{,x}} , \tag{2.16c}$$

and we have as well

$$\eta = - \frac{\partial \psi}{\partial \theta} , \tag{2.16d}$$

In eq.(2.16b) σ is the *macroscopic shear stress* (classical elasticity) associated with the shear deformation. The last term in eq.(2.16b) defined by eq.(2.16c) represents the *microstress* due to the *lattice curvature* $S_{,x}$. The microstress can be introduced from a more general theory of strain gradient elasticity [18-20], roughly speaking, such a theory tends to introduce the local curvature or the local rotation of the Cartesian frame embedded in the lattice. The effect of the local curvature on the lattice distorsion has been introduced in the lattice model through the non-central interaction potential. On the other hand, the strain gradient theory of elasticity can be recovered from the *weak nonlocal behaviour* of the so-called quasi-continuum theory derived from lattice theory [10]. Finally, eq.(2.16d) gives the entropy density. Physically, a surface energy density can be associated with the strain gradient this leads to the notion of correlation length. This means that the order parameter -shear strain- changes rapidly over a length of the order of the correlation length. The introduction of strain gradient becomes efficient when structures in domains and walls are implicated.

(ii)- *Ginzburg-Landau theory* . The next step consist in choosing a specific form for the elastic potential ψ. Within the framework of the Ginzburg-Landau theory of the first-order phase transition [14,22] and in agreement with our lattice model, the elastic potential density ψ takes on the form

$$\psi(S, S_{,x} ,\theta) = \psi_1 (S) + \psi_2 (S_{,x}) + \psi_0 (\theta) , \tag{2.17a}$$

and

$$\psi_1(S) = \frac{1}{2} a(\theta)\, S^2 - \frac{1}{4} S^4 + \frac{1}{6} S^6 \quad , \tag{2.17b}$$

$$\psi_2(S_{,x}) = \frac{1}{2} \gamma (S_{,x})^2 \quad , \tag{2.17c}$$

where $a=(C_T)^2$ in accordance with the notations used in the previous section and a satisfies the Curie-Weiss temperature law. ψ_o is a reference energy function only of temperature. The potential (2.17) is the simplest free-energy which fulfils all the conditions required by the Ginzburg-Landau theory of the first-order phase transition [22-24]. Inserting eqs(2.17a)-(2.17c) into eqs(2.15), (2.16a)-(2.16c) we recover the equation of motion (2.11) deduced from the lattice model, and this equation can be rewritten as

$$S_{,tt} = \Sigma_{,xx} \quad . \tag{2.18}$$

Now, owing to eqs(2.16b) and (2.17b) the shear stress is given by

$$\sigma = C_T^2\, S - S^3 + S^5 \quad , \tag{2.19}$$

and the microstress defined by eq.(2.16c) reads then as

$$\chi = \gamma\, S_{,x} \quad . \tag{2.20}$$

The entropy density is

$$\eta = - \frac{dC_T^2}{d\theta} S^2 - \frac{d\psi_0}{d\theta} \quad . \tag{2.21}$$

It is worth commenting the *stress-strain relation* (2.19). The latter is not univoque and this leads to the notion of hysteresis which characterizes ferroelastic materials. At hight temperature the material is *nearly linear* and for rather small strain the linear elasticity is valid, moreover the elastic potential (2.17b) has only one minimum and the curve is convex (see Fig.1.2). At lower temperature the material has a nonlinear behaviour and it is so-called *pseudoelastic* displaying unstable regions corresponding to the nonconvex parts of the elastic potential curves. On loading the specimen, we force the transition from the austenitic phase to the martensitic one and the process goes on until the austenitic phase disappears completely on the expense of the martensite phase. On unloading the strain-stress path is different because of the hysteresis phenomenon. At much lower temperature, the *hysteresis* is growing and a vanishing stress can be reached at nonzero strain, martensitic phases are created. The reverse transformation to austenitic phase can be obtained only by loading in the opposite direction and in this range of temperature the specimen is *ferroelastic* . In this case only the martensitic phase is stable and this corresponds to the minima of the elastic energy. Some schematic stress-strain relations are sketched in Fig.1.3. The elastic modulus is given by

$$C' = \frac{\partial\sigma}{\partial S} = 2\, C_T^2 - 3\, S^2 + 5\, S^4 \quad , \tag{2.22}$$

which is not a constant and it may be negative in the nonconvex regions of the elastic potential where the state of the material is unstable. This elastic modulus corresponds to the shear strain on a (IIO)-plane in the <IIO> direction, that is $C'=(C_{11}-C_{12})/2$ in crystallographic notation. In further section, we assume that the temperature changes during the characteristic period of dynamic process are not enough to induce heat conduction. Therefore, *adiabatic motion* with constant entropy density is assumed.

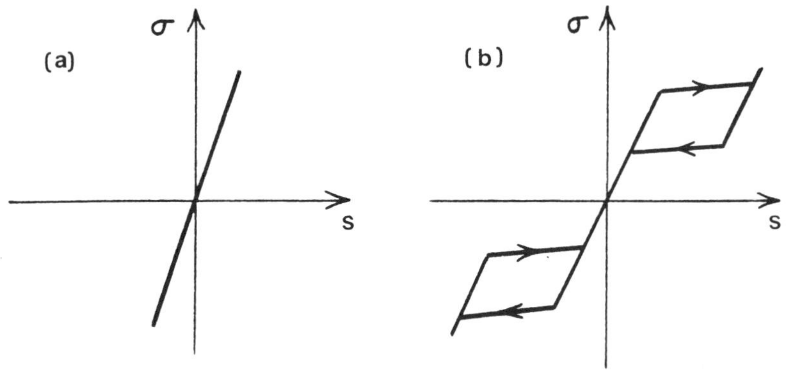

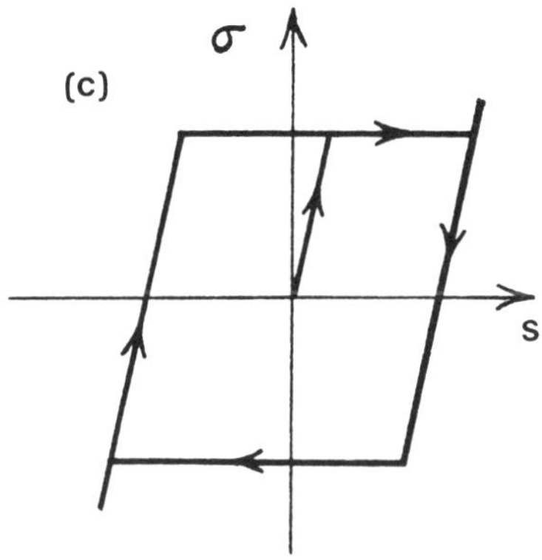

Fig. 1.3 *-Schematic stress-strain response curve (a) at high temperature: elasticity, (b) at intermediate temperature: pseudo-elasticity and (c) at low temperature: ferroelasticity.*

2.4- Linear case.

If the amplitude of the transverse displacements are sufficiently small, the nonlinear terms in eq.(2.10) can be neglected and harmonic sinusoidal wave solutions travelling along the stacking direction can be looked for. This yields the dispersion relation for the linear waves

$$\omega^2 = 4C_T^2 \sin^2(k/2) \left(1 + \frac{4\delta}{C_T^2} \sin^2(k/2)\right) \quad , \tag{2.23}$$

where ω is the angular frequency and k is the wave-number lying in the first Brillouin zone. The dispersion relation curves are depicted in Fig.1.4 for different values of the ratio $\rho=\delta/(C_T)^2$. For $\rho<-8$ the dispersion curves differ slightly from the classical case (see curves (a) and (b)). If $-8\leq \rho \leq -4$ the dispersion curves have a minimum at k_o (see curve (c) and (d)). Finally, for $\rho > -4$ ω is real only for $0\leq k \leq k_1$ and the wave becomes *unstable* for wave-number greater than k_1. Since the coefficient C_T^2 (or a) depends on temperature, at a given δ, on heating, the acoustic branch undergoes a *softening* at the limit of the Brillouin zone ($k=\pi$). This is also caused by the presence of the coefficient δ introduced by the non-central potential. Note that, since $C_T^2 > 0$, δ must be negative and this corresponds to the *competitive interaction* , nevertheless, the case a<0 can be envisaged and $\delta>0$ leads to the *cooperative interactions.*

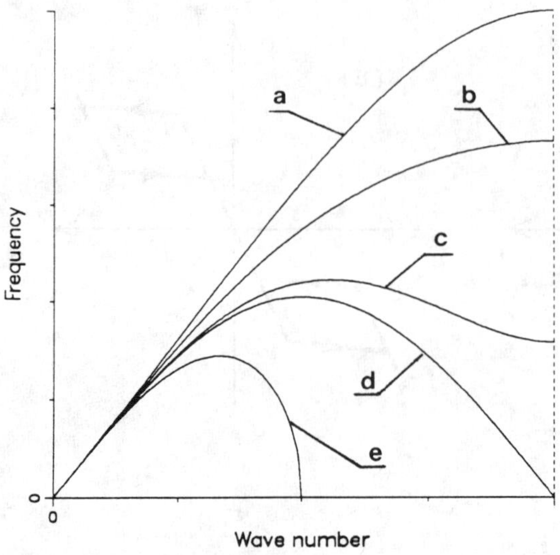

Fig. 1.4 *-Dispersion relation of the transverse acoustic waves. (a) and (b)* $C_T^2/\gamma < -8$, *(c)* $-8 < C_T^2 < -4$, *(d)* $C_T^2 = -8$ *and (e)* $C_T^2 > -8$ *(instability).*

2.5- Solitary wave solutions.

Our purpose is to examine the possible nonlinear excitations in such media on the basis of the present lattice model, the harmonic approximation is therefore no longer applicable.In order to close eq.(2.10) it is necessary to introduce appropriate boundary conditions. We choose periodic conditions at each end of the chain, these conditions are particularly useful for the numerical simulations of the discrete system (2.10) and furthermore, this is equivalent to suppose that the specimen is made of periodic arrangements of largely separated domain walls. In the mechanical point of view the boundary conditions can be written as

$$\sigma_1 + \sigma_2 = 0 \ , \qquad (2.24a)$$

$$\chi_1 = \chi_2 = 0 \ , \qquad (2.24b)$$

The first condition involving the macroscopic stress means that the atomic chain is at equilibrium while the second condition imposes vanishing microstress at each end of the chain. The number 1 denotes the first particle (or the first close-packed atomic layer) and N is the number of the last particle. These conditions can be transposed into the continuum stress σ and microstress χ by accounting for eqs(2.19) and (2.20) respectively and we can write

$$(C_T^2 S_1 - S_1^3 + S_1^5) + (C_T^2 S_N - S_N^3 + S_N^5) = 0 \ , \qquad (2.25a)$$

$$(S_{,x})_1 = (S_{,x})_N = 0 \ . \qquad (2.25b)$$

Now, we look for solutions of eq.(2.11) or (2.18) subject to the condition (2.24) or (2.25) function only of the phase variable ξ=x-Ct where C is a constant phase velocity. Inserting such a solution into eqs(2.18) and (2.16a) we obtain thus the equation

$$C^2 S'' = (\sigma - \chi'')'' \quad , \qquad (2.26)$$

where the prime stands for the derivative with respect to ξ. We integrate twice eq.(2.26) with respect to ξ, by accounting for the fact that the shear deformation is bounded for large ξ we obtain

$$C^2 S = \sigma - \chi'' + C_1 \quad , \qquad (2.27)$$

where C_1 is the remaining constant of integration which is determined from the boundary conditions for S_1 and S_N

$$C^2 S_N = \sigma_N + C_1 \quad . \qquad (2.28)$$

The same condition hold true for the particle 1 and the condition (2.24b) has been accounted for. At this stage we can explicite the possible and compatible boundary conditions on S_1 and S_N to allow for the propagation of solitary waves. We must first satisfy the conditions (2.24a) and (2.24b) (or (2.25a) and (2.25b)). The second condition is fulfilled if S_1 and S_N are constant, so that the first condition is satisfies if

$$S_N = -S_1 = S_0 \quad , \qquad (2.29a)$$

or

$$\sigma_1 = \sigma_N = 0 \quad , \qquad (2.29b)$$

The first condition, in turn, implies from eq.(2.28) that C_1=0. Then eq.(2.28) provides S_o

$$S_0 [(C_T^2 - C^2) - S_0^2 + S_0^4] = 0 \quad , \qquad (2.30)$$

which leads to two kinds of solution whether S_o=0 or $S_o \neq 0$. The second condition (eq.(2.29b)) gives a vanishing stress at each end of the chain. Accounting for eqs(2.19) and (2.29b) we obtain

$$S_0 (C_T^2 - S_0^2 + S_0^4) = 0 \quad , \qquad (2.31)$$

which provides either S_o=0 or $S_o \neq 0$. On substituting eq.(2.31) into eq.(2.28), we have

$$C^2 S_0 = C_1 \qquad (2.32)$$

which gives the constant C_1. We can remark that, for the last condition, a travelling wave is obtained either if $C_1 \neq 0$ or if $S_o = C_1 = 0$. The latter condition has been met in the first condition case. Now, we sum up all the possible conditions induced by the compatible boundary conditions: (i) martensitic solitary wave in an austenitic matrix $S_1 = S_N = S_o = 0$ and C_1=0, (ii) martensitic-martensitic kink $S_N = -S_1 = S_o \neq 0$, C_1=0 and S_o is given by the nonzero root of eq.(2.30) and (iii) martensitic solitary wave in a martensitic matrix $S_1 = S_N = S_o \neq 0$ and $C_1 = C^2 S_o$, S_o is provided by the nonzero root of eq.(2.31). In fact, in the solitary wave problem, we consider the amplitude of the wave as a free parameter, so that the wave

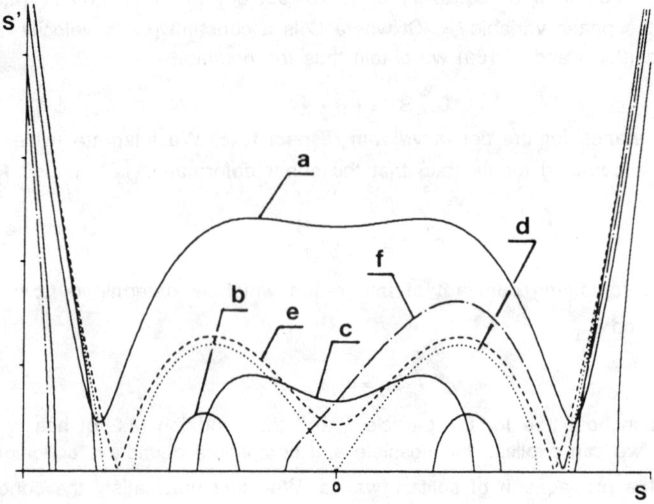

Fig. 1.5 *-Phase trajectories of eq.(2.33) in the phase plane (S',S) -(a) unbounded solution, (b) and (c) periodic solutions, (d) martensitic soliton moving in austenitic matrix (section 2.5a) (e) martensitic kink (section 2.5b) and (f) martensitic or austenitic soliton moving in martensitic matrix (section 2.5c).*

velocity C is function of the amplitude wave.

The first integral of eq.(2.27) can be obtained and it reads as

$$\frac{1}{2}\gamma(S')^2 = W(S) \tag{2.33a}$$

with

$$W(S) = \frac{1}{2}(C_T^2 - C^2)S^2 - \frac{1}{4}S^4 + \frac{1}{6}S^6 + C_1 S + C_2 \tag{2.33b}$$

where C_2 is an integration constant and S' is defined everywhere W is positive. The constant C_2 is determined from the boundary conditions on S. The discussion of the solutions of eq.(2.33a) can be qualitatively obtained on the basis of the phase trajectories plotted in the plane (S,S'). Fig.1.5 gives the trajectories for different values of C_1, C_2 and $(C_T{}^2 - C^2)$. According to the value of these constants we have unbounded solutions curve (a), periodic solutions or periodon-like solutions curves (b) and (c), special cases correspond to the solitary waves, the curve (d) passing through zero corresponds to the conditions (i) (see here-above), the curve (e) which is symmetric, is obtained from the conditions (ii) and the curve (f) corresponds to the conditions (iii), whereas the external prolongations of these curves lead to unbounded solutions. The periodic solutions can be expressed by means of Jacobian elliptic functions and correspond to arrays of solitons describing the structure of martensitic-austenitic walls [25]. Then the integration constant C_2 is determined so that the extrema of $S(\xi)$ coincide with the roots of W(S).

a - Solitary wave : case $S_o=0$.

For this case, the solution vanishes for large ξ (austenitic phase). Owing the boundary conditions given by eqs(2.29) and (2.32) we get $C_2=0$, so that the equation of motion (2.33) can be rewritten as

$$\gamma(S')^2 = S^2\left[(C_T^2 - C^2) - \frac{1}{2}S^2 + \frac{1}{3}S^4\right]. \quad (2.34)$$

An exact solution of eq.(2.34) is given by

$$S^2(\xi) = \frac{S_m^2}{1 + \left(1 - \dfrac{\frac{2}{3}S_m^2}{1 - \frac{2}{3}S_m^2}\right)\sinh^2(Q\xi)}, \quad (2.35a)$$

with

$$Q^2 = 2S_m^2\left(1 - \frac{2}{3}S_m^2\right)/\gamma, \quad (2.35b)$$

and S_m is the amplitude of the solitary wave which satisfies the equation $W(S_m)=0$. It turns out that the solitary wave velocity depends strongly on the wave amplitude and it is given by

$$C^2 = C_T^2 - \frac{1}{2}S_m^2\left(1 - \frac{2}{3}S_m^2\right). \quad (2.36)$$

Nevertheless the solution (2.35a) is subject to some constraints on the amplitude S_m:

(i) for $|C_T| \geq \sqrt{3}/4$ and $\gamma > 0$ then $|S_m| < \sqrt{3}/2$,

(ii) for any C_T and $\gamma < 0$ then $|S_m| > \sqrt{3/2}$,

(iii) for $|C_T| < \sqrt{3}/4$ and $\gamma > 0$ then $S_m^2 < \frac{3}{4}\left(1 - \sqrt{1 - \frac{16}{3}C_T^2}\right)$,

otherwise the wave does not travel and only a static solution exists. The case $\gamma>0$ corresponds to competitive interactions whereas $\gamma<0$ leads to cooperative interactions. Moreover, if $|S_m|>\sqrt{3/2}$ the wave is supersonic since $|C|>C_T$. When S_m^2 approaches 3/2 or 0 the solitary wave velocity is close to the acoustic transverse velocity and on the other hand if C_T is high and S_m is not too large, the velocity depends slightly on the amplitude and in this case the elastic potential can be approximated by a parabola (see Fig.1.2).

The numerical simulations of the present solution are given in Fig.1.5. The numerical scheme is directly provided by the discrete system (2.10) where the exact solution (2.35a) has been used for the initial conditions. The first illustration (Fig.1.5a) represents a small layers of martensitic phase moving in an austenitic matrix with a relatively large amplitude satisfying the condition (i). We can observe small oscillations which propagate on either side of the solitary wave, this is due to the discreteness effect of the lattice. In Fig.1.5b we have the same situation but the amplitude of the solitary wave is smaller and the wave spreads out a little more than in Fig.1.5a. These both solutions correspond, in fact, to the curve (d) of the phase trajectories plotted in Fig1.4. Furthermore, the austenitic phase is an absolute minimum of the elastic energy (see Fig.1.2), whereas the amplitude S_m is a metastable minimum. If $C_T=0$, S_m goes to zero if $\gamma>0$ the martensitic layer desappears at the expense of the austenitic phase. But if $\gamma<0$ (cooperative interactions) the solitary wave exists even if $C_T=0$.

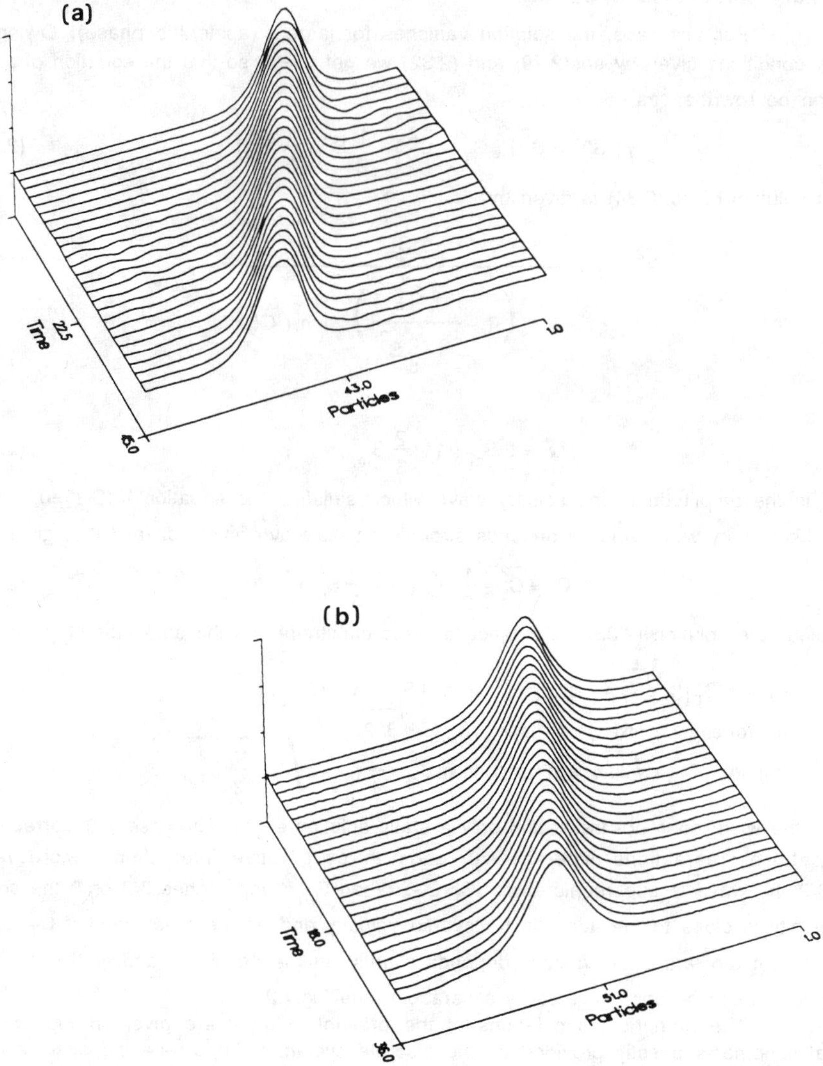

Fig. 1.6 - *Martensitic solitary wave moving in an austenitic matrix* ($S_o = 0$), *(a)* $C_T=0.806$ *and* $S_m = 0.85$ ($C= 0.68$ *and* $\gamma = 1.446$), *(b)* $C_T = 0.806$ *and* $S_m = 0.3$ ($C=0.78$ *and* $\gamma = 1.446$).

b- Solitary wave : case $S_1=-S_N=S_o$.

This case corresponds to a moving wall between two martenstic domains of different orientations. The strain state in the martensitic domain is given by the nonzero solution of eq.(2.30). However an alternative situation consists in giving the amplitude S_0 and then we deduce the velocity as function of S_0 from eq.(2.30)

$$C^2 = C_T^2 - S_0^2\,(1 - S_0^2) \quad . \tag{2.37}$$

On the other hand, in the present case $C_1=0$ and C_2 is chosen such as $W(S_0)=0$ and the equation of motion can be written as

$$\gamma\,(S')^2 = \frac{1}{3}\,(S^2 - S_0^2)^2\,(S^2 - 2\,S_0^2 - \frac{3}{2}) \quad . \tag{3.38}$$

This equation possesses the following solution

$$S\,(\xi) = \frac{S_0 \tanh\,(Q\xi)}{\left[1 + \dfrac{2S_0^2}{4S_0^2 - 3}\,(1 - \tanh^2(Q\xi\,))\right]^{1/2}} \tag{2.39a}$$

with

$$Q^2 = S_0^2\,(2\,S_0^2 - 1\,)/2\gamma \quad . \tag{2.39b}$$

The existence of such a solution implies some constraints on S_0 and C_T since the solution (2.39a) and the velocity defined by eq.(2.37) must be real. Then S_0 must fulfil either of the following conditions

(i) for $|\,C_T\,| > 1/2$ and $\gamma < 0$ we have $|\,S_0\,| < 1/\sqrt{2}$ and $C_T^2 - 1/4 < C^2 < C_T^2$,

(ii) for $|\,C_T\,| > 1/2$ and $\gamma > 0$ we have $|\,S_0\,| > \sqrt{3}/2$ and $C^2 > C_T^2 - 3/16$,

(iii) for $|\,C_T\,| < 1/2$ and $\gamma < 0$ we have $S_0^2 < \frac{1}{2}\,(1 - \sqrt{1 - 4C_T^2}\,)$ and $|\,C\,| < C_T$,

(iv) for $|\,C_T\,| < 1/2$ and $\gamma < 0$ we have $S_0^2 > \frac{1}{2}\,(1 + \sqrt{1 - 4C_T^2}\,)$ and no constraint on C .

We notice that the solitary wave is supersonic if $|S_0|>1$ this case may occur for the conditions (ii) and (iv). As in the previous case we have either competitive or cooperartive interactions. We also notice that if $S_0 \to 0$ the solitary wave approaches the transverse acoustic velocity and in addition, for $|C_T|>1/2$ there is a gap between $1/\sqrt{2}$ and $\sqrt{3}/2$ for the possible value of the wave amplitude.

Fig.1.7 collects the numerical simulations . As in the previous case the analytic solution (2.39a) is used for initializing the numerical scheme given by the discrete system (2.10) . Two sorts of solution can be displayed. The first one is presented in Fig.1.7a and

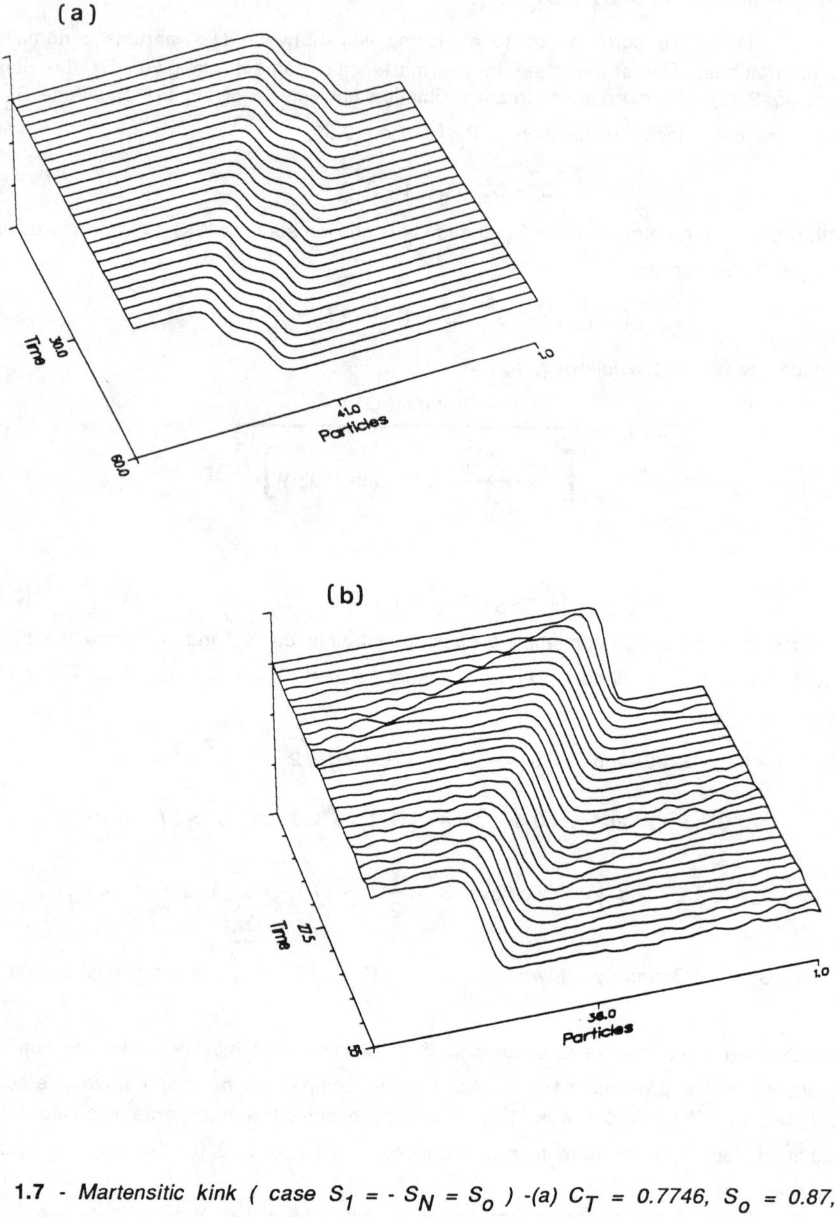

Fig. 1.7 - *Martensitic kink (case $S_1 = - S_N = S_o$) -(a) C_T = 0.7746, S_o = 0.87, C=0.645 and γ =1., (b) C_T = 0.6325, S_o = 1.02, C = 0.665 (supersonic wave) and γ = 2.*

describes the smooth transition from one martensitic domain to the other one with a different orientation. Moreover, the transition is not direct since the interface passes through a small austenitic zone which is metastable because, in this case, the austenitic phase is not an absolute minimum of the elastic energy (see Fig1.2). This small austenitic region exists if $\sqrt{3}/2<|S_0|<1$. the second situation is presented in Fig.1.7b, and the transition from one martensitic domain to the other one is direct. The latter case satisfies the condition (ii), but the wave is supersonic and the kink is narrower, we can observe some phonon radiations caused by the discreteness effects.

c- Solitary wave : $S_1=S_N=S_0\neq 0$.

This situation describes the propagation of a solitary wave of an austenitic or martensitic layer moving in a martensitic matrix. The strain state of the martensitic matrix is S_0, at infinity on both sides there exists the same martensitic phase and S_0 is one of the minima of the potential (2.17b) and it is given by the nonzero root of eq.(2.31), but the solution exists only for $|C_T|<1/2$. For intermediate values of ξ the strain is smaller then S_0, let S_m be the closest value of $S(\xi)$ to S_0 given by the equation $W(S_m)=0$. The amplitude of solitary wave is then S_0-S_m. the equation of motion can be written as

$$\gamma\,(S')^2 = \frac{1}{3}(S-S_0)^2(S-S_m)\{(S+S_0+S_m)[(S+S_0)S+(S_m+S_0)S_m + 2S_0^2 - 3/2] + S_0(2S_0^2 - 3/2)\} \tag{2.40}$$

and the solitary wave velocity is given by

$$C^2 = (S_m+S_0)^2(S_m^2 + 2S_0^2 - 3/2)/3 \ . \tag{2.41}$$

We fixe the amplitude of the wave, whence S_m, and the velocity is then determined as function of S_m. Nevertheless, such a solution leads to some restrictions on the amplitude S_m

$$\text{if } \sqrt{3}/4 < C_T < 1/2 \text{ then } \sqrt{3/2 - 2S_0^2} \le S_m \le S_0 \ ,$$

$$\text{if } 2/5 < C_T < \sqrt{3}/4 \text{ then } -S_0 \le S_m \le \left[-S_0 - 3\sqrt{2 - 5S_0^2/3}\right]/4 < 0$$

$$\text{or } \left[-S_0 + 3\sqrt{2 - 5S_0^2/3}\right]/4 < S_m < S_0 \ ,$$

$$\text{if } C_T < 2/5,\ \text{ then } -S_0 < S_m < S_0 \ .$$

We remark that, according to the amplitude S_m, we can have different types of solitary waves. For $C_T<\sqrt{3}/4$ we have a martensitic layer moving in the matrix of the other martensitic variant. If $C_T<2/5$ we can have an austenitic or martensitic layer moving in a martensitic matrix since S_m can approach the value $-S_0$. In the intermediate range $2/5<C_T<\sqrt{3}/4$ we can have both types of solitary wave but there is a gap in S_m, and we cannot pass continuously from one type to the other one. We notice that S_m approaching $-S_0$ the velocity goes to zero and the wave does not travel. In the present case the solitary wave is always subsonic. When the amplitude of the solitary wave tends to zero ($S_m\to S_0$) the velocity

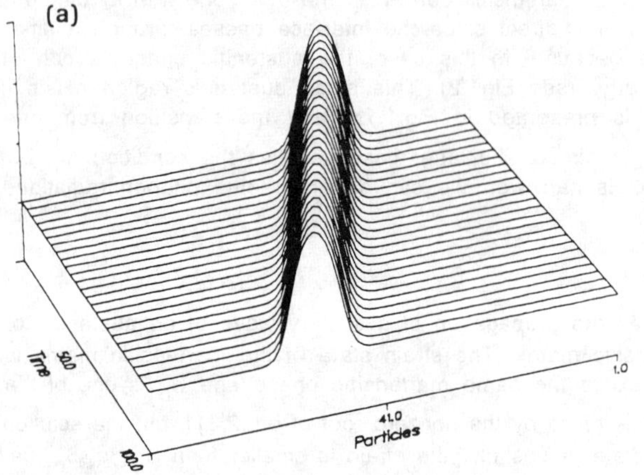

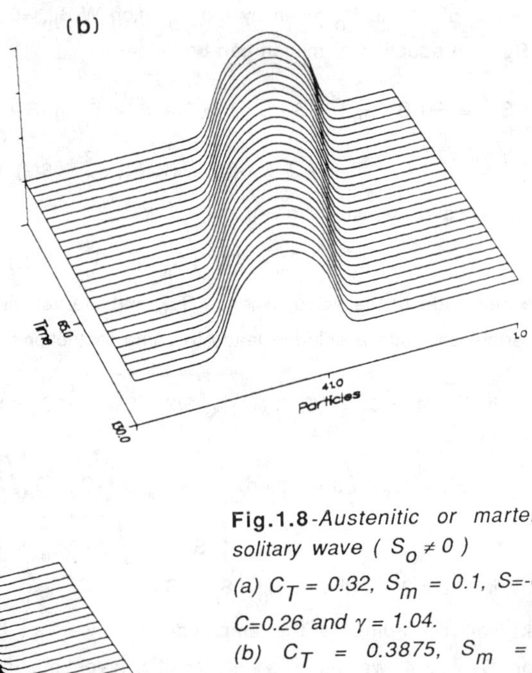

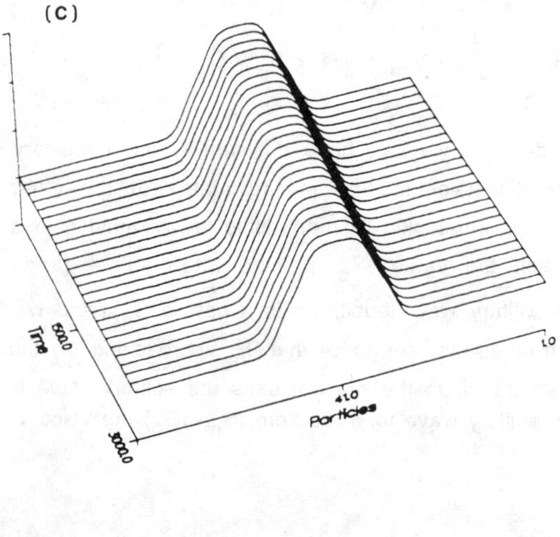

Fig.1.8-*Austenitic or martensitic solitary wave (* $S_o \neq 0$ *)*

(a) $C_T = 0.32$, $S_m = 0.1$, $S=-0.942$ $C=0.26$ *and* $\gamma = 1.04$.

(b) $C_T = 0.3875$, $S_m = 0.3$, $S_o=-0.9$, $C=0.164$ *and* $\gamma = 1.0375$.

(c) $C_T = 0.3873$, $S_m = 0.9$, $S_o=-0.9035$, $C= 0.002$ *and* $\gamma=1.0375$.

is nonvanishing and it is given by $C=2S_0(S_0^2-1/2)^{1/2}$, this velocity is very close to the transverse acoustic velocity in the austenitic phase, in fact, the amplitude is growing from S_0 to $-S_0$ the solitary wave velocity is then decreasing from C_T to 0.

In the present situation we cannot give an exact analytic solution, then eq.(2.40) is solved numerically and the solution serves as initial conditions for the numerical simulations. In Fig.1.8 we present different types of solitary waves according to S_m. The first plot (Fig.1.8a) describes the wave evolution for a small amplitude, the hump is relatively narrow. Next, in Fig.1.8b we have a solitary wave for an intermediate amplitude the hump becomes thicker whereas the velocity is smaller. Finally, the case where S_m approaches $-S_0$ is given in Fig.1.8c, the solitary wave is much thicker that means when S_m is very close to $-S_0$ the width of the wave becomes very large and the solitary wave decomposes into a kink-antikink pair, corresponding each to a martensitic kink. This can be interpreted as a nucleation process.

2.6- Conclusion.

In this section we have built a nonlinear lattice model for martensitic transformations involving both nonlinearities and competitive interactions, the latter, in fact, account for a refine description of the long-range interactions (three-body interaction). Nevertheless, this weak nonlocal behaviour may be improved by adding the harmonic interactions between the second nearest neighbours. The nonlinear part is deduced, in a naturel way, from the nonlinear interactions between the close-packed atomic planes and the energy fulfils all the physical requirements involving in martensitic transformations [12]. Passing to the continuum approximation we recover the Ginzburg-landau theory of the first-order phase transition from which we can obtain the shear stress state. On the basis of this model an exhaustive study of the nonlinear excitations is presented. We find three types of solitary wave which can travel. The first type (see section 2.5a) corresponds to a martensitic domain moving in an austenitic phase, the second type (see section 2.5b) represents a martensitic kink transiting from one martensitic domain to the other one with a different orientation, moreover, according to the amplitude of the kink, a small austenitic zone is included in this buffer zone. The last case (see section 2.5c) describes a layer of an austenitic or martensitic phase which travels in a martensitic matrix. In all cases we determine the solitary wave velocity as function of the amplitude. In its continuum version we recover a one-dimensional dynamic Ginzburg-Landau model for martensitic transformations in shape-memory alloys established by F. Falk [14,22]. On the basis of this model he studies the phenomenological theory of the mechanism of martensitic transformations and shape memory phenomena as well as the structure in domains and walls [25,26]. In the fully dynamic study we find three types of moving soliton whereas in [26] there are only two types, because in our study periodic boundary conditions have been considered which allows for an extra case. Starting from nonlinear theory of elasticity a rather general theory of thermoelasticity of phase transition can be developed with the view of describing structural phase transition and twinning [27-30] but the propagation of domain walls has not been envisaged. Typically , the order of magnitude of the parameters which are present in the model are $A_0=48J/cm^3K$ $(A=A_0(T-T_c))$ and $T_c=208°K$, $B=6\times105J/cm^3$, $C=4.5\times10^7J/cm^3$ and $D\approx2\times10^{-12}J/cm$ these coefficients correspond to those of eq.(2.5) for the example of the alloy $Au_{23}Cu_{30}Zn_{47}$ [17,22]. Moreover, the width of the solitary wave can be estimated for about 10 cell units in the case of Fig.1.4a and Fig.1.6a.

Nevertheless, the stability of the solitary waves is worth examining in order to test the behaviour of such waves for long times [31]. If the solution is unstable the width of the solitary wave growths and the solution can be interpreted as a nucleation phenomenon. On the other hand, the discreteness effects can become predominant, especially in the case of rather small thicknesses or amplitudes of the solitary waves and this can therefore contribute to the nucleation process.

3.- PART II - Model for deformable lattices equipped with rotatory microstructures.

3.1 - The model.

In the previous section we studied the effect of a finer description inherent to a weak nonlocal behaviour, but the degees-of- freedom of the lattice model were those of a classical deformable chain. In the present section we suppose that the particles cannot be reduced to a point endowed with a simple mass, but the particles display a *moment of inertia* so that *internal degrees-of-freedom* inherent to a finer description of the microstructured medium must be accounted for. A good picture of such media can be given by a deformable lattice equipped, at each of its nodes, with a molecule performing rigid-body rotational motions (which may have large amplitudes) about their mass center. This implies that *microgyration* of the molecules must be added to the usual lattice deformations. Moreover, since an inertia is associated with the molecules, we can thus define an angular momentum relating to the motions of the molecules and the *dynamics of coupled rotation-lattice deformation* can be envisaged. The interactions involved in the lattice model depend strongly on the micro-physics of the micro-system and may be related to some phase-transition phenomena in some cases. More precisely, the interactions of special interest are those which permit nonlinear excitations of the soliton type (especially competitive interactions often connected with a characteristic length of the system). Such nonlinear behaviours can be found in nematic liquid crystals [32], in long chains of polymers [33,34], in molecular ferroelectric crystals (e.g. Sodium Nitrite)[35-37], in elastic chains of macromolecules (e.g. DNA)[38].

Let us consider a one-dimensional monoatomic chain with N particles (but N is considered large enough in order to avoid any boundary effects) equipped, at each node, with a molecule (Fig.2.1). We suppose that each molecule is endowed with a microscopic electric dipole of constant length which indicates the molecule orientation. The force acting on the system result from (i) the ionic interactions between neighbouring ions (short and long-range interactions) as in classical ionic crystals, (ii) the mutual interactions between microscopic dipoles and (iii) the electrostatic interaction due to the external field **E**. The possible motions of the system are (i) the longitudinal displacements of the mass center of the molecules, (ii) the transverse displacements of the same mass center and (iii) the rigid-body rotational motions of the molecules or microscopic dipoles. We allow the dipoles to have two rotation angles (Fig.2.1b). the angle θ_n is the rotation of the plane Π about the z-axis and the angle φ_n corresponds to the rotation in the plane Π about an axis perpendicular to Π. We assume that the lattice displacements remain infinitesimal whereas the rotations may have large amplitudes in order to describe the possible reversal motion of the dipoles. Expanding the mutual electrostatic energy between dipoles up to the first order in lattice deformations we obtain the Hamiltonian describing the dynamic of our system [39,40]

$$H = K + W_{lat.} + W_d + W_{int.} + W_e \tag{3.1}$$

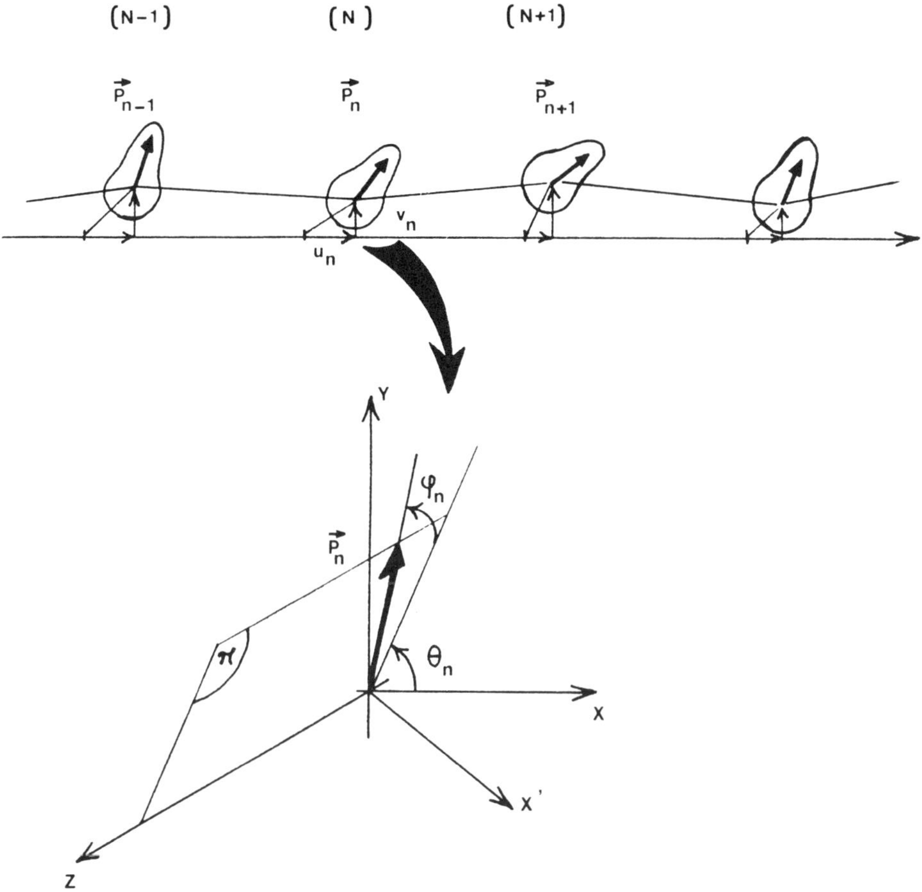

Fig. 2.1 - *Lattice model for deformable chain with rotatory microstructure. The deformation of the chain is given by* u_n *and* v_n *the longitudinal and transverse displacements. The microstructure is endowed with a microscropic dipole which rotates about the mass center by the angle* θ_n *about z-axis and* φ_n *about x'-axis.*

where we have defined

$$K = \sum_n \left[\frac{1}{2} M (\dot{u}_n^2 + \dot{v}_n^2) + \frac{1}{2} I_1 \dot{\theta}_n^2 + \frac{1}{2} I_2 \dot{\varphi}_n^2 \right] , \qquad (3.2a)$$

$$W_{la} = \sum_n \left[\frac{1}{2} k_L (u_{n+1} - u_n)^2 + \frac{1}{2} k_T (v_{n+1} - v_n)^2 \right] , \qquad (3.2b)$$

$$W_d = \sum_n \left[\varepsilon (\mathbf{p}_n \cdot \mathbf{p}_{n+1} - 3 p_n^x p_{n+1}^x) + \eta p_n^x p_{n+1}^x \right] , \qquad (3.2c)$$

$$W_{int} = -3\varepsilon \sum_n \left[(\mathbf{p}_n \cdot \mathbf{p}_{n+1} - 3 p_n^x p_{n+1}^x) \left(\frac{u_{n+1} - u_n}{a} \right) + (p_n^x p_{n+1}^y + p_n^y p_{n+1}^x) \left(\frac{v_{n+1} - v_n}{a} \right) \right] , \qquad (3.2d)$$

$$W_e = P_0 \sum_n \mathbf{E}.\mathbf{p}_n \qquad (3.2e)$$

Eq.(3.2a) represents the kinetic energy where M is the mass of the particle, I_1 and I_2 are the moments of inertia corresponding respectively to the rotations θ_n and φ_n. The lattice deformation energy is given by eq.(3.2b) where k_L and k_T are the longitudinal and transverse lattice constants. The expression (3.2c) denotes the mutual interaction between dipoles for the rigid chain where $\varepsilon = P_o^2/4\pi\varepsilon_o a^2$ and P_o is the modulus of the dipoles so that $\mathbf{P}_o = P_o \mathbf{p}_n$ and $\mathbf{p}_n$ is an unit vector (ε_o is the permitivity constant in vaccum). Moreover, in eq.(3.2c), anisotropic terms in the chain direction have been added where η is the anisotropic coefficient. Eq.(3.2d) represents the interaction energy between the dipoles and the lattice deformation. Finally, the energy due to the external field is given by eq.(3.2e) where **E** is the applied electric field which will be taken parallel to the y-direction. We note by p_n^x and p_n^y the projection of the dipole n onto the x-axis and y-axis, respectively. This situation is similar to that of the nonlinear excitations in the chains of spins, but the physics of this problem is different [41]. We note that the interactions between the lattice deformation and dipoles is equivalent to an electrostrictive coupling. In terms of rotation angles the unit vector $\mathbf{p}_n$ can be written as

$$\mathbf{p}_n = (\cos\varphi_n \cos\theta_n , \cos\varphi_n \sin\theta_n , \sin\varphi_n)^T \quad . \qquad (3.3)$$

At this stage, two configurations can be examined: (i) configuration (below referred to as A) where the dipole rotation occurs in the plane (x-y), in this case $\varphi_n = 0$ and (ii) configuration (below referred to as B) where the dipoles rotate about the chain axis and we have $\theta_n = \pi/2$.

3.2 - Configuration A - Equations of motion.

a- Discrete equations.

This case can be studied by setting φ_n equal to zero in the Hamiltonian (3.1) and eqs(3.2a-e). In terms of rotation angles the Hamiltonian of the special case reads as

$$H = \sum_n \frac{1}{2} (\dot{U}_n^2 + \dot{V}_n^2 + \dot{\Phi}_n^2) + \sum_n [\frac{1}{2} K_L (U_{n+1} - U_n)^2 + \frac{1}{2} K_T (V_{n+1} - V_n)^2]$$

$$- \sum_n [4\cos\frac{1}{2}(\Phi_{n+1} - \Phi_n) - \chi \cos\frac{1}{2}(\Phi_{n+1} + \Phi_n)]$$

$$- \frac{1}{2}\sum_n [\alpha (U_{n+1} - U_{n-1}) \cos\Phi_n - \beta (V_{n+1} - V_{n-1}) \sin \Phi_n]$$

$$+ \sum_n 2 F \sin(\Phi_n /2) \qquad (3.4)$$

after the following changes of variables and functions

$$\tau = \sqrt{ (\varepsilon - \eta)/ I_1 }\; t , \qquad 2 \theta_n = \Phi_n , \qquad (u_n , v_n) = \frac{1}{2} \sqrt{I_1 /M}\, (U_n , V_n) \quad (3.5a)$$

and where we have set

$$K_L = (I_1 /M (\varepsilon-\eta))\, k_L , \qquad K_T = (I_1 /M (\varepsilon-\eta))\, k_T ,$$

$$\alpha = \sqrt{(I_1 /M)}\, \frac{9\varepsilon}{a(\varepsilon-\eta)} , \qquad \beta = \sqrt{(I_1 /M)}\, \frac{6\varepsilon}{a(\varepsilon-\eta)} ,$$

$$\chi = 2(3\varepsilon-\eta)/(\varepsilon-\eta) , \qquad F = 2 P_0 E / (\varepsilon-\eta) . \qquad (3.5b)$$

Then, we deduce the equations of motion from the Hamiltonian (3.4), this yields

$$\ddot{U}_n = K_L (U_{n+1} + U_{n-1} - 2 U_n) - \frac{1}{2} \alpha (\cos\Phi_{n+1} - \cos\Phi_{n-1}) , \qquad (3.6a)$$

$$\ddot{V}_n = K_T (V_{n+1} + V_{n-1} - 2 V_n) + \frac{1}{2} \beta (\sin\Phi_{n+1} - \sin\Phi_{n-1}) , \qquad (3.6b)$$

$$\ddot{\Phi}_n = 2 [\sin\frac{1}{2}(\Phi_{n+1} - \Phi_n) - \sin\frac{1}{2}(\Phi_n - \Phi_{n-1})] + \frac{1}{2} \chi [\sin\frac{1}{2}(\Phi_n + \Phi_{n+1})$$

$$+ \sin\frac{1}{2}(\Phi_{n-1} + \Phi_n)] - \frac{1}{2} \alpha (U_{n+1} - U_{n-1}) \sin\Phi_n - \frac{1}{2} \beta (V_{n+1} - V_{n-1}) \cos\Phi_n$$

$$- F \cos(\Phi_n/2) . \qquad (3.6c)$$

These equations are valid for any kind of rotational motions, especially for very discrete cases when the dipole orientation undergoes rapid changes in space. The three equations (3.6) are coupled and are relatively involved. The solution of such equations cannot be reached analytically in the dynamic case, the continuum approximation is therefore required to proceed further. However, the case of a dynamically harmonic chain of dipoles is only tractable in the discrete situation [42-44].

b- Equations for the continuum model.

We assume that the discrete displacements and rotations are slowly varying over a crystalline spacing. This allows us to expand the trigonometric functions in eqs(3.6a-c) with respect to the finite differences of the type $\theta_{n+1}-\theta_n$ up to the first order. Furthermore, we limit our attention to the dynamic processes of spatially extended waves, where every variation in space and time is slow. Then, the functions U_n, V_n, Φ_n can be expanded in Taylor series around their value at x=na. After some manipulations we obtain the following system of partial differential equations for the continuous functions U, V and Φ

$$U_{,tt} - C_L^2 U_{,XX} = -\alpha (1 + \cos\Phi)_{,X} \quad , \tag{3.7a}$$

$$V_{,tt} - C_T^2 V_{,XX} = \beta (\sin\Phi)_{,X} \quad , \tag{3.7b}$$

$$\Phi_{,tt} - \Phi_{,XX} = \chi \sin\Phi - \alpha \sin\Phi\, U_{,X} - \beta \cos\Phi\, V_{,X} - F \cos(\Phi/2) \quad . \tag{3.7c}$$

In these equations the change of the space variable X=x/a has been considered and C_L and C_T are referred to the longitudinal and transverse elastic wave velocities, respectively. In the case of rotations Φ's of small amplitudes the linearized equations (3.7a-c) can be compared, on the one hand, to the linear equations deduced from a fully continuum approach to *ferroelectric crystals* [45,46] and, on the other hand, to the one-dimensional version of the *elastic micropolar media* [13,14,47]. However, here we have built up a continuum model for nonlinear micropolar media since the model is able to display large rotations. The structure of the sytem of dispersive nonlinear equations is of special interest because it is made of two wave equations for the elastic displacements (see eqs(3.7a) and (3.7b)) and a sine-Gordon equation for the dipole rotation, each elastic displacements being nonlinearly coupled to Φ through one of coupling coefficients α and β. The continuum version of the present microscopic model can be compared to the equations deduced from the nonlinear continuum of oriented media [48,49] or micropolar elastic crystals [50].

c- One-soliton solution (case F=0).

We confine now our attention on possible solutions of the system (3.7). Some obvious properties can be emphasized. Let (U,V,Φ) be a solution of (3.7), then (U,-V,-Φ) and (U,V,Φ+2kπ) are also solutions. The system (3.7) possesses uniform static elementary solutions

$$U = U_0 = \text{const.} \;, \qquad V = V_0 = \text{const.} \;, \qquad \Phi = k\pi \quad (k \in Z) \;. \tag{3.8}$$

Nevertheless, a more interesting solution can be looked for in the form of travelling waves, that is, U, V and Φ are functions of the single phase variable ξ=X-Cτ where C is the phase velocity. Then, eqs(3.8a-c) become

$$(C^2 - C_L^2)\, U_{,\xi\xi} = -\alpha (1 + \cos\Phi)_{,\xi} \quad , \tag{3.9a}$$

$$(C^2 - C_T^2)\, V_{,\xi\xi} = \beta (\sin\Phi)_{,\xi} \quad , \tag{3.9b}$$

$$(C^2 - 1)\, \phi_{,\xi\xi} = \chi \sin\Phi - \alpha \sin\Phi\, U_{,\xi} - \beta \cos\Phi\, V_{,\xi} \quad . \tag{3.9c}$$

Eqs(3.9a) and (3.9b) can be integrated once with respect to ξ, this gives

$$U_{,\xi} = \frac{\alpha}{C_L^2 - C^2} \cos\Phi \quad , \tag{3.10a}$$

$$V_{,\xi} = \frac{\beta}{C_T^2 - C^2} \sin\Phi \quad , \tag{3.10b}$$

where it has been assumed that the shear deformation $V_{,\chi}$ vanishes when Φ goes to $k\pi$ and C differs from $\pm C_L$ and $\pm C_T$. However, it is worth noting that the elongational deformation of the chain does not vanish when Φ goes to $k\pi$ but it remains a constsnt deformation

$$U_{,\chi}(\infty) = \frac{-\alpha}{C_L^2 - C^2} \quad , \tag{3.11}$$

that means, far from the localized distorsion (soliton) the dipoles are uniform and aligned and the dipoles, in turn, create internal strain $U_{,\chi}(\infty)$ it results a dilatation of the chain in its direction. Other conditions at infinity can be taken but this does not change the essence of the discussion. Now, on substituting eqs(3.10a-b) into eq.(3.9c), we obtain

$$(C^2 - 1) \Phi_{,\xi\xi} = \chi \left(\sin\Phi + \frac{1}{2} \delta(C) \sin 2\Phi \right) \quad , \tag{3.12a}$$

where we have set

$$\chi \, \delta(C) = \frac{\beta^2}{C_T^2 - C^2} - \frac{\alpha^2}{C_T^2 - C^2} \quad . \tag{3.12b}$$

We remark that the problem of the one-soliton solution of the somewhat complicated system (3.7) is equivalent to the solution of a nonlinear ordinary differential equation -eq.(3.12a)- which can be formally deduced from a *double sine-Gordon equation* when a one-soliton solution is sought [51-53]. Eq.(3.12) possesses a first integral which can be written as

$$\frac{1}{2} (C^2 - 1) (\Phi_{,\xi})^2 = V(\Phi) + E_0 \quad , \tag{3.13a}$$

$$V(\Phi) = \chi \left(\cos\Phi + \frac{1}{4} \delta(C) \cos 2\Phi \right) \quad , \tag{3.13b}$$

where E_0 is an integration constant which can be related to the total energy of the system. Now, the physical meaning of eq.(2.13a) becomes clear, it stands for the equation governing the motion of a particle of mass $(1-C^2)$in a periodic potential $V(\Phi)$ (note here the "mass" may be negative if $|C|>1$). Nevertheless, the qualitative discussion of the solutions of eq.(3.13) can be examined on the basis of the phase trajectories plotted in the plane $(F,F_{,\xi})$. Fig.2.2 gives the trajectories for different values of $\delta(C)$ and χ. According to the value of E_0 with respect to δ we have aperiodic solutions for Φ with a 2π pseudo-period (curves (a) and (b)). The curves (c) and (b) correspond to periodic solutions which can be expressed in terms of Jacobian elliptic functions [54]. Two particular values of E_0 lead to the curves (e) and (f) passing through the points $k\pi$ or Φ_0 and $2k\pi-\Phi_0$ which are stable and metastable points, respectively and for these cases the corresponding solutions are solitons. However, here we consider a solution of eq.(3.13) such that $\Phi \to k\pi$ at infinity and then $E_0=\chi(1-\delta/4)$, that means that the "particle" takes an infinite time to go from the point $-\pi$ to the point $+\pi$ (see Fig.2.2, curve (e)) or the

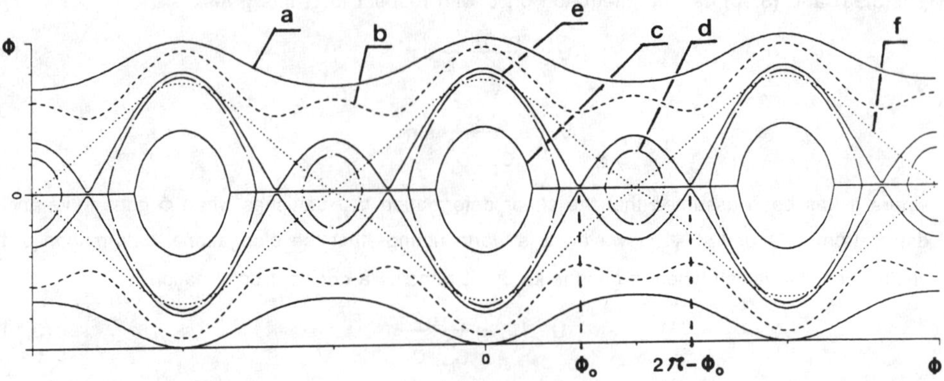

Fig. 2.2 *-Phase trajectories of the solutions of eq.(3.13) in the plane (Φ',Φ) -curves (a) aperiodic solutions, curves (c) and (d) periodic solutions, curve (e) $\delta > 1$, $E/\chi = \delta/4 + 1/2\delta$ soliton solution passing through Φ_o and $2\pi - \Phi_o$ (metastable points, $\cos\Phi_o = -\delta^{-1}$). and curve (f) $|\delta| < 1$, $E/\chi = 1 - \delta/4$ soliton solution (eq.(3.14a)).*

period of the wave is infinite. After a more or less lengthy calculation the solution to eq.(3.12) is then given by

$$\Phi = 2\tan^{-1}\left(e^{-Q\xi+\Delta}\right) - 2\tan^{-1}\left(e^{Q\xi+\Delta}\right) \quad , \tag{3.14a}$$

with

$$Q^2 = \chi(1-\delta)/(1-C^2) \qquad \text{and} \qquad \Delta = \cosh^{-1}\left(\sqrt{\chi(1-\delta)}\right) \quad . \tag{3.14b}$$

But such a solution requires that both $\chi(1\text{-}\delta)$ and $(1\text{-}C^2)$ are positive. We assume that $\chi>0$, the case $\chi<0$ will be discussed below. The second condition implies $|C|<1$ that means that the wave is subsonic and this leads to a stable solution [55,56]. It remains the condition $\delta<1$ which holds true since $|C|<1<C_T<C_L$. The solution (3.14a) represents the transition from the position $\Phi=\pi$ to the position $\Phi=-\pi$. Now, it is in order to give the strain state of the medium. Inserting eq.(3.14a) into eqs(3.10a) and (3.10b) we obtain

$$U_{,X} = \frac{\alpha}{C_L^2 - C^2}\left(\frac{1 + \sinh(\Delta + Q\xi)\,\sinh(\Delta - Q\xi)}{\sinh^2\Delta + \cosh^2 Q\xi}\right) \quad , \tag{3.15a}$$

$$V_{,X} = \frac{\beta}{C_T^2 - C^2}\left(\frac{\sinh(\Delta + Q\xi) - \sinh(\Delta - Q\xi)}{\sinh^2\Delta + \cosh^2 Q\xi}\right) \quad . \tag{3.15b}$$

It can be seen that eqs(3.15a) and (3.15b) are generated solitons in deformation corresponding to the soliton rotation given by eq.(3.14).

d- Comments: (i) In the form (3.14a), the one-soliton solution to the double sine-Gordon equation (3.12) decomposes into the sum of two one-soliton solutions, one varying from π to 0 and the other one from 0 to $-\pi$. Taken separatly, each of these is analogous to the one-soliton of the usual sine- Gordon equation for the rotation $\theta=2\Phi$. Furthermore, the decomposition leads to the same comments for the deformation field (3.15a) and (3.15b) where we can note the nonlinear decomposition characterized by a product in the expression of the elongational deformation -eq.(3.15a)- whereas in eq.(3.15b) for the shear deformation there is a sum. Other forms can be given for the solution of our problem [39,40].

(ii) *Acoustic phonons* : eqs(3.7a) and (3.7b) governing the elastic displacements are linear but nonhomogeous since the right hand sides depend on the solution Φ, however, a general solution to these can be decomposed into the solution of the system without right hand side and the particular solution generated by the solution Φ. The solutions of the homogeneous system are

$$U(X,t) = U_0\, e^{iq(X - C_L t)} \quad , \tag{3.16a}$$

$$V(X,t) = V_0\, e^{ik(X - C_T t)} \quad , \tag{3.16b}$$

which represent the longitudinal and transverse acoustic phonons, respectively. In fact, we should consider such solutions, but the latter would appear in the right hand side of the equation for the rotation Φ and this would add source terms in eq.(3.7c), leading then to a much more complicated problem. Nevertheless, in physical systems such acoustic phonons can be excited through thermal fluctuations which are not considered in the present work.

(iii) *Comments on the sign of* χ : In order to obtain the solution of eqs(3.7) we have assumed that $\chi>0$. In the opposite case ($\chi<0$) we then consider the following change of functions

$$\Phi = \pi + \Phi' \quad , \tag{3.17a}$$

$$U = - U' \quad , \tag{3.17b}$$

$$V = -V' \quad , \tag{3.17c}$$

These new functions are substituted into the system (3.7) so that the new functions Φ', U' and V' satisfy again the system (3.7), unless if $\chi=-\chi'$ and $\delta=-\delta'$ (with $\chi'>0$). The study of this case is then the same as in the case $\chi>0$. However, the topology of the solitons is different. Here, Φ decreases from 2π to 0 and the solitons in deformation have opposite signs. This means that the physical angle θ ($\theta=\Phi/2$) of the dipoles varies monotonously from π to 0. Fig.2.3a and Fig.2.3b show the numerical pictures of the dipole orientations when transisting from one state to the other one in the cases $\chi>0$ (Fig.2.3a) and $\chi<0$ (Fig.2.3b).

(iv) *Soliton thickness* : The thickness of the soliton is given by

$$\Delta = \pi \sqrt{(1-C^2)/\chi} \quad , \tag{3.18}$$

This is a remarkable result which shows that the thickness of the soliton (eq.(3.14a)) is not altered by the coupling coefficients. In the case of structures in domains and walls occuring in ferroelectric crystals (see for example the case of Sodium Nitrite [39]) Δ represents the wall thickness.

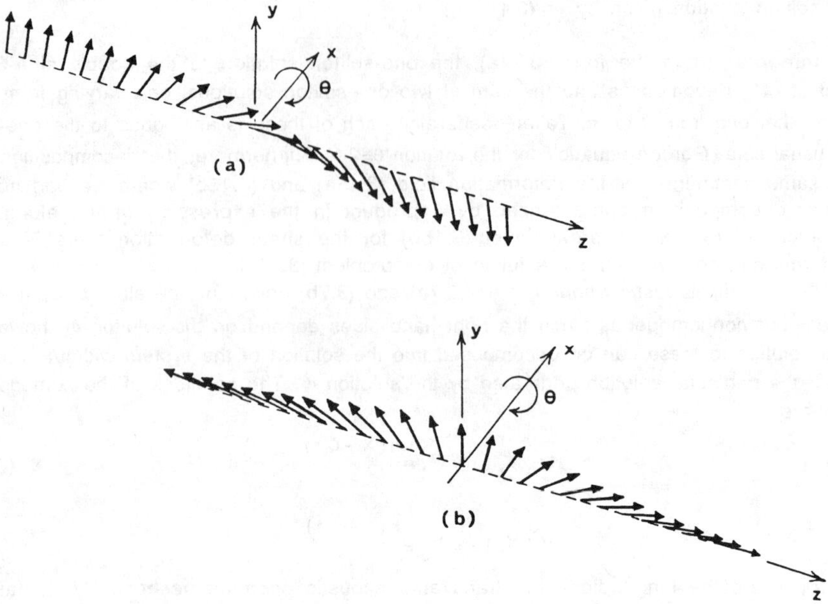

Fig. 2.3 *-Dipole orientation in the case of the configuration A -(a) $\chi > 0$, θ transits from $\pi/2$ to $-\pi/2$ and (b) $\chi < 0$, θ goes from π to 0.*

e- Numerical simulations.

The numerical scheme uses directly the discrete equations (3.6a-c). In order to ascertain the validity of the solution obtained in the continuum approximation, we consider the solution given by eqs(3.14), (3.15a) and (3.15b) as initial conditions for the discrete system (3.6). In Figs.2.4a-c we give numerical illustrations of this solution in space-time perspective representation. The graph in Fig.2.4a illustrates the rotational motion of the dipoles from the state $\theta=\pi/2$ for $X \to -\infty$ to the state $\theta=-\pi/2$ for $X \to +\infty$, here we have a so-called "*kink* ". In Fig.2.4b we have the elongational deformation generated through the coupling coefficient α by the soliton in rotation, this is a "*hump* " soliton which is essentially nonzero in the thickness of the kink. At last, the shear deformation is given in Fig.2.4c, this deformation changes sign when the angle Φ passes through zero. This a sort of "*double hump* " soliton generated by the soliton in rotation as well

The model presented here-above finds physical support in the case of ferroelectrics such as $NaNO_2$. the latter crystal possesses a central molecular group NO_2 equipped with a permanent dipole which may suffer large-amplitude rigid-body rotations about a crystallographic axis. The relative displacements of the different ions can be neglected with respect to the rotational motion of the molecular group and the mass-center

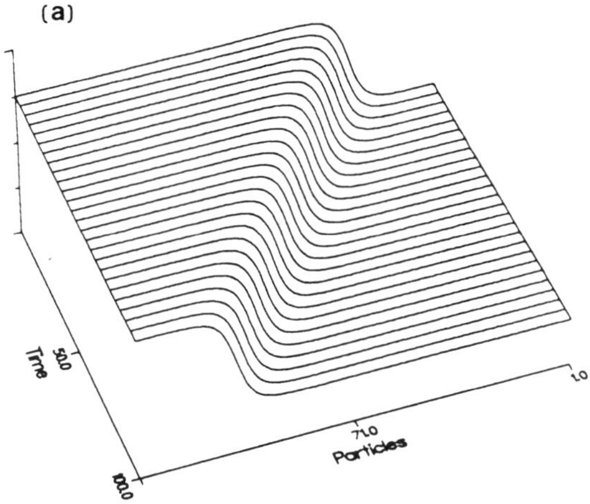

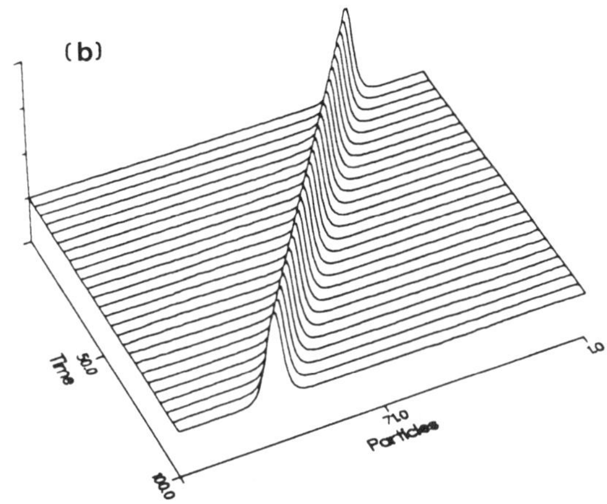

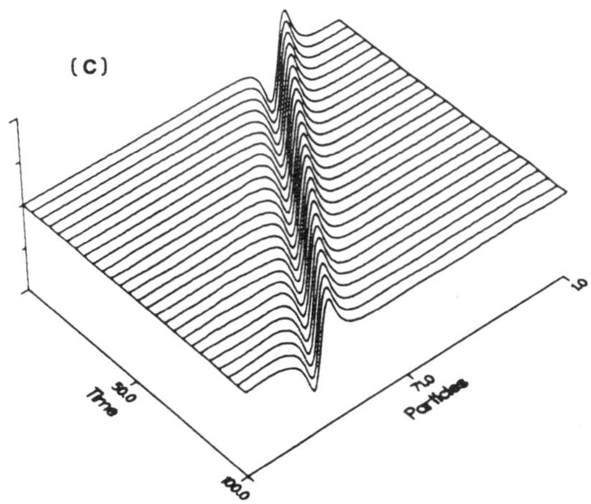

Fig. 2.4 *-Configuration A-solution of the system (3.6) - (a) Soliton in rotation, (b) Soliton in elongational deformation and (c) Soliton in shear deformation.*

displacements since they occur with relative high frequencies (10^{13}Hz in comparaison to 10^{9}-10^{10}Hz for the libration frequency of the molecular group). Therefore, such a crystal can be modelled by a monoatomic chain where each point is endowed with a mass (the mass of the crystalline cell) and an inertia associated with the molecular group NO_2. An estimate of the wall thickness gives $\Delta \approx 10$ lattice spacings in nondimensional notation which corresponds to $\delta \approx 5.6 \times 10^{-2} \mu m$ in physical notations. This is to be compared to the wall thickness of the order of 0.3-1.μm estimated by electronic microscopy [57-59]. Nevertheless, if we associate with the crystal-lattice distorsion induced by the rotation a thickness, that is, the wall thickness corresponding to the deformation field, we can estimate the latter: $\delta \approx 1.-2. \times 10^{-1} \mu m$. Moreover, this type of ferroelectrics exhibits incommensurate phase transition with a modulated structure [60] occuring in a rather small temperature interval between the paraelectric phase and the ferroelectric one.

3.3 - Influence of an applied field on the motion of a soliton.

a - Perturbation scheme.

The problem that we propose now to solve is the starting motion of a soliton from rest when a field F is suddently applied (at t_0) to the static solution of eqs(3.7). For sake of simplicity we suppose that $\alpha << \beta$ so that the coupling between the rotational motion and longitudinal displacement is then neglected. In this case the system (3.7) reduces to

$$V_{,tt} - C_T^2 V_{,XX} + \nu V_{,tXX} = \beta (\sin\Phi)_{,X} \quad , \tag{3.19a}$$

$$\Phi_{,tt} + \gamma \Phi_{,t} - \Phi_{,XX} = \chi \sin\Phi - \beta \cos\Phi \, V_{,X} - F \cos(\Phi/2) \quad , \tag{3.19b}$$

where the term depending on the external field has been considered and some dissipative processes have been added: (i) ν is the viscoelasticity coefficient corresponding to the transverse motion and (ii) γ is a damping constant associated with the rotational motion. The action of the field yields a forcing term in eq.(3.19b), which, however, depends on the rotation Φ. The new terms -external field and damping- may be considered as small perturbations when the field is applied to a static soliton, the latter does not stay at rest and it begins to move so that the soliton, which was initially static, depends now on time. The external field creates a torque acting on the dipoles and this field tends to rotate the dipoles over the soliton thickness. The *a priori* effects of the field are (i) to modulate in time the soliton location (translational mode), (ii) to alter the thickness of the soliton and (iii) to modify the shape of the soliton (phonon radiations). The perturbation scheme is based on the *adiabatic time variation hypothesis* [61,62] which consists of replacing the phase variable ξ by $\xi = X - W(t)$ with $dW/dt = C(t)$ in the solutions (3.14) and (3.15). Obviously, if F=0 there is no perturbation and C=const. The problem is then to determine W or C as function of time when the external field is applied. With this in mind, a perturbation scheme is developed in order to obtain the characteristic parameters of the evolving wave. The method is based on the fact that the solutions (3.14) and (3.15) with the new phase variable are forced to satisfy a variational principle associated with the Lagrangian of the system [63,64]. Since the form of the solution is known, we arrive at a new variational principle associated with the soliton location W(t) which leads to the Euler-Lagrange equation for W(t). Here, instead of developing the analysis of the perturbation scheme we prefer to give the numerical simulations of this problem.

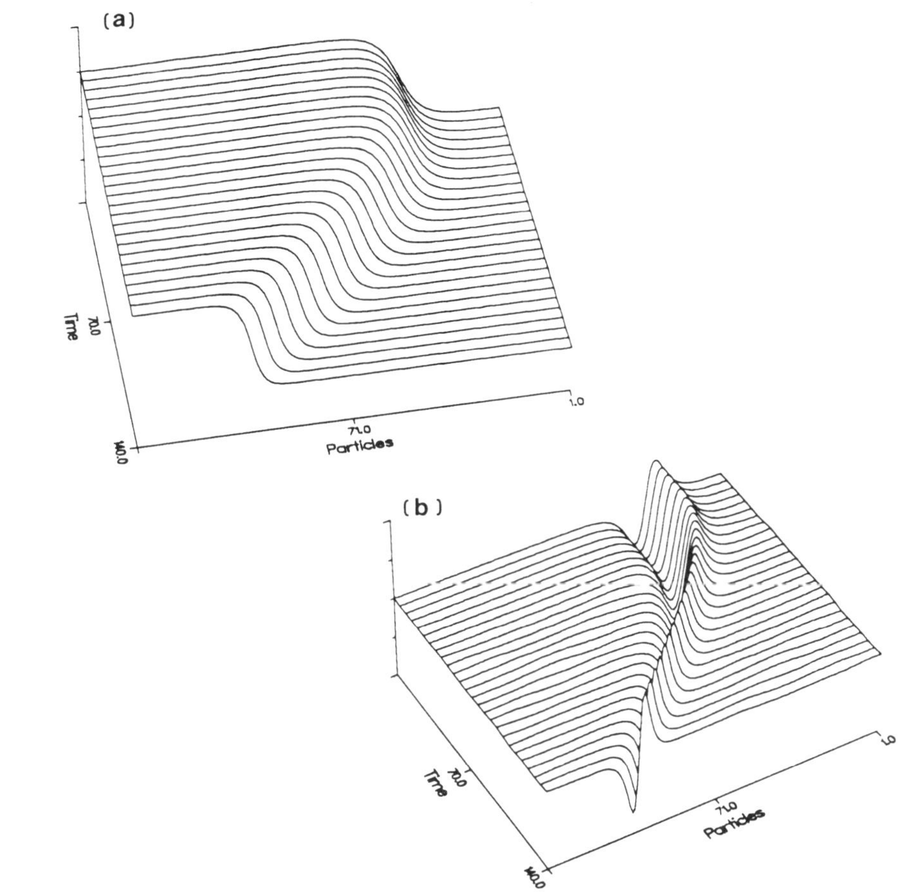

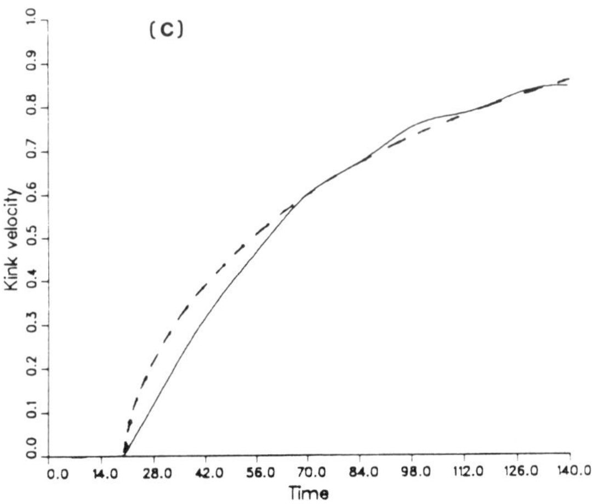

Fig. 2.5 -*Transient motion of a soliton under the action of an applied field: (a) soliton in rotation, (b) soliton in shear deformation and (c) variation in time of the soliton velocity, full line: numerical result, dash line: analytic result.*

b - Numerical results.

The problem of the starting motion of a soliton from rest can be simulated by means of the discrete system (3.6) (where U=0 has been set) with the static solutions (3.14) and (3.15) as initial conditions. Fig.2.5a gives the computed solution in rotation Φ. Between times $t=0$ and $t=t_o$ the soliton is at rest. The external field F is applied at $t=t_o$. For t close to t_o the kink begins to move and, progressively, gains a constant velocity. Fig.2.5b represents the corresponding shear deformation $V_{,X}$ which characterizes the lattice distorsion. In addition, we remark that the thickness of the wall becomes narrower as the velocity increases. From the computation of the solution in rotation it is numerically possible to reach the velocity of the soliton as function of time and compare it to that deduced from the perturbation method. The results are collected in Fig.2.5c. This ascertains first the validity of the perturbation method (average Lagrangian and adiabatic hypothesis) and secondly the validity of the continuum approximation. We notice that the phonon radiations due to the perturbations are hardly perceptible on the numerical illustrations which confirms the hypothesis.

Physically, the problem corresponds to the action of an electric field on a wall separating two ferroelectric domains. The field favours the growth of a domain at the expense of the other one. This is a means to change a structure in two domains into a single domain structure.

3.4 - Configuration B - Equations of motion.

a - Discrete equations.

Going back to the Hamiltonian (3.1) of the discrete system where $\theta_n=\pi/2$ has been set, we obtain the Hamiltonian in terms of rotation about the chain axis

$$H=\sum_n \frac{1}{2}(\dot{U}_n^2+\dot{V}_n^2+\dot{\varphi}_n^2)+\sum_n\left[\frac{1}{2}K_L(U_{n+1}-U_n)^2+\frac{1}{2}K_T(V_{n+1}-V_n)^2\right]$$

$$+\sum_n \cos(\varphi_{n+1}-\varphi_n)-\eta\sum_n(U_{n+1}-U_n)\cos(\varphi_{n+1}-\varphi_n)-\sum_n F(\cos\varphi_n-1)\,. \qquad (3.20)$$

The following changes of variables and functions have been considered

$$\tau=\sqrt{\varepsilon/I_2}\;t\;, \qquad (u_n\,,v_n)=\sqrt{I_2/M}\,(U_n\,,V_n)\;, \qquad (3.21a)$$

and we have set

$$(K_L\,,K_T)=(I_2/\varepsilon M)(k_L\,,k_T)\,,\qquad \eta=3a\sqrt{I_2/M}\quad\text{and}\quad F=P_0E/\varepsilon\,. \qquad (3.21b)$$

The equations of motion are deduced from the Hamiltonian (3.20) and read as

$$\ddot{U}_n=K_L(U_{n+1}+U_{n-1}-2U_n)-2\eta\sin\left(\frac{\varphi_{n+1}-\varphi_{n-1}}{2}\right)\sin\left(\frac{\varphi_{n+1}+\varphi_{n-1}-2\varphi_n}{2}\right)\,. \qquad (3.22a)$$

$$\ddot{V}_n=K_T(V_{n+1}+V_{n-1}-2V_n)\,, \qquad (3.22b)$$

$$\ddot{\varphi}_n=\sin(\varphi_{n+1}-\varphi_n)-\sin(\varphi_n-\varphi_{n-1})-\big[(U_{n+1}-U_n)\sin(\varphi_{n+1}-\varphi_n)$$

$$-(U_n-U_{n-1})\sin(\varphi_n-\varphi_{n-1})\big]-F\sin\varphi_n\,. \qquad (3.22c)$$

The transverse displacements do not, of course, play any role in this case, then it will not be considered further. The interesting set of equations consists then in an equation for the longitudinal displacement nonlinearly coupled to the equation governing the rotations φ_n. We notice that the equation for the rotation φ_n is strongly nonlinear. As in the case of the configuration A an analytic solution of this discrete system is not tractable, the continuum approximation is therefore required.

b - Equations for the continuum model.

The same hypothese```s used for the configuration A hold true in the present case. In fact, we consider dynamic processes of which the characteristic length is large as compared to the lattice spacing a. After some manipulations which amount to expanding the quantities U_n and φ_n about x=na, we obtain

$$U_{,tt} - C_L^2 U_{,XX} = -\frac{1}{2}\eta\,(\varphi_{,X}^2)_{,X} \quad , \tag{3.23a}$$

$$\varphi_{,tt} - \varphi_{,XX} = -\eta\,(U_{,X}\varphi_{,X})_{,X} - F\sin\varphi \quad , \tag{3.23b}$$

where we have set X=x/a and $C_L=\sqrt{K_L}$ is the longitudinal elastic velocity. Note that, in this configuration, the coupling terms do not involve trigonometric functions. On the other hand, if the coupling is neglected, it is necessary to have nonvanishing F to find nonlinear excitations and then eq.(3.23b) reduces to the usual sine-Gordon equation. In the linear case, for small rotations, the equations are uncoupled and this leads to the usual longitudinal acoustic mode and optic mode associated with the rotation φ or "libron". However, in its structure the equations differ from those of the configuration A because of the coupling terms.

c - One-soliton solution.

The system (3.31) has some obvious properties, let (U,φ) be a solution of (3.23), then $(U,\varphi+2k\pi)$ and $(U,-\varphi)$ are also solutions. The solutions

$$U = U_0 = \text{const.} \qquad \text{and} \qquad \varphi = k\pi \quad (k \in Z)$$

are elementary static solutions. We look for a solution of eqs(3.23a,b) in the form of travelling waves and U and φ are functions of the phase variable $\xi=X-C\tau$ where C is the phase velocity. Then, the system (3.23) transforms into

$$(C^2 - C_L^2)\,U_{,\xi\xi} = -\frac{1}{2}\eta\,(\varphi_{,\xi}^2)_{,\xi} \quad , \tag{3.24a}$$

$$(C^2 - 1)\,\varphi_{,\xi\xi} = -\eta\,(U_{,\xi}\varphi_{,\xi})_{,\xi} - F\sin\varphi \quad . \tag{3.24b}$$

The first equation can be integrated once with respect to ξ, and we obtain

$$U_{,\xi} = \frac{\eta}{2\,(C_L^2 - C^2)}\,(\varphi_{,\xi})^2 + A \quad , \tag{3.25}$$

where A is a constant of integration and we have supposed that $C \neq \pm C_L$. Now, on substituting eq.(3.25) into eq.(3.24b), we obtain an equation for the rotation φ. This equation possesses a first integral which reads as

$$(\varphi_{,\zeta})^4 - 2a(\varphi_{,\zeta})^2 + b - \cos\varphi = 0 \quad , \tag{3.26}$$

where we have set

$$\zeta = (F/\gamma)^{1/2}\,\xi \quad , \tag{3.27a}$$

$$a = -(\eta A + C^2 - 1)/4(\gamma F)^{1/2} \quad , \tag{3.27b}$$

$$\gamma(C) = 3\eta^2/8(C_L^2 - C^2) \quad , \tag{3.27c}$$

in eq.(3.34) a (or A) and b are arbitrary constants of integration. We notice that if F is changed into -F in eqs(3.24a-b) this is equivalent to shifting φ by π so that only the case F>0 is considered. Moreover, the existence of the solution implies that $|C|<C_L$, this is in contrast with the usual sine-Gordon system ($\eta=0$) for which we must have $|C|<1$. This means that the nonlinear excitations can travel with a velocity smaller than the longitudinal elastic velocity but greater than the libration wave velocity since $C_L>1$. Eq.(3.26) is generally very complicated to solve for any constants a and b. However, qualitative results can be discussed from the phase trajectories plotted in the plane $(\varphi,\varphi_{,\xi})$ as that has been done in the case of the configuration A and periodic or aperiodic solutions exist. But, here, we do not give the details of the discussion and we focus our attention on the special case b=1 and $a=\sqrt{2}$ for which eq.(3.26) is reduced to

$$(\varphi_{,\zeta})^2 = \sqrt{2}\,(1 - |\cos(\varphi/2)|) \quad , \tag{3.28}$$

this equation provides a solitary wave solution which transits from $\varphi=0$ to $\varphi=2\pi$. The solution can be written as

$$\varphi = -(1 + \mathrm{sign}(\xi))\,\pi + 8\tan^{-1}[\exp Q(\xi + \mathrm{sign}(\xi)\,\xi_0)] \quad , \tag{3.29a}$$

with

$$Q = \frac{1}{2}(F/2\gamma)^{1/4} \quad , \tag{3.29b}$$

$$\xi_0 = -\frac{1}{Q}\ln(\sqrt{2}-1) \quad . \tag{3.29c}$$

Next, on substituting eq.(3.37) into eq.(3.33) we obtain the elongational deformation

$$U_{,X} = 2\sqrt{\frac{F}{3(C_L^2 - C^2)}}\left[\frac{2}{\cosh^2 Q(|\xi| - \xi_0)} - 3\right] + (1 - C^2)/\eta \quad . \tag{3.30}$$

We can remark that $U_{,X}$ tends towards a nonvanishing value when $|\xi| \to \infty$, This means that, in each domain $\varphi=0$ and $\varphi=2\pi$ the electric dipoles create a nonzero deformation.

d - Comments.

i) If $\eta=0$, the system (3.24) reduces to the linear equation for the longitudinal displacement leading to the longitudinal elastic linear mode and to the classical sine-Gordon equation for the rotation of which the soliton solution is given by

$$\varphi = 4\tan^{-1}\left(\exp\sqrt{\frac{F}{1-C^2}}(X - C\tau)\right) \quad , \tag{3.31}$$

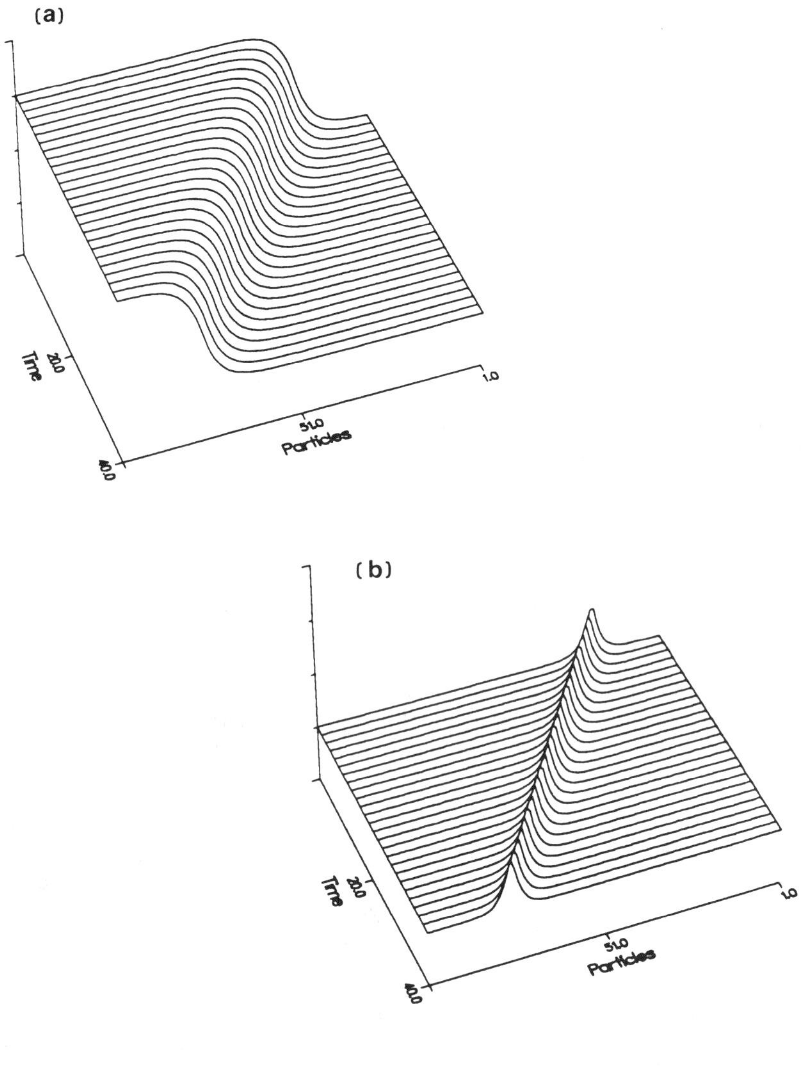

Fig. 2.6 *-Configuration B - Numerical simulation of the solution of eqs(2.30), (a) soliton in rotation and (b) solution in elongational deformation.*

here the same argument holds true if $F<0$ and we must have $|C|<1$ as in the classical sine-Gordon theory.

ii) It is worth calculating the solitary wave thickness, the latter reads as

$$\Delta = \pi /(\sqrt{2}\, Q) \quad , \tag{3.32}$$

and it depends strongly on the external force F and the coupling coefficient η as well as the velocities C_L and C (see eqs(3.27c) and (3.29b)).

e - Numerical results.

The numerical scheme is identical to that used for the configuration A. Starting from the initial conditions given by eqs(3.29) and (3.30) the discrete equations (3.22a-c) provide the numerical solutions and ascertain the validity of the continuum approximation. The results are shown in Fig.2.6. Fig.2.6a gives the rotation of the dipoles which consists in a kink travelling from the state $\varphi=0$ to the state $\varphi=2\pi$ while the associated elongational deformation is plotted in Fig.2.6b exhibiting a sharp hump within the thickness of the kink. Notice that the dipoles perform a revolution about the chain axis. This situation is similar to that of a compressible chain of dipoles [65] or compressible Heisenberg chains [41,66].

3. 5.- Conclusion.

The propagation of nonlinear excitations of the soliton type for two simple configurations has been examined on the basis of an anharmonic lattice model consisting in a deformable chain equipped with rotatory dipoles. The nonlinear phenomena are intimately connected with the additional degrees-of-freedom inherent to the rotation of the dipoles so that the model accounts for the nonlinear coupling between the rotational motion and elastic deformations. The propagation of *solitons in rotation* and *in (elongational and shear) elastic deformations* has been investigated. In the first configuration, referred to as A, both longitudinal and transverse elastic displacements polarized in the plane of rotation of the dipoles are coupled to the latter. In the second situation, configuration B, the dipoles rotate about the axis of propagation. This in fact is the same situation as in the Bloch wall in ferromagnetic crystals apart from the fact that in the latter the wave equations for the elastic displacements are not coupled to the equation governing the magnetic spin, whereas the configuration A is formally similar to that of the Neel wall in thin elastic ferromagnetic films [67]. Moreover, it has been shown that, in each configuration, the complete dynamic problem of the nonlinearly coupled system is equivalent to solving a *double sine-Gordon equation* or a *single sine-Gordon equation* in the case of one-soliton solutions according to the configuration. In the framework of the configuration A the influence of an external field on a soliton has been studied. This allows us to examine the transient motion of a soliton from rest by means of a perturbation method.

The nonlinear excitations of the soliton type studied in the present work can take place in various media when both the microstructure and internal degrees-of-freedom are considered. The underlying physics of the micro-motion corresponds to various media and the micro-system possesses both *nonlinearities* related to the internal degrees-of-freedom (rotations of large amplitudes) and *dispersion* (spatial nonuniformity in the dipole field). Thus the soliton propagation in our model seems to be similar to the study of nonlinear excitations in nematic liquid crystals [32]. The above-developed model can be applied to polymer media (poly-vinylidenefluoride) [33,34,68,69], or to the long elastic chains of macromolecules (biological materials, e.g. DNA) [38,70,71] where macromolecules undergo rotational motions of relatively large amplitudes. Furthermore, this model can be applied to describe the structure in domains and walls in molecular ferroelectric crystal of which Sodium Nitrite ($NaNO_2$) provides a good prototype [9,59]. The physical meaning of the soliton in rotation is then a moving wall coupled to the mechanical state and soliton in elongational and shear defromations are accordingly generated through the couplings. The interesting problem where the combination of both configurations A and B is considered can be envisaged and the relative stability of one configuration with respect to the other one can be studied. We meet the same situation in ferromagnetic materials [72-74]. Finally, our study is based essentially on the continuum approximation, however the highly discrete character of the models involving discreteness effects can give rise to particularly interesting phenomena which do not exist in the continuum version[75,76].

Aknowledgment: One part of the present work was supported by theme "Nonlinear Waves in Electromagnetic Elastic Materials" of A.T.P. "Mathematiques Appliquees et Methodes Numeriques Performantes" (M.P.B., C.N.R.S., Paris, France). The author wishes to thank G.A. MAUGIN for his continual encouragements. The author is also grateful to A. BISHOP and D.CAMPBELL for their hospitality at the Center for Nonlinear Studies and Theoretical Division of the Los Alomos National Laboratory where another part of this work has been done.

References.

[1] **A.C. Scott, F.Y.F Chu** and **D.W. McLaughlin**: A new concept in a applied science. *Proceeding of the IEEE 61*,1443-1483(1973).

[2] **A.C. Newell**: The history of the soliton. *Transaction of the ASME J. of Appl. Mech. 50*,1127-1138(1983).

[3] **G.L.Lamb,Jr.** and **D.W. McLaughlin**: Aspects of soliton physics.Topic in Current Physics 17, "*Solitons*", edited by R.K. Bullough and P.J. Caudrey, pp.65-106(Springer-Verlag, Berlin, 1980).

[4] **R.H. Enns, B.L. Jones, R.M. Miura** and **S.S. Rangnekar**: Nonlinear phenomena in physics and biology. *NATO Adv. Study Inst. Series75* (Plenum press, New York, 1981).

[5] **G.L. Lamb, Jr.**: *Element of soliton theory.* Pure and Applied Mathematic a Wiley-interscience series of texts, Monographs and Tracts (John Wiley & Sons, New York, 1980).

[6] **F. Galogero** and **A. Degasperis**: *Special transform and solitons I.* in Studies in Mathematics and its Applications 13 (North-Holand Publishing Company, Amsterdam, 1982).

[7] **M.J. Ablowitz** and **H. Segur**: *Soliton and the inverse scattering transform* (SIAM Studies in Applied Mathematics-Series of Monographs, Philadelphia, 1981).

[8] **R. Stojanovic**: *Mechanics of polar continua: Theory and applications.* (C.I.S.M. Lecture Notes, Udine, Italy, 1969).

[9] **C.B. Kafadar** and **A.C. Eringen**: Micropolar media I. The classical theory. *Int. J. Eng. Sci 9*, 271-305(1971).

[10]**I.A.Kunin**: *Elastic media with microstructure: One-dimensional model.* in Springer series in Solid-State Sciences, vol.26 (Springer-Verlag, Berlin, 1982).

[11]**J.M. Christian**: *The theory of transformations in metals and alloys.* (Pergamon Press, Oxford, 1975).

[12]**L.Delaey,R.V.krishnan,H.Tas** and **H.Warlimont**: Thermoelasticity, pseudoelasticity and the memory effects associated with martensitic transformations. *J. Mat. Science 9*, 1521-1555 (1974).

[13]**C.M. Wayman**: *Introduction to the theory of martensitic transformations.* (MacMillan, New York,1964).

[14]**F.Falk**: Landau theory and martensitic phase transitions, *J. de Physique Colloque C4, 43*, C4-3-C4-15 (1982).

[15]**C. Zener**: *Elasticity and anelasticity of metals* (University of Chicago Press, Chicago, 1965).

[16]**T.Suzuki**:Non-linear mechanical model for martensitic transformation.*J. Phys. Soc. Jpn. 45*, 860-868 (1978).

[17]**Y. Murakami**: Lattice softening, phase stability and elastic anomaly of the β-Au-Cu-Zn alloys. *J. Phys. Soc. Jpn .33*, 1350-1360 (1972).

[18]**R.A. Toupin**: Theory of elasticity with couple-stress. *Arch. Rational Mech. Anal. 17*, 85-112 (1964).

[19]**R.D. Mindlin** and **N.N. Eshel**: On the first strain gradient theories in linear elasticity. *Int. J. Solid Struct. 4*, 109-124 (1968).

[20] **P.Germain**: The method of virtual power in continuum mechanics.Part2: Microstructure. *SIAM J. Appl. Math. 25*, 556-575 (1973).

[21]**J. Laval**: L'elasticite du milieu cristallin. *J. Phys. Radium 18*, 247-369 (1957).

[22]**F.Falk**: One-dimensional model of shape memory alloys. *Arch. Mech. 3,* 63-84 (1983).

[23]**L. D. Landau** and **E. Lifshitz**: *Statistical Physics* (Pergamon Press, Oxford, 1980).

[24]**A. F. Devonshire**: Theory of ferroelectrics. *Adv. in Phys. 3*, 85-130, (1954).

[25]**F.Falk**:Ginzburg-Landau theory of static domainwalls in shape-memory alloys. *Z. Phys. B-Condensed Matter 51*, 177-185 (1983).

[26]**F. Falk**: Ginzburg-Landau theory and solitary waves in shape-memory alloys. *Z. Phys. B-Condensed Matter 53*, 159-167(1984).

[27]**J. L. Ericksen**: Some phase transition in crystal. *Arch. Rat. Mech. Anal. 73*, 99-124 (1980).

[28]**G.P.Parry**: Twinning in nonlinearly elastic monoatomic crystals. *Int. J. Solid Structures 16*, 275-281 (1980).

[29]**G.P.Parry**: The effect of stress on multiple bifurcations in ideal crystals. *J. Mech. Phys. Solids 32*, 133-148 (1984).

[30]**J.H. Maddocks** and **G.P. Parry**: A model for twinning. *J. of Elasticity 16*,113-133 (1986).

[31]**E.Magyari**: Stability of the martensitic phase boundaries in shape memory alloys. *J. Phys. F: Mat. Phys. 13*, L185-L188 (1983).

[32]**E Magyari**: The inertia mode of the mechanically generated solitons in nematic liquid crystals. *Z. Phys. B-Condensed Matter 56*, 1-3 (1984).

[33]**J.L. Skinner** and **R.G. Wolynes**: Transition state and Brownian motion theories of solitons. *J. Chem. Phys. 73*, 4015-4021 (1980).

[34]**K.J.Wahlstrand** and **P.G.Wolynes**: Simulation of a relativistic soliton model of crystalline polymer dynamics. *J. Chem. Phys. 82*, 5259-5263 (1983).

[35]**K. H. Michel** and **J. Naudts**: Dynamics of translations and rotations in molecular crystals. *J. Chem. Phys. 68*, 216-228 (1978).

[36]**K.H.Michel** and **E.Courtens**: Dynamics of translations and rotations in molecular crystals: Macroscopic and microscopic approaches. *Phys. RevB 23*, 513-522 (1981).

[37]**J.Naudts** and **E. Courtens**: Lattice dynamics of a partially disordered crystal: A simple model. *Phys. Rev. B 18*, 667-673 (1978).

[38]**S. Yomosa**: Solitary excitations in DNA double helices. in Springer Serie in Synergetics vol.30, "*Dynamical Problems in Soliton Systems*". Ed. by S.Takeno, pp.242-247, (Springer-Verlag, Berlin, 1985).

[39]**J.Pouget** and **G.A. Maugin**: Solitons and electroacoustics interactions in ferroelectric crystals. I: Single solitons and domain walls. *Phys. Rev. B 30*, 5306-5325 (1984).

[40]**G.A.Maugin** and **J.Pouget:** Solitons in microstructured elastic media: The physical and mechanical aspects. Proceeding of the 5th International Symposium on *Continuum Models of Discrete Systems*, Nottingham,U.K., 14-20 July 1985, Ed. A.J.M. Spencer (A.A. Balkema, Amsterdam, 1986).

[41]**M.Cieplak** and **L.A.Turski**: Magnetic solitons and elastic kink-like excitations in compressible Heisenberg chain. *J. Phys. C: Solid State Phys.13*, L777-L780 (1980).

[42]**A. Askar**: A model for coupled rotation-displacement mode of certain molecular crystals. Illustration for KNO_3. *J. Phys. Chem. Solids 34*, 1901-1907(1973).

[43]**J.Pouget, A.Askar** and **G.A. Maugin**: Lattice model for elastic ferroelectric crystals: Microscopic approach. *Phys. Rev. B 33*, 6320-6325 (1986).

[44]**J.Pouget, A.Askar** and **G.A. Maugin**: Lattice model for elastic ferroelectric crystals: Macroscopic approach. *Phys. Rev. B 33*, 6326-6331 (1986).

[45]**G.A. Maugin** and **J. Pouget:** Electroacoustic equations for one-domain ferroelectric bodies. *J. Acoust. Soc. Am. 68*, 575-587 (1980).

[46]**J. Pouget** and **G.A. Maugin**:Coupled acoustic-optic modes in deformable ferroelectrics. *J Acoust. Soc. Am. 68*, 588-601 (1980).

[47]**A.C Eringen**: Theory of micropolar elasticity. in *Fracture, vol. II*, ed. by H. Liebowitz, pp.621-729 (Academic, New York, 1968).

[48]**J. Pouget** and **G.A. Maugin**: Nonlinear dynamics of oriented elastic solid I. Basic equations. (submitted for publication to J. of Elasticity).

[49]**J. Pouget** and **G.A. Maugin**: Nonlinear dynamics of oriented elastic solid II. Propagation of solitons. (submitted for publication to J. of Elasticity).

[50]**G.A. Maugin** and **A. Miled**: Solitary waves in micropolar elastic crystals. *Int. J. Eng. Sci. 24*, 1477-1499 (1986).

[51]**C.A. Condat, R.A. Guyer** and **M.D. Miller**: Double sine-Gordon chain. *Phys. Rev. B 27*, 474-494 (1983).

[52]**R.K. Bullough, P.J. Caudrey** and **H.M. Gibbs**: The double sine-Gordon equations: A physically applicable system of equations. in Topic in Current Physics, vol.17,"*Solitons*", ed. R.K. Bullough and P.J. Caudrey, pp.107-141 (Springer-Verlag, Berlin, 1980).

[53]**D.K.Campbell, M. Peyrard** and **P. Sodano**: Kink-antikink interactions in the double sine-Gordon equation. *Physica 19D*, 165-205 (1986).

[54]**R.K. Dodd, J.C. Eilbeck, J.D. Gibbon** and H.C. Morris: *Solitons and nonlinear wave equations* (Academic Press, London, 1982).

[55]**A. Barone, F. Esposito, G.J. Magee** and **A.C. Scott**: Theory and application of the sine-Gordon equation. *Riv. del Naovo Cim. 1*, 227-267 (1971).

[56]**A.J. Callegar** and **E.L. Reiss**: Nonlinear stability problem for the sine-Gordon equation. *J. Math. Phys. 14*, 267-276 (1973).

[57]**S.Suzuki** and **M.Takagi**: Topographic study in ferroelectric $NaNO_2$ crystals.I. Structure of 180° domain wall. *J. Phys. Soc. Jpn. 30*, 188-202 (1971).

[58]**S. Suzuki** and **M. Takagi**: Topographic study on ferroelectric $NaNO_2$ crystals.II Mechanism of polarization reversal. *J. Phys. Soc. Jpn. 30*, 1302-1312 (1972).

[59]**W. Kinase, W. Makino** and **K. Takahashi**: Theory of the domain wall in $NaNO_2$ and the relation with the incommensurate structure. *Ferroelectrics 64*, 173-180 (1985).

[60]**Y.Ishibashi** and **H.Shiba**: Successive phase transitions in ferroelectric $NaNO_2$ and $Sc(NH_2)_2$. *J. Phys. Soc. Jpn. 45*, 409-413 (1978).

[61]**G.B. Whitham**: *Linear and nonlinear waves* (J. Wiley-Interscience,New York, 1974).

[62]**V.I. Karpman**: *Non-linear waves in dispersive media.* in International Series of Monographs in Natural Philosophys vol.71 (Pergamon Press, Oxford, 1975).

[63]**J. Pouget** and **G.A. Maugin**: Influence of an external electric field on the motion of a ferroelectric domain wall. *Phys. Lett. 109A*, 389-392 (1985).

[64]**J.Pouget**:Transient motion of a solitary wave in elastic ferroelectric in Lecture Notes in Physics 249. *Trends in Applications of Pure Mathematics to Mechanics*, ed. by E. Kroner and K. Kirchgassner, pp.156-163 (Springer-Verlag, Berlin, 1985).

[65]**M. Remoissenet**: Nonlinear excitations in a compressible chain of dipoles. *J. Phys. C: Solid State Physics 14*, L335-L338 (1981).

[66]**F.Fivez**:On the continuum limit of a classical compressible Heisenberg chain. *J. Phys. C: Solid State Phys. 15*, L641-643 (1982).

[67]**G.A.Maugin** and **A.Miled**:Solitary waves in elastic ferromagnets. *Phys. Rev. B 33*, 4830-4842 (1986).

[68]**H.D.Vey-Aharon, T.J.Sluckin, P.L.Taylor** and **A.J. Hopfinger** : Kink propagation as a model for poling in poly(vinylidenefluoride). *Phys. Rev. B 21*, 3700-3707 (1980).

[69]**P.L.Taylor** and**A.Barnerjea**: Incommensurate phase in a simple polymer. *Ferroelectric 66*, 135-142 (1986).

[70]**S.Takeno** and **S.Homma**: Topological solitons and modulated structure of bases in DNA double helices. *Prog. Theor. Phys. 70*, 308-311 (1983).

[71]**S. Yomosa**: Soliton excitations in deoxyribonacleic acid (DNA) double helices. *Phys. rev. A 27*, 2120-2125 (1983).

[72]**P.Kumar**: Soliton instability in an easy plane ferromagnet. *Physica 5D*, 359-369 (1982).

[73]**G.Wysin, A.R.Bishop** and **P.Kumar**: Soliton dynamics in an easy-plane ferromagnetic chain. *J. Phys. C: Solid State Physics 17*, 5975-5991 (1984).

[74]**G. Wysin, A.R. Bishop** and **J. Oitmaa**: Single-kink dynamics in an easy plane classical antiferromagnetic chain. *J. Phys. C: Solid State Phys.19*, 221-233 (1986).

[75]**N. Flytzanis, St. Pnevmaticos** and **M. Remoissenet**: Kink, breather and asymmetric envelope or dark solitons in nonlinear chains: I Monoatomic chain. *J. Phys. C: Solid state Phys. 18*, 4603-4629 (1985).

[76]**M. Peyrard** and **D. Kruskal**: Kink dynamics in hyghly discrete sine-Gordon system. *Physica 14D*, 88-102 (1984).

EXPERIMENTAL STUDIES OF MINERAL ENERGETICS

Alexandra Navrotsky
Dept. of Geological and Geophysical Sciences
Princeton University
Princeton, NJ 08544 USA

ABSTRACT. The energetics of mineral reactions can be studied by a variety of high temperature techniques. Studies of phase equilibria, vapor pressure, and oxidation-reduction reactions give values of free energies and activity composition relations. Calorimetric studies at high temperature, both by oxide-melt solution calorimetry and drop calorimetry, determine heats of formation, mixing, melting, phase transitions, and order-disorder processes. Application of thermochemical measurements is made to atmospheric and high pressure phases in the system $MgO-SiO_2$, with emphasis on the aspects of structure and bonding which stabilize olivine, pyroxene, spinelloid, spinel, garnet, ilmenite, and perovskite structures.

1. INTRODUCTION

The thermodynamics of mineral reactions can be studied by a variety of approaches, both experimental and theoretical. Because minerals are structurally complex, and their stability and phase relations are governed by a delicate balance of energetic, entropic, and volumetric factors which depend on temperature, pressure, and composition, such studies must pay careful attention to these details. Experimentally, one must be sure of sample purity, of the structural state of the samples studied, of careful experimental technique, and of equilibrium (or lack of it) in the experiments. In analyzing the data, one must make the punishment fit the crime by using thermodynamic models which are appropriate to the processes involved, especially when the possibility of order-disorder exists. These models must be complex enough to incorporate the essential physics and chemistry and yet simple enough to apply to data of somewhat limited accuracy. Thus although one might regard mineral thermodynamics as a sub-branch of solid state physics, the complexity of the structures and inherent "messiness" of natural mineral phases, and the difficulty with which definitive experiments can be made gives this field its own flavor. The purpose of this article is twofold: to summarize the state of present experimental techniques and to present an example of the interplay of

E. K. H. Salje (ed.), Physical Properties and Thermodynamic Behaviour of Minerals, 403–432.

thermodynamics, crystal chemistry, and theory in analyzing the behavior of mineral systems. This example describes the energetics of phase relations, mainly in the system $MgO-SiO_2$, at atmospheric and high pressure.

2. HEAT CAPACITIES, ENTROPIES, AND LATTICE VIBRATIONS

2.1. Basic Relations and Magnitudes

The heat capacity of a substance may be defined under conditions of constant pressure or constant volume:

$$C_V = (\partial E/\partial T)_V \tag{1}$$

$$C_P = (\partial H/\partial T)_P \tag{2}$$

where C_V and C_P are the heat capacities, E and H are internal energy and enthalpy, T is temperature, P is pressure, and V is volume. Because work must be done to expand a material during heating at constant pressure, C_P is larger than C_V, and

$$C_P - C_V = TV\alpha^2/\beta \tag{3}$$

where α and β are thermal expansivity and compressibility respectively. Thus

$$\alpha = V^{-1} (\partial V/\partial T)_P \tag{4}$$

$$\beta = -V^{-1} (\partial V/\partial P)_T \tag{5}$$

The reciprocal of the compressibility is the bulk modulus, K, used in the physics and geophysics literature.

For an ideal gas, $C_P - C_V$ is equal to the gas constant, R (8.314 JK^{-1}), α is 1/T (3.3×10^{-3} K^{-1} at room temperature), and β is 1/P (1 bar^{-1} at atmospheric pressure). For typical solids (also see Table (3)), α is of the order of 10^{-5} K^{-1}, β is of the order 10^{-6} bar^{-1}, and C_P-C_V is only a few percent of C_V, and that only at high temperature (>1000 K).

For a crystalline material, the thermal expansivity and compressibility reflect the response of bond lengths and bond angles to changes in T and P. Although single parameters, the volume expansion and volume compressibility, can be defined as above, much insight can be gained by looking at the response along individual crystallographic axes or of individual bond lengths. Thus, for example, the Mg-O bond in silicates such as Mg_2SiO_4 (olivine) expands with increasing temperature and contracts with increasing pressure to a much greater extent than does the Si-O bond. The response of individual coordination polyhedra to pressure and temperature, and its relation to bond character has been developed extensively (see Hazen and Finger, 1982, for a review).

At sufficiently high temperatures, when vibrations are fully excited, the heat capacity, C_V, approaches its classical value of 3nR per mole, where n is the number of atoms per formula unit (n = 7 for Mg_2SiO_4) and the factor of 3 arises because there are 3 independent vibrational degrees of freedom per atom. As an approximation to C_p at high temperature (since C_p-C_V is small), this generalization is known as the Law of Dulong and Petit. At low temperature, both C_p and C_V approach zero. For nonmetallic materials at sufficiently low T, $C_V = \alpha T^3$ (α a constant), while for metals, $C_V = \alpha T^3 + \gamma T$. The term linear in T, arising from electronic motion, predominates in metals at very low temperatures. The behavior of heat capacity at intermediate temperatures will be discussed further below.

2.2. Experimental Techniques

To obtain the standard entropy (S^o_{298}) of a phase, its heat capacity must be measured down to temperatures within the region where the T^3 law holds, and below any likely magnetic or other ordering phenomena. Adiabatic calorimeters exist in a number of laboratories to do this down to liquid helium temperatures and up to 350-400 K (Robie and Hemingway, 1972; Westrum et al., 1968). Especially for the lower temperature range, a relatively large amount of sample, one to ten grams, is required. Because of this, and because measurements are painstaking and time-consuming, calorimetric entropies are not routinely available for many substances, especially for exotic and interesting materials which can be synthesized in only small amounts. In the range 200-1000 K, differential scanning calorimetry (DSC) has made great headway in obtaining heat capacities with an accuracy of about 1% on samples as small as a few milligrams (Watanabe, 1982; Krupka et al., 1979; Stebbins et al., 1982). Above 1000 K, heat contents are generally measured by drop calorimetry, with the sample in a furnace at high temperature being dropped into a calorimeter at room temperature (Orr, 1953; Stebbins et al., 1982). Heat capacities, with an accuracy of about ±5-10% are obtained by taking the derivative $(\partial H/\partial T)_p$. These methods are summarized in Table 1.

Table 1. Means of Measuring Heat Capacities and Heat Contents

Method	Basic Measurement	Sample Required	Approx. Temp. Range
Adiabatic low temperature calorimetry	Temperature change for known input under adiabatic conditions	1 - 10 g	5 - 500 K
Differential scanning calorimetry	Temperature differences or power difference between sample and standard during controlled heating conditions	0.5 - 100 mg	200 - 1000 K
Drop calorimetry	Heat released when hot sample is dropped into calorimeter at room temperature	1 - 10 g	300 - 2400 K
Transposed temp. drop calorimetry	Heat absorbed when sample is dropped from room temperature into calorimeter at high temperature	20 - 500 mg	300 - 1800 K

As an example, the heat capacity, C_p, of Mg_2SiO_4 olivine (forsterite) is shown in Fig. 1. At 5-380 K, the data come from adiabatic calorimetry (Robie et al., 1982), at 300-1000 K both from DSC (Watanabe, 1982) and drop calorimetry (Orr, 1953), and above 1000 K, from drop calorimetry (Orr, 1953). The melting point of forsterite is

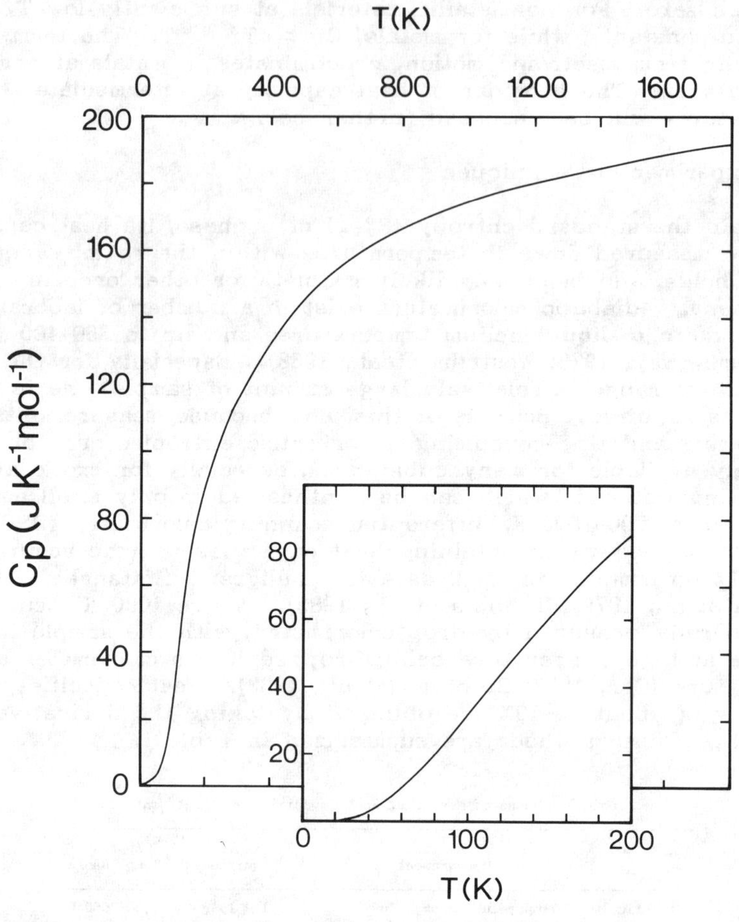

Fig. 1. Heat capacity of Mg_2SiO_4 olivine. Data from Robie et al. (1982), Orr (1953) and Robie et al. (1978). Arrow shows Dulong and Petit limit of 21R.

2163 K. Its heat of fusion has not yet been measured. Fig. 2 compares the heat capacities obtained by DSC (Watanabe, 1982) of the olivine (α), modified spinel (β) and spinel (γ) polymorphs of Mg_2SiO_4. The β and γ forms are high pressure phases which are generally available only in milligram amounts. These amounts are too small for adiabatic calorimetry, so values of C_p at low temperature must be

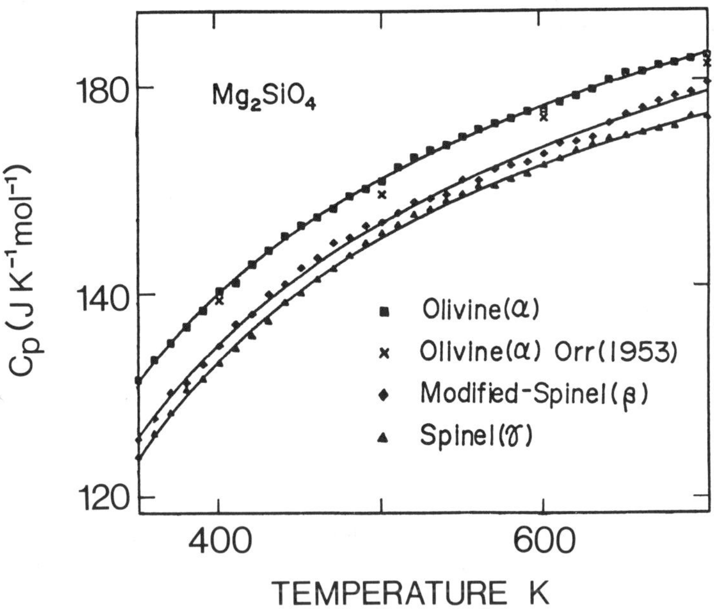

Fig. 2. Heat capacities of olivine (α), modified spinel (β) and spinel (γ) forms of Mg_2SiO_4 determined by DSC, from Watanabe (1982).

estimated by other means, see below. Above 1000 K at atmospheric pressure, the metastable β and γ polymorphs revert rapidly to the α form, which limits thermodynamic measurements.

2.3. The Interpretation of Lattice Heat Capacities

A crystal containing n atoms per formula unit has 3n vibrational degrees of freedom. For a discussion of basic principles, see Kieffer (1979, 1980). If one assumes that all these vibrations are independent and occur at the same frequency, the Einstein model of lattice vibrations results. Although it correctly predicts the low and high temperature limits of C_V, this model does not properly describe the behavior at low T, predicting C_V is proportional to T^2 rather than T^3, and at intermediate temperatures it does not allow a good fit to experimental data. The next level of complexity in approximating vibrational properties is the Debye model. Here one assumes that the density of vibrational states varies with the square of the vibrational frequency, with a maximum frequency or cutoff occurring when all degrees of freedom are accounted for. This maximum frequency is the Debye frequency, ν_D, and can be related to a Debye temperature, θ_D, with h Planck's constant and k Boltzmann's constant. The heat capacity at constant volume is given by

$$\theta_D = h\nu_D/k \quad (6)$$

$$C_V = 9nN_AkT^3/\theta_D^3 \int_0^{\theta_D/T} e^x x^4 (e^x-1)^2 \, dx = 3nN_Ak \, D(\theta/T) \quad (7)$$

where $x = h\nu_D/kT = \theta_D/T$. Thus the thermodynamic properties are a function of the ratio of the Debye temperature to the actual temperature, and the integral $D(\theta/T)$, known as the Debye function, is available in tabulations. The proper high temperature and low temperature limits obtain, and at low T

$$C_V = \frac{12}{5} nR\pi^4 (\frac{T}{\theta})^3 \quad (8)$$

In general θ_D increases with the strength of the bonding in the crystal and decreases with increasing molar volume for a given structure.

A test of how well the Debye model describes complex solids can be made by calculating the apparent Debye temperature from the heat capacity at various temperatures using Equation (7). If the Debye approximation were completely adequate, θ_D would be constant. As an example, data for NaCl and MgO (rocksalt), Mg_2SiO_4 (olivine) and SiO_2 (quartz) are shown in Fig. 3. One concludes (see Kieffer, 1979, 1980) that the Debye approximation works fairly well for solids of high symmetry containing only one kind of atom (Pb, C) or two atoms of similar masses and bonding characteristics (NaF, NaCl, KCl) but becomes increasingly inadequate as the bond type, coordination, and mass of different atoms becomes quite different (quartz, silicates). This reflects the fact that vibrational frequencies are not uniformly distributed, and that specific covalently bonded groups (e.g. SiO_4) retain their vibrational identity in crystals. For a few crystals, the vibrational density of states and its dispersion across the Brillouin zone are well characterized by infrared and Raman spectroscopy and by inelastic neutron scattering, and a rigorous calculation of heat capacity can be made. For most materials, this is not the case. Nevertheless, there are important features, not predictable from a Debye model, especially in complex solids, which need explanation. For example, heat capacities (and entropies) may "cross over." Thus pyrope garnet ($Mg_3Al_2Si_3O_{12}$) has a higher heat capacity than grossular ($Ca_3Al_2Si_3O_{12}$) at low temperature, but the reverse is true above 400 K. This leads to an anomalously high entropy for pyrope, which has been linked to the low frequency vibrations of Mg in the 8-coordinated site (Kieffer, 1980; Newton et al., 1977). Similarly, a number of high pressure phases (e.g. the perovskite form of $CdTiO_3$, $CdGeO_3$, and probably of $MgSiO_3$) have high vibrational entropies

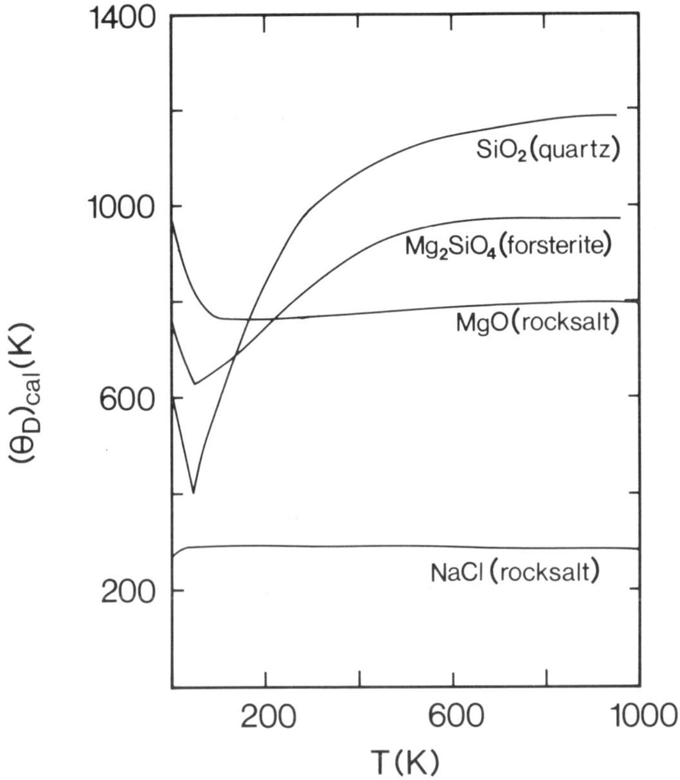

Fig. 3. Calorimetric Debye temperature, $(\theta_D)_{cal}$, for NaCl, MgO, SiO_2, Mg_2SiO_4. Simplified from Kieffer (1979, 1980).

(Navrotsky, 1980).

An approach to modelling heat capacities of complex materials has been developed by Kieffer (1979, 1980). A Debye-like approximation is retained for the low frequency acoustic modes, but the optic modes are described by two types of features: an optic continuum with a low and high frequency cutoff, and specific Einstein oscillators or narrow continua corresponding to well-defined vibrations, e.g. Si-O stretching. The approach is quite successful in modelling heat capacities and entropies of silicates (see below).

From an empirical point of view, the high temperature heat capacity, C_p, is often expressed by polynomials. To give the proper "knee" in C_p near the Debye temperature, various equations have been proposed. A form commonly used is

$$C_p = a + bT + cT^{-2} + dT^{-1/2} \tag{9}$$

As with all polynomials, one must be careful when extrapolating outside the range of the original data.

3. FREE ENERGIES OF MINERAL REACTIONS

3.1 General Principles

Consider the general reaction

$$aA + bB = cC + dD \tag{10}$$

where at least one reactant or product is a solid. For the reaction at atmospheric pressure and a given temperature

$$\Delta G^o = \Delta H^o - T\Delta S^o = -RT \ln K \tag{11}$$

with

$$K = (a_A)^a(a_B)^b/(a_C)^c(a_D)^d \tag{12}$$

The terms within parentheses are activities, related to chemical potentials, μ, by

$$RT \ln a_i = \mu_i \text{ (given state)} - \mu_i \text{ (reference state)} \tag{13}$$

At high pressure

$$\Delta G(P,T) = \Delta G^o(T) + \int_{1\ atm}^{P} \Delta V\, dP \tag{14}$$

where ΔV is the volume change for reaction (10). If two solid phase assemblages are at equilibrium at a given P, T, then $\Delta G(P,T) = 0$, which leads to the Clausius-Clapeyron Equation for the slope of a univariant phase boundary in P-T space

$$\left(\frac{dP}{dT}\right)_{equil} = \frac{\Delta S(P,T)}{\Delta V(P,T)} \tag{15}$$

If ΔS and ΔV can be approximated as constants independent of P and T, then dP/dT is constant and the phase boundary is a straight line is P,T space.

Thus to characterize the thermodynamics of solid state reactions, one needs to determine the following: (a) the enthalpy, entropy, and volume changes for reactions involving pure stoichiometric compounds and (b) the thermodynamic activities in phases of variable composition, generally substitutional solid solutions (e.g. $(Mg_xFe_{1-x})_2SiO_4$) or compounds with a homogeneity range (variable cation to anion ratio, e.g. $Fe_{1-x}O$). In general these parameters will be functions of temperature and pressure, though useful simplifying assumptions can often be made.

A further complication inherent to solids is the difficulty in attaining (and proving one has attained) equilibrium. This means that reactions between phases should ideally be "reversed," phase boundaries approached from high and low T, high and low P, and reactant and product phase assemblages. Also one must pay attention to possible equilibria within a single phase, e.g. cation distributions, nonstoichiometry, defects, each of which may affect thermodynamic properties. In general such problems of equilibrium and metastability are more serious for the solid state than for liquids and gases, both because of lower diffusion rates and of problems of nucleation. To overcome kinetic barriers, measurements must often be made at high temperature, especially when dealing with oxides.

3.2. Oxidation-reduction equilibria

For systems containing an easily reducible metal, equilibrium with the metal and an appropriate gas phase provides a useful means of measuring activities and free energies in a binary or ternary oxide (or chalcogenide) phase. Thus for the equilibrium

$$\text{"FeO"} \rightleftarrows Fe + 1/2\ O_2 \qquad (16)$$

$$\Delta G^0 = -RT \ln K = -RT \ln (pO_2)^{1/2}\ a_{Fe}/a_{\text{"FeO"}} \qquad (17)$$

For pure "FeO" in equilibrium with pure Fe, a knowledge of the equilibrium value of pO_2 at a given T gives ΔG^0 directly, while the temperature dependence of pO_2 can be used to derive values of ΔH^0 and ΔS^0. The value of pO_2 in equilibrium with both metal and metal oxide can be measured in two ways: by gas-mixing experiments or by a solid state electrochemical cell. In the former, a value of pO_2 is set by mixing oxygen with an inert gas for $pO_2 > 1$ to 10^{-4} atm, or, to attain lower values of pO_2, by using a subsidiary gas phase equilibrium such as

$$H_2O \rightleftarrows H_2 + 1/2\ O_2, \quad pO_2 = [K_{18}(pH_2O)(pH_2)^{-1}]^2 \qquad (18)$$

or

$$CO_2 \rightleftarrows CO + 1/2\ O_2, \quad pO_2 = [K_{19}(pCO_2)(pCO)^{-1}]^2 \qquad (19)$$

The ratio in which the major gases (CO and CO_2 or H_2O and H_2) are mixed then determines the partial pressure (fugacity) of oxygen. This ratio is varied until a value is found for which the solid sample equilibrated with the gas mixture contains both metal and oxide (Nafziger and Muan, 1967)

The electrochemical method depends on the use of a cell using a solid electrolyte (typically a CaO-stabilized zirconia) which conducts by the movement of oxide ions (Schmalzried, 1972). For the Fe/"FeO" equilibrium, such a cell may be written as

$$\mathrm{Pt|Fe,\text{"FeO"}||ZrO_2(CaO||air|Pt)} \tag{20}$$

The left side of the cell has its pO_2 fixed by the Fe/"FeO" equilibrium while the right side has $pO_2 = 0.21$ atm. The half cell reactions may be written as

$$\text{Left} \quad O^{=} + Fe \rightarrow \text{"FeO"} + 2e^{-} \tag{21}$$

$$\text{Right} \quad 1/2O_2 + 2e^{-} \rightarrow O^{=} \tag{22}$$

and the cell reaction

$$Fe + 1/2O_2 = \text{"FeO"} \tag{23}$$

with

$$\Delta G^{o} = -n\mathcal{F}E^{o} = -2\mathcal{F}E^{o} \tag{24}$$

where n is the number of electrons transferred, F the Faraday constant, E^{o} the measured cell voltage.

In the above reactions "FeO" has been written in quotation marks because this oxide is nonstoichiometric, containing a deficiency of cations and a virtually perfect anionic sublattice in the rocksalt structure. The defect equilibria and deviations from stoichiometry are functions of pressure, temperature, and oxygen fugacity. In silicates containing iron, the deviations from stoichiometry are much smaller than in "FeO" and, although the defect equilibria dominate transport and electrical properties, they are generally not major factors in the overall thermodynamics. However, in systems containing both Fe^{2+} and Fe^{3+}, e.g. the iron-titanium oxides, oxidation-reduction plays a major role.

Both the gas equilibration method and the solid cell method can be used to find the free energies of ternary compounds and activities in solid solutions. In general, easily reducible oxides include those of Fe, Ni, Co, Cu and the higher oxides of manganese. Careful studies of multicomponent systems (e.g. Mg-Fe-Si-O, Nafziger and Muan, 1967) can give free energy data for all end-member phases and activity-composition relations for their solid solutions. The temperature range of such studies is limited at low temperatures (typically ~1100 K for silicates) by sluggish equilibration and at high temperature by melting, vaporization, reaction with container material or other side reactions. Thus the temperature range over which equilibrium measurements can be made is not always large enough to derive accurate entropy and enthalpy values from the temperature dependence of free energy.

3.3. Vapor pressure measurements

If a constituent is relatively volatile, its vapor pressure can be used to probe its thermodynamic activity in the solid state. Measurement of vapor pressure fall into two classes: dynamic measurements in which one monitors weight loss or the amount of volatile species in a carrier gas stream as a function of time, and effusion measurements in which equilibrium in an almost closed system is attained, and the vapor is sampled through a pinhole. The detection and measurement of vapor species, by mass spectrometry, spectroscopic techniques, or chemical analysis, determines the sensitivity and accuracy of vapor pressure measurements.

Such measurements have proved especially useful in semiconductors, where one or both species in a binary system are often volatile, and the solid phases can show considerable homogeneity ranges. These methods have recently been applied quite successfully to measuring activities in aluminosilicate melts containing alkalis (Fraser and Rammensee, 1983).

3.4 High pressure phase equilibria

Table 2 lists the characteristics of several types of high pressure apparatus. In general, as P increases, the available sample volume decreases, and at pressures greater than about 100 kbar, one can seldom produce more than a few milligrams of sample, especially at high temperature. The appropriate calibration of T and P when both are high is a subject of ongoing development and controversy. The ability to perform physical measurements (X-ray diffraction, conductivity, spectroscopy) at high P and T is improving. In situ experiments are crucial to understanding the actual state of materials under these extreme conditions.

To properly constrain a phase boundary, it must be reversed, i.e. the low pressure phase transformed into the high pressure phase with increasing pressure, and the high pressure phase transformed into the low pressure one with decreasing pressure. Many if not most of the studies reported in the literature fail these criteria and represent synthesis diagrams rather than equilibrium phase diagrams. Although reactions at high P and T often occur rapidly, one must realize that such syntheses diagrams may occasionally represent reactions very far from equilibrium.

As an example of the calculation of thermodynamic parameters from a phase boundary, consider the reaction:

$$Mg_2SiO_4\ (\alpha) \rightarrow Mg_2SiO_4\ (\beta) \qquad (25)$$

Suito (1977) gives the boundary as

$$P\ (\text{kbar}) = 98 + 0.035\ T\ (\text{K}) \qquad (26)$$

Table 2. High Pressure Apparatus

Apparatus	Sample Vol. (cm^3)(approx.)	Pressure Range (kbar)	Temp. Range (K)(approx.)
autoclaves	100 - 10000	0 - 1.0	200 - 600
externally heated cold seal pressure vessels	10 - 500	0 - 5	300 - 1200
internally heated gas vessels	5 - 100	0 - 15	300 - 1800
piston-cylinder solid media apparatus	0.01 - 0.1	0 - 40	300 - 2000
multiple anvil solid media apparatus	10^{-3} - 10^{-2}	0 - 250	300 - 2000
diamond anvil cell (laser heated)	10^{-6} - 10^{-3}	0 - 1000 (up to 3000 at lower T)	200 - 3000
shock apparatus (dynamic methods, pressure for very short time only)	variable	0 to 2000	300 - 3000

First, if one ignores the difference in heat capacity, thermal expansion, and compressibility between olivine (α) and modified spinel (β), one can assume that the volume change, ΔV, for reaction (25) is a constant, with a value, obtained from lattice parameters under ambient conditions, of -3.13 cc mol^{-1}. At a point on the phase boundary, 1273 K, 143 kbar, the free energy ΔG (1273 K, 143 kbar), is zero. The free energy change at the temperature and atmospheric pressure, ΔG^0 (1273 K, 1 bar), is given by

$$\Delta G^0 = -\int_{1\ \mathrm{bar}}^{143\ \mathrm{kbar}} \Delta V\, dP = -\Delta V \times P_{trans} \tag{27}$$

$$\Delta G^0 = (-3.13) \times 143 = 447.59 \text{ cc kbar mol}^{-1} = 45361 \text{ J mol}^{-1}$$

From the Clausius-Clapeyron Equation (Equation (15)), ΔS at the phase boundary is given by

$$\Delta S = (\frac{dP}{dT})\Delta V = 0.035\ K^{-1}\ kbar \times -3.13\ cc\ mol^{-1}$$
$$= -0.110\ cc\ kbar\ K^{-1} = -11.15\ JK^{-1}\ mol^{-1} \qquad (28)$$

Neglecting $(\partial\Delta V/\partial P)_T$ and $(\partial\Delta V/\partial T)_P$ means that $\Delta S = \Delta S^o$ and

$$\Delta H^o = \Delta G^o + T\Delta S^o = 45361 - 11.15 \times 1273 =$$
$$31167\ J\ mol^{-1}\ (at\ 1273\ K) \qquad (29)$$

However, the olivine and modified spinel phases do not have identical compressibilities and thermal expansions; the denser phase generally has lower values of these parameters (see Table (3)). In

Table 3. Parameters used for phase boundary calculations for (α,β,γ) Mg_2SiO_4 polymorphs (from Akaogi et al., 1984)

Compound	V^o_{298} (cm^3mol^{-1})	$\alpha(K^{-1})$	K_o (kbar)	$_oK'$	$C_p = a+bT+cT^{-2}J\ mol^{-1}K^{-1})^j$ a	b$\times10^3$	c $\times 10^{-5}$
α-Mg_2SiO_4	43.67	3.7×10^{-5}	1290^g	5.2^g	155.8	22.2	-40.6
β-Mg_2SiO_4	40.54	3.0×10^{-5}	1660^h	4^i	151.8	23.1	-43.2
γ-Mg_2SiO_4	39.65	2.6×10^{-5}	2130^h	4^i	156.5	14.5	-47.8

general one must know the simultaneous effect of pressure and temperature on volume, but a simplification to the equation of state can be made by assuming that the effects of P and T can be separated. An approximate equation of state for the volume of a phase, is

$$V = V^o_{298}\ [1+\alpha(T-298)]\ [K'K^{-1}P+1]^{-1/K'} \qquad (30)$$

The term in the first brackets arises from assuming a constant coefficient of volume expansion, α. The term in the second brackets describes the compression according to the Birch-Murnaghan equation of state, which is generally adequate to pressures of 200 kbar. The integral $\int^P_{1atm} VdP$ at constant temperature is then given by, for a given phase

$$\int^P_{1atm} VdP = V\quad [1-\alpha(T-298]\ (1-K/K')^{-1}(K/K')[(PK'/K+1)^{1-1/K'}-1] \qquad (31)$$

An analogous calculation of thermodynamic parameters can then be done, which yields quite similar values of ΔS°_{1273} and ΔH°_{1273} to those above. The calorimetrically measured value (see below) of ΔH°_{973} is 30.1 ± 2.9 kJ mol^{-1}, which agrees within experimental error with the value calculated above from the Clausius-Clapeyron equation.

4. ENTHALPIES OF MINERAL REACTIONS

Heat capacity measurements, discussed in section (2) above, permit the characterization of the thermal properties of a single substance. The measurement of the enthalpy of a chemical reaction can be used to relate the thermodynamic properties of a given substance to those of other substances, and thus to obtain heats of formation, of mixing, and of solid state reactions. These calorimetric methods can be classified into two groups: direct methods in which the reaction studied by calorimetry is the one of interest, and indirect methods in which a thermochemical cycle is used to obtain the enthalpy of interest. In the former category is combustion calorimetry, in which the enthalpy of formation of oxides is obtained by burning the metals in oxygen (Charlu and Kleppa, 1973), or the heat of formation of fluorides is measured by burning the metals in a fluorine atmosphere (Kybett and Margrave, 1966). The absorption or release of oxygen by nonstoichiometric oxides (e.g. "FeO") can also be studied (Marucco et al., 1970). By combining such calorimetric measurements with free energy measurements and a knowledge of the composition of the phase as a function of the partial pressure of oxygen, the partial molar entropy of solution of the gas in the nonstoichiometric oxide can be obtained with fairly high accuracy. This lends insight into solubility mechanisms and defect chemistry.

The heats of mixing in molten salts, some oxide melts and metal alloys can be measured directly by calorimetry at high temperature (Hersh and Kleppa, 1965; Holm and Kleppa, 1967; Topor and Kleppa, 1983). Viscous melts, glasses, and many crystalline solid solutions or compounds cannot be studied by direct reaction calorimetry because they react far too slowly. Instead, indirect methods, such as solution calorimetry, are used. The solvent in such measurements can be an aqueous solution (hydrofluoric acid for oxides and silicates) (Robie and Hemingway, 1972), an acidified brine for carbonates (Capobianco and Navrotsky, 1987), or a molten salt or oxide melt (lead borate, alkali borate, or alkali molybdate or tungstate for oxides and silicates (Navrotsky, 1977)). Two important recent advances in high temperature solution calorimetry are the use of controlled fO_2 over an alkali borate melt for the study of Fe^{2+}-bearing phases (C. Chatillon-Colinet et al., 1983) and the development of solution calorimetric techniques for sulfides (Cemie and Kleppa, 1986). In addition, several approaches to high temperature calorimetry of hydrous systems have been made (Kiseleva and Ogorodova, 1984; Clemens et al., 1987; Clemens and Navrotsky, 1987; Navrotsky, 1987).

Thermochemical cycles can be devised to deal with phases which decompose if left to equilibrate in a hot (~1000 K) calorimeter prior to dissolution. One method (transposed temperature drop calorimetry) takes advantage of the rapid decomposition itself, directly measuring its enthalpy by dropping a sample from room temperature into a hot calorimeter with no solvent (Ito and Navrotsky, 1985). A second method (drop solution calorimetry) drops the metastable sample into the flux, where it dissolves (Akaogi and Navrotsky, 1984). These various calorimetric methods are summarized in Table (4).

As an example, the heats of solution in molten $2PbO \cdot B_2O_3$ near 1000 K of a number of oxides are given in Table (5). Using these values we can construct the following thermodynamic cycle:

$$SiO_2 \text{ (quartz)} \rightarrow \text{dilute solution}, \ \Delta H = -3.3 \text{ kJ mol}^{-1} \quad (32)$$

$$Mg_3SiO_4 \text{ (olivine)} \rightarrow \text{dilute solution}, \ \Delta H = 67.3 \text{ kJ mol}^{-1} \quad (33)$$

$$\text{dilute solution} \rightarrow 2MgSiO_3 \text{ (orthoenstatite)}, \ \Delta H = -70.0 \text{ kJ mol}^{-1} \quad (34)$$

Therefore for

$$Mg_2SiO_4 \text{ (olivine)} + SiO_2 \text{ (quartz)} = 2MgSiO_3 \text{ (orthoenstatite)}, \quad \Delta H = -3.3 + 67.3 - 70.0 = -6.0 \text{ kJ mol}^{-1} \quad (35)$$

Similarly, for the phase transformation

$$Mg_2SiO_4 \text{ (olivine)} = Mg_2SiO_4 \text{ (modified spinel)}, \quad \Delta H = 67.3 - 37.2 = 30.1 \text{ kJ mol}^{-1} \quad (36)$$

The first calculation confirms the observation, known from the phase diagram, that $MgSiO_3$ pyroxene is stable relative to a mixture of Mg_2SiO_4 (olivine) and SiO_2 (quartz) at atmospheric pressure (since the ΔS of reaction (35) is expected to be small). The second calculation shows that the modified spinel structure is indeed energetically metastable at atmospheric pressure, and supports the high pressure data discussed above.

5. ENERGETICS OF HIGH PRESSURE PHASE TRANSITIONS, WITH EMPHASIS ON THE MAGNESIUM SILICATES

5.1. Relations among olivine (α), modified spinel (β) and spinel (γ) phases

The structural features of the α, β, and γ phases are compared in Table (6). Thermodynamic data are summarized in Table (7). The enthalpies of transition have been measured by high temperature

Table 4. Methods of High Temperature Reaction Calorimetry

Method	Process	Application
Solution Calorimetry	Solute (T) + Solvent (T) = Dilute Solution (T)	sample must be able to equilibrate at calorimetric T
lead borate (700 °C)		oxides containing MgO, CaO, alkali, alkaline earth, Fe_2O_3, CoO, NiO, ZnO, GeO_2, Ga_2O_3
alkali molybdate or tungstate (700 °C)		titanates, vanadates
alkali borate (775-800 °C)		titanates, zirconates, stannates, Fe^{2+}-compounds under controlled fO_2
nickel sulfide (900 °C)		sulfides
Transposed Temperature Drop Calorimetry	Substance (25 °C) → Transformed substance (T)	heats of fusion, rapid phase transition, annealing, decomposition (process must be fast, <~30 min.)
Drop-Solution Calorimetry	Solute (25 °C) + Solvent (T) → Dilute Solution (T)	substances which would decompose to an ill-defined state at calorimetric T, hydrous phases, high P phase transitions
Gas-Solid Reaction	Substance (T) ± Gas (T) → Product (T)	oxidation-reduction, metal-hydrogen reaction

Table 5. Enthalpies of Solution in Molten $2PbO \cdot B_2O_3$ near 700 °C

oxide	ΔH_{sol} (kJ mol^{-1})
SiO_2 (quartz)	-3.3 ± 0.2
$MgSiO_3$ (orthoenstatite)	35.0 ± 0.8
Mg_2SiO_4 (olivine)	67.3 ± 2.3
Mg_2SiO_4 (modified spinel)	37.2 ± 1.7

solution calorimetry, the free energies of transition were obtained by phase equilibria, and $\Delta S°$ comes from $\Delta G° = \Delta H° - T\Delta S°$. The phase relations are shown in Figure (4).

Several points are evident from the data. The $\Delta H°$, $\Delta S°$, $\Delta V°$ of a given transition (e.g. olivine → spinel) have very different values for different compositions. The contribution to the free energy of the $\Delta H°$, $T\Delta S°$ and $P\Delta V$ terms are of comparable magnitude. Thus the location of

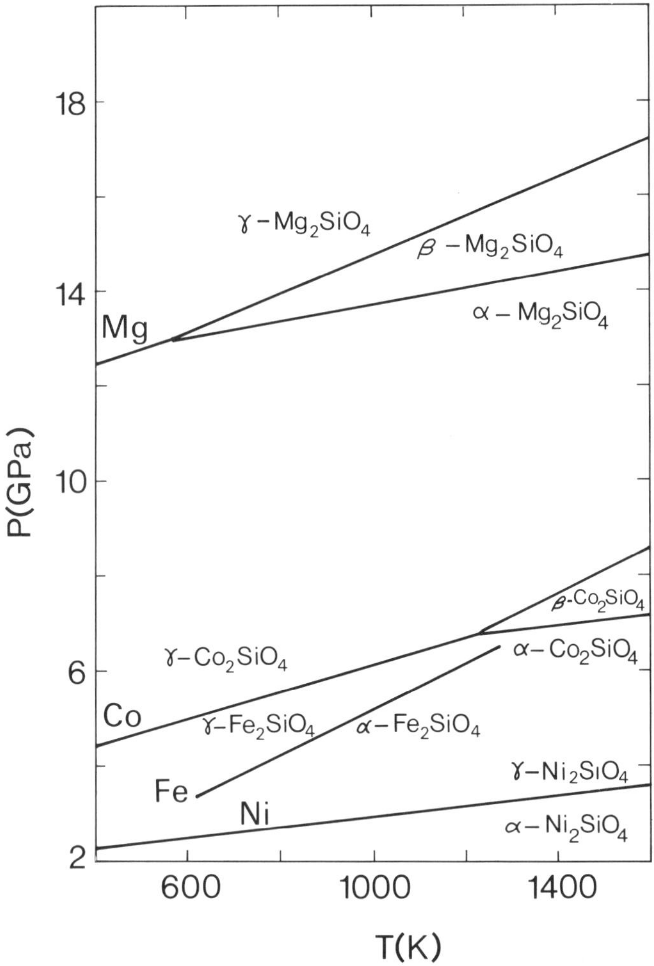

Fig. 4. Phase relations among olivine (α), spinel (γ), and modified spinel (β) for silicates of Mg, Co, Fe, Ni. From Navrotsky, 1987b.

Table 6. Structural Features of Some AB_2O_4 Phases

	Olivine (α)	Modified Spinel (β)	Spinel (β)
Oxygen packing	hcp	ccp	ccp
Si_2O_7 groups	no	yes	no
Mg_2SiO_4			
molar volume (cm^3)	43.67	40.54	39.65
average Mg–O (Å)	2.112	2.081	2.070
average Si–O (Å)	1.636	1.651	1.655
Co_2SiO_4			
molar volume (cm^3)	44.529	41.628	40.613
average Co–O (Å)	2.134	2.114	2.104
average Si–O (Å)	1.627	1.642	1.646

All crystallographic parameters refer to ambient conditions.

Table 7. Thermochemical Parameters for Phase Transitions in Silicates and Germanates of AB_2O_4 Stoichiometry

		$\Delta G°$ (kJ mol^{-1})	$\Delta H°$ (kJ mol^{-1})	$\Delta S°$ (J $mol^{-1}K^{-1}$)	$\Delta V°_{298}$ (cm^3 mol^{-1})
$\alpha \to \gamma$	Mg_2SiO_4	55.4	38.7	-16.7	-4.13
	Co_2SiO_4	24.4	11.3	-13.1	-3.92
	Fe_2SiO_4	20.5	3.9	-17.6	-4.35
	Ni_2SiO_4	11.8	6.0	-5.8	-3.42
	Mg_2GeO_4	0	-12.7	-11.7	-3.52
	Fe_2GeO_4	-29.7			-4.07
	Co_2GeO_4	-14.6			-3.89
	Ni_2GeO_4	-34.3			-3.12
$\alpha \to \beta$					
	Mg_2SiO_4	42.3	31.8	-10.5	-3.24
	Co_2SiO_4	18.0	9.0	-9.0	-2.90
	Fe_2SiO_4	27.9	16.4	-11.6	-3.31
	Mn_2GeO_4	-15.9			-3.61
$\beta \to \gamma$					
	Mg_2SiO_4	13.1	6.8	-6.3	-0.89
	Co_2SiO_4	6.4	2.3	-4.1	-1.02
	Fe_2SiO_4	-7.4	-13.4	-6.0	-1.04

Uncertainties in $\Delta G°$, $\Delta H°$ typically about ±1000 to ±2000 $J \cdot mol^{-1}$, in $\Delta S°$ about ±2 $J \cdot mol^{-1} \cdot K^{-1}$, in $\Delta V°$ ±0.05 $cm^3 \cdot mol^{-1}$. $\Delta G°$, $\Delta H°$, $\Delta S°$ values refer to temperatures near 1000 K.

a phase boundary in P-T space cannot be approximated from $\Delta H°$ alone. For example, although $\Delta H°$ for the olivine-spinel transition increases in the series Fe_2SiO_4, Ni_2SiO_4, Co_2SiO_4, Mg_2SiO_4, the pressure at which the spinel is first formed (near 1273 K) increases in the series Ni_2SiO_4 (near 30 kbar), Fe_2SiO_4 (near 60 kbar), Co_2SiO_4 (near 70 kbar), Mg_2SiO_4 (near 135 kbar).

For the olivine-spinel transition, $\Delta S°$ and $\Delta V°$ are roughly linearly related, see Fig. (5). The largest $\Delta V°$ and $\Delta S°$ (and $\Delta H°$ as well) are seen for Mg_2SiO_4, the smallest $\Delta V°$ and $\Delta S°$ for Ni_2SiO_4, despite the

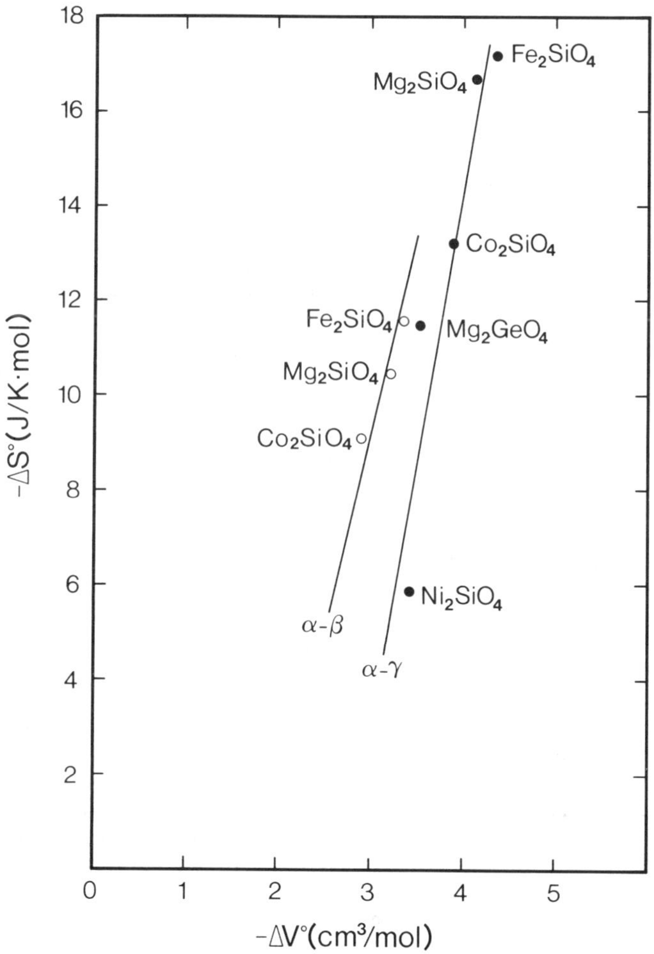

Fig. 5. Entropy and volume changes for olivine-spinel (α-γ) and olivine-modified spinel (α-β) phase transitions. From Navrotsky, 1987b.

fact that the ionic radii of Ni^{2+} and Mg^{2+} are very similar. If one attributes this difference between Ni and Mg transitions to crystal field effects, then the effect of the non-spherically symmetrical electron distribution is as important in determining entropy and volume as in its effect on energy.

The β-phase occurs between olivine and spinel; its stability field is limited to high temperatures by the α, β, γ triple point (see Fig. (4)). ΔV°, ΔH° and ΔS° for the α-β transition in Mg_2SiO_4 are 78, 81, and 63%, respectively, of the corresponding values for the α-γ transition. Similar relations hold for Co_2SiO_4. These results suggest that the β-phase is more similar in thermodynamic properties to the spinel than to the olivine. The α-β-γ transitions do not involve changes in the coordination numbers of any cations. However, the structures of β-phase and spinel are based on a slightly distorted cubic close packing of oxygen atoms, while the olivine structure is based on distorted hexagonal close packing. Recent structural studies on β-Mg_2SiO_4 and γ-Mg_2SiO_4 reveal that a small increase of mean Si-O bond length and a decrease of average Mg-O distance are observed in the $\alpha \rightarrow \beta \rightarrow \gamma$ transitions and that these changes are larger in the $\alpha \rightarrow \beta$ transition (See Table (6)). Similar relations hold for the $\alpha \rightarrow \beta \rightarrow \gamma$ transitions in Co_2SiO_4. Therefore, in terms of structural data, as well as in terms of the thermochemical data presented here, the β-phase is more similar to spinel than to olivine for both Mg_2SiO_4 and Co_2SiO_4.

The modified spinel (β-phase) has been observed in Mg_2SiO_4, Co_2SiO_4, Mn_2GeO_4, and Li_2WO_4 but not in the other silicates and germanates. A phase of similar structure is found at intermediate compositions in the systems Ni_2SiO_4-$NiAl_2O_4$ and Mg_2GeO_4-$MgGa_2O_4$ but not for those end-members. The β-phase is one member of a family of structures called spinelloids, which appear to be higher in enthalpy and volume than the spinels and which appear to owe their stability at high temperature to vibrational and/or configurational entropy (Akaogi and Navrotsky, 1984b). Why the β-phase is stable for Co_2SiO_4 and Mg_2SiO_4 but not for Fe_2SiO_4 and Ni_2SiO_4 is not clear.

Application of Kieffer's vibrational models to the Mg_2SiO_4 polymorphs (Akaogi et al., 1984) leads to the following conclusions. A number of vibrational models can be constructed which are consistent with the infrared and Raman spectra of each phase. These models predict entropies of transition which are in good agreement with those calculated from combining high pressure phase equilibria and calorimetry. The negative ΔS° values for $\alpha \rightarrow \beta$ and $\beta \rightarrow \gamma$ are based on rather subtle changes in the vibrational density of states. The negative ΔS° values for the transitions largely result from changes in the region of the density of states treated as an optic continuum. In particular, the low frequency cutoff for this continuum, consistent with spectra, thermodynamics, and models for dispersion of low frequency modes, is 130-140 cm^{-1} for olivine, 165-180 cm^{-1} for modified spinel, and 180-250 cm^{-1} for spinel. It is tempting to associate this shift to higher frequency with increasing regularity of the MgO_6 octahedron in the series α,β,γ. This regularity is in turn related to decreasing deviation from ideal close packing of the oxygens (hcp in α, ccp in β and γ) and changes in the pattern of occupancies of octahedral and tetrahedral interstices, which also result in increasing density in the sequence α,β,γ. O'Keeffe and Hyde (1981) pointed out that although the olivine-spinel transition does not involve any increase in the number of anions around a cation (conventional coordination number), it does involve an increase of the number of next nearest cations around a given cation, $SiMg_9$ to $SiMg_{12}$. They suggest that cation-cation nonbonded interactions may make an important energetic contribution. It appears reasonable that such an overall increase in packing efficiency results in additional vibrational constraints which eliminate some of the lowest frequency modes.

5.2. "Post-spinel" phases

At pressures considerably above those for the olivine-spinel transition, silicates and germanates either decompose to mixtures of MO (rocksalt) and SiO_2 or GeO_2 (rutile) or form new structures (garnet, ilmenite, perovskite) in which some or all of the Si or Ge is octahedrally coordinated. The large volume decrease associated with this increase in coordination number is the driving force for these transitions to "post spinel" phases. The wealth of phases observed is evidence for very closely balanced enthalpy, entropy, and volume factors. Table (8)

Table 8. Parameters in Vibrational Models

		PX	GT	IL	PV
Volume (cm^3/mol)		31.395	28.347	26.411	24.393
K (GPa)		108	154	211	262
α (const x 10^{-5}(°K^{-1})		2.203	2.57	2.60	4.00
Directionally averaged velocities (km sec^{-1}),	K_1	4.67	3.58	5.30	5.75
	K_2	5.09	5.05	5.80	6.30
	K_3	8.07	8.96	9.80	10.40
q (fraction of modes)		0.0125	0.0125	0.100	0.050
Optic continua					
First	ν_L(cm^{-1})	140	150	288	174
	ν_u(cm^{-1})	545	630	800	800
	q (fraction of modes)	0.6775	0.7875	0.900	0.950
Second	ν_L	650	850	---	---
	ν_u	760	1050	---	---
	q	0.1900	0.200	---	---
Third	ν_L	940	---	---	---
	ν_u	1045	---	---	---
	q	0.1200	---	---	---

lists available thermochemical data for $MgSiO_3$ phases and Fig. (6) compares the enthalpies, relative to the phase stable at ambient conditions, of a number of silicates and germanates (Navrotsky, 1987b). In the germanates, the wollastonite and garnet phases are

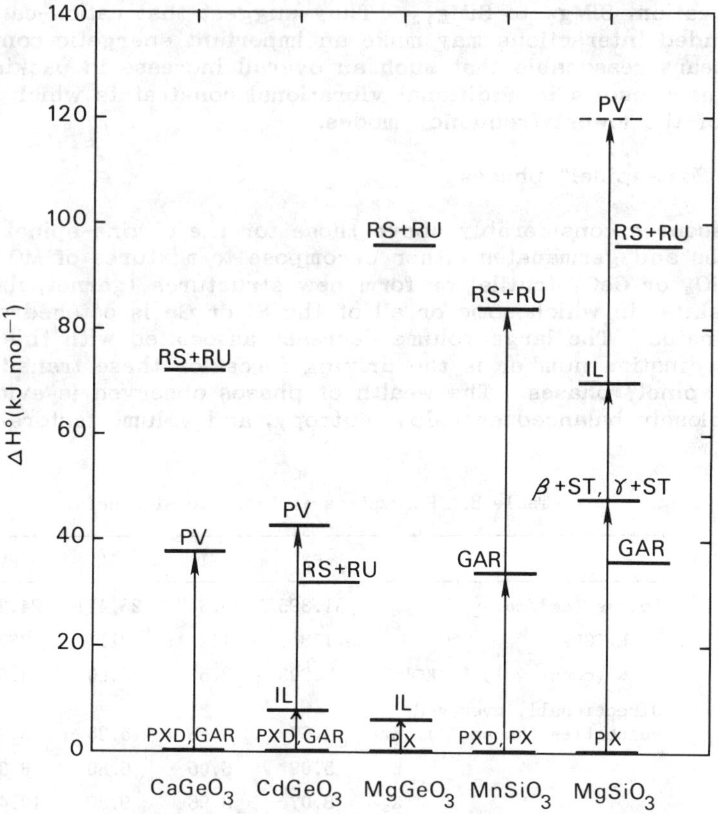

Fig. 6. Enthalpies of high pressure ABO_3 polymorphs relative to phases stable at atmospheric pressure. Solid lines indicate relatively well-constrained data; dashed lines indicate estimates. From Navrotsky, 1987b.

very similar in enthalpy; the garnet is a high pressure phase not because of significantly unfavorable energetics but because of its lower entropy and volume. $MnSiO_3$ and $MgSiO_3$ garnets, in contrast, are energetically much higher than the corresponding pyroxenes, with $MgSiO_3$ garnet estimated to be comparable in enthalpy to the assemblage spinel plus stishovite. $MgSiO_3$ ilmenite is much higher in enthalpy relative to pyroxene than is $MgGeO_3$ ilmenite. The perovskites $CaGeO_3$, $CdGeO_3$ and $MgSiO_3$ are all energetically rather unfavorable phases.

5.3. Lattice vibrational models for post-spinel phase transitions

The Kieffer approach is particularly useful for these high pressure phases, which certainly will not be available in quantities suitable for low temperature Cp measurements in the foreseeable future. The input data needed to constrain the model are space group, molar volume, elastic constants, bulk modulus, thermal expansion, and vibrational (Infrared and Raman) spectra. All these data are obtainable for ultrahigh pressure phases and indeed are of interest to geophysics for other reasons. The vibrational models work best when comparing phases with quite different crystal structures and vibrational spectra. Then, although one does not need to assign all vibrational bands, one can see how changes in the spectra (reflecting structural changes) influence the entropy. For hypothetical high pressure phases, or those not readily quenched to ambient conditions for detailed measurements, one can judiciously invent or "synthesize" a likely vibrational density of states based on crystal chemical systematics. One can use this to calculate an entropy and then to deduce whether this phase is likely to appear under certain conditions (for example, both high temperature and high pressure).

Vibrational calculations based on actual spectra have been quite successful for the $CaGeO_3$ polymorphs (Ross et al., 1986). Here, for illustrative purposes, the $MgSiO_3$ polymorphs (orthopyroxene, ilmenite, garnet, perovskite) are compared. For each of the phases, a representative vibrational model is shown in Fig. 7. The pyroxene is modelled by three optic continua; it could equally well be described by the first (lowest frequency) continuum and two Einstein oscillators, one near 700 cm^{-1}, one near 1000 cm^{-1}. The highest frequency bands correspond to tetrahedral Si-O stretching modes. Similar models have been described by Kieffer (1979). Kato and Kumazawa (1985) reported an infrared spectrum for $MgSiO_3$ garnet. The model in Fig. 7 is consistent with that data and with data and models for $Mg_3Al_2Si_3O_{12}$ garnet (pyrope) discussed by Kieffer (1979). Note that a smaller number of modes are located above 650 cm^{-1} (0.20 for garnet, 0.31 for pyroxene). This reflects the difference in structure (orthosilicate versus chain silicate) and the fact that in $MgSiO_3$ garnet half the silicon is in octahedral coordination. $MgSiO_3$ ilmenite has all its silicon in octahedral coordination. Its Raman spectrum (Ross and McMillan, 1984) shows no peaks below 300 cm^{-1}, consistent with a structure in which all octahedra are quite regular and strongly linked by edge-sharing and none above 800 cm^{-1}, consistent with the absence of tetrahedral silicon. A vibrational model consisting of a single optic continuum has been proposed by Ross (1985) and Ross and McMillan (1987) is shown here. A further constraint is provided in that this model fits Cp data at 175-550 K measured by DSC by Ashida (unpublished). For perovskite, vibrational spectra are currently unavailable though such measurements may appear shortly. A model for perovskite was invented as follows, based in part on analogy with $CaGeO_3$. A single optic continuum was chosen. Its low frequency cutoff was chosen such that it bore an analogous relation to the low frequency cutoffs for $CaGeO_3$ garnet and perovskite, i.e.

$(\nu_1)MgSiO_3(PV) = \nu_1 MgSiO_3(GT) \times \frac{\nu_1 CaGeO_3(PV)}{\nu_1 CaGeO_3(GT)}$. The high frequency cutoff was taken the same as for $MgSiO_3$(IL) corresponding to the absence of tetrahedral silicon.

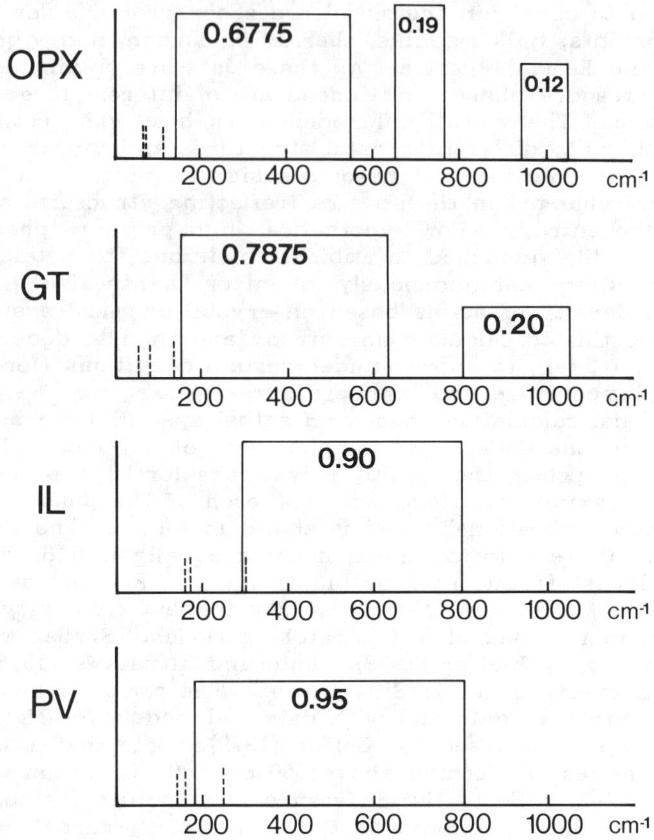

Fig. 7. Vibrational models for $MgSiO_3$ polymorphs. OPX = orthopyroxene, GT = garnet, IL = ilmenite, PV = perovskite. Dashed lines represent cutoff frequencies of Debye-like] acoustic modes. Solid "boxes" represent optic continua; the fraction of modes in each is indicated by a number.

The parameters used for the vibrational entropy calculations are summarized in Table 8. The acoustic modes, compressibilities, and thermal expansions were chosen to be consistent with available data and estimated when unavailable. C_V was calculated by statistical mechanics from the vibrational density of states (Kieffer, 1979; Ross et al., 1986). Anharmonicity effects are incorporated in converting C_V to C_p by Eq. (3). The calculated standard entropies of the phases involved are shown in Table 9, and entropies of transition in Table 10.

The calculated results suggest the following. Pyroxene is the phase of highest entropy, ilmenite of lowest, with garnet and perovskite intermediate. Above 1000 K, perovskite has a slightly higher entropy than garnet. The IL→PV transition has a definitely positive $\Delta S°$ and a negative dP/dT. Above 1000 K, $\Delta S°$ of phase transitions is virtually independent of temperature.

One should caution that the vibrational modelling used here has certain limitations. First, the vibrational density of states is only a very simplified approximation. Second, dispersion across the Brillouin zone is hard to predict, adding uncertainty to the location of the low frequency cutoff. Third, if some group of frequencies is very temperature dependent (i.e. soft modes), this cannot be readily incorporated. Thus distortional transitions in perovskite cannot be

Table 9. Entropies from Vibrational Calculations ($J\ K^{-1}\ mol^{-1}$)

	PX	GT	IL	PV
$S°_{298}$	64.78	62.35	51.72	60.44
$S°_{500}$	113.02	110.33	99.01	109.53
$S°_{1000}$	190.97	188.14	176.94	188.42
$S°_{1500}$	239.81	236.94	225.83	237.52
$S°_{2000}$	275.06	272.19	261.11	272.91

Table 10. Entropies of Transition ($J\ K^{-1} mol^{-1}$)

	PX→GT	GT→IL	IL→PV
298 K	-2.43	-10.63	+ 8.72
500	-2.69	-11.32	+10.52
1000	-2.83	-11.20	+11.48
1500	-2.87	-11.11	+11.69
2000	-2.87	-11.08	+11.80

modelled. Nevertheless, models like those above appear to be very useful in making semiquantitative predictions of heat capacities and entropies.

It is possible to construct a phase diagram for $MgSiO_3$ using entropies of transition consistent with these vibrational parameters.

Tentative values of enthalpies of transition, generally consistent with calorimetry, but in some cases a little different from published values, are given in Table 11, as are volumes of transition. Because this work is currently in progress (Ashida and Navrotsky, in preparation), these values should be considered as very provisional, illustrative and pedagogic in nature, and subject to revision. The phase diagram, calculated neglecting differences in thermal expansion, compressibility, and heat capacity, is shown in Fig. 8. This calculated diagram reproduces most of the essential topology of the experimental phase diagrams (see Akaogi et al., 1984; Ito and Navrotsky, 1985; Kato and Kumazawa, 1985). In particular, the transitions γ+ST = 2IL, β+ST = 2PV, and IL=PV have negative P-T slopes and garnet appears only as a very high temperature phase.

From the above calculations and from experimental observations,

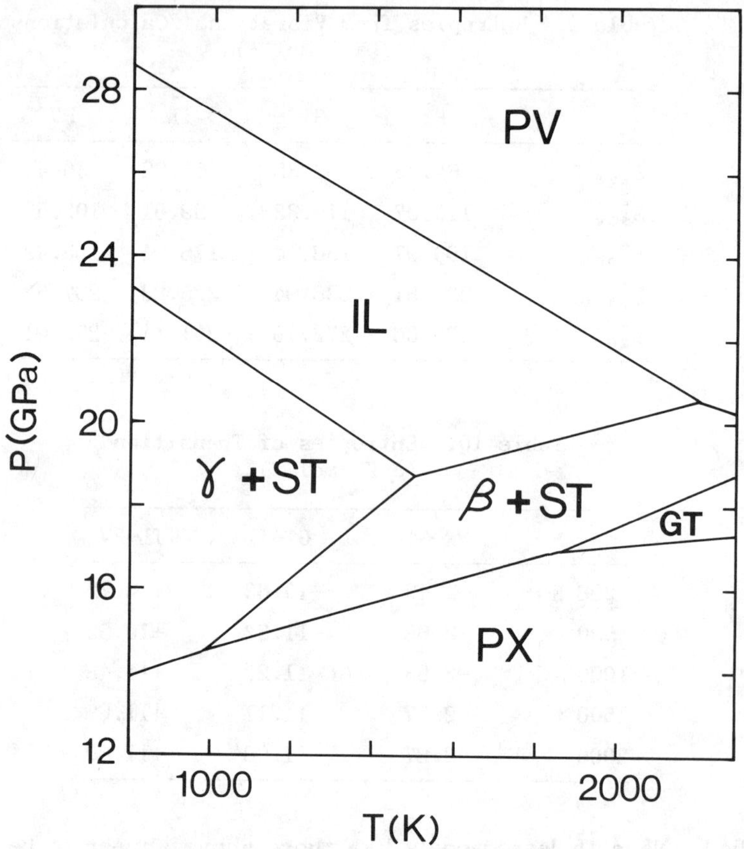

Fig. 8. $MgSiO_3$ phase diagram calculated using parameters in Table 11. PX = pyroxene, β = modified spinel, γ = spinel, ST = stishovite, GT = garnet, IL = ilmenite, PV = perovskite.

Table 11. Thermochemical Parameters Used in $MgSiO_3$ Phase Diagram Calculations - Very Provisional Values

Reaction	$\Delta H°$ ($J\ mol^{-1}$)	$\Delta S°$ ($J\ K^{-1}mol^{-1}$)	$\Delta V°$ (cm^3mol^{-1})
$\beta=\gamma$	6820	-9.33	-1.10
2PX=β+ST	93052	-23.56	-7.99
2PX=γ+ST	99872	-32.89	-9.09
PX=IL	61050	-14.04	-4.94
β+ST=2IL	29048	-4.51	-1.89
γ+ST=2IL	22228	+4.82	-0.79
PX=GT	46856	-2.86	-3.05
2GT=β+ST	46196	-20.70	-4.94
IL=PV	67896	+11.67	-2.02
β+ST=2PV	164840	+18.83	-5.93

one can draw the following general conclusions. For phase transitions involving no changes in cation coordination numbers, ΔS generally scales with ΔV and phase transitions have positive P-T slopes. The lower entropies of denser phases generally result from small shifts in the vibrational density of states, with the denser phase showing fewer low frequency modes. When a change in coordination occurs, much more pronounced changes occur in the vibrational spectrum. Since an increase in coordination number lengthens and weakens the bonds in the first coordination sphere (while increasing their number), this change generally implies a shift in the vibrational density of states toward lower frequencies. This shift is especially pronounced when tetrahedral Si-O bonds are converted to octahedral Si-O linkages. The result is that entropy-volume systematics break down, and the denser phase has a substantially higher entropy (at 298-2000 K) than otherwise expected and, in extreme cases, has an entropy higher than that of the lower pressure polymorph (e.g. the ilmenite-perovskite transition). Similar anomalies in the relation of other physical properties (elastic constants, thermal expansion, compressibility) to density may also occur in phase transitions involving coordination number changes. Such effects, though especially pronounced in silicates, can be expected to occur for many solid-state systems undergoing phase transitions with changes in nearest neighbor coordination numbers.

Acknowledgements

A. Navrotsky's work has been supported for the past 15 years by the Solid State Chemistry Program, Division of Materials Research, U.S. National Science Foundation.

References

Akaogi, M. and Navrotsky, A. (1984a) 'The quartz-coesite-stishovite transformations: new calorimetric measurements and calculation of phase diagrams.' *Phys. Earth Planet. Int.* **56**, 124-134.

Akaogi, M. and Navrotsky, A. (1984b) 'Calorimetric study of the stability of spinelloids in the system $NiAl_2O_4$-Ni_2SiO_4.' *Phys. Chem. Min.* **10**, 166-172.

Akaogi, M., Ross, N. L., McMillan, P. F. and Navrotsky, A. (1984) 'The Mg_2SiO_4 polymorphs (olivine, modified spinel, and spinel) - thermodynamic properties from oxide melt calorimetry, phase relations, and models of lattice vibrations.' *Amer. Mineral.* **69**, 499-512.

Capobianco, C. and Navrotsky, A. (1987) 'Solid solution thermodynamics in $CaCO_3$-$MnCO_3$.' *Amer. Mineral.* (in press).

Cemic, L. and Kleppa, O. J. (1986) 'High temperature calorimetry of sulfide systems. I. Thermochemistry of liquid and solid phases of Ni + S.' *Geochim. Cosmochim. Acta* **50**, 1633-1641.

Charlu, T. V. and Kleppa, O. J. (1973) 'High temperature combustion calorimetry, I. Enthalpy of formation of tungsten oxides.' *Jour. Chem. Thermo.* **5**, 325-330.

Chatillon-Colinet, C., Newton, R. C., Perkins, D. and Kleppa, O. J. (1983) 'Thermochemistry of (Fe^{2+},Mg) orthopyroxene.' *Geochim. Cosmochim. Acta* **47**, 1597-1603.

Clemens, J. D. and Navrotsky, A. (1987) 'Mixing properties of $NaAlSi_3O_8$ melt-H_2O: new calorimetric data and some geological implications.' *Jour. Geol.* (in press).

Clemens, J. D., Circone, S., Navrotsky, A. and McMillan, P. F. (1987) 'Phlogopite: high temperature solution calorimetry, thermodynamic properties, Al-Si and stacking disorder, and phase equilibria.' *Geochim. Cosmochim. Acta* (submitted).

Fraser, D. G. and Rammensee, R. H. (1983) 'The mixing properties of melts in the system $NaAlSi_2O_6$-$KAlSi_2O_6$ determined by Knudsen cell mass spectrometry.' *Bull. Min.* **106**, 111-117.

Hazen, R. M. and Finger, L. W. (1982) *Comparative Crystal Chemistry*, John Wiley and Sons, N.Y.

Hersh, L. S. and Kleppa, O. J. (1965) 'Enthalpies of mixing in some binary liquid halide mixtures.' *Jour. Chem. Phys.* **42**, 1309-1322.

Holm, J. L. and Kleppa, O. J. (1967) 'Thermochemistry of the liquid system lead (II) oxide - boron oxide at 800°.' *Inorg. Chem.* **6**, 645-648.

Ito, E. and Navrotsky, A. (1985) '$MgSiO_3$ ilmenite - calorimetry, phase equilibria, and decomposition at atmospheric pressure.' *Amer. Mineral.* **70**, 1020-1025.

Kato, T. and Kumazawa, M. (1985) 'Garnet phase of $MgSiO_3$ filling the

pyroxene-ilmenite gap at very high temperature.' *Nature* **316**, 803-805.
Kieffer, S. W. (1979) 'Thermodynamics and lattice vibrations of minerals, I-III.' *Rev. Geophys. Space Phys.* **17**, 1-19, 20-34, 35-59.
Kieffer, S. W. (1980) 'Thermodynamics and lattice vibration of minerals, IV.' *Rev. Geophys. Space Phys.* **18**, 862-886.
Kiseleva, I. A. and Ogorodova, L. P. (1984) 'High temperature solution calorimetry for determining the enthalpies of formation for hydroxyl-containing minerals such as talc and tremolite.' *Geochim. Internat.* **12**, 36-46.
Krupka, K. M., Robie, R. A., and Hemingway, B. S. (1979) 'High-temperature heat capacities of corundum, periclase, anorthite, $CaAl_2Si_2O_8$ glass, muscovite, pyrophyllite, $KAlSi_3O_8$ glass, grossular, and $NaAlSi_3O_8$ glass.' *Amer. Mineral.* **64**, 86-101.
Kybett, B. D. and Margrave, J. L. (1966) 'Two chambered combustion bomb for fluorine calorimetry.' *Rev. Sci. Instr.* **37**, 675-676.
Marucco, J. F., Gerdanian, P. and Dode, M. (1970) 'Determination des grandeurs molaires partielles de melange de l'oxygene dans le protoxyde de fer a 1075°C.' *Jour. Chimie Physique* **67**, 904-913, 914-916.
McMillan, P. and Ross, N. L. (1987) 'Heat capacity calculations for Al_2O_3 corundum and $MgSiO_3$ silicate. *Phys. Chem. Min.* **14**, 225-234.
Nafziger, R. and Muan, A. (1967) 'Equilibrium phase compositions and thermodynamic properties of olivines and pyroxenes in the system MgO-"FeO"-SiO_2.' *Amer. Mineral.* **52**, 1364-1385.
Navrotsky, A. (1977) 'Recent progress and new directions in high temperature calorimetry.' *Phys. Chem. Min.* **2**, 89-104.
Navrotsky, A. (1980) 'Lower mantle phase transitions may generally have negative pressure-temperature slopes.' *Geophys. Res. Lett.* **7**, 709-711.
Navrotsky, A. (1987a) 'Calorimetric study of melts, crystals, and glasses, especially in hydrous systems.' in *Magmatic Processes: Physiochemical Principles*, B. O. Mysen, Ed. The Geochemical Society Spec. Pub. No. 1, 411-422.
Navrotsky, A. (1987b) 'High pressure transitions in silicates and related compounds.' *Prog. Solid State Chem.* (in press).
Newton, R. C., Charlu, T. V., and Kleppa, O. J. (1977) 'Thermochemistry of high pressure garnets and clinopyroxenes in the system $CaO-MgO-Al_2O_3-SiO_2$.' *Geochim. Cosmochim. Acta* **41**, 369-377.
O'Keeffe, M. and Hyde, B. G. (1981) 'Why olivine transforms to spinel at high pressure.' *Nature* **293**, 727-728.
Orr, R. L. (1953) 'High temperature heat contents of magnesium orthosilicate and ferrous orthosilicate.' *Jour. Amer. Chem. Soc.* **75**, 528-529.
Robie, R. A. and Hemingway, B. S. (1972) 'Calorimeters for heat of solution and low temperature heat capacity measurements.' U. S. Geol. Survey Prof. Paper 755.
Robie, R. A., Hemingway, B. S. and Fisher, J. R. (1978) 'Thermodynamic properties of minerals and related substances at 298.15 K and 1 bar pressure and at higher temperatures.' U. S. Geol. Survey Bull. 1452.

Robie, R. A., Hemingway, B. S. and Takei, H. (1982) 'Heat capacities and entropies of Mg_2SiO_4, Mn_2SiO_4, and Co_2SiO_4 between 5 and 380 K.' *Amer. Mineral.* **67**, 470-482.

Ross, N. L. (1985) 'A thermochemical and lattice vibrational study of high pressure phase transitions in silicates and germanates.' Ph.D. Thesis, Arizona State Univ., Tempe, AZ 85287, U.S.A. (Geology Dept.).

Ross, N. L., and McMillan, P. (1984) 'The Raman spectrum of $MgSiO_3$ ilmenite.' *Amer. Mineral.* **69**, 719-721.

Ross, N. L., Akaogi, M., Navrotsky, A., Sasaki, J., and McMillan, P. (1986) 'Phase relations among the $CaGeO_3$ polymorphs (wollastonite, garnet, and perovskite structures): studies by high pressure synthesis, high temperature calorimetry, and vibrational spectroscopy and calculations.' *Jour. Geophys. Res.* **91**, 4685-4696.

Schmalzried, H. (1972) 'Two aspects of solid state thermodynamics at elevated temperatures: point defects and solid state galvanic cells.' *Ann. Rev. Mat. Sci.* **2**, 143-180.

Stebbins, J. F., Weill, D. F. and Carmichael, I. S. E. (1982) 'High temperature heat contents and heat capacities of liquids and glasses in the system $NaAlSi_3O_8$-$CaAl_2Si_2O_8$.' *Contrib. Min. Pet.* **80**, 276-284.

Suito, K. (1977) 'Phase relations of pure Mg_2SiO_4 up to 200 kilobars.' In *High Pressure Research - Applications to Geophysics*", M. H. Manghnani and S. Akimoto, Eds., Acad. Press, NY, pp. 255-266.

Topor, L. and Kleppa, O. J. (1983) 'Thermochemistry of binary liquid gold alloys: the systems Au-Ni, Au-Co, Au-Fe, and Au-Mn.' *Metal. Trans.* **15B**, 573-580.

Watanabe, H. (1982) 'Thermochemical properties of synthetic high-pressure compounds relevant to the earth's mantle.' In *High Pressure Research in Geophysics*, S. Akimoto and M. H. Manghnani, Eds., Center for Acad. Pub., Tokyo, Japan, p. 441-464.

Westrum, E. F. Jr., Furukawa, G. T., and McCullough, J. P. (1968) 'Adiabatic low temperature calorimetry.' In *Experimental Thermodynamics*, J. P. McCullough and D. W. Scott, Eds., Butterworths and Co., London, Vol. I, pp. 133-214.

HEAT CAPACITY OF SOLIDS

M.J. Tello and A. López-Echarri
Departamento de Física. Facultad de Ciencias. Universidad del Pais Vasco. Apdo. 644. Bilbao (Spain).

ABSTRACT. The importance of specific heat measurements is emphasized in relation with Solid-Solid phase transitions. The performances of two types of high accurate calorimeters are analized as well as different problems related with the automatization in both adiabatic and AC techniques. Some interesting examples such as lattice specific heat calculus, classical and non classical Landau systems, Order-disorder phenomena, critical fluctuations, etc, are used in order to stress the importance of this type of measurements.

I. INTRODUCTION

The first definition of heat capacity together with the first table of specific heat values were published by J.M. de Magellan[1] in a short work in the spring of 1780. J.M. de Magellan was a writer about scientific instruments and a disseminator of scientific information. The specific heat data in his table were supplied by Richard Kirwan who probably obtained the experimental data using the mixtures method. This first heat capacity definition in Magellan's paper was also probably due to Kirwan, who gave his own definition some years later[2]. This author was also the first to express consistently all heat capacities relative to water as.

Although the first written reference was the one published by Magellan, it is probable that some years earlier J. Black in his lectures at Glasgow made the first serious contributions to the science of heat, introducing concepts such as: heat capacity, latent heat, etc. An interesting analysis about the early history of specific heat is given by E.L. Scott[3].

From the beginning of the heat capacity history, a long way has led to the modern high precision experimental methods as well as to theoretical predictions in relation with microscopic models. We also know at present that the knowledge of the thermodynamic parameters of a material is absolutely necessary for both, scientific studies and technological applications.

In relation with the experimental measurements a question is open now: Is it better in the future to buy or to build one's own calorimeter?. This question has not a simple response. If we want standard measurements with a very good accuracy, the actual commercial continuous heating calorimeters such as differential scanning calorimeters are enough. The problems appear when one wants to use the

E. K. H. Salje (ed.), Physical Properties and Thermodynamic Behaviour of Minerals, 433–458.

experimental results from this apparatus in the same way than the results from an equilibrium calorimetry. In this lecture this point will become clear. In any case, actually both, commercial and hand made calorimeters have their place with different functions.

If we make reference to the equilibrium cryogenic calorimetry we find that ten years ago a specific heat curve in the 5K-300K temperature range could be the work of 20 or 30 days, depending on the equilibrium times in each experimental specific heat point. At present the automatization enables to load the calorimeter, press the button of the computer and pick up the disk with more information than that obtained by manual measurements. The automatization systems have made possible new advances in the classical calorimeters as well as the apparition of new techniques which allow to extend the precission of the heat capacity measurements to newer kinds of materials and to use different experimental conditions such as: Measurements under pressure, electric fields, uniaxial stress, etc. A list of some calorimetric methods as well as different reviews of this area can be found in references[4-9].

This lecture is divided into three parts. In the first we make some emphasis on the importance of the specific heat measurements. In the second we show two high accuracy measuring methods and in the third we show some interesting examples of specific heat results in solid state physics.

2. IMPORTANCE OF THE SPECIFIC HEAT MEASUREMENTS

Ruhemann[10] suggested that the high accuracy heat capacity measurements have more fundamental significance than any other investigation, at least at low temperatures. Without having this extremal position, we can show a lot of reasons that prove the importance of carrying out accurate specific heat measurements.

The theoretical description of the specific heat of solids plays an important role in the introduction of the quantum ideas. The first contribution was the Einstein's model in 1907, where the quantum harmonic oscillator is introduced with a unique vibration frequency corresponding to a characteristic temperature θ_E. In the following years different approaches were proposed: The continuous model of Debye in 1912, the use of several Einstein characteristic temperature values to achieve a better specific heat fit by Nernst and Lindemann in 1911, or the lattice dynamics approach by Born-Von Karman in 1912. See for example references 11 to 13. Within these primitive models the Born-Von Karman formalism allows a very accurate estimation of the lattice heat capacity; however the great number of experimental parameters (obtained from different spectroscopies) as well as the number of independent parameters involved, make this formalism quite impractical for the usual analyses of the specific heat data.

From these early years many attempts have been made to attain a reasonably simple, accurate and useful model of phonon distribution. With such a model we would get a more accurate analysis of the thermodynamic data in the phase transitions, magnetic systems, Schottky heat-capacities, etc., the prediction of the thermodynamic properties and the possibility to use the specific heat data to obtain distribution functions. For this purpose the main difficulty arises from the slow convergence of the basic Einstein function which represents the elemental contribution to the specific heat at fixed phonon frequency.This fact leads to great difficulties when the specific heat results are used to determine the phonon spectra[14]. It is observed that the specific heat curve is very insensitive to significant changes in the distribution function. This means that from the first step of this attempt, accurate experimental data are necessary for a successful result.

Some progress has been established through different manipulations of the original specific heat data: the use of high order derivatives of the C_v curve, which show a better convergence at high temperatures[15]; the expansion of the distribution function as a Fourier series[16], schematic and simple trial spectra to fit the specific heat data[17] and the use of the frequency moments of the phonon spectra[18] did not attempt to surmount the remaining basic slow convergence of the derived functions. In the following we briefly show three recent models which present a practical interest.

The first model due to Kieffer[19] was principally used in mineralogy. In this model the distribution function is divided into three frequency groups. The first group comprises the acoustical modes, the second is made of densely packed optical modes and the third group is formed by those optical phonons with frequencies lying outside the last group. The phonon distribution of the acoustical modes is obtained using a sinusoidal dispersion relation for a one-dimensional lattice to the three-dimensional vibrations. In this case the acoustic data determine three cut-off frequencies. The first group of optical modes is approximated by a constant phonon density using infrarred and Raman experimental data for the low-end and high-end frequencies. The second group of optical modes (usually belonging to internal branches) are approximated by a Dirac δ-function. Two important problems are present in this interesting approach: The first is that we need a lot of elastic and spectroscopic data in order to obtain the predicted heat capacity. Whith this method it is almost impossible to fit the experimental heat capacity curve, but a comparison between predicted and calculated specific heats is possible instead. In this way very accurate specific heat data are absolutely necessary if we want to obtain some interesting results out of this comparison. The adiabatic calorimetry becomes an important experimental technique for these purposes.

The most recent single parameter model for a phonon distribution (see reference 20 and references cited therein) is characterized by a single characteristic temperature as fitting parameter. The two transverse acoustical modes are considered to be degenerated with the same dispersion relation than the logitudinal mode. The (3nZ-3) optical modes are also divide into (2nZ-2) transverse modes and (nZ-1) longitudinal ones (n= number of atoms in a formula unit and Z= number of formula units in a primitive cell). Evidently these assumptions strictly hold only for phonons which propagate in highly symmetric directions. The frequency range shows a maximun value for the acoustical branch at the edge of the first Brillouin zone ($w_m = (2k/\mathbf{M})^{-1/2}$ in the average mass approximation). In the same approximation the maximum frequency for optical modes can be estimated as $w_m=(D/d)w_m$, where D and d are magnitudes related with the primitive cell size ($\sim V^{1/3}$) and the smallest interaction distance respectively. If X-ray data are not available the smallest sum of the ionic radii may be a good approximation.

The transverse and longitudinal optical modes will be distributed in a first approximation from w_{mT} to $(V^{1/3}/d)w_{mT}$ and from w_{mL} to $(V^{1/3}/d)w_{mL}$ respectively. This frequency range is too narrow but can be elongated using correction factors which take into account the influence in both, lowest and highest frequencies of the heaviest and lightest atoms respectively. The form of the distribution function is assumed to follow the Debye behaviour under the continuous approximation with the incorporation of the discreteness given by the atomic arrangement in a real crystal. Finally the internal mode frequencies are considered the same as in the free ion state. Dirac δ-functions at these frequencies (for example: OH^-, NH_4, etc) represent the phonon density for the internal modes. Of course, a better fit would be obtained if the internal frequencies were measured in the crystal. It must be noticed that if we have extremely small or large force constants in the crystal, related with external modes, their frequency can also be represented by a Dirac δ-function.

The molar lattice specific heat at constant volume is therefore, in the harmonic approximation, the sum of five contributions: Transversal and longitudinal optical and acoustical modes respectively as well as the internal modes contribution.. A general and schematic distribution function for this model is shown in figure 1. In this figure w_i accounts for the degeneration of each internal mode, θ is the characteristic temperature of this model which is used as a reference parameter in the formulation of the molar isometric heat capacity, and K_B the Boltzmann constant.

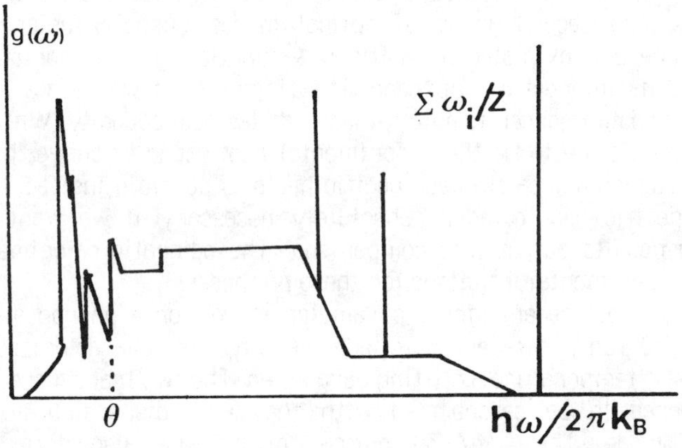

Figure 1. Schematic distribution function according to ref. 20

The model was applied to different minerals. Figure 2 shows the behaviour of both, apparent characteristic temperature and the Debye one, for a scapolite sample[20] in the temperature range from 8K to 350K. This figure shows that the apparent θ-T curve is insensitive to temperature, which provides an interesting method to extrapolate lattice specific heat into an unmeasured temperature region. Other applications to the resolution of broad transitions in deerite, grunerite, disilicate crystals, etc, are announced in the same reference.

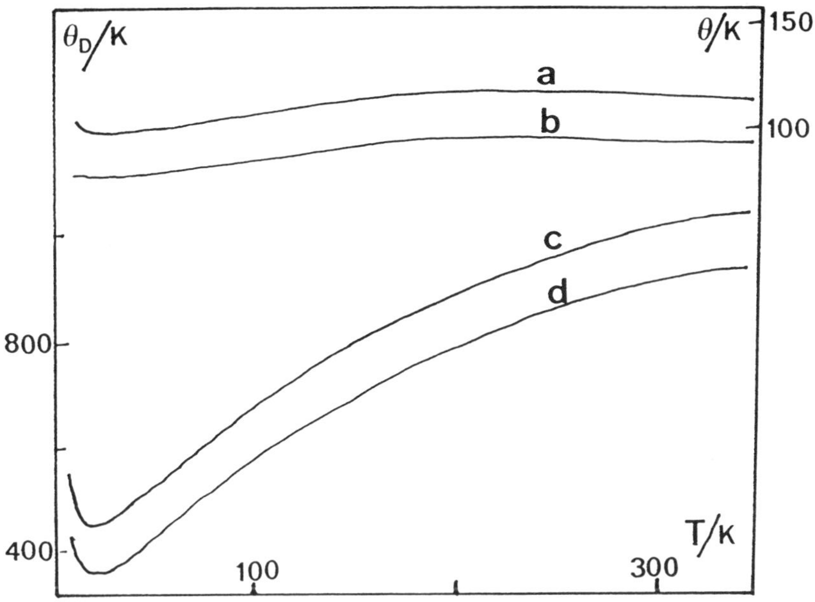

Figure 2. Temperature dependence of the characteristic temperature (a and b) and the Debye temperature (c and d) for two scapolite samples[20].

A recent and very promising approach to this problem has been recently published[21]. A different specific heat transformation which shows a convenient convergence at high temperatures is proposed and has been sucessfully applied to some simple physical situations. The obtained phonon spectra is claimed to be similar to that found by incoherent inelastic neutron scattering. Progress in the computed data processing also with the continuous accuracy increase of the original specific heat data would improve the initial success of this technique. In any case, it seems that only the simplest solid systems can be discussed within this framework. Resolution around the abrupt peaks of the distribution function is really poor although previous and/or incomplete knowledge of some details of the phonon spectra can drastically change this situation.

As a result of this brief review of the various techniques described above we can conclude that at present important problems still remain unsolved and lie on the following points:

a) The present accuracy of the specific heat measurements should be highly improved. This is not a simple task attainable in the near future, mainly due to the peculiar characteristics of this physical magnitude. If for example the present experimental error could be reduced to 0.01% (in the best case one order of magnitude better than the actually estimated) the phonon spectra shape could be reasonably resolved. We cannot forget the univoque relation between the specific heat and the solid frequency distribution, which reduces the problem to a mathematical exercise when the quality of the experimental data is guaranteed.

b) The anharmonicity contribution adds another difficulty to the correct estimation of the specific heat at constant volume. The knowledge of both, elastic constants and thermal expansion coefficients as a function of temperature is necessary in order to find a pure harmonic specific heat contribution. In some cases the Nernst-Lindemann relation is successfully used. The electronic contribution is only relevant at low temperatures and in any case less important than the preceding one. However, the technique described by Loram[21], seems to account for an indirect determination of these unpleasant contributions to the total specific heat.

Actually, the knowledge of experimental specific heat curves is absolutely necessary to characterize the experimental behaviour of any physical system. On the other side we know at present the sensibility of the specific heat data in order to check the theoretical models which describe the behaviour of the real systems around the phase transitions. In this way it is interesting to notice the sensibility of the specific heat measurements in the study of the critical behaviour of such systems. Throughout this lecture we show some examples related with these points.

3. EXPERIMENTAL SYSTEMS

3.1. Adiabatic Calorimetry

3.1.1. Introduction

This calorimetric technique is considered as one of the most accurate methods for specific heat measurements. The remarkable accuracy of these types of calorimeters lies fundamentally on two special characteristics of the experimental method: a) measurements are made between equilibrium states and b) the heat leaks are carefully minimized. This last feature highly improves the experimental results when long thermal equilibrium times are present. For this reason, adiabatic calorimetry has been widely used in high precision studies of solid-solid phase transitions. The temperature range which can be conveniently covered by adiabatic calorimeters goes from the liquid helium range up to 800 K. However, serious limitations appear at high temperatures where the thermal radiation becomes the main source of errors.

For measurements below room temperature the most commonly described cryostats are derived from the one described by Nernst[22]. Along the century and specially in the last three decades, continuous improvements of the electronics and automatization have enormously increased the accuracy and resolution of this type of calorimeters. A typical experimental precision of 0.1% is commonly attained below room temperature at this moment.

3.1.2. Calorimeter

A detailed and wide review on theoretical and practical aspects of adiabatic calorimetry can be found in reference 4. The general features of the cryostats as well as the measuring techniques described therein continue being valid at present. The fundamental changes are basically related with the development of the instrumentation and with the automatic data acquisition and control. A typical scheme of a modern automatic adiabatic calorimeter is shown in figure 3. In the following we refer only to some technical aspects which are more interesting from an experimental point of view.

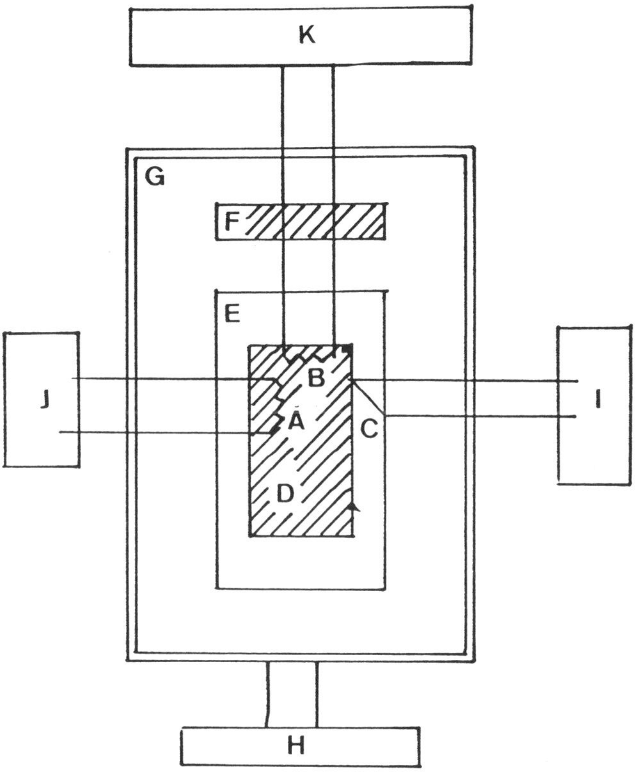

Figure 3. Schematic representation of an adiabatic calorimeter. A=Pt Thermometer; B= heater; C= thermocouples; D= sample vassel; E= adiabatic shield; F= ring; G= cryostat; H= vacuum system; I= adiabatic control circuit (4-chanals); J= temperature measuring circuit and K= Energy supply and measuring.

a) **Sample uessel** includes the thermometers and heater as well as the solid or liquid sample. The sample vessel size and consequently the sample weight required are principally determined by the temperature sensor size, usually a Pt-resistence thermometer. In the last years the availability of miniature platinum thermometers have allowed to work with smaller sample weights (1gr. or less). In any case, together with the sample the vessel is filled with low pressure helium gas (10 cm Hg) in order to reduce the equilibrium times. The thermal isolation of the vessel is provided by a high vacuum ($\sim 10^{-6}$ mm Hg). Evidently a small heat conduction along the signal wires cannot be completely avoided.

b) The **Adiabatic shield** is a temperature controlled shield which is the most characteristic element in the adiabatic calorimetry. The sample vassel is inside this adiabatic shield and its temperature is mantained, in every moment, at the same value of that of the external surface of the sample vessel. With a fine temperature control between the shield and the vessel the thermal leaks by radiation of the sample are minimized. The accuracy of this experimental technique depend on this control. The maximum efficiency is achieved in the low temperature region. For this temperature control, PID amplifiers have been extensively used in the last decades. With these amplifiers a typical thermal imbalance of 10^{-3}K between the sample and the shield is usually attained during the equilibrium periods as well as at the heating pulses.

The adiabatic shield is surrounded by a set of different non controlled or "floating" shields which reduce the thermal exchange between the calorimeter and the external medium. These shields have a great importance in high temperature regions where two temperature controlled shields (adiabatic shields) are recommended.

3.1.3. Automation

The resolution and accuracy of the adiabatic calorimetry are mainly limited by the design and instrumentation employed in each case, but other aspects such as the increasing amount of experimental data, the adequacy of the system control, the time spent in manual operation, etc, are greatly facilitated by a small computer. The automatic data acquisition also permits experimental studies which are unaccessible to the manual mode (i.e: thermal relaxation, etc).

The detailed characteristics of each particular automatic system are strongly dependent on both, the peculiarities of design of the experimental system where the automatization must be implemented and the preferences of the experimentalist. In the last few years some different automatic calorimeters have been described in the literature. References 23 to 26 are some representative examples of different automatization ways. In the following, as an example, we briefly show some special characteristics as well as different measuring possibilities of one of such models[26].

The temperature of the sample cell is measured by a Pt-resistance thermometer (25 Ω nominal) using a dc-voltmeter with a sensibility of 0.1 μV, which represents 0.001K at room temperature. The signal is directly driven to the voltmeter because any other intermediate system would introduce noise higher than the signal to be measured. For example a multiplexer introduces up to 3 μV of noise. Each temperature measurement is made in the following way: five sequencially independent resistance measurements are performed with a 2 sec. interval and immediately converted into temperature units by the computer program. The actual cell temperature is the average among these values and is assigned to the mean time of this set of measurements. With this procedure we eliminate the noise of the dc-voltimeter last digit and we observe an important improvement of the sensibility. The experimental specific heat data can be obtained using two different experimental methods: pulse method and thermogram technique.

a) In the **pulse method** a continuous test of the sample cell temperature is realized during the equilibrium periods. Once the temperature drift drops to the imposed value (tipically 10^{-6}K/sec), the thermal equilibrium is considered to be obtained and a heating period starts. Power input is supplied by a programmable digital-analogic converter and continuous measurements of the heating current (each 20 sec. approx) are made during the whole heating period. The adiabatic shield control is achieved by three independent PID temperature controls which allow for a maximum temperature imbalance of 1mK. After each equilibrium period, the sample specific heat is determined by the integrated power input and by the corresponding temperature step. The temperature is determined by the temperature drifts before and after the heating pulse extrapolated to the mean time of the heating period. The program allows the molar specific heat values as well as all the relevant data to be preserved in a diskette. In any moment a graphic program permits to plot the measured specific points in a window, as well as more complementary useful information without interferring the measurement operation. A simple scheme of this measuring method is showed in the figure 4.

b) In the **thermogram technique** a continuous cooling or heating process is used to measure the specific heat of the sample[27]. This experiment is recommended when solid-solid phase transitions take place. The resolution of the specific heat anomalies is greatly improved and a lot of significant points are easily obtained by the computer. In this case we perform a continuous measurement of the sample temperature as a function of time with a control of the heating source in every moment.

c)**Thermal relaxation** studies are another possibility of an automatic adiabatic calorimeter. In this case we plot the temperature as a function of time after a heating pulse. After our experience the experimental data fit adequately as follows:

$$T = T_{eq} + Ae^{-t/\tau_1} + Be^{-t/\tau_2} \qquad (1)$$

These two exponential terms are related with the thermal conductivities of the sample vessel (high) and sample (usually low) respectively. When the sample undergoes some kind of thermal anomaly which produces a change in its thermal conductivity, one observes that τ_2 accounts for those anomalies whereas τ_1 is insensitive to this change. In this way we can say that τ_2 is an indirect measurement of the thermal conductivity of the sample. Although absolute measurements of this magnitude are not allowed by this procedure, we find that the automatic adiabatic calorimetry can provide interesting information about the relative evolution of the thermal conductivity as a function of temperature.

3.2. AC Calorimetry

It is a new high resolution method developped in the last twenty years. The main features of this new technique make it possible to work with extremely small samples (down to 10 mg) and also to avoid problems involved with the sample adiabatic isolation. For these reasons, the temperature range of this AC calorimetry

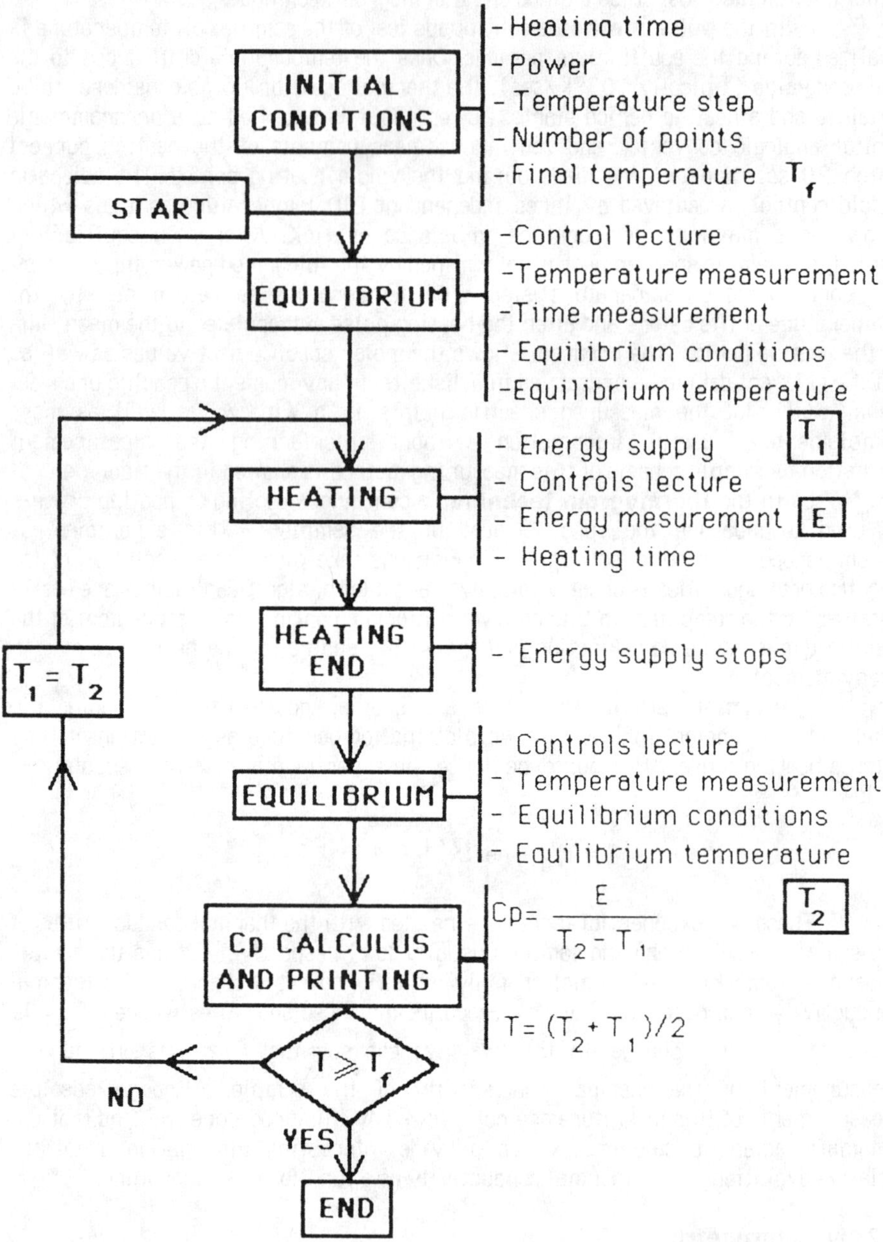

Figure 4. Simple scheme of the automatic measuring method.

can be very wide (Specific heat data along the range 1K to 1500K have been reported). The resolution of this experimental technique around the specific heat anomalies has been widely demonstrated since its first applications in phase transition studies and critical exponent determinations[28-30].

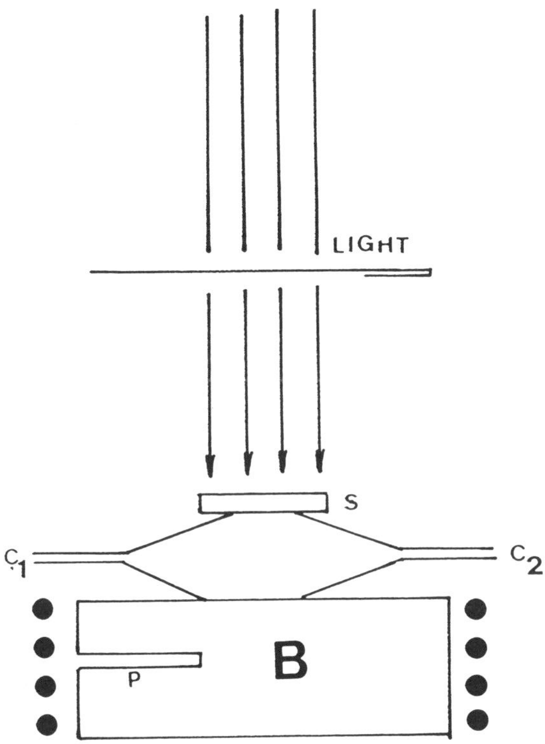

Figure 5. Schematic representation of an AC calorimeter

A typical AC calorimetric set up is shown in figure 5(More detailed descriptions can be seen in references 31-41). The slab shaped sample **S** (which can be a solid or a liquid specimen) of depth **h** and area **a** is heated on one side with a periodic source **P** (chopped light or electrical heater). Two independent thermocouples c_1 and c_2 are attached to the opposite face of the sample and to a temperature controlled copper block **B** which acts as an isothermal heat reservoir.The temperature of this block (T_B) can be measured either by a thermocouple or by a Pt resistance thermometer (**p**). Consequently, the actual sample temperature is:

$$T_S = T_{DC} + T_B \tag{2}$$

where T_{DC} is the mean temperature difference between the sample and the block measured by the thermocouple c_1 and the nanovoltmeter U. The periodic heating source produces a thermal oscillation T_{AC} between the sample and the block. The amplitude and phase of this signal is measured by means of the thermocouple c_2 using a lock-in amplifier (L) with the heating source frequency as reference signal.

If the heating source power is given by:

$$P = P_0 \cos(w/2)t \tag{3}$$

the amplitude of the thermal response is given by the following expression:

$$T_{AC} = \frac{P_0}{2wC}\left[1 + \frac{1}{w^2\tau_1^2} + w^2\tau_2^2 + \frac{2K_b}{3K_s}\right]^{-1/2} \tag{4}$$

where C is the specific heat of the sample plus adenda, τ_1 is the thermal relaxation time between the sample and the reservoir, τ_2 the thermal relaxation time of the sample plus adenda and K_b and K_s the thermal conductances between sample and block and the sample itself respectively. Typical values of this set of parameters make the last term under the square root in equation (4) negligible. In addition, an appropriate choice of the working frequency may satisfy the condition:

$$w\tau_1 \ll 1 \ll w\tau_2^2 \tag{5}$$

In this case, equation (3) takes the simpler form:

$$T_{AC} = \frac{P_0}{2wC} \tag{6}$$

which permits a direct determination of the sample specific heat.

The heating of the sample is actually accomplished by means of an electrical heater or chopped light. In the first case we have serious problems when the sample size is very small and in any case this heater increases the adenda contribution to the specific heat. Another difficulty with this heating source appears when it works at low frequencies ($w<2Hz$)[42]. In this case the sensibility of this experimental technique is perturbed by the presence of second harmonic components due to the non ideality of the sinusoidal heating signal. In this case the use of a square wave signal avoids most part of these problems[42]. A square wave heating signal is employed, in a natural way, when chopped light is used as heating source. This procedure simplifies the sample holder but recquires the depot of an absorbing material (evaporated carbon or similar) over the sample in order to facilitate the heat transfer to it.

We shall now briefly discuss about condition (5) which permits to simplify equation (4) and therefore to remove a lot of parameters in the final specific heat calculation. For excessively thick samples or/and with very poor thermal conductivity, the optimum work frequency in order to satisfy (5) can be very low. In this case problems may arise with the one Hz limit of the lock-in amplifier. In any case, a previous signal amplification by means of a low noise transformer which also matches the impedance between the AC thermocouple signal and the lock-in amplifier is recommended. With frequency values as low as 0.1-0.01 Hz a fast microvoltmeter with automatic data acquisition will succesfully substitute the lock-in amplifier as signal detector.

In the following we show some typical values: Dielectric crystals between 0.1 to 0.5 mm thick usually attain the optimum frequency around 1 Hz. Very thin films (around 0.01 mm) of very good thermal conductors (metals, alloys, etc) have optimum work frequencies up to 50 Hz. In the opposite case, very poor thermal conductors or too high sample thickness can only satisfy equation (5) in a very low frequency range. This is the case of liquid crystals.

A priori theoretical calculation of the frequency range in which equation (5) is accomplished could be a difficult task. It needs the previous knowledge of the thermal relaxation times τ_1 and τ_2 which are also temperature dependent. Fortunately two different experimental techniques can be used to determine the optimum frequency range. The first one is based on the linear behaviour of $\mathbf{T}_{AC}$ vs. w at constant C (equation (6)). In this case a previous and tedious work is necessary along the whole temperature interval in order to establish the appropriate frequency range from the plot of the last mentioned function. The second and more direct method uses another point of view. When the condition given by equation (5) is attained, a phase shift of $\pi/4$ between the period of the heating source and $\mathbf{T}_{AC}$ takes place. This situation can be easily checked with the lock-in amplifier. Figure 6 shows this phase shift when the sample thermal relaxation time is very much faster than the thermal relaxation time between the sample and the heater reservoir. In this situation the sample is always in thermal equilibrium.

Specific heat measurements can be performed in two different ways. The first one is a $\mathbf{T}_{AC}$ average over several periods at each fixed temperature. This experimental situation can be achieved with a programmable temperature controller (PID or similar) which stabilizes in sequential steps the copper block temperature before each measurement. The step of the DC temperature increments can be as low as necessary because it does not affect the specific heat determination. This characteristic establishes another interesting difference against adiabatic calorimetry for which small temperature increments always increase the noise to signal ratio and explains the good specific heat resolution of the AC calorimetric technique. In any case, a practical limit is attained when the step size approximates the T_{AC} values (2-10 mK). The second procedure permits a direct plot of the continuously variable block temperature against T_{AC}. This can be easily done with an X-Y recorder or computer. Temperature rates as low as 0.1K/h have been reported in the literature. Both experimental measuring methods can be used either on heating or cooling sequences with the same experimental resolution. This is an additional advantage against other calorimetric techniques (i.e.:adiabatic calorimetry).

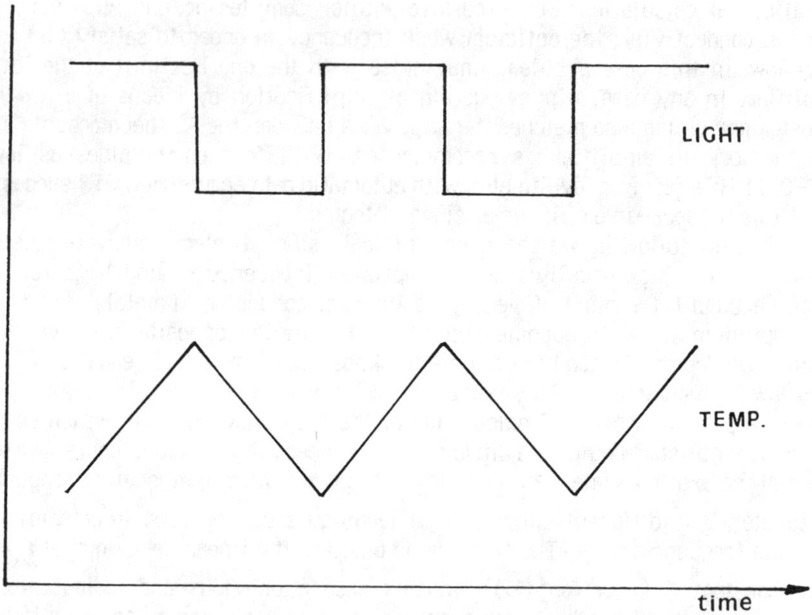

Figure 6. Phase shift when the sample thermal relaxation time is very much faster than the thermal relaxation time between the sample and the heater reservoir.

An important difference between the AC calorimetry and the adiabatic one is that the first is not constrained to the adiabatic isolation. This circumstance makes the AC calorimetry very useful as an accurate experimental technique for specific heat measurements under external fields. Measurements under electric or magnetic fields, pressure, etc have been reported. Figures 7 shows an example of this type of measurements under an electric field. In the case of uniaxial or hydrostatic pressure this technique allows to determine not only the pressure-temperature diagrams, but also to characterize the phase transition thermodynamics when pressure is applied. Indeed, it can be employed to obtain the equation of state of solids and provides interesting information about the influence of pressure on the vibration spectra.

The main disadvantage of the AC calorimetry lies on the difficulty to estimate absolute values of specific heats. Although a careful calibration of the adenda contribution should allow to obtain the sample specific heat, absolute

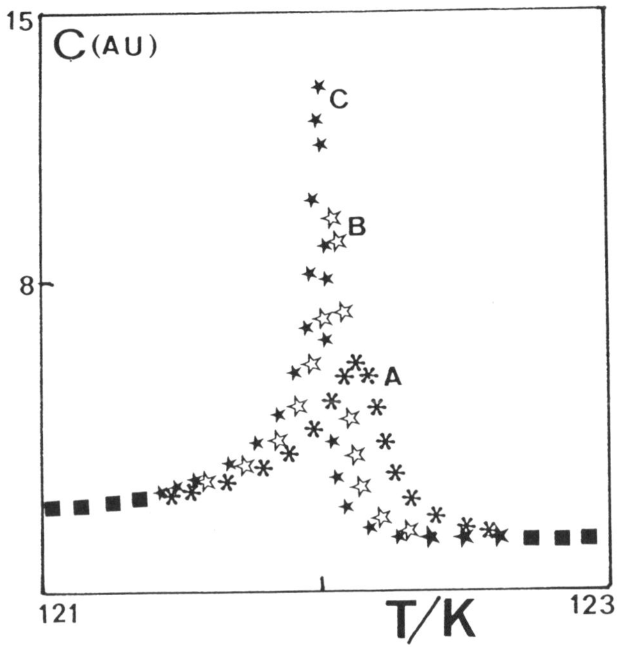

Figure 7. Specific heat measurement of KDP under electric field[77]. A (E= 4.77 esu), B (E= 2.40 esu) and C (E= 1.20 esu).

accuracy better than 1% seems to be unpracticable, mostly due to the size of the sample and to the unpredictible adenda variation in different measurements. It should be finally noted that a direct measurement of the enthalpy is not allowed. As a consequence, latent heats of first order phase transitions cannot be directly determined.

In spite of these important inconveniences, the interest of the AC calorimetric technique has increased in the last years. It has been succesfully applied in the field of the solid-solid phase transitions as the relative resolution (better than 0.1 %) allows to test the validity of the phase transition models and to study specific heat critical behaviours. Studies on ferroelectric or antiferroelectric crystals, magnetic and structural phase transitions as well as on metal and alloys have been published. A brief review on these topics can be found in reference 43. Finally we should notice the interesting application of the AC calorimetry on the thermodynamic study of the liquid crystal phase transitions even with applied pressure. For this purpose special miniature sample cells should be designed.

4. EXPERIMENTAL RESULTS

One of the most interesting applications of both, adiabatic and AC caloriemtric techniques is the study of the anomalies of the specific heat in the phase transitions. During the last three decades much experimental work was carried out with adiabatic calorimetry and more recently with the AC-calorimeters. This experimental technique provides experimental data on the critical behaviour much more accurately than those which were previously available. With this new experimental results a lot of old problems become interesting again because of the discrepancies between the theoretical predictions and these experimental results. In the following we show some examples which were mainly chosen to reinforce the importance of the accurate specific heat measurements in the study of phase transitions in solids.

4.1. Solid to Solid Phase Transitions within the Landau Framework

The experimental results of a great number of crystals that undergo solid to solid phase transitions can be described, almost in a first approximation, by the phenomenological Landau Theory. In the following we show some examples related with specific heat measurements carried out using adiabatic or AC calorimeters.

In the Landau theory of phase transitions the free energy is expanded in powers of the order parameter. In our discussion we consider a usual and general expresion given by:

$$F = F_0 + (a/2)(T-T_0)\mathbf{q}^2 + (b/4)\mathbf{q}^4 + (c/6)\mathbf{q}^6 \qquad (7)$$

where in a first approximation a, b and c are weakly temperature dependent parameters and $\mathbf{q}$ is the order parameter. Discussions about order parameter dimensionality, symmetry, as well as the thermodynamic constraints of the free energy expansion, can be seen in references 44 to 47. Equation (7) accounts for both, first ($b<0$) and second ($b>0$) order phase transitions respectively; therefore T_0 has a different meaning in both cases. The other two coefficients are positive in any case. In the particular case of $b=0$, equation (7) accounts for tricritical points.

By minimising the free energy with respect to the order parameter and using the standard thermodynamic calculus we obtain the temperature dependence of the thermodynamic magnitudes. The final expressions are given by:

$$\Delta S = -a\mathbf{q}^2 \qquad (8)$$

and

$$\Delta C_E = \begin{cases} T(a^{3/2}/4c^{1/2})(T_d-T)^{-1/2} & (T<T_c) \\ 0 & (T>T_c) \end{cases} \quad (9)$$

where T_c is the temperature for which the experimental specific heat curve has the maximum value and T_d is the metaestability limit on cooling. If we name T_0 the metaestability limit on heating the theory gives:

$$T_d-T_0 = b^2/4ac \quad (10)$$

On the other side the theory shows that T_d and T_0 temperatures are related with T_c as follows: $T_0=T_c$ for second order phase transitions ($b>0$) and $T_c=T_0+3b^2/16ac$ for first order ones ($b<0$). Evidently in the tricritical points $T_d=T_c=T_0$.

Equation (9) gives all specific heat contributions which exceed the normal lattice contribution (for example, magnetic, ferroelectric, etc.). Therefore the fit of the measured specific heat must be made using the following equation:

$$C = C_L + \Delta C^{\varphi}_{E\epsilon} \quad (5)$$

where C_L is the lattice contribution (we see here an example where a very good determination of the lattice contribution becomes important) and $\Delta C^{\varphi}_{E\epsilon}$ is given by equation (9). φ and ϵ show the thermodynamic constraints.

If the experimental excess specific heat below T_c shows much higher values than above T_c, we can look for equation (9) in order to describe the main contribution of the excess specific heat in the low temperature phase. Evidently a plot $(T/\Delta C_E)^2$ vs T must be a straight line for $T<T_c$. However in many experimental systems additional contributions with different origins appear which make the curve bend off from the Landau line near T_c. These additional contributions can be caused by fluctuation effects[48,49] and/or smearing of T_c, etc., and can be written as:

$$\Delta C_A \sim M|t|^{-\alpha} \quad (12)$$

where $t = (T-T_c)/T_c$ and is an effective exponent including corrections to scaling. In the case of critical fluctuations this temperature dependence is nearly symmetric around T_c.

Finally we must cite that in the last years evidence of logarithmic corrections in uniaxial ferroelectric and ferromagnetic systems has been obtained by different experimental techniques. For these systems the Larkin-Khmel'nitskii

(LK)[50] theory predicts a cross-over region where the logarithmic contribution becomes important. The size of this transition region, where the fluctuations are the most important contribution, was calculated for different uniaxial systems[51,52] with values which go from 0.5K for $LiTbF_4$ to 10^{-3} for TGS. It must be noticed than most part of the characteristic parameters involved in the model are very sensitive to the non-critical background and therefore to the accuracy of the experimental measurements, purity of the sample and quality of the crystal .In this case the AC calorimetry is an important experimental technique.

In the following we show some experimental examples of phase transitions which can be described within the Landau framework. It must be noticed that the choice of these examples is completely arbitrary throghout a great number of papers in the literature.

4.1.1. Uniaxial systems:

Among the uniaxial ferroelectric crystals we choose two compounds which exhibit a ferro-paraelectric second order phase transition: Triglycine sulphate (TGS) and Tris-sarcosine calcium chloride (TSSC). In both compounds the first measurements (refs 53 and 54for TGS and refs 55 and 56 for TSSC) show that the transition obeys a simple Landau theory although announcement of deviations from the classical behaviour nearer the transition temperature were made. More accurate specific heat measurements confirm that both, TGS[54] and TSSC[52] fit the Landau predictions logarithmically corrected in the $10^{-2} < |t| < 10^{-5}$ temperature range. In this case, adiabatic calorimetry was used for TSSC and AC calorimetry for TGS. Figure 8 shows the typical experimental curves with the classical Landau predictions as a continuous line. In this case we must remark that the specific heat measurements allow the observation of this region closer to the transition temperature with no external field perturbation. In this way accurate specific heat experimental data clarify discrepancies derived from the experimental technique employed.

SbSI is also an uniaxial system with a para-ferroelectric phase transition showing a tricritical point as detected by non calorimetric experimental techniques[57]. The character of this phase transition changes from first to second order when the hydrostatic pressure is applied. Accurate specific heat measurements[58] (AC-calorimetry) confirm the first order character of this phase transition. The experimental data can be fit below T_c using equation (9) with $b > 0$. This fit was also carried out with a free exponent for $(T_d - T)$ in equation (9). The mean exponent value between those found in heating and cooling specific heat curves were in complete agreement with the Landau prediction (equation 9)[58]. Above T_c the experimental specific heat shows a short tail which was fit using a power law (equation 12) with $0,8 < \alpha < 1.5$ depending on the sample history. These high values cannot be explained using the critical exponent predictions of the theoretical models and were attributed to sample impurities[58]. In any case critical fluctuations seem to be present also in this case, however their contributions seem to be much lower than those due to defects or impurities.

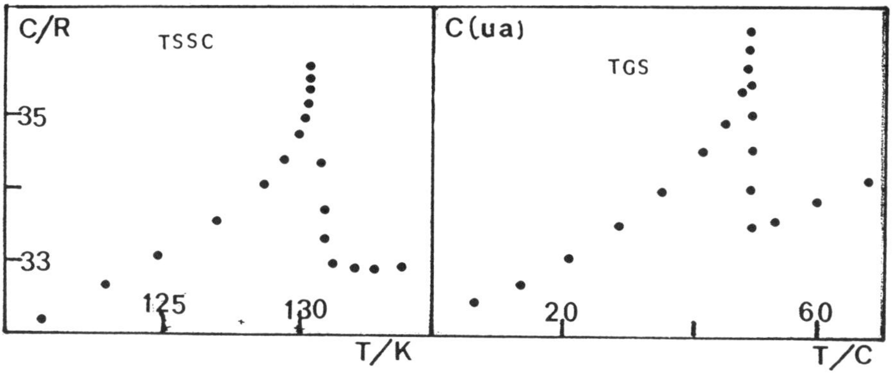

Figure 8. Specific heat measurements of TSSC[52] and TGS[54].

Among the compounds discussed here, the dipolar Ising ferromagnet $LiTbF_4$ shows a great interest . This is also a uniaxial system (spin dimensionality n=1) with dipolar interactions, and therefore the Landau specific heat near T_C must be affected by logarithmic corrections in agreement with the LK-predictions. Experimental specific heat measurements[59] carried out with a high precision calorimeter[60] show again that is easier to obtain quantitative results for the logarithmic contributions with this experimental technique than other properties with strong leading singularities such as the susceptibility or the magnetization. It must be noticed that in this case T_C=2.885K and that 18 experimental points were measured in the $0<t<10^{-3}$ temperature interval.

The experimental points near T_C can be fit by means of a power law as (12) at both sides of the transition temperature. Within a reasonable range of t ($10^{-2}>t>10^{-3}$) the value $\alpha=\alpha'=-0.15$ was obtained. This value implies Heisenberg-like behaviour with n=3, whereas the $LiTbF_4$ is uniaxial with n=1. This example shows how the power law fit must be used with a simultaneous analysis of the microscopic characteristics of the system if we want to find any physical meaning to the found exponent values.

4.1.2. **Structural phase transitions**:

A great number of compounds exhibit one or more structural phase transitions in the temperature range from 0K to the fusion or decomposition temperature. A nearly complete classification can be seen in references 61 and 62 and in the references cited therein. In this case the analysis of the phase transitions is made using the two phase space groups in order to find the order parameter symmetry. We show here three examples where accurate specific heat measurements cannot fit using only equation (9).

Bis-penthylammonium tetrachlorozincate exhibits a Pnma $-P2_12_12_1$ first order phase transition at 234.3K. The experimental measurements of the macroscopic order parameter (gyrotropic tensor component) can be fit using the Landau theory[63] but the specific heat data cannot be fit using equation (9). This is a typical case where a combination between equations (9) and (12) can be useful to fit the experimental results (figure 9). One should notice the rounding around T_c. In this case the high specific heat peak value together with the high value of the α exponent are probably related with a coupled order-disorder mechanism of the organic chains due to hindered rotations of the NH_3 groups among different equilibrium positions.

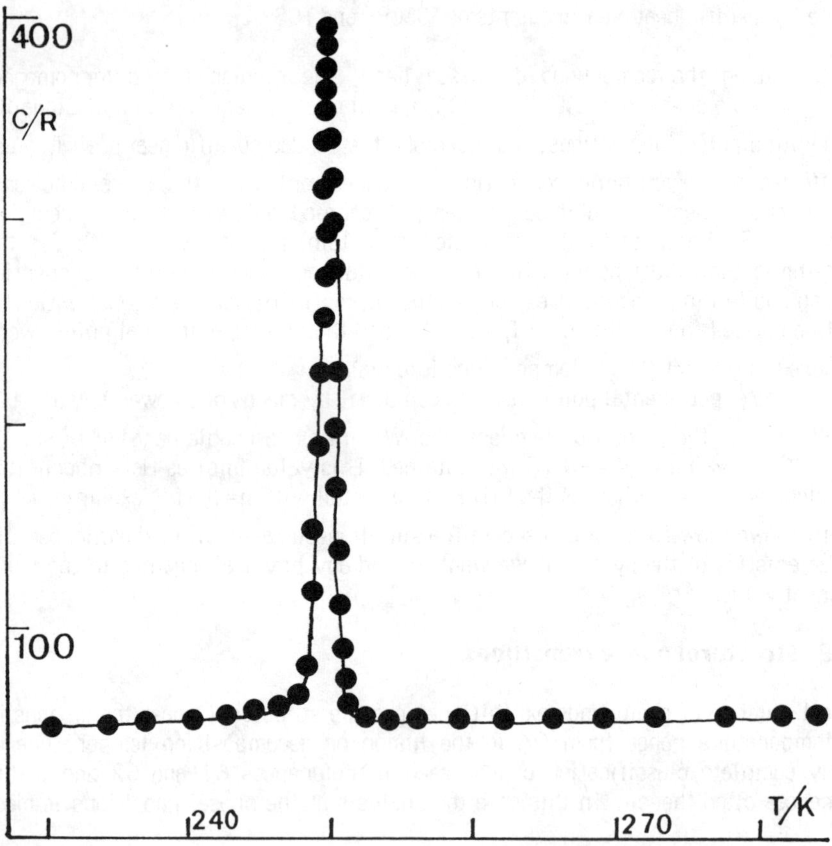

Figure 9. Theoretical fit of the specific heat experimental data of Bis-penthylammonium tetrachlorozincate acording to equations (9) and (12).

Another interesting example refers to those cubic parent phase perovskites which exhibit a cubic-tetragonal phase transition (for example: $CsPbCl_3$, $SrTiO_3$, $TlCdF_3$, $KMnF_3$, etc). All these compounds have been extensively studied from both, experimental and theoretical points of view. In the following we shall refer only to specific heat results in three cases where the Landau approach can be used. In a great part of the halogen compounds the total specific heat[49,64,65] excess above Tc is much higher than below, which suggests that equation (9) could described fairly well the main contribution to the specific heat in the low temperature phase. In all the cases the experimental data can be fit using equations (9) and (12) below Tc and using only equation (12) above Tc. The α-exponents at both sides of Tc range between 0.3 to 1.37. These values are far from those given by the renormalization group theory for one, two or three component systems. This means that the main contribution to the specific heat exceeding the Landau curve is caused by non critical fluctuation contributions. The fluctuations drive the first order character in $KMNF_3$ which was confirmed by the observation of tricritical points in specific heat measurements under uniaxial pressures along the (100) and (110) directions. One should notice that among the halogen compounds cited above, $KMnF_3$ also shows the lowest T_d-T_c values, which is in agreement with the constraint T_c=T_d in the tricrital point. The non critical fluctuation contributions (defects, inhomogeneities, impurities, etc) cause a smearing of the Landau step in all the cases cited.

In the case of $CsPbCl_3$ two more phase transition are present (D_{4h}-D_{2h}-D_{2h}) which have also been fit using the same equations[49]. Very near T_c the experimental curves do not follow pure laws, which is related with the limitations imposed by the noise of the experiemtnal systems. In the case of $KMnF_3$, specific heat studies under uniaxial pressure have shown the presence of two consecutive tricritical points[64] joined by a second order line. When the pressure is applied along (110) the specific heat curve shows the appearance of two new phase transitions. We want to remark here the possibilities of the AC-calorimetry in order to study critical phenomena as well as phase transition studies under external fields.

4.2. Order-Disorder Systems

Among the systems which exhibit order-disorder changes, the ammonium salts take a special place because the microscopic mechanisms are usually due to rotational energy levels of the ammonium ions which can be computed by quantunmechanics theoretical calculus. From an experimental point of view these systems represent an interesting problem in order to show the difficulties that are involved in the separation of the rotational contribution in the measured specific heat curves. The theoretical models in the case of cubic crystals[66] work with constant volume specific heat whereas the experimental measurements are carried out at constant pressure conditions. This problem is more complicated in the case of non-cubic systems, where the theoretical models work on the assumption of "constant geometry" conditions and, therefore the measured specific heat values must be corrected in this case to constant geometry. On the other side we must make the correction of the contribution of the thermal expansion to the heat capacity at constant pressure. In the case of solids the condition of constant pressure means, in a good approximation, constant stress (C_ε), and therefore in a rigurous analysis of the experimental data we must use the following relationship[67]:

$$Cp - C_\varepsilon = VT\sum \mathbf{C}_{ijkl}\, \varepsilon_{ij}\, \varepsilon_{kl} \qquad (13)$$

where C_ε = constant strain specific heat, $\mathbf{C}_{ijkl}$ = isothermal elastic constant tensor and ε_{ij} =thermal expansion tensor. The measured specific heat is given by:

$$C_m = (C_p - C_\varepsilon) + C_{int} + C_{lat} + C_{rot} + C_{ext} \qquad (14)$$

where C_{int}, C_{lat} and C_{rot} are the respective contributions from the internal vibration of the ions, lattice vibrations and rotational or librational movements of the ammonium ion. C_{ext} is the contribution from any abnormal source.

Equation (14) was applied to the calculation of the rotational specific heat in the ammonium sheelites[68,69]. NH_4ReO_4 has the C_{44} space group with four formula units per crystallographyc unit cell which gives 60 unit cell vibrations. Most part of the frequencies of the Raman and IR active modes were measured at 77K or lower. Below 100K the ammonium ion shows librational movements with two different librational frequencies. With all these frequencies and using Einstein and Debye functions the different contributions can be calculated. The non calculated contribution of the specific heat will be the anisotropic thermal expansion given by (13) and the rotational motion of the ammonium ion. Therefore a rigorous calculus of the rotational contribution requires the evaluation of equation (13). Figure 10 shows the results on NH_4ReO_4.

The most usual method employed up to now for these calculus was to use an alkali metal salt as a reference material. However this technique can only be used when the thermal expansion coefficients of the ammonium salt are similar to those of the alkali metal salt. Two things become important from our discussion: the first is related with the difficulties to fit theoretical models of rotational motion to experiment and secondly, the importance of the elastic constant measurements for crystals that are rotationally disordered.

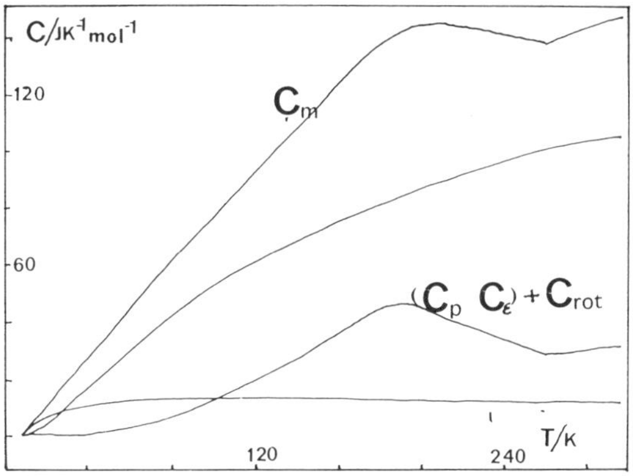

Figure 10. Rotational specific heat contribution of NH_4ReO_4

The group of the ammonium halide salts (for ex: $NH_4Br_xCl_{1-x}$, etc) also shows order-disorder phase transitions from a parent disordered cubic phase (CD) to a distorted one characterized as ferro-ordered cubic (CF) or antiferro-ordered tetragonal (AFT). The order-disorder microscopic mechanism in this case involves a competition between direct interactions due to octopole-octopole terms which favour parallel NH_4 orientaions (CF-phase) and indirect interactions arising from coupling between the pseudospin and transverse phonons which stabilize antiparallel chains of NH_4 ions (TAF-phase)[70,71]. In this case the high resolution of the ac-calorimetric technique has allowed to analyse different theoretical approaches from specific heat measurements in function of x and T in terms of the critical exponents predictions[72]. An interesting point in this analysis, which can be extended to another similar analysis cited in this lecture, is related with the theoretical approach in order to take into account the smearing of the specific heat data.This effect is due to a distribution of the transition temperatures which produces different fit of the available experimental data. In this case the excess specific heat can be obtained from equation (9) with a normalized distribution of local transition temperature values (g(Tc)). The general expression is given by:

$$C_p = \int Cp(T,T_c)g(T_c)dT_c \qquad (15)$$

where $g(T_c)$ in a first approximation can be assumed to be a Gaussian distribution centered at T_c. This discussion opens an interesting problem related with the analysis of the experimental results: Are the fits unique? or have we different values in function of the fitting procedure?. In any case we observe that very accurate experimental data of different physical parameters are neccessary in order to define a theoretical fit which can definitely prove a theoretical approach.

At present an interesting order-disorder problem is the one related with the glassy phases that appear in some compounds, for example: N_2/Ar/Co solid alloy system, H_2, D_2, $(KBr)_{1-x}(KCN)_x$, etc. In the case of N_2/Ar solid alloy system, as well as in the solid H_2 or D_2 quadrupolar glass phases, the diatomic molecules have an electric quadrupole-quadrupole orientational interaction which leads to a quenching of the molecular rotation at low temperature. This coupling leads to a molecular rotationally ordered phase through a hcp-fcc structuralphase[73,74] transition The phase transition temperature is strongly dependent on the quadrupolar concentrations and below some critical values of this component the hcp phase is stable up to T=0K. However, in this case below certain temperature the system exhibits a quadrupolar glass state.

The specific heat data of these systems can be fit as[75,76]:

$$C= AT + BT^3 + LT^3 \qquad (16)$$

where LT^3 is the typical lattice Debye low temperature term and the other two account for the anomalous specific heat contribution attributed to 180º flips of $(CN)^-$ ions or CO-dipoles in the case of $(KBr)_{1-x}(KCN)_x$ alloy and N_2/Ar/CO alloys respectively. In both cases as well as in the ortho-H_2 or para-D_2 the phase diagram points correspond to the specific heat peaks for each concentration. The specific heat data give complementary and valuable information about the modifications of the basic properties in function of the partial substitution. For example, in the case of N_2/CO system the specific heat data show that the CO molecule is a good substitute for the N_2 molecule in the solid lattice and does not seriously affect the nature of the rotationally ordered phase. We have e new clear example of the sensibility of the specific heat measurements.

REFERENCES

1. J.H. de Magellan, "Essai sur la nouvelle theorie du feu elementaire, et de la chaleur des corps" London 1780
2. R. Kirwan, "An estimate of the temperature of different latitudes". London 1787
3. E.L. Scott, Annals of Science, **38** (1981) 141
4. J.P. Mc Cullough and D.W. Scott (eds). "Experimental Thermodnamics" Vol. 1. Butterworths. London 1968.
5. O.V. Lounasma, "Experimental principles and methods below 1K" Academic Press, London 1974.
6. E. Gmelin, Thermocimica Acta **29** (1979) 1
7. D. Gonzalez, J. Calorim. Anal. Thermo. **11** (1980) 101
8. S.T. Lakshmikumar and E.S.R. Gopal, Indian Inst. Sci. **63A**(1981)277.
9. E.F. Westrum, J. of Thermal Analysis **14** (1978) 5.
10. M.B. Ruhemann, "Low Temperature Physics" Cambridge Univ. Press. 1937.
11. T.L. Hill, "An introduction to Statistical Thermodynamics". Adison Wesley. London 1962.

12. L.E. Reichl, "A Modern Course in Statistical Physics" Edward Arnold Ltd. 1980.
13. M. Born and K. Huang "Dynamical Theory of Crystal Lattices" Osford, Clarendon Press 1954.
14. E.W. Montroll, J. Chem. Phys. **10**(1942)218
15. R.G. Chambers, Proc. Phys. Soc. **78**(1961)941
16. H. Grayson-Smith and J.P. Stanley, J. Chem. Phys. **18**(1950)236
17. T.H.K. Barron and J.A. Morrison, Proc. R. Soc. **A256**(1960)427
18. J.L. Hwang, J. Chem. Phys. **22**(1954)154
19. S.W.Kieffer in S.W. Kieffer and A. Navotsky(eds); Microscopic to Macroscopic, Reviews in Mineralogy, **14**(1985)
20. E.F.Westrum and N. Komada, Thermochimica Acta 109 (1986)11
21. J.W. Loranm, J. Phys. C: Solid State Phys. **19**(1986)6113
22. W. Nernst, Ann. Phys. **36**(1911)395
23. E. Gmelin and P. Rodhammer, J. Phys.E: Sci. Inst. **14**(1981)223
24. P.R. Clayton, L.A.K. Staveley and R.D. Weir, J. Chem. Phys **75** (1981)5464
25. F.L. Oetting and E.D. West, J. Chem. Therm. **14**(1982)107
26 J.A. Zubillaga, A. López-Echarri and M.J. Tello, Thermochim. Acta **92**(1985)283
27. R. Burriel, Private Commun.
28. P. Handler, D.E. Mapother and M. Rayl, Phys. Rev. Lett. 1**9** (1967)356
29. P. Sullivan and G. Seidel, Phys. Lett. **25A**(1967)229
30. P. Sullivan and G. Seidel, Phys. Rev. **173**(1968)679
31. C.W. Chu and G.S. Knapp, Phys. Lett. **46A**(1973)33
32. J.D. Baloga and C.W. Garland, Rev. Sci. Inst. **48**(1977)105
33. M. Yoshizawa and T. Fujimura, Bull. Res. Inst. Sci. Meas. Tohoku Univ. **26**(1977)135
34. S. Ikeda and Y. Ishikawa, Jpn. J. Appl. Phys. **18**(1979)1367
35. M. Yoshizawa and T. Fujimura, Bull. Res. Inst. Sci. Meas. Tohoku Univ. **28**(1979)35
36. A. Eichler and W. Gey, Rev. Sci. Inst. **50**(1979)1445
37. C.R. Wolfe, Mol. Cryst. Liq. Cryst. **85**(1982)337
38. T. Suzuki, T. Tsuboi and H. Tanaki, Jpn. J. Appl. Phys. **21** (1982)368
39. S. Stokka and K. Fossheim, J. Phys. E:Sci. Inst. **15**(1982)123
40. J.M. Viner, D. Lamey, C.C. Huang, R. Pindak and J.W. Goodby, Phys. Rev **28A**(1983)2433
41. S. Imaizumi, K, Suzuki and I. Hatta, Rev. Sci. Inst. **54** (1983)1180
42. X.C. Jin, P.H. Hor, M.K. Wu and C.W. Chu, Rev. Sci. Inst. **55** (1984)993
43. I. Hatta and A.J. Ikushima, Jpn. J. Appl. Phys. **20**(1981)1995
44. L.D. Landau and E.M. Lifshitz, Statistical Physics. Pergamon Press. 1969
45. E. Fatuzo and W.J. Mertz,"Ferroelectricity"North-Holland 1967

46. J. Kocinski,"Theory of Symmetry Changes at Continuous Phase Transitions" Elsevier 1983.
47. J.C. Toledano and P. Toledano, "Landau Theory of Phase Transi tions" Scient. World Publ. (to appear).
48. A. Aharony and P.C. Hohemberg, Phys. Rev. **13B** (1976)3081
49. S. Stokka, F. Fossheim, T. Johanses and J. Feder, J. Phys. C: Solid State Phys. **15** (1982) 3053
50. A.I. Larkin and D.E. Khmelnitzkii, Soviet Phys. JEPT **29**(1969) 1123
51. Th. Nattermann, Phys. Status Solidi **B85**(1978)291
52. M.J. Tello, M.A. Perez-Jubindo, A. López-Echarri and C. Socias, Solid State Commun. **50**(1984)957
53. B.A. Strukov, S.A. Taraskin K.A. Minolva and V.A. Fedorikhin, Ferroelectrics **25**(1980)399
54. R. Tobon and J.E. Gordon, Ferroelectrics, **17**(1977)409
55. A. López-Echarri and M.J. Tello, J. Phys. D:Appl. Phys.**14** (1981)71.
56. E. Sandvol and E. Courtens, Phys. Rev. **27B**(1983)5660
57. G.A. Samara, Phys. Rev. **17B**(1978)3020
58. S. Stokka, K. Fossheim and S. Ziolkiewicz, Phys. Rev. **24B** 24B(1981)2807.
59. G. Ahlers, A. Kornblit and H,J, Guggenheim, Phys. Rev. Lett. **34** (1975)1227.
60. A. Kornblit and G. Ahlers, Phys. Rev. **8B**(1973)5163
61. J.C. Toledano, J. Sol. State Chem. **27**(1979)41
62. P. Toledano and J.C. Toledano, Phys. Rev. **16B**(1977)386
63. A. Gómez-Cuevas, J.M. Perez-Mato, M.J. Tello, G. Madariaga, J. Fernandez and A. López-Echarri, Phys. Rev. **29B**(1984)2655
64. S.Stokka and K. Fossheim, J. Phys. C: Solid State Phys. **15**(1982)1161
65. M. Chabin, F, Gilletta and C. Ridou, Phys. Status Solidi **48**(1978)67.
66. R. Burriel, PhD Zaragoza Univ. 1979
67. D.C. Wallace, Thermodynamics of crystals, J. Willey 1972
68. R.D. Weir and L.A.K. Staveley, J. Chem. Phys. **73**(1980)1386
69. R.J.C. Brown, J.E. Callanan, R.D. Weir and E.F. Westrum, J. Chem. Phys. **85**(1986)5963
70. Y. Yamada, M. Mori and Y. Noda, J. Phys. Soc. Jpn. **32**(1972)1565
71. C.W. Garland, K.J. Lushington and R.C. Leung, J. Chem. Phys. **71**(1979)3165
72. K.J. Lushington and C.W. Garland, J. Chem. Phys. **72**(1980)5752
73. L.G. Ward, A.M. Saleh and D.G. Haase, Phys. Rev. **27B**(1983)1832
74. C.S. Barret and L. Meyer, J. Chem. Phys. **42**(1965)107
75. M. Meissner et al. Phys. Rev. **32B**(1986)609
76. D.G. Haase, C.I. Nicholls and M.S. Conradi (to be published)
77. E. Sandvold and K. Fossheim, J. Phys. C: Solid State Phys. **19**(1986)1481

MAGNETIC ORDERING AND THERMODYNAMICS IN SILICATES

J.M.D. Coey
Department of Pure and Applied Physics, Trinity College, Dublin 2, Ireland
and S. Ghose
Department of Geological Sciences, University of Washington, Seattle WA 98195, USA.

ABSTRACT. An elementary account is given of the properties of the magnetic ions most commonly found in silicate minerals, Fe^{2+} and Fe^{3+}, including the effect of the crystal field on the former. Collective magnetic behaviour is discussed, mainly in term of molecular field theory, and the influence of cation disorder is considered. There follows a survey of collective magnetic order of end-member iron minerals belonging to the garnet, pyroxene, amphibole, serpentine, mica and olivine families. They usually have complex antiferromangetic structures with Néel points in the range 1 - 100 K, but disorder may induce spin glass states. Finally, the thermodynamic consequences of magnetic order are considered, in relation to the enthalpy and entropy of iron minerals

I INTRODUCTION

1.1 Background

Silicates are the main constituent of the Earth's crust and mantle. Abundances of major elements, and the 3d and 4f transition elements in the crust are shown in figure 1. Nine major elements constitute 99% of the atoms in the crust, which is composed of oxides of cations (principally Si^{4+}) with an average valence $M^{3.15+}$. With the exception of iron, which represents just less than 2% of the atoms in the crust, all of these ions have a filled-shell configuration in their normal oxidation state ($1s^2 2s^2 2p^6$ for Si^{4+}, Al^{3+}, Mg^{2+}, Na^+ and O^{2-}). Iron is singular in that it has a partly-filled d shell with two common charge states $Fe^{3+}(3d^5)$ and $Fe^{2+}(3d^6)$. These ions are generally found in a high-spin state in oxides, and therefore carry a magnetic moment. Abundances of other transition elements which form magnetic ions, shown in figure 1b, indicate that iron is forty times as abundant as all the rest of the magnetic elements put together. Any broad discussion of the magnetic properties of natural minerals must inevitably focus on iron oxides and iron silicates.

E. K. H. Salje (ed.), Physical Properties and Thermodynamic Behaviour of Minerals, 459–499.

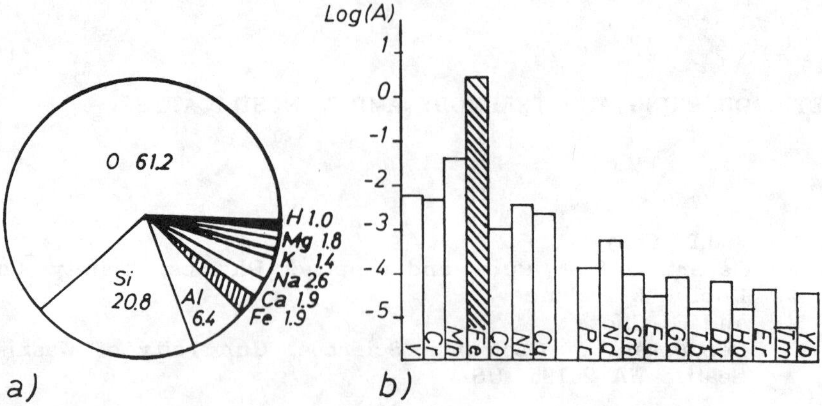

Figure 1. a) Major elements in the Earth's crust, expressed as atom %, b)magnetic elements, on a logarithmic scale. (Ti and Ce are assumed to be quadrivalent, hence nonmagnetic)

Study of the magnetism of natural iron oxides dates back to the discovery of lodestone in antiquity. The common minerals magnetite (Fe_3O_4) and haematite (Fe_2O_3) have the remarkable property of exhibiting spontaneous magnetization σ_s at ambient temperature. Values for the pure oxides are 100 J/T/kg and 0.5 J/T/kg†. A physical explanation of the origin of this magnetization (ferrimagnetism and canted antiferromagnetism, respectively) together with an understanding of the concepts of superparamagnetism and natural remanence in terms of an ensemble of two-level systems led to a theory of the magnetism of rocks [1]. In due course, rock magnetism and paeleomagnetism combined to provide convincing proof of the theory of global plate tectonics, based on reversals of the Earth's magnetic field. The work on natural iron oxides, which spanned the years 1940-1975, was a magnificent contribution of Physics to Earth Science. There are many texts on the subject [2-5].

Here we will be concerned with more recent work on the magnetism of iron silicates and its thermodynamic consequences. No silicate is magnetically ordered at ambient temperature; if they order

---

† We adopt the modern, S1 system of units. The normal magnetic field is B_o (= $\mu_o H$), measured in Tesla; abbreviated as T. The induction in a medium is $B = B_o + \mu_o M$ where M is the magnetic moment per m^3 and μ_o is the permeability of free space ($4\pi \times 10^{-7}$ T.m/A). Magnetic energy of a piece of material of volume V placed in a field B_o is $- V M.B_o$, hence the units of M are $J/T/m^3$. In practice it is the magnetization per kg which is normally measured, $\sigma = M/\rho$ in J/T/kg where ρ is the density. Susceptibility is defined as $\kappa = (dM/dB_o)$ or $\chi = (d\sigma/dB_o)$. Accordingly, we tend to use M and κ in the theoretical sections but σ and χ when referring to experimental data. The old cgs system is still in widespread use in the research literature. There $\mu_o = 1$, B_o is in Gauss (1 Gauss = 10^{-4} Tesla), $B = H + 4\pi M$ and magnetic moment is normally measured in emu/g (1 emu/g = 1 J/T/kg).

TABLE 1

Crystal chemistry of common cations

Ion	Configuration	Ionic radius (Å)	Coordination number
Si^{4+}	Ne	0.39	4
Al^{3+}	Ne	0.51	6,4
Fe^{3+}	$Ar3s^2 3p^6 3d^5$	0.64	6(4)
Mg^{2+}	Ne	0.66	6
Fe^{2+}	$Ar3s^2 3p^6 3d^6$	0.74	6(8)
Na^+	Ne	0.97	8
Ca^{2+}	Ar	0.99	8
K^+	Ar	1.33	8 - 12

at all, it is at cryogenic temperatures and generally without any net spontaneous magnetization. Nevertheless, these materials exhibit a variety of interesting magnetic properties which often reflect peculiarities of the silicate structures. Besides, the possibility of magnetic order must be taken into account when calculating the enthalpy and entropy of iron minerals.

The structures of oxides, including silicates, may be considered as more or less close-packed arrays of large O^{2-} ions (r = 1.40 Å) with smaller, positively-charged cations in the interstices. 61% of the atoms, but 94% of the volume of the crust is oxygen. The sizes of the spherical ions which can just be accommodated in tetrahedral (4-coordinated), octahedral (6-coordinated) and cubic (8-coordinated) sites are 0.31, 0.58 and 1.02 Å respectively. These values should be compared with the ionic radii of the major cations, shown in table 1. Silicon fits well in 4-coordination, and SiO_4 tetrahedra are the characteristic feature of all silicate structures (except the recently-discovered high-pressure phases with the perovskite structure). Al^{3+} may substitute for Si^{4+} in tetrahedral sites, but is more commonly found in octahedral coordination. Fe^{3+} may substitute isomorphically for Al^{3+} in octahedral sites but it is rarely found in tetrahedral coordination. Mg^{2+} is an octahedral cation, which may be substituted by Fe^{2+}. Na^+, Ca^{2+} and K^+ occupy larger sites with a higher coordination number. On account of this chemical interchangeability, pure end-member iron silicates are rather uncommon,

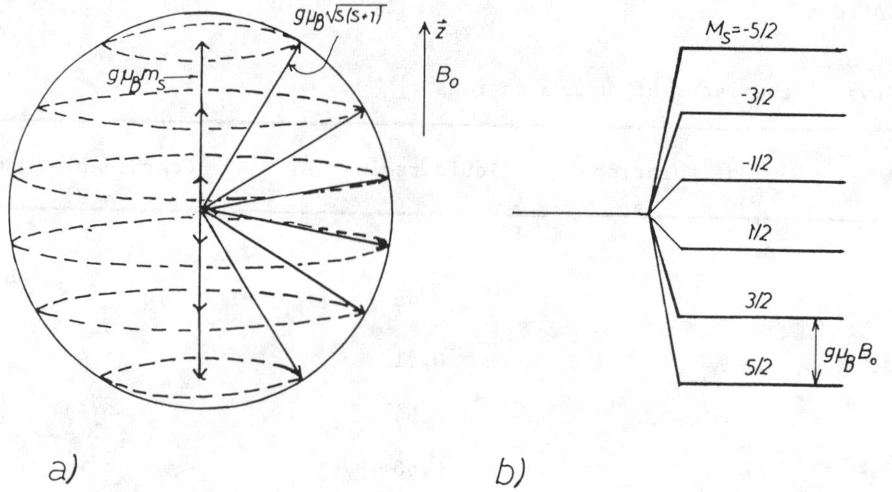

Figure 2. a) Vector model of the Fe^{3+} ion showing the six possible projections of the magnetic moment along the direction of a magnetic field B_o, b) Zeeman splitting of the six energy levels.

although they provide a key to understanding the magnetic properties of natural minerals. The typical iron-bearing silicate is rather a random solid solution where magnetic and nonmagnetic cations, $Al^{3+} \leftrightarrow Fe^{3+}$, $Mg^{2+} \leftrightarrow Fe^{2+}$ or different magnetic cations, $Fe^{3+} \leftrightarrow Fe^{2+}$ occupy crystallographically-equivalent sites more or less at random. The resulting magnetic dilution and disorder have a profound effect on the magnetism.

1.2 Properties of noninteracting ions

a) *Fe^{3+}*: Magnetically, Fe^{3+} is the simpler ion. The half-filled d-shell $3d^5$ is spherically symmetric and there is no orbital angular momentum. The spectroscopic term for the free iron is $^6S_{5/2}$; its spin angular momentum has magnitude $\sqrt{S(S+1)}\hbar$ where $\hbar$ is Plank's constant and the spin quantum number S = 5/2. The corresponding magnetic moment has magnitude $g\mu_B\sqrt{S(S+1)}$ where the g-factor is 2 for the spin moment of the electrons. μ_B is the Bohr magneton, 9.274 x 10^{-24} J/T. When subject to an external or internal field B_o, the projection of the magnetic moment along the field direction, conventionally the z-axis, can take only six discreet values $m_i = g\mu_B M_s$ where the magnetic quantum number M_s = 5/2, 3/2, 1/2, -1/2, -3/2 or -5/2. Then

$$(1) \quad E_i = -g\mu_B M_s B_o$$

The Zeeman splitting of two adjacent energy levels is $g\mu_B B_o$, as illustrated in figure 2. The lowest energy state has a moment of $5\mu_B$ but it will be solely populated, (magnetic saturation) only at very low temperatures or in enormous magnetic fields. Populations of the

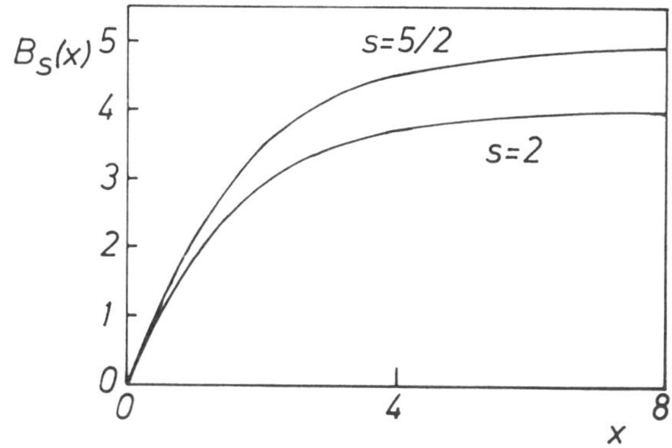

Figure 3. Magnetization curves of ions with S=2 and S=5/2, given by the Brillouin function, equation 5. The susceptibility is the initial slope.

Zeeman levels are proportional to the Boltzmann factor exp $(-E_i/kT)$, so the condition for saturation is $g\mu_B B_o \gg kT$. In the opposite limit, when the levels are populated with equal probability, there is essentially no net moment and the full entropy of disorder k ln(2S+1) = k ln6 is developed. A useful numerical result is $\mu_B B_o/k$ = 0.67 K, so that a field of 1 Tesla (which could be produced by a large electromagnet) acting on a moment of 1 μ_B corresponds to an energy equivalent to only 0.67 K. The susceptibility and magnetization may be calculated by taking the thermodynamic average

(2) $$\langle m\rangle = \sum_{1}^{6} m_i \exp(-E_i/kT) / \sum_{1}^{6} \exp(-E_i/kT)$$

If $g\mu_B B_o \ll kT$, $\exp(-x) \simeq (1 - x)$ and equations 1 and 2 give $\langle m\rangle = g^2\mu_B{}^2 S(S + 1)\ B_o/3\ kT$. Hence the Curie law expression for the susceptibility $\kappa = N\langle m\rangle/B_o$ is

(3) $$\kappa = Ng^2\mu_B{}^2\ S(S + 1)\ /\ 3\ kT = C/T$$

where N is the number of ions per m^3. (χ or χ_m is obtained if N is defined as the number of ions per kg or per mole, respectively). In the general case, $g\mu_B Bo \not\ll kT$, and it is necessary to evaluate equation 2 without any approximation. The result is

(4) $$\langle m\rangle = g\mu_B S\ B_s(x)$$

where x is the dimensionless variable $g\mu_B SB_o/kT$ and $B_s(x)$ is the Brillouin function of argument S, namely

(5) $$B_s(x)=(2S+1)/2S)\coth((2S+1)/2S)x - (1/2S)\coth(x/2S)$$

This function is tabulated, for instance by Smart [7]. It is plotted in figure 3 for S = 2 and S = 5/2.

The magnetic properties of the Fe^{3+} ion in a crystal are approximately those of the free ion. In particular, the susceptibility and magnetization are isotropic in the sense that they do not depend

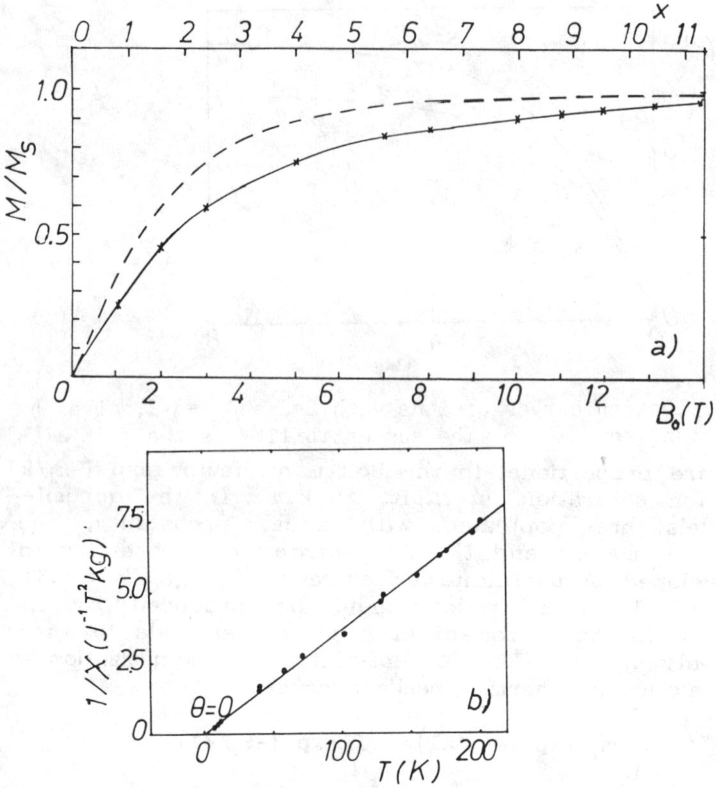

Figure 4. a) Curie law variation of the susceptibility of a vermiculite crystal which contains 2.6 wt% Fe^{3+}, b) Magnetization curve at 4.2 K. $B_{5/2}(x)$ is indicated by the dashed line.

on the orientation of the field relative to the crystal axes. Some data on the sheet silicate vermiculite which contains ferric impurities are shown in figure 4. Deviation from the Brillouin function $B_{5/2}(x)$, shown dashed, is due to weak interactions between ferric ions, on an energy scale order 1 K.

b) *Fe^{2+}*: High-spin Fe^{2+}, with electronic configuration $3d^6$, has an extra electron with spin opposite to the other five in the half-filled shell. According to Hund's rules, the spectroscopic term of the free ion is $^5D_4$; the spin quantum number of the ion is S = 2, and the orbital quantum number associated with the extra electron is L = 2. Spin-orbit coupling would link L and S into a state described by the total quantum number J = L + S = 4. but the situation is different if the Fe^{2+} ion is located in a crystal. A strong electrostatic field created by the neighbouring ions interacts directly with the anisotropic charge distribution of the sixth d-electron's orbit. To calculate the magnetic properties of the ion, we must now consider the Hamiltonian

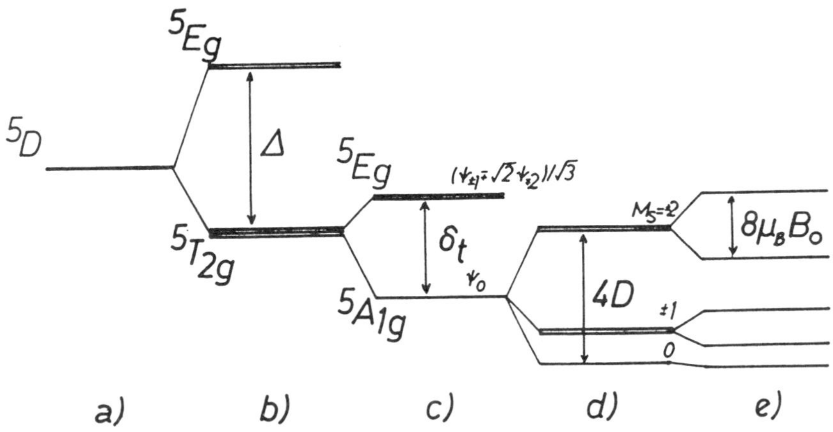

Figure 5. Energy-level scheme for the Fe^{2+} ion, not to scale, showing a) the free ion, b) cubic crystal-field splitting, c) trigonal crystal-field splitting, d) spin-orbit splitting and e) Zeeman splitting. The orders of magnitude of the interactions are respectively 10^4, 10^3, 10 and 1 K.

(6) $\quad H = H_{cf}+H_{so}+H_Z$

(For Fe^{3+}, the crystal field and spin orbit terms, H_{cf} and H_{so}, were negligible because the ion had L = 0; eigenvalues of H_z were given by equation 1).

The effect of the crystal field interaction is to split the five-fold degenerate (2L + 1) energy levels of the free ion, which correspond to the five different orbitals that can be occupied by the sixth d-electron, by energies of order 10^4 K. In an octahedral site, the primary effect is to split the one-electron states into a group of two (e_g) and a group of three (t_{2g}). The former have lobes of the d-wave functions pointing towards the oxygen ligands, and are therefore much higher in energy than the latter, where the lobes point between the ligands. Distortion of the octahedron will further raise the degeneracy, as illustrated in figure 5 for the case of a trigonal extension. The ground state is stabilized by an energy δ, typically or order 10^3 K. Crystal-field effects in minerals are discussed further by Burns [8].

It turns out that the eigenfunctions of H_{cf} are linear combinations of the basis set Ψ_{ml} (M_l = 0, ±1, ±2) for which l_z = 0. The orbital angular momentum of the ion is <u>quenched</u>, and it therefore behaves as if S, not J, is the appropriate quantum number. The magnetic moment again has magnitude $g\mu_B\sqrt{S(S+1)}$ with g = 2, and the maximum projection along the z-axis is $g\mu_B S = 4\mu_B$. However, it is now necessary to consider the effect of the second term in equation 6. The spin-orbit interaction is represented by a term λ L·S where λ = -150 K for Fe^{2+}. It is therefore a perturbation on H_{cf}, the effect of which is to bring back a small orbital moment to the eigenstates. Although S remains a good quantum number, the g factor may be

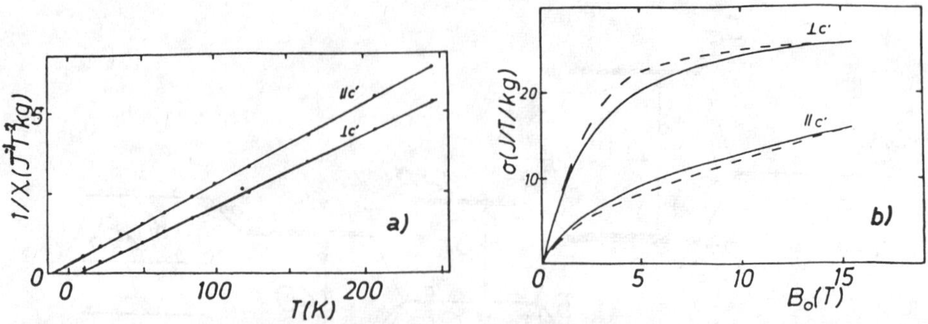

Figure 6. a) Inverse susceptibility of a chlorite crystal containing 5 wt% of Fe^{2+}, b) magnetization curves at 4.2 K measured parallel and perpendicular to the c' axis, normal to the layers. Dashed lines are calculated from equation 6.

different from 2, and it can be quite anisotropic, which means that the susceptibility and magnetization will depend on the relative orientation of the applied field and the crystal axes. As an illustration, we show in figure 6 the magnetization and susceptibility of another sheet silicate, a platelet of chlorite containing Fe^{2+}, with the field applied parallel or perpendicular to the normal to the plane of the sheets. The latter is the 'easy' direction. Note that the susceptibility becomes more and more anisotropic as the temperature is lowered.

Calculation of the susceptibility or magnetization involves diagonalizing the 25 x 25 matrix corresponding to equation 6, with elements in the (M_L,M_S) basis. The crystal field is frequently expressed as a sum of terms $\Sigma\ B_n^m\hat{O}_n^m$, where Bn^m are crystal-field parameters depending on the ion involved and on the crystal structure, and $\hat{O}n^m$ are the Stevens operators involving angular momentum operators $\hat{L}^2$, $\hat{L}_x$ $\hat{L}_y$ and $\hat{L}_z$ which are tabulated by Hutchings [9]. The spin orbit term was given above, and the general Zeeman term is $-\ \mu_B\ B.\ (L\ +\ 2S)$. In certain cases, such as the trigonal distortion illustrated in figure 5, the structure of the lowest energy levels produced by the spin-orbit interaction can be represented by a spin Hamiltonian. In that case $-\ DS_z{}^2$, with $D \simeq 10$ K. The lowest energy level has $S_z = 0$.

There will obviously be specific heat and entropy associated with energy-level schemes such as that of figure 5. The complete entropy associated with the (2L + 1) orbital states could only be developed well beyond the melting point, but some of the entropy associated with the lower orbital triplet, kln3, will appear at room temperatures and above, in silicates in the solid state, in addition to the kln5 associated with spin disorder. In general, with a set of n energy levels with degeneacy g_i, the total energy of a system of N noninteracting ions is:

$$(7)\quad E = N \sum_{o}^{n} E_i g_i \exp(-Ei/kT) / \sum_{o}^{n} g_i \exp(-E_i/kT)$$

The heat capacity is obtained by calculating dE/dT. The simple case of two levels with $g_1=g_2=1$ illustrates the result. The heat capacity

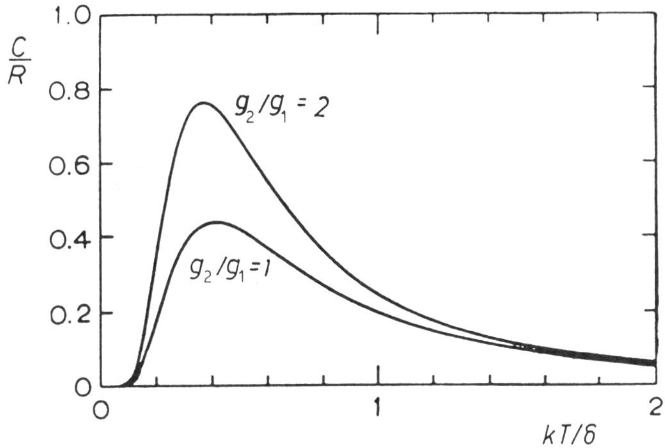

Figure 7. Specific heat of a two-level system (Schottky anomaly). Curves are drawn for $g_1=g_2=1$ and $g_1=1$ & $g_2=2$.

associated with single-ion excitations

$$(8) \quad C = \frac{N\delta}{kT^2} \; \frac{\exp(\delta/kT)}{[1 + \exp(\delta/kT)]^2}$$

gives the 'Schottky anomaly' where δ is the separation of the two energy levels. A Schottky anomaly is illustrated in figure 7; it is a broad peak centred at 0.4 δ, with the heat capcity varying as exp $(-\delta/kT)$ at low temperatures, and tailing off as $1/T^2$ when $kT \gg \delta$. Such contributions are to be expected in the specific heat of silicates containing small or moderate amounts of Fe^{2+}, with δ in the range 10 - 10^3 K.

Finally, it is worth mentioning another contribution to the specific heat which is expected in certain mixed-valence minerals containing Fe^{2+} and Fe^{3+} on equivalent sites. At high temperatures, the sixth 3d electron may hop from site to site by thermal activation, yielding a disordered array of the two cations, but at lower temperatures charge ordering usually takes place with a lowering of the symmetry and elimination of the entropy of disorder. The process occurs much more readily than order-disorder transitions involving cations of different elements, because all that is needed is movement of electrons, not ions. Examples of collective charge order transitions (sometimes known as Verwey or Wigner transitions) in minerals can be found in magnetite, ilvaite and cronstedtite.

1.3 Magnetic Interactions and Collective Behaviour

Collective magnetic behaviour such as ferromagnetism or antiferromagnetism results from interactions between the magnetic moments of different ions. The principal coupling mechanism is not the familiar magnetic dipolar interaction, but the <u>exchange</u> interaction which is electrostatic and quantum-mechanical in nature. The interaction between the spins of two ions at sites i and j usually tends to couple them parallel or antiparallel. It can be represented by the Heisenberg hamiltonian for the system

$$(9) \quad H_H = -2\Sigma J_{ij} S_i . S_j$$

J_{ij} is the exchange constant which may be positive (ferromagnetic) or negative (antiferromagnetic). It falls off rapidly with the distance r_{ij}, as it depends on direct overlap of the 3d wavefunctions of the ions (direct exchange) or on the overlap with a common ligand (superexchange). Superexchange via an O^{2-} or an OH^- ion is the usual magnetic interaction in silicates. J is usually negative for an Fe^{3+}-O-Fe^{3+} bond, and varies as $\cos^2\psi$ where ψ is the bond angle. It also depends sensitively on the bond length. The tendency is therefore for spins of neighbouring ferric ions to align antiparallel. For ferrous ions, the interaction is negative when $\psi \approx 180^o$, but it is weakly positive for $\psi \approx 90^o$. First-principles calculations of J are impractical, but empirical rules giving the magnitude and sign of superexchange interactions in ionic compounds were provided by Goodenough [10]. Another, phenomenological interaction term, which represents a tendency for the ions to couple perpendicular, was given by Dzialoshinskii and Moriya [11]

$$(10) \quad H_{DM} = \sum_{ij} D_{ij} . S_i \wedge S_j$$

However, $|D|$ is usually much less than $|J|$, and it is identically zero in certain symmetry.

An appropriate way of treating the Heisenberg hamiltonian is to represent the interaction by an effective field B_o^i which is supposed to be proportional to the average moment $\langle m \rangle$

$$(11) \quad B_o^i = \lambda \langle m \rangle$$

Equation 9 becomes

$$(12) \quad H_{mf} = -g\mu_B \Sigma S_{zi} B_o^i$$

so that the approximation involves replacing $S_i . S_j$ by an average $S_i . \langle S_j \rangle$. Since all the ions are taken into account in the average $\langle S_j \rangle$, the theory does not reproduce local fluctuations; it is equivalent to a model where the interactions have infinite range. The relation between J and λ is $J = g^2\mu_B{}^2\lambda/2Z$ where Z is the magnetic coordination number, provided only the nearest-neighbour interaction $J_{ij} = J$ need be considered. The magnetization in a small external field is then $\langle m \rangle = g^2\mu_B{}^2 \; S(S + 1) \; (B_o + \lambda\langle m \rangle)/3kT$, and equation 3 for the susceptibility must be replaced by the Curie-Weiss law.

$$(13) \quad \kappa = Ng^2\mu_B{}^2 \; S(S+1)/3k(T-\theta) = C/(T-\theta)$$

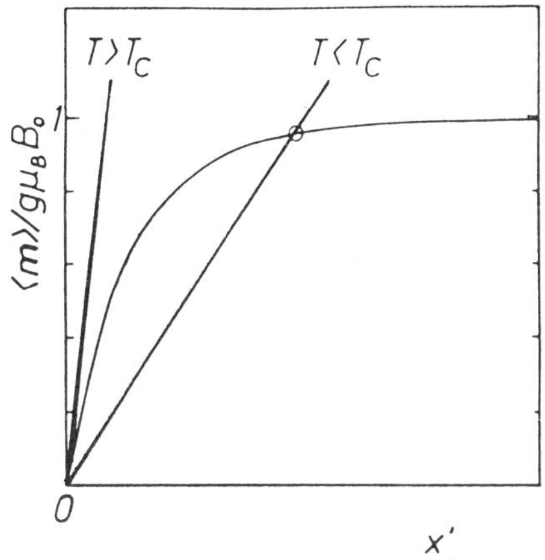

Figure 8. Graphical solution of equations 4 & 11, described in the text, showing the development of spontaneous magnetization when $T < T_c$.

where $\theta = g^2\mu_B{}^2\lambda\ S(S+1)/3k = 2ZJS(S+1)/3k$ is the paramagnetic Curie temperature, obtained experimentally by plotting $1/\chi$ against T. If more than one type of exchange bond is present, θ gives the weighted average $\Sigma Z_i.J_i$.

If J is positive, the susceptibility diverges at a critical temperature T_c, which is equal to θ when only the one exchange constant need be taken into account. Below T_c, the material acquires a spontaneous ferromagnetic magnetization $M_s = N\langle m\rangle$, even in the absence of an applied field. M_s is obtained by simultaneous solution of equation 11 and equation 4, with x replaced by $x' = g\mu_B SB_o^i/kT$. Graphical solution for S = 5/2 is illustrated in figure 8.

The divergence of χ at T_c is an example of the singularities which appear in derivatives of the thermodynamic functions† at a magnetic phase transition. The variations of spontaneous magnetization M_s, susceptibility and heat capacity as a function of T/T_c when $(1-T/T_c) \to 0$, and of $M(B_o)$ at T_c when $B_o \to 0$ all show power-law behaviour.

$$(14)\quad M_s \simeq (1 - T/Tc)^{\beta} \qquad M \simeq B_o{}^{1/\delta}\ (T = T_c)$$

$$\kappa \simeq (T/Tc - 1)^{-\gamma} \qquad C \simeq |1-T/Tc|^{\alpha}$$

where the indices $\alpha,\beta,\gamma\ \delta$ are known as static critical exponents. The variation of M, M_s, χ, and C calculated from mean field theory are shown in figure 9. α is zero, which means that the specific heat exhibits a simple discontinuity rather than a divergence at T_c, β is ½,

† The expression embodying the first and second laws for a magnetic system is $dU = TdS + B_o dM$. The gibbs free energy is $G = U-TS-BM$, hence $dG = -SdT - MdB_o$. Magnetization and entropy are given by the first derivatives of the Gibbs function, susceptibility and specific heat by the second derivatives.

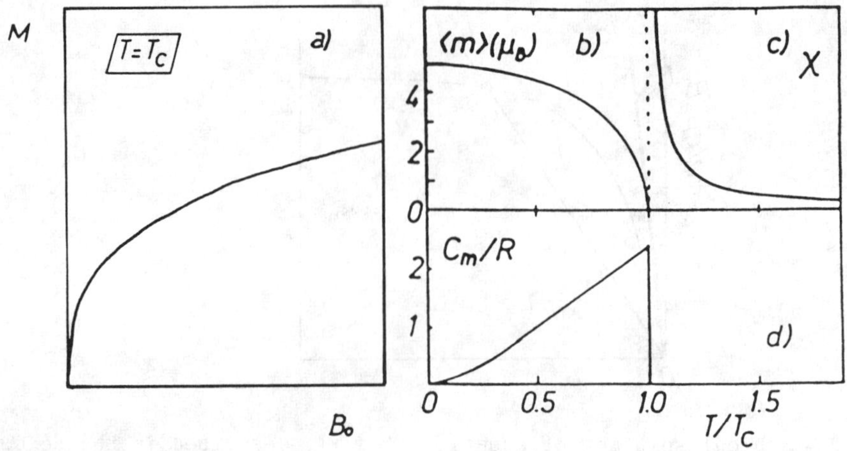

Figure 9. Variation of a) magnetization at T_c, b) spontaneous magnetization, c) susceptibility and d) specific heat calculated from molecular field theory for a ferromagnet with S=5/2.

γ is 1 (c.f. equation 13) and $\delta = 3$. Other physical quantities which exhibit a singularity at T_c and can also be described by power laws [13] include the ferromagnetic correlation length ξ, which varies as $(T/Tc-1)^{-\nu}$, where $\nu = ½$. Actually only two of the critical exponents are independent; they are related by equalities such as $\alpha + 2\beta + \gamma = 2$ and $\gamma = \beta\ (\delta-1)$.

The essential structure of the ferromagnetic phase transition, summarized by the critical exponents, can be recovered from a quite different starting point. If following Landau, one assumes that the free energy F can be expanded in a power series involving even powers of the magnetization (since $M \rightarrow -M$ must leave F unchanged).

$$(15) \quad F = A_o + A_2(T)\ M^2 + A_4(T)\ M^4 + A_6(T)\ M^6 + \ \ldots$$

then the relations of equation 14 can be obtained with the <u>same</u> exponents as before by making use of results such as $B_o = (\partial F/\partial M)_T$, $\chi = (\partial M/\partial B_o)_T$, $C_M = -T(\partial^2F/\partial T^2)_M$ [13].

In a ferromagnet, the magnetization M plays the role of <u>order parameter</u>; it is the order parameter which exhibits as essential singularity at the phase transition (figure 9b) and which together with its conjugate field (B_o) is central to the thermodynamic description of the phenomenon.

Landau theory and molecular field theory are both examples of the class of <u>mean field theories</u>, which all yield the same critical behaviour. The snag is that they are not in good agreement with what happens at the Curie point. In nickel, for example, power law divergences are found, but $\alpha = -0.10$, $\beta = 0.33$, $\gamma = 1.32$, $\delta = 4.2$. These values, however, can be reproduced by the modern theory of phase transitions which is based on the renormalization group [14].

A useful concept for classification of phase transitions is the <u>universality class</u>, defined by just two numbers (d,n) the dimensionality of the space and order parameter of the system. The space dimensionality of crystals is usually three, but if they have sheets or chains of interacting cations with negligibly weak interactions between these units, then the space dimensionality d can be effectively reduced from three to two or one. Some silicate structures could be of reduced dimensionality in this sense, since the magnetic exchange interactions are short-ranged. The dimensionality of the order parameter is equivalent to the spin dimensionality for a magnetic system. At first sight this too would seem to be three, since the Heisenberg hamiltonian, equation 9, involves $S_x^i S_x^j + S_y^i S_y^j + S_z^i S_z^j$. However, single-ion anisotropy may effectively reduce the number of spin components active at low temperature. In the energy-level scheme of figure 5d) the singlet ground state has $S_z = 0$, so only the x and y components can interact magnetically. With ferromagnetic exchange, the spontaneous magnetization would be constrained to lie in the xy plane, so that n = 2. The xy hamiltonian is $H = -2\Sigma\ J(S_x^i S_x^j + S_y^i S_y^j)$. if D were opposite in sign, the ground state would be a doublet with $M_s = \pm 2$. Only the z-component is then non-zero, n = 1, and the resulting, Ising hamiltonian is $H = -2\Sigma J S_z^i S_z^j$.

The microscopic anisotropy energy resulting from the sum of the single-ion contributions in a magnetically-ordered material is known as the magnetocrystalline anisotropy. In uniaxial systems, it is represented by a term $-K_1\cos^2\theta$ (J/m^3), where θ is the angle between the magnetization and the easy axis. For the Fe^{2+} ion of figure 5d), $K_1 = 4DN$. A related quantity is the anisotropy field B_a required to saturate the magnetization in the hard direction: $B_a = 2DS^2/g\mu_B S$. Each universality class (n,d) is thought to be characterized by a particular set of critical exponents regardless of the details of the interactions. Mean field theory would be correct in the physically inaccessible region d = 4. When d = 1, it can be shown by an elementary argument that no phase transition at all can occur at any nonzero temperature†. Conditions in which a phase transition can be expected are summarized in table 2.

Besides the inadequate description of the phase transition, molecular field theory also fails to reproduce exactly the temperature-dependence of the magnetization at lower temperatures. This is because of collective quantum excitations in the ground state known as spin waves or magnons. Each magnon consists of an excitation of a spin from its ground state $M_s = S$ to $M_s = (S - 1)$, which is delocalized and spread out over the crystal in the form of a wave of wave-vector q. For small q, the waves, which are bosons, obey the dispersion relation

$$E(q) = D' q^p \qquad (16)$$

with p = 2 for a ferromagnet in three dimensions. The number of these excitations at temperature T is $1/(\exp(E(q)/kT)-1)$. It follows

†Consider a row of N spins. Invert all the spins to the right of any point in the chain. The change in free energy $\Delta F = 4JS^2 - kT\ln N$. As $N \to \infty$, ΔF is negative at arbitarily small temperatures hence there is no stable state with long-range order and no phase transition.

TABLE 2.

Magnetic phase transitions as a function of space and spin dimensionality

d	n = 1(Ising)	2(xy)	3(Heisenberg)
1	No	No	No
2	Yes	(Yes)	No
3	Yes	Yes	Yes

that the temperature variation of the magnetization is $M(T) = M(O)(1 - B'T^{3/p}$ the Bloch $T^{3/2}$ law for a feromagnet, where $B' \propto D'^{3/2}$ Furthermore the low-temperature specific heat of magnetic C_m origin will also vary as $T^{3/p}$. At higher temperatures, spin wave interactions become important, and correction terms involving higher powers of T must be taken into account in the expressions for the magnetisation and specific heat [15].

So far, we have considered the ferromagnet as the model of collective magnetic order, because the order parameter and its conjugate field have a straightforward interpretation. But in fact no silicates order ferromagnetically; they exhibit a rich variety of other sorts of magnetic order, discussed in Part 2, which come about because some or all of the exchange interactions J_i are negative. A dissymmetry exists between the two signs of J. If J is positive, the result will be collinear ferromagnetic order regardless of crystal structure; even noncrystalline solids can order ferromagnetically [16]. The possibilities when J is negative are more interesting because they depend on the topology of the lattice [17]. If the lattice can be divided into two interpenetrating sublattices so that the cations on one sublattice interact only with neighbours on the other, (bcc or NaCl cation structures, for example) then the molecular field theory (eg equations 11 - 13) can be simply carried over to the antiferromagnetic case [7, 12]. M_s is now the sublattice magnetization, and it is the staggered susceptibility that diverges at the magnetic ordering temperature T_N (the Néel point). The normal susceptibility just shows a cusp-like maximum there; θ in equation 13 is negative, and equal to $-T_N$. The power in the dispersion curve for antiferromagnetic magnons, however, is $p = 1$ (like phonons) so the sublattice magnetization and specific heat at low temperatures both vary as T^3. In fact there is a small gap at $q = 0$ due to magnetocrystalline anisotropy, which leads to expontential behaviour at the lowest temperatures.

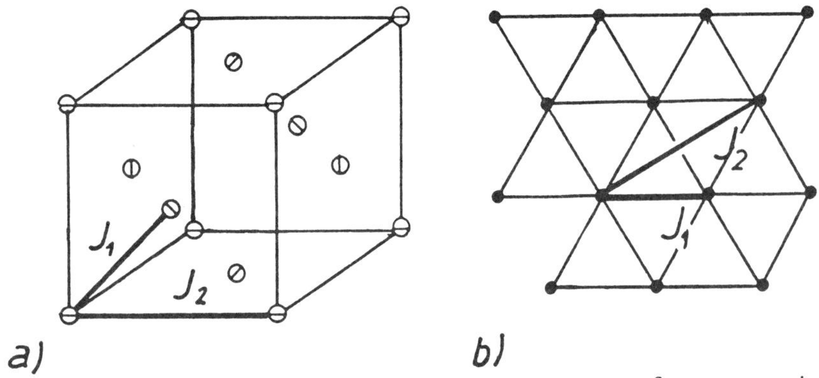

Figure 10. Examples of simple structures where antiferromagnetic exchange bonds are frustrated. a) fcc, b) 2d triangular. Nearest neighbour and next-nearest neighbour exchange paths are shown.

However, for other lattices, barely more complex such as the fcc or triangular lattices, it may be impossible to find a spin structure which will satisfy all the nearest-neighbour exchange bonds simultaneously (Figure 10). In such cases, some of the interactions are frustrated, and there are typically a number of degenerate possible ground states, with the same energy. The system may adopt a noncollinear ground state with more than two sublattices; examples of three and four sublattice structures are shown in figure 11). Noncollinear spin structures in ionic materials have recently been reviewed by Coey [17]. Weak longer-range interactions may help stabilize one of the degenerate structures, as for example in MnO (fcc lattice). Furthermore, the Néel point of a system with frustrated antiferromagnetic interactions will be much lower than that expected from the strength of the individual bonds. For the fcc lattice, the molecular field theory adapted to take account of the four interpenetrating cubic sublattices, predicts a collinear structure with $T_N = -\theta/3$; each magnetic sublattice consists of two crystalline sublattices, so that only two thirds of the neighbours of a given ion can have their spins aligned antiparallel to it; the remaining one third are parallel at an energy cost of $4JS^2$ per bond.

Still more varied magnetic ground states arise when there are ferromagnetic and antiferromagnetic interactions with different shells of neighbours [17]. Consider the planar structure illustrated in figure 12, where there is a ferromagnetic interaction J_1 with Z_1 neighbours within the planes and antiferromagnetic coupling J_2 with Z_2 neighbours in adjacent planes. The result is a planar antiferromagnetic structure with the moments of alternate ferromagnetic planes aligned in opposite directions. If J_2 is positive, but there is a weak third-neighbour interaction J_3 such that $0 < Z_2J_2/Z_3J_3 < -4$, then a helical structure will result with twist angle Φ from plane to plane given by $\cos\Phi = -Z_2J_2/4Z_3J_3$. Besides competing exchange, noncollinear structures may also arise when single-ion anisotropy favors different easy directions at different crystallographic sites in a structure.

Finally we consider what happens when an external magnetic field

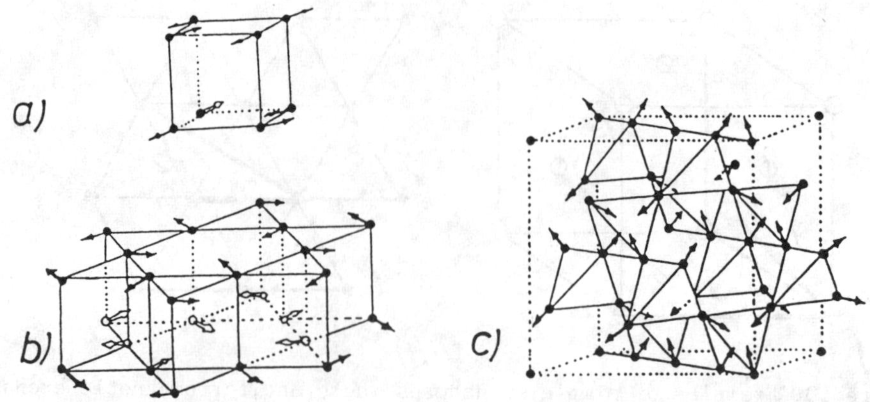

Figure 11. Magnetic structures of three crystalline polymorphs of FeF_3. Interactions in a) are unfrustrated, whereas in b) and c) there is increasing frustration. a) is a two-sublattice collinear antiferromagnet; b) and c) are three and four sublattice noncollinear structures. (after G. Ferey)

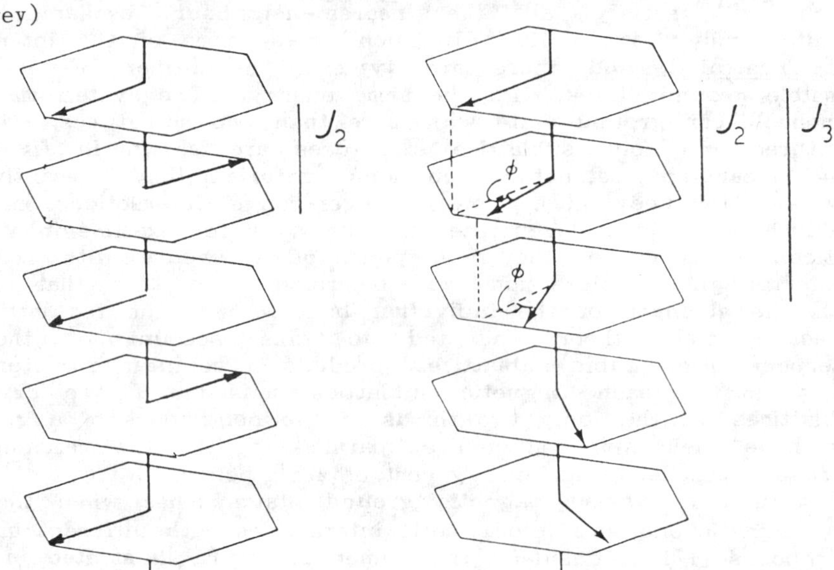

Figure 12. Some magnetic structures which can arise from the simultaneous presence of ferromagnetic and antiferromagnetic interactions: a) planar antiferromagnet, b) proper helix.

is applied perpendicular to the c-axis of the planar antiferromagnet of figure 12, supposing there is some in-plane anisotropy, represented by an anisotropy field B_a'. Depending on the relative strength of B_a' and the molecular field $B_0^{1\prime}$ representing the interplane exchange J_2 there can be an abrupt transition as a function of applied field, where the magnetization of both sublattices turns to lie along, or close to the

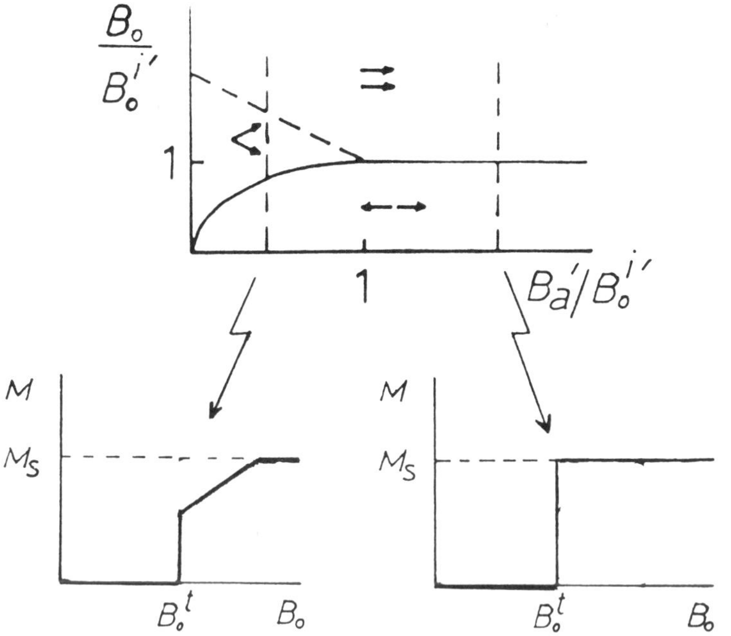

Figure 13. Magnetic phase diagram for a metamagnet in a field parallel to the easy direction of magnetization. Magnetization curves are shown at a) and b).

direction of the applied field. Two possible $M(B_o)$ curves are shown in figure 13. If $B_a' > B_o^{i'}$, the threshold field for the discontinuous transition is $B_o^t = B_o^{i'}$; otherwise $(B_o^t)^2 = (2B_o^{i'} - B_a)B_a$. The former case, where the jump is from an antiferromagnetic to a collinear ferromagnetic structure is a metamagnetic transition, whereas the latter is also known as a spin flip. It is possible to deduce the interplane exchange coupling J_2 from the value of B_o^t.

1.4 Cation disorder

A feature of natural silicates which is bound to effect their magnetic properties is chemical disorder on equivalent lattice sites. Since none of the major cations which can substitute for Fe^{2+} or Fe^{3+} is magnetic, chemical disorder amounts to magnetic dilution, removing the sites occupied by Mg^{2+}, Al^{3+}... from the magnetically-interacting network. A small amount of nonmagnetic substitution in an iron end-member, (Fe_xM_{1-x}) with $x \lesssim 1$ will, reduce the magnetic ordering temperature in proportion to x; the expression for the Curie temperature given after equation 13 becomes $T_c = 2(xN)JS(S+1)/3k$ where the magnetic coordination number Z is replaced by xN, N being the number of neighbouring cation sites. Proportionality of T_c and x cannot be maintained at small values of the iron concentration because isolated iron sites will appear, and ultimately the iron will be found in small groups - single ions, dimers, trimers - which cannot interact

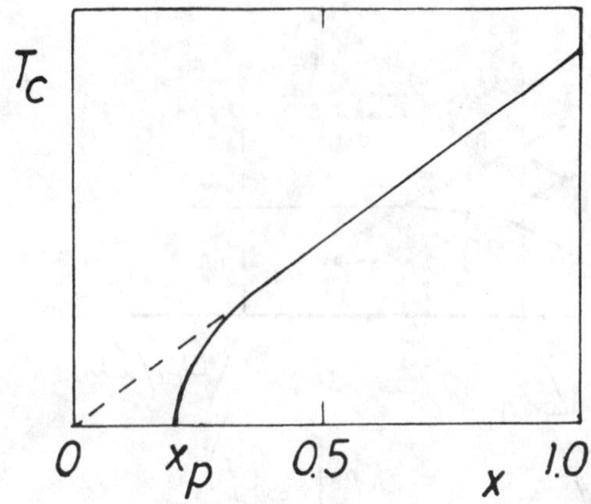

Figure 14. Variation of the magnetic ordering temperature of a solid solution of magnetic (x) and nonmagnetic ions.

with each other via nearest-neighbour exchange coupling. The critical concentration at which an infinite cluster appears is known as the percolation threshold x_p. It depends on the structure of the crystal lattice, but an approximate expression is $x_p \simeq 2/N$. It is only when $x > x_p$ that a collective phase transition can occur; the variation of T_c with x is sketched in figure 14 (A thermodynamic phase transition requires an infinite system, because it is only possible to generate a singularity in the free energy from an infinite sum of non-singular terms). Percolation is discussed in some detail by Zallen [18].

Magnetic interactions in small clusters of ions will modify the susceptibility and specific heat, although there cannot be any spontaneous magnetization. Consider, for example, a ferric dimer with hamiltonian $H = -2JS_1.S_2$ and negative J. The two spins are coupled to form a total spin $\tilde{S} = 5$, and there is a set of 11 energy levels with a nonmagnetic ground state having $M_{\tilde{S}} = 0$. The magnetization curve at T = 0 will be a staircase and the specific heat will be a series of superposed Schottky anomalies. As the number of interacting ions in the cluster increases, so the specific peak becomes sharper, but it shows a singularity only for an infinite cluster.

Returning to the influence of cation disorder on the magnetic ordering transition, a question of interest is how much inhomogeneity can the system support and yet exhibit a single phase transition? One extreme would be when the system is composed of grains of significantly different iron contents. Each one would then have a transition at a temperature T_c proportional to x, and the transition for the whole sample would appear smeared out. However, if we suppose that the disorder is quite statistical, the variation in T_c from one region to another will be $\Delta T_c \simeq x(\xi/a)^{3/2}$, where ξ is the size of the region and a is the interatomic spacing. However, at a phase transition the correlation length diverges as $(T/T_c-1)^{-\nu}$. If the material is to show a single transition, the correlation length must diverge without mismatch, so that ΔT_c for the two adjacent regions must be less than $|T - T_c|$. Hence $\nu > 2/3$ [19]. The correction length exponent is related to the specific heat exponent via the scaling relation $\alpha = 2-d\nu$ (which, incidently, shows that the mean field theory exponents given in section 1.3 are only consistent with $d = 4$). Hence the phase transition for the system will not be destroyed by statistical cation disorder provided $\alpha < 0$, which is the case (just) for the 3d Heisenberg model.

Effects of disorder become more severe when the exchange interactions in the system are already frustrated. What happens then is typically appearance of a random, noncollinear magnetic structure [17]. Many examples are found among the ferrites - ternary iron oxides with the spinel structure, such as $[Mn]\{Fe_2\}O_4$. There are both tetrahedral [A] and octahedral {B} cation sites, in the ratio 1:2. The principal exchange interactions J_{AB} and J_{BB} are both negative, but the former is dominant so the A and B sublattices are coupled antiparallel in a ferrimagnetic spin structure. However in $[Mn_{1-x}Zn_x]\{Fe_2\}O_4$, for example, the A-B interactions experienced by a cation at a particular B site are weakened by the number of nonmagnetic Zn A neighbours. If $Z_{AB}J_{AB} \sim Z_{BB}J_{BB}$, the energy of the cation may be minimized by canting its moment away from the ferrimagnetic axis. The net magnetization is thereby reduced. The canting angle depends essentially on the number of magnetic A neighbours, and it is uncorrelated from site to site.

Randomly canted ferrimagnets are an example of a much larger class of disordered magnets with competing exchange interactions, the spin glasses [16]. A distribution of exchange interactions including positive and negative values of J with equal probability leads to a magnetically 'ordered' state where the spins are frozen in random directions with, on average, no correlations whatsoever between them, i.e. $\langle S_i(0).S_j(r)\rangle = 0$. The system enters such an ideal spin-glass state from the paramagentic state on cooling below the spin freezing temperature T_f. Spin dynamics in the neighbourhood of T_f are quite different and much slower than those near an ordinary Curie or Néel point and the magnetization depends on the thermal history of the sample. Despite a vast amount of theoretical and experimental work on the subject, the nature of the transition to the spin glass state is still a matter of controversy. Part of the problem experimentally is to reach equilibrium, which may take years a degree or two below T_f. The following views find their partisans:

i) There is no phase transition, but the spin freezing simply reflects the sluggish dynamics of the system (rather like the glass transition in a polymer).

ii) There would be a phase transition at $T = 0$, but it is inaccessible for the same reason.

iii) There is a thermodynamic phase transition at T_f, but it belongs to a quite different universality class and has a different order parameter from any of the systems discussed in the previous section. For example, mean field theory is thought to be exact in six dimensions for spin glasses, rather than four.

There has also been much interest in the passage from a normal ferro-, antiferro- or ferrimagnetic state to spin glass, with increasing disorder in the distribution of exchange interactions. The diagram in figure 15 shows what happens [17].

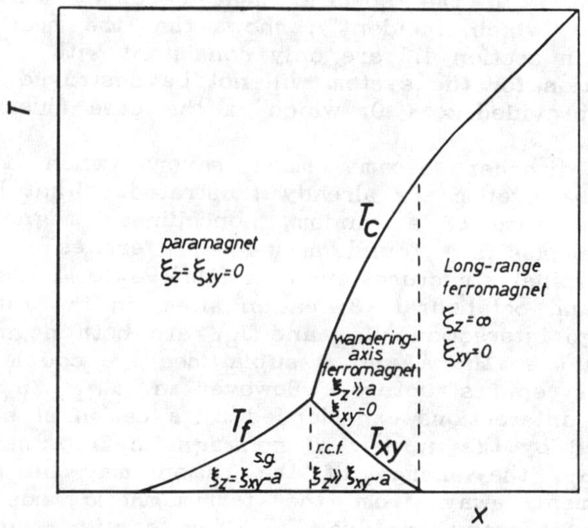

Figure 15. Magnetic phase diagram for a diluted , random ferromagnet with antiferromagnetic second-neighbour interactions. A similar diagram is obtained for a frustrated diluted antiferromagnet. rcf and sg are randomly canted and isotropic spin glass states.

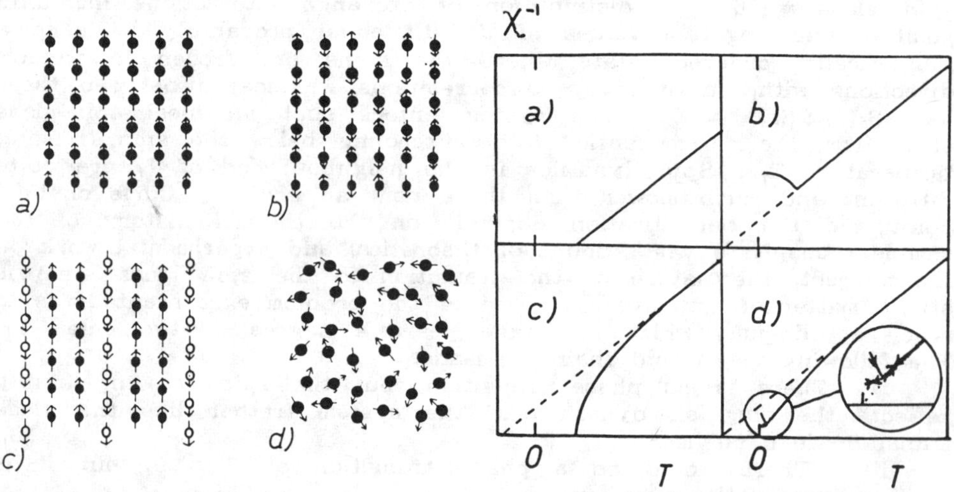

Figure 16. Some magnetic structures with the corresponding temperature-dependence of the inverse susceptibility: a) ferromagnet, b) antiferromagnet, c) ferrimagnet and d) speromagnet (amorphous spin glass).

Finally, we summarize in figure 16 some of the representative magnetic structures encountered in this part, and the corresponding variation of the susceptibility.

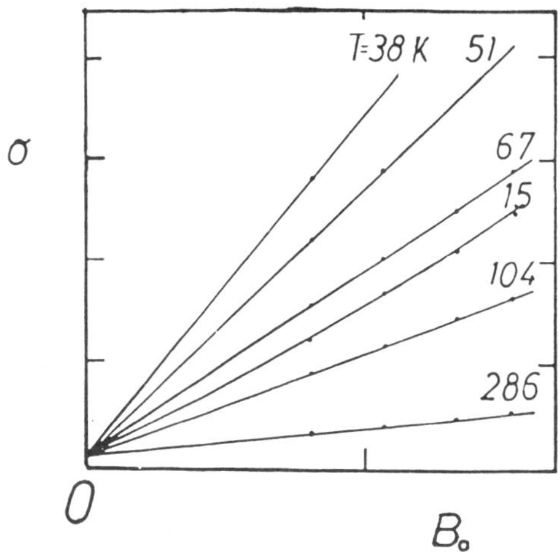

Figure 17. Magnetization curves of hedenbergite, showing the presence of a trace of magnetite impurity.

1.5 Experimental methods

Methods of investigating magnetic order in minerals are the following:

i) Magnetization and susceptibility measurements. Large fields (>10 T) are required to approach saturation at low temperatures, especially for ferrous minerals with large single-ion anisotropy. Susceptibility measurements should be made in at least two fields in order to separate the true susceptibility, which is linear in field, from the contribution of traces of any readily-saturated ferrimagnetic component (generally magnetite in natural samples). See figure 17.

ii) Magnetic neutron diffraction [20]. This is the method, *parexcellence* for determining magnetic structures of powders and single crystals. It requires access to neutron beams (ILL, Brookhaven,...)

iii) Mössbauer spectroscopy of ^{57}Fe [21] can provide a variety of information on site occupancy, second-order crystal field parameters, orbital ground state and level scheme of Fe^{2+} and magnetic order via the hyperfine interactions and their temperature dependence.

iv) Specific heat measurements [55] (corrected for phonon contributions) generally show λ anomalies at magnetic transitions (but not at the spin freezing temperature of spin glasses), and a characteristic power-law variation at low temperatures due to magnons. Broad Schottky anomalies may also appear due to crystal-field splitting of the Fe^{2+} energy levels, or magnetic interactions within isolated groups of ions.

v) Other techniques capable of throwing some light on magnetic order in silicates include optical spectroscopy [8], esr, ultrasonic measurements (for the elastic constants) and measurements of bulk and linear magnetostriction.

2. MAGNETIC ORDER IN SILICATES

A Iron End Members

We consider first silicates where the effects of cation disorder on the magnetic properties are slight; these are natural or synthetic samples which approach the composition of the ideal iron end-member. For magnetic studies, it makes sense to adopt a structural classification based on the connectivity of the iron sites (usually octahedral), rather than the silica tetrahedra as is traditional in mineralogy. Many silicates fall in the same category whichever scheme is used, but there are exceptions such as olivine, which is a mineral with isolated SiO_4 groups, but a framework structure of connected octahedra.

2.1 Group Structures

Silicates which contain iron in isolated sites (i.e. with no common oxygen ligand) include the garnets almandine $\langle Fe_3^{2+}\rangle\{Al_2^{3+}\}[Si_2]O_{12}$ and andradite $\langle Ca_3\rangle\{Fe_2^{3+}\}[Si_3]O_{12}$. We adopt the convention that [ ], { } and ⟨ ⟩ refer to cations with 4-fold, 6-fold and greater then 6-fold oxygen coordination. Part of the garnet structure is illustrated in figure 18. The iron in almandine occupies the c-sites which have distorted eight-fold coordination, and are connected by SiO_4 tetrahedra. From the discussion of the short-range character of superexchange interactions in section 1.3, almandine might not be expected to order magnetically. In fact it does, with a Néel point of 7.5 K [22], and a large λ-anomaly in the specific heat near this temperature [23]. Andradite also orders antiferromangetically, at 11.5 K [24], and there too the specific heat exhibits a large anomaly [25]. These ordering temperatures indicate that the strength of superexchange interactions mediated by the SiO_4 tetrahedron can be significant. The bonding of the SiO_4 group is ionic and covalent, in roughly equal measure.

2.2 Chain Structures

We will take the minerals here, mainly pyroxenes and amphiboles, in order of increasing width of the chains of magnetic ions.

a) *Pyroxenes.* The pyroxene structure is indicated in figure 19. In hedenbergite, $\langle Ca\rangle\{Fe^{2+}\}[Si_2]O_6$, the M1 sites are occupied by Fe and M1 sites by Ca. Acmite, $\langle Na\rangle\{Fe^{3+}\}[Fe_2]O_6$, is the ferric analogue. Susceptibilities of the two compounds are compared in figure 20, from which it can be seen that a) the principal exchange interaction is <u>positive</u> in hedenbegite (θ = 35 K), but <u>negative</u> in acmite (θ = -46 K). Despite the principally-ferromagnetic exchange due to the near - 90° Fe^{2+} -O-Fe^{2+} bond angle for the edge - sharing M2 octahedra, hedenbegite, actually orders antiferromagnetically, as shown by the cusp in the susceptibility. There is a corresponding λ-anomaly in the specific heat at 35 K [28]. The obvious magnetic structure is one with ferromagnetic M1 chains, coupled antiferromagnetically by weaker,

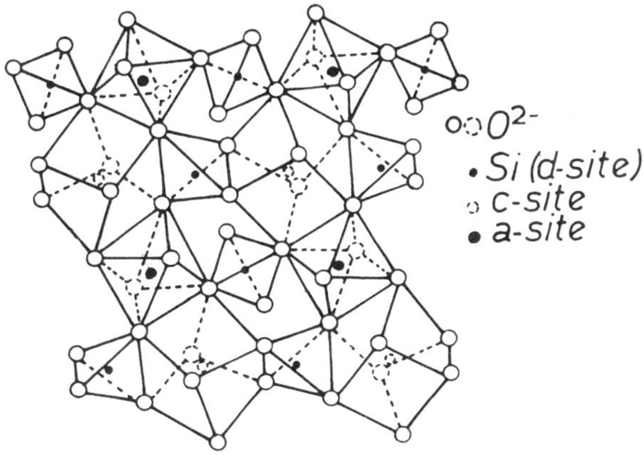

Figure 18. Part of the garnet structure projected onto (001).

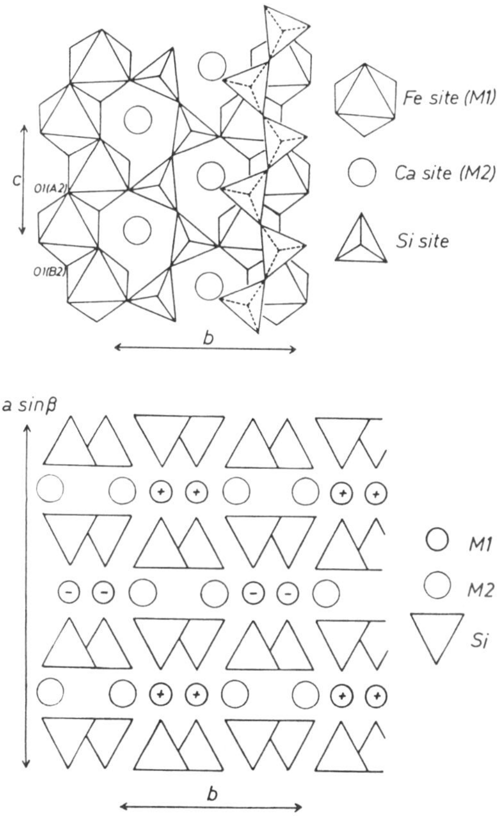

Figure 19. Structure of monoclinic pyroxene in an approximate projection onto (100). The lower parrt shows how the ribbons are packed, and indicates the magnetic structure of hedenbergite

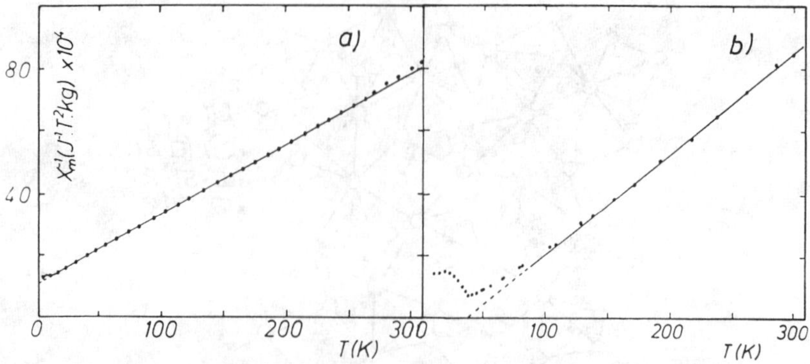

Figure 20. Inverse susceptibilities of synthetic samples of acmite (Fe^{3+}) and hedenbergite (Fe^{2+}). [26,27]

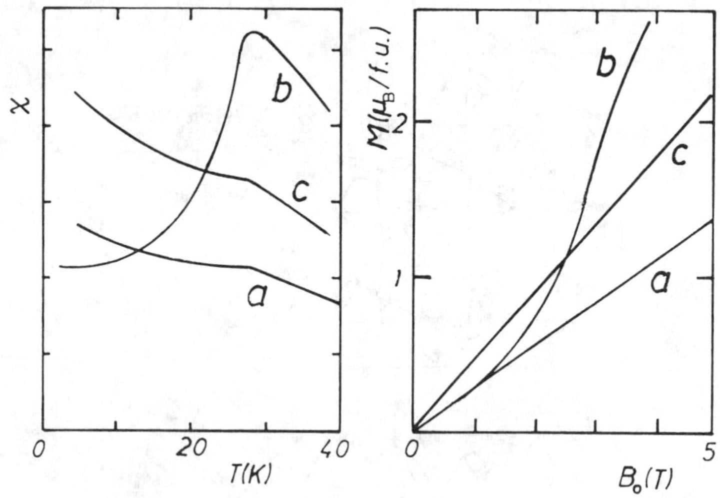

Figure 21 . Susceptibility and magnetization curves for a small crystal of natural orthoferrosilite [31].

negative interchain interactions [27]. Recent neutron diffraction studies on natural [29] and synthetic [30] samples have confirmed this structure, and the moments are found to lie in the ac plane at 135° to $\vec{c}$.

When both M1 and M2 sites are occupied by Fe^{2+}, the pyroxene structure becomes orthorhombic; orthoferrosilite $\{Fe_2^{2+}\}[Si_2]O_6$ is the ferrous end member. As in hedenbergite, the exchange is predominately ferromagnetic (θ = 30 K), and there is a linear antiferromagnetic structure, but moments lie along $\vec{b}$ [31,32] below the Néel point of 40 K. Below 8 K however the spins at both M1 and M2 sites are canted away from the $\vec{b}$ axis [32]. This transition, which is absent in hedenbergite, is most likely due to competing directions of single-ion anisotropy at M1 and M2 sites. The anisotropy of the highly-distorted M2 site is the greater, and favours the $\vec{b}$ axis, whereas that of the M1 site is less and favours a perpendicular direction. The levels issuing from the $^5T_{2g}$ triplet for M1 (figure 5) are almost degenerate [33]. Other interesting features of the magnetization are the strong anisotropy of the susceptibility of a natural crystal (figure 21a), which also indicates that $\vec{b}$ is the antiferromagnetic axis. There is an upturn in the magnetization measured $\parallel \vec{b}$ in fields greater than 2 T (figure 21b), which signals a spin flip transition. Also the temperature-dependence of the M1 sublattice magnetization can be fitted to $M = M_0\ (1-T/T_N)^{\beta}$, where β = 0.16. If this exponent carries over into the critical region, it indicates reduced effective magnetic dimensionality (β for the 2d Ising model is ⅛).

b) *Amphiboles.* The structures of amphiboles closely resemble those of pyroxenes, except that the octahedral ribbons are wider (figure 22) with four cation sites, in the ratio 2:2:1:2; M4 is the equivalent of the pyroxene M2.

In crocidolite (blue asbestos) and the related iron-rich riebeckite minerals, M4 is occupied by Na. Charge balance is maintained by putting Fe^{3+} on M2 and keeping Fe^{2+} on M1 and M3 sites at the centre of the ribbon. Crocidolite orders at T_N = 30 K in a structure with ferromagnetic ribbons coupled antiferromagnetically to their neighbours. The spin direction at 5 K lies at 36° to $\vec{b}$ [34].

The ferrous end-member grunerite has Fe^{2+} in all four sites. As usual in ferrous chain structures of this type, the exchange is principally ferromagnetic (θ = 67 K, figure 23). Each ribbon is ferromagnetic, with moments along $\vec{b}$, but couples antiferromagnetically to its four neighbours. As in orthoferrosilite, there is a spin canting below 8 K involving M1 and M4 sites, and a second peak in the susceptibility at this temperature. The spin canting is shown clearly by the appearance of an 010 magnetic reflection in neutron diffraction [37].

Both grunerite and crocidolite exhibit spin flip transitions with an onset at about 1.5 T at 4.2 K (figure 23b). Analysis of this transition in pyroxenes and ampiboles leads to the result that the ratio of intra-ribbon exchange coupling to inter-ribbon exchange coupling is of order 5 [31,34]. This is too small for the materials to show clear effects of reduced spatial dimensionality in their critical behaviour, and it indicates again that significant interactions can propagate via SiO_4 groups.

c) *Other Chain Structures.* Ilvaite is the only other chain silicate whose magnetic properties have been examined in detail [38]. It is a mixed valence ribbon silicate with ideal formula $\langle Ca\rangle\{Fe_2^{2+}Fe^{3+}\}[Si_2]O_8(OH)$. The double chains of edge-sharing octahedral A-sites are occupied by a mixture of Fe^{2+} and Fe^{3+}, whereas the Fe^{2+} B-site octahedra are attached above and below the chains (figure 24a). The B-site charges are disordered above about 370 K, where the structure is orthohombic and there is evidence of charge relaxation (electron hopping) in Mössbauer spectra with an activation energy of 0.1eV [39]. A specific heat anomaly around 370 K

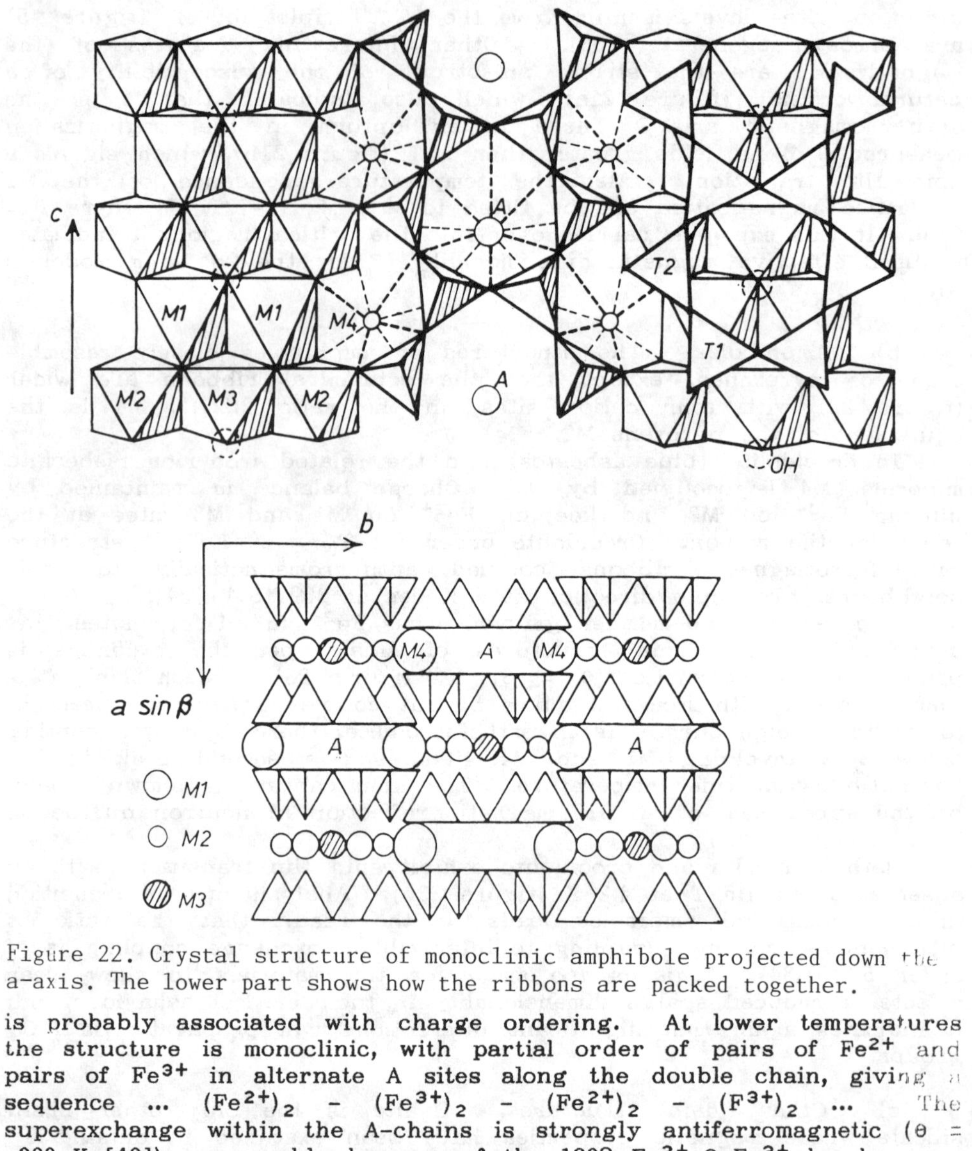

Figure 22. Crystal structure of monoclinic amphibole projected down the a-axis. The lower part shows how the ribbons are packed together.

is probably associated with charge ordering. At lower temperatures the structure is monoclinic, with partial order of pairs of Fe^{2+} and pairs of Fe^{3+} in alternate A sites along the double chain, giving a sequence ... $(Fe^{2+})_2$ – $(Fe^{3+})_2$ – $(Fe^{2+})_2$ – $(F^{3+})_2$ ... The superexchange within the A-chains is strongly antiferromagnetic (θ = -300 K [40]), presumably because of the 180° Fe^{2+}-O-Fe^{3+} bond across the chain. The Néel point is 116 K, with a magnetic structure composed of two lines of oppositely-directed spins along the double chains, but long-range magnetic ordering of B sites is only complete below 36 K. The B sites, if they were symmetrically positioned above the A chains, would not be expected to order at all because they interact with two Fe^{2+} and two Fe^{3+} ions, each pair having opposite spin. This is an example of a *'spin fou'* or idle spin. In fact a slight distortion of the structure permits the B sites to order at low

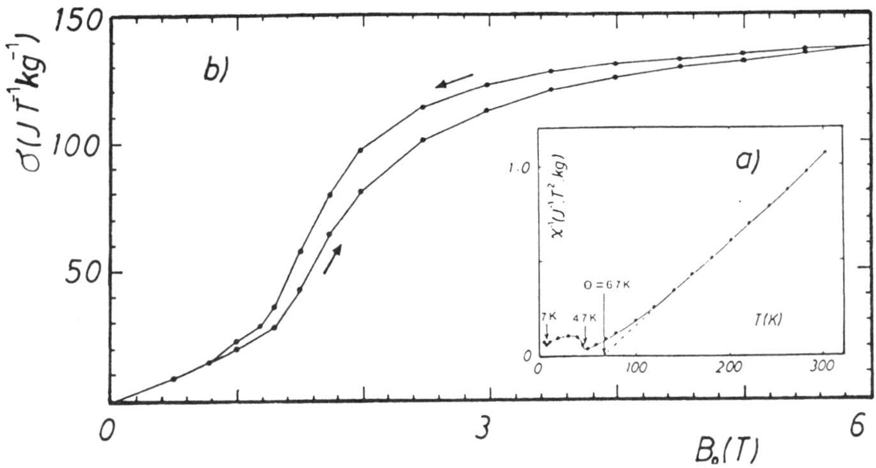

Figure 23. Magnetic properties of grunerite: a) inverse susceptibility and b) magnetization curve at 4.2 K [37].

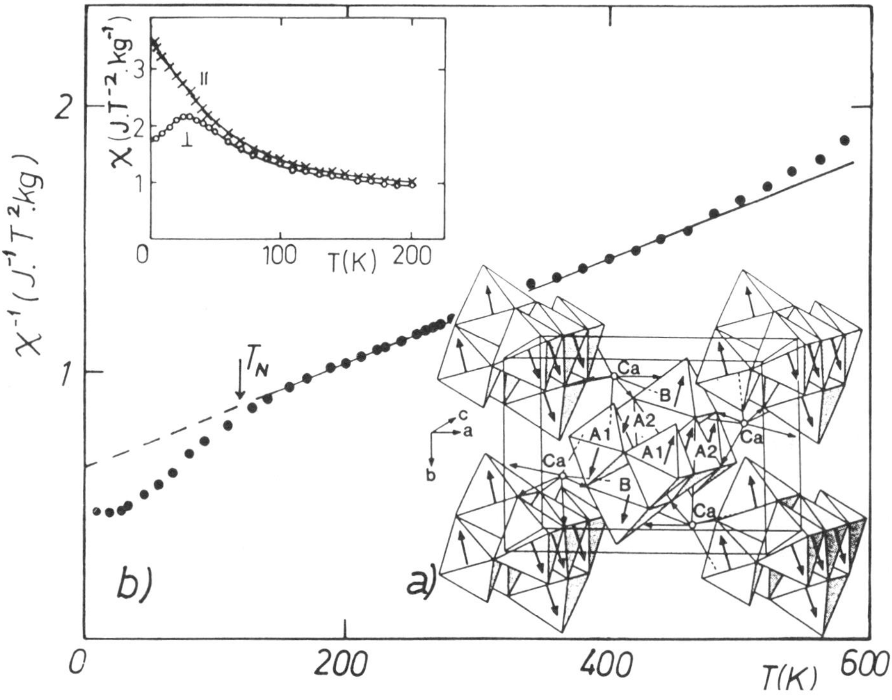

Figure 24. a) Crystal and magnetic structure of ilvaite and b) its susceptibility

temperatures, also with their moments along the b-axis.

Other chain silicates which order magnetically include ferrobustamite, which is another structural variant of $FeSiO_3$, and deerite which has broad bands of cations including nine distinct octahedral sites.

2.3 Sheet Structures

The magnetic properties of natural sheet silicates have been quite extensively investigated using magnetization and susceptibility measurements, Mössbauer spectropscopy and neutron diffraction [41-44], and there are recent reviews of the data on these minerals [6, 45].

The two main structural groups are based either on a 1 : 1 layer associating a sheet of silica tetrahedra and a sheet of edge-sharing metal octahedra (serpentine group) or a 2:1 layer where the octrahedral sheet is sandwiched between two tetrahedral sheets (mica group). They are illustrated in figure 25. There is a close relation, both structurally and magnetically, with amphiboles and pyroxenes since the ribbons in the chain silicates can be considered as slices of the 2:1 layers. In either group the octahedral sites may be all occupied - the dioctahedral case - or 1/3 of them may be vacant with the vacancies ordered or random - the trioctahedral case. Three different possible cation arrangements are shown in figure 25c). The separation between octahedral sheets (7 - 16 Å is much greater than the cation-cation distance within them (2.7 Å) so the exchange coupling might be expected to impart a quasi-two-dimensional magnetic character to the minerals. Structures of sheet silicates are reviewed by Bailey [46].

The ferrous end-members of the 1:1 and 2:1 structural families are greenalite $\{Fe_3{}^{2+}\}[Si_2]O_5(OH)_4$ and minnesotaite. $\{Fe_3^{2+}\}$ $[Si_4]$ O_{10} $(OH)_2$. Minnesotaite does not have exactly the same structure as talc [47] but from a magnetic point of view, the differences are unimportant. Both minerals have rather similar magnetic properties [41,48]. A peak in the susceptibility at low temperatures suggests antiferromagnetic order, yet samples show ferromagnetic hysteresis at 4.2 K (figure 26). The susceptibility of ferrous sheet silicates is quite anisotropic, χ being greatest in the plane of the sheets and least in the perpendicular c′ direction. The anisotropy and related properties can be adequately explained by the crystal field model with a trigonal component parallel to c′ which stabilizes the orbital singlet ground state (figure 5). The powder average susceptibility however is almost independent of crystal field, and gives θ = 24 K for greenalite and 38 K minnesotoite. Exchange interactions within octahedral sheets are ferromagnetic, with $J \simeq 1$-2 K. The Neél temperatures are 17 and 28 K, respectively.

Understanding of the reason for the paradoxical combination of symptoms of ferromagnetic and antiferromagnetic order comes from the neutron diffraction patterns. The magnetic scattering, obtained as the difference between the diffraction patterns above and well below the Néel point shows a sequence of reflections of the type 00½,

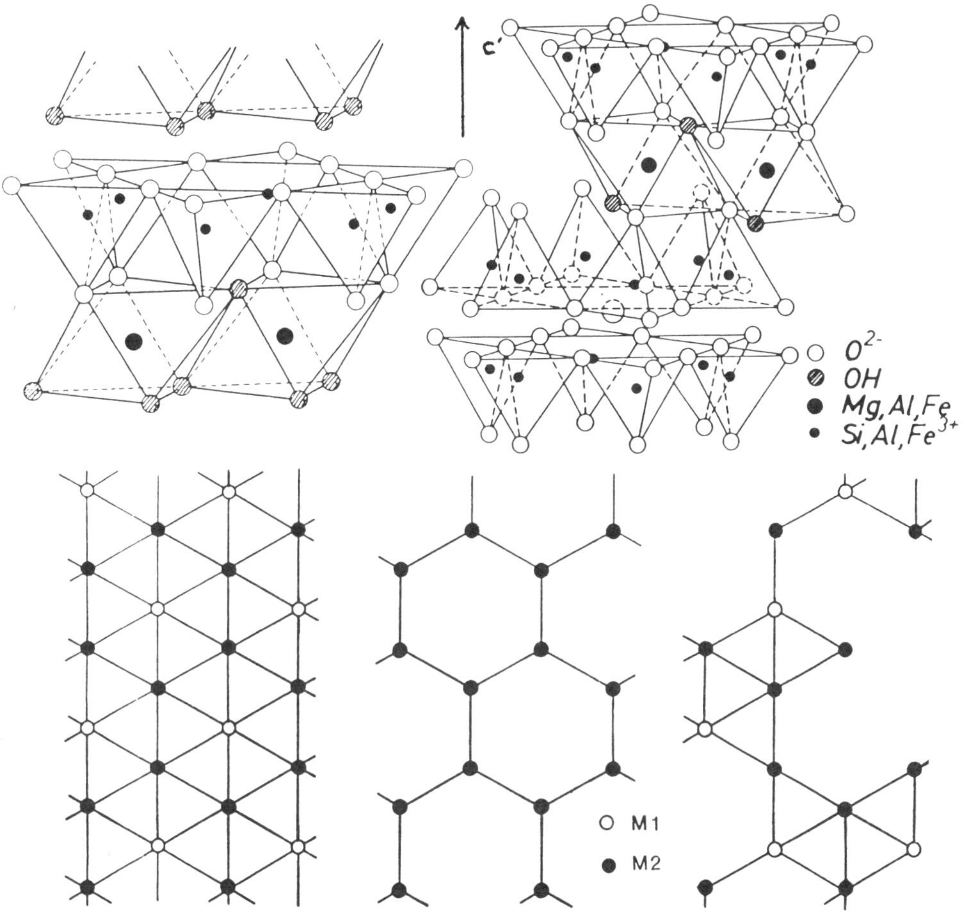

Figure 25. Schematic structure of sheet silicates: a) 1:1 layer, b) 2:1 layer, c) possible cation occupancy schemes of the octahedral sheet.

003/2, ... which show that the magnetic cell is doubled along c´. The ferrous sheet silicates are planar antiferromagnets (figure 12a) with the magnetization of each ferromagnetic octahedral sheet lying in its plane, and antiferromagnetic coupling from sheet to sheet. The spin flip transistion occurs in such a small applied field (≃0.05 T) that ferromagnetic hysteresis appears. The transition is spread out in field in a powder sample, but a detailed neutron study in an applied field shows that once the planar antiferromagnetic state is broken up, the sample never returns to it at any point around the hysteresis loop [48]. Only by heating above T_N and cooling in zero field can the mineral recover its planar antiferromagnetic structure. Analysis of the threshold field shows that the intersheet coupling is 10-50 times weaker than intrasheet coupling [41-45], so the sheet silicates are better approximations to low-dimensional magnets than the chain silicates. Some characteristic low-angle scattering due to short-range magnetic correlations has been seen in minnesotaite over an extended

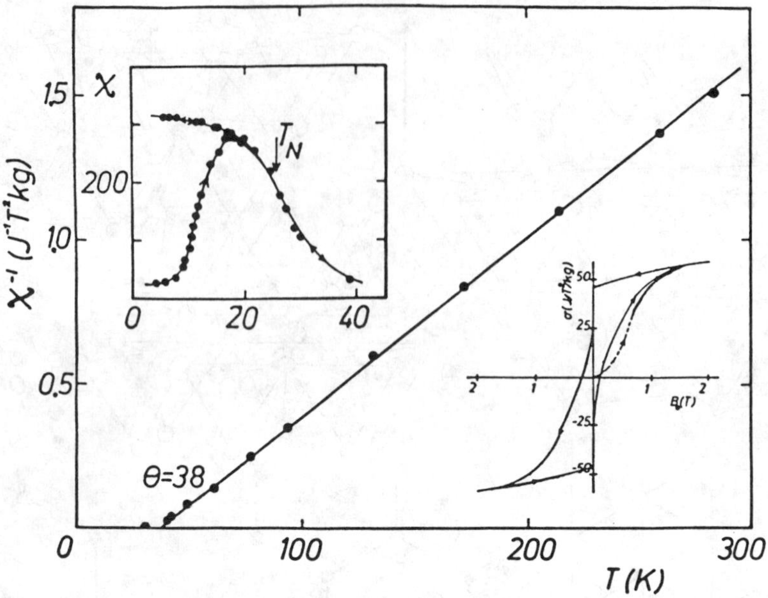

Figure 26. Magnetic properties of minnesotaite.

range of temperature on either side of T_N [48].

Cronstedtite is an unusual 1:1 layer trioctahedral mineral. The ideal formula $\{Fe_2^{2+}Fe^{3+}\}$ $[Si\ Fe^{3+}]$ O_5 $(OH)_4$ shows half the tetrahedral silicon replaced by ferric iron. It is the silicate with the highest electrical conductivity due to $Fe^{2+} \leftrightarrow Fe^{3+}$ charge hopping, and has a charge ordering transistion in the liquid nitrogen range. It orders magnetically below 12 K in a structure with antiferromagnetic octahedral sheets [49].

Ferripyrophyllite is the trioctrahedral ferric end member, with Fe^{3+} ordered on M2 sites in the octahedral sheet, giving an unfrustrated honeycomb magnetic lattice. The principal exchange coupling is antiferromagnetic, and ferripyrophyllite orders antiferromagnetically below 18 K [44].

2.4 Framework structures.

The olivine structure is a framework from the point of view of the metal ions. There are chains of edge-sharing octahedral M1 sites running parallel to $\vec{b}$ which are connected to sheets of corner-sharing octahedral M2 sites perpendicular to $\vec{a}$ as shown in figure 28. The magnetic properties of olivines have recently been reviewed by Fuess et al [50]. The ferrous mineral fayalite $\{Fe_2\}[Si]O_4$ orders antiferromagnetically at 65 K where there is a large specific heat anomaly [51]. The magnetic ground state is noncollinear, with M2 moments and the net sublattice magnetizations along $\vec{b}$ but M1 moments are canted away from $\vec{b}$ and lie in directions nearly pependicular to $\vec{a}$. The structure of each M1 chain is canted ferromagnetic, whereas that

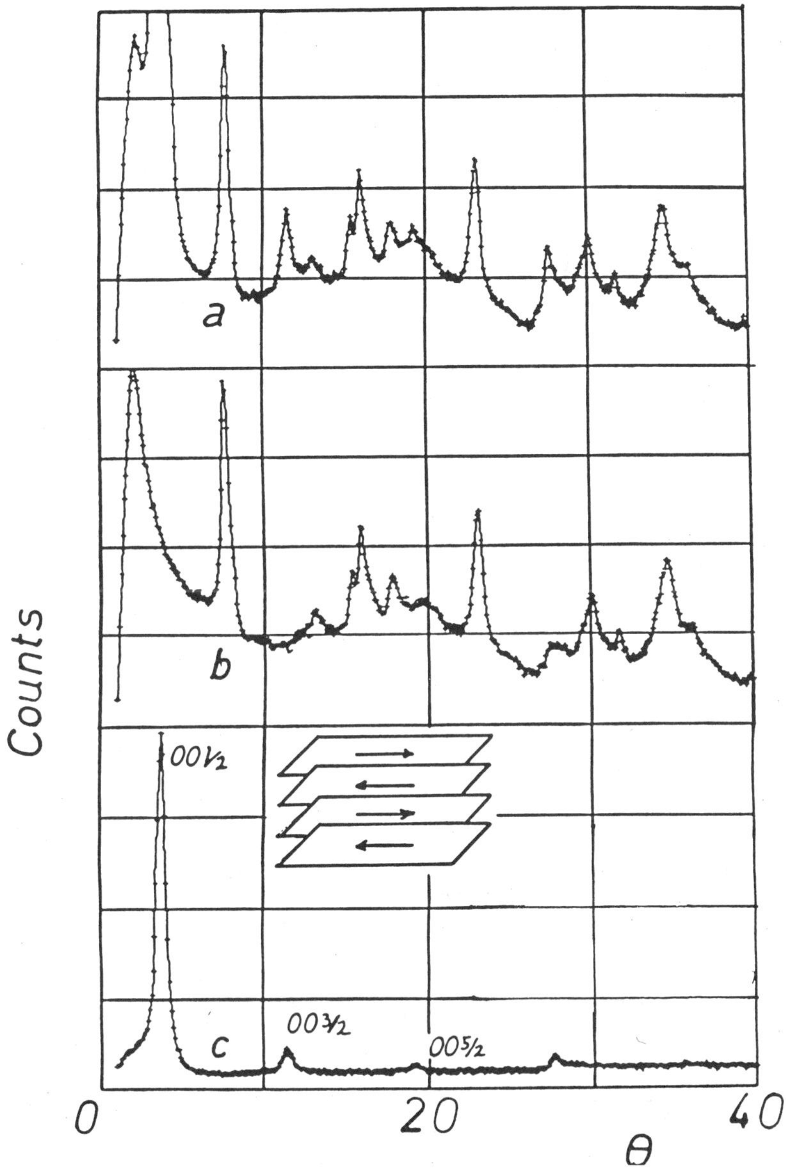

Figure 27. Neutron diffraction patterns from minnesotaite; a) and b) are above and below the magnetic ordering temperature, c) is the difference diagram showing the coherent magnetic scattering. The planar antiferromagnetic structure is shown in the insert.

of the M2 sheets is collinear antiferromagnetic. It was originally thought that the canting sets in only below a temperature $T_k \simeq 23$ K where there is a maximum in the powder susceptibility due to a maximum in the c-axis component (figure 29), but recent neutron and

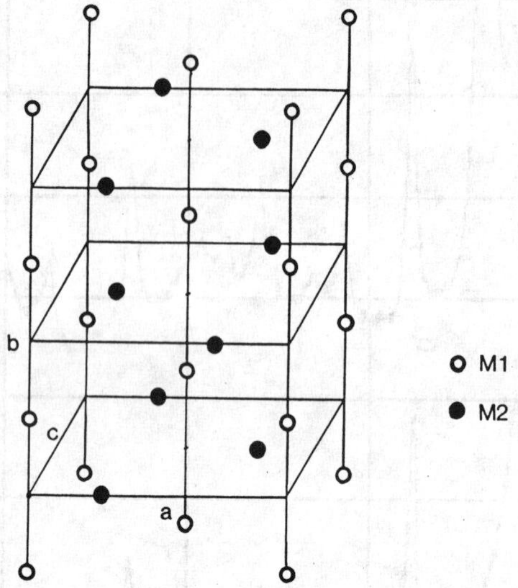

Figure 28. The cation structure of olivine.

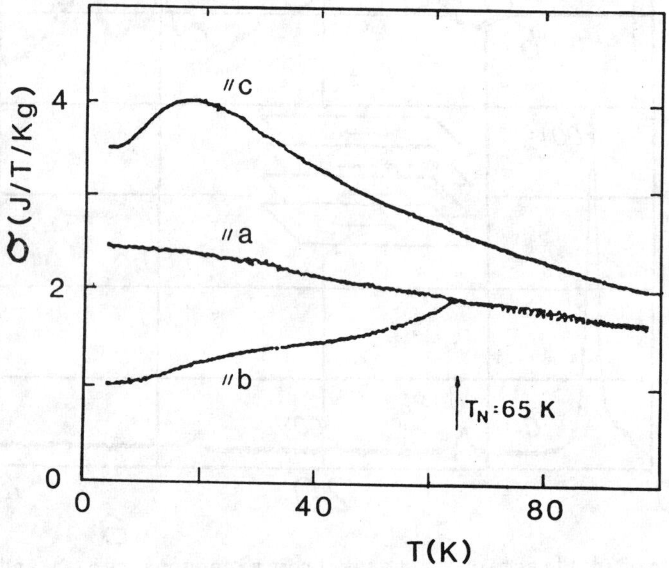

Figure 29. Magnetic susceptibility of a fayalite crystal [50].

Mössbauer studies have established, that the M2 site canting persists at all temperatures up to T_N so there is no second magnetic phase transition at T_k [50]. The likely cause of the noncollinear structure is competing single-ion anisotropy, the crystal field favoring the c-axis as easy direction for M1 sites and the b-axis for M2 sites.

The iron-deficient ferric olivine laihunite is a special case magnetically [52]. Half the M1 sites in each chain are vacant, and M2 sites are occupied by Fe^{3+} in the ideal formula $\{Fe^{2+}_{0.5}\,\square_{0.5},Fe^{3+}\}[Si]O_4$. The ferric iron in the M2 sheets orders antiferromagnetically below about 160 K, in such a way that coupling with the M1 iron exactly cancels (another example of a '*spin fou*'). The magnetic modes of M1 and M2 sites are decoupled. Magnetic ordering of M1 iron at lower temperatures, ≃ 80 K, may be associated with a structural distortion.

The mineral kirschsteinite $\{Ca,Fe\}[Si]O_4$ has M1 sites occupied by ferrous iron, with Ca in M2. It is actually a chain silicate from the magnetic point of view. Surprisingly, it shows no sign of magnetic order down to 1.6 K which indicates that M1-M1 exchange is very weak for Fe^{2+}. By contrast, the Co and Mn analogous order antiferromagneticaly above liquid helium temperature with ferromagnetic and antiferromagnetic order along the chains, respectively.

Table 3 summarises the magnetic properties of the iron and member minerals discussed above.

B Cation Disorder

It should be noted that many of the end members for which magnetic structures have been determined are synthetic samples. Natural iron-rich silicates generally involve some cation disorder, and it is now necessary to consider what effect this can have on the magnetic ground state.

The effect of nonmagnetic substitution for iron on the Curie or Néel point of a simple magnetic system was shown in figure 14. However, it is clear from table 3 that almost none of the silicates whose properties we have reviewed are simple antiferromagnets in the sense that $T_N \simeq -\theta$. While the Fe^{3+}-Fe^{3+} interactions are generally negative, the Fe^{2+}-Fe^{2+} interaction in most sheet and chain silicates is actually positive. Fe^{2+}-Fe^{3+} interactions may be of either sign. Substitution of nonmagnetic ions, or iron ions of the other valence, can disrupt the local balance of exchange, and even provoke noncollinear spin structures [17].

Another point is that magnetic order in one or two dimensions is much more vulnerable to disruption by nonmagnetic impurities than it is in three dimensions. Long-range magnetic order in a linear chain, assuming that it existed, could be disrupted by a single impurity since the configurations ++□++ and ++□-- both have the same energy.

Most results on the effect of cation disorder on the magnetism of minerals are available for sheet silicates, but there is a little data on other systems. A natural orthopyroxene $\{Fe_{0.87}Mg_{0.13}\}[Si]O_3$, where Mg substitutes preferentially on M1 sites, has a Néel point of 26 K (as opposed to 38 K for the end member) although θ increases slightly [31]. The implication of the disproportionate reduction in T_N is that the M1 sites contribute more to exchange interactions consistent with the linear antiferromagnetic ground state than do M2 sites. An interesting result is that a natural Mg-substituted hedenbegite $\langle Ca\rangle\{Fe_xMg_{1-x}\}[Si_2]O_6$, with $x = 0.8$ was found to order as a linear

Table 3. Selected References to Magnetic Order in Iron End-Member Silicate Minerals

	Mineral	Ideal formula	T_N (K)	θ_p (K)	Magnetic Structure	Comments	References
Group Structures	Almandine	$\langle Fe_3^{2+}\rangle\{Al_2\}[Si_3]O_{12}$	7.5	-	antiferromagnetic	-	22.23
	Andradite	$\langle Ca_3\rangle\{Fe_2^{3+}\}[Si_3]O_{12}$	11.5	-	(antiferromagnetic)	-	24.25
Chain Structures	Hedenbergite	$\langle Ca\rangle\{Fe^{2+}\}[Si_2]O_6$	38	35	linear antiferromagnet	-	27.30
	Acmite	$\langle Na\rangle\{Fe^{3+}\}[Fe_2]O_6$	8	-46	(antiferromagnet)	-	26
	Orthoferrosilite	$\{Fe_2^{2+}\}[Si_2]O_6$	38	30	linear antiferromagnet	Spins cant below 10K	31.32
	Crocidolite	$\langle Na_2\rangle\{Fe_3^{2+}Fe_2^{3+}\}[Si_8]O_{22}(OH)_2$	30	27	linear antiferromagnet	-	34.35
	Grunerite	$\{Fe_7^{2+}\}[Si_8]O_{22}(OH)_2$	47	67	linear antiferromagnet	Spins cant below 8K	36.37
	Ilvaite	$\langle Ca\rangle\{Fe_2^{2+}Fe^{3+}\}[Si_2]O_8(OH)_2$	116	-300	antiferromagnetic double chains	B sites order collinearly below 36K	38
Sheet Structures	Greenalite	$\{Fe_3^{2+}\}[Si_2]O_5(OH)_4$	17	24	planar antiferromagnet	ferromagnetic hysteresis	41
	Cronstedtite	$\{Fe_2^{2+}Fe^{3+}\}[SiFe]O_5(OH)_4$	12	18	antiferromagnet		49
	Minnesotaite	$\{Fe_3^{2+}\}[Si_4]O_{10}(OH)_2$	28	38	planar antiferromagnet	ferromagnetic hysteresis	48
	Ferripyrophyllite	$\{Fe_2^{3+}\}[Si_4]O_{10}(OH)_2$	18	-22	antiferromagnet		44
Framework Structures	Fayalite	$\{Fe_2^{2+}\}[Si]O_4$	65	-37	canted antiferromagnet	Spins are canted from T_N	50, 51
	Laihunite	$\{Fe^{2+}Fe_2^{3+}\}[Si_2]O_8$	160	-	antiferromagnetic M2 sheets	M1 orders only below 80K	52

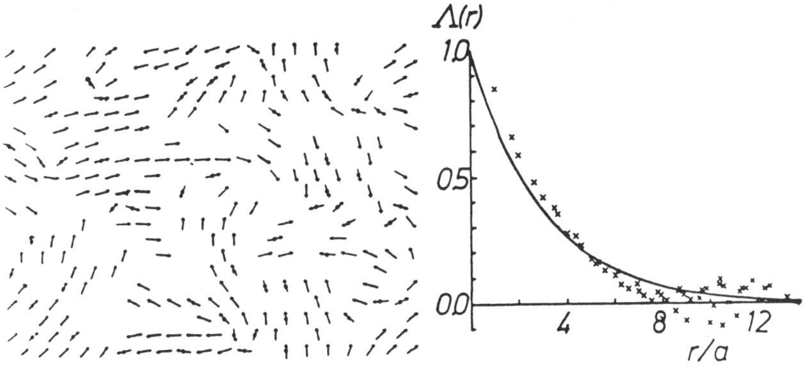

Figure 30. Results of Monte Carlo simulations of the ground state of a triangular lattice representing a trioctahedral ferrous mineral with ferric and magnesium impurities. The short-range character of the ferromagnetic spin correlations is illustrated [43].

antiferromagnet at T_N = 35 K, which is good evidence that pyroxenes are not magnetically one-dimensional. If they were, T_N would fall very rapidly when x decreases from 1.

The common iron-rich sheet silicates biotite and thurnigite (iron-rich chlorite) appear at first sight to behave analogously to greenalite and minnesotaite. They show a similar peak in susceptibility in the range 5-10 K, and exhibit ferromagnetic hysteresis at lower temperature. Nevertheless, magnetic neutron diffraction has revealed no trace of any magnetic Bragg peak down to 1.5 K [43,45] (although ferromagnetic reflections can be readily induced in a small external field). The reason for the absence of long-range magnetic order in these minerals which contain a great deal of iron appears to be the simultaneous presence of Fe^{2+} and Fe^{3+} in the octahedral sheets. Antiferromagnetically-coupled Fe^{3+} - Fe^{3+} pairs will appear in the normally ferromagnetic octahedral sheets. Monte-Carlo simulations with vector spins, an example of which is shown in figure 30, indicate that a statistical distrubution of Fe^{3+} and other, nonmagnetic ions leads to ferromagnetic short-range correlations [43]. The correlation length is only a few interactomic spacings, and there is no long-range order within the plane, or from plane to plane. The Fe^{3+} pairs turn the planar antiferrromagnet into a sort of two-dimensional spin glass.

Cation disorder may also be relvant in explaining why nontronite shows no sign of magnetic order, although the iron composition is practically the same as that of ferripyrophyllite. A random dioctahedral distribution, like that illustrated in figure 25c, would lead to frustration of the antiferromagnetic exchange and a greatly reduced spin freezing transition, $T_f \sim -\theta/10$ [16]. Mössbauer hyperfine splitting at 1.5 K, where there is no coherent magnetic Bragg scattering, suggests a randomly frozen spin structure.

Mössbauer spectra of olivines belonging to the fayalite-fosterite series $\{Fe_xMg_{1-x}\}_2[Si]O_4$ with $0.1<x<0.7$ have been interpreted in terms of a cluster model, where the effective size of the iron-rich regions was in the range 20 - 40Å, and increased with increasing x [53].

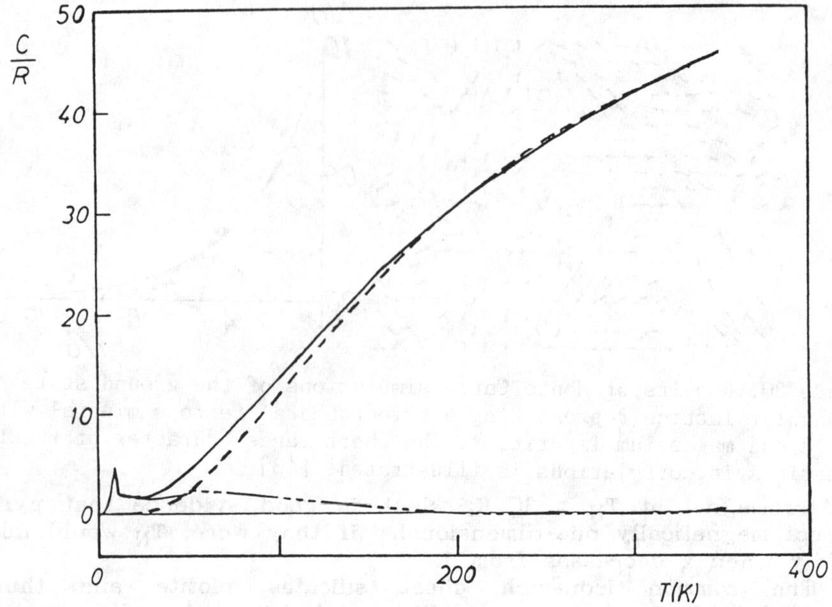

Figure 31. Specific heat of almandine garnet, compared with than of grossular, Anomalies due to magnetic ordering at 11.5 K and crystal-field splitting of the Fe^{2+} energy levels are visible [23].

Cluster models have been extensively applied to spin glasses, and the disappearance of the normal antiferromagnetic transition with cation disorder is characteristic of frustrated exchange interactions.

3. THERMODYNAMIC CONSEQUENCES

It has long been known that the entropy and enthalpy of iron minerals are significantly greater than those of their magnesium or aluminium counterparts. Excesses are found in iron end-members as well as in solid solutions. As an example, we show in figure 31 a comparison of the low-temperature specific heat of almandine (T_N = 11.5 K), and that of grossular. The magnetic and crystal-field contributions were discussed in principle in section 1.2b). Here we will give estimates these contributions for some iron silicates discussed above.

We have seen how the iron end-members order antiferromagnetically, typically with Néel points in the range 1 - 100 K. While some short-range dynamic spin correlations persist at temperatures above T_N, shown by the tail at high temperatures on the λ-anomaly in the specific heat [23], the full magnetic entropy is developed by $2T_N$ [54]. It is therefore safe to assume that at ambient temperature and above the spin disorder is complete. The magnetic contribution to the entropy S_m is readily calculated as R ln(2S+1) or 13.4 J/mole of Fe^{2+} (S=2) and 14.9 J/mole of Fe^{3+} (S=5/2). In order to calculate the corresponding value of the nonmagnetic enthalpy, H_m,

Table 4. Contributions to Entropy and Enthalpy due to Iron

	T_N (K)	S_m	S_{cf} J/mole/K	S_{tot}	H_m	H_{cf} kJ/mole	H_{tot}
Almandine	7.5	40.2	10.5	339.7	0.1	2.1	52.7
Andradite	11.5	29.8	-	316.4	0.1	-	53.5
Hedenbergite	38	13.4	3.5	174.2	0.1	0.7	28.2
Acmite	8	14.9	-	(170)	0.0	-	(27)
Ferrosilite	38	26.8	7.0	(170)	0.3	1.4	(27)
Grunerite	47	93.8	24.5	(663)	1.3	4.9	(105)
Annite	8	40.2	10.5	(340)	0.1	2.1	(54)
Fayalite	65	26.8	7.0	(149)	0.5	1.4	(22.4)

Estimates of total entropy and enthalpy of minerals are in brackets

it is necessary to have a model for the sublattice magnetization. The simplest model is provided by mean field theory [7]. Here the magnetic enthalpy $H_m = C_m\, dT$ is constant at any temperature above T_N, since the model takes no account of short-range correlations. C_m is the contribution to the specific heat of magnetic origin. The estimate for Fe^{2+} is $H_m = 3.9\ T_N$ J/mole, whereas for Fe^{3+} it is 4.0 T_N J/mole

The crystal-field term for Fe^{2+} is more difficult to estimate in general terms, since it depends on the distortion of the iron site from cubic symmetry which gives rise to splitting of the T_{2g} orbital triplet (the ferrous site in almandine is distorted 8-coordination, so the splitting of the ^5D orbital levels is rather different to that shown in figure 5). These splittings may be deduced from the temperature-dependence of the Mössbauer quadrupole splitting [21], or from optical spectra [8]; they can be quite different at inequivalent sites in the same mineral. For simplicity, we assume here a distorted octahedral site with a singlet ground state separated from a doublet excited state by $\delta \simeq 1000$ K. At ambient temperature the enthalpy is about 700 J/mole, and the entropy is approximately 3.5 J/mole/K. These values are used in Table 4, where the magnetic and crystal field contributions are compared with the total entropy and enthalpy derived from specific heat measurements. When no complete specific heat data are available for a particular mineral, rough estimates of entropy and enthalpy are given, usually based on 17 J/mole/K or 2.7 kJ/mole per atom in the formula. The object of the exercise is just to show the relative importance of the contributions which are specific to iron. The ideal formulae for minerals listed in table 4 were given in table 3, except annite (biotite), <K> $\{Fe^{2+}_3\}[Si_3Al]O_{10}(OH)_2$.

It is clear from table 4 that the magnetic entropy is a significant proportion, 10-20%, of the total but the crystal-field term for Fe^{2+} is only about a quarter of this. By comparison, the magnetic enthalpy is never more than about 2% of the total, whereas the crystal-field term is about 5%.

In solid solutions, the extra entropy will scale straightforewardly with the iron concentration x, but the entropy will scale with x^2, assuming $T_N \propto x$. Below the percolation threshold $x_p \sim 0.2$, the extra entropy will be quite negligible.

Finally, we remark that the other region of interest in specific heat measurements on magnetic materials is the region of temperature well below the ordering temperature, say 1 - 10 K for the silicates we have been considering. There, a power-law variation is expected, which will reflect the character of the magnons excited. There are no published data yet for silicates, and there have been no measurements of spin wave dispersion by inelastic neutron scattering on single crystals. This is an area for future work.

4. CONCLUSIONS

There is now a fairly complete picture of magnetic order in common iron end-member silicates. All of them order at relatively low temperatures, in the range 1-100 K, in antiferromagnetic structures with no net magnetization. The magnetic structures of ferrous pyroxenes, amphiboles and micas reflect the dominant ferromagnetic Fe^{2+} - Fe^{2+} exchange, and crystal-field effects on the Fe^{2+} ions.

The real challenge offered by silicates is to understand the properties of the materials as they are provided by Nature. In this respect, the work on iron end-members and synthetic minerals provides a basis from which the rich and complex effects of disorder on magnetic properties can be evaluated. We have seen examples of how disorder - induced random frustration of the magnetic interactions can lead to canted or random noncollinear structures (nontronite, biotite). It is in characterizing and understanding the magnetism of real, disordered systems that most remains to be done. Computer experiments seem destined to play an increasing part in this field.

ACKNOWLEDGEMENTS

We are grateful to Kevin Donnelly, Marcella Maughan, Brenda McDonald and Fergal Supple for help in preparing this paper.

REFERENCES

[1] L. Néel *Ann Géophys* 5 99 (1949).

[2] T. Nagata, *Rock Magnetism* Maruzen 1961.

[3] W. O'Reilly *Rock and Mineral Magnetism* Blackie 1984.

[4] M.W. McElhinny, *Palaeomagnetism and Plate Tectonics*, Cambridge 1973.

[5] A. Hallam, *A Revolution in the Earth Sciences*, Oxford 1973.

[6] J.M.D. Coey, 'Magnetic properties of iron in soil oxides and clay minerals' in *Iron in Soil Oxides and Clay Minerals*. J.W. Stucki and B.A. Goodman (editors) Reidel, Dordrecht, 1987, Ch 14.

[7] J.S. Smart *Effective Field Theories of Magnetism*, Saunders 1966.

[8] R.G. Burns *Mineralogical Applications of Crystal Field Theory*, Cambridge 1970.

[9] M.T. Hutchings, *Solid State Physics*, 16 227 (1964).

[10]J.B. Goodenough, *Magnetism and the Chemical Bond*, Interscience 1963.

[11]T. Moriya in *Magnetism*, G.T. Rado and H Suhl (editors) Academic Press 1963, vol 1 ch 3.

[12]A. Herpin, *Théorie du Magnétisme*, Presses Universitaires de France 1968.

[13]M.E. Stanley, *Introduction to Phase Transitions and Critical Phenomena*, Oxford 1971.

[14]G. Toulouse and P. Pfeuty, *Introduction au Groupe de Renormalisation et ses Applications*, Presses Universitaires de Grenoble 1975.

[15]P.D. Loly, *Can. J. Phys.* (1987).

[16]K. Moorjani and J.M.D. Coey, *Magnetic Glasses*, Elsevier 1984.

[17]J.M.D. Coey, *Can. J. Phys.* (1987).

[18]R. Zallen, *The Physics of Amorphous Solids*, Wiley, New York 1983.

[19]A.B. Harris, *J. Phys.*, C7 1671 (1974).

[20]G.E. Bacon, *Neutron Diffraction in Experimental Magnetism*, G.M. Kalvius and R.S. Tebble, (editors). Wiley, Chichester 1979.

[21]J.M.D. Coey, *Silicate Minerals in Mössbauer Spectroscopy Applied to Inorganic Chemistry*, G.J. Long, (editor) Plenum, New York 1984, voll, ch 14.

[22]W. Prandl, *Z. Kristallogr.*, **134** 334 (1971).

[23]G.W. Metz, S.R. Bohlen, E.J. Essene, L.M. Anovitz, E.F. Westrum, B.S. Hemingway and M.D. Barton (to be published).

[24]E. Murad, *Amer. Mineral.* 69 722 (1984).

[25]R.A. Robie, Z. Bin, B.S. Hemingway and M.D. Barton (to be published).

[26]O. Ballet, J.M.D. Coey, G. Fillion, S. Ghose and J.R. Reynard, (to be published).

[27]J.M.D. Coey and S. Ghose, *Solid State Commun.,* 53 143 (1985).

[28]H.T. Haselton, R.A. Robie and B.S. Hemingway, *Geochin et Cosmochin Acta* (to be published).

[29]A. Wiedenmann and J.R. Regnard, *Solid State Commun.* 57 499 (1986).

[30]S. Ghose, A.W. Hewat and J.R. Wiedner (to be published).

[31]A. Wiedermann, J.R. Regard, G. Fillion and S.S. Hafner, *J. Phys.,* C 19 3683 (1986).

[32]S. Ghose, A.W. Hewat and N.V. Dang, *Phys. Chem. Minerals* (to be published).

[33]D.S. Goldman and G.R. Rossman, *Amer. Mineral* 62 151 (1977).

[34]A. Moukarika, J.M.D. Coey and N.V. Dang, *Phys. Chem. Minerals* 9 269 (1983).

[35]R.J. Borg and I.Y. Borg, *Phys. Chem. Minerals* 5 219 (1980).

[36]J. Linares, J.R. Regnard and N.V. Dang, *J. Magn. Magn. Mat,* 31 - 34 715 (1983).

[37]S. Ghose, D.E. Cox and N.V. Dang, (to be published).

[38]S. Ghose, Electron ordering and associated crystallographic and magnetic phase transitions in ilvaite, *Structural and Magnetic Phase Transitions in Minerals,* S. Ghose, J.M.D. Coey and E. Salje, editors Springer (1987).

[39]F.J. Litterst and G. Amthauer, *Phys. Chem Minerals* 10 250 (1986).

[40]J.M.D. Coey, J. Allan, Kan Xuemin, N.V. Dang and S. Ghose, *J. Appl. Phys.* 55 1963 (1984).

[41]J.M.D. Coey, O. Ballet, A. Moukarika and J.L. Soubeyroux, *Phys. Chem. Minerals,* 7 141 (1981).

[42]O. Ballet and J.M.D. Coey, *Phys. Chem. Minerals,* 8 218 (1982).

[43]O. Ballet, J.M.D. Coey and K.J. Burke, *Phys. Chem. Minerals* 12 370.

[44]J.M.D. Coey, F.V. Chukhrov and B.B. Zvyagin, *Clays and Clay Minerals* 32 198 (1984).

[45]J.M.D. Coey in *Proceedings of the International Clay Conference 1985*, L.G. Schultz, H. van Olphen and F.A. Mumpton, editors, Clay Minerals Society 1987, p. 261.

[46]S.W. Bailey, Structures of Layer Silicates in *Crystal Structures of Clay Minerals and their X-ray Identification.* G.W. Bridley and G. Brown, editors, Mineralogical Society (1980) p1.

[47]S. Guggenheim and R.A. Eggleton (to be published).

[48]O. Ballet, J.M.D. Coey, P. Mangin and M.G. Townsend, *Solid State Comm.* 55 787 (1985).

[49]J.M.D. Coey, A. Moukarika and C.M.McDonagh, *Solid State Comm* 50 125 (1982).

[50]H. Fuess, O. Ballet and W. Lottermoser 'Magnetic order and phase transitions in olivines' in *Structural and Magnetic Phase Transitions in Minerals*, S. Ghose, J.M.D. Coey and E. Salje (editors) Springer 1987.

[51]R.A. Robie, C.B. Finch and B.S. Hemingway, *Amer. Mineral.* 67 463 (1982).

[52]Kan Xuemin and J.M.D. Coey, *Amer. Mineral.* 70 576 (1985).

[53]V.U.S. Rao, F.E. Huggins and G.P. Huffman, *J. Appl. Phys.* 50 2408 (1979).

[54]R.A. Robie, B.S. Hemingway and H. Takei, *Amer. Mineral.* 67 470 (1982).

[55]E.S.R. Gopal, *Low Temperature Specific Heats,* Plenum, New York, 1972.

MOLECULAR DYNAMICS SIMULATIONS IN THE SOLID STATE SCIENCES.

Martin Dove,
Department of Earth Sciences,
University of Cambridge,
Downing Street,
Cambridge
CB2 3EQ

ABSTRACT. Molecular Dynamics Simulation is a computer modelling technique that has proved to be a very fruitful tool in the general study of condensed matter, but it has yet to be exploited in the field of earth sciences. In this paper the basic ideas of the technique are described, and the technical details aşsociated with the use of the method are discussed. The wide range of information that can be obtained from these simulations is described in some detail. Some recent work on orientationally disordered crystals and incommensurate systems is described as representative examples of the use of this simulation technique. Potential applications of the molecular dynamics simulation technique to problems in minerals science are discussed.

1 COMPUTER SIMULATIONS

1.1. INTRODUCTION

Computer simulations are becoming a more frequent part of twentieth century life as the relative cost of computer technology lowers to the point where is cheaper to simulate some aspect of life than it is to experience it in reality. Thus air pilots are trained using simulated cockpits, with realistic graphics displays that model the view from the cockpit; drugs are developed on computers by modelling their interactions with biological molecules as this is a cheaper way to sort through the almost unlimited variety of available chemicals than to perform an equally limitless number of experiments; the response of structures such as cars to catastrophic events, like crashes, are simulated on computers as a simulated crash is cheaper to stage than a real one. Over the past two decades or so scientists have come to realise that they too can be served by the technique of computer simulation. The motivation in their case is not so much cost-effectiveness, but the recognition that computer simulations as "experiments" with nature's equations can provide information in many cases when the theoretical models are inadequate and when experimental data are ambiguous. This is often the case in the condensed matter

E. K. H. Salje (ed.), Physical Properties and Thermodynamic Behaviour of Minerals, 501–590.

sciences, where one is continually faced with the intractability of the "many body problem".

At the start of this paper, therefore, we should deal with the question as to why anyone working in the earth sciences should want to use computer simulations, and at the same time consider how the expanding area of computational science fits into the more familiar relationship between experiment and theory. One of our aims is to attempt to explain the properties of materials on the basis of the forces between atoms and the interactions with and between the elementary excitations at atomic level. This has traditionally been the domain of the theoretician, who has relied upon the work of the experimentalist both to provide many ofthe parameter values that are required and to test the predictions of the theory. We can illustrate the picture with a very simple example. A material that undergoes a phase transition will have a broken symmetry that is characterized by an order parameter Q. If it is found experimentally that the transition is accompanied by a spontaneous strain η that varies quadratically with the order parameter, it is possible to describe the situation using an appropriate Landau free energy expansion:

$$F = \tfrac{1}{2}a(T-T_0)Q^2+\tfrac{1}{4}bQ^4+\tfrac{1}{6}cQ^6+\tfrac{1}{2}C\eta^2+\tfrac{1}{2}d\eta Q^2 \qquad [1.1.1]$$

This expression has been discussed by Cowley (1980). The terms that have been included in the expansion have the appropriate symmetry for the particular system. It is generally assumed that all the coefficients are temperature independent. From this free energy expansion one can calculate the form of various quantities that can be measured experimentally (e.g. the order parameter, susceptibility, specific heat, elastic constants, spontaneous strain), and from these measurements it will be possible to determine the values of the coefficients a,b,c, etc. It will then be possible to use this free energy expression to calculate other thermodynamic quantities, such as the entropy, that are of interest to the geologist. As a description of the thermodynamic behaviour of a material equation 1.1.1 may be all that a geologist will be interested in to enable his studies to progress. However, as minerals physicists we would be most unhappy to stop at this point, for our equation 1.1.1 as it stands is more descriptive than explanatory. Why is this equation valid? What determines the values of the coefficients? Why is the system unstable with respect to the particular form of the order parameter Q? What determines the strength of the coupling to the strain? What determines the symmetry of the low temperature phase? These questions are probing the underlying microscopic mechanisms that drive the transition and that provide the justification for the use of a free energy expansion such as equation 1.1.1. Unfortunately the very success of Landau's theory in correlating a wide range of data with just a few adjustable parameters has meant that it is difficult to distinguish between different microscopic models for particular phase transitions, as has been discussedby Cowley (1980), since most theories will finally end up with a similar free energy to a Landau free energy, with microscopic parameters that will also need to be determined by fitting to experimental data. Thus there may be a need to obtain additional information that can only be obtained by large scale calculations, using any of the variety of simulations techniques that are available. For example, given a model Hamiltonian it may be possible to calculate the values of a number of the coupling parameters that are important for the existence of the phase transitions, or even to provide information about the origin of the values of these parameters.

As far as minerals are concerned, any microscopic model will involve wave vector dependent variables that either relate to ordering schemes or phonon amplitudes. These variables will be coupled by harmonic and anharmonic interactions, and will be free to fluctuate. From an experimental viewpoint one can study these microscopic variables, and how their fluctuation amplitudes depend on wave vector. From a theoretical approach the free energy expansion can be shown to correspond to a microscopic model such as a renormalised phonon model, but the large numbers of interactions involved mean that it will be extremely difficult to calculate the transition temperature or the expansion coefficients. Much of this information can be extracted more easily from a realistic computer simulation calculation.

This example has illustrated the situation when different experimental data are capable of being explained by different theories, making it hard to determine which theory is correct. There are other situations where the experimental data are so featureless that they provide no hints as to even the best route for interpretation; examples are systems that contain a high degree of disorder, where experimental spectra contain so few detailed features that they can be fitted by more-or-less any functional form. In these cases, the results ofsimulations can be decisive in the development of a good theoretical understanding of the case in hand.

Thus computer simulations are becoming an increasingly useful tool in the theoretical analysis of experimental data, particularly in those caseswhere an exact theory cannot be obtained, and where even any approximationmay be untestable. On the other hand, a good simulation may be able to help when the appropriate experimental data are unavailable, either because there is no experimental probe of the behaviour in question, or because the experiments are just not possible. For example, the interest in any particular system may be at temperatures or pressures that are unattainable in the laboratory; examples should be familiar to an earth scientist!

For these reasons computer modelling is an expanding area of science, not just in the earth sciences but also in the solid and liquid state sciences, statistical physics, elementary particle physics, cosmology and astronomy, and the biological and biochemical sciences. In this article I will consider one method that has not been used much in earth sciences but which has been extensively and fruitfully used in the other branches of science that I have mentioned, namely the *Molecular Dynamics Simulation* (MDS) technique. I will contrast this method with other methods that are more commonly used (and which will be described in other articles in this volume), I will give technical details to the extent to which my experience will stretch, and will then describe the use of this technique to some problems in solid state physics. Finally I will air some thoughts on the applications (more future than present) of the MDS technique to earth science problems.

1.2. SIMULATION MODELS.

There are a number of techniques that are thought of as computer simulations. As we have just said, the main technique that will be described in detail in this paper is the Molecular Dynamics Simulation technique, but there are three other techniques that I will briefly describe in this section in order to give a general context to the rest of this paper. The simplest simulation method (although it is by no means simple!) is the "static lattice energy" calculation. This is used to find the lowest energy configuration of a given set of atoms in a crystal that is related to a given starting configuration. A model for the interaction energy between all of the atoms is assumed, and the program then calculates the structure that has the lowest lattice energy by an iterative procedure (such as a steepest-descent algorithm), working within user-supplied constraints such as the lattice symmetry or the size of the unit cell. With this method one can study the influence of the form of the potential model on the structure in order to be able to calculate best effective potentials, and one can also study the response of the equilibrium structure to various externally applied forces, such as pressure, or to forced internal distortions. Static quantities such as elastic constants and dielectric constants can be readily calculated, and as well as using the comparison with experimental data to improve model potentials, when reliable potentials are available these quantities can be calculated for use when experimental data are not available. Static simulations can also be used for calculating other coupling constants (such as between two separate distortion mechanisms), and they have been extensively used for calculating the energy of defects and activation energies. This method does not incorporate temperature, and so it is effectively a simulation at zero temperature, and because of this these simulations will only study the region of phase space that is close to a potential minimum. However, a considerable amount can be learned from such a simulation, and for many applications these calculations will not be costly in computer time.

The second technique, which is a simple development from the static lattice energy calculation, is the "lattice dynamics" calculation. Using the forces and structures from the lattice energy calculations the phonon dispersion curves are calculated within the harmonic approximation. This technique provides additional means of comparing the predictions of a given atomic force model with experimental data (such as the long-wavelength mode frequencies from Raman and Infra-red spectroscopy, or the dispersion curves and mode eigenvectors obtained from inelastic neutron scattering experiments). It is possible to use the results of a standard lattice dynamics calculation to give the harmonic (predominant) contribution to properties such as temperature factors, specific heat capacity and diffuse scattering, all of which vary with temperature due to the increasing population of levels rather than through anharmonic effects. These calculations give additional tests of the validity of a given force model, and, as for the static calculations, when the force model is reliable they can give useful information when experimental data are unavailable. The response of the lattice dynamics to external forces can also be calculated, a simulation that is particularly relevant for studying displacive phase transitions that occur under pressure, where the external force can cause a softening of a phonon branch. From these calculations the dominant internal contribution to an instability can be identified. To some extent the effect of temperature can be simulated via the modification of one of more parameters; this is known as the "quasi-harmonic" approximation. For example, the calculations can be performed using different values for the unit cell parameters that correspond to their values at different temperatures, thereby simulating the effect of temperature by assuming that any changes in the phonon spectra are due to the changes in the unit cell size with temperature. Again, one might be able to study any mode instabilities within this approximation. However, as for the static lattice calculation, the direct effects of temperature are not explicitly included, and certainly any effects of anharmonic interactions other than those that cause thermal expansion are neglected. Therefore there may be many cases for which this approach is unable to provide useful information, for example, when a displacive phase transition is caused by anharmonic interactions that are not directly related to the thermal expansion, or when the high temperature phase contains disorder. These two techniques will be described in more detail in other papers in this volume.

One problem associated with the fact that these two techniques do not include the effects of temperature is that although in some cases they may be able to give information about any structural instabilities that might arise on heating and that could lead to phase transitions, it is not possible to study the properties of the high temperature (higher symmetry) phases, as many quantities of interest will be found to be unstable (e.g. elastic constants, phonon frequencies). This is because these methods have not included the anharmonic interactions that give the stability of these phases at high temperatures, and which thereby renormalise the effective interactions. This is particularly true if the anharmonic interactions provide a stability in the presence of disorder. A second disadvantage of these two techniques is that they do not provide a search of the configurational phase space for other stable configurations, nor do they sample phase space for other configurations that arise through thermal fluctuations. The two other simulation techniques have been developed in order to include temperature directly so that these disadvantages can be overcome. As we have said, the rest of this paper will be devoted to the MDS technique. We briefly discuss here the other technique, called the "Monte Carlo" (MC) simulation technique.

The name of this technique reflects the fact that the method relies on a random sampling approach. In this method an ensemble of particles is defined, each of which has it's own set of variables (e.g. position, spin orientation), and these are altered in any update step.

Specifically, in each update step a particle is chosen from the ensemble at random. Any of its variables are changed by a random fraction of a pre-determined step. For example, if the particle is an atom, its position may be changed by a small random displacement, or if the particle is a spin then its orientation will be changed (of course, if it is an Ising spin the spin direction will be completely reversed). After the change has taken place,the new energy of the ensemble is calculated. If the energy is lowered, the change is considered to be a good one, the update step is said to be completed, and the simulation moves on to the next update step. If the energy is raised by the variable change, the update is not discarded (otherwise the simulation would end up in a potential minimum, of which there will be a large number depending on the start configuration), but is retained depending on the result of a probability test. A random number between 0 and 1 is obtained and compared with the quantity exp(-energy change/temperature). The variable change is retained if the random number is the lower of the two numbers, but is discarded otherwise. The simulation then moves on to the next update step. This way there is a non-zero probability of the ensemble jumping out of any potential minimum, and when the simulation is run for a large number of update steps (usually several millions in the case of Ising spins) the set of configurations generated represent an ergodic pathway through the multi-dimensional phase space of the particle variables. The simulations therefore effectively give a statistical sampling of the configurational phase space with the correct thermal weighting of exp(-energy/temperature). If the simulation is run for enough time steps, the statistical sampling will give a good approximation to the partition function, which for most cases will be too complicated to calculate analytically. From the set of configurations it is then possible to calculate thermodynamic quantities at a number of different temperatures. This technique is often used in the study of spin systems where exact analytical solutions of the appropriate Hamiltonian are not available (e.g. the 3-d Ising model or random field models), and can yield information about critical temperatures, exponents etc. in addition to other thermodynamic quantities. However, as this technique only gives a sample of instantaneous configurations and not any dynamic information, it is not so often used in the study of interactions between atoms in condensed phases (In the early days, however, the MC method was often applied to such problems, but now the MDS technique has become the routine method for such applications). The great advantage that a MC simulation gives over the previous two techniques is that it includes the effects of temperature directly, and in problems such as Ising spin systems, where the dynamics are of little interest, the ability of the method to give an accurate statistical sampling of phase space makes it an extremely useful tool. Extensive details of the use of MC simulations arefound in a book by Binder (1983).

1.3. THE MOLECULAR DYNAMICS SIMULATION TECHNIQUE.

The basic idea of a molecular dynamics simulation calculation is very simple. The computer stores the positions and velocities of a number of particles in a fictitious "box" at a given time, and from a given model for the forces between the different particles it solves Newton's classical equations of motion in order to predict the changes in the set of positions and velocities. Newton's equations are continuous in time, but the simulation uses an integration algorithm to calculate the discrete changes in these dynamic variables over a small given time interval. This process is repeated iteratively for several thousand time steps, and effectively the simulation generates the trajectories of the particles in time. For small enough time steps, the method gives a good approximation of the continuous evolution of the system with time. Hence, starting from a given configuration of particle coordinates and velocities, the simulation generates the trajectory of the entire ensemble through the phase space spanned by the dynamic variables of all the particles in a completely deterministic way. Conditions such as the density of the system and the temperature can easily be imposed during an initial equilibration stage of the simulation. This simple idea makes this kind of simulation intuitively the most satisfying of the different computer simulation techniques.

Within the current limits of available computational power, it is usual for the MDS calculations to use 100-10000 particles and, at least for atomic systems, for the simulations to run for 10-100 ps. The range of systems that can be studied by this approach is very wide indeed. At the atomic level, provided that good models for the interactions between the particles exist, it is possible to study any particle systems such as solids, fluids, glasses, solutions etc., and the simulations can provide fundamental information about a very wide range of properties such as structures, dynamics, lattice excitations, and transport phenomena. At the other extreme, MDS techniques can be used to study the classical evolution of cosmological aggregations of matter such as the stars in galaxies.

In addition to the physical appeal of the method and its conceptual simplicity, the MDS technique has a number of very useful properties. It is a simulation technique that includes temperature directly in the equations of motion rather than via a statistical sampling algorithm. This is essentially equivalent to the point that the simulations give information about the dynamic properties of the system at a non-zero temperature. The simulations also include the effects of anharmonic interactions directly. Similarly, it is just as easy to study systems that contain considerable disorder (e.g. liquids) as it is to study highly ordered systems. However, despite the simplicity of the method, the calculations involved in performing a reasonable simulation are considerably greater than for the other techniques, and so a MDS calculation will not always be the automatic choice. There will be many applications for which neither temperature nor anharmonic interactions are important considerations, and in these cases the static lattice energy and harmonic lattice dynamics simulations will be the ideal tools. If temperature is important, but the aim is only to calculate statistical averages, then the MC method will be used, particularly if it is important to perform calculations under a number of different conditions. But if the interest is in the dynamic behaviour of complex systems, then the effort involved in performing a good MDS calculation will usually give high dividends.

In the next section we will discuss the MDS method in more detail. As this section contains mostly technical information, the reader may wish to be selective about which parts are worth reading, and appropriate parts can be consulted when required. The five sub-sections discuss the technical details of performing a MDS calculation, the different ensembles that can be simulated, the anharmonic force models that are frequently used, the range of information that can be extracted from a MDS calculation, and the computer requirements respectively. The third section gives examples of the application of the MDS method, which are aimed at illustrating the wide range of information that can be obtained from the simulations as discussed in section 2.4. The final section discusses applications and aspects of the technique that are based more in the future than in the present, particularly those that relate to the minerals sciences.

2.1. TECHNICAL DETAILS FOR MOLECULAR DYNAMICS SIMULATIONS.

2.1.1. Historical development. Some remarks about the historical development of the use of MDS calculations might serve to introduce this section on the technical details. The first interest in these simulation calculations was triggered by the fact that they provided means of giving information about the microscopic properties of liquids for well-defined systems. The first calculations were performed using infinitely hard particles (e.g. hard spheres) by Alder and Wainwright (1957, 1959). The techniques for this kind of problem, which uses discontinuous potentials and impact forces, are different from the techniques for continuous potentials, and so we will not discuss this case, except to note that the simulations on these apparently unrealistic models reproduce many of the features of real fluids surprisingly well. The first simulations in which continuous potentials were used were performed by Rahman (1964, 1966), using models appropriate for fluid argon, and these calculations were extended by Verlet and Levesque (Verlet, 1967, 1968; Levesque and Verlet, 1970). Once it had been established that the simulations showed the qualitative features of true fluids and that the results they gave were in reasonable agreement with corresponding experimental results, further calculations were performed on ionic fluids (Woodcock, 1971, 1972; Jacucciet al, 1976), molecular fluids such as Cl_2 (Murphy et al, 1983), F_2 (Singer et al, 1977), N_2 (Cheung and Powles, 1975, 1976; Weis and Levesque, 1976), CS_2 (Steinhauser and Neumann, 1979; Tildesley and Madden, 1981), CH_4 (Murad et al, 1979), CCl_4 (McDonald et al, 1982) and water (Rahman and Stillinger, 1971, 1974), and simple fluids with metallic potentials (Schiff, 1969).

In addition to the work on liquids, the MDS technique has been applied to the study of simple solids, and successful simulations have been performed on systems such as solid argon (Hansen and Klein, 1976), NaCl (Jacucci et al, 1975, 1976) and solid nitrogen (Weis and Klein, 1973; Klein and Weis, 1977). However, the value of MDS calculations is that they can give direct information about systems that are not well understood, and the fact that simple crystals, unlike simple fluids, are reasonably well understood has meant that the application of the technique to solids is not as well developed as for liquids. The exceptions are the cases where there is disorder in the solids, such as mobile ions in fast-ion conductors or molecular crystals with orientational disorder. In general, however, the ability of MDS calculations to predict anharmonic effects and to correlate the observed behaviour with the anharmonic atomic forces has not been exploited.

In the rest of this section we will discuss the development of MDS techniques following in the tradition of these earlier efforts, which, as we will discuss, are simulations of the systems in a microcanonical ensemble. A recent small revolution has led to the introduction of formalisms that allow simulations to be performed in a whole range of thermodynamic ensembles, and we will discuss these developments below in section 2.2.

2.1.2. Equations of motion. It is traditional practice for the simulation to have a fixed number of particles N and a fixed sample size and shape of volume V, and the equations of motion have been developed for the particles moving and interacting within these constraints. For translational motions the only equations that need to be solved have the well-known form:

$$\frac{d}{dt}\underline{r}_i(t) = m_i\,\underline{F}_i(t) \qquad [2.1.1]$$

where $\underline{F}_i(t)$ is the instantaneous force acting on the i-th particle at a position $\underline{r}_i(t)$, and m_i is the mass of the particle. In order to approximate this continuous equation of motion by a discrete integral form we express the position of the i-th atom at a time t+Δt as a Taylor expansion:

$$\underline{r}_i(t+\Delta t) = \underline{r}_i(t)+\Delta t\,\dot{\underline{r}}_i(t)+\frac{(\Delta t)^2}{2!}\ddot{\underline{r}}_i(t)+\frac{(\Delta t)^3}{3!}\dddot{\underline{r}}_i(t)+\ldots \qquad [2.1.2]$$

If we add to this the corresponding equation for $\underline{r}_i(t-\Delta t)$ and neglect the terms of order $(\Delta t)^4$ and higher, we readily obtain an expression for the calculation of our new particle position:

$$\underline{r}_i(t+\Delta t) \approx 2\underline{r}_i(t)-\underline{r}_i(t-\Delta t)+(\Delta t)^2\,\underline{F}_i(t)/m_i \qquad [2.1.3]$$

On the other hand, if we subtract the equation for $\underline{r}_i(t-\Delta t)$ from the equation for $\underline{r}_i(t+\Delta t)$, we obtain an expression for the velocity of the particle at time t:

$$\dot{\underline{r}}_i(t+\Delta t) \approx \frac{1}{2\Delta t}\left[\underline{r}_i(t+\Delta t)-\underline{r}_i(t-\Delta t)\right] \qquad [2.1.4]$$

These equations are known as the "Verlet algorithm", after their inventor (Verlet, 1967). The error in the expression for the velocity is of the order of $(\Delta t)^3$, whereas the error in the expression for the position is of the order $(\Delta t)^4$. However, the different accuracies of these expressions is unimportant since this algorithm does not use the velocity in the determination of the particle's position at the next time step. Although this might appear to be a rather simple method of integrating the equation of motion, it turns out that this algorithm is sufficiently stable to be of general practical use, and attempts at improving upon it are only marginally successful without paying a high price in increased computational effort.

The Verlet equations of motion are often used in an alternative form known as the "leapfrog algorithm". In their sequence of use, the equations for the leapfrog algorithm are:

$$\dot{\underline{r}}_i(t+\Delta t/2) \approx \dot{\underline{r}}_i(t-\ t/2)+\Delta t\ \underline{F}_i(t)/m_i \quad [2.1.5a]$$

$$\underline{r}_i(t+\Delta t) \approx \underline{r}_i(t)+\Delta t\ \dot{\underline{r}}_i(t+\Delta t/2) \quad [2.1.5b]$$

$$\dot{\underline{r}}_i(t) \approx [\dot{\underline{r}}_i(t-\Delta t/2)+\dot{\underline{r}}_i(t+\Delta t)]/2 \quad [2.1.5c]$$

where

$$\dot{\underline{r}}_i(t-\Delta t/2) \approx [\underline{r}_i(t)-\underline{r}_i(t-\Delta t)]/\Delta t \quad [2.1.5d]$$

An alternative algorithm due to Beeman (1976) uses an approximation for the higher order terms in the expansion 2.1.2:

$$\underline{r}_i(t+\Delta t)\approx\underline{r}_i(t)+\Delta t\dot{\underline{r}}_i(t)+\frac{(\Delta t)^2}{6}[4\ddot{\underline{r}}_i(t)-\ddot{\underline{r}}_i(t-\Delta t)] \quad [2.1.6a]$$

$$\dot{\underline{r}}_i(t+\Delta t)\approx \dot{\underline{r}}_i(t)+\frac{\Delta t}{6}\ [2\ddot{\underline{r}}_i(t+\Delta t)+5\ddot{\underline{r}}_i(t)-\ddot{\underline{r}}_i(t-\Delta t)] \quad [2.1.6b]$$

In fact it is easy to show that the Beeman algorithm and the Verlet algorithm generate identical trajectories. The operation of the Verlet algorithm is clear: at each time step, t, the forces $\underline{F}_i(t)$ are calculated and the positions are then updated. When the Beeman algorithm is used, the new positions are calculated at the start of each iteration, the forces are then calculated, and finally the new velocities are calculated.

These algorithms are only appropriate when the forces depend only upon relative positions. However, there are occasions when the forces in the equations of motion are also dependent upon the instantaneous velocities, particularly when the variables are not linear positions but are rotational or generalised coordinates. In these cases it will be necessary to use predictor-corrector methods, in which the velocity is first predicted to use in the equations of motion, and then corrected in a further iteration step. General details are given by Gear (1971) and summarised by Rahman and Stillinger (1971). It is possible to add a corrector stage to the Beeman algorithm, which involves including a prediction of the particle velocities after the calculation of the new positions:

$$\dot{\underline{r}}_i(t+\Delta t) \approx \dot{\underline{r}}_i(t)+\frac{\Delta t}{2}[3\ddot{\underline{r}}_i(t)-\ddot{\underline{r}}_i(t-\Delta t)] \quad [2.1.6c]$$

The predicted velocities can be used to calculate the new forces, and the correct velocities are then calculated by equation 2.1.6b. If necessary steps 2.1.6c and 2.1.6b can be repeated until convergence is reached, which need not be a very time consuming process if the position-dependent forces are held in memory. However, for suitably sized time steps only one iterative step may be required.

We have noted that the variables in the equations of motion need not be the particles' positions and velocities, as there may be other dynamic variables in the simulations such as orientations, generalised molecular coordinates or sample variables (section 2.2). The use of generalised coordinates in MDS calculations has been described by Refson (1985) and Refson and Pawley (1987).

The equations of motion presented here are for isolated systems and thus have the property that the total energy and linear momentum are conserved (the initial total linear momentum will usually be zero by choice); the conditions of constant energy, constant volume and constant number of particles implies that the equations of motion are those of a microcanonical ensemble. The requirement of the conservation of energy thus provides a good internal test of the equations used in a simulation. It is found that the Verlet algorithm gives good conservation of energy, which essentially provides part of the justification for its use. It should be noted that the use of the approximations in the integration scheme and the existence of rounding errors on the computer will mean that the energy will not be conserved *exactly*, but the requirements of any stable algorithm are that the fluctuations in the energy should be minimised and that there should be no systematic change in the energy over the duration of the simulation. As a rule of thumb the fluctuations in the total energy should be no greater than a few percent of the fluctuations of the potential energy.

It is this stability criterion that determines the size of the time steps to be used. There is a compromise between the greater accuracy that small time steps bring and the heavier computational demands that accompany the use of smaller time steps. In general it is preferable to use time steps that are as large as possible within the stability conditions on the total energy described above. Typical sizes are of the order of 0.001-0.01ps, depending on the system under study. It is clear that smaller time steps are needed at higher temperatures (with respect to the potential energy scale) and for systems with small masses.

We should also address the question of the reproducibility of the particle trajectories in the simulations. Tests have shown that when different runs are performed starting from the same initial configuration but using different time steps, the trajectories of the individual particles in the different simulations begin to diverge after about 3ps, and are distinct after only another 1-2ps. This behaviour has not been explained, but suggests that the trajectories are only reliable over a time scale of 3ps or so. Thus time dependent quantities become unreliable for longer times than this, but we will see below that there are additional reasons for considering longer times than about 3ps to be susceptible to error. Other tests have shown, though, that a number of static and dynamic quantities are fairly insensitive to the size of the time step, so that it appears that even when individual trajectories contain propagating errors, their interactions with their neighbours as revealed in the different analysis do not propagate the same errors.

2.1.3. Finite size samples and periodic boundaries. The limits of computer resources limit the number of particles within a sample that can reasonable be simulated. Typical simulations can use as few as 100 particles, and rarely are sample sizes of greater than 10000 particles used. With such small samples it would be extremely likely that the existence of free surfaces would drastically affect the behaviour of the ensembles. Therefore all simulations of the bulk properties of condensed matter use periodic boundaries to remove the effects of the free surfaces. In general the repeating sample for liquids will be a cube, but it has been suggested that the efficiency of the system can be improved by using sample shapes such as dodecahedra that are more "rounded". For solids the shape of the sample will be determined by the unit cell shape, and if the three unit cell axes are described by the vectors $\underline{a}_1$, $\underline{a}_2$, and $\underline{a}_3$, the sample shape will be described by the sample vectors $\underline{A}_2 = \underline{n}_2.\underline{a}_2$ etc., where n_2 is an integer.

One of the features of the use of finite sized samples is that the number of allowed wave-vectors is limited. Specifically, the allowed wave-vectors are integral multiples of the reciprocal lattice vectors of the sample cell $\underline{A}_2^*$ etc., given as:

$$\underline{A}_1^* = \underline{a}_2 \times \underline{a}_3 / V \qquad [2.1.7a]$$

$$\underline{A}_2^* = \underline{a}_3 \times \underline{a}_1 / V \qquad [2.1.7b]$$

$$\underline{A}_3^* = \underline{a}_1 \times \underline{a}_2 / V \qquad [2.1.7c]$$

where V is the sample volume. Thus any allowed wave-vector $\underline{k}$ is given as

$$\underline{k} = m_1\underline{A}_1^* + m_2\underline{A}_2^* + m_3\underline{A}_3^* \qquad [2.1.8]$$

where m_1, m_2 and m_3 are integers. This limitation on the range of allowed wave-vectors is one of the main disadvantages of the use of small sample sizes. For example, if a sample has a size of 4x4x4 unit cells, the wave-vector resolution is only 1/4 of a reciprocal lattice vector. Since many single-particle properties are determined by the collective dynamic properties (e.g. normal mode lattice vibrations), it might be expected that the sizes of these quantities will be affected by the wave-vector resolution, as indeed is found to be the case with smaller sample sizes. In addition, since many properties that we will be interested in are wave-vector dependent, the study of these properties will demand a good resolution in wave-vector. Typical experiments will work with a resolution of about 1/10-th of a reciprocal lattice vector, which suggests that a useful sample size would be 1000 unit cells. At this point, the simulation will become a very large calculation indeed! It should also be noted that if the sample axes are not along the directions of the unit cell axes, there may not be allowed wave-vectors that lie along the symmetry directions in reciprocal space. In general there should be no need to construct samples that do not have this property, but recent work that will be described in chapter 3 has used a large parallel processor that has necessitated using odd sample shapes. However, in these cases, the number of primitive unit cells (4096) has been large enough that there has always been a set of wave-vectors that lie close enough to the symmetry directions that this problem has not been of practical importance.

A related disadvantage of working with small samples is that we are limited to the study of systems that have relatively short correlation lengths. This then precludes the study of critical behaviour at phase transitions, where the correlation length can diverge to sizes of many thousands of unit cell lengths, and systems with important small domain structures, where the existence of the domains is the crucial aspect that determines many of the properties of these systems.

In addition to the problems inherent in the use of small samples, there is a further problem associated with the use of periodic boundary conditions. A propagating fluctuation (e.g. a sound wave) can travel through the boundaries and back into the sample, eventually reaching the same point of disturbance that gave rise to the propagation in the first instance. There is no way of telling what will happen, but it is clear that it is possible for the fluctuation to interact with itself again, and this feedback process may set up a resonance behaviour. The time taken for a propagating mode to travel back to its origin, τ_{rec}, is called the "recurrence time", and if the fastest sound velocity is v and the linear dimensions of the sample cell are of the order L, the recurrence time is simply given as:

$$\tau_{rec} = L/v \qquad [2.1.9]$$

For typical values of v and L, τ_{rec} will be of the order of 2-3ps, which is of the same order as the time scale for the divergence of the particle trajectories discussed above. τ_{rec} will be equivalent to the period of the longest wavelength longitudinal acoustic mode.

Despite these problems associated with sample size effects, it is found in practice that with sample sizes greater than 100 unit cells most properties are fairly insensitive to the size of the sample, and a considerable amount of information can be obtained with reliability from such calculations. Provided that the computational resources are available, sample sizes of about 1000 unit cells would seem to be a practical choice. As we shall discuss later, when even larger systems are used additional problems associated with statistical averaging begin to emerge in force!

2.1.4. Inter-atomic forces. We shall discuss in detail in section 2.3 the particular models that can be used in simulation calculations. Here we just want to comment on the way that forces are used in MDS calculations. In the study of liquids or other systems with mobile ions, where the local environment of an atom is constantly changing in time, it is commonpractice to take a fictitious sphere around each atom (with a radius of up to half the box length) and explicitly include only the interactions with the other particles lying in this sphere. This defines a "cut-off radius" that we will discuss in section 2.3. There may be problems with the conservation of energy with this approach, in that an interacting neighbour will appear to *disappear* as it moves outside of the cut-off surface, and immediately the energy (and force) will change discontinuously. This may not make a practical difference, but it is possible to redefine the energy so that the energy at the cut-off distance is zero by simply subtracting an appropriate amount. This will not affect the equations of motion at all. Also the model for the energy can be modified so that the derivative with respect to distance (and hence the force) goes continuously to zero at the cut-off distance, for example by multiplying the model inter-atomic energy function by some other functionthat does not modify the energy very much at most particle separations but does change the behaviour in the vicinity of the cut-off radius.

When dealing with solids, the diffusion of atoms will be on a timescale that is orders of magnitude slower than the time scales used in a simulation, and so it will be extremely unlikely that a particle will be found to move way from its lattice site. Therefore it will be possible to establish a list of interacting neighbours at the start of the simulation, so that the program will only ever have to bother to calculate the interactions between these neighbours.

We should also note that in a MDS calculation, the calculation of the forces for the equations of motion is the most time consuming part of the simulation, and any optimization of this part of the simulation program will help reduce the computational requirements of any calculation. For example, careful thought of the best neighbour lists to use can speed up the execution of the program, and even faster version of the square root function can make a useful difference to the execution time.

2.1.5. A typical simulation! As we have said, a typical simulation will have a sample containing 100-1000 particles, with each particle interacting with its neighbours via an assumed force model. At the start of the simulation, all the particles are given random positions (or, if it is a crystal, random small displacements from their lattice positions) and random velocities. The equations of motion are then solved using the algorithms discussed above. In the early stage, of course, the sample is far from equilibrium, and so the first stage of a simulation is just used to equilibrate the sample. The sample may start to heat or cool down as it equilibrates, and so some temperature rescaling will be required. This will usually involve multiplying all of the particle velocities by the square root of the ratio of the desired temperature and the calculated temperature (section 2.4). This rescaling does not have a physical basis, but in the equilibration stage this doesn't matter; a more physical method of controlling the temperature is to use a canonical ensemble, and the method for this is outlined in the next section. Once the steady temperature is established the rescaling can stop and the sample can be allowed to equilibrate a bit longer. The aim is to establish a steady state for the temperature, and also to allow for equipartition of the temperature over all the particles and over the different degrees of freedom of motion. It is not possible to establish the equipartition over the normal modes of vibrations of the lattice as a separate process. One possible test is to check that the amplitude of the temperature fluctuations settles to a roughly constant value once the systematic drifts have ceased. In some cases, particularly if there are some harmonic force constants in use, it may be necessary to use a more sophisticated method for controlling the temperature, and so the constant temperature algorithm discussed in the next section can be used.

Once the sample has been fully equilibrated, the following runs will be used to generate data that will be analyzed along some of the lines discussed in section 2.4. Depending on the information required from the simulations, this stage of the runs can last up to 100ps. If a time step of 0.005ps is being used, this will correspond to running the simulations for 20000 time steps. A typical sample using 500 particles on an average computer will use about 5s of computer time for each time step. Thus the total run will take 28hr of computer time, which is clearly a large use of resources. Since each run uses so much computing time, there is little flexibility to perform several runs under different conditions. The eventual output will be either just the particle trajectories or other calculated quantities, which will be stored for further analysis.

2.1.6. Rotational motions. Our discussion so far has assumed that all the particles are simple atoms or ions, interacting via central isotropic forces only. However, many systems of interest contain molecules or molecular ions, with the constituent atoms held together by forces that are much stronger than the inter-molecular forces. In these cases it will be common to assume that the molecules can be treated as rigid bodies, and each rigid body will have three translational degrees of freedom and three rotational degrees of freedom. The equations of motion for rigid bodies are well known, and can be found in any book on classical mechanics (e.g.Goldstein, 1980). If the molecule has a number of force centres (corresponding say to the constituent atomic sites) the torque, N, experienced by the rigid body is easily related to the forces experienced by each atom:

$$\underline{N} = \sum_j \underline{r}_j \times \underline{F}_j \qquad [2.1.10]$$

where the sum is over all atoms in the rigid body. For torques that arise from anisotropic interactions (e.g. electrostatic multipole interactions) additional expressions will be required; some of these have been given by Price et al (1984) (see section 2.3).

There are a number of ways of handling the rotational equations of motion. The well-known Euler angle formalism can be used, and the relationship between the time derivatives of the Euler angles and the components of the angular velocity are straightforward. However, the Euler angle method has a serious drawback in that when the angle for the second rotation Θ, as defined by Goldstein (1980), is zero, the first

and third angles are no longer uniquely defined. This leads to a singularity in the equations of motion as discussed by Evans (1977). So instead of using the Euler angle formalism to describe the rotational motions of rigid bodies it is common to define the orientation of a rigid body using a set of related variables known as quaternions. There are four of these variables, and they are related to the Euler angles by the relations:

$$Q_0 = \cos(\Theta/2)\cos(\psi+\varphi)/2 \qquad [2.1.11a]$$

$$Q_1 = \sin(\Theta/2)\cos(\psi-\varphi)/2 \qquad [2.1.11b]$$

$$Q_2 = \sin(\Theta/2)\sin(\psi-\varphi)/2 \qquad [2.1.11c]$$

$$Q_3 = \cos(\Theta/2)\sin(\psi+\varphi)/2 \qquad [2.1.11d]$$

and the rotation matrix that converts from space axes to body axes is simply given as:

$$A=\begin{pmatrix} Q_0^2+Q_1^2-Q_2^2-Q_3^2 & 2(Q_1Q_2-Q_0Q_3) & 2(Q_1Q_3+Q_0Q_2) \\ 2(Q_1Q_2+Q_0Q_3) & Q_0^2-Q_1^2+Q_2^2-Q_3^2 & 2(Q_2Q_3-Q_0Q_1) \\ 2(Q_1Q_3-Q_0Q_2) & 2(Q_2Q_3+Q_0Q_1) & Q_0^2-Q_1^2-Q_2^2+Q_3^2 \end{pmatrix} \qquad [2.1.12]$$

Note that the definitions used here follow those given by Pawley (1981); they differ from those given by Evans (1977) only in the signs of Q_1 and Q_3. The set of quaternions is defined to be normalised, such that:

$$Q_0^2 + Q_1^2 + Q_2^2 + Q_3^2 = 1 \qquad [2.1.13]$$

so that there are only three independent values as we would expect. The set of quaternions define a point on the surface of a hypersphere in four-dimensional space. The existence of the set of quaternions follows simply as an extension to higher dimensions of the representation of the two polar axes of a vector in the space of the three normalised cartesian coordinates x,y,z, or of the representation of the orientation of a vector in a plane by the two coordinates x,y instead of by one angle.

Although the quaternion equations are discussed in detail by Evans (1977), who was essentially the one who introduced the use of these variables into molecular dynamics simulations, we should give the essential properties as relate to the equations of motion. Additional information can be found in the paper by Pawley (1981). The rules of addition and multiplication for two quaternions $\underline{Q}$ and $\underline{Q}'$ are straightforward, where we define the quaternion to have a scalar component Q_0 and a vector component $\underline{q}=(Q_1,Q_2,Q_3)$ so that $\underline{Q}=(Q_0,\underline{q})$:

$$\underline{Q} + \underline{Q}' = (Q_0+Q_0', \underline{q}+q') \qquad [2.1.14]$$

$$\underline{Q}.\underline{Q}' = (Q_0Q_0'-\underline{q}.\underline{q}', Q_0\underline{q}'+Q_0'\underline{q}+\underline{q}\times\underline{q}') \qquad [2.1.15]$$

The quaternion $\underline{P} = \underline{Q}.\underline{Q}'$ can be easily constructed in matrix form following this equation:

$$\underline{P} = \begin{pmatrix} P_0 \\ P_1 \\ P_2 \\ P_3 \end{pmatrix} = \begin{pmatrix} Q_0 & -Q_1 & -Q_2 & -Q_3 \\ Q_1 & Q_0 & -Q_3 & Q_2 \\ Q_2 & Q_3 & Q_0 & -Q_1 \\ Q_3 & -Q_2 & Q_1 & Q_0 \end{pmatrix} \begin{pmatrix} Q_0' \\ Q_1' \\ Q_2' \\ Q_3' \end{pmatrix} = \begin{pmatrix} Q_0' & -Q_1' & -Q_2' & -Q_3' \\ Q_1' & Q_0' & Q_3' & -Q_2' \\ Q_2' & -Q_3' & Q_0' & Q_1' \\ Q_3' & Q_2' & -Q_1' & Q_0' \end{pmatrix} \begin{pmatrix} Q_0 \\ Q_1 \\ Q_2 \\ Q_3 \end{pmatrix} \qquad [2.1.16]$$

If we treat any normalised vector as a quaternion with the scalar component set to zero, the rotation of a vector $\underline{u}$ by a quaternion $\underline{Q}$ is given as:

$$\underline{u}' = \underline{Q}.\underline{u}.\underline{Q}^{-1} \qquad [2.1.17]$$

where

$$\underline{Q}^{-1} = (Q_0, -\underline{q}) \qquad [2.1.18]$$

The matrix 2.1.12 follows in a straightforward fashion from equations 2.1.15 to 2.1.17.

The time derivative of a quaternion can be shown to be related to the angular velocity via the matrix equation (Pawley, 1981; Evans, 1977):

$$\begin{pmatrix} \dot{Q}_0 \\ \dot{Q}_1 \\ \dot{Q}_2 \\ \dot{Q}_3 \end{pmatrix} = \frac{1}{2} \begin{pmatrix} Q_0 & -Q_1 & -Q_2 & -Q_3 \\ Q_1 & Q_0 & Q_3 & -Q_2 \\ Q_2 & -Q_3 & Q_0 & Q_1 \\ Q_3 & Q_2 & -Q_1 & Q_0 \end{pmatrix} \begin{pmatrix} 0 \\ \omega_1 \\ \omega_2 \\ \omega_3 \end{pmatrix} \qquad [2.1.19]$$

Since this expression is of the form of equation 2.1.16, we can easily use the expression for the inverse quaternion, equation 2.1.18, to obtain the reverse relation:

$$\begin{pmatrix} 0 \\ \omega_1 \\ \omega_2 \\ \omega_3 \end{pmatrix} = 2 \begin{pmatrix} Q_0 & Q_1 & Q_2 & Q_3 \\ -Q_1 & Q_0 & -Q_3 & Q_2 \\ -Q_2 & Q_3 & Q_0 & -Q_1 \\ -Q_3 & -Q_2 & Q_1 & Q_0 \end{pmatrix} \begin{pmatrix} \dot{Q}_0 \\ \dot{Q}_1 \\ \dot{Q}_2 \\ \dot{Q}_3 \end{pmatrix} \qquad [2.1.20]$$

The final equation that we need is for the second time derivative of the quaternion variables. Again, this can be easily obtained from the equations we have discussed so far, namely by combining equations 2.1.19 and 2.1.20 to give:

$$\begin{pmatrix} \ddot{Q}_0 \\ \ddot{Q}_1 \\ \ddot{Q}_2 \\ \ddot{Q}_3 \end{pmatrix} = \frac{1}{2} \begin{pmatrix} Q_0 & -Q_1 & -Q_2 & -Q_3 \\ Q_1 & Q_0 & Q_3 & -Q_2 \\ Q_2 & -Q_3 & Q_0 & Q_1 \\ Q_3 & Q_2 & -Q_1 & Q_0 \end{pmatrix} \begin{pmatrix} -2 \sum_{i=0}^{3} \dot{Q}_i^2 \\ \dot{\omega}_1 \\ \dot{\omega}_2 \\ \dot{\omega}_3 \end{pmatrix} \qquad [2.1.21]$$

The equations for the two time derivatives of the quaternions will therefore be obtained in the simulations from the calculated angular velocities and the torques. These variables can then be used in the normal time step algorithms just as for translational motions. The rotational equations of motion that link the torque, $\underline{N}$, to the changes in the angular velocity are easily expressed in the frame of the rigid body for which the inertia tensor is diagonal (Goldstein, 1980):

$$\begin{aligned} I_1\dot{\omega}_1 &= N_1 + (I_2-I_3)\omega_2\omega_3 \\ I_2\dot{\omega}_2 &= N_2 + (I_3-I_1)\omega_3\omega_1 \\ I_3\dot{\omega}_3 &= N_3 + (I_1-I_2)\omega_1\omega_2 \end{aligned} \qquad [2.1.22]$$

where I_1, I_2 and I_3 are the principle moments of inertia. If the principle components of the inertia tensor are not identical, the last term in equation 2.1.22 is non-zero, so that equation 2.1.22 represents a case of the use of velocity-dependent forces as discussed in section 2.1.2, thus necessitating the use of predictor-corrector steps in the integration algorithm.

There is one other technique that is used for rotational motions, namely the method of constraints (Ciccotti et al, 1982). However, for simple systems the quaternion method is sufficiently robust that it can beconsidered to be the automatic choice, particularly since it is useful to be able to use the quaternions in the analysis stage. Therefore we simply refer readers who might be interested in the constraint method to the quoted reference.

2.2. ENSEMBLES.

2.2.1. Microcanonical ensemble. The use of equations of motion that lead to the microcanonical ensemble (called the NVE ensemble, reflecting the constant number of particles, N, volume, V, and energy, E) has been the traditional approach to molecular dynamics simulation, and these equations formed the basis of the technical details presented in the last chapter. For many purposes the microcanonical ensemble is satisfactory to use. However, the interest in phase transitions in solids has recently prompted interest in the use of other ensembles for molecular dynamics simulations. There are probably two main reasons for this. The first is that one is interested in the equilibrium structures formed when the thermodynamic conditions are changed, for example, by changing the temperature or the pressure. In order to observe changes in the equilibrium structures it is important that the size and shape of the unit cell are not constrained to fixed values but that the simulation should be free to find the equilibrium unit cell. This prompted the development first of a constant pressure ensemble for liquids where the shape of the sample remained constant but the size of the sample was allowed to relax (Andersen, 1980), and secondly of a constant pressure ensemble for solids where the shape of the unit cell was also allowed to relax (Parrinello and Rahman, 1980). By using this kind of ensemble, the changes in the shape of the crystalline lattice that occur at a phase transition can be easily simulated; prior to the development of the constant pressure ensemble phase transitions could be simulated in the microcanonical ensemble only by using samples that were sufficiently large that the new phase obtained on cooling was produced as small crystallites within the sample (Pawley and Thomas, 1982).

The second reason for developing new methods of simulating different ensembles is that in order to be able to compare the results of MDS calculations with experimental and theoretical results, it would be useful to be able to perform the simulations under the same thermodynamic conditions as the experiments are carried out under and for which the theoretical results are appropriate. These conditions are usually at constant temperature.

We will see that it is useful to express the equations of motion in the Lagrangian formulation. The Lagrangian for the microcanonical ensemble is given as:

$$L = \frac{1}{2}\sum_i m_i|\dot{\underline{r}}_i|^2 - \frac{1}{2}\sum_{ij}\varphi_{ij}(\underline{r}_i,\underline{r}_j) \qquad [2.2.1]$$

where $\varphi_{ij}(\underline{r}_i,\underline{r}_j)$ is the potential energy of interaction between the i-th and j-th particles with respective instantaneous positions $\underline{r}_i$ and $\underline{r}_j$. The equation of motion of any variable Q in the Lagrangian is obtained from the Lagrangian equation of motion (Goldstein, 1980):

$$\frac{d}{dt}\left(\frac{\partial L}{\partial \dot{Q}}\right) = \frac{\partial L}{\partial Q} \qquad [2.2.2]$$

so that the equations of motion for the position variables are simply obtained as:

$$m_i\ddot{\underline{r}}_i = -\sum_j \partial\varphi_{ij}(\underline{r}_i,\underline{r}_j)/\partial\underline{r}_i = \underline{F}_i \qquad [2.2.3]$$

2.2.2 Constant pressure ensembles. Andersen (1980) first developed a method for simulating a constant-pressure ensemble for a liquid, in which the shape of the sample remained constant but the volume was treated as a dynamic variable and was therefore allowed to fluctuate. For solids it is important that the shape of the sample should be allowed to vary as well. In order to model an ensemble that includes the response of the sample size and shape to an applied external pressure P_{ex}, Parrinello and Rahman (1980) postulated the use of the following Lagrangian as an extension of equation 2.2.3:

$$L = \frac{1}{2}\sum_i m_i\dot{\underline{s}}_i{}^t\underline{\underline{h}}^t\underline{\underline{h}}\,\dot{\underline{s}}_i - \frac{1}{2}\sum_{ij}\varphi_{ij}(\underline{r}_i,\underline{r}_j) + \frac{1}{2}W\sum_{k\ell}\dot{h}_{k\ell}^2 - P_{ex}V \qquad [2.2.4]$$

where the matrix $\underline{\underline{h}}$ defines the size and shape of the simulation sample, and the variables $\underline{s}_i$ are now scaled coordinates, so that the components of $\underline{s}_i$ will lie between 0 and 1. The real space coordinates of the i-th molecule are then written as:

$$\underline{r}_i = \underline{\underline{h}}\cdot\underline{s}_i \qquad [2.2.5]$$

The parameter W has the dimensions of mass. The volume of the sample is simply equal to det($\underline{\underline{h}}$). It is useful to introduce the tensor $\underline{\underline{G}} = \underline{\underline{h}}^t\cdot\underline{\underline{h}}$. The equations of motion for the scaled coordinates $\underline{s}$ are readily obtained for the constant pressure case from equations 2.2.2 and 2.2.4:

$$m_i\ddot{\underline{s}}_i = \underline{\underline{h}}^{-1}\underline{F}_i - m_i\underline{\underline{G}}^{-1}\dot{\underline{\underline{G}}}\,\dot{\underline{s}}_i \qquad [2.2.6]$$

which represents the usual equation for Newtonian motion (equation 2.2.3) with a correction for the changing shape of the sample. The equations of motion for the sample vectors $h_{k\ell}$ are also readily obtained from the Lagrangian:

$$W\ddot{h}_{k\ell} = \sum_i m_i(\underline{\underline{h}}\,\dot{\underline{s}}_i)^t_k\,(\dot{\underline{s}}_i)_\ell - \qquad [2.2.7]$$

$$- \frac{1}{2}\sum_{ij}\frac{\partial\varphi_{ij}}{\partial r_{ij}}\,\frac{(\underline{r}_{ij})_k}{r_{ij}}(\underline{s}_i - \underline{s}_j)_\ell - P_{ex}\frac{\partial V}{\partial h_{k\ell}}$$

This equation can be expressed in a slightly different way by defining an internal stress tensor $\underline{\Pi}$ with components given as:

$$\Pi_{k\ell} = \frac{1}{V}\Big[\sum_i m_i(\underline{\underline{h}}\,\dot{\underline{s}}_i)^t_k(\underline{\underline{h}}\,\dot{\underline{s}}_i)_\ell + \qquad [2.2.8]$$

$$+ \frac{1}{2}\sum_{ij}(\underline{F}_{ij})_k(\underline{\underline{h}}(\underline{s}_i - \underline{s}_j))_\ell\Big]$$

where the force acting between the i-th and j-th particles is $\underline{F}_{ij}$ $(=(\partial\varphi/\partial r_{ij})\,\underline{r}_{ij}/r_{ij})$. Our equation of motion for h_{kl} is therefore given as:

$$W\ddot{h}_{k\ell} = \sum_m \left(\Pi_{km} - P_{ex}\delta_{km}\right)\sigma_{m\ell} \qquad [2.2.9]$$

where

$$\sigma_{k\ell} = \frac{\partial V}{\partial h_{k\ell}} = V(\underline{\underline{h}}^{-1})_{\ell k} \qquad [2.2.10]$$

This way of expressing the equation of motion for $\underline{h}$ highlights the fact that the sample cell changes its size and shape in response to fluctuations in the internal stress. In addition, we can also see how the method can be extended by using an external stress tensor $\underline{S}$ instead of a uniform pressure. The only change in the dynamical equations is that 2.2.9 is replaced by the general equation:

$$W\ddot{h}_{k\ell} = \sum\left(\Pi_{km} - S_{km}\right)\sigma_{m\ell} \qquad [2.2.11]$$

This has been used to study the effects of uniaxial stress in simple materials (Parrinello and Rahman, 1981).

The Lagrangian 2.2.4 conserves the momentum and Hamiltonian exactly. There is a weaker conservation of enthalpy, H, which is exact in the limit of large N. The internal pressure (or stress) does fluctuate, but its mean value will be equal to the external pressure. Thus this constant-pressure method is called an NPH ensemble.

Many technical details associated with the NPH ensemble have been discussed by Nose and Klein (1983a). In particular, they show how the use of long range forces can be incorporated into the method (section 2.3 of the present paper discusses the general use of long range forces). They also discuss one way of reducing the number of free variables in the $\underline{\underline{h}}$ tensor to six (three of the variables essentially only define the orientation of the sample). The fact that the choice of the term in the Lagrangian for the kinetic energy associated with the sample shape variables is arbitrary has also been discussed by these authors. This term can be modified so that the kinetic energy is instead given as:

$$\frac{1}{2}\,\mathrm{Trace}\,(\dot{\underline{\underline{h}}}^t\underline{\underline{B}}^t\underline{\underline{B}}\,\dot{\underline{\underline{h}}}) \qquad [2.2.12]$$

and Ray (1983) has suggested that a good choice for $\underline{\underline{B}}$ would be $\underline{\underline{G}}$. This modified form of the Lagrangian gives better consistency with Andersen's' constant pressure method, and gives a result for the average pressure that is consistent with the virial equation. On the other hand, the equations of motion for the sample vectors generated by the modification 2.2.12 are more complicated than the equations given above (2.2.7-2.2.11), and since both the original and modified Lagrangian formulations give NPH ensembles, there is little advantage to be gained in using the modified form. The choice of the mass variable W is also arbitrary. The size of W controls the frequency of the fluctuations in $\underline{h}$. It has been found that values of the order of a few atomic masses are useful; it has been suggested that the size of W can be related to the velocity of sound in simple systems (Andersen, 1980; Nose and Klein, 1983a). It has been formally demonstrated that the size of W does not affect the calculation of static averages, and it appears also that the calculation of dynamic quantities (section 2.4) is unaffected by the choice of W (Andersen, 1980; Nose and Klein, 1983a).

Finally, it should be noted that the equations of motion have velocity dependent forces; as discussed in section 2.1.2 this dictates the choice of integration algorithm employed.

2.2.3 The canonical ensemble. We have pointed out that it is possible to control the temperature by simply rescaling the velocities at each time step (section 2.1.5). However, it is not clear that this simple method generates a true thermodynamic ensemble. For this reason Nose (1984) has developed a constant temperature algorithm which, for the coordinates and momenta, generates trajectories for a canonical ensemble. The equations of motion for a canonical ensemble are obtained by postulating a Lagrangian of the form:

$$L = \frac{1}{2}\sum_i m_i s^2 |\dot{\underline{r}}_i|^2 - \frac{1}{2}\sum_{ij}\varphi_{ij}(\underline{r}_i,\underline{r}_j) + \frac{1}{2}Q\dot{s}^2 - (f+1)k_B T_{eq}\ln(s) \qquad [2.2.13]$$

where a new dynamic variable, s, and an associated mass variable, Q, have been introduced. T_{eq} is the externally chosen temperature for the sample, and f is the number of degrees of freedom of the physical system. The equations of motion for this Lagrangian are readily obtained:

$$m\underline{\ddot{r}}_i = \frac{1}{s^2}\underline{F}_i - \frac{2\dot{s}m\underline{\dot{r}}_i}{s} \qquad [2.2.14]$$

$$Q\ddot{s} = \sum_i m_i s|\underline{\dot{r}}_i|^2 - \frac{(f+1)k_B T eq}{s} \qquad [2.2.15]$$

Nose has proved that that these equations generate configurations that correspond to a canonical ensemble at a temperature of T_{eq}. The simulations using this method have the Hamiltonian, momentum and angular momentum conserved, and the temperature fluctuates about its mean value of T_{eq} without systematically deviating from this mean. Therefore this ensemble is essentially an NVT ensemble. Generalisation to the NPT ensemble is straightforward.

The variable s that appears in the dynamic equations has been interpreted by Nose as a scaling factor for time. In principle each time step should be divided by the instantaneous value of s in order to obtain a value for the real step. In practice it may be preferable to multiply each time step by the average $\langle 1/s \rangle$, particularly if the dynamic properties are of interest. The frequency of the fluctuations in s is determined by the value of Q.

2.2.4. Summary. The one question that we are left with is to decide the appropriate choice of ensemble conditions for any simulation that we wish to undertake. As a general rule, the NPH or NPT ensembles are most useful when the aim is to calculate an equilibrium structure at a given temperature, for example when studying phase transitions. Examples of the use of the NPH ensemble for this purpose given by Parrinello and Rahman (1980), Parrinello et al (1983), and Nose and Klein (1983b,c). As we stated above, the NVE ensemble will only be of any use for this when the sample is large enough to be able to break up into small crystallites ofthe lower-symmetry phase. However, the constant pressure ensembles also have constraints on the numbers of particles and their general arrangements, and there may well be a mismatch between the arrangement ofa number of particles in a single domain of one phase and a single domain of a related phase. Thus it may be necessary to know the structure of the equilibrium phases in advance and set up the simulation with the right sample dimensions and the right number of interactions in a neighbour list in order to reproduce any transition found in nature. For this reason it cannot be concluded that any structure calculated in the simulation is a true equilibrium structure, and experimentation with the conditions of the simulations will always be necessary.

The NVT ensemble is probably the best choice for calculating static quantities as comparison with experimental results and theoretical predictions is probably easier when using the canonical ensemble. However, the NVE ensemble will probably be the best ensemble to use when studying dynamic properties. The fluctuations in the sample size in the constant pressure ensemble are much larger than would be expected in nature, and the effects of these fluctuations are unclear. In addition, the fluctuations of s in the constant temperature ensemble may also have unexpected effects, so that the NVE ensemble represents a good safe choice. If the computational effort that the study of dynamic properties demands is to be put into the use of the NVE ensemble, this ensemble might as well be used for the simultaneous calculations of static quantities.

2.3. POTENTIAL MODELS.

2.3.1. Model functions. It is common in simulations to use four kinds of potentials: long range electrostatic potentials, short range attractive dispersive interactions, short range repulsive interactions, and covalent bonded potentials. The electrostatic interactions are the familiar multipole-multipole terms, of which the most common is the Coulomb monopole-monopole interaction that varies as 1/distance. In molecular systems (containing molecules or molecular ions) it may be necessary to include higher order interactions (e.g. involving dipole or quadrupole moments). Price et al (1984) have tabulated the equations for energies, forces and torques for the longer-ranged multipolar interactions. However, in many cases, the multipolar expansion for the electrostatic interaction between two molecules may give only a slow convergence, particularly when the two molecules are close together; the use of the expansion is only valid when the separation distance is larger than the molecular size. Then it is necessary to partition the molecule into a number of smaller units (e.g.the constituent atoms themselves) which have their own associated multipole moments, and then calculate the net interactions as sums over the interactions between these units. Then the individual multipolar expansions will be quickly convergent, as the size of the units will be much smaller than the distances between interacting units. It is possible to obtain a reasonable estimate of the sizes of the moments on the units from calculations of the molecular wave functions. Stone (1981) has devised a method of calculating the multipole moments on a number of sites from the calculated wave functions, which is called a "Distributed Multipole Analysis" (DMA). The user is free to choose what sites he likes for the location of the multipole moments, although it is usual practice to associate the moments with the atomic sites. The DMA will then calculate the multipole moments associated with each site. It is usually adequate just to use the charges given by this analysis, as the multipole expansion is usually accurate enough with only the lowest order terms included, but it may sometimes be necessary to include additional moments. For example, it was found necessary to include a local quadrupole moment in a recent calculation of the hydrogen bonding in thiourea, an important model incommensurate system (Dove and Lynden-Bell, 1986; see section 3.3), and calculations for malononitrile, which has an important sequence of reentrant phase transitions (Rae and Dove, 1982), showed that a significant part of the molecular dipole moment comes from the local atomic dipole moments and not only from the charge distribution (unpublished results by the present author). The accuracy of this approach depends on the accuracy of the wave function calculation; typically the use of SCF wave functions will give an over-estimation of the individual multipole moments by up to 20%. The moments can be scaled if there are experimental determinations of some of the net molecular moments (e.g. dipole moments), and the scaling factor, or even the individual moment distribution, can be refined at a later stage (see below).

The dispersive interactions also follow a series expansion, with terms with a distance-dependence of $(1/r)^n$, where n>6 and where some of the terms have zero coefficients. Because of the short-range nature of these interactions, it is usual to only include the lowest order (n=6) term. In addition, it is common to consider the dispersive interaction in conjunction with the repulsive term. This is because of the fact that although the coefficients of the dispersive terms are in principle calculable either from a knowledge of the electronic wave functions or the electronic polarizabilities and excitation energies, it is considerably easier to obtain appropriate values via a fitting procedure; this will always be performed at the same time as fitting a model repulsive interaction, and the values obtained will always be correlated with the form of the repulsive potential employed. The method of obtaining potential parameters by a fitting procedure will be described below. There is as yet no justifiable model for the repulsive interaction, and so it is common to use either of two phenomenological representations for the repulsive interaction taken with the dispersive interaction:

$$\varphi(r) = -Ar^{-6} + B\exp(-Cr) \qquad [2.3.1]$$

$$\varphi(r) = -4\varepsilon\left[(\sigma/r)^6 - (\sigma/r)^{12}\right] \qquad [2.3.2]$$

where $\varphi(r)$ is the total energy between two atoms separated by a distance r. The coefficients in the Lennard-Jones model can easily be seen to be related to the depth of the potential well and the equilibrium separation. It is possible to rewrite the equation for the Buckingham model in terms of new parameters that have similar physical significance (Califano et al, 1981).

Although we are forced to use a purely phenomenological model for the repulsive energy, in practice the range of atomic separation distances for which the repulsive interaction is significant is so small that there is very little sensitivity to the exact form of the potential model used. In fact, it is generally found that the values of the two coefficients B and C in the Buckingham model are so highly correlated that independent values for these parameters cannot be obtained.

When we are dealing with molecules it is common practice to consider each atom to be an interaction site, so that the net molecule-molecule interaction is simply a sum of the atom-atom interactions. However, the atom-atom interactions given above are isotropic, but it is unlikely that the electron density around an atom in a molecule will also be isotropic. For many cases the effects of anisotropy may be negligible, but for some atom types (such as nitrogen and chlorine) anisotropic atom-atom models may be necessary. Price and Stone (1982, 1984) have investigated the effects of anisotropy in some simple systems (including solid chlorine) by working with models in which the parameter C in the Buckingham representation is taken to be dependent upon the orientation of the molecules. A pragmatic alternative may well be to continue to use an isotropic model (which is computationally far more convenient), but to move the interaction centre along the covalent bond, so as either to flatten or elongate the effective potential.

The models that we have described so far are appropriate for interactions between non-bonded units. When we are considering systems containing molecular units it is usual to assume that the covalent bonds within a molecule are much stronger than the interactions between units so that vibrations involving bond stretching or bond bending will be decoupled from vibrations involving the intermolecular forces, and thus itis possible to forget about the internal motions altogether and simply treat the molecule as a rigid unit; we have described the equations for handling rigid bodies in section 2.1.6. If however it is still wished to allow for molecular flexibility it will usually be adequate to treat the covalent bonds as harmonic oscillators, with bond stretching and bond bending force constants that can be obtained empirically. But if we are simulating a covalent framework structure (e.g. diamond-structures or silicates) the use of harmonic interactions only will be inadequate, since the simulations will not allow energy to flow between different harmonic modes of motion. One common representation of the single covalent bond is the Morse interaction:

$$\varphi(r) = E_o[\exp(2\alpha(r_o-r)) - 2\exp(\alpha(r_o-r))] \quad [2.3.3]$$

where E_o, r_o and α are coefficients. This clearly allows for anharmonic as well as harmonic vibrations. As for the other empirical models, the coefficients of the Morse potential will need to be obtained from fitting to experimental data. In addition, a function will be required for bending of bonds involving a chain of three or more atoms, and it is usual to just simply invoke a harmonic potential model for the bond angles. The explicit use of bond-bending potentials will necessitate the use of a constraint algorithm for the equations of motion. If it appears that this discussion of the covalent bonds appears much more crude than for the electrostatic and dispersive interactions, it should be noted that there have been very few simulations in which the covalent bonds are the most important source of atomic interaction, and so there is still a considerable amount of work that need to be done in this respect. Recent uses of *ab initio* quantum mechanical calculations of binding energies in silicate structures look to be a promising route towards the development of justifiable anharmonic potential models of the covalent bond.

There are two further features that we have not yet mentioned. The first is the neglect of induction effects, in which additional energies can be obtained by polarisation of the atoms due to local electrostatic fields. These induction effects are usually neglected, although they can be important. For example, it has long been recognised that it is necessary to include atomic polarizabilities in lattice dynamics calculations of ionic crystals, and the shell model is used routinely to handle this interaction. However, polarization effects are very time consuming to incorporate into a simulation as they involve the use of self-consistency iterations, and so unless there is a demonstrated need to include these interactions they are generally neglected. Nevertheless, some simulations have been performed with polarisation effects included and the effects of these additional interactions have been shown to be not insignificant (Jacucci et al, 1976). The second feature is to do with the fact that our discussion has implicitly assumed that the many-body interactions can be separated into sums of pair interactions. This assumption is justifiable for electrostatic interactions, repulsive interactions and dispersive interactions for n=6, but is not true for covalent interactions and higher order dispersive interactions. The polarisation effects discussed above are also examples of many-body interactions. The calculations of three-body (and even higher order) interactions are again not trivial, and for the most part they are neglected.

This completes our survey of the commonly used potential models. Most simulations to date have been for ionic systems, in which the electrostatic Coulomb interactions and repulsive interactions are most important, and for molecular systems, in which the higher order multipolar electrostatic interactions, dispersive interactions and repulsive interactions have roughly equal weight. For these systems it has been found that the models we have discussed give adequate representations ofnature, although there are still areas in which the models can be refined (e.g. the use of anisotropic atom-atom interactions). On the other hand,there is still a considerable amount of work that needs to be done in order to obtain a satisfactory representation of covalent interactions. Since covalent bonds are an essential feature of many minerals (e.g. the framework silicates), progress will need to be achieved before simulations can be applied routinely to mineral systems.

Finally, it is also worth reminding ourselves in this section that the role of simulations is not so much to simply reproduce reality as it is to help explain reality. Therefore, the simulator must be prepared to look for a compromise in the choice of potential model; a model that yields the overall behaviour of interest can give considerable useful information even if there are deficiencies in the fine detail, and the additional effort and computational time required to refine the model to reproduce the fine detail will probably not give any more physical information. It is clearly important to take a pragmatic approach in this area!

2.3.2. Fitting potential parameters. The semi-empirical models that we have discussed will all contain a number of coefficients whose values will need to be determined. This aspect of the development of a simulation model may well prove to be one of the most time-consuming parts of the enterprise! There are several ways of determining these parameters. The electrostatic multipole moments can be obtained from the distributed multipole analysis described above, perhaps leaving an overall scale factor to be determined separately. Some of the parameters for the Buckingham or Lennard-Jones interactions can be taken from other studies, e.g. a model for a system that contains sodium ions can incorporate the parameters involving sodium taken from a separate study of, say, sodium chloride. This assumes the transferability of atom-atom potential parameters for the same atom type from one system to another. This assumption is often used in the study of molecular crystals, and general parameters for the common atoms C, H, F, N etc. have been obtained that appear to be applicable to a large range of different molecules (e.g.Williams and Houpt, 1986). The number of independent coefficients in a model can be reduced by constraining the coefficients for interactions involving unlike atoms to be related to the coefficients for interactions involving the corresponding like atoms. Thus, for the Buckingham and Lennard-Jones interaction, the coefficients for interactions involving unlike atom types i and j can be constrained as:

$$A_{ij} = [A_{ii}A_{jj}]^{\frac{1}{2}} \; ; \; B_{ij} = [B_{ii}B_{jj}]^{\frac{1}{2}} \; ;$$
$$C_{ii} = \tfrac{1}{2}[C_{ii}+C_{jj}] \qquad [2.3.4]$$
$$\varepsilon_{ij} = [\varepsilon_{ii}\varepsilon_{jj}]^{\frac{1}{2}} \; ; \; \sigma_{ij} = [\sigma_{ii}+\sigma_{jj}]$$

By using previous calculations, transferable parameters and constraints the number of coefficient values that will need to be determined can be considerably reduced. The remaining parameters can then be obtained using experimental data as a source of information. One useful set of data is the crystal structure, particularly if the structure is known at very low temperatures, as it is possible to use static lattice energy calculations at this point. Many programs (e.g. WMIN; Busing,1981) are able to work in two modes of operation. In the common mode, the program will change the calculated structural parameters in order to obtain the lowest energy structure, and this is the mode that corresponds to a static lattice simulation (section 1.2). The second mode is more useful for the purposes of model development. In this the program will adjust the undetermined potential parameters in order to minimised a residual M that is given by the derivatives of the lattice energy Φ with respect to the set of structural parameters p_j:

$$M = \sum_j \left(\frac{\partial\Phi}{\partial p_j}\right)^2 \qquad [2.3.5]$$

Essentially the program is fitting the potential parameters to give the best agreement between the calculated and observed crystal structures. This approach can work well, but it requires there to be more structural parameters than undetermined potential coefficients. This criteria can easily be met for low-symmetry structures, but clearly this approach will not work for high symmetry structures such as the perovskite structure, which in an extreme case will give just one structural parameter for three atom types! In this situation it will be necessary to use additional experimental data such as lattice vibrational frequencies and elastic constants. Programs for fitting of potential parameters in lattice dynamics calculations are not as available as the lattice energy programs, and at this point it may be necessary to use more of a trial-and-error approach. Indeed, it may even be necessary to use MDS calculations themselves in the refinement of models, using information such as thermal expansion.

2.3.3 Long range effects. As we mentioned in section 2.1.4, it is usual practice to only include the interactions between particles that lie within a certain distance of each other. However, some of the model interactions that we have discussed are long-range interactions, and the long range contributions that are neglected due to the use of a cut-off may be not be insignificant compared to the shorter distance interactions. There are a number of computational techniques for handling the long range part of these interactions.

If the form of the interaction between two particles is such that the average over all orientations of the particles is non-zero, and provided that it is not so long range that the lattice sums remain convergent, it is possible to use an integration averaging for the long range effects, using the approximation that at large distances the discrete atomic sites can be represented as a uniform density. A typical case would be an interaction of the form Ar^{-n}, where $n>3$. For any atom pairs, the long range average interaction energy $\varphi^{(d)}$ would then be simply given as:

$$\varphi^{(d)} = \rho\int_{r_0}^{\infty} 4\Pi r^2 dr.Ar^{-n} = \frac{4\Pi\rho A}{(n-3)r_0^{n-3}} \qquad [2.3.6]$$

where ρ is the density of interaction sites. r_0 is an effective cut-off radius, and is simply calculated from the number of direct short range interactions that have been included (N):

$$r_0^3 = 3N/4\Pi\rho \qquad [2.3.7]$$

It can be useful in some of the cases that involve orientation dependent interactions to extend this analysis to include integrations over angles. This extension is not difficult, and the relevant equations for the energies, forces and torques for the multipolar interactions that might necessitate this extension have been given by Price et al (1984).

It is more common though, when the simulation models involve long range multipolar interactions, to exploit the use of the periodic boundary conditions by employing standard methods in lattice calculations. The most popular method is the Ewald summation. This is best illustrated by the example of the use of Coulomb interactions. The method uses the mathematical identity for r^{-1}:

$$\frac{1}{r} = \frac{2}{\sqrt{\Pi}}\int_0^{\infty} \exp(-r^2\rho^2)d\rho \qquad [2.3.8]$$

The integral can be separated into two parts, so that a lattice sum can be given as:

$$\sum \frac{1}{r} = \frac{2}{\sqrt{\pi}} \sum \int_0^G \exp(-r^2\rho^2)d\rho + \frac{2}{\sqrt{\pi}} \sum \int_G^\infty \exp(-r^2\rho^2)d\rho \qquad [2.3.9]$$

For convenience, we assume the sums to be over all lattice sites in the crystal, and over the periodic repetitions of the sample. The second sum converges fairly quickly, and apart from expressing it in terms of the complementary error function (the second term in equation 2.3.11) it is left alone. As our sample is a periodically repeating unit, we can use Fourier methods to perform the first sum in equation 2.3.9: a long range function in real space will be a short ranged function in reciprocal space. Specifically, the sum over repeating origins $\underline{\ell}$ can be expressed in terms of the reciprocal lattice vectors $\underline{k}$ that have been given in equation 2.1.8.

$$\frac{2}{\sqrt{\pi}} \sum_{\underline{\ell}} \exp(-|\underline{\ell}+\underline{r}_j-\underline{r}_i|^2\rho^2) = \frac{2\pi}{V} \sum_{\underline{k}} \frac{1}{\rho^2} \exp(-k^2/4\rho^2)\exp(i\underline{k}.(\underline{r}_j-\underline{r}_i)) \qquad [2.3.10]$$

where $\underline{r}_i$ and $\underline{r}_j$ are the positions of two atoms in the sample. After integration of equation 2.3.10, and by expressing the real space sum as a complimentary error function, the lattice sum in equation 2.3.9 can be re-written as:

$$\phi = \frac{1}{2} \sum_{\underline{\ell},ij} q_i q_j |\underline{\ell}+\underline{r}_j-\underline{r}_i|^{-1} = \frac{1}{2} \sum_{ij} q_i q_j \left[\frac{4\pi}{V} \sum_{\underline{k}} k^{-2}\exp(-k^2/4G^2)\exp(i\underline{k}.(\underline{r}_j-r_i)) + \mathrm{erfc}(G|\underline{r}_j-\underline{r}_i|)\,|\underline{r}_j-\underline{r}_i|^{-1}\right] \qquad [2.3.11]$$

where all the summations now converge quickly. It should be noted that the $\underline{k}=0$ contribution is not included in the reciprocal space summation as it is zero. Because of the fast convergence of the real space sum, it effectively only includes the $\underline{\ell}=0$ contributions, and will only include the set of i,j for distances smaller than some cut-off distance. The real space sum in equation 2.3.11 includes the interactions between the charges in any molecule, which we will not want to be included in the lattice energy, and also the real space sum includes the interactions of the charges with themselves. Therefore we need to subtract these self interactions. The self interactions are therefore included to give the terms for $\underline{\ell}=0$

$$\Phi_{self} = \frac{1}{2}\sum_{ij}{}' q_i q_j \left[\mathrm{erfc}(G|\underline{r}_j - \underline{r}_i|)|\underline{r}_j - \underline{r}_i|^{-1} - |\underline{r}_j - \underline{r}_i|^{-1} \right] \quad [2.3.12]$$

$$= -\frac{1}{2}\sum_{ij}{}' q_i q_j \mathrm{erf}(G|\underline{r}_j - \underline{r}_i|)|\underline{r}_j - \underline{r}_i|^{-1} \quad [2.3.13]$$

where the prime indicates that the sum is over all the atoms in a single molecule. The small argument limit of erf(x) is simply:

$$\mathrm{erf}(x) = 2x/\sqrt{\pi} \quad [2.3.14]$$

so that the self interaction simply becomes equal to:

$$\Phi_{self} = -\frac{2G}{\sqrt{\pi}}\sum_i q_i^2 - \frac{1}{2}\sum_{ij}{}' q_i q_j \ \mathrm{erf}(G|\underline{r}_j - \underline{r}_i|)|\underline{r}_j - \underline{r}_i|^{-1} \quad [2.3.15]$$

where the sum now excludes the terms for i=j since they are given in the first term. This interaction must be added to equation 2.3.11, and equation 2.3.11 is then modified to exclude the real space contributions when $\underline{\ell}$ =0 for i and j in the same molecule and for i=j. Although the periodic boundary conditions give an artificial aspect to the simulations,this method is still often used and would seem to be a reasonable solution to the problem of the long range interactions. The use of this method though increases the computational time. In particular, the size of the sample determines the density of points in reciprocal space, so that the larger the sample the more terms there are in the reciprocal space sum. The number of terms in the real space sum is proportional to N when each atom is surrounded by a fixed cut-off radius, but the number in the reciprocal space sum will vary as N^3. The relative convergence of the two sums is controlled by the parameter G. In static lattice calculations typical values of G are of the order $\pi^{1/2}/V^{1/3}$, where V is the volume of the repeating unit, but in the case ofa simulation sample it is more appropriate to chose a larger value of G in order to accelerate the convergence of the real space sum. In many cases values of G can be chosen so that the reciprocal space sum is much smaller than the real space sum whilst still giving fast convergence of the real space sum. Then it will be possible simply to ignore the reciprocal space contribution since it is shielded by the modifications to the short range interactions. There are two other techniques that are much less commonly used. One is a Fourier technique that replaces the long range interactions by an interaction with a general background that is approximated by a mesh. This model therefore involves particle-particle and particle-mesh interactions, giving it the name PPPM. This technique is described in detail by Hockney and Eastwood (1981). The other technique is called the "reaction field" method, and allows the sample to react to an additional field related to the long wavelength fluctuations in the simulation sample. Various aspects of this method have been described by Van Gunsteren et al (1978).

2.4. ANALYSIS OF RESULTS FROM A SIMULATION.

A molecular dynamics simulation calculation essentially generates the trajectories of an ensemble of particles through the phase space spanned by the set of single particle dynamical variables. These trajectories or any instantaneous configuration can either be visually inspected or formally analyzed, and in this chapter we will describe a number of ways by which information that will be of interest in the study of the solid state can be extracted. I will assume, following the discussion in section 2.2, that the microcanonical ensemble will have been used.

2.4.1. Visual representation. Although formal analysis will be expected to yield the quantitative information about the system of interest, there may well be occasions when there is the need for a qualitative description of the state of the system, and for this it may be useful to use the simulations to give a visual representation of the system. By way of comparison, the end product of a crystal structure analysis may be a list of atomic coordinates, but a diagram showing a projection of the structure will usually be more useful. So, if in the course of the simulation there has been a change in the mean configuration of the ensemble (e.g. with anatomic rearrangement or equilibriation, or even a phase transition), or if some essential features need to be recognised (e.g. the rotational motions of a molecule), then some sort of visual representation may be required before the formal analysis can proceed. It is of course possible to use standard molecular drawing programs to plot sections (e.g. crystal layers) of instantaneous configurations (Pawley and Thomas, 1982; Pawley and Dove, 1983). However, it is now possible to go beyond this approach to make movie films of selected portions of the ensemble.

The basic method and the potential uses of this approach have been described by Dove et al (1986a). Instantaneous configurations of a subset of the whole ensemble (e.g. one layer containing say 25 unit cells) can be stored at regular time intervals, and by using high resolution raster graphics devices each configuration can be displayed on a screen and used as a single or double frame in the movie. Each atom will be represented by a coloured sphere, and these devices are able to give good colour contrast to help distinguish individual atoms and are able to use illumination shading to create the impression of depth, as well as being able to use a number of other features to improve the fine detail if so required. At the present time some effort is required in order to produce such a movie. A typical display image might take about 2 minutes to be produced and would be used for two frames. For a standard film speed of 24 frames per second, a film lasting a minute would take 24 hours to make! For motions with a power spectrum in the 1THz range, a scale of 1ps of simulation time to 1s of film time has been found to be useful. Although this is a time consuming process, it can provide a considerable amount of information when the dynamic properties of the system cannot be described in the language of normal mode excitations, e.g. when there is a considerable degree of disorder. But it is hoped that the rapid development in computer technology will one day enable simulations to give a real-time display whilst proceeding, so that this qualitative element will become more widely used.

2.4.2. Averages and fluctuations. Many quantities of interest will be of the form of thermal averages of real space quantities, for example the mean position of an atom in a unit cell. These averages are easy to calculate in a simulation, and involve simply taking the mean over all the particles in the ensemble, and then extending the calculation of this mean using a large number of configurations over a suitable time period. This latter average is necessary because the conditions of the ensemble (e.g.temperature) fluctuate in time. Real space averages are fairly accurate quantities, particularly when there is a large number of particles in the ensemble. Other quantities may be related to fluctuations. (We define the fluctuation in a quantity x to be given as $\langle x^2\rangle - \langle x\rangle^2$, where the brackets denote configurational and time averages.) Single particle fluctuations (e.g. the mean square atomic displacement) will be as easy to calculate and as accurate as the mean. However, the fluctuations in ensemble properties (e.g. temperature, or collective coordinates as discussed below) will be harder to calculate with the same high accuracy, as there is a loss of a factor of N when we cannot make a configurational average. In addition, the size of any fluctuation will in general scale inversely with the number of particles.

2.4.3. Thermodynamic quantities. We can consider the usual thermodynamic quantities to be illustrative examples of the previous discussion. The mean temperature is simply obtained from the particle velocities summed over all particles and averaged over time:

$$\langle T\rangle = \frac{1}{3Nk_B}\sum_i m_i \langle V_i^2\rangle \qquad [2.4.1]$$

The potential energy is given as an average of the particle potential energies, again averaged over a number of different configurations, and adding the long-range integrated contribution. The total energy (the standard internal energy E) is simply given as the sum of potential and kinetic energies. In the microcanonical ensemble this is a conserved quantity and so should not fluctuate. It is possible to calculate the specific heat capacity C_v from the temperature fluctuations by a result given by Lebowitz et al (1967):

$$C_V = \frac{3}{2}Nk_B\left[1 - \frac{3N\langle(\delta T)^2\rangle}{T^2}\right] \qquad [2.4.2]$$

Note that this is the heat capacity at constant volume and entropy, and that it differs from the usual equation for the specific heat capacity in the canonical ensemble, which involves the fluctuations in energy. Note also that there is now no condition that requires the sign of the heat capacity to be positive, which is a peculiar feature of the microcanonical ensemble! For the reasons given above, it is not usually possible to calculate the specific heat to an accuracy greater than a few percent.

As far as the author is aware, it has not yet been shown how to calculate either the entropy or the free energy from a MDS calculation using the microcanonical ensemble, although this may be due to the fact that physicists are less interested in the absolute values of these quantities than are minerals scientists. On the other hand, it has recently been shown (Branka and Parrinello, 1986) that the free energy can be calculated directly using the constant temperature ensemble of Nose (1984).

2.4.4. Pressure and stress. The pressure is readily calculated in a simulation using the standard virial expansion and including the effects of the integral correction to the long range interactions:

$$P = \rho k_B T - \frac{1}{3V}\sum_i \underline{F}_i \cdot \underline{r}_i - \frac{\partial\varphi^{(d)}}{\partial V} \qquad [2.4.3]$$

where $\underline{F}_i$ is the force acting on the i-th particle at position $\underline{r}_i$, $\varphi^{(d)}$ is the distant correction to the potential (e.g. equation 2.3.6) and ρ is the density. Stress is also easily calculated, and any stress component $\underline{\sigma}_{ij}$ is given in terms of the corresponding strain components e_{ij}:

$$\sigma_{ij} = \rho k_B T \delta_{ij} - \frac{1}{V}\frac{\partial\varphi}{\partial e_{ij}} \qquad [2.4.4]$$

where δ_{ij} is the Kronecker delta function.

It is possible to use the fluctuations in these quantities to obtain the corresponding susceptibilities. Cheung (1977) has derived an expression for the compressibility in the microcanonical ensemble:

$$\beta_S = \left[\frac{2(P+\rho k_BT)}{3} + \frac{1}{9V}\sum_{ij} \langle \underline{r}_i \cdot (\underline{\nabla}.\,{}^t\underline{\nabla}\varphi)\underline{r}_j{}^t \rangle - \frac{N\langle(\delta P)^2\rangle}{\rho k_BT}\right]^{-1} \qquad [2.4.5]$$

This approach can be readily modified to calculate the elastic constants in the microcanonical ensemble, using instead the fluctuations in the strain components (unpublished result):

$$C_{ijk\ell} = \frac{V}{k_BT} \langle \delta\sigma_{ij}\delta\sigma_{k\ell} \rangle - \rho k_BT\left[\delta_{i\ell}\delta_{jk} + \frac{2}{3}\delta_{ij}\delta_{k\ell}\right] - \frac{1}{V}\left\langle \frac{\partial^2\varphi}{\partial e_{ij}\partial e_{k\ell}} \right\rangle \qquad [2.4.6]$$

The elastic constant tensor determines the strain fluctuations in the constant pressure ensemble, and in this case the elastic constants are simply given as

$$(\underline{\underline{C}})^{-1}_{ijk\ell} = \frac{\langle V \rangle}{k_BT} \langle \delta e_{ij}\delta e_{k\ell} \rangle \qquad [2.4.7]$$

and the compressibility is given as

$$\beta_S = \langle (\delta V)^2 \rangle / Vk_BT \qquad [2.4.8]$$

It should be noted, however, that as for any fluctuations in ensemble properties (as opposed to single particle properties), the accuracy of the calculated susceptibilities will be of the order of a few percent.

2.4.5. Dielectric constant. It turns out that the dielectric constant is a very difficult quantity to calculate with any accuracy. Within a rigid ion model the sole contribution to the dielectric constant will be due to the motions of charge particles or the reorientations of dipole moments. The interactions between these particles will be long-ranged, and so the simulations will be using the techniques that have been discussed insection 2.3. Because of these problems with long range effects, it turns out that the calculated values of the dielectric constant are often affected by the shape of the simulation sample. There is still research being carried out into the calculation of dielectric properties. That said, it should be clear that the dielectric constant will be related to fluctuations in the dielectric polarisation of the ensemble.

2.4.7. Distribution functions. Since atoms are not held rigidly on their lattice sites, it may often be useful to know the probability of finding an atom in a different position at any time. In this case, we can define $h(\underline{r})dV(\underline{r})$ as the probability of finding the atomic centre within a volume element $dV(\underline{r})$ at a position $\underline{r}$ with respect to some origin (e.g. its lattice position). This example may not be too useful unless the atom is fairly mobile with a wide anisotropic distribution (as in a fast ion conductor), but it is clear that probability distribution functions of this form are not at all difficult to calculate. More useful may be the distribution function for the orientation of a molecule, as there are many examples where the orientation can be disordered. We call this an "orientational distribution function" (odf). The odf $h(\Theta, \varphi)$ for a single bond in the molecule can be defined such that $h(\Theta, \varphi)\sin\Theta\, d\Theta\, d\varphi$ gives the probability of the bond lying within the solid angle element $\sin\Theta\, d\Theta\, d\varphi$ about the direction defined by the two polar angles Θ and φ. Thus for any instantaneous configuration of N bonds the number lying within this solid angle element will be $Nh(\Theta, \varphi)\sin\Theta d\Theta d\varphi$. For an isotropically uniform distribution $h(\Theta, \varphi)=1/4\pi$. The calculation of these distribution functions can be readily performed by using simple binning algorithms.

In addition to describing the distribution of a single particle, we are often interested in the way in which the positions or orientations of two molecules are related. The simplest example is the pair distribution function $g(r)$, which gives information about the local environment of an atom. The probability of finding an atom in a volume element $dV(r)$ at a *distance* r from a reference atom is equal to $\rho g(r)dV(r)$, where ρ is the density. In many cases we are not so much interested in the angular dependence of the probability as in the radial dependence. Thus the number of atoms lying between r and r+dr of the reference atom is simply $4\pi r^2 \rho g(r)dr$. At small distances $g(r)=0$ as the atoms cannot get closer than a certain minimum separation. At very large distances the positions of atoms are unrelated, so that $g(r)$ is simply given as 1, since the number of atoms within a shell at large distances will simply be related to the mean density. The pair distribution function $g(r)$ is particularly useful in the study of disordered systems where the local environment is not known; examples are for fluids, systems with substitutional disorder, systems with local random strains, or systems undergoing phase transitions. Many theories of disordered systems will result in a calculation of an appropriate $g(r)$, and in these cases simulations are able to provide excellent tests.

2.4.6. Crystallographic temperature factors. Our last example of real space averages and fluctuations is the calculation of the rigid body temperature factors that are determined in a standard crystal structure analysis. The set of translational parameters are easily calculated, and are given as:

$$T_{ij} = \langle u_i u_j \rangle \qquad [2.4.9]$$

where the subscripts i,j denote motions along any two components of the orthogonal crystal axes. The librational temperature factors can be calculated from the fluctuations and mean values of the quaternions. If the instantaneous orientation for a molecule, described by the quaternion Q, is a small rotation away from the average orientation, that has the average quaternion $\langle \underline{Q} \rangle$, we can write the relationship between these two quaternions as:

$$\underline{Q} = \underline{q} \cdot \langle \underline{Q} \rangle \qquad [2.4.10]$$

The small rotation is thus given by the quaternion $q=(1,\varphi_1/2,\varphi_2/2,\varphi_3/2)$, where the angles $\varphi_1,\varphi_2,\varphi_3$ are the rotation angles about the orthogonal crystal axes. Following the rules of quaternion multiplication (equation 2.1.16), and noting the definition of the inverse of a quaternion (equation 2.1.18), we obtain values for the rotation angles:

$$\underline{q} = \begin{pmatrix} 1 \\ \varphi_1/2 \\ \varphi_2/2 \\ \varphi_3/2 \end{pmatrix} = Q.\langle Q \rangle^{-1} = \begin{pmatrix} Q_0 & -Q_1 & -Q_2 & -Q_3 \\ Q_1 & Q_0 & -Q_3 & Q_2 \\ Q_2 & Q_3 & Q_0 & -Q_1 \\ Q_3 & -Q_2 & -Q_1 & Q_0 \end{pmatrix} \begin{pmatrix} \langle Q_0 \rangle \\ -\langle Q_1 \rangle \\ -\langle Q_2 \rangle \\ -\langle Q_3 \rangle \end{pmatrix} \qquad [2.4.11]$$

The components of the libration tensor $\underline{L}$ are given as:

$$L_{ij} = \langle \varphi_i \varphi_j \rangle \qquad [2.1.12]$$

and can therefore be easily calculated using the result given in equation 2.4.11. The quantities that need to be calculated in the simulation are therefore the average quaternion components and the means of products of components. The translation-libration tensor $\underline{S}$ can also be calculated using equation 2.4.11; the components are given as:

$$S_{ij} = \langle u_i \varphi_j \rangle \qquad [2.1.13]$$

This method for the calculation of the $\underline{L}$ tensor form the quaternion averages has been given by Della Valle and Pawley (1984), who have used these calculations in a large MDS study of naphthalene in order to compare the predictions of a number of different Buckingham-type potential models with corresponding experimental data at a number of different temperatures.

A typical curve for g(r) for a liquid is given in figure 2.4.1. The

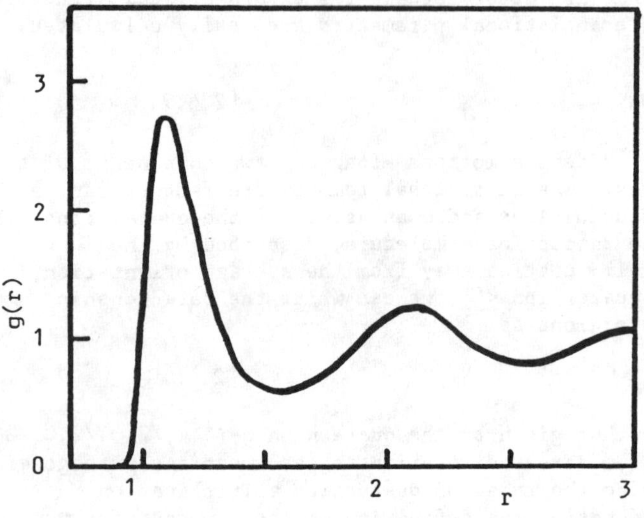

fig. 2.4.1 Representative form of g(r) for monatomic fluid. The unit of distance r is approximate atomic diameter.

strong peak is the signature of a surrounding shell of atoms, and the number of atoms in this shell is simply given as:

$$n = \int_{r_1}^{r_2} g(r)\rho 4\pi r^2 \, dv \qquad [2.4.14]$$

where the limits on the integral are the values of r either side of the peak. g(r) can be defined for specific types of atoms, like or unlike, and is not at all difficult to calculate. Formally, g(r) for two atom types i and j is defined in terms of the local densities $\rho_i(r)$ and $\rho_j(r)$:

$$g_{ij}(\underline{r}_i-\underline{r}_j) = \langle \rho_i(\underline{r}_i-\underline{r}_j)\rho_j(o)\rangle / \rho_i\rho_j \qquad [2.4.15]$$

where the brackets denote space and time averages. This form of g(r) will be useful later on, where we will discuss the fact that a scattering experiment will give information about the spatial Fourier transform of equation 2.4.12. This point leads us into the next section on correlation functions.

2.4.8. Space and time correlation functions. It is clear that any experiment will not provide information about instantaneous positions, velocities etc., but will give information about average values of quantities and the average evolution of these quantities in time and space. In addition, any theory will also not predict the absolute instantaneous configurations but will again predict average values and, hopefully, the average evolution of these quantities. Hence the important calculations in a simulation will not be instantaneous quantities but the average behaviour. We have introduced the calculations of mean quantities and the single particle probability or distribution functions, which will tell us about the range of values that any quantity will take. We are also interested in the the relationship between quantities separated in space and time, which introduces the idea of a correlation function. A correlation function describes the average relationship between a quantity at one point in time and space and at any other point. The function will be zero when the quantities at the two points are statistically independent, and will generally be normalised to unity when the quantities are completely related to each other (e.g. when the two quantities have zero time and spatial separation). To illustrate this, we will consider the pair distribution function g(r) that we discussed above. The relevant correlation function is formally given as h(r)=g(r)-1. This has the feature that h(r) is zero at large distances, and is equal to -1 at very small distances. h(r) provides information about the extent to which local fluctuations in density are correlated over different distances. The form of h(r) tells us that local fluctuations in two regions of space at large distances are completely unrelated, but at smaller distances the microscopic fluctuations in the local density associated with the motion of a single atom are reflected in the corresponding motions of the neighbouring shells. The complete inverse correlation at small distances is associated with the correlations due to the atomic size.

Time correlation functions tell us how various quantities change with time, and in particular give information about the extent to which the quantity values at different times are related. They can therefore give information about single particle dynamic properties, such as diffusion constants, relaxation times and power spectra. The simplest example might be the correlation function associated with the orientation of a dipole moment $\underline{p}(t)$. Thus we can define the self correlation function:

$$P(t) = \langle \underline{p}(0).\underline{p}(t)\rangle / \langle p^2\rangle \qquad [2.4.16]$$

In many cases P(t) will vary as $\exp(-t/\tau)$, where τ is a relaxation time and gives a measure of the reorientation time of the dipole moment, although at very short times the relaxation will be "Gaussian-like", with $dP(t=0)/dt=0$. The relaxation time τ essentially gives the time scale over which the value of $\underline{p}$ at any time is determined by its value at a previous time. In the case where $\underline{p}$ can be interpreted as a pseudo-Ising spin, the relaxation time will give the average time that the spins point in any direction before they flip to the opposite direction. It is often interesting to consider the time Fourier transform of correlation functions such as equation 2.4.16; in the case of the dipole moment correlation function, where there is only a weak coupling between neighbouring dipoles, the power spectrum that is obtained by Fourier transformation of the correlation function is the quantity that is measured in an infra-red absorption experiment (In the case of strong coupling between neighbouring dipoles, the relevant correlation function for Infra-red absorption is for the total dipole moment of the sample.). However, in many other cases the power spectrum obtained from the correlation function will not have a corresponding experimental realisation, or else an experiment may measure several different power spectra simultaneously so that any one cannot be unambiguously extracted from the data, and it is this ability of the simulations to calculate quantities that cannot be measured experimentally that makes it such an important technique.

Time correlation functions are not very difficult to calculate. For the self correlation function X(t) obtained from the variable x(t), the averaging process is simply explained in the equation:

$$X(t) = \frac{1}{T}\sum_{i}\int_{0}^{T} x_i(t')x_i(t'+t)dt', \quad T\to\infty \qquad [2.4.17]$$

where the sum is over all atoms, i, in the sample and the integral is replaced by a sum over successive configurations. The calculation of X(t) is not difficult in principle, but may be computationally time consuming in some cases since, in addition to the summations over all particles, for each value of t there is the summation over all configurations from the start of the run to the end of the run. Therefore if the total simulation has n timesteps, and the correlation function is calculated for m timesteps, the number of calculations that go into the calculation of X(t)is proportional to mxn. This will be particularly time consuming if the configurations are stored on magnetic tape, since the tape will need to be read, rewound and reread many times! It is possible to perform these calculations in a computationally faster way using a well-known Fourier transform technique:

$$a_i(\omega) = \int_{-\infty}^{\infty} x_i(t)e^{-i\omega t}dt \qquad [2.4.18]$$

$$X(t) = \sum_{i}\int |a_i(\omega)|^2 e^{i\omega t}dt \qquad [2.4.19]$$

These equations are a form of the Wiener-Khintchine theorem, and will be found in most books that discuss power spectra (e.g. McQuarry, 1976). The equations 2.4.18 and 2.4.19 can be evaluated fairly quickly using standard Fast Fourier Transform routines; the use of this method for the analysis of MDS data has been described by Futrelle and McGinty (1971). In general single particle correlation functions can be calculated reasonably accurately, with typical errors of the order of 0.01. This level of accuracy is more than adequate for most purposes.

One time correlation function that is of considerable importance is the velocity self-correlation function, C(t), given as:

$$C(t) = \langle \underline{v}(0).\underline{v}(t)\rangle / \langle \underline{v}(0).\underline{v}(0)\rangle \qquad [2.4.20]$$

A typical example of this function for a liquid is shown in figure 2.4.2. The structure in this function can be correlated to the actual

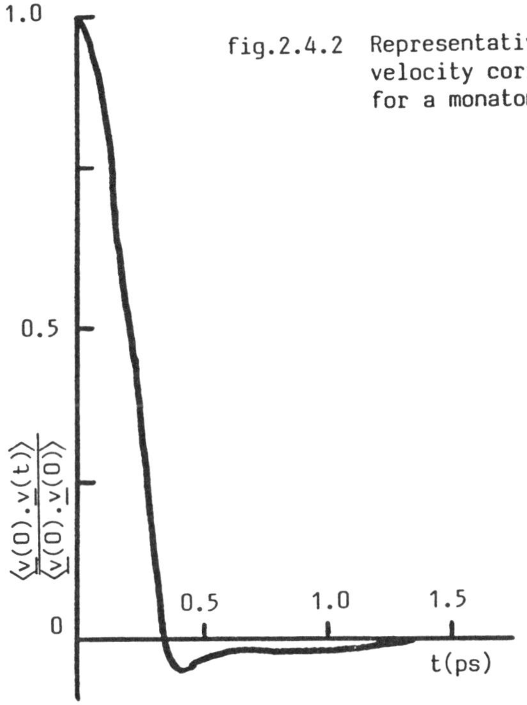

fig.2.4.2 Representative form of the velocity correlation function for a monatomic fluid.

motions of the atoms in their local environment. The fact that the function falls below zero implies that the atoms change their directions of motion, which implies that they undergo vibrations within a local cage of neighbouring atoms. This is clearly to be expected for a crystal, but the extent ofthis cage effect for mobile atoms in a disordered medium (e.g. a liquid, or for mobile ions in a fast ion conductor) is an important piece of information. The power spectrum of C(t) for a crystal can easily be shown to be proportional to the quasi-harmonic phonon density of states, and so provides a quick way of looking at the overall dynamical properties of the system. The integral of C(t) over time, which is the zero frequency value of the power spectrum, gives a relaxation time τ , which in turn is related to the atomic diffusion constant D at a temperature T:

$$D = k_B T\tau/m \qquad [2.4.21]$$

where m is the atomic mass. This will only be calculable for systems that have mobile atoms, such as liquids. However, the components of the time correlation function 2.4.20 can be the time derivatives of any quantity, and the resultant diffusion constant will define the diffusive behaviour of this variable in the corresponding phase space. For example, if we form a correlation function from the angular velocities, then we can calculate a rotational diffusion constant that will describe the behaviour of the molecular reorientational motions in the rotational phase space.

2.4.9. Collective variables. The results that we have described so far are associated with the properties of individual particles. In many cases, however, we will be interested instead in the behaviour of the whole sample, which is more conveniently described by the use of the spatial Fourier components that give the collective variables. We can define a set of collective variables $X(\underline{k})$ as the spatial Fourier transforms of a corresponding set of variables $x(\underline{R}_i)$ associated with each of the equilibrium atomic or molecular positions $\underline{R}_i$. For N particles $X(\underline{k})$ is given as:

$$X(\underline{k}) = N^{-1/2} \sum_i x(\underline{R}_i)\exp(i\underline{k}.\underline{R}_i) \qquad [2.4.22]$$

where the sum is over all the atoms in the sample (i=1,N), and the only allowed values of $\underline{k}$ are those that are determined by the size and shape of the molecular dynamics ensemble (section 2.1.3). It is easily seen that the individual values of $x(\underline{R}_i)$ will be given as the inverse transform of the set of $X(\underline{k})$. The sum in 2.4.22 is an instantaneous sum, so strictly speaking we should consider the variables $X(\underline{k},t)$ and $x(\underline{R}_i,t)$.

The collective variables are of interest as they describe the microscopic properties of the whole ensemble and they determine the behaviour of the individual particles. For example, the fluctuating displacements of single atoms from their mean positions are determined by the full phonon spectrum rather than by a mean local potential. In many studies using MDS we are interested in phase transitions involving ordering processes. There may be a number of possible (or even competing) ordering schemes that are defined by particular wave vectors. These will be seen as fluctuations in the collective variables, with amplitudes that may be strongly dependent upon $\underline{k}$. Some ordering schemes may involve the same ordering occurring in each unit cell, so that critical fluctuations will occur at $\underline{k}=0$, but other ordering schemes may involve say odd or even ordering in adjacent unit cells, with large amplitude fluctuations when $\underline{k}$ is at the appropriate Brillouin zone boundary point. Thus it can be seen why in many cases a collective variable description may be more informative than a single particle description. Moreover, whereas resonance and absorption experiments give information about single particle properties, the large class of scattering experiments that are often performed are usually more sensitive to the collective properties than to the single particle properties (section 2.4.11).

The fluctuations in the collective variables are conveniently analyzed by static correlation functions of the form:

$$M_{xx}(\underline{k}) = \langle X(\underline{k})X(-\underline{k})\rangle \qquad [2.4.23]$$

where the average is over time. The use of $\underline{k}$ and $-\underline{k}$ in this thermal average is due to the fact that fluctuations obey the requirements of conservation of wave-vector. For any number of quantities in a product the sum of the wave-vectors of the individual components of the product must equal zero or, for an Umklapp term, a reciprocal lattice vector. Note that as the collective variables are ensemble quantities, the correlation functions $M_{xx}(\underline{k})$ are not as accurate as single particle correlation functions, as we have already discussed earlier. Correlation functions such as $M_{xx}(\underline{k})$ are closely related to a corresponding susceptibility $\chi_{xx}(\underline{k})$:

$$\chi(\underline{k}) = M_{xx}(\underline{k})/k_BT \qquad [2.4.24]$$

For example, if the variables x that give $X(\underline{k})$ are proportional to a local dipole moment, the susceptibility 2.4.24 will be the $\underline{k}$-dependent dielectric susceptibility, and the $\underline{k}=0$ susceptibility will be the general susceptibility as described in section 2.4.5. We can also calculate cross correlation functions for two collective variables $A(\underline{k})$ and $B(\underline{k})$:

$$M_{AB}(\underline{k}) = \langle A(\underline{k})B(-\underline{k})\rangle \qquad [2.4.25]$$

In addition to the statistical errors on such functions, which, as we have noted, are larger for collective correlation functions than for single particle correlation function, there may be further errors associated with residual non-equipartition. When the MDS sample is equilibrated, the tests for equilibrium and equipartition are usually performed on the single particle velocities. Thus it is possible that the single particle velocities can show equipartition whilst there is not a corresponding equipartition of the thermal energy in the corresponding vibrational collective modes. If the crystal is sufficiently anharmonic the equipartition in the collective variables will be established through the anharmonic interactions, but for modes than are only very weakly anharmonic (e.g. the low-$\underline{k}$ acoustic modes) these problems may remain. We can however test for such errors by calculating the real and imaginary components of our correlation functions. For a correlation function ofthe form given in equation 2.4.23, the imaginary component will always be rigorously zero. However, for cross correlation functions such as 2.4.25 there is no such requirement, but instead general theory will predict that either the real or imaginary component should be zero for modes with full equipartition. If we say that our correlation function should be real, the residual errors will be revealed through the fact that in practice the imaginary component of $M_{AB}(\underline{k})$ will not be statistically averaged to zero, but will have a value that is hopefully small compared to the value of the real component. The ratio of the values of the real and imaginary components will give a reasonable estimate of the errors on the calculated values of $M_{AB}(\underline{k})$, but this estimate should not be taken point-by-point, and instead all the values of the imaginary components of $M_{AB}(\underline{k})$ should be taken together to indicate the accuracy of $M_{AB}(\underline{k})$ over the whole range of $\underline{k}$. The effect of the non-equipartition errors can be distinguished from the normal statistical errors since the size of the imaginary component will not fall to zero on increasing the averaging time (For random errors the amplitude of the imaginary component will fall with the square root of the averaging time.). This non-equipartition problem is identical to the statement that the ensemble fails in practice to sample phase space in an ergodic fashion in the time scale of the simulation, as has been discussed by Dove et al (1986b). However, in spite of the gloomy note of this discussion, it is found in practice that the accuracy of the calculated correlation functions such as given by equations 2.4.23 and 2.4.25 will be sufficient for most purposes.

One can also construct the time correlation functions, which in general will be of the form:

$$M_{AB}(\underline{k},t) \;=\; \langle A(\underline{k},0)B(-\underline{k},t)\rangle \qquad [2.4.26]$$

These correlation functions are calculated in the same way as the single particle time correlation functions (equation 2.4.17), and the discussion of the statistical and non-ergodic errors apply here also; when A=B, the imaginary part will only vanish at t=0. The time correlation functions for these collective variables will correspond to either propagating or relaxational modes, and in many cases of interest, the nature of the correlation function will change with $\underline{k}$, so that a mode that propagates for some wave-vectors will be only relaxational at others, and there will be a range of wave-vectors over which the behaviour changes. The time correlation functions 2.4.26 are affected by the existence of the recurrence time discussed in section 2.1.3. Therefore, they are only to be considered as reliable over the time interval from zero to the recurrence time, for beyond this the forms of the correlation functions include contributions from modes that have travelled the length of the sample and through the periodic boundaries, and even if the correlation function is relaxational there will still be interactions with the fast travelling waves. For example, it is commonly found that a propagating mode that decays in time will increase in amplitude again after further time rather than eventually decaying to zero. As the collective correlation functions can be long-lived, calculating their time Fourier transforms in order to obtain the power spectra may not be as straightforward as for the single particle time correlation functions; in general the latter functions are are short lived, since they are composed of a sum over a large number of modes with different frequencies. The transform of a truncated function will contain spurious features associated with the cut-off since it will contain the convolution with the transform from a top-hat function, which will give ripples in the resultant power spectra. There are three ways around this. The first is to multiply the correlation function by a short life-time function, such as a Gaussian, which will eliminate the cut-off problems at the expense of broadening all the features in the power spectrum (due to the convolution of the power spectrum with the transformed Gaussian). As the correlation function has the heaviest weighting at short times, it is possible to go beyond the recurrence time in order to reduce the line broadening in the transform. The second approach is to use a reasonable model for the time correlation function in order to extrapolate the correlation function past the recurrence time until it has decayed to a satisfactorily small amplitude, and then Fourier transform this extrapolated function. The third method is to "fit" a power spectrum that has been back-transformed against the correlation function, using an algorithm such as the "maximum entropy" method (e.g. Tildesley and Lynden-Bell, 1986). This procedure can work very well. However, although any peaks in the power spectrum will be correctly calculated, it is not clear that their widths will be accurate. In any case, the accuracy of these time correlation functions will probably only be good enough to enable reliable quantitative conclusions to be drawn in special cases, although the overall qualitative behaviour will in general be reasonable. If the only interest is in the peak positions in the power spectra rather than the the line widths, the Fast Fourier Transform method of equations 2.4.18 and 2.4.19 can be used, noting that for power spectra the second transform does not need to be performed.

2.4.10. Normal mode analysis. In the previous discussion we did not attempt to identify the particular real-space variables that were transformed to give the collective variables. Usually the variables will be atomic displacements, so that the time correlation functions will define the normal modes of vibrations (phonons) of the system. For many cases the anharmonic interactions that are implicit in the MDS calculation will, as in nature, lead to renormalizations of the harmonic frequencies to give a quasi-harmonic description of the lattice excitations, so that we can continue to work in terms of normal mode coordinates. Only in the very simple cases (i.e. the uninteresting ones!) will the normal modes correspond to the simple vibrations that have polarisations along the ensemble axes. Therefore we need a general method to obtain the atomic motions that characterize each normal mode. For n atoms in the unit cell there will be 3n normal modes for each wave-vector. If we define three orthogonal axes, which we label 1,2,3, and define the displacement of the μ-th atom of the $\underline{\ell}$-th unit cell along the j-th direction as $u_j(\mu,\underline{\ell})$, we can define the corresponding collective variable $U_j(\mu,\underline{k})$ as:

$$U_j(\mu,\underline{k}) = (m_\mu/N)^{1/2} \sum_{\ell} u_j(\mu,\ell)\exp(i\underline{k}.\underline{R}(\mu,\ell)) \quad [2.4.27]$$

where m_μ is the mass of the atom, and $\underline{R}(\mu,\underline{\ell})$ is the equilibrium position of each atom. Note that for future application we have defined a mass-weighted collective variable. We can gather all of the separate collective variables $U_j(\mu,\underline{k})$ for each wave-vector into a 3n component column vector $\underline{T}(\underline{k})$. The components of $\underline{T}(\underline{k})$ are not in general the normal mode coordinates as they have not been defined to be orthogonal. Instead we can define the normal mode coordinates to be orthogonal linear combinations of the components of $\underline{T}(\underline{k})$, and we denote the resultant 3n component column vector as $\underline{Q}(\underline{k})$. The transform from $\underline{T}(\underline{k})$ to $\underline{Q}(\underline{k})$ therefore involves a 3nx3n matrix $\underline{\underline{A}}(\underline{k})$:

$$\underline{Q}(\underline{k}) = \underline{\underline{A}}(\underline{k}).\underline{T}(\underline{k}) \qquad [2.4.28]$$

whose components need to be determined. The static thermal averages of the normal mode coordinates can easily be related to the corresponding averages of the original collective variables:

$$\langle \underline{Q}(\underline{k}).\underline{Q}^t(-\underline{k})\rangle = \underline{\underline{A}}(\underline{k}).\langle \underline{T}(\underline{k}).\underline{T}^t(-\underline{k})\rangle.\underline{\underline{A}}^t(-\underline{k}) \qquad [2.4.29]$$

where the superscript t denotes a transposed matrix. Since the normal mode coordinates are orthogonal, this equation represents the diagonalisation of the matrix $\underline{S}(\underline{k})=\langle\underline{T}(\underline{k}).\underline{T}^t(-\underline{k})\rangle$. Therefore the eigenvalues of $\underline{S}(\underline{k})$ give directly the amplitudes of the normal mode coordinates. Equally important, though, is the fact that the eigenvectors of $\underline{S}(\underline{k})$, which can be complex, give the components of the matrix $\underline{A}(\underline{k})$ which transforms the original collective variables to the normal mode coordinates. This matrix can be used at each time step to calculate the instantaneous value of each of the normal mode coordinates $\underline{Q}(\underline{k},t)$ from the set of original collective variables $\underline{T}(\underline{k},t)$ that are calculated in the simulations directly:

$$\underline{Q}(\underline{k},t) = \underline{\underline{A}}(\underline{k}).\underline{T}(\underline{k},t) \qquad [2.4.30]$$

The subsequent analysis of a MDS calculation will use the 3n components of $\underline{Q}(\underline{k},t)$ in a correlation function analysis of the form that we have already discussed. If the system contains rigid molecules or molecular ions, the analysis will be extended to include the rotation angles about the orthogonal axes in the set of dynamical variables, weighted by their respective moments of inertia. These angles can be readily calculated from the instantaneous values of the quaternions (equation 2.4.11). A preliminary account of this method has been given by Dove and Lynden-Bell (1986b) but in that study the computational effort involved in the diagonalisation of the matrix $\underline{S}(\underline{k})$ was reduced by taking account of the symmetry. By noting the symmetries of all the possible normal modes, smaller matrices could be constructed for each symmetry type. This approach however is only feasible when the unit cell is relatively simple; when there is a large number of atoms in a unit cell of low symmetry (as found in many minerals) the use of the full matrices is a simpler procedure.

The normal mode coordinates calculated in this way can be used to construct time correlation functions. In general these will represent propagating vibrational modes with single frequencies. The corresponding power spectra will therefore contain single peaks with widths that indicate the extent of damping of the vibrations. The peak positions will clearly correspond to the vibrational frequencies of the normal mode vibrations. There is a standard result for quasi-harmonic lattice dynamics that relates the square of the vibrational frequency of a normal mode to the inverse susceptibility (Cowley, 1980). From equation 2.4.24 we can therefore obtain the quasiharmonic mode frequency from the static correlation function for a normal mode $\omega_i(\underline{k})$:

$$\omega_i^2(\underline{k}) = k_B T \langle Q_i(\underline{k}) Q_i(-\underline{k}) \rangle \qquad [2.4.31]$$

2.4.11. Scattering theory. In this section I will relate the discussion of correlation functions to the results of scattering experiments; specifically I will discuss neutron coherent scattering experiments, for these experiments are able to give information in both the wave-vector and frequency domains, and unlike X-ray scattering experiments they do not have form factors that vary with the scattering vector. More detailed discussion of the theory can be found in the book by Marshall and Lovesey (1971). The intensity of a scattered neutron beam is proportional to the quantity $S(\underline{Q}, \omega)$, called the scattering function, which contains information about the collective density fluctuations in the sample:

$$S(\underline{Q},\omega) = \sum_{ij} b_i b_j \iint \langle \rho_i(o,o) \rho_j(\underline{r}_{ij},t) \rangle \times \exp(i[\underline{Q}.\underline{r}_{ij} - \omega t]) dt d\underline{r}_{ij} \qquad [2.4.32]$$

where $\rho_i(\underline{r}, t)$ is the instantaneous density of the i-th atom type, b_i is the corresponding scattering length, ω is the change in frequency of the scattered neutrons and $\underline{Q}$, the scattering vector, is the change in wave-vector of the neutron beam. It is more common to write the equation for $S(\underline{Q}, \omega)$ as the time Fourier transform of the intermediate scattering function $F(\underline{Q},t)$:

$$S(\underline{Q},\omega) = \int F(\underline{Q},t) e^{-i\omega t} dt \qquad [2.4.33]$$

where:

$$F(\underline{Q},t) = \langle \tilde{\rho}(\underline{Q},o) \tilde{\rho}(-\underline{Q},t) \rangle \qquad [2.4.34]$$

and the density operator $\tilde{\rho}(\underline{Q},t)$ is given as:

$$\tilde{\rho}(\underline{Q},t) = \sum_i b_i \exp(i\underline{Q}.\underline{r}_i(t)) \qquad [2.4.35]$$

It is clear that equation 2.4.35 gives the spatial Fourier transform of the instantaneous density, with each atom having the weighting of its neutron coherent scattering length b:

$$\tilde{\rho}(\underline{Q},t) = \sum_i b_i \int \rho_i(\underline{r},t)\exp(i\underline{Q}.\underline{r})d\underline{r}$$

$$= \sum_i b \int \delta(\underline{r}-\underline{r}_i)\exp(i\underline{Q}.\underline{r})d\underline{r} \qquad [2.4.36]$$

As well as performing experiments that give the intensity at different frequencies and wave-vectors, it is possible to perform an experiment at a fixed value of $\underline{Q}$ in which the measurement is of the intensity integrated over all frequencies. This defines the total scattering function S($\underline{Q}$), as would be observed in an x-ray scattering experiment:

$$S(\underline{Q}) = \int S(\underline{Q},\omega)d\omega = \sum_{ij} b_i b_j \langle \rho_i(0)\rho_j(\underline{r}_{ij})\rangle \, d\underline{r}_{ij}$$

$$= \sum_{ij} b_i b_j \langle \exp(i\underline{Q}.\underline{r}_{ij})\rangle \qquad [2.4.37]$$

The connection with the pair distribution function (equation 2.4.15) should be noted.

The correlation functions in the equations for the scattering function are all relatively easy to calculate in a MDS analysis, and the calculated scattering function can then be compared with experimental measurements. A direct comparison will necessitate the convolution of the calculated line profile with the experimental resolution function. In addition, it is also known that there is a difference between the classical scattering function and the quantum mechanical scattering function, which is related to the requirements of fulfilling detailed balance. The classical scattering function is an even function of frequency, whereas the quantum mechanical function has the property:

$$S(\underline{Q},\omega) = S(\underline{Q},-\omega)\exp(\hbar\omega/k_BT) \qquad [2.4.38]$$

Schofield (1960) has suggested that the classical scattering function should be multiplied by $\exp(h\omega/2k_BT)$ in order to reproduce this property, and it is common practice to follow this pragmatic advice.

There are a number of different ways that the scattering equations can be handled. If one is interested only in the motions of a subset of atoms, the scattering lengths of all other atoms can be chosen to be zero, so that the calculation probes only the motions that one is interested in. This is particularly useful when trying to explain experimental data, as the choice of scattering lengths may help to distinguish particular contributions to the full scattering profile. Alternatively, it is possible to perform the familiar n-phonon scattering expansion of the density-density correlation function 2.4.32:

$$\exp(i\underline{Q}.[\underline{u}(o)-\underline{u}(\underline{R},t)]) = \\ = \exp(-\langle\underline{Q}.\underline{u}]^2\rangle)\sum_m \frac{1}{m!}\left(\langle\underline{Q}.\underline{u}(o)][\underline{Q}.\underline{u}(\underline{R},t)]\right)^m \qquad [2.4.39]$$

where $u_i(\underline{k},t)$ is the collective displacement variable. The n=0 term is simply the Bragg scattering (elastic) term. The n=1 term is for scattering from one phonon (normal mode), and is formally equivalent to the correlation functions for the normal mode coordinates (section 2.4.10). Since most of the information in a scattering experiment relates to one-phonon scattering (multi-phonon scattering processes give almost featureless scattering), when the purpose of the simulation is to explain experimental data it is often better to work with these correlation functions than to calculate the full scattering function, particularly as any interesting feature may be related to only one correlation function. The full one-phonon scattering function can then be constructed from these correlation functions if required. Another advantage of using the normal mode coordinate correlation functions is that the set of correlation functions need only be calculated for one set of wave-vectors. In contrast, the scattering function contains peaks for each normal mode, and the height of any peak is proportional to $(\underline{Q}\cdot\underline{e})^2$, where $\underline{e}$ is the corresponding mode eigenvector as given in equation 2.4.29. In order to obtain power spectra that contain only a few peaks of characterized eigenvectors, the calculations need to be performed for a wide range of carefully chosen scattering vectors, just as in the corresponding experiments.

It is also possible to separate the total scattering function $S(\underline{Q})$ into the components due to single-particle disorder and those due to correlated motions. For atomic systems (i.e. no molecules) the scattering can be resolved into three components, namely the Bragg scattering $S_{Bragg}(\underline{Q})$, the single particle term $S_{self}(\underline{Q})$ and the correlated scattering term $S_c(\underline{Q})$:

$$S_{Bragg}(\underline{Q}) = \left|\sum_i b_i\langle\exp(i\underline{Q}.\underline{r}_i)\rangle\right|^2 \qquad [2.4.40]$$

$$S_{self}(\underline{Q}) = \sum_i b_i^2(1-|\langle\exp(i\underline{Q}.\underline{r}_i)\rangle|^2) \qquad [2.4.41]$$

$$S_c(\underline{Q}) = \sum_{ij} b_i b_j\Big[\langle\exp(i\underline{Q}.\underline{r}_i)\exp(-i\underline{Q}.\underline{r}_j)\rangle - \\ - \langle\exp(i\underline{Q}.\underline{r}_i)\rangle\langle\exp(-i\underline{Q}.\underline{r}_j)\rangle\Big] \qquad [2.4.42]$$

Additional terms will occur if we have molecules as well, for then there is scattering from single particle rotational motions, correlated rotational motions, and coupled translation-rotation scattering. These addition terms have been given by Dove and Pawley (1984). The great advantage of this approach is that the simulations can be used to give information concerning the origin of diffuse scattering in an experiment.

2.5.COMPUTERS.

To close this chapter on technical details, we should discuss briefly the hardware issue, since in many cases the one factor that inhibits progress is the availability of computing time. The sizes of the samples that can be used, the length of the simulations, the number of different sample characteristics (e.g. temperatures) that can be examined, and the complexity of the models are all determined by the problem of the hardware capabilities. The standard university mainframe computer may well be adequate for studies of simple model systems with small sample sizes (ofthe order of 250 ions). Unfortunately, minerals generally have fairly complicated unit cells, and many problems of interest will require the use of large samples (of the order of 1000 unit cells) in order to give an adequate wave-vector resolution, and for these problems the use of a general purpose shared mainframe will not be practical. There are three main solutions to the computational problem. The first is the use of a special dedicated processor that will have been custom built for the purpose of running MDS calculations. This route has been explored successfully by a few groups, but clearly it requires considerably more effort and technical expertise than can be given by very many of us. However, with the development of the transputer (a revolutionary specially designed small processor that can easily be connected to other transputers to form an array), it will become easier to design and build special purpose processors from off-the-shelf components.

The second solution is to use supercomputers, such as the CRAY series. These machines are continuing to become more powerful, as well as more available within the scientific community. However, even these machines will prove to have limitations, as increases in power do not come in order-of-magnitude steps, so that eventually the increase in demand will overtake the increase in power. An attractive advantage of this solution is that it is not a difficult matter to make standard MDS programs (which are usually written in FORTRAN) run on a supercomputer. Further increases in speed can be obtained by optimising the program to take account of the hardware features of the supercomputer, but this will not be too difficult a task and will certainly not require a major revision of the program's design.

The third solution is to use parallel array processors (Hockney and

Jesshope, 1981). Current machines have computational power that is equivalent to that of supercomputers, but have the potential for order-of-magnitude improvements. The philosophy behind the design of such systems is that considerable gain in power can be obtained when the computer does many tasks at once, rather than just one thing at a time as in a computer with a more common architecture. An example is the ICL Distributed Array Processor (DAP). The architecture of the DAP has 4096 separate processors connected on a square 64x64 grid, with hardware connections between nearest neighbour processors, so that each processor can transfer information to its neighbours. The connections between neighbours also exist across the boundaries of the grid, so that the DAP has built-in periodic boundary conditions. Each processor also has its own associated memory. The operation of the individual processors is controlled by a master controller, and a single instruction will cause each processor to perform an identical operation on data stored in its memory. For example, each processor might add together two numbers, or it might transfer a number to its neighbour on the left and receive another number from its neighbour on the right. Thus the DAP performs 4096 separate but identical operations simultaneously, and so it can easily be seen how the DAP achieves its computing speed. This kind of architecture is known as a Single Instruction Multiple Data (SIMD) machine. Although the SIMD architecture might sound somewhat restrictive, in practice it is ideal for many lattice calculations such as MDS calculations for solids. Each processor with its associated memory can correspond either to one unit cell or to one molecule, and it is a relatively easy matter to map a 3-d lattice onto the 2-d grid of the DAP whilst still making full use of the periodic boundary connections. The simulations that are discussed in the next chapter were all performed using the DAP with 4096 molecules. It should be noted that very few other workers are able to use such large samples at the present time.

The DAP is a relatively old machine, and by current standards the processors are fairly slow. For example, they operate as single bit serial processors (rather than as 32-bit processors), although this feature is of considerable advantage for MC simulations of Ising models. Therefore it should be noted that the power of the DAP could be considerably increased simply by using present day electronics.

The programming language for the DAP is an adapted version of FORTRAN, so that writing programs is not a difficult skill to learn. What is more difficult, though, is the fact that any program needs to be written taking account of the special parallel architecture of the DAP. Therefore it is not possible to simply alter an existing FORTRAN program, but a radical rethink of the whole approach is demanded. This drawback may well deter many potential users, but if the major advances in the use of computer simulations (particularly MDS) in the mineral sciences are to occur, they will probably only be achieved with the use of parallel processors. The DAP is not the only parallel array processor available on the market. For example, several companies are marketing parallel array processors that are made from the transputer. The specific array designs can be varied to suit the applications, and go beyond the SIMD structure to an architecture whereby the computer can perform different operations on different data simultaneously; such an architecture is known as Multiple Instructions Multiple Data (MIMD). The transputers are programmed in their own language, known as OCCAM; it is clear that any revolutionary new computer architecture will necessitate the use of new languages and new algorithms

Finally, we should also remark on the fact that the very large simulations that will be feasible in the future will generate enormous quantities of data. In the past it has been common practice to store the sample configurations every few time steps, and to perform the analysis after the simulations have been completed. However, this procedure is impractical with large simulations; as well as the vast volume of storage space required, it is in fact quicker to regenerate the data and perform the analysis during the simulation than it is to manipulate and move around the stored configurations.

3. ILLUSTRATIVE EXAMPLES.

3.1. INTRODUCTION.

In this section I will give two examples of the use of the MDS techniques to tackle problems in solid state physics in which the meeting point between theory and experiment has not yet been established. The first example is of systems that have orientational disorder, and the second is the problem of the incommensurate and lock-in phase transitions. Although the systems discussed are not minerals, the scientific principles have obvious relevance to minerals physics, since similar phenomena are observed in naturally occurring minerals.

However, it is also important to note that although the physical mechanisms discussed in this section may not be directly relevant to many minerals of interest, the types of interactions that are discussed in the section will almost certainly have analogues in the interactions that give rise to phase transitions in minerals. In these cases the analysis methods discussed in this section may well have direct relevance to the study of minerals.

The simulations that are discussed in this section have all been performed on the ICL Distributed Array Processor (section 2.5). This has allowed the use of 4096 molecules in each case, giving samples that were large enough to be able to study collective properties as well as real space properties.

3.2. ORIENTATIONALLY DISORDERED CRYSTALS.

3.2.1. Background. One field of study in which computer simulations have made a major contribution is the study of orientationally disordered (OD) crystals. These are systems containing molecules or molecular ions in which the molecules have their positions ordered on a lattice but have their orientations disordered. The disorder can be either a site occupancy type disorder, with the molecular orientations pointing in any of two or more directions (like an Ising spin) and occasionally jumping between the different possible orientations, or a free rotation, or else the disorder can be (and usually is) described as intermediate between these extremes. It is reasonable to think of OD systems as being intermediate stages between true liquids and true crystals, with the positions being crystal-like and the orientations being liquid-like, with the rotational melting temperature being lower than the translational melting temperature. In general, the molecules are highly symmetric with only weak rotational interactions. Examples are more common than one might expect, and include N_2, O_2, H_2, H_2O, CH_4, CBr_4, NH_3, HCl, NaOH,NH_4Cl, $NaNO_2$, and KCN. Sherwood (1979) gives more detail about these and other systems. Ice is of course a very important example for Earth Sciences in general, and some of the other examples are important in the study of the geology of other light planets. It is also believed that the important mineral calcite has an OD phase at high temperatures, with disorder of the carbonate molecular ions, and that several important mineralogical properties (e.g. anomalous thermal expansion) are directly related to the existence of the disorder and the ordering phase transition. It is not unlikely that other minerals will be found to be affected by OD phases at high temperatures, and because of this a good understanding of the general properties of OD phases is highly desirable. Moreover, these systems provide excellent model systems for general theories of phase transitions in which some kind of disorder that cannot be described as a normal mode displacement will be strongly coupled with the crystal phonon bath. For example, it has recently been recognised that some OD systems (particularly when diluted with ordered crystals, e.g. $(KCN)_x(KBr)_{(1-x)}$ (Michel, 1986a) display orientational glass phases that are similar to the magnetic spin glasses but which also show the transport properties of true isotropic glasses. An understanding of these orientational glasses, in which the existence of an underlying lattice considerably aids the theoretical interpretation of such states, may well lead to an understanding of the properties of the more common isotropic glasses. For all of these reasons a good understanding of the general features of OD systems is highly desirable.

There are a number of questions that any study of OD systems is faced with. How are the molecules oriented on average? How do the individual molecules move? What effect does the disorder have on the collective dynamics, especially the acoustic phonons? How does the onset of order couple to a lattice distortion? Indeed, why are the molecules disordered at all? Although some features of the answers to these questions can be obtained from suitable experiments (e.g. correlation times from relaxation experiments, phonon line-widths), the kinds of results that experiments give (e.g. power spectra) are usually featureless and hence can be described by any number of models. Therefore, although experiments can give a description of the properties of OD phases, they are less successful in aiding an explanation of these same properties and intesting theoretical predictions. For this reason MDS calculations performed on a number of systems, e.g. NaCN (Lynden-Bell et al, 1983), NH_4Cl (Klein et al, 1983), CH_4 (Bounds et al, 1980), CBr_4 (present paper), SF_6 (present paper), and adamantane (Meyer et al, 1986), have enabled progress to be made. In passing, it is worth reminding ourselves that these problems in the use of experimental data to provide explanations and tests for theories exist for most disordered systems, and similarly MDS calculations have been exploited in the study of the disorder in liquids and fast-ion conductors, to name two areas.

The work that I wish to describe here as illustrative of the power ofthe MDS technique concerns three OD systems: SF_6, CBr_4 and C_2F_6. A number of properties of these systems are given in table 3.2.1. The reason for studying these three systems is that they are representative of specific symmetric shapes of molecules: CBr_4 is tetrahedral (like

TABLE 3.2.1

Summary of the properties of the systems

molecule:	SF_6	CBr_4	C_2F_6
range of OD phase:	96-223K	320-365K	~50-173K
crystal lattice:	bcc	fcc	bcc
site symmetry:	O_h	O_h	O_h
molecular symmetry:	O_h	T_d	D_{3d}, O_h*
lowest ℓ value:	4	3	2,4*

*first value is for the total molecule, second is for the approximate F_6 octahedron

many molecular ions found in minerals), SF_6 is octahedral, and C_2F_6 is similar in shape to SF_6 except that the octahedron is slightly elongated along one of the 3-fold axes. Other shapes are dumbells (e.g. N_2 and O_2, and the CN ion is effectively a dumbell shape), asymmetric tops (e.g. NH_3, $C(CH_3)_3CN$), triangles (e.g. CO_3, NO_3) and arrow heads (e.g. NO_2), and several of these have been studied by other workers. The point of this emphasis on the molecular shape is that the properties of OD systems (including the nature of the ordering transition) will often be determined by the general shape (i.e. symmetry) of the molecules.

Returning to table 3.2.1, it can be seen that these materials have cubic structures with one molecule in the primitive cell. The temperature regimes for the existence of the OD phases are reasonably large, particularly for SF_6. We will comment on other information given in this table below. The discussion of these examples will be restricted to the aspects that demonstrate the range of properties that can be studied by MDS calculations; for more details the reader is referred to the papers by Dove and Pawley (1983, 1984), Dove et al (1986a,b), Dove and Lynden-Bell (1986a, 1987), and Dove (1986).

3.2.2. Dynamic variables. Before proceeding with a detailed description of the simulations and the results they have given, we should for a moment consider the problem of the formal description of the dynamic variables. As in an ordered crystal, we can define the translational variables as small displacements of the position of the molecular centre of mass from the equilibrium position; thus the instantaneous position of the centre of mass of a molecule whose average position is at $\underline{R}$ is:

$$\underline{r}(\underline{R},t) = \underline{R}+\underline{u}(R,t) \qquad [3.2.1]$$

where $\underline{u}(\underline{R},t)$ is the instantaneous displacement. On the other hand, we are not able to think of the molecular orientation as corresponding to a small rotation from an equilibrium orientation (as one might do for an ordered molecule). Instead we need to work with a set of variables that define the absolute orientation. The set of orientational variables that are generally used are symmetry adapted linear combinations of spherical harmonics obtained from the unit vector $\underline{x}_i$ pointing towards the instantaneous position of the i-th atom from the position of the molecular centre of mass. Thus for each value of ℓ we obtain $(2\ell+1)$ variables $B_{lm'}$:

$$B_{\ell m'} = \sum_i \sum_m C_{\ell mm'} Y_{\ell m}(\underline{x}_i) \qquad [3.2.2]$$

The set of coefficients $C_{lmm'}$ are chosen to give an appropriate set of variables such that they have symmetries consistent both with the molecular and site symmetries. The coefficients are also scaled to give the normalisation condition:

$$\langle B_{\ell m'}^2 \rangle = \frac{1}{2\ell+1} \text{ or } \frac{1}{4\pi(2\ell+1)} \text{ for all } \ell m' \qquad [3.2.3]$$

where the angle brackets denote an average over all orientations. In addition, the orientational variables have a zero mean, i.e.

$$\langle B_{\ell m'} \rangle = 0 \qquad [3.2.4]$$

The symmetry of the situation determines the orientational variables that are of interest. Assuming here the common case of cubic lattices, so that the molecules are on sites of O_h symmetry, the lowest values of l for which non-zero orientational variables can be defined are l=2 for symmetric dumbells, l=3 for tetrahedons, and l=4 for octahedrons. It is not usually necessary to work with higher order variables than the lowest allowed ones. The sets of variables for O_h site symmetries and l=2,3,4 are given in Table 3.2.2. For some of these variables we can associate physical meaning. For dumbells, the E_g variables correspond to the molecules lying along the <100> axes and the T_{2g} variables correspond to orientations along the <110> and <111> directions. For tetrahedrons the A_{2u} variable corresponds to the standard orientation with the bond vectors lying along <111> directions (the two possible settings of this orientation giving the sign of the variable), and the T_{2u} variables correspond to the six orientations obtained from the A_{2u} orientations by rotation of the molecules by $\pi/4$ about any of the

Table 3.2.2. Orientational variables for ℓ=2,3,4, constructed from spherical harmonics $Y_{\ell m}$. For clarity we do not give the relative normalization factors. We define $Y_{\ell mc}=(Y_{\ell,m}+Y_{\ell,-m})$ and $Y_{\ell ms}= -i(Y_{\ell,m} - Y_{\ell,-m})$. For additional details the reader is referred to Altman and Cracknell (1965).

	ℓ = 2	ℓ = 3	ℓ = 4
A_{1g}			$B_1=\sqrt{14}\ Y_{40}+\sqrt{5}Y_{44c}$
A_{2u}		$B_1=-Y_{32s}$	
Eg	$B_1=Y_{20}$		$B_2=\sqrt{10}\ Y_{40}-\sqrt{7}Y_{44c}$
	$B_2=Y_{22c}$		$B_3=\ \ -Y_{22c}$
T_{1g}			$B_7=\sqrt{7}Y_{11s}-Y_{33s}$
			$B_8=\sqrt{7}Y_{11c}-Y_{33c}$
			$B_9=\ \ Y_{44s}$
T_{2g}	$B_3=Y_{22s}$		$B_4=Y_{11s}-\sqrt{7}Y_{33s}$
	$B_4=Y_{21s}$		$B_5=Y_{11c}+\sqrt{7}Y_{33c}$
	$B_5=Y_{21c}$		$B_6= -Y_{22s}$
T_{1u}		$B_2= Y_{30}$	
		$B_3=\sqrt{5}iY_{33s}-\sqrt{3}iY_{31s}$	
		$B_4=-\sqrt{5}iY_{33c}-\sqrt{3}iY_{31c}$	
T_{2u}		$B_5= Y_{32c}$	
		$B_6=-\sqrt{3}iY_{33s}-\sqrt{5}iY_{31s}$	
		$B_7=-\sqrt{}Y_{33c}+\sqrt{5}iY_{31c}$	

<001> axes. In fact, in CBr_4 the orientations of the molecules in the low temperature ordered phase (monoclinic C2/c with 32 molecules in the non-primitive cell) closely correspond to the set of T_{2u} orientations. For octahedrons the A_{1g} variable corresponds to the molecule having all bonds along the <100> axes. For SF_6 the important variables are those for l=4, for CBr_4 they are for l=3, and for C_2F_6 we describe the C-C bond by the l=2 variables and the F_6 cage by the l=4 variables.

In fact, the A_{1g} variables (which are unchanged under all symmetry operations of the site symmetry group) form an expansion in l that can describe the average orientation of a molecule and from which the bond orientation distribution function (odf; see section 2.4.7) can be constructed. For O_h site symmetry these are the well-known Kubic harmonics K_l. The bond odf is given as:

$$g(\Omega) = \sum_{\ell} c_{\ell} K_{\ell}(\Omega) \qquad [3.2.5]$$

where the coefficients $c_l = \langle K_l \rangle$, and where now the angle brackets denote the ensemble and time average rather than the rotational average. Of course, we also need the l=0 term in the expansion. This expansion is expected to convergence quickly so that only the lowest order terms need to be included (l=0,4,6,8 for cubic lattices); this condition is appropriate when the molecules have considerable rotational freedom. The coefficients in the expansion can be determined without too much difficulty from neutron diffraction data (Press and Huller, 1973).

3.2.3. Intermolecular potentials. For each of the three systems we have used Lennard-Jones potentials (equation 2.3.2) for the atom-atom interactions involving the halogen atoms, and have neglected the interactions involving the central atoms. The parameters for SF_6 were obtained in an approximate way by obtaining the value of σ from the known unit cell length in a static lattice energy calculation, and a corresponding value of ε was estimated from typical values of lattice energies (Pawley, 1981). Although this seems rather crude, at the time the structure of the low temperature phase had not been determined, so that it was not possible to use that information, and the origin of the disorder (which would perhaps discourage this approach) had not been identified (see below). Nevertheless, the model used has been spectacularly successful in reproducing and explaining many of the properties of SF_6, in particular predicting correctly the structure of the ordered phase before it could be extracted from the experimental data (Pawley and Thomas, 1982; Powell et al, 1987). However, it was recognised that the model potential was slightly softer than the true potential (as was expected in view of the way it was obtained), and our recent determination of the structure of the low temperature phase has enabled us to suggest an improved value of σ (Powell et al, 1987). In addition, we were able to justify the neglect of the interactions involving the sulphur atom, as inclusion of these interactions in a static energy model for the ordered phase failed to yield a realistic improvement. We have used both values of σ in our calculations for C_2F_6.

The case of CBr_4 was slightly more problematic. We used the simulations to obtain values for the model potential parameters, by scanning the range of values in order to find the values which gave the best agreement for $\langle u^2 \rangle$ and the cell length at zero pressure and 345K. It turned out that no combination of parameter values were able to give reasonable agreement, which we interpreted as evidence for anisotropy in the real potentials. Our pragmatic solution was to move the interaction centres in along the Br-C bond, which is quite reasonable to do if, as is often the case, the real interactions are flattened normal to the bonds (section 2.3.1). We found that by using this procedure our models could then reproduce the experimental values of $\langle u^2 \rangle$ and the cell length whilst maintaining the overall molecular size. However, for other quantities the agreement between the MDS results and experiments was not as good as for SF_6, but the overall behaviour observed in the simulations (particularly the ordering instabilities observed) were in good agreement with reality. Therefore we can have confidence in the science that the simulations have produced.

3.2.4. Statistical disorder. In the disordered phase of CBr_4 the site symmetry (O_h) is higher than the molecular symmetry (T_d), so that the orientational disorder must arise as a statistical feature with the odf having the full site symmetry. It has been suggested (More et al, 1980; Coulon and Descamps, 1980) that the disordered phase has each molecule statistically distributed in any of the six T_{2u} orientations, occasionally jumping between different orientations. As expected, the simulations have shown that this is an oversimplified picture, but nevertheless there is a preference for the molecules to have T_{2u} orientations and to stay for awhile in these orientations, but the jumps between orientations are slow so that there will always be a large number of molecules with different orientations.

In addition, steric hindrance tends to preclude certain relative orientations (so that in fact the perfect picture corresponds only to the ordered structure), and so some of the molecules will be forced into different orientations. In the disordered phase of C_2F_6 there will also be similar statistical disorder of the C-C bonds. Neutron diffraction studies of C_2F_6 (Powell et al, 1984) and MDS calculations have shown that the C-C bonds lie along any of the <111> directions, with the F atoms lying on average close to the cell axes. It was generally thought that this statistical disorder accounted for most of the properties of the disordered phase, but the MDS results have shown that in fact this statistical disorder is only a small factor; most properties are most affected by the disordering mechanism found in SF_6 (next section).

3.2.5. Orientational Frustration. The interesting feature of SF_6 is that the molecular symmetry is identical to the site symmetry, so that there will not be any statistical orientational disorder. Neutron diffraction measurements have given the S-F bond odf (by refining values of the Kubic harmonic coefficients; Dolling et al, 1979), which shows that the average molecular orientation has the S-F bonds lying along the <001> directions, with a rms angle between the bond and the axis direction of about 20°. At face value therefore it appears that we simply have an ordered crystal with a large librational amplitude. However, other results have been obtained (from experiments as well as from simulations) that are very characteristic of disordered systems, and so the questions arise as to the nature and origin of the orientational disorder. Figure 3.2.1 shows the mean-square of the angle between a S-F bond and its closest unit cell axis and the mean-square displacement $\langle u^2 \rangle$ as functions of temperature as calculated in the simulations. The displacement fluctuation can be seen to increase approximately linearly with temperature, as would be expected for thermal fluctuations. However, the rotation fluctuation only slightly decreases on cooling, so that it appears that it will never decrease to zero, even at 0K. Thus the rotational amplitude is not a thermal fluctuation effect, which essentially means that the orientational disorder is not simply a kinetic effect.

The advantage of MDS studies is that we have the interaction potentials that result in the properties of interest available to us from the start; the experimentalist would not have this information. Calculations of the interactions between molecules of different relative

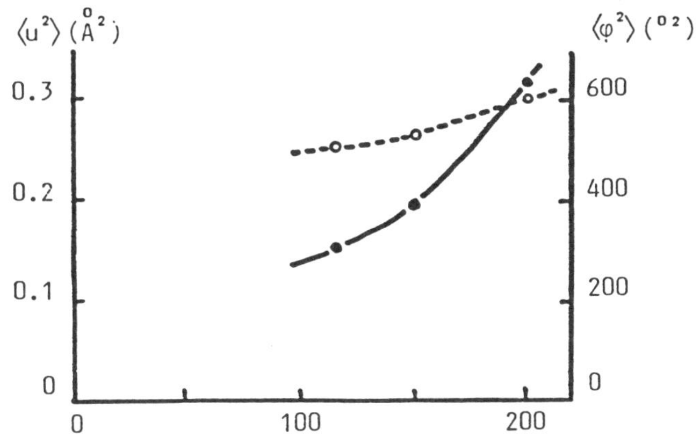

Figure 3.2.1. Mean square displacement $\langle u^2 \rangle$ (full circles and curve) and mean square angle of S-F bond (open circles and broken curves) for SF_6 as functions of temperature.

orientations showed that the interaction between nearest neighbour molecules with their ideal orientations (recall that the crystal structure is bcc, so that the nearest neighbours are along <111>) is attractive and strong, whereas the interaction between next-nearest neighbours (along the cell edge) is repulsive, because of the close overlap of F atoms. Thus there is a competition of interactions, with the nearest neighbour interactions favouring orientational order and the next-nearest neighbour interactions opposing the order. This competition, which we have called "orientational frustration", is relieved dynamically at high temperatures, but when the kinetic energy is no longer enough to balance this frustration the crystal structure transforms to give a low-symmetry phase. The essential features of this model have been confirmed by calculations of the pair distribution function for the closest F...F contact distances, which show that the F atoms tend to be further apart than would be expected if their positions were uncorrelated, with a mean separation distance that is close to the equilibrium distance given by the model potential. Additional calculations of the distribution function for relative orientations of the closest S-F bonds of two next-nearest neighbour molecules along the cube axis show that the bonds are almost never both simultaneously aligned along the axis, and instead their orientations are usually rotated by about 20° with respect to each other. In fact, the full distribution function was consistent with a situation in which one bond might lie along the cell axis with the other precessing about it. Comparison with the results expected for uncorrelated motions confirm the features of the orientational frustration model. This result has been confirmed from movie films produced from the simulations (Dove et al 1986a).

The orientational frustration model is able to explain why many properties of the disordered phase of SF_6 have only a weak temperature dependence. For example, the mean square rotation angle that is shown in figure 3.2.1 will be determined mainly by the orientational frustration, and any increase at high temperatures due to thermal fluctuations might be expected to be countered by the reduction in the angle due to the thermal expansion of the unit cell length.

A number of tests in the MDS calculations for C_2F_6 have shown that there is a strong orientational frustration component of the orientational disorder in that system, due to the fact that the fluorine cage is only a slightly distorted octahedron and is thus similar to the fluorine cage in SF_6. Thus it is highly likely that many of the properties of C_2F_6 will be determined not by the statistical disorder of the C-C bond (which will really only affect the dynamics via the moment-of-inertia disorder) but by the orientational frustration. We will see later on that the correlation times associated with the different disordering mechanisms differ by a factor of 5. This simple argument has not been noted elsewhere, and indeed would not have been obtained other than by the MDS technique.

3.2.6. <u>Structural properties</u>. MDS calculations are able to give very good structural information, and our results for SF_6 are able to illustrate this. We used a constant pressure algorithm (although not one as sophisticated as the method described in section 2.2.2) to observe the structure as a function of temperature. The OD phase was stable over the range between about 80-220K. Figure 3.2.2 shows the variation of the unit cell length with temperature, and figure 3.2.3

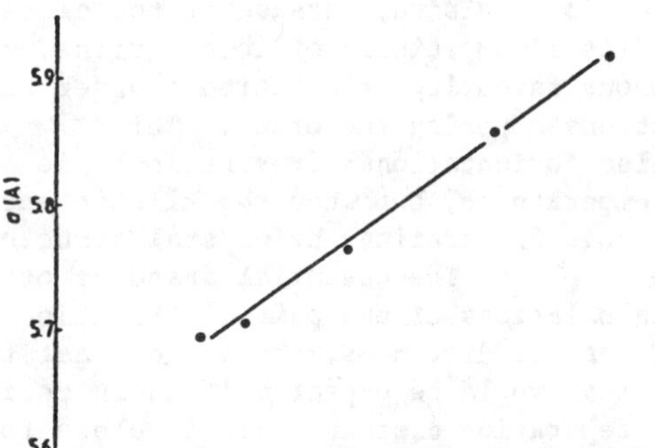

Figure 3.2.2. Temperature dependence of the unit cell length a as determined in the MDS study of SF_6.

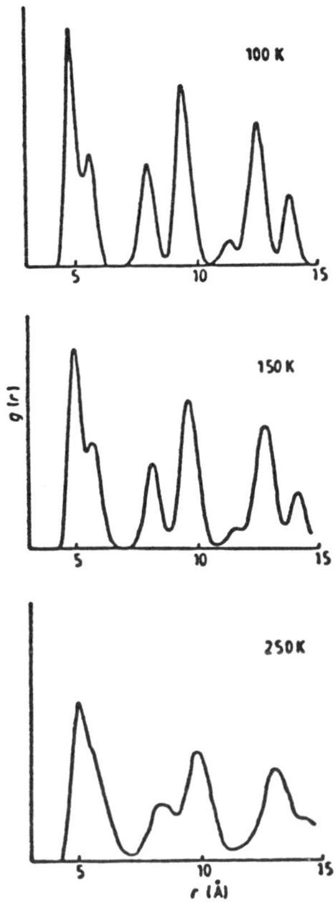

Figure 3.2.3. The radial distribution functions g(r) for S-S distances r for several temperatures of the MDS sample, of SF_6.

shows the variation of the S...S pair correlation function with distance. The obvious effect of temperature on the pair distribution function is that the peaks are broadened. Of greater interest is the fact that when the simulations were run at lower temperatures (80K and 25K) ordered structures appeared as small crystallites in the sample; these are shown in figure 3.2.4. At 80K the structure appears to be hexagonal, with 1/3 of the molecules remaining disordered (Pawley and Dove, 1983), and at 25K the structure is triclinic and fully ordered (Pawley and Thomas, 1982). At the time the simulationswere performed there were no structural data available concerning the low temperature structures of SF_6; our

recent neutron diffraction results have essentially confirmed the prediction of the triclinic structure, and have shown that the partially-ordered hexagonal phase is probably only metastable but that it nevertheless is the intermediate step in the cubic-triclinic transition pathway (Powell et al, 1987). the metastable existence of this hexagonal phase has been observed by electron diffraction (Raynerd et al, 1980).

a)

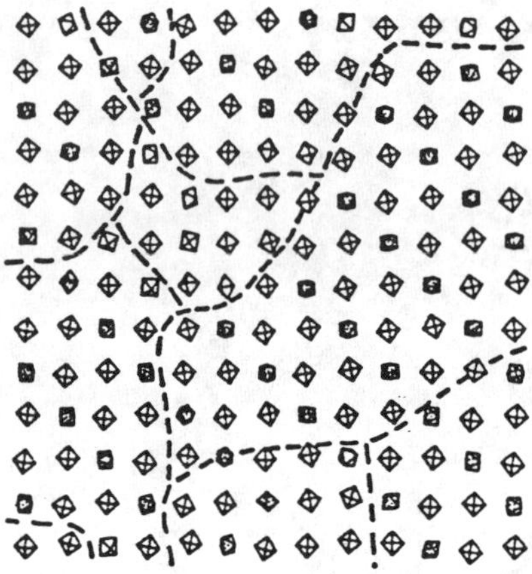

b)

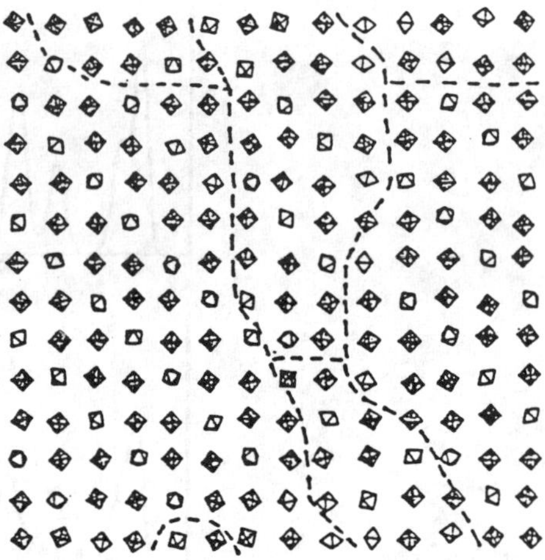

Fig.3.2.4 - Sections through the SF_6 MD cell for (a) the 25K triclinic phase and (b) the 80K trigonal partially plastic phase. In both cases the crystallites are outlined by dashed lines. In (a) the molecules which are stippled are those positioned in an orientation quite different from the bcc orientation, and these become disordered in the intermediate phase (b) where they are represented unstippled.

3.2.7. Single particle dynamics. The correlation functions of the form 2.4.16 have been calculated for all the orientational variables defined in table 3.2.2 for the three systems of interest. The results for CBr_4 are shown in figure 3.2.5; the results for the other two systems are similar. The interesting point concerning the results for CBr_4 is that the correlation function with the slowest decay is for the T_{2u} variables, which is consistent with the ordering of these variables in the low temperature structure. From the correlation functions it was

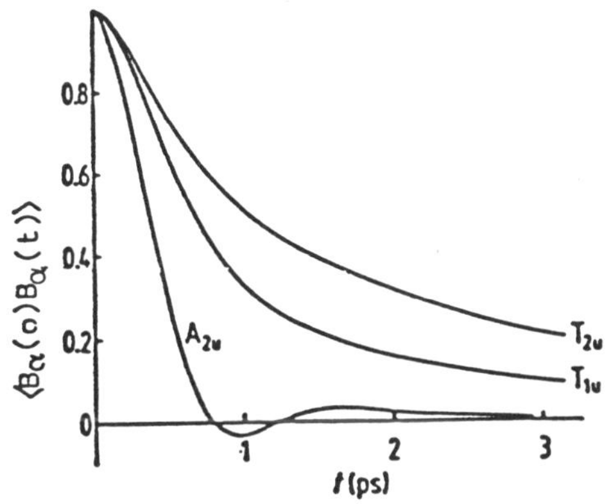

Fig.3.2.5. Auto-correlation functions for the orientational variables in CBr_4. The three curves are for the $A_{2u}(\alpha=1)$, $T_{1u}(\alpha=2,3,4)$ and $T2u(\alpha=5,6,7)$ variables.

possible to obtain relaxation times (see the discussion following equation 2.4.16), and these are given in table 3.2.3. As expected, the relaxation times for the SF_6 variables are shorter at higher

TABLE 3.2.3.
Approximate relaxation times (in ps) for the orientational correlation functions $\langle B(0)B(t)\rangle$

	SF_6 (115K)	SF_6 (150K)	SF_6 (200K)	CBr_4 (345K)	$C_2F_6(F_6)$ (115K)	$C_2F_6(C_2)$ (115K)
A_{1g}	0.9	0.6	0.5		1.3	
A_{2u}				0.5		
E_g	0.8	0.5	0.4		1.0	1.4
T_{1g}	0.4	0.3	0.2		0.9	
T_{2g}	0.4	0.3	0.2		0.7	5.4
T_{1u}				1.9		
T_{2u}				2.5		

temperatures. The relaxation times for SF_6 are generally fairly short, particularly when compared with the time for complete reorientation (about 5.5ps). The relaxation times for C_2F_6 are consistent with the suggestion that there are the two disordering processes involved in the dynamic properties (statistical disorder and orientational frustration). The 1-4 correlation functions decay on the same time scale as the corresponding relaxation times for SF_6, and the relaxation times for the C-C bond dynamics (the 1-2 variables) are considerably longer than the relaxation times for the F_6 cage. The general features of these systems are that the long relaxation times are indicative of tendency towards particular orientations. However, the lack of librational character in the other correlation functions suggests that the molecules do not librate in the potential wells before reorienting, but that instead the motions are considerably more complex.

The velocity and angular velocity correlation functions are similar in form for each of these systems, and so we focus our discussion on the correlation functions for SF_6 and their power spectra that are shown in figure 3.2.6. The velocity correlation function falls below zero in a

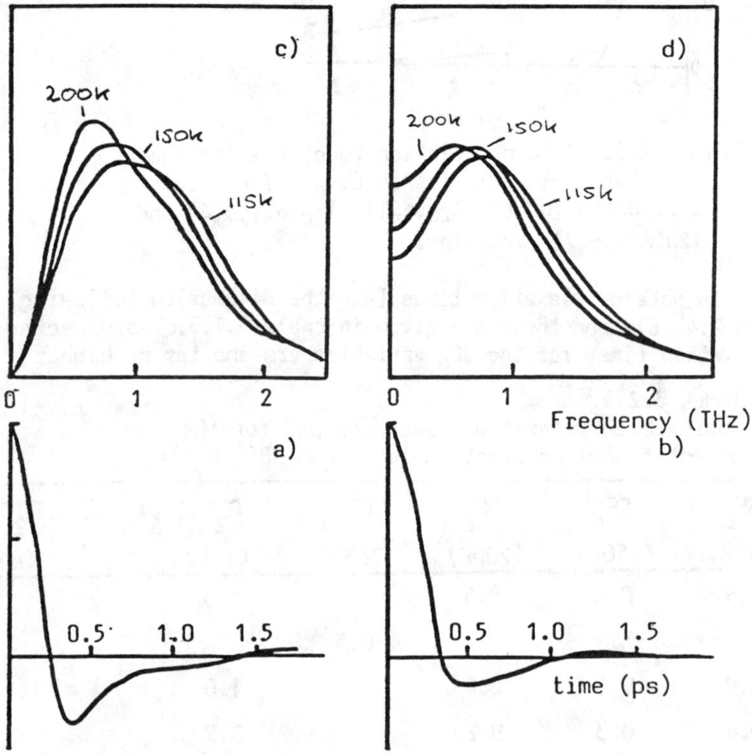

Figure 3.2.6. a) linear velocity correlation of SF_6 at 150K. b) rotational velocity correlation of SF_6 at 150K. c) translational power spectra of SF_6 at different temperatures. d) rotational power spectra of SF_6 at different temperatures.

manner that is consistent with the molecules bouncing back and forth in the cage of the local environment. The same effect is also seen in the angular velocity correlation function. This means that the rotational motions of the molecules are constantly changing direction; if the molecule was rotating with only a light damping of the motion, the angular velocity correlation function would be relaxational and would not drop below zero. Thus the form of the angular velocity correlation function is able to provide useful qualitative information about the nature of the rotational dynamics. The observed power spectrum of the velocity correlation function is similar to the acoustic phonon density of states for an ordered crystal, as one might expect. The shift in the peak position with temperature shows the anharmonic shifts in vibrational frequencies with temperature. The fact that the power spectra fall to zero at zero frequency is consistent with the absence of translational diffusion. On the other hand, the power spectra for the rotational motions show large values at zero frequency, which suggest that the rotational motions can be said to be diffusion-like. As we have discussed in section 2.4.8, the rotational diffusion has an associated diffusion coefficient, D, that is related to the zero-frequency value of the power spectrum. In the case of SF_6 we have a variation of this diffusion constant with temperature. It is generally assumed that the diffusion coefficient follows an Arrhenius behaviour:

$$D(T) = D_0\exp(-E_a/k_BT) \qquad [3.2.6]$$

where E_a is an activation energy that will be related to the height of the energy barrier against molecular reorientation. Using the data presented in figure 3.2.6 a value for E_a for SF_6 of 140K is obtained. This energy corresponds roughly to the mid-temperature of the stability range of the OD phase, and suggests therefore that the molecules do not simply occasionally reorient by jumping over an energy barrier, but that instead the molecules rotate as a result of the collapse of any potential barriers to reorientation. This conclusion is consistent with visual observations (Dove et al, 1986a) and with the absence of a librational character of the orientational variable time correlation functions, and is anticipated inthe orientational frustration model. It should be noted that the rotational diffusion and associated activation energy can be measured in hydrogenous OD systems by incoherent neutron scattering, and it is often found that the activation energy is much higher than the maximum temperature of the OD phase. Hence we have been able to show that the case of SF_6 is different from many other OD systems, as one might expect from the absence of any statistical disorder.

3.2.8. <u>Diffuse scattering in SF_6.</u> Bragg scattering essentially only gives information about the average particle distribution functions (in this case, the odf), yet the orientational frustration model would suggest that there should be important short range correlations of the orientations of neighbouring molecules. Such correlations would usually be expected to be seen in a neutron diffuse scattering experiment, in experiments to measure $S(\underline{Q})$ (equation 2.4.37). In addition, as we will see below, most collective excitations have very short correlation times, so that phonon scattering would not be expected to be a prominent contribution to the total diffuse scattering. The components of the expansion of $S(\underline{Q})$, which in this case included the rotational single-particle term, the rotational cross-term and the rotation-translation cross-term in addition to the expressions given in section 2.3.11 (equations 2.4.41 and 2.4.42; the additional terms are given in Dove and Pawley, 1984) were calculated in the simulations performed at 150K. The results for the total scattering and the two single particle terms are presented in figure 3.2.7. The correlated contributions could not be calculated with the same accuracy, but their combined effects can be noted from this figure. The general observation, however, is that most

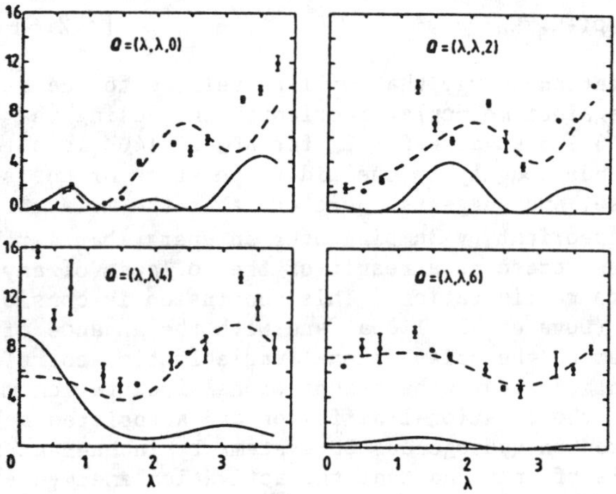

Figure 3.2.7. The single-molecule contributions to the total scattering $S(\underline{Q})$ in SF_6. The full and broken curves correspond to displacement and orientational terms respectively. The data points with error bars are the total scattering amplitudes and are given so that it is possible qualitatively to observe the effects of neighbour correlations. The errors on the single-particle distributions are considerably smaller than for the total scattering and are thus not shown.

of the scattering appears to be due to the the single-particle scattering. The general oscillatory nature of the scattering has been observed experimentally in a number of systems (Dolling et al, 1979; Powell et al, 1984), where it has been assigned to single-particle rotation scattering. However, it is clear that the single-particle translation scattering has a similar form and is just as important. The scattering intensity shown in figure 3.2.7 has also been observed in a measurement of S($\underline{Q}$) from a single crystal of SF_6 (Powell, unpublished data). The only large effect on the total S($\underline{Q}$) from correlated motions appears in the region from $\underline{Q}$ =(112) to $\underline{Q}$ =(222), where the effects of the correlated motions are both constructive and destructive. It is clear that the calculation of diffuse scattering by MDS calculations can give very useful information in the interpretation of experimental data, particularly with the help of appropriate an expansion of S($\underline{Q}$).

3.2.9. Collective fluctuations. Following the formal description given for collective properties in section 2.4.9, we define collective variables for the displacements and orientational variables:

$$u(\underline{k}) = N^{-1/2} \sum_i u(\underline{R}_i)\exp(i\underline{k}.\underline{R}_i) \qquad [3.2.7a]$$

$$B(\underline{k}) = N^{-1/2} \sum_i B(\underline{R}_i)\exp(i\underline{k}.\underline{R}_i) \qquad [3.2.7b]$$

Since the systems under discussion have cubic structures, the orthogonal linear combinations of these two types of variables are easily constructed for symmetry directions in reciprocal space (for example, along [110] the three displacement variables are u_z, u_x+u_y and u_x-u_y, neglecting normalisation prefactors). We do not yet consider the modes formed by mixing the two types of motion. It is readily seen from equation 2.4.31 that the acoustic mode frequency is given in the quasi-harmonic approximation by:

$$\omega_i^2 = \frac{k_B T}{m\langle u_i(\underline{k})u_i(-\underline{k})\rangle} \qquad [3.2.8]$$

Results for SF_6 and CBr_4 are given in figure 3.2.8. In one of the more successful theories of the dynamics of DD crystals, developed by Michel

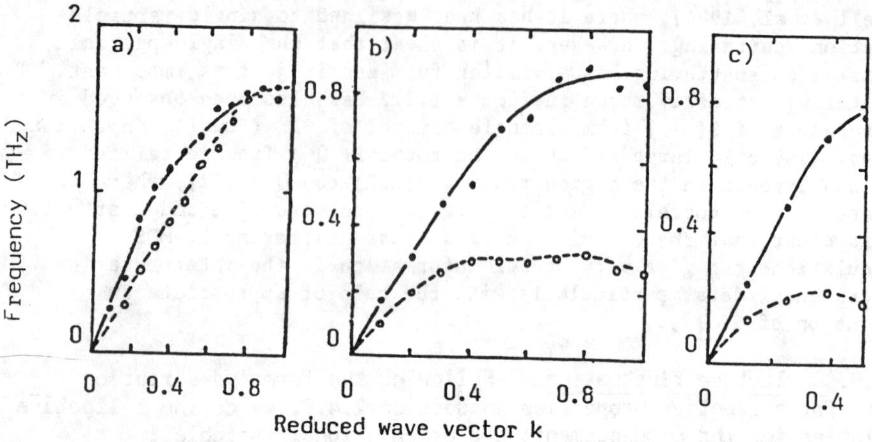

Fig.3.2.8. Acoustic phonon frequency dispension curves in a) SF_6 at 200K for k along [001]; b) CBr_4 at 345K for k along [001]; c) CBr_4 at 345K for k along [111]. Full circles are for longitudinal modes, and open circles are for transverse modes. Curves are drawn as guides to the eye.

and co-workers (Michel and Naudts, 1978; Michel, 1986b) and references cited there) there is an important bilinear interaction between the orientational variables and the displacement variables, the so-called "rotation-translation coupling". The Hamiltonian in the lowest order approximation is given as:

$$H = \frac{1}{2}\sum_{\underline{\underline{k}}}\{\underline{u}(\underline{k})\underline{\underline{M}}(\underline{k})\underline{u}(-\underline{k}) + \underline{u}(\underline{k})\underline{\underline{v}}(\underline{k})\underline{B}(-\underline{k}) + \underline{B}(\underline{k})\underline{\underline{J}}'(\underline{k})\underline{B}(-\underline{k})\} \qquad [3.2.9]$$

Michel and his co-workers have shown how the various coefficients in the Hamiltonian determine the form of the static correlation functions (as obtained from equations 2.4.23 and 2.4.25) for the displacement and orientational variables, and also how the various correlation functions are related. In addition, the $\underline{k}$-dependence of the coefficients has been shown to be related to the molecular symmetry, and it is this dependence that determines the nature of the ordering phase transition. For example, the bilinear rotation-translation coupling leads to a softening of some of the acoustic modes, so that if the instability is a zone centre ordering, the phase transition will be accompanied by a softening of one or more of the elastic constants. The extent of this coupling can be determined through calculations of the cross correlation function:

$$\lambda_{\alpha\beta}(\underline{k}) = \frac{\langle u_\alpha(\underline{k})B_\beta(-k)\rangle}{\langle u_\alpha(\underline{k})u_\alpha(-\underline{k})\rangle^{1/2}\langle B_\beta(\underline{k})B_\beta(-\underline{k})\rangle^{1/2}} \qquad [3.2.10]$$

where we have included the normalisation in the denominator so that complete coupling is indicated by $|\lambda(\underline{k})| \to 1$. At this point the OD phase is unstable against ordering, and the phase transition occurs. In addition, the acoustic mode is also made unstable at this point. In figure 3.2.9 the form of $\lambda(\underline{k})$ for CBr_4 and SF_6 is shown for several different modes and temperatures. There are a number of points that are revealed by this figure. The first is that the coupling only exists between displacement and orientational variables of the same symmetry. The second point is that $\lambda(\underline{k})$ is real for CBr_4 but imaginary for SF_6; this follows from a general result first noted by Pawley and Cochran (1964) that the rotational and displacement motions are exactly out of phase for a centro-symmetric molecule. The third point is that the maximum and minimum values of the coupling parameters for SF_6 and CBr_4 are at opposite points in the Brillouin zone: the coupling vanishes at the zone centre in CBr_4 but at the zone boundary in SF_6. This is consistent with a result first suggested by Michel and Rowe (1985), that

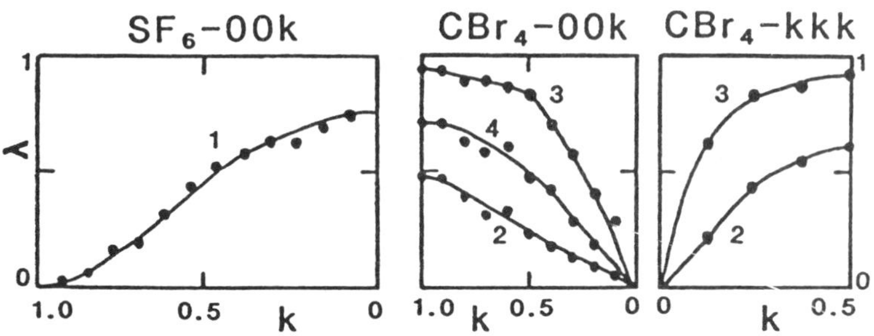

Figure 3.2.9. Rotation-translation coupling parameter λ as function of reduced wave vector k for SF_6 and CBr_4. Coupled models are 1. TA-T_{2g}; 2. TA-T_{1u}; 3. TA-T_{2u}; 4. LA-T_{1u}.

the form of the coupling at the zone centre is determined by whether the molecule has a centre of symmetry. The reason for this is that in the case of centro-symmetric molecules (where the lowest allowed value of l is even) the coupling is between the orientational mode and a shear mode (the coefficient $v(\underline{k})$ in the Hamiltonian is proportional to $\underline{k}$), which can be non-zero at $\underline{k}=0$, but in the case of the non-centrosymmetric molecule the coupling is between the orientational variable and a displacement mode, which must necessarily vanish in the very-long wavelength limit. The fourth point is that the coupling between the T_{2u} orientational variables and the transverse acoustic modes in CBr_4 is close to unity at the zone boundary points. These orientational variables, as we recall, are the variables that describe the ordering in the low-temperature phase. The ordering transition is expected to be associated with instabilities at the zone boundary point along the [111] direction and at a point half way along this direction. It is clear from figure 3.2.9 that these instabilities are shown in the simulations. However, the simulations also predict the existence of incipient instabilities associated with the orderings of the same orientational variables at other zone boundary points, which is a prediction that is not easy to confirm experimentally. Thus we can conclude that the ordering phase transition in CBr_4 is driven by a strong coupling between the orientational and transverse displacement motions. The softening of the transverse acoustic modes at the zone boundary points due to the coupling is clearly seen in figure 3.2.8. The fifth point is that the coupling in SF_6 is only very weakly temperature dependent, and we therefore conclude that the coupling does not play a role in the ordering transition in SF_6, at least to lowest order ($l=4$). This is consistent with calculations of the elastic constants of SF_6, which do not soften on cooling.

3.2.10 Collective dynamics. To conclude our review of the use of the MDS technique in the investigation of the properties of OD systems, we consider the effect of the disorder and the rotation-translation coupling on the collective dynamics of SF_6 and CBr_4. Our first use of MDS to tackle this problem illustrates the role of the technique in interpreting experimental data. Figure 3.2.10 shows the scattering function $S(\underline{Q}, \omega)$ (equations 2.4.32-2.4.38) for SF_6 at 200K measured in a neutron scattering experiment together with the results of MDS calculations. The agreement between the experimental results and the calculations is reasonable, in that the spectra given by both techniques are generally featureless over the region 0-4THz. The experimental results and the simulations show similar behaviour for all temperatures and for all scattering vectors away from the Bragg vectors, and no evidence was found for the existence of propagating modes at these scattering vectors. By calculating the scattering function with the scattering length for fluorine set to zero the centre-of-mass motion could be studied in isolation, and these calculations showed that the experiments were indeed probing non-propagating acoustic modes rather than orientational modes. The corresponding time correlation functions (equation 2.3.34) are shown in figure 3.2.11, where it is seen that the correlation time is less than 1ps. It should be noted that this

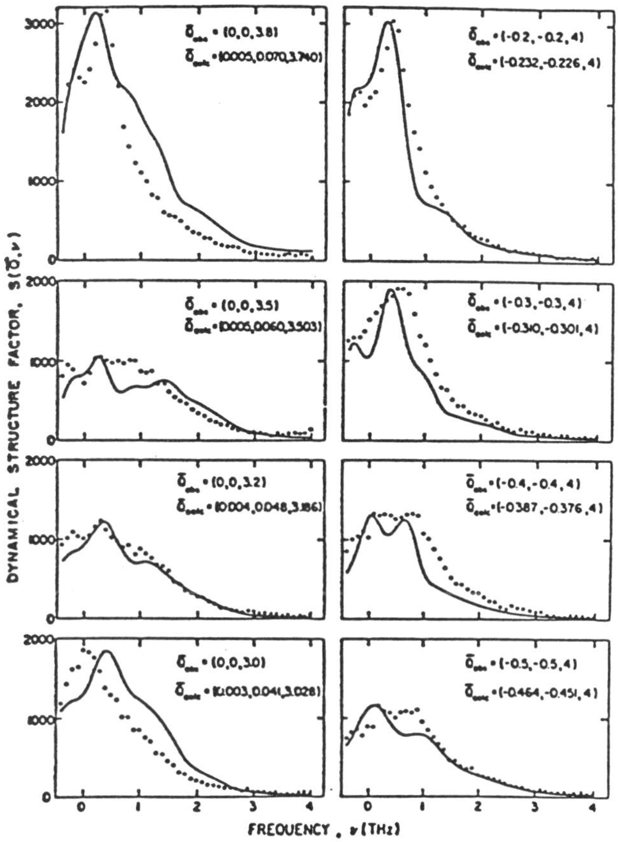

Fig.3.2.10. Comparison of observed and calculated dynamical structure factors for scattering vectors along the symmetry directions [001] and [110] near the reciprocal lattice point (004). The filled circles show the background subtracted data. The line shows the calculated function after convolution with the experimental resolution. Q_{obs} is the experimental momentum transfer while Q_{calc} is that for which the calculation is made.

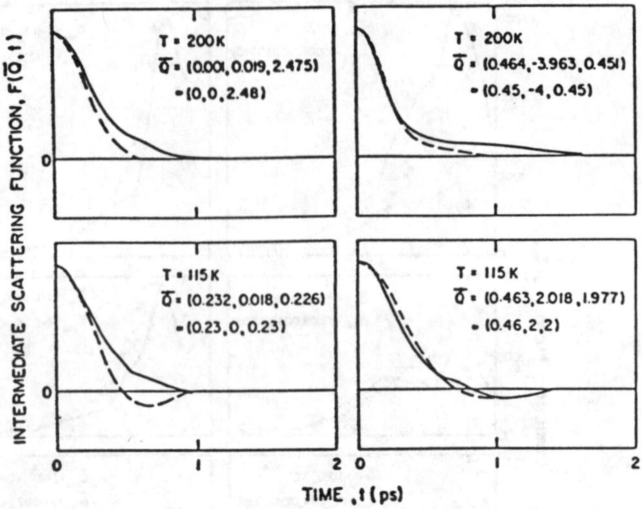

Fig.3.2.11. Intermediate scattering function F(Q,t), calculated as a function of time by MDS at temperatures of 200K and 115K. The components of the scattering vector, Q, are in units of 2π/a. The symbol = shows the scattering vector for which the simulation was actually made while the symbol ≈ shows the nearest momentum transfer lies on one of the symmetry directions. The solid line shows the total F(Q,t) while the dashed line shows its centre of mass component $F_{cm}(Q,t)$.

behaviour is not observed to such an extent in most other OD crystals for which a comparable study (experimental or simulation) has been performed. However, the simulations suggested that propagating excitations should exist at very small wave-vectors, and a recent inelastic neutron scattering experiment has confirmed this prediction (Dove and Powell, 1987).

As has been discussed in section 2.4.11, it is more convenient to study collective modes via the time correlation functions formed from the collective variables given by equation 3.2.7 than by calculating the scattering function directly. The most interesting results are for CBr_4; the correlation functions for SF_6 are all similar to the form shown in figure 3.2.11. Sample results for the time correlation functions formed from the transverse acoustic modes and coupled T_{2u} and T_{1u} orientational variables are shown in figure 3.2.12. The general feature is that on increasing wave-vector the acoustic mode changes in character from a propagating mode to a non-propagating mode similar to the behaviour observed in SF_6. However, as the wave-vector further increases, the bilinear rotation-translation coupling also increases, and the acoustic mode becomes relaxational with a very long relaxation time.

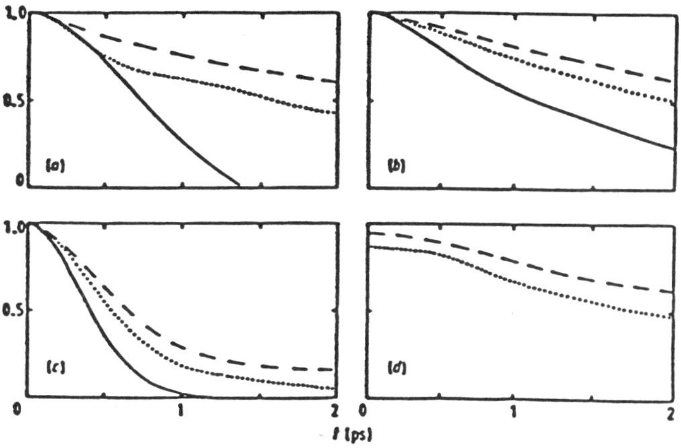

Fig.3.2.12. Correlation functions for TA orientation coupling along the 001 direction. (a) The displacement-displacement correlation functions. (b) The T_{2u} orientation-orientation correlation function. (c) The T_{1u} orientation-orientation correlation function. (d) The T_{2u} cross-correlation function. Full, dotted and broken curves refer to the reduced wave-vectors (0,0,0.3),(0,0,0.6) and (0,0,1) respectively.

In order to understand the behaviour of the collective dynamics of these systems, we have interpreted the results of the simulations in terms of a simple phenomenological model with values of adjustable parameters that we have obtained from the simulations directly or from fitting the model to the calculated time correlation functions. The model gives an equation of motion for a set of collective variables V_i, where the vector $\underline{V}$ (t) is assumed to follow the equation:

$$\frac{d}{dt}\underline{V}(t) = (\underline{\underline{\Omega}} - \underline{\underline{R}})\underline{V}(t) \qquad [3.2.11]$$

The components of the vector $\underline{V}$ are defined so that they are orthogonal and normalised. For the case where a displacement variable couples to two orientational modes (the usual case for our systems), V_1 is a displacement variable, V_2 is the corresponding momentum variable, and V_3 and V4 are linear combinations of the two orientational variables and the displacement variable, with coefficients chosen to satisfy the orthogonality condition (In the absence of any coupling these last twovariables are purely orientational variables.). The restoring force matrix $\underline{\Omega}$ in equation 3.2.11 is simply constructed from the calculated acoustic mode frequencies (equation 3.2.9) together with appropriate coupling constants; for more technical details the reader is referred to the paper by Dove and Lynden-Bell (1986). The relaxation matrix $\underline{R}$ is assumed to have all components except R_{22}, R_{33} and R_{44} zero, and the values of these non-zero components are obtained by fitting. R_{22} essentially models the anharmonic damping of the acoustic mode, and R_{33} and R_{44} give the relaxation frequencies of the two orientational modes.

The time correlation functions that have been calculated in the simulations can readily be constructed from the products of the form $V_i(0)V_j(t)$. The restoring force and relaxation matrices are taken to be wave-vector dependent. This equation-of-motion phenomenology is related to the theoretical work of Michel, and was first applied to the analysis of the results of MDS calculations by Lynden-Bell et al (1983). The general results are the same as obtained by Michel and Naudts (1978) for their model with $R_{22}=0$, and it was generally found that the model gave results that were in very good agreement with the correlation functions calculated in the simulations.

In the absence of the rotation-translation coupling, this model gives a propagating acoustic mode and two relaxational acoustic modes. Application to the case of CBr_4 showed that the effect of the strong coupling observed at the zone boundary points was to change the character of the acoustic mode completely from being a propagating mode to a relaxational mode with a long decay-time, consistent with the simulation calculations. In the case of SF_6, where the coupling is strongest at the zone centre, the model shows that the effect of the coupling is different for the two cases where the acoustic mode frequency is smaller or greater than the orientational relaxation frequency. As the system crosses over into the second regime, the acoustic modes begin to dampen so that the scattering function changes from having narrow peaks at the phonon frequencies to have broad peaks with a broad central peak (Michel and Naudts, 1978).

From fitting the model time correlation functions to the calculated time correlation functions values of the relaxation coefficients could readily be obtained; these are shown for both SF_6 and CBr_4 in figure 3.2.13. The main point that is revealed by these data are that the

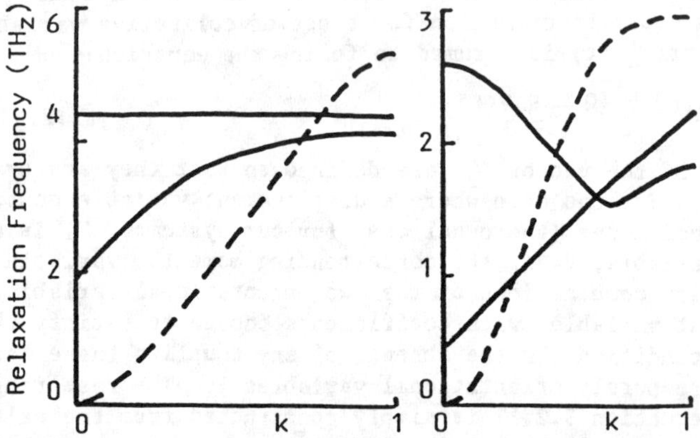

Fig.3.2.13. Relaxation frequencies for acoustic (TA) and two coupled rotational modes in the [001] direction for a) SF_6 and b) CBr_4. Full curves are for the rotational modes: broken curves are for the acoustic modes. The curves follow the patterns of the data points, which are not included in this figure in order to give clarity.

relaxation frequencies for the acoustic modes, R_{22}, which arise from anharmonic processes only and not from the bilinear rotation-translation coupling, vary as k^2 in the small wave-vector limit, and beyond this limit the acoustic mode relaxation frequencies are of similar size to the orientational relaxation frequencies, R_{33} and R_{44}. The large relative size of R_{22} in these cases contrasts with the much smaller effect observed in NaCN (Lynden-Bell et al, 1983), and is probably associated with the absence of ionic forces.

To summarise the qualitative description of the acoustic mode collective dynamics of these two OD systems, we can correlate the observed behaviour with both the effects of the rotation-translation coupling and the anharmonic relaxation effects. In the case of SF_6, the coupling is strongest at the zone centre. At very small wave-vectors, the acoustic phonon frequencies are smaller than the orientational relaxation frequencies, and the acoustic modes are able to propagate with little damping. As the wave-vector increases, the acoustic mode frequency increases and the effects of the rotation-translation coupling cause a damping of the acoustic modes. On further increasing the wave-vector, the rotation-translation coupling becomes less important, but the anharmonic interactions as seen in the form of R_{22} become important, so that the acoustic modes remain damped even to the zone boundary where the rotation-translation coupling vanishes. In CBr_4, the order of these effects is reversed. On increasing the wave-vector the rotation-translation coupling is initially small, so that the first effect to come into play is the anharmonic damping. On further increasing the wave-vector, although the effects of the anharmonic damping become even more important, they are superseded by the effects of the very strong rotation-translation coupling interactions which cause the acoustic mode to become relaxational with a very long life time.

We have used our model to generate the time correlation functions beyond the recurrence time (section 2.1.3) for CBr_4, so that we could calculate the acoustic mode scattering function (equation 2.4.39) in order to compare with experimental results. The results for the scattering function for wave-vectors along the [111] direction are shown in figure 3.2.14. On increasing wave-vector, the main scattering is found to be in the central peak that becomes as narrow as 0.03THz at the zone boundary. This looks exactly like the scattering that would be expected for a rotational mode, as expected with strong rotation-translation coupling. This narrow quasi-elastic scattering has been observed in a coherent inelastic neutron scattering study of CBr_4 (More et al, 1984), but the identification of the scattering with acoustic motions rather than rotational motions could not be made on the basis of the experimental data alone.

Finally, the existence of the strong anharmonic acoustic mode relaxation is not yet fully understood. Michel(1986b) has suggested that this behaviour might be obtained by using a Hamiltonian of the form of equation 3.2.9 with additional third order terms in the displacements and orientational variables. He has extended the analysis of the basic Hamiltonian to derive expressions for the power spectra of the time correlation functions in terms of the fundamental coupling coefficients. However, these results have not yet been tested. Work is currently in

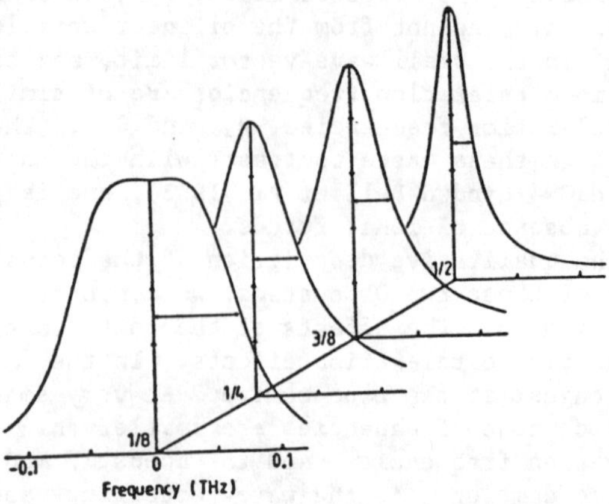

Fig.3.2.14. Fourier transforms of the TA displacement-displacement correlation functions (calculated from the phenomological theory presented in the text) along the 111 direction. The values 1/8, 1/4, 3/8, 1/2 are to indicate the reduced wave-vector.

progress to assess these theoretical results for the acoustic modes in SF_6 in the light of the simulation results and our recent neutron scattering results. This is an example of the three-way interplay between pure theory, pure experiments, and computer simulations.

3.2.11. Summary. It should be clear from this section that the application of the MDS technique to the problem of orientationally disordered crystals has been very fruitful. In particular, it is hoped that the application of a range of analysis methods has been well-illustrated by this example. It should be stressed that the information that the MDS calculation has provided has in several areas been decisive, in that the information could not be obtained by other approaches. The simulations have shown that the origin of the disorder in SF_6 is due to orientational frustration interactions; the simulations have enabled neutron scattering data to be interpreted and have successfully predicted the existence of well-defined collective acoustic excitations at very small wave-vectors; the role of rotation-translation coupling in both CBr_4 and SF_6 has been elucidated, particularly with regard to the ordering phase transitions in these systems, and the results have shown that the collective acoustic excitations in CBr_4 become relaxational modes with long life-times in the limit of strong coupling; the existence of both orientational frustration and statistical disorder in C_2F_6 has been identified, showing that many features are this system are determined primarily by the orientational frustration than the statistical disorder.

Among the results from the application of the MDS techniques to other OD systems by other workers, we should note the studies of the alkali cyanides (Lynden-Bell et al, 1983) and $NaNO_2$ (Lynden-Bell et al, 1984; Heine et al, 1984). The alkali cyanides have become the main model system for both experimental and theoretical investigations of OD phenomena. An extensive study of NaCN (Lynden-Bell et al, 1983) has shown that the theory of rotation-translation coupling of Michel and Naudts (1978) gives a good description of these systems. Additional calculations have shown how the detailed behaviour of the alkali cyanides is very sensitive to the form of the model inter-atomic potentials, and in particular it has been shown that small changes in the CN^- quadrupole moment can have drastic effects on the ordering processes (Bounds et al, 1981). Thus these studies have given crucial information about the role of the inter-atomic forces in the ordering transitions in these systems, and have added considerable detail to the bare theories of rotation-translation coupling.

In the case of $NaNO_2$, the orientational ordering transition gives rise to the existence of an incommensurate structure over a small temperature interval. The simulation calculations have given information concerning the appropriate rotation-translation coupling interactions as functions of wave-vector, showing that the interaction is strongest at the same incommensurate wave-vector observed in nature (Lynden-Bell et al, 1984). The studies have provided useful information about the structural distortions that accompany the incommensurate transition (Heine et al, 1984).

It is hopefully also clear that the MDS technique can in principle be just as usefully applied to a range of other solid systems; collective excitations, coupling interactions and scattering cross-sections can be calculated for any system. One of the main features of OD systems is the existence of the bilinear coupling interaction, and analogous coupling interactions associated with phase transitions can be expected to be found in other systems. Such interactions, which are not often directly accessible to experiments, can be conveniently studied by MDS calculations, by the evaluation of appropriate fluctuation amplitudes. The wave-vector and temperature dependence of these fluctuations can becorrelated with the main characteristics of the associated phase transitions, thereby providing information about the driving mechanisms behind the transitions. By taking this kind of approach, appropriate simulations can be used to test theories of particular phase transitions by performing explicit calculations to test any approximations or assumptions. The use of MDS calculations will also provide the crucial link between the inter-atomic forces and the observed behaviour.

3.3. SIMULATIONS OF THIOUREA

3.3.1. Incommensurate behaviour in thiourea. Our MDS study of thiourea was motivated by the recognition that thiourea represents a model system in the study of incommensurate phase transitions. The reason for this is that with changing temperature and pressure there are a number of stability regions that correspond to lock-in phases with different lock-in wave-vectors ($\underline{b}^*/3$, $\underline{b}^*/7$, $\underline{b}^*/8$, $\underline{b}^*/9$) in addition to the existence of a ferroelectric ($\underline{k}$=0) ground state. Although thiourea is a molecular crystal and structurally somewhat simpler than many minerals, it is a good model system for minerals physics since it is now recognised that a number of minerals have modulated phases. If it is possible to understand in detail the behaviour of a relevant model system, we will be in a considerably better position to understand similar phenomena in minerals. There are two main questions to be asked about the transitions in thiourea: what is the origin of the incommensurate phase transition, and what gives the stability of the lock-in phases? It is possible to write out a phenomenological free energy that gives an incommensurate instability, but it is not clear that the inherent mechanism is captured in this approach. The lock-in transitions with wave-vectors of $\underline{b}^*/n$ may arise from n-th order Umklapp terms, but they may also arise from third order interactions involving phonon modes with different wave-vectors. It would therefore be useful if a realistic microscopic model such as can be given in a MDS calculation could differentiate between different models. In addition, the crystal chemistry of this system needed understanding; in particular an apparent hydrogen bonded framework has often been pointed out. As the purpose of this section is only to outline the application ofthe MDS method, some of the technical details will be glossed over, andthe interested reader is referred to our recent paper (Dove and Lynden-Bell, 1986). It should be noted though that this simulation represents one of the most ambitious projects undertaken with this technique.

3.3.2. Intermolecular potential. This example gives a good illustration of the effort that is required to obtain an adequate intermolecular potential before simulation calculations can be performed. Thiourea has the chemical formula $SC(NH_2)_2$, and is a planar molecule with C_{2v} symmetry about the S-C bond. The low-temperature structure is ferroelectric, and both the high and low temperature structures are orthorhombic. The structures of these two phases are shown in figure 3.3.1. The incommensurate wave vector lies along the [010] direction, and the important molecular motions consist of rotations about the b axis and displacements in the a-c plane. The polarisation of the ferroelectric phase lies along the [100] direction.

The charge distribution of thiourea was obtained from a DMA calculation (section 2.3.1), assuming that the charges are located at the atomic sites. It was assumed that the atom-atom dispersive and repulsive interactions could be modelled by Lennard-Jones interactions (equation 2.3.2), and approximate values for the parameters (assuming combining laws of the form given in equation 2.3.4) were gleaned from the literature. This starting model was then refined by fitting a calculated static structure against the experimental low-temperature structure using the static lattice energy program WMIN (Busing, 1980; section 2.3.2). It was found that this simple model was unable to give a good fit, and in particular the molecular orientations could not be

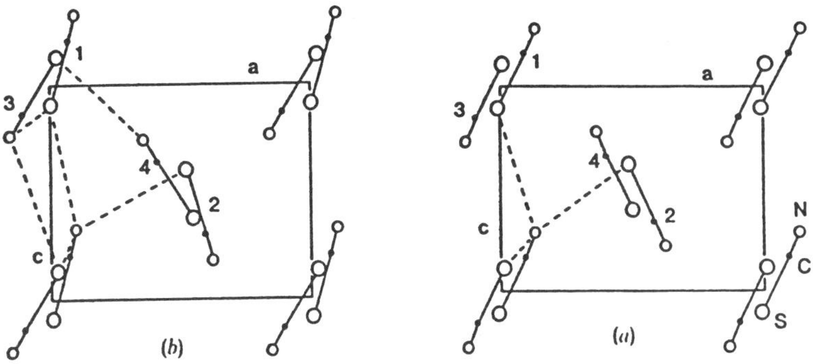

Fig.3.3.1. The crystal structure of the paraelectric (a) the ferroelectric (b) phases of thiourea, as determined by Elcombe and Taylor (1968) viewed down the crystal b axis. The network of hydrogen bonds is shown by the dashed lines, and the numbers label the four molecules in the unit cell. Molecules 1 and 2 lie in one plane, and molecules 3 and 4 lie in another plane at a height of b/2.

reproduced by this model. The existence of a hydrogen bond network involving the sulphur atoms had been pointed out by Elcombe and Taylor (1967), and this is shown in figure 3.3.1. It was clear that our simple model did not take account of this hydrogen bonding. Our attempts to model the hydrogen bonds using specific functions (e.g. Morse potentials, equation 2.3.3), which implicitly assume that the hydrogen bonds are chemical bonds, were unsuccessful. Instead, the DMA calculations suggested that there should be a large quadrupole moment on the sulphur atom, and this was modelled by placing two charges on either side of the atom along the direction of the C-S bond. This refinement of the model was able to reproduce the main features of the structure. The dispersive r^{-6} interaction between the sulphur and hydrogen atoms was found to be negligible in comparison to the repulsive r^{-12} interaction; the electrostatic interactions provide most of the attraction within the hydrogen bond interaction. It was also found that the structure was insensitive to the Lennard-Jones interactions involving the carbon atoms and the hydrogen atoms. Thus the final refinement of the potential model was achieved using just the charge distribution plus the quadrupole moment on the sulphur atom, the H..S repulsion for close contacts, and the Lennard-Jones interactions for the sulphur and nitrogen atoms. From the good agreement between the calculated and observed structures we were able to conclude that the hydrogen bonds observed in thiourea are not true chemical bonds, but arise due to anisotropic electrostatic multipole interactions.

Despite the success of the model in reproducing the overall features of the crystal structure, the model was unsuccessful in calculating a large energy difference between the ferroelectric and paraelectric phases, and in fact when the long range part of the electrostatic interactions were neglected the paraelectric phase had the lower energy. We believe that the important component of the model for the phase transitions that is missing is an induction interaction; in particular we expect that polarisation of the C-S bond due to local fields will play a significant role in the phase transition. Nevertheless, as we will see, the model as it stands reproduces many features associated with the phase transition. Therefore the simulations were performed using this basic model in order to find the extent to which our basic model captured the main features of the phase transitions, as might be revealed in precurser fluctuations.

3.3.3. Simulations. The MDS calculations were performed using a sampleof 1024 unit cells that was elongated along the b axis (16 unit cells long) in order to increase the number of allowed wave-vectors along this direction. Each molecule was allowed to interact with its 22 nearest neighbours, and the long-range electrostatic interactions were neglected. This meant that about 15% of the total electrostatic energy was not included. Our continued calculations, the preliminary results of which are not discussed here, are using the Ewald summation with a value for G (equation 2.3.9) chosen so that the neglect of the reciprocal space sum accounts for only a very small part of the total electrostatic energy. We are also attempting to model the induction effects by allowing the dipole moment along the C-S bond to vary in response to instantaneous electric fields. Since the local electric field experienced by any molecule depends on the induced polarisation of all its neighbours, at each time step it is necessary to perform a self consistent iteration to obtain all the induced dipole moments in the sample.

3.3.4. Single particle dynamics and collective fluctuations. The main aim of the study is to identify the role of phonon-phonon interactions in the incommensurate and lock-in transitions. A preliminary insight into the phonon spectra can be obtained by calculating the power spectra from the translational and rotational velocity time correlation functions (section 2.4.8). The power spectra for thiourea for simulations performed at 85K are shown in figure 3.3.2, where we also show the power spectrum obtained from the time correlation function of the orientation of the molecule about the crystal b axis. This function is the phonon density of states for the corresponding motions weighted by a factor of ω^{-2}, thereby accentuating the low frequency regime of the density of states. Of particular importance is the peak at about 0.7 THz in the latter power spectrum, which can also be seen in the rotational density of states. This corresponds to the soft optic phonon that is associated with the phase transitions.

The collective variables calculated in the simulations were converted to normal mode coordinates by the method outlined in section 2.4.10. However, use was made of the symmetry of the different motions

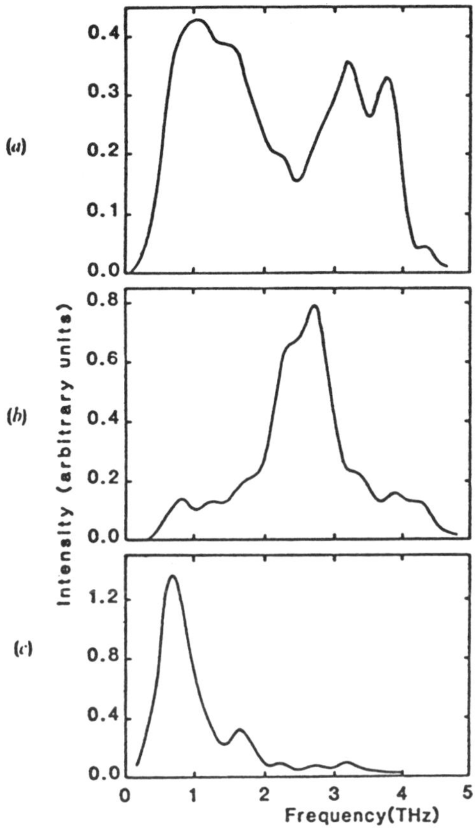

Fig.3.3.2. Power spectra obtained from the correlation functions for a) velocities, b) angular velocities, c) molecular orientation about the b axis.

to minimise the number of calculations required (see Dove and Lynden-Bell, 1986, for a fuller description). The frequencies associated with each normal mode were calculated using the standard quasi-harmonic result (equation 2.4.31). Of particular note is the fact that one of the normal modes of the known symmetry of the soft mode had a low frequency (~0.7THz). In figure 3.3.3 we show the harmonic phonon dispersion curves (section 1.2) for all three high symmetry directions, together with the dispersion of the soft mode branch calculated in the simulations and some experimental data obtained by neutron scattering (MacKenzie, 1975). The figure is arranged in an extended zone scheme, where the branches of the four different symmetry types are displayed together. It is clear that our model has given good agreement between the low-frequency branches and the experimental measurements, and in

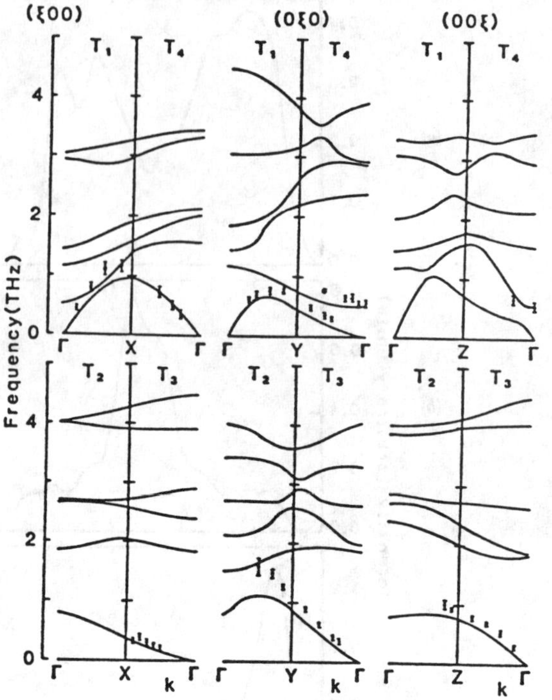

Figure 3.3.3. Harmonic phonon dispersion curves for the paraelectric phase of thiourea calculated from our model. The points with error bars represent experimental data obtained with neutron scattering (MacKenzie 1975). The branch in the b^* direction marked with an asterisk is the soft branch that provides the mechanism for the phase transitions.

particular has given a soft mode of the dorrect symmetry together with some fairly soft acoustic modes. The dispersion curves show that the softness of the optic branch spreads across the zone in the [010] direction, but that the branch quickly rises in frequency with wave-vector in other directions. This is consistent with the tendency towards an incommensurate instability. It was also found that the normal mode frequencies calculated from the simulations are close to the frequencies calculated by the harmonic lattice dynamics calculation. We can therefore conclude that the anharmonic effects are not very large.

The model of Michel and Parlinski (1984) and Parlinski (1985) assumes that the lock-in phase arise through Umklapp terms involving the soft modes of T_2 and T_2 symmetry, whose normal modes we denote as $(Q_2(\underline{k})$ and $Q_4(\underline{k})$ respectively. For example, it is expected that terms of the form $Q_m(\underline{b}^*-2\underline{k})Q\,(\underline{k})Q_n(\underline{k})$, where m,n are either 1 or 4, will be important. We calculated a number of these terms as thermal averages, but to within the level of noise on the calculations the thermal averages were zero. We calculated some higher order terms, but found that although there were some dramatic peaks at incommensurate wave-vectors, these could be understood within quasi-harmonic approximation as reflecting the underlying soft mode (since terms of the order $\langle Q_m(\underline{k})Q_m(-\underline{k})\ldots\rangle$ are approximately equal to $\langle Q_m(\underline{k})Q_m(-\underline{k})\rangle\langle\ldots\rangle$), and Umklapp terms were always found to be negligibly small.

3.3.5. Conclusion. There is one main conclusion that arises from this work so far, and that is that although the model that we have used for thiourea can reproduce many of the features of nature and also displays a tendency towards the phase transitions observed in this system, the model lacks the one small characteristic that will produce the phase transitions. For apart from this one aspect, the analysis that we have performed on this system has not shown any precurser effects apart from the soft direction in reciprocal space. Our guess is that this one important characteristic is the induction interaction, and we are currently using a model that includes this interaction.

However, there is one observation that comes out of this conclusion, and that is that the driving force for a displacive phase transition may well be a very small interaction when compared to all the other interactions in the system. This is highlighted by the fact that even the soft mode does not apparently have any significant anharmonic effects. It would seem that the main interactions produce only a low resistance to any particular distortion (so that the nature of the phase transition can be understood on the basis of our model), but that a small additional interaction is required to actually drive the system unstable with respect to this barrier. This conclusion is similar to some observation on the phase transitions in the perovskite family of structures given by Cowley (1980). This contrast with the previous example of orientation disordered crystals, where the driving mechanism of the phase transition is anentropic effect, and the existence of a phase transition from a high symmetry (disordered) phase is guaranteed by the second law of thermodynamics! It is clear that the realistic simulation of displacive phase transitions is considerably harder to achieve than of order-disorder transitions.

4. CONCLUSIONS: THE OUTLOOK FOR THE USE OF MOLECULAR DYNAMICS SIMULATIONS IN MINERALS PHYSICS.

4.1 GENERAL SUMMARY

The two main areas of discussion in this paper have been the technical methods and analysis techniques for molecular dynamics simulations, and case studies employing these techniques. The illustrations have shown how appropriate simulations can provide information about systems for which theories are necessarily too complicated either to be reliably tested or to be able to give definite information, and for which experiments are unable to give sufficient unambiguous results to allow for a detailed understanding of the basic phenomena. The main characteristics of these systems that make MDS calculations so valuable are the existence of extreme disorder and the crucial importance of specific anharmonic interactions. These systems therefore lie outside of the range of systems for which conventional calculation techniques can play a role (e.g. harmonic lattice dynamics), and these characteristics are those for which the connection between the specific details of microscopic interactions and general theories are too difficult to evaluate. This connection is specifically provided by the MDS technique.

It should therefore be clear that the MDS techniques are able to provide information about the structural and thermodynamic properties ofthe system under study, as well as information about the single particle and collective dynamics and interactions between these elementary excitations. The paper has emphasised the role of static and dynamic correlation functions, which provide the link between experiment and theory but which in any system may not be directly accessible to experiments, or which may not be calculable from microscopic theories for reasons such as the lack of information concerning atomic interactions or because of difficulties associated with renormalization of force constants due to other interactions. In providing the connection between experiments and theory, via the use of specific models for atomic forces, the MDS technique is actually able to provide information about the atomic forces found in real systems and their role in determining the behaviour of these systems. The simulations therefore give the opportunity of fine-tuning the parameters in any model.

The fact that the discussion has focussed on the use of the NVE ensemble should not detract from the fact that considerable information has also been obtained about particular systems using different ensembles. In particular, the new constant pressure ensembles can be used to study the response of structures to changes in their environments in order to predict equilibrium properties; the NVE ensembles can then be used to study the dynamic and thermodynamic properties of these new equilibrium structures if they are of interest. For example, constant stress methods have been used to study the response of single crystals to uniaxial loading (Parrinello and Rahman, 1981). Furthermore, we have not touched on the use of the class of non-equilibrium MDS techniques, which probe the response of samples to non-equilibrium perturbations such as shear stresses. We have also not considered the use of the MDS technique to study imperfect crystals. It should hopefully be apparent that there is a very wide range of potential applications of the MDS techniques.

The main point, though, is that the MDS technique has become a relatively refined technique, so that it can take its place along-side other more familiar experimental and theoretical techniques in the armoury of the physicist, which are themselves becoming more widely used in the study of minerals. In particular, during the past decade the technique has been used in the study of solids to the point at which it can now be said to be almost a routine technique (Klein, 1987). We should therefore consider the ways in which the MDS technique can be applied to the study of minerals, taking note of any advances that might be required ahead of these applications.

4.2. APPLICATION IN THE FIELD OF MINERALS PHYSICS: GENERAL OUTLOOK.

4.2.1. Potential areas of application. In general, the MDS technique has been applied to systems that have simple constituents but that have "complicated" behaviour, the obvious example being simple monatomic and diatomic fluids. The aims of these simulations have been to explain the observed behaviour and correlate it with the known atomic forces. The case of minerals is somewhat to the opposite extreme. The basic constituents are themselves complex, particularly in terms of the large numbers of atoms in the unit cells and the forms of the interatomic forces. However, the motivation for using simulations in the study of minerals will be the same, but the fact that minerals are complex and also that it is possible to make progress with experiments and phenomenological theories has meant that so far MDS calculations have not been widely used in the study of minerals. I suggest, though, that there are at least three areas of study to which the MDS technique might be usefully applied.

The first area is the study of equilibrium structures of systems under extreme conditions, such as very high temperatures and pressures, that cannot be reached in the laboratory. Probably the only direct information will come from computer simulations. Although static lattice energy calculations can play a very important role here, the fact that many of these structures will undergo phase transitions at high temperatures will mean that full anharmonic simulations will be required. Starting from a potential model for a particular system that has been shown to be reliable, the equilibrium structures formed under the extreme conditions can be predicted using NPH or NPT MDS ensembles (section 2.2). Such predictions will clearly play a very valuable role in Earth Science.

The second area is related to the first area, and this is that MDS calculations can be used to study properties of systems that for any reason cannot be studied in the laboratory. The cases for which this might be so include structures that exist under only extreme conditions, structures that cannot be obtained as single crystals, or structures that are formed by annealing for times much longer than laboratory times. In these cases realistic simulations might give information about thermodynamic, structural, elastic or dynamic properties, such as thermal expansion. As before, the MDS technique will be essential when static calculations or harmonic lattice dynamics calculations are inappropriate (e.g. when there are phase transitions, disorder or important anharmonic interactions).

The third area is the study of phase transitions in minerals, particularly displacive or order-disorder, rather than reconstructive transitions. The simulations can give information about the important interactions that drive the transition. They will also be able to provide the link between specific features of the interatomic forces and the observed behaviour. As was stressed in the introduction to this paper, our goal is an understanding of these phase transitions rather than just a description or characterization, and this understanding will include the prediction of the values of parameters that enter any phenomenological model.

4.2.2. Necessary developments. Although it is possible to be optimistic about the application of the MDS technique, there are at least two main areas on the technical side that need major advances to occur before this application becomes routine or even reliable. But we are now in the position where we can expect such advances to be made, and we should therefore be encouraging them!

The first necessary advance is the development of appropriate potential models. The most successful use of simulations have been for systems for which the interatomic forces can be modelled with some reliability, e.g. molecular or ionic systems. In the case of minerals, there are probably two types of potentials that need to be considered. The first type is for common ions, such as Mg^{2+}, Ca^{2+}, Na^{+} and K^{+}, which interact via Coulomb, dispersive and repulsive interactions only, as described in section 2.3. Essentially what is required is a set of parameterized models for ion-ion interactions and for interactions between these different ions and other major constituents of minerals, e.g. Si, Al and O atoms, that will be transferable between different systems so that we can use these models for any mineral that we are interested in simulating. This advance has been made in the study of organic molecular physics, where the range of systems of molecules containing C, H, N, O, F and Cl atoms can now be easily simulated used reliable tested potential models. The second type of potential is for covalent bonds. Analytical forms for the covalent interactions for specific atom types such as Si-Si, Si-O, and Al-O need to be developed, and we will also need information about the transferability of these functions between different environments (e.g. different chain or framework structures), including the effect of environment on the residual atomic charges. It will be essential to be able to model the anharmonic parts of the potentials correctly as well as the harmonic contributions if we are to be able to study phase transitions, as in many cases the existence of phase transitions are very sensitive to subtle changes in the anharmonic interactions. This may sound like an idealistic dream, particularly the wish for transferable potentials, but unless it is fulfilled it is unlikely that simulation methods will become routine tools in minerals physics. That the development of such potentials was forthcoming in other branches of solid state physics gives us optimism in this regard! The current use of quantum chemistry methods to calculated *ab initio* covalent potentials for silicates may well give the necessary breakthrough in this area.

The second necessary advance is the increase of the power of the computers we use. The current availability of computing resources generally limits the sizes of systems that can be simulated, which is another reason beside the problem of potential models why the MDS technique has only be applied to systems with simple constituents. As we pointed out in section 2.1, the study of wave-vector dependent quantities requires the use of simulation samples that contain as many as 1000 unit cells. In addition, the use of large samples will be required in the study of systems that have long range effective interactions, as might be mediated by elastic strain fields for example. On the other hand, many minerals can contain as many as 200 atoms in the unit cell, so that a useful mineral simulation will have up to 2×10^5 atoms in the ensemble. This is two or three orders of magnitude larger than many of today's simulations! One answer to this problem is to just wait for similar increases in the power of our large central mainframe computers, but such an increase may be a long time coming. An alternative answer is to use the powerful parallel array processors that are discussed in section 2.5. The simulations for thiourea that were described in section 3.3, which were performed on the ICL Distributed Array Processor, used samples of 1024 unit cells with 40 interaction sites per unit cell. This illustrates the current power of parallel array processors. It is likely that there will be incredible advances in the power of these machines, and it is also likely that because of the relative low cost and small sizes of such machines they will only need to be shared by a few workers. Examples are the Transputer arrays and the next model of the ICL Distributed Array Processor.

There is very little that we can contribute to the development of appropriate computing resources for simulations of minerals. But we can be alert to these trends, so that when it does become possible to perform the necessarily large simulations as a matter of routine the problems of the model potentials will have been largely overcome.

Acknowledgements

A large part of the work reported in this paper has been carried out in collaboration with Stuart Pawley (Edinburgh), Ruth Lynden-Bell (Cambridge), Brian Powell (Chalk River) and David Fincham (York), whose contributions are gratefully acknowledged. I am also grateful to Pat Hancock, who rescued the text from the word processor for me! Finally, I thank my wife, Kate, who helped in checking the text and preparing some of the figures.

REFERENCES

Alder, B.J. and Wainwright, T.W., 1957, J. Chem. Phys. 27, 1208
------------------------------ 1959, J. Chem. Phys. 31, 459
Altmann, S.L. and Cracknell, A.P., 1965, Rev. Mod. Phys. 37, 19
Andersen, H.C., 1980, J. Chem. Phys. 72, 2384
Beeman, D., 1976, J. Comput. Phys. 20, 130
Binder, K., 1983. 'Monte Carlo Methods In Statistical Physics' (Springer-Verlag, Berlin)
Bounds, D.G., Klein, M.L. and Patey, G.N., 1980, J. Chem. Phys. 72, 5348
Bounds, D.G., Klein, M.L. and McDonald, I.R., 1981, Phys. Rev. Lett. 46, 1682
Branka, A. and Parrinello, M., 1986, Mol. Phys. 58, 989
Busing, W.R., W.R., 1981, Oak Ridge National Laboratory Report ORNL-5747
Califano, S., Schettino, V. and Neto, N., 1981, 'Lattice Dynamics of Molecular Crystals.' (Springer-Verlag, Berlin)
Cheung, P.S.V., 1977, Mol. Phys. 33, 519
Cheung, P.S.Y. and Powles, J.G., 1975, Mol. Phys. 30, 921
- - - - - - - - - - - - - - - 1976, Mol. Phys. 32, 1983
Ciccotti, G., Ferrario, M. and Rychaert, J-P., 1982, Mol. Phys. 47, 1252

Coulon, G. and Descamps, M., 1980, J. Phys. C: Solid State Phys. 13, 2947
Cowley, R.A., 1980, Adv. Phys. 29, 1
Della Valle, R.G. and Pawley, G.S., 1984, Acta Cryst. A40, 297
Dolling, G., Powell, B.M., and Sears, V. F., 1979, Mol. Phys. 37, 1859
Dove, M.T., 1986, J. Phys. C: Solid State Phys. 19, 3325
Dove, M.T., Fincham, D. and Hubbard, R., 1986a, J. Mol. Graph. 4, 79
Dove, M.T., Pawley, G.S., Dolling, G. and Powell, B.M., 1986b, Mol. Phys. 57, 865
Dove, M.T. and Lynden-Bell, R.M., 1986a, J. Phys. C: Solid State Phys. 19, 3342
\- - - - - - - - - - - - 1986b, Phil. Mag. B54, 443
\- - - - - - - - - - - - 1987, in 'Dynamics of Molecular Crystals' (ed. J. Lascombe; Elsevire, Amsterdam) p.705
Dove, M.T. and Pawley, G.S., 1983, J. Phys. C: Solid State Phys. 16, 5969
\- - - - - - - - - - - - 1984, J. Phys. C: Solid State Phys. 17, 6581
Dove, M.T. and Powell, B.M., 1987, in preparation
Elcombe, M. and Taylor, J.C., 1968, Acta Cryst. A24, 410
Evans, D.J., 1977, Mol. Phys. 34, 317
Futrelle, R.P. and McGinty, D.J., 1971, Chem. Phys. Lett. 12, 285
Gear, G.W., 1971, ' Numerical Initial Value Problems in Ordinary Differential Equations' (Prentice Hall)
Goldstein, H., 1980, 'Classical Mechanics' (Addison-Wesley, Massachusetts)
Hansen, J.P. and Klein, M.L., 1976, Solid State Comm. 20, 771
Heine, V., Lynden-Bell, R.M., McConnell, J.D.C. and MacDonald, I.R., 1984, Z. Phys. B56, 229
Hockney, R.W. and Eastwood, J.W., 1981, 'Computer Simulation Using Particles' (McGraw-Hill, New York)
Hockney, R.W. and Jesshope, C.R., 1981, 'Parallel Computers: Architecture, Programming and Algorithms' (Adam Hilger, Bristol)
Jacucci, G., Klein, M.L. and McDonald, I.R., 1975, J. Phys. Lett. 36, L97
Jacucci, G.., McDonald, I.R. and Rahman, A., 1976, Phys. Rev. A13, 1581
Klein, M.L., 1987, in 'Dynamics of Molecular Crystals' (ed. J. Lascombe; Elsevier, Amsterdam), p.659
Klein, M.L. and Weis, J.J., 1977, J. Chem. Phys. 67, 217
Klein, M.L., McDonald, I.R. and Ozaki, Y., 1983, J. Chem. Phys. 79, 5579
Lebowitz, J.L., Percus, J.K. and Verlet, L., 1967, Phys. Rev. 153, 250
Levesque, D. and Verlet, L., 1970, Phys. Rev. A2, 2514
Lynden-Bell, R.M., Klein, M.L. and McDonald, I.R., 1983, Mol. Phys. 48, 1093
Lynden-Bell, R.M., Klein, M.L. and McDonald, I.R., 1984, Z. Phys. B54, 325
MacKenzie, D.R., 1975, J. Phys. C: Solid State Phys. 8, 1607
Marshall, W. and Lovesey, S.W., 1971, 'Theory of Thermal Neutron Scattering' (Oxford University Press, London)
McDonald, I.R., Bounds, D.G. and Klein, M.L., 1982, Mol. Phys. 45, 521
McQuarrie, D.A., 1976, 'Státistical Mechanics' Harper and Row, New York)
Meyer, M., Marhic, C. and Ciccotti, G., 1986, Mol. Phys. 58, 723
Michel, K.H., 1986a, Phys. Lett. 57, 2188

Michel, K.H., 1986b, J. Chem. Phys. 84, 3451
Michel, K.H. and Naudts, J., 1978, J. Chem. Phys. 68, 216
Michel, K.H. and Parlinski, K., 1984, Phys. Rev. B29, 396
Michel, K.H. and Rowe, J.M., 1985, Phys. Rev. B32, 5818
More, M., Lefebvre, J., Hennion, B., Powell, B.M. and Zeyen, C.M.E., 1980 J. Phys. C: Solid State Phys. 13, 2833
More, M., Lefebvre, J. and Hennion, B., 1984, J. de Physique 45, 303
Murad, S., Evans, D.J., Gubbins, K.E., Streett, W.B. and Tildesley, D.J., 1979, Mol. Phys. 37, 725
Murphy, C.S., Singer, K. and Vallauri, R. 1983, Mol. Phys. 49, 803
Nose, S., 1984, Mol. Phys. 52, 255
Nose, S. and Klein, M.L., 1983a, Mol. Phys. 50, 1055
- - - - - - - - - - 1983b, J. Chem. Phys. 78, 6938
- - - - - - - - - - 1983c, Phys. Rev. Lett. 50, 1207
Parlinski, K., 1985, J. Phys. C: Solid State Phys. 18, 5667
Parrinello, M. and Rahman, A., 1980, Phys. Rev. Lett. 45, 1196
- - - - - - - - - - 1981, J. Appl. Phys. 52, 7182
- - - - - - - - - - 1982, J. Chem. Phys. 76, 2662
Parrinello, M., Rahman, A. and Vashista, P., 1983, Phys. Rev. Lett. 50, 1073
Pawley, G.S., 1981, Mol. Phys. 43, 1321
Pawley, G.S. and Dove, M.T., 1983, Chem. Phys. Lett. 99, 45
Pawley, G.S. and Thomas, G., 1982, Phys. Rev. Lett. 48, 410
Powell, B.M., Press, W., Dolling, G. and Sears, V.F., 1984, Mol. Phys. 53, 941
Powell, B.M., Dove, M.T., Pawley, G.S. and Bartell, L.S. 1987, Mol. Phys. (in press)
Press, W. and Huller, A., 1973, Acta Cryst. A29, 252
Price, S.L., Stone, A.J. and Alderton, M., 1984, Mol. Phys. 52, 987
Price, S.L. and Stone, A.J., 1982, Mol. Phys. 47, 1457
- - - - - - - - - - - - 1984, Mol. Phys. 51, 569
Rae, A.I.M. and Dove, M.T., 1982, J. Phys. C: Solid State Phys. 16, 3233
Ray, J.R., 1983, J. Chem. Phys. 79, 5128
Raynerd, G., Tatlock, G.J. and Venables, J.A., 1980, Acta Cryst. B38, 1896
Refson, K., 1985, Physica, 131B, 256
Refson, K. and Pawley, G.S., 1987, in preparation
Rahman, A., 1964, Phys. Rev. 110, A405
Rahman, A., 1966, J. Chem. Phys. 45, 258
Rahman, A. and Stillinger, F.H. 1971, J. Chem. Phys. 55, 3336
- - - - - - - - - - - - 1974, Phys. Rev. A10, 368
Schiff, D., 1969, Phys. Rev. 186, 151
Schofield, P., 1960, Phys. Rev. Lett. 4, 239
Sherwood, J.N. (ed), 1979,'The Plastically Crystalline State' (Wiley, Chichester)
Singer, K., Taylor, A. and Singer, J.V.L., 1977, Mol. Phys. 33, 1757
Steinhauser, O. and Neumann, M., 1979, Mol. Phys. 37, 1921
Stone, A.J., 1981, Chem. Phys. Lett. 83, 233
Tildesley, D.J., and Madden, P.A., 1981, Mol. Phys. 42, 1137
Tildesley, D.J., and Lynden-Bell, R.M., 1986, J. Chem. Soc. Faraday Trans. 2 82, 1605

Van Gunsteren, W.F., Berendsen, H.J.C. and Rullmann, J.A.C., 1978, Faraday Discuss. 66, 58
Verlet, L., 1967, Phys. Rev. 159, 98
- - - - - 1968, Phys. Rev. 163, 201
Weis, J.J. and Bevesque, D., 1976, Phys. Rev. A13, 450
Weis, J.J. and Klein, M.L., 1973, J. Chem. Phys. 63, 2869
Williams, D.E., and Haupt, D.J., 1986, Acta Cryst. B42, 286
Woodcock, L.V., 1971, Chem. Phys. Lett. 10, 257
- - - - - - - 1972, Proc. Roy. Soc. A328, 83

THE COMPUTER SIMULATION OF THE LATTICE DYNAMICS OF SILICATES

Geoffrey D. Price
Department of Geological Sciences
University College London
Gower St
London WC1E 6BT
UK

Stephen C. Parker
School of Chemistry
University of Bath
Bath, BA2 7AY
UK

ABSTRACT. We use an approach based upon the Born model of solids, in which potential functions represent the interactions between atoms in a structure, to calculate the phonon dispersion of forsterite and the lattice dynamical behaviour of the beta-phase and spinel polymorphs of Mg_2SiO_4. The potential used (THB1) was derived largely using data from simple binary oxides, and includes 'bond bending' terms, that model the directionality of the Si-O bond, in addition to the pair-wise additive Coulombic and short range terms. The infrared and Raman behaviour of forsterite are successfully model by the potential, and the phonon dispersion relationships of the Mg_2SiO_4 polymorphs predicted by THB1 are used to calculate the heat capacities, entropies, thermal expansion coefficients and Gruneisen parameters of these phases. The predicted heat capacities and entropies are in outstandingly good agreement with those determined experimentally. The predicted thermodynamic data of these phases is used to construct a phase diagram for this system. The success that we have in predicting the lattice dynamical and thermodynamic properties of the Mg_2SiO_4 polymorphs shows that the potential THB1 provides the previously elusive quantitative link between the microscopic or atomistic behaviour of a mineral and its macroscopic properties.

1. INTRODUCTION

One of the fundamental aims of mineralogical research is to determine which minerals exist under given pressure, temperature and compositional conditions, and to understand why those specific minerals are more stable than any others. Such an understanding of the macroscopic behaviour of minerals can only be obtained, however, from a detailed

E. K. H. Salje (ed.), Physical Properties and Thermodynamic Behaviour of Minerals, 591–618.

knowledge of the microscopic or atomistic nature of the minerals involved. This microscopic understanding is required because it is the response at the atomic level to changes in pressure and temperature that eventually determines the bulk properties of a material. The link between the microscopic properties of materials and their macroscopic behaviour is best made via their vibrational spectra, which are determined purely by interatomic interactions, and to which many bulk thermodynamic properties such as heat capacity, entropy, etc. are directly related. If, therefore, we wish to understand the behaviour and properties of a mineral from first principles, we must be able to predict and rationalise its lattice vibrations - or equivalently its infrared and Raman spectra and phonon dispersion curves.

The frequencies of atomic vibrations within a crystal are determined by the strength and nature of the bonding, which holds the atoms in that crystal together. More generally, these bonding forces can be described in terms of interatomic potentials, which not only determine the vibrational characteristics of a crystal but also its structure and physical properties, such as its elastic and dielectric behaviour. Because the vibrational behaviour of a material provides this vital link between its internal bonding and thermodynamic behaviour, considerable effort has been put into the study of the vibrational behaviour and lattice dynamics of the major Earth-forming minerals (e.g. Kieffer, 1979 a, b, c). In addition, attempts have been made to develop models which reproduce not only the vibration behaviour of silicates but also their structural and physical properties (e.g. Devarajan and Funck 1974; Iishi, 1978). However none of these previous efforts have been fully successful, and the quantitative link between the atomic forces which determine silicate structures and their vibrational and bulk properties has not been made. This shortcoming led McMillan (1985) to conclude that the lack of suitable models for the internal interactions in silicates represents one of the major obstacles to our achieving a full understanding of their macroscopic behaviour. Our recent work (discussed at greater length in Price et al., 1987 a, b) has involved the computer simulation of the lattice dynamics of the major mantle-forming mineral forsterite (Mg_2SiO_4) and its beta-phase and spinel structure polymorphs. Our model is among the first successfully to predict both the vibrational behaviour of these silicates, and the atomic interactions that determine their structural and elastic behaviour. The potentials used in our studies were derived purely from the known structural and physical properties of simple binary oxides, and their ability to predict correctly the lattice dynamics of forsterite and its polymorphs represents a fundamental advance over previous, force-constant models (e.g. Iishi, 1978), which were simply derived by fitting to the spectrascopic data that they aimed to model.

The computer model adopted in our study of the lattice dynamics of Mg_2SiO_4 uses the atomistic approach based upon the classical Born model of solids, in which potential functions represent the interactions between ions or atoms in the structure. The theory of this approach is described in greater detail by Prof Catlow elsewhere in this volume. It has been widely shown that this type of computer simulation may successfully predict both the structures of minerals and their elastic, dielectric and defect properties (e.g. Price and Parker, 1984; Price et al., 1985; Catlow et al., 1986). Early studies concentrated upon the use of pair-wise additive interatomic potentials, and ignored the role of many body interactions. Recent successes, however, in modelling more complex silicates by using bond-bending terms to simulate some aspects of three-body interactions (Matsui and Busing, 1984; Sanders et al., 1984; Catlow et al., 1986) have provided the impetus for our study of the polymorphs of Mg_2SiO_4, in which we have used potentials which also include O-Si-O bond-bending terms to simulate the directional bonding thought to be important in silicate phases. The details of the potential (referred to throughout as THB1) used in this study are given in Table I and discussed further in Price et al. (1987 a, b). Table II shows how successfully potential THB1 predicts the static properties of the Mg_2SiO_4 polymorphs. In the following sections we outline how our potential model was used to calculate the vibrational behaviour of forsterite and its polymorphs, and their associated thermodynamic properties.

TABLE I

The potential model and the parameters used in THB1

$$U = 1/2 \sum e^2 q_i q_j r_{ij}^{-1} + A_{ij} \exp(-r_{ij}/B_{ij}) - C_{ij} r_{ij}^{-6} + $$
$$+ 1/2 \sum k_{ijk} (\theta_{ijk} - \theta_o)^2 + 1/2 \sum k_i r_i^2$$

q_{Mg}	+2.0	$q_{O\text{-}Shell}$	-2.8480
q_{Si}	+4.0	$q_{O\text{-}Core}$	+0.8480
$A_{Mg\text{-}O}$	1428.5	$B_{Mg\text{-}O}$	0.2945
$A_{Si\text{-}O}$	1283.9	$B_{Si\text{-}O}$	0.3205
$A_{O\text{-}O}$	22764.3	$B_{O\text{-}O}$	0.1490
$C_{Si\text{-}O}$	10.7	$C_{O\text{-}O}$	27.88
$k_{O\text{-}Shell}$	74.9	$k_{O\text{-}Si\text{-}O}$	2.09

Units: A_{ij} in eV, B_{ij} in Å, C_{ij} in eVÅ^6, $k_{O\text{-}Shell}$ in eVA^{-2}, $k_{O\text{-}Si\text{-}O}$ in eVrad^{-2}. Short range cut off = 7.5Å

TABLE II

Predicted and observed properties of the Mg_2SiO_4 polymorphs

	Forsterite		Beta-Phase		Spinel	
	Obs	THB1	Obs	THB1	Obs	THB1
a	4.754	4.784	5.698	5.654	8.065	8.020
b	10.194	10.261	11.438	11.397		
c	5.981	5.991	8.257	8.281		
V	43.67	44.26	40.54	40.15	39.65	38.82
K	1.25	1.54	1.74	2.20	1.84	2.50
μ	0.81	0.78	1.14	1.17	1.19	1.26
β_{1000}	0.38	0.23	0.33	0.17	0.28	0.15
γ_{1000}	1.12	0.90	1.26	0.90	1.37	0.88

Units: a,b and c in Å, the molar volume (V) in $cm^3 mol^{-1}$, K and μ in Mbars, and $\beta \times 10^{-4}$ in K^{-1}. Observed data quoted in Price and Parker (1984), Price et al., (1985), Watanabe (1982) and Weidner et al. (1984).

2. ATOMISTIC SIMULATION TECHNIQUES

The relationship between the interatomic potential (U) and the normal modes of vibration of a structure has been described in detail in a variety of standard texts (e.g. Born and Huang, 1954; Ziman, 1964; Cochran, 1973). In brief however, the vibrations of the atoms in a crystal consisting of N unit cells, each containing n atoms, can be described by 3nN Newtonian equations of motion. If U is the interatomic potential, then if any atom i in the crystal is displaced by an amount u_i from its equilibrium position, it will experience a restoring force F_i;

$$F_i = \partial U/\partial u_i \qquad \text{.........(1).}$$

Thus the equation of motion for each atom i, of mass m_i, can be expressed as;

$$m_i \frac{\partial^2 u_i}{\partial t^2} = - \frac{\partial U}{\partial u_i} \qquad \text{..........(2).}$$

From the theory of small displacements (e.g. Ziman, 1964) it can be shown;

$$\frac{\partial U}{\partial u_i} = \sum \frac{\partial^2 U}{\partial u_i \partial u_j} u_j \qquad \text{..........(3),}$$

which can be interpreted as meaning that the sum on the right handside of equation (3) is the force acting on the ith atom due to the displacement u_j of all the other j atoms in the crystal. In accordance with Block's theorem (e.g. Ziman, 1964) however, equations (2) and (3) must be translationally invariant, so we may write;

$$u = e(q)\exp(iq.R - w(q)t) \qquad \text{..........(4),}$$

where q is the reciprocal lattice wave-vector for lattice waves, w(q) is the frequency of the vibrational mode, R is the interatomic separation, and e(q) is the polarization vector describing the atomic displacements involved in the vibration. This treatment reduces the number of equations that now need to be considered from 3nN to 3n, as the equivalence of all unit cells means that we need only consider one of them in order to calculate the dynamics of the whole assembly. Substitution of equations (3) and (4) into equation (2) yields;

$$mw^2(q)e(q) = D(q)e(q) \qquad \text{..........(5),}$$

where D(q) is the dynamical matrix;

$$D(q) = \sum \frac{\partial^2 U}{\partial u_i \partial u_j} \exp(iq.R) \qquad \text{..........(6).}$$

For a given value of q, solution of equation (5) yields 3n eigenvalues which are the squared frequencies ($w^2(q)$) of each of the normal modes of the crystal, and 3n sets of eigenvectors ($e_x(q)$, $e_y(q)$ and $e_z(q)$) which describe the pattern of atomic displacements for each normal mode.

In our study, we use a suite of programs such as CASCADE (the details of which are given in Catlow and Mackrodt, 1982), and the related programs THBREL and THBPHON, which are supported by the SERC at their Daresbury Laboratory. In the following sections we present the THB1 predicted normal modes of forsterite at values close to q=0, which correspond to the modes excited by the relatively long wavelength radiation used in infrared and Raman spectroscopy.

3. THE q=0 LATTICE VIBRATIONS OF FORSTERITE

Forsterite has four formula units, and consequently 28 atoms, per unit cell, which give rise to 84 normal modes of vibration. Detailed analysis of the symmetry of these modes has been presented in a variety of works (see McMillan, 1985), but in brief the sum of the irreducible representation of the forsterite space group (Pbnm) can be written as;

$$\Gamma^{total} = 11A_g + 11B_{1g} + 7B_{2g} + 7B_{3g} + + 10A_u + 10B_{1u} + 14B_{2u} + 14B_{3u} \quad ....(7).$$

At the centre of the Brillouin zone (q=0), three of these 84 modes are purely translational;

$$\Gamma^{trans} = B_{1u} + B_{2u} + B_{3u} \quad ....(8).$$

The remaining 81 modes, however, are truly vibrational in nature. Of these, those which are symmetric with respect to inversion (grade or g) are Raman active, 10 with A_u symmetry are spectroscopically inactive, and the remaining 35 modes with B_{1u}, B_{2u} and B_{3u} symmetry are infrared active.

The infrared and Raman spectra of forsterite have been extensively studied (e.g. Servoin and Piriou, 1973; Paques-Ledent and Tarte, 1973; Iishi, 1978), and there is general agreement on the frequencies of the characteristic vibrations (Table III). Most infrared and Raman bands have been unambiguously identified, but there is still some controversy over some of the weaker or more poorly resolved peaks (see McMillan, 1985). The spectroscopically active vibrational modes have frequencies that lie in the range between 1100 cm^{-1} and 150cm^{-1}, but modes with frequencies greater than 800 cm^{-1} are clearly separated by a 150cm^{-1} wide gap from those with frequencies below approximately 650cm^{-1}. It is generally agreed that the high frequency modes are the result of Si-O stretching motions, while modes with frequencies below 650cm^{-1} are thought to be the result of vibrations involving more complex displacements of Si, O and Mg. Considerable uncertainty remains, however, concerning the detailed nature of the atomic motions in many vibrational modes, despite the efforts which have been made to elucidate them by using isotopic substitution techniques (e.g. Paques-Ledent and Tarte, 1973).

In addition to these experimental studies, attempts have been made to model empirically the lattice dynamics of forsterite. These studies by Devarajan and Funck (1974) and by Iishi (1978) involved the development of force-constant models to describe the vibrational behaviour of Mg_2SiO_4. In such an approach interactions between atoms are described by a simple harmonic spring constant, and occasionally include Coulombic effects. The numerical values of the spring constants are determined purely by fitting to the observed vibrational frequencies, and are therefore in no way predictive. Neither are they able to be used in other atomistic studies, such as point defect or diffusion modelling. Devarajan and Funck (1974) fitted 24 parameters to 62 observed vibrational frequencies, while Iishi (1978) developed models with up to 14 parameters fitted to forsterite spectra. The success and accuracy of

TABLE III

Observed and calculated q=0 vibration frequencies of forsterite.

Obs	THB1	I	DF	Obs	THB1	I	DF
Ag				B1g			
960-966	943	962	962	972-976	963	962	965
854-856	851	886	867	863-866	864	888	870
822-826	807	839	827	835-839	821	839	826
606-609	630	606	625	626-632	667	616	626
541-545	580	537	556	577-583	617	519	526
420-424	427	430	435	428-434	454	418	423
334-340	360	350	347	418	428	394	379
325-329	344	318	329	314-318	337	333	334
304-305	312	269	289	260-265	325	287	294
221-227	221	213	208	215-224	269	217	231
181-183	184	165	167	149-192	231	206	192
B2g				B3g			
880-884	901	894	887	917-922	960	894	893
583-588	611	563	570	588-595	606	548	556
436-441	433	467	457	407-412	417	424	426
354-368	365	373	383	371-376	390	362	351
324	345	312	306	312-318	327	323	334
237-244	207	268	276	272	250	269	276
142-150	149	157	153	226	133	242	240
Au				B1u			
.	938	.	890	885-994	879-1028	893-903	884
.	519	.	552	502-585	513-598	512-512	544
.	479	.	496	483-489	476-502	482-494	508
.	452	.	435	423-459	450-475	439-444	456
.	391	.	397	365-371	375-390	391-391	404
.	340	.	346	296-318	347-362	336-350	366
.	271	.	262	274-278	313-321	240-244	237
.	250	.	233	224	250-251	209-209	229
.	166	.	178	201	176-177	185-186	170
.	58	.	112	0	0	0	0
B2u				B3u			
987-993	995-1008	958-959	958	980-1086	995-1084	957-967	954
882-979	874-980	900-906	863	957-963	960-991	901-901	873
838-843	815-817	834-834	825	838-845	812-815	832-835	823
....	671-672	....	622	601-645	657-658	569-577	617
537-597	558-605	581-581	559	562-566	573-638	529-528	545
510-516	538-547	542-543	507	498-544	520-571	473-473	503
465-493	488-524	467-479	450	438-469	482-516	....	448
421-446	448-488	427-430	418	403-438	437-471	421-421	434
400-412	431-444	364-364	355	378-386	376-394	359-376	342
352-376	355-390	338-354	333	320-323	349-349	327-332	312
294-313	298-308	280-284	265	293-298	307-313	280-280	...
280-283	265-290	250-250	...	274-276	293-293	242-241	259
244	....	....	219	224	....	218-219	212
144	146-146	137-138	128	201	193-197	190-191	189

these fitting exercises can be determined by inspection of Table III. Devarajan and Funck (1974) calculated frequencies (column DF in Table III) that are in excellent accord with their set of observed values, with only 11 frequencies in error by more than 30cm^{-1}. Their model included no Coulombic effects, so that they were not able to reproduce the TO-LO splitting of the infrared active modes. In contrast, Iishi's (1978) most successful model did contain a Coulombic component, but as it has fewer overall parameters than that of Devarajan and Funck (1974) it is not surprising that the fit that Iishi achieved with his model is slightly inferior to that of Devarajan and Funck. Nevertheless, of the calculated values obtained by Iishi (column I in Table III) only 16 are in error by more than 30cm^{-1}. It should be noted, however, that there are some minor differences between the sets of frequencies used by these workers in deriving their force-constant models, which makes it difficult to compare them directly.

In Table III, we also present the vibrational frequencies of the normal modes of forsterite predicted by the potential THB1. This potential, derived purely from MgO (periclase) and SiO_2 (quartz), predicts vibrational frequencies for forsterite which are in excellent agreement with those observed, and have accuracies which compare favourably with those achieved by the models of Devarajan and Funck (1974) and Iishi (1978). Of the 71 infrared and Raman active vibrational modes predicted by the THB1 potential, 61 had frequencies within 30cm^{-1} of unambiguously observed modes, and only 4 were in error by more than 50cm^{-1}. There appears to be some experimental uncertainty over the exact frequencies of some of the infrared active modes, which makes the exact quality of the prediction of these modes difficult to assess. The average error in the predicted Raman frequencies is only 20cm^{-1}, and the predicted magnitudes of the TO-LO splittings of the infrared active modes are generally comparable with those observed, in contrast to those calculated by Iishi (1978). Until more detailed inelastic neutron scattering data are available (e.g. Rao et al, 1987), there is no way to assess the accuracy of the predicted frequencies of the 10 spectroscopically inactive A_u modes, however it appears that the values predicted by the potential THB1 and those calculated by Devarajan and Funck (1974) are broadly in agreement. In the next section we consider in more detail some of the outstanding problems of the lattice dynamics of forsterite, and assess how the predictions made by the potential THB1 can be used to resolve some of these ambiguities.

4. THE CRYSTAL DYNAMICS OF FORSTERITE

As discussed in the previous section, the frequencies of both the infrared and Raman spectra of forsterite are relatively well known. In contrast however, the actual vibrations or atomic displacements which give rise to the adsorption bands in these spectra of forsterite are far less well determined. It was briefly mentioned that the general principles of spectroscopy and studies of molecular vibrational spectra have led to the deduction that the high frequency bands (1100-800 cm^{-1})

are the result of Si-O stretching motions. Indeed, the interpretation of the general vibrational behaviour of forsterite can be greatly simplified by considering, in the first instance, the constituent SiO_4 groups to be dynamically isolated. Under these circumstances, it would be expected that the higher frequency vibrational modes in forsterite would be associated with internal vibrations within the SiO_4 tetrahedron, while the lower frequency modes would be the result of external or lattice vibrations, involving rotation or libration of the SiO_4 tetrahedra as well as the translation of the SiO_4 group and the Mg cations. It should be noted, however, that both the classification of crystal vibrations as internal or external, and the subdivision of these latter into rotational and translational vibrations should be viewed with considerable caution. In the general case, lattice vibrations are not purely translational or rotational, and the separation of internal vibrations from lattice vibrations is only approximate, even in molecular crystals (Lazarev, 1974). Nevertheless, as will be shown below, significant progress towards understanding the lattice dynamics of forsterite can be made if this simple view of its vibrational behaviour is adopted.

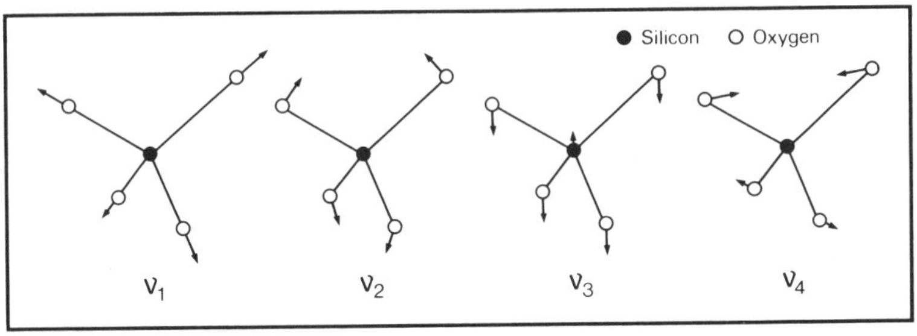

Figure 1. The vibrational modes of a tetrahedral molecule.

An isolated tetrahedral molecule is known to have four distinct vibrational modes. The higher frequency modes involve symmetric (v_1) and anti-symmetric (v_3) stretching, while the symmetric (v_2) and anti-symmetric (v_4) bending modes are generally of lower frequency (Figure 1). Conventional symmetry analysis (e.g. McMillan, 1985) has shown that the simple v_1 and v_3 modes of the isolated tetrahedron would give rise to 15 active modes in forsterite, which correspond exactly to the number of bands identified in the high frequency (1100-800cm^{-1}) region of its infrared and Raman spectra. The inference that these high frequency modes are essentially due to internal Si-O stretching is supported by the unpolarised infrared and Raman studies performed by Paques-Ledent and Tarte (1973) on forsterites containing varying $^{28}Si/^{30}Si$ and $^{24}Mg/^{26}Mg$ isotope ratios. These workers noted that bands in the range 1100-800cm^{-1} were unaffected by the Mg isotope content, indicating that little or no Mg translation was involved in these high frequency modes. They found, however, that bands in the range 991cm^{-1}, 960-950cm^{-1} and 900-885cm^{-1} were sensitive to variations in the Si isotope content, and inferred that they were related to v_3, anti-symmetric stretching modes, in which significant Si displacement is to be expected.

Paques-Ledent and Tarte (1973) also found that bands in the range 650-475cm^{-1} showed a frequency dependence on both Si and Mg isotope content. They were unable to infer the exact nature of the vibrations involved, however it is generally accepted that the vibrations in this region of the forsterite spectra are associated with internal Si-O bending vibrations. Obviously, from the observations of Paques-Ledent and Tarte (1973) these vibrations must be of significantly mixed nature. Below 475cm^{-1}, these workers found infrared peaks at 466, 425, 415, 364, 320, 300 and 277cm^{-1} which showed large frequency shifts with Mg isotope variation, but which had little Si frequency dependence (McMillan, 1985). From there observations Paques-Ledent and Tarte (1973) inferred that the bands were associated with large Mg translations and possibly SiO_4 rotations. All the remaining bands observed in their study appeared to involve both Mg and Si displacements.

Various other experimental investigations have been performed (see McMillan, 1985) in an attempt to elucidate the general features of the atomic motions involved in the vibrations which give rise to the infrared and Raman spectra of forsterite. However, by themselves such studies will be incapable of describing the detailed nature of the atomic motions associated with each vibrational mode, since these exact considerations can only be obtained from atomistic models of the lattice dynamics of forsterite. As described above, the solution of equation (5) yields not only the eigenvalues of each of the normal modes, from which the frequencies of the vibrations are derived, but also the eigenvectors which describe the pattern and relative magnitude of the atomic displacements for each mode.

In Figure 2, we present projections of the forsterite structure on to which are plotted some of the eigenvectors calculated using potential THB1 for each of the forsterite normal modes. In Table IV we have attempted to summarise the dominant vibrational components in each of the modes, in terms of the internal vibrations of a SiO_4 tetrahedron (v_1, v_2, v_3 and v_4), SiO_4 group rotations (R) about the x, y and z crystallographic axes, and of translations (T) of the SiO_4 group and Mg1 (at the centre of symmetry) and Mg2 (on the mirror plane) cations parallel to the x, y or z axes. Detailed study of the calculated eigenvectors shows that most of the modes involve complex patterns of atomic motion, and therefore exhibit evidence of considerable mixing of the various simple modes discussed above. As a result, Table IV should be viewed only as an approximate description of the forsterite lattice vibrations. However, we can conclude that the general features of the

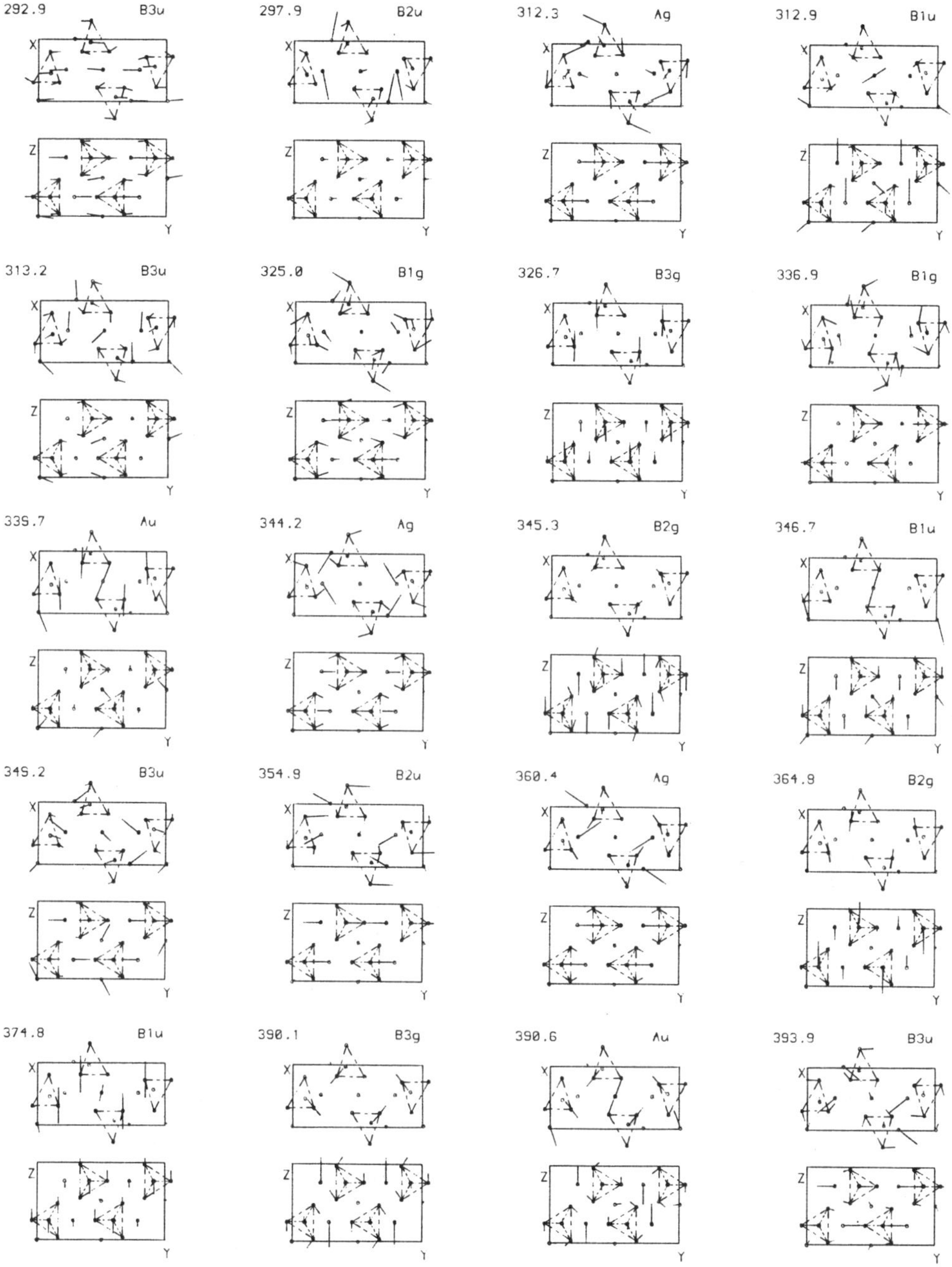

Figure 2. A selection of vibrational modes of forsterite predicted by THB1.

TABLE IV

Assignments of the vibrational modes predicted by potential THB1.

Ag	943	v3	B1g	963	v3
	851	v3		864	v3
	807	v1 + Ty(Si)		821	v1 + Txy(Si)
	630	v4 + Tx(Si)		667	v4 + Tx(Si) + Ty(Mg)
	580	v4 + Ty(Si,Mg2)		617	v4 + Ty(Si) + Ty(Mg)
	427	v2 + Txy(Si,Mg2)		454	v2 + Tx(Mg2)
	360	Rz(SiO_4) + Txy(Mg2)		428	Rz(SiO_4) + Ty(Mg2)
	344	Rz(SiO_4) + Txy(Mg2)		337	Rz(SiO_4) + Tx(Mg2)
	312	Ty(SiO_4,Mg2)		325	Rz(SiO_4) + Txy(Mg2)
	221	Ty(SiO_4,Mg2)		269	Txy(SiO_4,Mg2)
	184	Rz(SiO_4) + Tx(Mg2)		231	Ty(SiO_4) + Tx(Mg2)
B2g	901	v3	B3g	960	v3 + Tz(Mg2)
	611	v4 + Tz(Si)		606	v4 + Tz(Si,Mg2)
	433	v2		417	v2 + Tz(Si)
	365	Ry(SiO_4) + Tz(Mg2)		390	Ry(SiO_4) + Tz(Mg2)
	345	Tz(SiO_4,Mg2)		327	Tz(SiO_4,Mg2)
	207	Rx(SiO_4) + Tz(Mg2)		250	Rx(SiO_4) + Tz(Si,Mg2)
	149	Rx(SiO_4) + Tz(Mg2)		133	Rx(SiO_4) + Tz(Si,Mg2)
Au	938	v3	B1u	879	v3
	519	v4 + Tz(Si,Mg2)		513	v4 + Tz(Si)
	479	v2 + Tz(Si)		476	v2 + Tz(Si)
	452	Txyz(Mg1) + Tz(SiO_4,Mg2)		450	Txyz(Mg1) + Ry(SiO_4)
	391	Ry(SiO_4)+Tz(Mg2)+Tx(Mg1)		375	Tx(Mg1) + Ry(SiO_4)
	340	Txyz(Mg1) + Ry(SiO_4)		347	Txy(Mg1) + Rx(SiO_4) + Tz(Mg2)
	271	Txz(Mg1) + Rx(SiO_4)		313	Txyz(Mg1) + Rx(SiO_4) + Tz(Mg2)
	250	Txyz(Mg1)+Rx(SiO_4)+Tz(Mg2)		250	Rx(SiO_4)
	166	Ty(Mg1) + Rx(SiO_4)		176	Ty(Mg1)
	58	Txyz(M1) + Tz(SiO_4,Mg2)		0	Tz(Mg1,SiO_4,Mg2)
B2u	995	v3	B3u	1084	v3
	874	v3		991	v3
	815	v1 + Txy(Si)		815	v1
	671	v4 + Tz(Mg1,Si,Mg2)		658	v4 + Tx(Si)
	558	v4 + Ty(Mg1,Si,Mg2)		638	v4 + Tx(Mg1,Si,Mg2)
	538	v2 + Tz(Mg1)		571	v2 + Txy(Mg1,Si,Mg2)
	488	v2 + Txyz(Mg1)		516	Tyz(Mg1) + Ty(SiO_4,Mg2)
	448	Tx(Mg1) + Rz(SiO_4)		471	v2 + Tz(Mg1)
	431	Txz(Mg1) + Rz(SiO_4)		394	Rz(SiO_4) + Txy(Mg2)
	355	Ty(Mg2) + Rz(SiO_4)		350	Txyz(Mg1) + Ty(SiO_4,Mg2)
	298	Ty(Mg1,SiO_4) + Tx(Mg2)		313	Tx(Mg2) + Rz(SiO_4) + Txy(Mg1)
	265	Ty(Mg1) + Rz(SiO_4) + Tx(Mg2)		293	Ty(Mg1,SiO_4,Mg2)
	146	Tx(Mg1,SiO_4)		193	Ty(Mg1,SiO_4)
	0	Ty(Mg1,SiO_4,Mg2)		0	Tx(Mg1,SiO_4,Mg2)

lattice dynamics of forsterite, which are predicted by potential THB1, are in excellent agreement with those previously inferred from experiment. Figure 2 and Table IV show that the high frequency vibrations (1100-800cm^{-1}) are predicted only to involve SiO_4 internal v_1 and v_3 modes, while the bands in the range 650-400cm^{-1} are largely due to v_2 and v_4 Si-O stretching modes but also involve significant Si and Mg displacements, as suggested by the isotopic studies of Paques-Ledent and Tarte (1973). Bands with frequencies below approximately 400cm^{-1} are predicted to be external modes of generally mixed character. Further discussion of these calculations can be found in Price et al. (1987a).

In potential THB1, we have a model of the effective interatomic interactions in forsterite that is of sufficient accuracy as to reproduce not only the major structural features of forsterite, but also to predict its lattice dynamics as successfully as any previously developed model. This capability is made even more impressive when it is recalled that the empirically derived parts of the potential were obtained purely by fitting to the structural and physical properties of MgO and SiO_2, and that they contain no input derived from a previous knowledge of the behaviour of forsterite. Its success is a powerful testimony to the philosophy of potential transferability, which holds that if a potential is presented as being able to describe the real nature of effective interatomic interactions within a solid, then it should be transferable from one structure to another. We believe that the bond-bending or quasi-three-body terms in potential THB1 model the real effect of bond directionality in silicates, and that the inclusion of such terms in silicate potentials is both a major and vital development. The fact that the potential THB1 reproduces the known q=0 lattice dynamics of forsterite leads us to expect that it may also accurately predict the phonon dispersion, heat capacity, Gruneisen parameters, thermal expansion coefficients and the other bulk thermodynamic properties of forsterite and its polymorphs, which are investigated in the following sections.

5. LATTICE DYNAMICS AND THERMODYNAMIC PROPERTIES

The dependence of w (the vibrational frequency) on q (inversely related to the wavelength of the lattice vibration) gives rise to the widely described phonon dispersion relationships. As discussed for example by McMillan (1985) and Ghose (1985), the fact that the energy, or frequency of a lattice vibrational mode or phonon depends on the wavelength of the vibration within the crystal, reflects the effect that the phase of the vibrational mode in adjacent unit cells has upon the frequency of the mode. Because of the periodic nature of crystals, phonon dispersion can be fully described by simply considering the behaviour of phonons within the first Brillouin zone (the Wigner-Seitz reciprocal lattice unit cell, Figure 3). Phonons with a wavelength large in comparison with the unit cell size (such as those stimulated by infrared radiation) have a q-vector which approaches zero, and so appear close to the Brillouin zone centre (the Γ point, Figure 3), while those with a wavelength close to that of the unit cell, will fall near to the edge of the zone (e.g. points X,Y and Z in Figure 3).

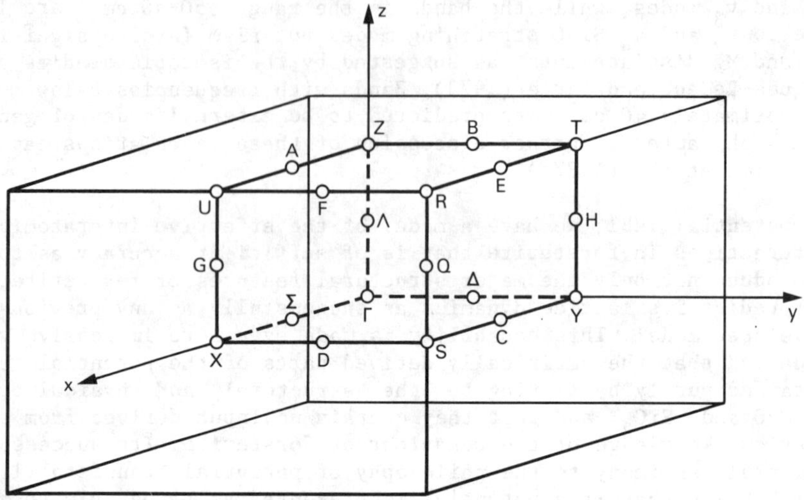

Figure 3. The Brillouin zone for an orthorhombic crystal.

The link between the vibrational behaviour of a mineral and its thermodynamic behaviour can most conveniently be made if we first define a density of states function, G(w), such that G(w)dw is the number of modes, when summed over the entire Brillouin zone, having frequencies in the range w to w+dw. The overall energy of a crystal can now be defined in terms of G(w) (e.g. Cochran, 1973; Salje and Werneke, 1982) thus:

$$E = U + \int_0^{\omega_m} hw\{1/2 + [\exp(hw/kT) - 1]^{-1}\}G(w)dw \qquad \text{........(9)},$$

where U is the static lattice energy, h is Planck's constant, k is Boltzmann's constant, T is temperature (in K), and w_m is the maximum vibrational frequency of the crystal. By differentiation of equation (9), the heat capacity of a crystal at constant volume, C_v, is given by:

$$C_v = k\int_0^{\omega_m} (hw/kT)^2 \exp(hw/kT)\{\exp(hw/kT) - 1\}^{-2} G(w)dw \qquad \text{.......(10)}.$$

The other major thermodynamic functions of a mineral, such as its entropy (S), enthalpy (H), Gibbs free energy (G) and heat capacity at constant pressure (C_p), can only be calculated from E or C_v if the molar volume (V), the coefficient of volume expansion (β) and isothermal bulk modulus (K_T) of the mineral are known, since:

$$C_p = C_v + \beta^2 TVK_T \quad \text{........(11)},$$

and

$$S = \int C_p T^{-1} dT \quad \text{........(12)}.$$

As discussed above, in the atomistic approach to the modelling of crystal structures, the adiabatic bulk modulus (K_S) of a material can be simply calculated from the elastic constants derived from the second derivatives of the interatomic potentials used in the simulation. The thermal expansion coefficient of a mineral is, however, somewhat more difficult to determine theoretically, but can best be calculated from the following relationship:

$$\beta = \gamma_{th} C_v / K_T V = \gamma_{th} C_p / K_S V \quad \text{........(13)},$$

where (γ_{th}) is the thermal Gruneisen parameter of the mineral. There is an extensive literature on the nature and definition of Gruneisen parameters (e.g. Born and Huang, 1954; Barron et al., 1980; Kieffer, 1982; Poirier and Liebermann, 1984; Barton and Stacey, 1985). However, in brief, in addition to the thermal Gruneisen parameter defined above, we can also define a related function, the mode Gruneisen parameter (γ_i), thus:

$$\gamma_i(q) = d\ln(w_i(q))/d\ln V \quad \text{.......(14)}.$$

The mode Gruneisen parameter is a measure of the change in vibrational frequency of a normal mode (w_i) as a function of the change in molar volume. As w is a function of q, there are in effect an infinite number of γ_i's for any one mineral. We can, however, define the average mode Gruneisen parameter ($\overline{\gamma}$), thus:

$$\overline{\gamma} = C_v^{-1} \sum C_{vi} \gamma_i \quad \text{.......(15)},$$

where C_{vi} is the contribution to the total heat capacity, C_v, made by each mode i. The vital link between the thermodynamic and lattice dynamic definitions of the Gruneisen parameter, can be made as a result of Slater's (1939) proof that, within the quasi-harmonic approximation (in which γ_i's are assumed to be volume independent), $\overline{\gamma}$ is equivalent to the thermal Gruneisen parameter, γ_{th}. Thus by using interatomic pair potentials to calculate not only the lattice vibrational frequencies at ambient pressures, but also at a variety of molar volumes (or pressures), it is possible to evaluate not only C_v, but if the quasi-harmonic approximation is assumed, the average mode Gruneisen parameter, the thermal expansion coefficient and all the other thermodynamic properties of a mineral can also be determined. In the following sections, we will describe how successful our potential model THB1 is in predicting the thermodynamic properties of Mg_2SiO_4 forsterite, and show how it can be used to predict the phase relations between the Mg_2SiO_4 polymorphs.

6. THE THERMODYNAMIC PROPERTIES OF FORSTERITE

The first stage in predicting the thermodynamic properties of forsterite involves calculating its phonon dispersion relations and the resulting density of states, G(w). Figure 4 shows the forsterite phonon dispersion relations predicted by potential THB1 for a circuit ΓXRΓ within the forsterite Brillouin zone (Figure 3). Each segment in the circuit (i.e ΓX, XR, etc) is plotted from the vibrational frequencies calculated at 35 equally spaced points along the line joining the endpoints of the segment. The calculation of the dispersion relations in each segment took approximately 3 minutes of cpu time on the University of London Cray 1s. To our knowledge there are no experimentally derived data on the phonon dispersion in forsterite, against which these predictions can be tested. Given the 84 (3 x 28 atoms per cell) modes to be found in forsterite and their relatively complex dispersion, it is hardly surprising that systematic inelastic neutron scattering studies of Mg_2SiO_4 have not been made. Thus, rather than spending a considerable amount of computing time calculating other unverifiable dispersion relations along further circuits in the forsterite Brillouin zone, it was felt that it would be more efficient to obtain a density of states function by calculating the vibrational behaviour of the forsterite lattice at a grid of points within its Brillouin zone.

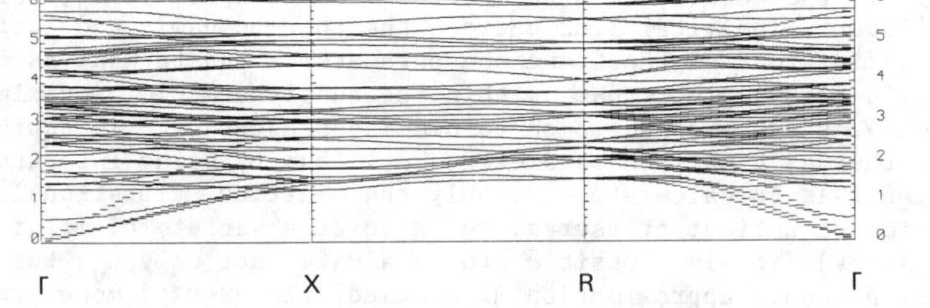

Figure 4. A calculated set of phonon dispersion curves for forsterite.

Considerable efforts have been made in the past to define an efficient way in which to obtain an accurate description of the density of states of a material from a limited sampling of its phonon dispersion spectrum (e.g. Baldereschi, 1972; Chadi and Cohen, 1973). Indeed, Baldereschi (1972) defined a series of 'mean-value points' for some high symmetry structures, at which the vibrational behaviour of the lattice represents an excellent approximation to the average behaviour when summed over the Brillouin zone as a whole. However, rather than confining our calculations to a single, mean-value point, we calculated the vibrational behaviour at a set of points in an MxMxM orthogonal grid (M = 1,2,3) within the forsterite Brillouin zone. The grid was centred at the Γ point and the axes of the grid coincided with the crystallographic axes of the lattice. The grid points were regularly spaced at $(0.5)^{M-1}$ reciprocal lattice units, but because of the symmetry of forsterite, vibrational frequencies only needed to be calculated at points, the coordinates (q_x,q_y,q_z) of which fell within the constraints $0 \leqslant q \leqslant 0.5$. Thus the first grid considered (M = 1) was the trivial case of a single point at the Brillouin zone centre (0,0,0); the second grid generated eight points with coordinates of the type (0,0,0), (0.5,0,0), (0.5,0.5,0), etc.; and the third grid considered contained 27 points with coordinates such as (0.25,0,0), (0.5,0.25,0), etc. The total number of modes evaluated for each grid was $84xM^3$, and the frequencies thus calculated were used to construct a histogram of the density of states of forsterite. The contribution of each point in the grid to the density of states function was weighted by the multiplicity of the star of lattice vectors to which that point belonged. The resulting normalized density of states for forsterite calculated from the 3x3x3 grid is shown in Figure 5. The most striking feature of this density of states plot is the $100cm^{-1}$ wide gap between the high frequency, internal Si-O stretching modes and the lower frequency lattice modes. In her development of a model density of states function for forsterite, Kieffer (e.g. 1985) found that a separation of internal and external modes, similar to the gap predicted by our potential, was most important if accurate descriptions of its thermodynamics of were to be obtained. Our predicted density of states is also vary similar to that recently reported by Rao et al. (1987). The heat capacities at constant volume, C_v, calculated using equation (10) and the density of states obtained from the data calculated on the three grids of reciprocal lattice points discussed above, are shown in Figure 6. The values of C_v calculated using G(w) functions evaluated from frequencies calculated for the 8 and 27 point grids are virtually identical, and are in excellent agreement with values inferred from experimental studies (see Kieffer, 1979 a, b, c and references therein). As might be expected, the heat capacity calculated when only the frequencies appropriate to the Γ point are used is in poorer agreement with that inferred from experiment, thus emphasising the need to consider dispersion effects when calculating thermodynamic properties. It appears, however, that an 8 point grid allows sufficient sampling of the Brillouin zone of a material like forsterite, to enable accurate calculations of its thermodynamic properties to be made. It is perhaps appropriate at this stage to point out that it is a significant

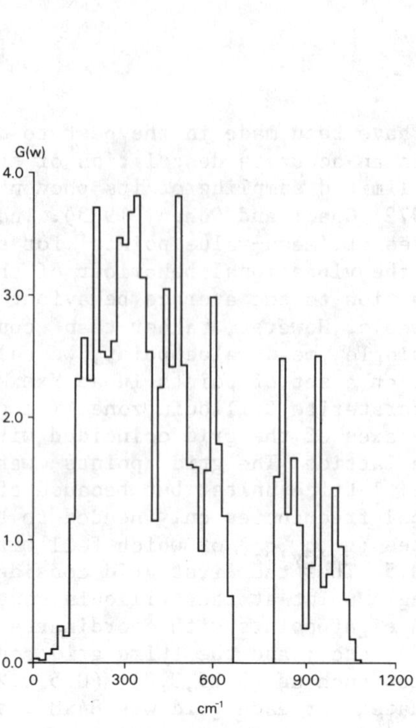

Figure 5. The THB1 predicted density of states for forsterite.

achievement for an atomistic model to be able to predict so accurately the basic thermodynamic data of a mineral as complex as forsterite, and to our knowledge THB1 is the first silicate potential model, determined independently from binary oxides, capable of being used in such a quantitative fashion. It should be noted, however, that although Kieffer's continuum model calculations (e.g. 1985 and references therein) were also able to simulate to a very large extent the heat capacity of forsterite and other minerals, their construction requires a significant input concerning the spectroscopic behaviour of the mineral of interest. In contrast, our model is truly predictive, and therefore represents a considerable advance in our ability to link microscopic and macroscopic phenomena.

In Figure 7, we present the potential-predicted, average mode Gruneisen parameter for forsterite, calculated as a function of temperature using equations (14) and (15). $\overline{\gamma}$ was calculated from the 672 γ_i's of the modes sampled on the 8 point grid, which were themselves obtained from the difference between the simulated vibrational behaviour of forsterite at 1bar and at 25kbars. It can be seen that the general nature of the predicted variation of with temperature is very similar to that recently determined experimentally by White et al. (1985). It appears, however, that, except at very low temperatures, our model produces a systematic underestimate of $\overline{\gamma}$, which becomes most marked at high temperature where the predicted value of 0.9 is almost 20% less than the observed value of 1.1. This underestimate of the high temperature Gruneisen parameter occurs because under these conditions the

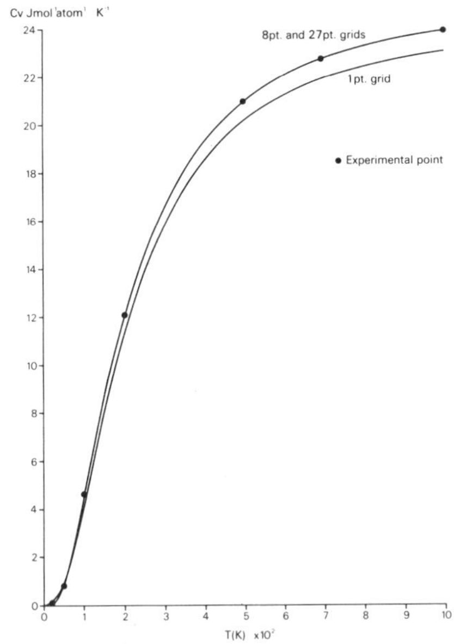

Figure 6. The THB1 predicted C_v for forsterite calculated using frequencies from 1, 8 and 27 point grids. Experimental data from Robie et al. (1982) and Berman and Brown (1985).

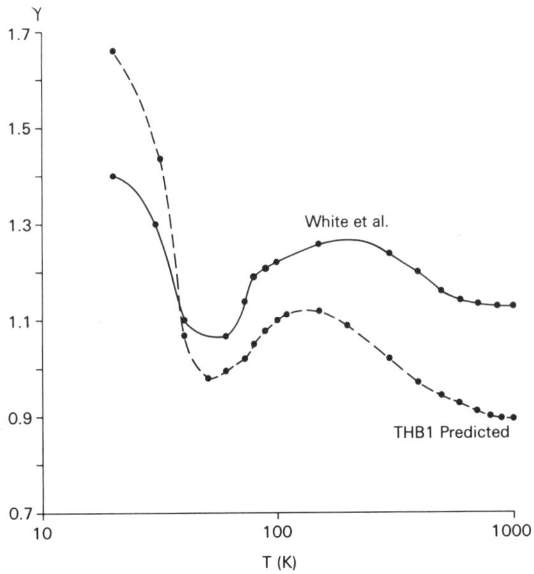

Figure 7. The predicted and observed mean Gruneisen parameter of forsterite as a function of temperature. The experimental data comes from White et al. (1985).

quasi-harmonic approximation does not hold, and the anharmonic effects, neglected in our analysis, should be include to obtain a full description of the lattice dynamics of a crystal.

Figure 8 shows the predicted coefficient of volume expansion for forsterite obtained from equation (13), using the previously calculated Gruneisen parameters and elastic constants (assumed to be temperature independent). It can be seen that although the experimentally observed (White et al., 1985) form, and the three orders of magnitude variation of the thermal expansion coefficient with temperature is well reproduced, the errors associated with the underestimated Gruneisen parameter propagate to give an underestimate in the thermal expansion coefficient of about 35% of the observed value. Nevertheless, within the limits of the quasi-harmonic approximation, a prediction of even this accuracy is still an outstandingly fine achievement for an atomistic model. In Table V, we present the C_p and entropy (S) data calculated using the above equations (11) and (12). The errors in γ and β, have very little effect on these major thermodynamic properties, and the predicted heat capacity and entropy data at temperatures above 100K are within 2% of experimental values (Robie et al., 1982), and even at 10K the agreement with experiment is excellent. We conclude, therefore, that even within the quasi-harmonic approximation the atomistic model THB1, although not perfect, can predict the thermodynamic properties of forsterite most satisfactorily. In the next section, we show how effectively this potential also models the properties of the high-density, spinel and beta-phase polymorphs of Mg_2SiO_4, and show how this approach can be used to describe the phase relations in the Mg_2SiO_4 system as a whole.

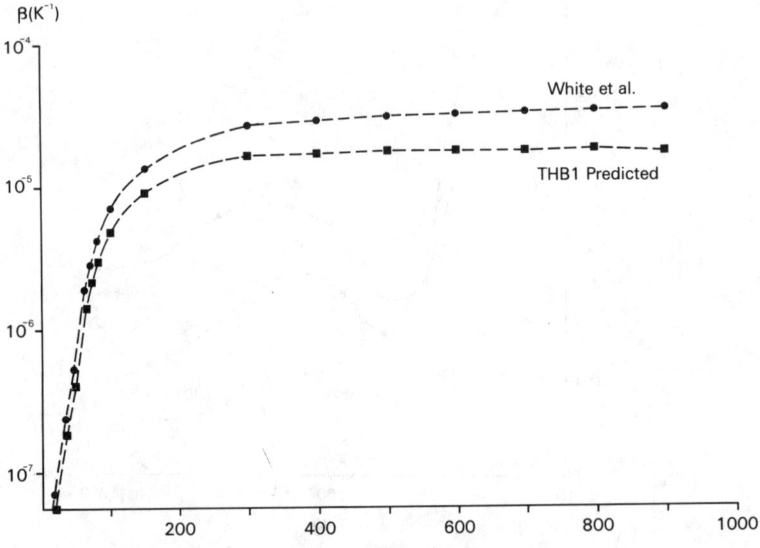

Figure 8. The observed and predicted thermal expansion coefficient of forsterite.

TABLE V

Predicted and observed C_p and S for forsterite

T	C_p Obs	C_p THB1	S Obs	S THB1
10	0.028	0.023	0.009	0.011
20	0.275	0.218	0.083	0.077
30	1.059	0.953	0.317	0.291
40	2.843	2.809	0.839	0.801
50	5.787	5.903	1.768	1.742
60	9.804	10.04	3.163	3.169
80	20.20	20.39	7.347	7.417
100	32.26	32.05	13.13	13.20
150	61.81	60.33	31.94	31.64
200	85.95	83.63	53.17	52.31
250	104.6	101.8	74.44	73.00
300	119.1	115.8	94.85	92.86
350	130.2	126.6	114.1	111.6
400	137.7	135.1	132.0	129.0
500	148.3	147.0	163.9	160.6
600	155.8	154.7	191.6	188.1
700	161.7	160.0	216.1	212.4
800	166.8	163.7	238.1	234.0
900	171.2	166.5	258.0	253.4
1000	175.1	168.9	276.2	272.9

Observed data from Robie et al.(1982). Units:T in K, C_p and S in $Jmol^{-1}K^{-1}$.

7. THE THERMODYNAMIC PROPERTIES OF THE Mg_2SiO_4 POLYMORPHS

Potential THB1 predicts the unit cell and elastic properties of the beta-phase and spinel structure polymorphs of Mg_2SiO_4 approximately as well as it it does those of forsterite (see Table II). The predicted spinel and beta-phase molar volumes are within 2% of those measured, and the calculated shear moduli are within 5% of those observed. The predicted bulk moduli are, however, up to 35% too stiff. Although known in general outline (e.g. Akaogi et al., 1984; Williams et al., 1986), the infrared and Raman behaviour of the high-density Mg_2SiO_4 polymorphs have not been studied to the same extent as the spectroscopic behaviour of forsterite. Thus, rather than presenting a detailed analysis of the Γ point vibrational behaviour of the spinel and beta-phase polymorphs, we chose to assess the success that potential THB1 has in modelling the lattice dynamics of these phases by comparing the observed thermodynamic properties of these phase with those predicted by our model. The phonon frequencies of the spinel and beta-phase structures were therefore evaluated at a set of points on a 2x2x2 grid, and their thermodynamic properties calculated using the approach discussed above.

Estimates of the thermal Gruneisen parameters of the beta-phase and spinel polymorphs are given by Watanabe (1982), while recently Williams et al. (1986) have measured the mode Gruneisen parameters for some of the infrared active modes of the beta-phase structure. Watanabe calculates the high temperature γ_{th}'s of the beta and spinel phases to be approximately 1.26 and 1.37 respectively. Williams et al. were unable to calculate an exact value for $\bar{\gamma}$ from their data, because they did not know how γ_i varied as a function of V, nor did they know the correct weightings to assign the individual γ_i's in calculating $\bar{\gamma}$. However, from their data they conclude that the $\bar{\gamma}$ for beta-Mg_2SiO_4 probably lies between 1.22 and 1.34. In comparison our predicted $\bar{\gamma}$ values of 0.90 and 0.88, for the beta-phase and spinel structures respectively, again appear to be underestimated. As was the case for forsterite, the origin for this underestimate stems from our neglect of anharmonic effects.

As was seen in the previous section, an underestimation of the magnitude of the Gruneisen parameter propagates to give a low calculated value of the thermal expansion coefficient. Thus, the predicted high temperature thermal expansion coefficients for the beta-phase and spinel polymorphs (Table II) are between 40 and 50% lower than those measured. These errors do not appear to propagate into the calculation of the heat capacity, however, and the calculated values of C_p as a function of temperature for these phases (Table VI) are in excellent agreement with those reported by Watanabe (1982), differing as they do by only a few percent. In addition, the calculated entropies of these phases are in very good agreement with those inferred by Watanabe (1982), with standard entropies (S_{298}) differing by no more than 5 $Jmol^{-1}K^{-1}$.

In order to assess the usefulness of the concept of the 'mean-value point' discussed above, the phonon vibrational behaviour of Mg_2SiO_4 spinel was evaluated at q = (0.311, 0.148, 0.0), the point in the Brillouin zone calculated by Baldereschi (1972) to be the 'mean-value point' for materials with a face-centred cubic, real-space lattice. The resulting values for C_p and S are also given in Table VI, and are virtually identical to those calculated using an 8-point grid, thereby giving considerable support to Baldereschi's original suggestion that accurate thermodynamic data can be obtained from limited, but highly selective sampling of the phonon dispersion spectra of a material.

Although the errors in the predicted entropies of the three Mg_2SiO_4 polymorphs may be small in comparison with their molar entropies, the magnitude of these errors may be significant when compared with the differences in the entropies (ΔS) between the three polymorphs. Any significant errors in the predicted values of ΔS will severely limit the usefulness of any thermodynamic calculations made using potential THB1. Good estimates exist for the ΔS's of the possible polymorphic phase transformations in the Mg_2SiO_4 system. These have been determined

TABLE VI

Predicted and observed C_p and S data for the beta-phase and spinel polymorphs of Mg_2SiO_4.

	Beta-Mg_2SiO_4			Spinel-Mg_2SiO_4				
T	Cp Obs	Cp 8pt	S grid	Cp Obs	Cp 8pt	S grid	Cp Mean	S point
10		0.005	0.002		0.000	0.000	0.001	0.001
20		0.179	0.050		0.013	0.003	0.155	0.040
30		0.749	0.219		0.244	0.047	0.679	0.192
40		1.984	0.592		1.141	0.231	1.667	0.513
50		4.146	1.255		2.966	0.669	3.370	1.060
60		7.299	2.277		5.740	1.445	5.945	1.892
80		16.09	5.511		13.71	4.118	13.62	4.581
100		26.92	10.23		23.94	8.242	23.76	8.674
150		55.44	26.58		52.28	23.29	52.19	23.66
200		80.02	46.01		77.62	41.91	77.62	42.27
250		99.34	66.03		97.75	61.48	97.79	61.85
300	110.8	114.1	85.52	107.7	113.2	80.73	113.2	81.11
350	124.7	125.6	104.0	122.5	124.9	99.10	125.0	99.49
400	134.1	134.3	121.4	132.4	133.9	116.4	134.0	116.8
500	146.1	146.5	152.8	144.6	146.4	147.7	146.5	148.2
600	153.7	154.3	180.2	151.9	154.2	175.1	154.3	175.6
700	159.2	159.5	204.4	156.9	159.5	199.4	159.6	199.8
800	163.6	163.2	225.9	160.6	163.1	220.9	163.2	221.4
900	167.3	165.9	245.3	163.6	165.8	240.3	165.9	240.8
1000	170.6	168.1	264.8	166.2	167.8	259.8	167.9	259.9

Observed data from Watanabe (1982). Units: C_p and S in $Jmol^{-1}K^{-1}$.

calorimetrically by Akaogi et al. (1984), and from accurate phase relation studies by, for example, Sawamoto (1986). In Table VII, we compare our predicted values of ΔS_{1000} with a number of experimentally determined or inferred values. These predicted values of ΔS_{1000} are in outstandingly good agreement with those obtained experimentally, and lie well within the spread of experimental data.

Our predicted values of ΔS can be combined with our predicted structural data to obtain values for $\Delta S/\Delta V$, or the Clausius-Clapeyron slopes of the phase boundaries in the Mg_2SiO_4 system. A similar exercise was been carried out by Akaogi et al. (1984) to allow them to compare the results of their calorimetric and Kieffer model studies with experiment. In order to be able to compare our predictions with their data, we calculated $\Delta S/\Delta V$ at T = 1000K and P = 140kbars. The high pressure, high temperature molar volumes of the polymorphs were estimated using a Murnaghan equation of the type:

TABLE VII

Predicted and observed values of ΔS_{1000} and $\Delta S/\Delta V$ for phase transitions between the Mg_2SiO_4 polymorphs.

	ΔS_{1000}	$\Delta S/\Delta V_{1000,140}$
Olivine/Beta-phase		
THB1	-8.1	2.7
Akaogi et al.	-10.5 +/-2.0	3.3+/-0.7
Kawada		6.5
Suito		3.5
Watanabe	-13.0 +/-2.0	4.1+/-0.7
Beta-phase/Spinel		
THB1	-5.0	4.5
Akaogi et al.	-6.3 +/-3.7	7.0+/-2.8
Kawada		10.0
Ohtani	-4.4	4.3
Sawamoto	-6.1 +/-0.4	6.1+/-0.5
Suito	-5.5	5.5
Watanabe	-3.8 +/-1.6	8.9+/-1.2
Olivine/Spinel		
THB1	-13.1	2.9
Akaogi et al.	-16.7 +/-3.0	4.1+/-0.8
Watanabe	-16.6 +/-2.0	4.0+/-0.5

Units: ΔS in $Jmol^{-1}K^{-1}$ and $\Delta S/\Delta V$ in $MPaK^{-1}$.

$$V_{T,P} = V_{0,0}(1 + \beta T)(PK'/K + 1)^{-1/K'} \quad ....(16),$$

in which we used the commonly accepted value of K' = 4, but THB1 predicted values for $V_{0,0}$, β and K (Table II). The resulting predicted Clausius-Clapeyron slopes are given in Table VII, along with those found by direct, petrological experiment or by calorimetry. The best experimentally studied boundary in this system is that between the beta-phase and spinel polymorphs (see for example Sawamoto, 1986), for which reliable experimental estimates of $\Delta S/\Delta V$ range from 0.043 to 0.061 kbar/K. In comparison our potential model THB1 predicts a value of 0.045 kbar/K. A somewhat wider range of values is reported in the literature for the slope of the forsterite/beta-phase boundary. Estimates range from between 0.025 kbar/K (Akaogi et al., 1984) and 0.068 kbar/K (Kawada, 1979), against which our predicted value of 0.027 again compares well. The accuracy of our predictions may be somewhat fortuitous, but nevertheless the fact that an atomistic model can be used to obtain such accurate predictions of these subtle, yet vital thermodynamic relationships must be considered a major achievement.

If we make the reasonable assumption that our predicted Clausius-Clapeyron slopes are not strongly pressure and temperature dependent, then all that is required to generate the predicted phase diagram for the Mg_2SiO_4 system is the determination of two fixed points in PT space, at which the forsterite and beta-phase, and the beta-phase and spinel polymorphs have equal free energies. The THB1 predicted free energies of these phases were calculated at T = OK, using THBREL to calculated the static lattice energies of the three polymorphs at a variety of simulated pressures, and THBPHON to calculate the their zero point energies. The phase diagram which can be derived using the predicted, zero-temperature transformation pressures and the calculated Clausius-Clapeyron slopes is shown in Figure 9.

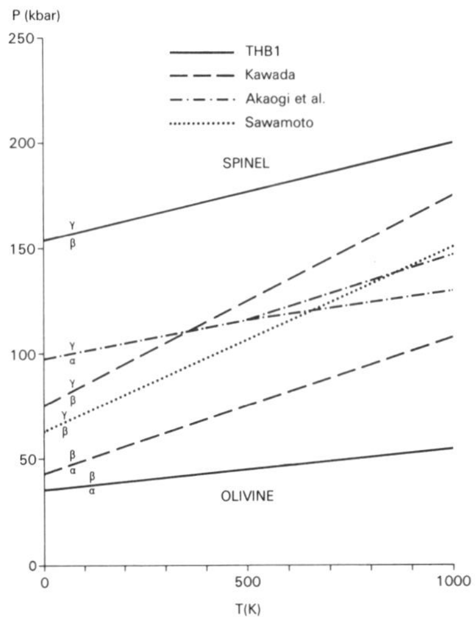

Figure 9. The predicted and observed phase boundaries in the Mg_2SiO_4 system.

The zero-temperature transformation pressure for the forsterite to beta-phase reaction has been inferred to occur at between 43 and 98 kbars (see Akaogi et al., 1984), with greater credibility being given to the upper value. Sawamoto (1986) recently reported experiments which suggest that that the beta-phase to spinel transformation would occur at 79 kbars at T = OK, thereby supporting the inference that a triple point exists in this system at finite pressures and temperatures (Akaogi et al., 1984). The THB1 predicted transformation pressures of 35 and 155 kbar, appear to be about 45 kbars too low and 70 kbars too high respectively. This poor agreement with the experimental data is disappointing after the previous successes of the potential, but the source of the error can be identified as resulting from the overestimate of the magnitude of the static lattice energy of the beta-phase. Indeed, the accuracy of our predicted phase diagram is extremely sensitive to the lattice energy differences between the polymorphs, and a 0.1% reduction in the calculated lattice energy of the beta-phase polymorph is all that is required to remove the 45-70kbars errors in our predicted transformation pressures. As discussed by Price et al. (1985), the beta-phase is considerably more difficult to model than spinel or olivine, because of the presence in its structure of Si_2O_7 groups. It is probable that in order to model the beta-phase structure satisfactorily it will be necessary to include Si-O-Si bond bending terms. The inclusion of such terms could easily bring about the necessary 0.1% change in the beta-phase lattice energy, and thereby produce a significantly more accurate prediction of the transformation pressures in this system.

8. CONCLUSION

Despite the few short comings of the potential THB1 discussed above, it must be stressed that it is by far the best microscopic or atomistic model of magnesium silicates ever developed. Its thermodynamic predictions are at least as successful as any Kieffer model used to describe the properties of the Mg_2SiO_4 polymorphs (Akaogi et al., 1984), and unlike these Kieffer models, no previous knowledge of the lattice dynamics of the phases of interest were used in its development. But however successful, there is always scope for improvement in any potential model, and we believe that in the future it will be necessary to pay more attention to the effect of many body interactions upon the properties and behaviour of silicates. It will be necessary to consider not only Si-O-Si interactions, but also those involving Mg. Preliminary studies by Weidner and Price (1987) have already shown that the non-central forces associated with the Mg-O octahedra that are responsible for the Cauchy violations in MgO, are also at work in forsterite, and that if we wish to be able to model the detailed behaviour of magnesium silicates it will be necessary to include such terms in our potential models. As well as improving our potential models, however, it may also be necessary to increase the sophistication of our lattice dynamical studies. Thus, for example, it will be necessary to include the effects of anharmonicity upon our calculations, and it is desirable that we should develop techniques to simulate the effect of temperature, as well as pressure on the structure and properties of the phases that we model.

The approach that we adopt in the modelling of silicates could possibly be criticised on the grounds that our potentials are largely derived empirically. Such criticisms may be based on aesthetic arguments, or on the more acceptable grounds that it is dangerous to used empirically derived potentials to model systems under conditions of pressure, temperature or interatomic separation not sampled in the original fitting of the potential parameters. This latter argument would carry more weight, however, if there were an acceptable, theory-based alternative method of developing interatomic potentials or calculating interatomic forces in structures as complicated as those in the Mg_2SiO_4 system. Progress is undoubtedly being made in deriving interatomic potentials from quantum mechanical calculations on silicate clusters (e.g. Lasaga and Gibbs, 1987), and similarly the modelling of silicates by using modified electron gas (MEG) calculations has been greatly advanced by recent workers such as Hemley et al. (1987). However, even these most recent MEG studies are not capable of successfully modelling the lattice dynamics of silicates, and have a tendency to over estimate the vibrational frequencies of lattice modes by several hundred wavenumbers (e.g. Hemley et al., 1987). As a result, these MEG calculations can, as yet, play only a minor role in the study of the thermodynamic behaviour of silicates. The limitations of ab initio, quantum mechanical or MEG approaches to the modelling of silicates will, no doubt, be overcome in the future, particularly with the advent of the next generation of super-computer. However, in the meantime, we believe that atomistic models such as ours will play an important role in the study of both the perfect lattice and defect properties of silicates.

9. ACKNOWLEDGEMENTS

The authors would like to thank Richard Catlow and Maurice Leslie for their help and guidance. GDP would like to thank the Royal Society for the receipt of a 1983 University Research Fellowship.

10.REFERENCES

Akaogi M, Ross NL, McMillan P, Navrotsky A (1984) Amer Mineral 69: 499-512.
Baldereschi A (1973) Phys Rev B7: 5212-5215.
Barron THK, Collins JG, White GK (1980) Advances in Physics 20: 609-730.
Barton MA, Stacey FD (1985) Phys Earth Planet Interiors 39: 167-177.
Berman RG, Brown TH (1985) Cont Mineral Petrol 89: 168-183.
Born M, Huang K (1954) Dynamical theory of crystal lattices. Clarendon Press, Oxford.
Catlow CRA, Mackrodt WC (1982) In : Computer simulation of Solids. Lecture notes in Physics 166: 3-20 (Editors: Catlow CRA and Mackrodt WC). Springer-Verlag, Berlin.
Catlow CRA, Doherty M, Price GD, Sanders MJ, Parker SC (1986) Mat Sci Forum 7: 163-176.
Chadi DJ, Cohen ML (1973) Phys Rev B8: 5747-5753.
Cochran W (1973) The dynamics of atoms in crystals. Edward Arnold, London.
Devarajan V, Funck E (1974) J Chem Phys 62: 3406-3411.
Ghose S (1985) In: Microscopic to Macroscopic. Reviews in Mineralogy 14: 127-163. (Editors: Kieffer SW and Navrotsky A). Mineralogical Soc America, Washington.

Hemley RJ, Jackson MD, Gordon RG (1987) Phys Chem Minerals 14: 2-12.
Iishi K (1978) Amer Mineral 63: 1198-1208.
Kawada K (1977) PhD Thesis, University of Tokyo.
Kieffer SW (1979a) Rev Geophys Space Phys 17: 1-19.
Kieffer SW (1979b) Rev Geophys Space Phys 17: 20-34.
Kieffer SW (1979c) Rev Geophys Space Phys 17: 35-58.
Kieffer SW (1982) Rev Geophys Space Phys 20: 827-849.
Kieffer SW (1985) In: Microscopic to Macroscopic. Reviews in Mineralogy 14: 65-126. (Editors: Kieffer SW and Navrotsky A). Mineralogical Soc America, Washington.
Lasage A, Gibbs JV (1987) Phys Chem Minerals 14: 107-117.
Lazarev AN (1974) In: The infrared spectra of Minerals (Editor: Farmer VC). Mineralogical Society, London.
Matsui M, Busing WR (1984) Amer Mineral 69: 1090-1095.
McMillan P (1985) In: Microscopic to Macroscopic. Reviews in Mineralogy 14: 9-64. (Editors: Kieffer SW and Navrotsky A). Mineralogical Soc America, Washington.
Ohtani E (1981) J Phys Earth 27: 189-208.
Paques-Ledent MT, Tarte P (1973) Spectrochim Acta A29: 1007-1016.
Poirier JP, Liebermann RC (1984) Phys Earth Planet Interiors 35: 283-293.
Price GD, Parker SC (1984) Phys Chem Minerals 10: 209-216.
Price GD, Parker SC, Leslie M (1987a) Mineral Mag 51: 157-170.
Price GD, Parker SC, Leslie M (1987b) Phys Chem Minerals: In Press.
Price GD, Parker SC, Yeomans J (1985) Acta Cryst B41: 231-239.
Rao KR, Chaplot SL, Choudhury N, Ghose S, Price DL (1987) Science 236: 64-65.
Robie RA, Hemingway BS, Takei H (1982) Amer Mineral 67: 470-482.
Salje E, Werneke Ch (1982) In: High-pressure researches in geosciences (Editor: Schreyer W). E. Schweizerbart'sche Verlagsbuchhandlung, Stuttgart.
Sanders MJ, Leslie M, Catlow CRA (1984) J Chem Soc Chem Comm 1271-1273.
Sawamoto H (1986) Phys Chem Minerals 13: 1-10.
Servoin JL, Piriou B (1973) Phys Status Solidi B55: 677-686.
Slater JC (1939) Introduction to chemical physics. McGraw-Hill, NewYork.
Suito K (1977) In: High-pressure research applications to geophysics (Editors: Manghnani MH and Akimoto S). Academic Press, New York.
Watanabe H (1982) In: High pressure research in geophysics (Editors: Akimoto S and Manghnani MH). Centre for Academic Publications, Tokyo.
Weidner DJ, Price GD (1987) Phys Chem Minerals: In Press.
Weidner DJ, Sawamoto H, Sasaki S (1984) J Geophys Res 89: 7852-7860.
White GK, Roberts RB, Collins JG (1985) High Temp High Press 17: 61-66.
Williams Q, Jeanloz R, Akaogi M (1986) Phys Chem Minerals 13: 141-145.
Ziman JM (1964) The principles of the theory of solids. CUP, Cambridge.

COMPUTER MODELLING OF SILICATES

C.R.A. Catlow,
Department of Chemistry,
University of Keele,
Keele,
STAFFS, ST5 5BG,
U.K.

ABSTRACT. We describe recent work on the modelling of SiO_2 and silicates using empirically derived effective potentials. We discuss briefly questions of methodology and review the different types of potential models that have been investigated. The greatest success appears to be enjoyed by shell model potentials which incorporate a 'bond-bending' term. Applications to the modelling of structural properties of zeolites, and to sorption in these materials are described; and we conclude with a summary of recent work on the defect structures of Mg_2SiO_4.

1. INTRODUCTION

The application of modelling methods to minerals has developed and expanded rapidly in the last five years. In this article we discuss the present status of studies using 'effective potentials'. We have reviewed the field briefly on previous occasions (Catlow and Parker, 1985; Catlow et al., 1985a); and in the present article we concentrate on the latest developments. We hope that this chapter and that of Price show that modelling methods can now be applied to most classes of mineral structure and to a wide range of problems in mineral physics.

2. AIMS AND METHODOLOGY

Simulation studies start from the specification of an interatomic potential model which essentially expresses the potential energy of the system as a function of atomic coordinates and may include a representation of electronic polarisation. The model is usually broken down into pair potential and many-body terms as discussed below. Having specified the model, we then aim to calculate, or more ambitiously to predict, the following:

A Structures,	i.e. the equilibrium cell dimensions and atomic coordinates of unit cells. In more restricted studies the former may be held fixed and only the latter allowed to vary.
B Properties,	in which we include elastic, dielectric, piezoelectric and lattice dynamical properties.
C Defects,	i.e. the energies and entropies of defect formation, migration and interaction.

E. K. H. Salje (ed.), Physical Properties and Thermodynamic Behaviour of Minerals, 619–638.

D Dynamics, in which we are concerned with the detailed mechanisms for ion migration which effect conductivity and diffusion in solids.

The justification of such studies must lie ultimately in their predictive value, that is in that ability to model correctly the behaviour of new materials and the behaviour of materials under extreme conditions of temperature and pressure – a topic of obvious importance to the mineralogist.

The techniques used, which have been amply reviewed elsewhere (Catlow and Mackrodt, 1982; Catlow, 1986) can be classified into three types.

(i) Energy minimisation studies of perfect structures. These methods entail (i) lattice summations (employing the Ewald technique (Tosi, 1964) for the Coulomb terms) (ii) energy minimisation, i.e. adjustment of unit cell coordinates and cell dimensions until the minimum energy structure has been generated. (iii) calculation of the dynamical matrix, (i.e. the matrix of 2nd derivatives of energy with respect to atomic coordinates) and of the derivatives of the energy with respect to strains, from which the range of properties listed in (B) above can be calculated. Such calculation can now be undertaken in a rapid and automated manner for large complex unit-cells containing several hundred atoms or ions. A number of computer codes are available including our own METAPOCS code and the WMIN program (Busing, 1979).

Important recent progress has been made by Parker (1987) who has developed free energy minimisation methods. In these techniques, entropies are obtained from the calculated dispersion curves by sampling out the Brilloiun zone; these are then combined with the lattice energy term to give the free energy, which is then minimised with respect to structural variables. An automated code, PARAPOCS, is available for such calculations which are discussed in greater detail in the chapter of Price.

(ii) Molecular Dynamics Simulations, which include kinetic energy explicitly in contrast to the 'static' simulation techniques described above. A detailed discussion is given in the chapter of Dove in this book. The procedure is however, simple. An ensemble of particles is specified to which periodic boundary conditions are applied; a start-up configuration is generated by assigning all particles positions and velocities; the equations of motion of the system are then solved in an iterative manner by specifying a time step, t, (which must be less than the characteristic time of any important process, e.g. an atomic vibration, in the system): the time steps are successively applied and after the elapse of each, coordinates and velocities are updated according to the classical equations of motion. In the early stages of the simulation the system equilibrates, i.e. attains a Maxwellian distribution of velocities and equipartition between potential and kinetic energies. During this stage, velocities may be periodically 'scaled' in order to achieve the target temperature. After equilibration has been achieved, the simulation is run usually for several thousand time steps, and the coordinates and velocities of each step are stored; they provide a record of the time evolution of the systems.

The technique is computationally expensive; runs with several hundred particles may require several hours on a modern supercomputer. Moreover if the effects of the electronic polarisation of atoms or ions are included in the potential the time required becomes prohibitively large, which is a serious weakness of the technique. The method is, however, very powerful as it provides information directly on structural and dynamical properties of the system. Diffusion coefficients may be obtained as may the mechanisms effecting migration. A whole range of correlation functions (of which the simplest is the radial distribution function) may be calculated and compared with experiment. Most applications to silicates have concerned the modelling of glasses (see e.g. the contribution of Angell (1987) to these proceedings) and silicate melts. The technique does, however, have a major rôle to play in the study of higher temperature crystalline silicates, especially those where high rates of diffusion might be expected.

C Mott–Littleton Calculations of Defect Energies

The basis of the Mott–Littleton (1938) procedure, which has been known for nearly fifty years, is the embedding of the defect in a region of crystal which in modern calculations contains typically 100–500 atoms and which is treated atomistically by adjusting all atomic coordinates (and dipoles) until all atoms are at zero force. This atomistic treatment must be used in the region close to the defect, where the defect forces are strong, if the relaxation of the lattice around the defect is to be correctly modelled. In contrast the more distant, weak field regions outside the inner region may be modelled by a quasi–continuum approach which calculates the polarisation, P(r), at a distanced r from the defect, according to formula:

$$\underline{P}(r) = \frac{1}{4\pi} \frac{q\underline{r}}{r^3} (1-\epsilon_0^{-1}), \qquad .....(1)$$

where q is the effective charge of the defect and ϵ_0 the static dielectric constant. In practice the two regions (I and II) are interfaced by a region IIa.

There are now several accounts (Catlow et al., 1982; Mackrodt, 1984; Catlow, 1986) of the methodology and achievement of defect calculations, which have been widely and successfully applied to an extensive range of oxide and halide crystals. This work has established the reliability of the techniques, given good interatomic potentials. The techniques are 'static' in the sense that kinetic energy is not included. For the majority of defect processes this is not an important restriction; but for materials with very high atomic or ionic mobilities e.g. 'superionic conductors', it is a serious omission and recourse must be made to the dynamical techniques discussed above.

Recently, it has become possible to calculate defect entropy as well as energy terms. Configurational contributions to the entropy arising from orientational or site degeneracy are easily evaluated. More difficult is the vibrational term whose origin is the perturbation of the phonon frequencies by the defect. This may be written as:

$$S_{vib} = -k\ln \left[\prod_{i=1}^{3N'} \omega'_i \Big/ \prod_{i=1}^{3N} \omega i \right] + 3K(N'-N)(1-\ln\hbar KT) \qquad ......(2)$$

where the primes indicate perturbed frequencies. Gillan, Jacobs and Harding have recently developed embedded crystallite methods which consider the perturbation of the frequencies in a region surrounding the defect (see Gillan & Jacobs, 1983; Harding, 1985). Alternative super–cell procedures will be discussed below.

Defect calculations can examine migration as well as formation processes, provided migration is effected by a jump mechanism. We have discussed elsewhere (Catlow, 1983; Catlow, 1986) the meaning of the latter term and the conditions under which it is an appropriate model, (which is the case for most materials). The absolute rate theory approach of Vineyard (1957) may then be applied, i.e. the frequency, ν, of defect jumps is written as:

$$\nu = \nu_o \exp(-G_{ACT}/KT) \quad \ldots\ldots\ldots(3)$$

where G_{ACT} is the free energy of activation. i.e. the difference between the free energy of the 'ground state' of the defect and the saddle point for the migration process. Once the latter has been identified, the component enthalpy and entropy terms can be calculated by the techniques outlined above.

We should note that the calculations we have just considered refer to constant volume processes; however, the corresponding constant pressure quantities are readily evaluated (Jacobs, 1983; Corish et al., 1981). In addition, the calculations all refer to an isolated defect (or defect cluster) embedded in a perfect lattice. An alternative approach is to use a periodically repeating array of defects or defect super-cell, which is then treated by the perfect lattice techniques described above. The resulting calculated energies or entropies will of course include a term arising from defect interactions, which will, however, be small if the super-cell is sufficiently large. The technique is receiving considerable attention in entropy calculations following early work of Harding and Stoneham (1981).

This completes our brief account of the methodology of current simulations. We now turn our attention to the fundamental input to the simulations, i.e. the potential model.

3. INTERATOMIC POTENTIALS

Most simulation studies of inorganic solids have the following features in common:

(a) The use of Born model description of cohesion employing ions with formal or partial changes.

(b) The description of the short-range terms due to overlap repulsion and dispersion using pair-potentials which may be numerical or analytical; in the latter case, the well-known Buckingham potential is commonly used, i.e. we write:

$$V(r) = Ae^{-r/\rho} - Cr^{-6} \quad \ldots\ldots.(4)$$

(c) The inclusion in some potentials of a shell-model (Dick and Overhauser, 1958) description of polarisability which describes the polarisable ion in terms of a core coupled to a mass-less shell by an harmonic spring. The shell simulates the polarisable valence shell electrons, and displacement of the shell from the core describes the development of a dipole moment

(d) In our recent model of SiO_2, discussed in greater detail below, we have found it necessary to include bond-bending terms about the O-Si-O bond angles, i.e. we add an extra energy term, $E(\theta)$:

$$E(\theta) = \frac{1}{2} K_B(\theta-\theta_o)^2, \quad \ldots\ldots.(5)$$

where θ_o is the tetrahedral angle. The term is intended to describe the tendency of the covalent bond to retain 'tetrahedrality' of the SiO_4 groups.

Potential models need, of course, to be parameterised. There are two broad strategies for achieving this. First, parameters can be adjusted via least squares fitting routines in order to achieve the best possible agreement between calculated and observed properties. This empirical procedure, has been widely used by ourselves and others; it has the inherent limitations first of requiring that suitable model compounds are available and secondly of giving information on the potential only at interatomic spacings corresponding to those observed in the model compounds. These objections are overcome by theoretical methods for calculating potentials directly. Both electron-gas and ab-initio methods have been used: the former by Gordon (1987) have been used in developing a potential model for SiO_2 and the latter by Lasaga and Gibbs (1987), as will be discussed below. Theoretical and empirical methods of parameterisation often are, of course, complementary and where possible both techniques should be used.

3.1 Potential models for Silicates.

Several groups have in recent years studied the development of Born model potentials for silicates, and a recent review is available from Burnham (1985). In the present chapter we will concentrate our own recent attempts to develop empirical potential models for silicates which are instructive in the way they reveal the strengths and limitations of different approaches. We can identify three distinct phases in the development of our empirical models: first rigid ion models with formal charges; second, the investigation of partial charges; third the inclusion of bond-bending terms and a shell model description of polarisability.

A Rigid-Ion, Formal-Charge Models

Models of this type were developed by Parker (1982) and Parker et al (1984) by fitting to structural properties of a range of ortho- and meta-silicates. The models were then investigated in energy minimisation studies of structures outside the range used in the parameterisation. A number of structures were simulated with reasonable accuracy. However, it became clear that such models have severe limitations. First, for a number of cases, (e.g. Mg_2SiO_4) inaccurate bond lengths are calculated; secondly, the physical properties (e.g. the dielectric and elastic constants) predicted by such potentials often gave poor agreement with experiment; thirdly the potentials fail totally when applied to layer and framework structured silicates. Nevertheless, the development of these models was of value as it showed that Born model potentials could be used in simulating structural properties at least of certain types of silicate.

B Partial Charge Models

These were thoroughly investigated in an important paper of Price and Parker (1984) who studied models in which the effective charge on Si varied from ~1.4 to 2.1. They found that such models were capable of giving considerable improvements both in calculated bond-lengths (which for the case of Mg_2SiO_4 were correct within 0.004 A) and physical properties. The models are, however, still unsuitable for framework structured silicates.

C Bond-Bending, Shell Model Potentials

The failure of the two-body potentials to model framework structured silicates provided the incentive to study the bond-bending terms described in section (3) above. We also included a shell model treatment of polarisation on the oxygen atoms. Formal charges were employed. To parameterise the model we fitted to the structural, elastic and dielectric properties of α–SiO_2. Potential parameters are reported in table (1a) and a comparison of experimental and fitted values given in table (1b), which also reports the best fits for a simple two–body model.

Table (1)

Table 1a
Potential parameters and oxygen core/shell displacements

Parameter	Si-O potential	O-O potential
A/eV	1283.9	22764.3
ρ/Å	0.3205	0.149
C/(eV/Å^6)	10.6616	27.88
Oxygen core charge/|e|	+0.8482	
Oxygen shell charge/|e|	-2.8482	
Core/shell spring constant/(eV/Å^2)	74.9204	
X Shell displacement/Å*	+0.0997	
Y Shell displacement/Å	-0.0302	
Z Shell displacement/Å	-0.0666	
Bond-bending constants/ (eV/radians^2)	2.097	

*components of shell displacements relative to core

Table 1b
Observed and calculated crystal properties of α-quartz

	Experimental	Bond bending potential	2-body potential
Elastic constants (10^{11} dyn cm^{-2})			
C_{11}	8.683	8.815	6.204
C_{33}	10.498	10.605	7.466
C_{44}	5.826	5.296	3.301
C_{66}	3.987	4.269	2.737
C_{14}	-1.8064	-1.666	-1.012
C_{13}	1.193	1.151	1.629
Static dielectric constants			
E_{11}	4.520	4.452	5.513
E_{33}	4.640	4.812	6.086
High frequency dielectric constants			
	2.4	2.04	2.069

It is clear from the latter table that inclusion of the bond–bending term improves considerably the performance of the model in simulating the properties of α–SiO_2. How well does the model perform, however, when applied to properties outside the data range used in the parameterisation? Three tests have been applied to date.

First phonon dispersion curves were calculated and compared with experiment (Dorner et al., 1980). The agreement is reasonable as shown in figure (1) for the ($\xi\xi$O) branches. Secondly we studied the extent to which our calculations could reproduce the variation of the Si–O–Si bond angle with pressure; the calculations imposed the cell dimensions measured for each pressure. The calculated Si–O–Si angle is compared with experiment in table (2). The agreement is good. The third and possibly most demanding test was to examine the structures of a number of polymorphs of SiO_2, namely cristobalite, coesite and tridymite. We showed, by energy minimisation techniques that these were in equilibrium with structures very close to the observed ones, with cell dimensions correct within 1% and atomic coordinates within 0.1 Å.

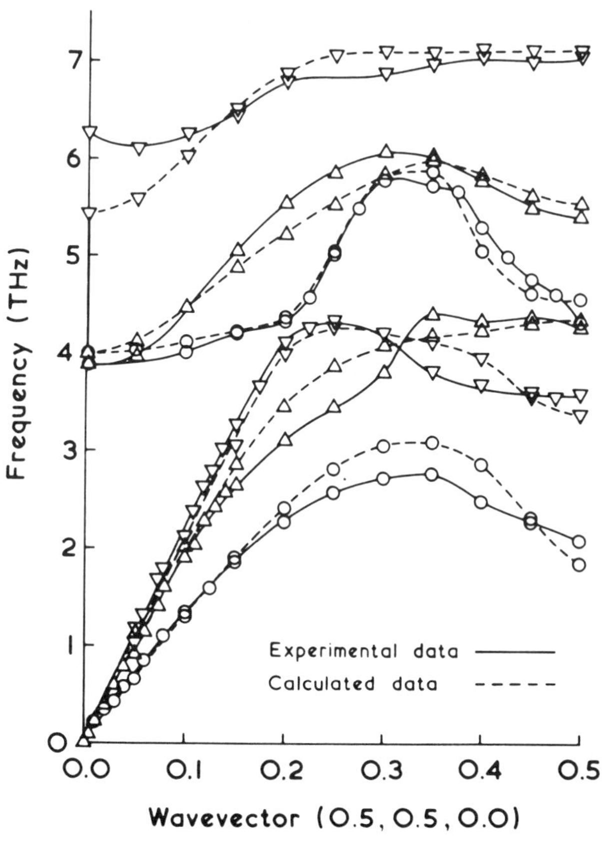

(1) Phonon Dispersion Curves for α–SiO_2.

Table (2)

Observed and calculated change of Si-O-Si angle with pressure

Pressure (kbar)	Experimental Si–O–Si angle (°)	Calculated Si–O–Si angle (°)
0.0	144.1	143.9
10.6	141.5	141.2
21.1	139.4	139.1
28.2	138.7	137.9

The results (which are discussed further by Catlow et al. 1985b) are clearly encouraging and suggest that framework structured silicates may be accurately modelled using effective potentials. How generally applicable are the potentials? Recent work of Parker (1987) has shown that both the olivine and spinel polymorphs of Mg_2SiO_4 may be reliably modelled by potentials which use the parameters in table (1a) together with Mg...O potentials developed from the study of crystalline MgO. Results are summarised in table (3). Preliminary studies of layer structured silicates including clays are also emerging. Section (4) will consider applications to zeolites, which have complex porous framework structures.

Table (3)

Modelling of Mg_2SiO_4 for both olivine and spinel structures

		Experiment	2-body fully ionic	2-body partially ionic	bond bending
Average Si...O Distance (Å)	olivine	1.63	1.48	1.63	1.64
	spinel	1.66	1.58	1.66	1.66
Unit cell Volume (Å^3	olivine	289	288	296	294
	spinel	263	258	262	258
Bulk Modulus (Mbar)	olivine	1.4	1.6	1.5	1.6
	spinel	2.1	2.1	2.1	2.5
Static Dielectric Constant	olivine	6	17	4	6
	spinel	-	-	-	-

At this stage we should consider the widely debated question of ionicity in silicates. First we should emphasise that the use of 'ionic model' potentials does not imply the assumption that the valence shell electron distribution corresponds to the fully ionic model (although the latter distribution is a very uncertain quantity as discussed elsewhere in this conference). It merely implies that interaction between atoms can be modelled by a potential comprising a long range R^{-1} term plus a short range repulsive term. Ionic model potentials do imply that the dipoles due to atomic displacements of δr are given by $q\delta r$ but the validity of this assumption is tested at last partially by our ability to model dielectric properties. We have discussed these and other points related to ionicity in solids in greater detail elsewhere (Catlow and Stoneham, 1983).

As noted earlier, the inherent limitations of the empirical fitting procedures provide a powerful incentive for developing theoretical methods for calculating interatomic potentials. In this context we should note the recent work of Lasaga and Gibbs (1987) who used Hartree–Fock calculations on both the $Si(OH)_4$ and $Si_2O(OH)_6$ molecules in order to derive interatomic potentials for silicates. Several calculations were performed in which the geometry of the molecules is varied: the calculated energies are then fitted to an interatomic potential function. The results of the calculations are encouraging. The resulting calculated Si....O potential, shows a minimum close to 1.6Å which agrees well with the normally observed Si...O bond length in silicates. Moreover, on comparing the results of the calculation with the empirically derived Si...O potentials a reasonable measure of agreement is obtained. Further calculations of this type using larger silicate molecules or clusters would clearly be desirable.

Finally we should emphasise that potential models whether derived empirically or from theoretical methods, should be tested by reference to as wide as possible a range of properties. The ability to model structure alone is not sufficient: a good, general potential should be able to model elastic, dielectric and lattice dynamical properties as well. In some cases, this may not be possible with presently available models. An example is the rutile phase of TiO_2 for which it does not seem possible to model both structural and dielectric properties (Catlow et al., 1985b; Mostoller and Wang, 1985). In such cases, more sophisticated models are required.

4. APPLICATIONS

In this final section we draw attention to some of the more notable recent applications of the simulation methods. The chapter of Price in this book describes the studies of the thermodynamics of mantle minerals based on simulation methods using the potentials described above. Here we concentrate on recent applications first to structure and sorption in zeolites and secondly to the defect properties of magnesium olivine. These studies illustrate the range and complexity of problems that can now be investigated using simulation techniques.

4.1 Simulation Studies of Zeolites

Our work on zeolites is motivated both by the intrinsic interest of the problems raised by their structural complexity and secondly by their importance as heterogeneous catalysts. Our work has investigated three types of problem: first the structural properties of the extra–framework cations which charge compensate the replacement of Si by Al in the framework; secondly the structures of the frameworks themselves; and thirdly the important problem of the sites occupied by sorbed molecules in zeolites – a problem at the heart of our understanding of how these porous structures function as catalysts.

A Cation Distribution in Zeolites

During the last three years we have examined two systems in detail: first Sr zeolite A (Sanders et al., 1984) where we showed that for Si/Al ratios of 1:1 Sr^{2+} cations are confined to 6–ring sites in accordance with the latest experimental data of Pluth and Smith (1982). The cations are distributed over two types of 6–rings which project both into and out of the aluminosilicate cubo–octahedra. The details of the calculated distribution agree well with experiment; and the calculations helped to settle a controversy concerning the possibility of occupancy of 8–ring sites which our work suggests is not feasible on energetic grounds.

Our second study (Sanders and Catlow, 1985) concerned the cation distribution in K^+–faujasite, where we considered both the site distribution of the potassium ions and the dependence of this distribution on that of the aluminium ions over the framework tetrahedral sites. We showed that the 6–ring sites which project outwards into the supercage (The S_{II} sites) are the most favoured energetically; while the S_I and S_I' sites which are six ring sites adjacent to the bridging hexagonal prisms that project, respectively, into and out of the cubo–octahedra, are similar energetically, for which we therefore predict similar occupancies. Again, the calculations are in good agreement with experiment.

B Inclusion of Framework Relaxations

A far more demanding test of available potentials is provided by their ability to reproduce the structures of the aluminosilicate frameworks which were held fixed in the calculations discussed in section (A) above. To date we have investigated framework relaxation in both zeolite A (with Sr^{2+} and Na^+ as extra–framework cations) (Sanders et al., 1984; Jackson and Catlow, 1987; Sanders, 1984) and in the pentasil structured zeolite, silicalite (Hope, 1985; Jackson and Catlow, 1987), discussed further below). In both cases the calculations in which framework relaxation was included were encouraging, as the structure after equilibration was close to that determined experimentally. The calculations used bond–bending potentials of the type discussed in section (3), and as an illustration of the present status of the field we will consider in some detail the case of Na zeolite A, for which we have performed a detailed comparison of the energy minimised structure for the compound with the Si/Al ratio of 1:1. We considered the following potential models which are tabulated in table (4), and which are discussed in greater detail by Jackson and Catlow (1987).

model (a) rigid ion O^{2-} + $Na^+ \ldots O^{2-}$ potential 1
model (b) rigid ion O^{2-} + $Na^+ \ldots O^{2-}$ potential 2
model (c) shell model O^{2-} + $Na^+ \ldots O^{2-}$ potential 1
model (d) shell model O^{2-} + $Na^+ \ldots O^{2-}$ potential 2
model (e) as (d) but with the alternative 3 body O-Al-O potential

Table 4
Potential parameters for zeolite frameworks

(i) 3 body model, rigid ions

interaction	A (eV)	ρ(Å)	C (eVÅ^6)
$Si^{4+}...O^{2-}$	1584.167	0.32962	52.64511
$Al^{3+}...O^{2-}$	1460.3	0.29912	0.0
$O^{2-}...O^{2-}$	22764.0	0.149	27.88

3 body potential for Si^{4+}, Al^{3+}: $k = 4.5815$ eV/rad^2
$\theta_o = 109.47^\circ$

(ii) 3 body model, shell model

interaction	A (eV)	ρ (eV)	C (eVÅ^6)
$Si^{4+}...O^{2-}$	1283.907	0.32052	10.66158
$Al^{3+}...O^{2-}$	1460.3	0.29912	0.0
$O^{2-}...O^{2-}$	22764.0	0.149	27.88

harmonic potential for O^{2-}: $Y^- = -2.86902\ |e|$, $k = 74.92$ eVÅ^6

3 body potential for Si^{4+}, Al^{3+}: $k = 2.09724$ eV/rad^2
$\theta_o = 109.47^\circ$

(iii) alternative 3 body potential for $Al^{3+}...O^{2-}$

A (eV)	ρ (Å)	C (eVÅ^6)
498.602	0.38528	0.0

3 body potential : $k = 2.0705$ eV/rad^2, $\theta_o = 109.47^\circ$

(iv) Potential parameters for non-framework cations

Model	A (eV)	ρ (Å)	C (eVÅ^6)
(1)	1226.84	0.3065	0.0
(2)	5836.84	0.2387	0.0

note on cutoffs
for all short range potentials a cutoff of -~10Å was used

The structure was then minimised to constant volume using each of the above potential models. Calculated lattice energies are given in table (5). Bond lengths and bond angles obtained are compared with the experiment in tables (6)–(8).

Table (5)

Calculated lattice energies for Na^{+} zeolite A

model	lattice energies, eV	
	after relaxation to constant volume	after relaxation to constant pressure
(a)	-5162.32	-5163.96
(b)	-5179.02	-5179.66
(c)	-5297.73	-5297.93
(d)	-5314.46	-5314.51
(e)	-5203.58	-5203.65

TABLE (6)

Comparison of experimental and calculated Si-O bonds and O-Si-O angles

bond length or angle	experiment*	model (a)	model (b)	model (c)	model (d)	model (e)
Si-O(1)	1.595	1.537	1.493	1.525	1.523	1.531
Si-O(2)	1.586	1.541	1.562	1.551	1.554	1.562
Si-O(3)	1.604	1.579	1.602	1.548	1.550	1.554
mean	1.597	1.552	1.552	1.541	1.542	1.549
O(1)-Si-O(2)	108.8	109.5	110.4	107.7	106.3	106.2
O(1)-Si-O(3)	111.3	109.3	109.7	110.0	110.9	110.9
O(2)-Si-O(3)	107.2	109.2	108.3	108.7	107.9	107.7

*Experimental data taken from Pluth and Smith (1980)

TABLE (7)

Comparison of experimental and calculated Si-O bonds and O-Si-O angles

bond length or angle	experiment	model (a)	model (b)	model (c)	model (d)	model (e)
Al-O(1)	1.723	1.731	1.734	1.740	1.747	1.718
Al-O(2)	1.717	1.718	1.721	1.761	1.761	1.747
Al-O(3)	1.741	1.736	1.739	1.777	1.784	1.806
mean	1.731	1.728	1.731	1.759	1.764	1.757
O(1)-Al-O(2)	108.1	109.5	107.1	114.1	117.0	118.0
O(1)-Al-O(3)	112.3	108.8	109.9	108.4	108.2	108.4
O(2)-Al-O(3)	106.0	109.7	109.7	107.4	105.6	105.2

TABLE (8)

Comparison of experimental and calculated
Si-O bonds and O-Si-O angles

bond length or angle	experiment	model (a)	model (b)	model (c)	model (d)	model (e)
T-O(1)-T	142.2	142.4	152.1	137.1	141.3	142.2
T-O(2)-T	164.7	169.7	159.8	171.6	165.4	165.1
T-O(3)-T	144.8	153.3	148.3	151.0	147.5	144.6
Na(1)-3O(3)	2.3227	2.4707	2.3596	2.4768	2.3735	2.3776
Na(1)-3O(2)	2.9146	2.8995	2.9407	2.9672	2.9911	2.9926
Na(2)-O(1)	2.557	2.6643	2.3832	2.7069	2.5131	2.5115
Na(2)-o(1)	2.673	2.8116	2.4332	2.8307	2.4043	2.3985

From these tables, the following general observations can be made:

(i) all potentials reproduce the experimental structure to broadly the same accuracy.

(ii) all potentials tend slightly to underestimate the Si–O bond lengths.

(iii) the most difficult angles to reproduce are those involving two tetrahedral atoms and a bridging atom. However, the most detailed model, model (e), reproduces these angles reasonably well.

(iv) The worst agreement in bond lengths concerns the Na(2) atoms which are located in the 8–rings. The main reason for this is that each 8–ring has 4 possible positions for occupation by Na^+ ions, and in the calculation a definite choice had to be made, which might not correspond with the experiment.

(iv) all potentials used were obtained by fitting to other structures (usually α quartz), except the $Na^+ \ldots O^{2-}$ potentials which were calculated by the electron gas method. It is gratifying that such good overall agreement was obtained for a material other than that used to fit the potentials. To improve the agreement (especially for the Si–O bond lengths) some refitting of the potentials may be needed, but it is clear that the types of potential we have derived are capable of reproducing detailed structural properties of complex silicates.

C Sorption in Zeolites

If we are to understand the details of how the architecture of the zeolite framework controls the nature of the reactions of sorbed molecules, it is clearly necessary to obtain detailed information on the sites at which sorption occurs. Recent work has shown that it may be possible in certain cases to derive such information from diffraction techniques (Fitch et al., 1986; Wright et al., 1985). There is clearly, however, an incentive to devise reliable theoretical methods for predicting sorption sites and energies. To this end, we recently undertook a detailed study of a relatively simple problem: we investigated the sorption of krypton into silicate (discussed above), a 'pentasil' structured zeolite (see fig. 2) – so called as it contains 5 and 10 rings. Silicalite and the closely related ZSM–5 zeolite have high Si/Al ratios (>200) and have attracted considerable attention as they catalyse the methanol–gasoline conversion.

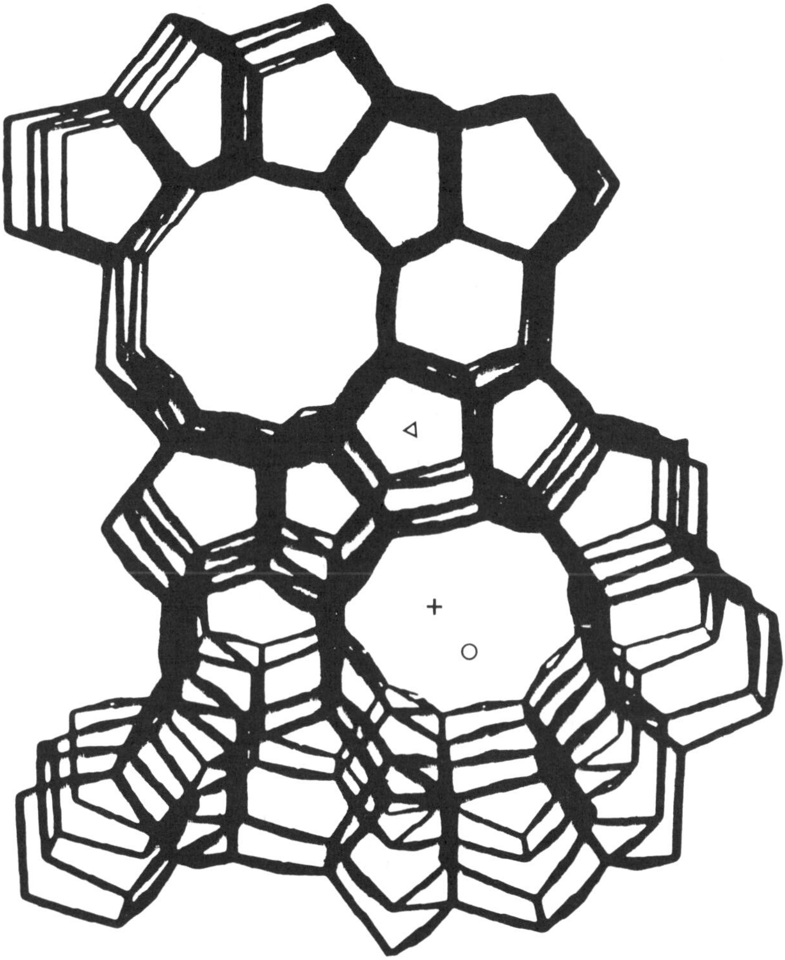

(2) The structure of silicalite with sorption sites marked.

We examined a wide range of sites for sorption of the Kr atoms. The potentials used were derived and are discussed by Hope (1985) and Hope et al. (1984). Three sites were clearly identified. Their positions are shown in fig. (4). The sorption energies are given in table (9). We should stress that these calculations included effects of framework relaxation around the sorbed atom, which was indeed handled by treating the atom as a 'defect' around which the structure was relaxed using the techniques summarised in section (2). The results of including these effects are appreciable; the energy is changed by ~1 kJ/mole. Much larger effects could be anticipated for large complex molecules.

In a companion study, Hope (1985) and Hope et al. (1987) undertook experimental sorption studies of the same problems using low pressure, low temperature volumetric techniques. It is gratifying that this work again identified three sites whose energies are compared with the calculations in table (4). The discrepancy between theory and experimental (of ~6kJ/mole) is due to a combination of factors. Omission of thermal energies in the calculations will lead to an overestimation of the sorption energy probably by 1-2 kJ/mole. There is also some uncertainty in the 'zero' of the energy scale for the experimental results. And, of course, there will be some error due to inadequacies in the Kr... framework atom potentials. The important feature of the calculations is, however, that they show that it is possible using simulations, to identify sorption sites.

Table (9)

Experimental and Calculated Energy Values in Krypton Sorbed in Silicate (All values are in KJ mol^{-1})

SITES*	$-U_{CALC}$ ($KjMOL^{-1}$)	Δ_{CALC}	$-U_{EXPT}$	ΔU_{EXPT}
1(red)	20.9		12.8	
		2.0		0.7
2(blue)	18.9		12.0	
		1.6		0.7
3(green)	17.3		11.3	

*Colours refer to fig. (2).

Work has also been reported on more complex sorption problems, notably the recent study of Wright et al. (1985) on the sorption of pyridine in zeolite L. Their study showed that it was possible to identify accurately sites occupied by large molecules in zeolites by energy minimisation techniques using effective potentials.

4.2 Defect Energies in Mg_2SiO_4

Since mass transport in this important mantle mineral may be controlled by intrinsic disorder the values of the formation energies of intrinsic defects are of considerable importance. Our study (Doherty et al., 1987) has used the Mott–Littleton methodology discussed earlier in this paper to examine Mg^{2+} and O^{2-} Frenkel and a variety of Schottky disorder reactions. In addition, we have examined the activation energies for magnesium vacancy and interstitial migration. Our results, summarised in table (10), indicate that Mg^{2+} Frenkel pairs are energetically most favourable, although the difference is small compared with the Schottky Mg^{2+} ... O^{2-} vacancy pair. The results suggest Arrhenius energies for intrinsically controlled Mg^{2+} diffusion of ~3 eV, while in vacancy excess material (produced by, for example substitution of Mg^{2+} by Fe^{3+}) Arrhenius energies of ~2 eV would be expected. The experimental results are still not firmly established, but the above values agree reasonably with most data.

Table (10)

Calculated Defect Energies in Mg_2SiO_4

Mg^{2+} Frenkel Energy : 5.3 eV (Mg(1) vacancy + tetrahedral interstitial)

O^{2-} Frenkel : 6.7 eV (O(3) vacancy + tetrahedral interstitial)

Mg^{2+} vacancy .... O^{2-} vacancy Schottky pair : 5.5 eV

Mg^{2+} Interstitialcy motion : 0.3 eV

Mg^{2+} Vacancy motion : 1.8 eV

CONCLUSIONS

It is clear from our work and that of others, that computer modelling techniques can be successfully applied to several types of problem in mineral physics. The key to further progress lies in the development of improved interatomic potential, where there is need for further input from theoretical methods both electron–gas and ab–initio, and for further testing of potentials by reference to a wider range of physical properties.

ACKNOWLEDGEMENTS

I would like to thank R.A. Jackson, S.C. Parker, G.D. Price and A.N. Cormack for many helpful discussions, and to acknowledge their contributions to the work summarised in this chapter.

REFERENCES

Angell, A. (1987), Phys. Chem. Min - in press.
Burnham C.W., (1985), Reviews in Mineralogy 14, 347.
Busing W.R., Trans Am. Crystallogr. Assoc. (1970) 6, 57.
Catlow C.R.A. (1983), Solid State Ionics, 8, 89.
Catlow C.R.A. (1986) in 'Modern Techniques in Defect Physics', (eds A.V. Chadwick and M. Tarenzi) (Plenum Press).
Catlow C.R.A. (1986), Ann. Rev. Mat. Sci., 16, 517.
Catlow C.R.A. and Mackrodt W.C. (1982) (eds), 'Computer Simulation of Solids', Lecture Notes in Physics, vol. 166 (Springer-Berlin).
Catlow C.R.A. and Stoneham A.M., (1983), J. Phys. C. 16 4321.
Catlow C.R.A. and Parker S.C. (1985) in 'Point Defects in Minerals' (ed R.N. Schock), Geophysical Monograph 31, 26.
Catlow C.R.A., Doherty M., Parker S.C. and Sanders M.J. (1985a), Materials Science Forum 7 118.
Catlow C.R.A., Freeman S.C. and Royle R.L., (1985b) Physica B. 131, 1.
Catlow C.R.A., Corish J., Jacobs P.W.M. and Lidiard A.B. (1981), J. Phys. C. 14, L121.
Dick B.G. and Overhauser A.W., (1958) Phys. Rev. 112, 90.
Doherty M.D., Catlow C.R.A. and Price G.D. (1987).- to be published.
Dorner B., Grimm H. and Rzany H. (1980) J. Phys. C. 13, 6607.
Fitch A.N., Jobic H. and Renouprez A. (1986) J. Phys. Chem. 90, 1311.
Gillan M.J. and Jacobs P.W.M. (1983) Phys. Rev. B28, 759.
Gordon R.G., (1987) Phys. Chem. Min. - in press.
Harding J.H., (1985) Physica B. 131, 13.
Harding J.H. and Stoneham A.M. (1981) Phil. Mag. A, 43, 705.
Hope A.T.J. (1985) PhD Thesis, University of London.
Hope A.T.J., Catlow C.R.A. and Leng C.A. (1987) - to be published.
Jackson R.A. and Catlow C.R.A., Molecular Simulation - in press.
Jacobs P.W.M. (1983) in 'Mass Transport in Solids', (eds F. Bénière and C.R.A. Catlow) (Plenum Press)
Lasaga A.C. and Gibbs G.V. (1987), Phys. Chem. Min. - in press.
Mackrodt W.C. (1984) in 'Transport in Non-Stoichiometric Compounds', (eds G. Petot-Ervas, Hj Matzke and C. Monty) (North Holland).
Mostoller M. and Wang J.C. (1985) Phys. Rev. B32, 6773.
Mott N.F. and Littleton M.J. (1938) Trans Faraday Soc., 34, 485.
Parker S.C. (1982) PhD Thesis, University of London.
Parker S.C., Cormack A.N. and Catlow C.R.A. (1984) Acta. Cryst. B40, 200.
Parker S.C. (1987) - to be published.
Pluth J.J. and Smith J.V. (1980), J. Am. Chem. Soc., 102, 4704 (1980).
Pluth J.J. and Smith J.V., (1982) J. Am. Chem. Soc. 104, 6977.
Sanders M.J. (1984) PhD Thesis, University of London.
Sanders M.J., Catlow C.R.A. and Smith J.V. (1984) J. Phys. Chem. 88, 2796.
Sanders M.J., and Catlow C.R.A. (1985) Proc. 6th Int. Zeolite Conference (ed D. Olsen anad A. Bisio) (Butterworths, London) p131.
Tosi M. (1964) Solid State Physics (eds F. Seitz and S. Turnbull) Vol 16, p1.
Vineyard G., (1957) J. Phys. Chem. Solids 3, 157.
Wright P.A., Thomas J.M., Cheetham A.K. and Nowak A.K. (1985) Nature, 318, 611.

UV TO NIR SPECTRA OF SILICATE MINERALS OBTAINED BY MICROSCOPE SPECTROMETRY AND THEIR USE IN MINERAL THERMODYNAMICS AND KINETICS

K.Langer
Technical University Berlin
Institute of Mineraloy and Crystallography
Ernst Reuter Platz 1
D 1000 Berlin 12

ABSTRACT. Microscope spectrometry, MISPE, allows to measure polarized spectra with areal resolution down to 3 µm of single crystals at normal and high pressures. Such spectra of synthetic single crystals, grown under controlled thermodynamic conditions to contain only one type of $3d^N$-ion, yield unambiguous band assignments which are the basis for determining the site geometry and crystal field stabilization energy of the $3d^N$-ion in the respective structural matrix.

The latter property contributes to the Gibbs free energy ΔG (enthalpy ΔH or inner energy ΔU) of reactions, involving $3d^N$-ions, such as phase transitions, formation of $3d^N$-ion bearing solid solutions, or $3d^N$-ion partitioning between coexisting phases.

Due to the high areal resolution of MISPE, $3d^N$-concentration profiles can be obtained including such ions which cannot be determined with the microprobe as Fe^{3+} or Mn^{3+}. Such profiles are needed to evaluate the kinetics of the formation of $3d^N$-ion bearing point defects in silicates.

1. INTRODUCTION

Thirty years ago, Holmes and McClure (1957) have shown that subtracting the crystal field stabilization energy, $CFSE_{3d^N}$, from experimentally determined heats of hydration of gaseous $3d^N$-ions, $\Delta_{h,m}$,

$$\Delta H_{h,m} - CFSE_{3d^N} = \Delta H_{corr} \tag{1}$$

yields enthalpy values, ΔH_{corr}, which fit, within the limits of experimental error, a straight line $\Delta_{h,t}$ = const · Z. Here, Z is the atomic number of the $3d^N$-ion and $\Delta_{h,t}$ the heat of hydration, theoretically predicted by pure electrostatic theory based on spherical charge distribution as e.g. in Mg^{2+}, Ca^{2+}, Al^{3+} etc. Furthermore, plots of the lattice energies of dihalides,monoxides and -sulfides of $3d^N$-ions as a function of Z (George and McClure, 1959) show nearly the same shape as the $\Delta_{h,m}$ = f(Z) plots, which proves a contribution of the crystal field stabilization energy also to the lattice energies.

Since that time many geoscientists measured and used crystal field

E. K. H. Salje (ed.), Physical Properties and Thermodynamic Behaviour of Minerals, 639–685.

stabilization energies, $CFSE_{3d^N}$, to solve problems such as partitioning of $3d^N$-ions between melt and crystals (e.g. Burns and Fyfe 1964, Curtis 1964), distribution of $3d^N$-ions between coexisting metamorphic minerals (Schwarcz 1967), thermodynamics of the formation of solid solutions (e.g. Strens 1968, Ganguly 1976), intracrystalline $3d^N$-ion partitioning (e.g. Burns 1970a), or phase transitions (e.g. Burns and Sung 1978). The work of geoscientists in this field was greatly encouraged by the pioneering work of Burns (1970b).

The required $CFSE_{3d^N}$-values in the respective phases may be extracted from optical spectra, polarized spectra in case of non-cubic minerals, on the basis of crystal field or ligand field theory (e.g. Ballhausen 1962). Problems to obtain the required data are:

(i) Proper band assignment. This problem arises from the presence of more than one type of $3d^N$-ion in chemically complex natural minerals. Furthermore, in natural minerals excitation processes other than dd transitions may cause additional bands.

(ii) The size of crystals is often too small to obtain single crystal spectra using conventional spectrometers.

To overcome the second problem, Burns (1966) developed a microscope-spectrometric method, which was further improved by Mao and Bell (1971, 1973). These methods allow to measure spectra from spots with minimum diameters of approximately 200 μm (Burns 1966) or 30 μm (Mao and Bell 1971, 1973).

The solution of the first problem is to measure synthetic single crystals obtained in high pressure experiments under defined thermodynamic conditions of p, T, fO_2 (f_{H_2O}) to contain only one type of $3d^N$-ion (Langer and Abu-Eid 1977). Normally, in such experiments crystals with maximum dimensions of some 10 to 100 μm are obtained.

Therefore, microscope spectrometric methods were developed capable to measure spectra, up to 40,000 cm^{-1} (250 nm) in the UV, with an areal resolution of about 3 μm (Langer and Frentrup 1979). These methods allow to record polarized spectra of very small synthetic and natural silicate crystals. $CFSE_{3d^N}$ may be evaluated from these spectra. Furthermore, since spectral features depend on the number of absorbing species per unit volume of the crystal under study, local concentrations of absorbing species in larger crystals may be determined with the just mentioned areal resolution. This opens the possibility to use the methods in kinetic studies, where concentration profiles are needed.

It is the aim of the present paper to discuss recent results in the field outlined in the last paragraph. Before doing so, a brief theoretical introduction may be useful.

2. SOME THEORETICAL ASPECTS

2.1 Spectroscopy

As in all matter, resonant absorption of electromagnetic radiation in minerals occurs at discrete energies

$$\Delta E = h \cdot \nu \qquad (2)$$

with h Planck's constant and ν frequency. Therefore, absorption spectra of minerals

$$\left.\begin{array}{l} T = (I/I_o) \\ A = \log\,(1/T) \\ \alpha = A/t \end{array}\right\} = \left\{\begin{array}{l} f\,(\lambda) \\ f\,(\tilde{\nu}) \end{array}\right. \qquad (3)$$

exhibit absorption edges and bands as schematically shown for two garnets in Figure 1.

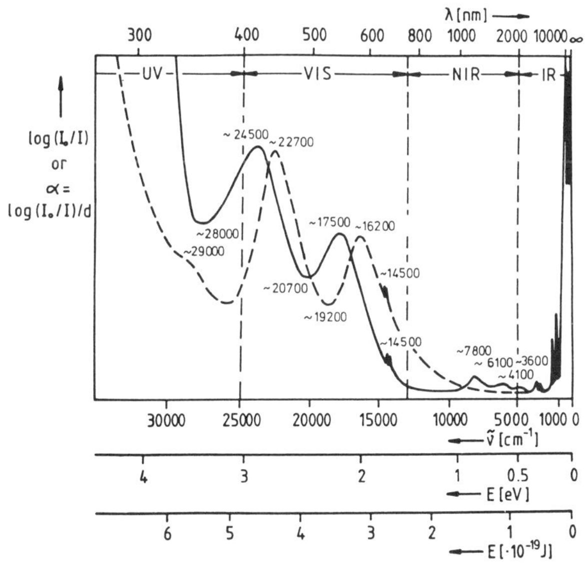

Figure 1. Schematical absorption spectra in the UV to IR spectral ranges of pyrope from Meronitz, Bohemia (with 5.5 wt% Cr_2O_3 (UV|VIS|NIR: Amthauer 1976), solid line, and of uwarovite from Outokumpu, Finland, with 16.9 wt% Cr_2O_3 (Amthauer 1976, Langer 1984), broken line.

In equations (3), I_o and I are the intensities of incident and transmitted radiation, respectively, T is the transmittance of the mineral crystal, A the absorbance - sometimes called "optical density" D -,t is the crystal thickness in [cm], α the linear absorption coeffcient in [cm^{-1}], λ the wave length, and $\tilde{\nu} = 1/\lambda$ in [cm^{-1}] the wavenumber. This latter quantity is directly related to the energy of the radiation by

$$\Delta E = h \cdot c_o \cdot 1/\lambda \tag{4}$$

where c_o is light velocity in vacuum. Hence, the wavenumber is the quantity appropriate as abscissa in absorption spectra (cf. Figure 1). Wavenumbers are recalculated to other energy units

$$1\ cm^{-1} = 1.2397\text{-}10^{-4}\ eV = 4.7442 \cdot 10^{-24}\ cal = 1.9863 \cdot 10^{-23}\ J \tag{5}$$

To calculate these energies per g-atom or per mole, the conversion factors are to be multiplied by Avogadro's number.

The absorption bands in minerals originate from transitions between the energetic ground state and various energetically excited states of the electronic or vibronic system of the mineral structure

$$E\Psi_g + \Delta E = (\Delta E + E)\ \Psi_e^* \tag{6}$$

with Ψ wave function. Mechanisms of excitation are detailed in Table 1.

Peak hights of bands, i.e. $\alpha_{\tilde{\nu}}$ max, or linear absorption coefficients at high or low energy wings of a band, $\alpha_{\tilde{\nu}}$, are related to the number of excited species per unit volume in the respective mineral structure, i.e. to their concentration c in [g-atom $\cdot\ 1^{-1}$]

$$\alpha_{\tilde{\nu}} = f\ (c) \tag{7}.$$

The function f in equation (7) is often a constant, especially in diluted systems, such that

$$\alpha_{\tilde{\nu}} = \varepsilon_{\tilde{\nu}} \cdot c \tag{8}$$

with $\varepsilon_{\tilde{\nu}}$ the molar absorptivity in [$1 \cdot$ g-atom$^{-1} \cdot cm^{-1}$]. Provided, proper calibrational crystals have been measured, $\varepsilon_{\tilde{\nu}}$, can be obtained from

$$\varepsilon_{\tilde{\nu}} = (\alpha_{\tilde{\nu}} \cdot V_{el} \cdot N_L) \mid (1000 \cdot X_{ion} \cdot z) \tag{9}$$

with V_{el} unit cell volume of the mineral structure in cm^3, N_L Avogadro's number, X_{ion} number of absorbing species per formula unit, and z number of formula units per unit cell.

Equations (7) to (9), or the respective formalism using integral intensities

$$\int_{\tilde{\nu}_1}^{\tilde{\nu}_2} \alpha d\tilde{\nu}$$

form the basis for spectroscopic concentration determinations, provided the correct band assignments are known. Thus, Hålenius and Langer (1980) designed a microscope-spectrometric method for Fe^{2+}-Fe^{3+}-determinations in chloritoids, which was based on band assignments of Hålenius et al. (1981). Goldman and Rossman (1977) determined the

Table 1. Excitation processes in silicate minerals in the UV to IR spectral range.

Excitation	Designation	Characteristics of bands: spectral range	selection rules (a)	intensity (ε in $l \cdot mol^{-1} \cdot cm^{-1}$)	$\Delta\tilde{\nu}_{1/2}$ (cm^{-1})	Theory
1. Transitions between e-states predominantly centered at the oxygen ligands of coordination polyhedra and those predominantly centered at the central ion	Ligand-metal charge transfer LM-CT	UV (>30000 cm^{-1}, low energy wings down into the VIS)	no	extremely high (10^4 to 10^6)	ca. 5000	MO-theory (SCF-X_α) band theory
2. Electron hopping between $3d^n$ cations, M^{x+} with different valence x in edge- or face-connected coordination polyhedra ("electron delocalization")	Metal-metal charge transfer MM-CT	NIR-VIS (ca. 9000-ca. 24000 cm^{-1} depending on R_{M-M}, type of M, polyhedral polymerization)	no	very high along MM vector, depending on concentration product $[M^x][M^{x+1}]$	3000 - 6000	Exchange theory (Hush)
3. Electronic transitions between crystal field split d states, localized at M^{x+} cations	dd transitions	NIR-UV (ca. 4000-ca. 30000 cm^{-1})				Crystal field theory
a. spin-allowed (without change of M = (2S+1))			parity forbidden (e.g. Cr^{3+}, Mn^{3+}, Fe^{2+}, Ni^{2+})	weak (10^0 to $3 \cdot 10^2$) (b) depending on concentration $[M^{x+}]$	ca. 1000 - 3000	
b. spin-forbidden (with change of M = (2S+1))			parity and spin-forbidden (e.g. Mn^{2+}, Fe^{3+})	very weak (10^{-1} to 10^1) (b) depending on concentration $[M^{x+}]$	ca. 20 - 600	
4. Excitation of energetically higher states of F-centers (electrons allocated at structure defects)	F-centers	only in some cases proven in silicates (e.g. rose quartz, smoke quartz, rubellite)	no	similar as 1. and 2.	similar as 1. and 2.	
5. Fundamental vibrations of groups of atoms (e.g. OH^-, CO_3^{2-}) or molecules (e.g. H_2O, CO_2)		IR (ca. 4000 - 1200 cm^{-1})	no	very strong, depending on concentration	ca. 4 - 100 cm^{-1}	Vibrational theory of isolated groups (e.g. Siebert)
6. Lattice vibrations		IR (< 1200 cm^{-1})	no	very strong	similar as 5.	Dynamic lattice theory (Born and Huang)

a) Except symmetry related selection rules which apply to all sorts of transitions in case of crystal symmetries lower than cubic

b) Numbers hold for single ions, ion pairing or larger clustering yield intensity enhancement

intercrystalline Fe^{2+}-Mg distribution in natural orthopyroxene and Steffen et al. (1987) in synthetic orthopyroxenes grown under controlled fO_2.

The value of $\varepsilon_{\tilde{\nu}}$ depends on the transition probability of the respective transition which, in turn, is governed by various selection rules (Table 1). In case of dd transitions, which will be predominantly discussed in this paper, the Laporte rule forbids transitions in which the wave function does not change sign on $\Psi_g \rightarrow \Psi_e^*$. This is the case when ions d^1 to d^4 and d^6 to d^9 are located in sites with an inversion center. Hence, $\varepsilon_{\tilde{\nu}}$ of such transitions is low but may increase in polyhedra lacking an inversion center (static lifting of Laporte's rule), or on coupling with vibrations abolishing the inversion center (dynamic lifting of Laporte's rule). The spin selection rule forbids transitions in which the multiplicity M = (2S+1) is changed on transition. Such dd-transitions occur but are very low in intensity because both, the Laporte and spin selection rules, exclude them. In case of $3d^5$-ions (Mn^{2+}, Fe^{3+}) transitions of such type are the only ones which are possible.

In addition, all transisitons mentioned in Table 1 underly symmetry related selection rules when the mineral under study has lower symmetry than cubic. This causes the band intensities to depend on the orientation of the crystal with respect to the elctric or magnetic wave vector of the radiation, an effect that causes the pleochroism of minerals.

Half widths of bands are greatly influenced by the half life time τ of the excited states, which are in the order of 10^{-12} sec for all transitions of Table 1 except spin-forbidden dd-transitions where τ is two to three orders of magnitude higher.

Ground and excited states of a $3d^N$-ion, e.g. of Cr^{3+} (N=3), Mn^{3+} (N=4), Fe^{2+} (N=6), in the field of a coordination polyhedron in a mineral structure originate from a lifting of d-orbital degeneracy, with respect to energy and symmetry of spectroscopic states, when the respective free $3d^N$-ion is transferred from the gaseous state into the polyhedron. This crystal field splitting of the spectroscopic ground state for an octahedral environment with point symmetry O_h is shown for the example of Cr^{3+} in Figure 2 and for the spectroscopic ground states of all $3d^N$-ions in Figure 3. Transitions between ground and excited crystal field states give rise to the dd bands in the optical spectra of minerals such as in Figure 1.

From Figure 2 and 3 it is obvious that the ground state in octahedral coordination is lower in energy than that of a free ion in a spherical field. This energy gain of the ion in octahedral environment relative to the free ion in a spherical field is the crystal field stabilization energy.

From Figure 2, it is further directly evident that in octahedral fields, point group O_h, this quantity is given by

$$- \text{CFSE}_{oct} = [4n - 6\,(N-n)] \cdot Dq \qquad (10),$$

where N is the total number of d-electrons and n the number of electrons allocated in the t_{2g} orbitals (cf. Figure 2, left). For a given $3d^N$-ion, the crystal field parameter Δ = 10 Dq depends strongly on the mean ligand-to-metal distance R

$$\Delta = 10\,Dq = 5q\,\overline{r^4} \cdot \frac{1}{R^5} \qquad (11)$$

as obtained from crystal field theoretical considerations, based on the point charge model by Jœrgensen (1963). Here, q is the

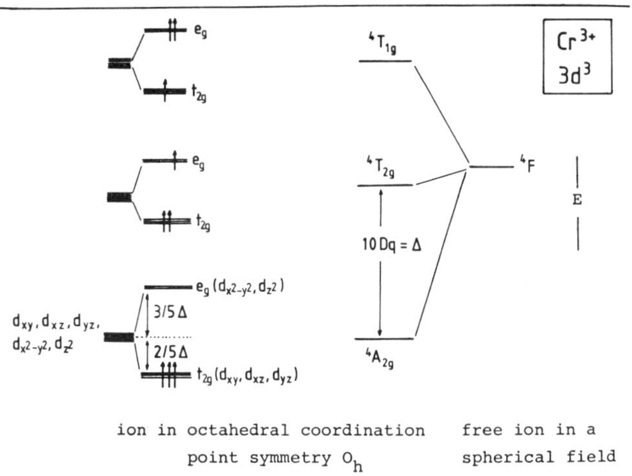

Crystal field stabilization energy: $CFSE_{Cr^{3+},oct} = -3\cdot(\frac{2}{5}\cdot 10Dq)$

for any transition ion ($d\varepsilon^n d\gamma^{N-n}$): $CFSE_{oct} = -[4n - 6(N-n)]Dq$

Figure 2. Crystal field splitting of the ^4F spectroscopic ground state of Cr^{3+} in octahedral field. Left: representation of states in the one electron orbital scheme. Right: designation of crystal field split terms by symbols of the irreducible representations of symmetry groups of the respective multielectron arrangement. Definition of the crystal field stabilization energy is also given. $d\varepsilon$ are the t_{2g} orbitals, $d\gamma$ the e_g orbitals.

d^1		d^2		d^3		d^4		d^5		d^6		d^7		d^8		d^9	
Ti^{3+}		V^{3+}		Cr^{3+}		Mn^{3+}		Fe^{3+}		Co^{3+}							
				V^{2+}		Cr^{2+}		Mn^{2+}		Fe^{2+}		Co^{2+}		Ni^{2+}		Cu^{2+}	
^2D		^3F		^4F		^5D		^6S		^5D		^4F		^3F		^2D	

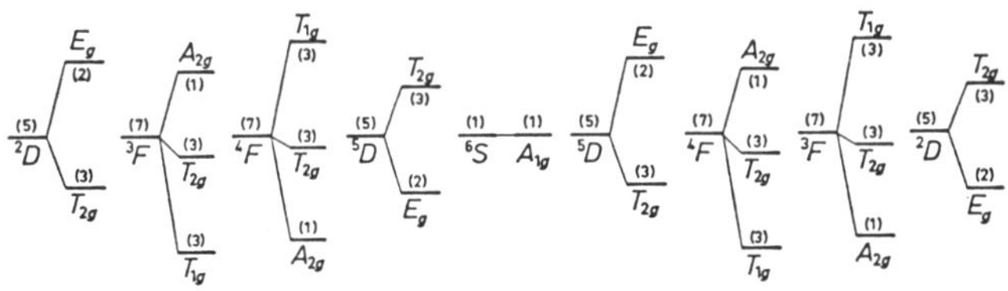

$-CFSE_{oct}$ 4Dq 6Dq 12Dq 6Dq 0 4Dq 6Dq 12Dq 6Dq

Figure 3. Crystal field splitting of the spectroscopic ground states and crystal field stabilization $(2S+1)L_J$ of the whole series of $3d^N$-ions in an octahedral environment, point group O_h (after: Schläfer and Gliemann 1967).

effective charge on the ligands and $\overline{r^4}$ is the mean of the fourth power of the d-electron-core distance. For structures with oxygen ligands, e.g. silicates, and for transition ions of the first long period, the $3d^N$-ions, the first term in equn. (11), $5q\overline{r^4}$, is with fair approximation a constant.

In tetrahedral fields, point group T_d, and cubal fields, point group O_h, the sequence of crystal field split terms in Figures 2 and 3 is reversed. The crystal field splitting parameters $\Delta = 10Dq$ are related by

$$\Delta_{oct}:\Delta_{cub}:\Delta_{tetr} = 1:(-8/9):(-4/9) \tag{12}$$

The crystal field stabilization energies in tetrahedral and cubal fields are

$$-CFSE_{tetr,cub} = [6n - 4(N - n)]\cdot Dq_{tetr,cub} \tag{13}$$

where N again is the total number of d-electrons and n the number of electrons allocated in the e (cubal e_g) orbitals, which are now at lower energies than the barycenter (cf. Figure 2) compared to octahedral coordination as indicated by the minus signs in equn. (12).

The energies of all crystal field split states can be calculated by the Schrödinger equation

$$\Psi_{n,l,m_l}(r,\theta,\phi) = E\Psi_{n,l,m_l}(r,\theta,\phi) \tag{14}$$

with respect to the radial and angular parts of the wave function Ψ with d-character, i.e. $l = 2$

$$\begin{aligned}\Psi_{n,l,m_l}(r,\theta,\phi) &= R_{n,l}(r)\cdot Y_{l,m_l}(\theta,\phi)\\ &= R_{n,l}(r)\cdot\Theta_{l,m_l}(\theta)\cdot\Phi_{m_l}(\phi)\end{aligned} \tag{15}$$

(e.g. Schläfer and Gliemann 1961). By this, the energies of the spectroscopic states of the free $3d^N$-ions are derived.. The influence of the field of the ligands is then calculated as a perturbation of the free ion wave function, where the perturbation results in energy changes ΔE of the originally degenerate systems. These energy changes are given as the square root of a secular determinant of the form

$$\begin{vmatrix} H_{11}-\Delta E & H_{12} & \cdots & H_{1r}\\ H_{21} & H_{22}-\Delta E & \cdots & H_{2r}\\ \vdots & \vdots & & \vdots\\ H_{r1} & H_{r2} & \cdots & H_{rr}-\Delta E \end{vmatrix} = 0 \tag{16}$$

where the matrix elements H_{ij} are integrals of the form

$$\int_0^{\infty}\int_0^{\pi}\int_0^{2\pi} R_{n,l}(r)\Theta_{l,m_l}(\theta)\Phi_{m_l}(\phi)\ V\ R_{n,l'}(r)\Theta_{l',m_l'}(\theta)\Phi_{m_l'}(\phi) r^2 \sin\theta\ dr\ d\theta\ d\phi \qquad (17)$$

(Dunn et al. 1965, cf. also Schläfer and Gliemann 1961), where V is the potential at any point around the $3d^N$-ion.

In solving the secular determinant, crystal field theory which is based on the point charge model includes the crystal field parameter 10Dq. Ligand field theory extends this approach by including also the interelectronic repulsion of d-electrons in the form of Racah parameters B and C ($C/B \cong 4.7$, Sugano et al. 1970). Tanabe and Sugano (1954 a, b) calculated, thus, energy level diagrams for all $3d^N$-ions by solving the secular determinant in terms of 10Dq, B, and C and plotted the results in $E/B = f(10Dq/B)$ diagrams. As an example, the Tanabe-Sugano diagram for the $3d^3$ systems (V^{2+}, Cr^{3+}, Mn^{4+}) is shown in Figure 4. Such diagrams allow to graphically determine 10Dq and/or B from measured spectra as schematically shown in the right part of Figure 4. Solutions of the secular determinant for energy differences between ligand field split states of octahederal $3d^N$-ions are presented in Table 2. These can be used to calculate the just mentioned crystal field parameters from properly assigned spectra.

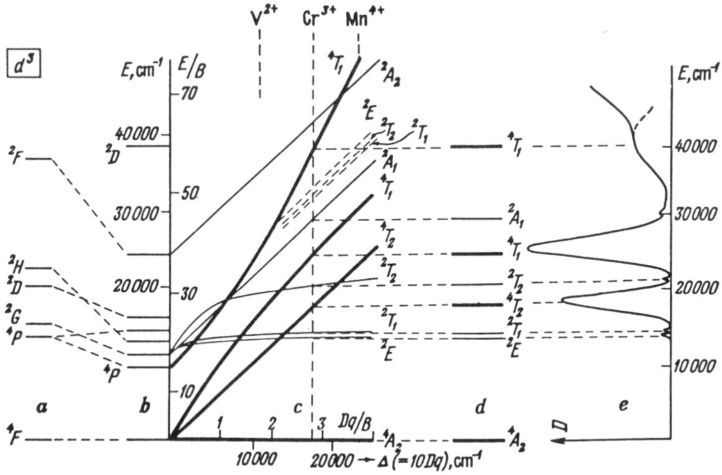

Figure 4. Tanabe-Sugano diagram (Tanabe and Sugano 1954 a,b) for octahedral $3d^3$-ions calculated in terms of 10Dq and Racah interelectronic repulsion parameters B and C. Left: free ion terms calculated for $B = 918$ cm^{-1} and $C = 3{,}600$ cm^{-1} (a.) and for $B = 650$ cm^{-1} and C $3{,}210$ cm^{-1} (b.). Right: unpolarized spectrum of ruby $Al_2O_3:Cr^{3+}$ (from: Marfunin 1979)

Table 2. Energy differences between ground and excited states of $3d^N$-ions in octahedral fields, point symmetry O_h. Energies of electronic transitions are derived by solving the secular determinant in terms of 10Dq, B, and C, i.e. the ligand field approach is used.

Number of 3d electrons	Ion	Ground State	Excited State	Energy of Electronic Transition	References
d^1	Ti^{3+}	$^2T_{2g}$	^2E_g	10Dq	Dunn et al. (1965), Hutchings (1964), Figgis (1966)
d^2	V^{3+}	$^3T_{1g}$	$^3T_{2g}$	$1/2(10Dq-15B)+1/2[(10Dq+15B)^2-12B.10Dq]^{1/2}$	Lever (1968), König (1971), Tanabe and Sugano (1954a,b)
			$^3T_{1g}$	$[(10Dq+15B)^2-12B.10Dq]^{1/2}$	
			$^3A_{2g}$	$1/2(30Dq-15B)+1/2[(10Dq+15B)^2-12B.10Dq]^{1/2}$	
d^3	Cr^{3+}	$^4A_{2g}$	$^4T_{2g}$(F)	10Dq	Tanabe and Sugano (1954a,b), Poole (1964), König (1971), Lever (1968), Reinen (1969)
			$^4T_{1g}$(F)	$1/2(15B+30Dq)-1/2[(15B-10Dq)^2+12B.10Dq]^{1/2}$	
			$^4T_{1g}$(P)	$1/2(15B+30Dq)+1/2[(15B-10Dq)^2+12B.10Dq]^{1/2}$	
d^4	Mn^{3+}	^5E_g	$^5T_{2g}$	10Dq	Lever (1968), Orgel (1966), Dunn <u>et al.</u> (1965)
d^5	Fe^{3+}, Mn^{2+}	$^6A_{1g}$	$^4T_{1g}$	$-10Dq + 34B - 26B^2/10Dq$	Lever (1968), Ballhausen (1962), Tanabe and Sugano (1954a,b), Figgis (1966)
			$^4T_{2g}$	$-10Dq + 42B - 39B^2/10Dq$	
			$^4A_{1g}$, ^4E_g (G)	10B + 5C	
			^4E_g(I)	17B + 5C	
d^6	Fe^{2+}	$^5T_{2g}$	^5E_g	10Dq	Figgis (1966), Dunn et al. (1965), Lever (1968), McClure (1959), Hutchings (1964)
			$^1A_{1g}$	$-20Dq+5B+8C-120B^2/10Dq$	
			$^3T_{1g}$	$-10Dq+5B+5C-70B^2/10Dq$	
			$^3T_{2g}$	$-10Dq+13B+5C-106B^2/10Dq$	
			$^1T_{1g}$	$-10Dq+13B+7C-34B^2/10Dq$	
			$^1T_{2g}$	$-10Dq+21B+7C-118B^2/10Dq$	
d^7	Co^{2+}	$^4T_{1g}$(F)	$^4T_{2g}$(F)	$1/2(10Dq-15B)+1/2[(10Dq+15B)^2-12B.10Dq]^{1/2}$	König (1971), Lever (1968), Tanabe and Sugano (1954a,b), Ballhausen (1962)
			$^4A_{1g}$(F)	$1/2(30Dq-15B)+1/2[(10Dq+15B)^2-12B.10Dq]^{1/2}$	
			$^4T_{1g}$(P)	$[(10Dq+15B)^2-12B.10Dq]^{1/2}$	
d^8	Ni^{2+}	$^3A_{2g}$	$^3T_{2g}$(F)	10Dq	König (1971), Lever (1968), Tanabe and Sugano (1954a,b)
			$^3T_{1g}$(F)	$1/2(15B+30Dq)-1/2[(15B-10Dq)^2+12B.10Dq]^{1/2}$	
			$^3T_{1g}$(P)	$1/2(15B+30Dq)+1/2[(15B-10Dq)^2+12B.10Dq]^{1/2}$	
d^9	Cu^{2+}	^2E_g	$^2T_{2g}$	10Dq	Dunn <u>et al.</u> (1965). Ballhausen (1962)

It should be noted that the interelectronic repulsion parameter B decreases strongly if a $3d^N$-ion is brough from the free ion state into a crystal field and further decreases as oxygen metal orbital-overlap, i.e. the degree of covalence of the M-O bond increases. Therefore, the ratio

$$\beta = B^{3d^N}_{field} / B^{3d^N}_{free} \qquad (18)$$

is a relative measure of covalency.

From the discussion so far it is clear

(i) from properly assigned mineral spectra, we obtain the crystal field parameters

(ii) from these, we may calculate $CFSE_{3d^N}$ in the respective structural matrix

(iii) $CFSE_{3d^N}$ depends strongly on the type of $3d^N$ -ion centered polyhedra and on the mean $3d^N$-ion oxygen distance.

A third factor which bears on the value of $CFSE_{3d^N}$ is polyhedral distortion, which lowers the point groups of the afore mentioned polyhedra. Symmetry reduction lifts the degeneracies of E- and T-terms. This results in a a further splitting of the respective ground- and/or excited states (Figure 3). It is directly obvious from Figure 3, considering also Figure 2, that any further ground state splitting will enhance $CFSE_{3d^N}$ in case of the ions Ti^{3+}, V^{3+}, Cr^{2+}, Mn^{3+}, Fe^{2+}, Co^{2+}, and Cu^{2+}. Therefore, these ions gain higher stabilization energy in distorted than in regular octahedra with the same mean M-O distance. On the other hand, Cr^{3+} and Ni^{2+} with their nondegenerate ground states (Figure 3) experience no such additional stabilization in distorted octahedra. The lifting of ground state degeneracy is known as Jahn-Teller effect of transition metal ions. From theoretical considerations, Jahn and Teller (1937) concluded that ions and molecules with degenerate ground state are unstable in highly symmetric environment and distort it so as to lift the degeneracy. Strong Jahn-Teller effect is encountered spectroscopically in case of Ti^{3+}, Cr^{2+}, Mn^{3+}, and Cu^{2+}.

The splitting and change of symmetry of crystal field states on descending point symmetry of the $3d^N$-ion bearing polyhedron may be obtained from correlation tables (e.g. Wilson et al. 1955, Marfunin 1979). A characteristic feature of transitions in low symmetry environments is that they occur only when the electrical dipole moment vector $\vec{P} = P_x + P_y + P_z$ has distinct symmetry relations with respect to those of the wave functions of ground and excited states, Ψ_g and Ψ_e^*. To obtain these, Ψ_g and Ψ_e^* states as well as the components P_x, P_y, and P_z along x, y, and z of the transition moment vector $\vec{P}$ are transformed to their irreducible representation Γ for the respective point group. Then, if the direct products $\Gamma\psi_g \dot{x} \Gamma_{P_x} \dot{x}\ \Gamma\psi_e^*$, $\Gamma\psi_g\ \dot{x}\ \ \Gamma_{P_y}\ \dot{x}\ \Gamma\psi_e^*$, and $\Gamma\psi_g \times \Gamma_{P_z} \times \Gamma\psi_e^*$ are totally symmetric A_1 (or A_{1g}, except *e.g. for* D_{4h}), the transition is symmetry-allowed in the respective direction x, y, or z of the polyhedron. Thus, the analysis of the polarization behaviour of properly assigned dd bands in silicate spectra yield the symmetry of the $3d^N$-ion bearing sites in the mineral structure. Often, the point group, thus obtained, is higher than the site symmetry according to the space group of the structure ("pseudo-symmetry").

An approach to theoretically calculate, on the basis of the point charge model (Ballhausen 1954), the energies of d-orbitals in distorted coordination polyhedra has been worked out by Wood and Strens (1972). However, this holds for those transition elements only, whose cations have one effective d-electron (Ballhausen 1954). Energy level diagrams, equivalent to the Tanabe-Sugano diagrams, have been calculated by König and Kremer (1977) for low symmetry fields by superimposing on 10Dq appropriate distortion parameters characterizing the deviation of the low symmetry fields from regular octahedral fields.

As a summary, from properly assigned, polarized single crystal spectra of silicate minerals we may obtain

(i) Type of $3d^N$-ion (ions) present in the case of well identified dd transitions.

(ii) $3d^N$-ion distribution in multisite structures. The same ion on different sites with different geometry experiences different 10Dq and, hence, creates the respective sets of bands.

(iii) $CFSE_{3d^N}$ in the respective site or sites of the mineral structure.

(iv) Effective site symmetry or symmetries (pseudo-symmetry) of the $3d^N$-ion containing octahedra.

2.2 How does $CFSE_{3d^N}$ enter thermodynamic funtions?

In the preceeding section it was stated that the crystal field stabilization energy, $CFSE_{3d^N}$, is defined with respect to the energy of the free, gaseous ion in a spherical field (cf. also Schläfer and Gliemann 1961). Therefore, values of $CFSE_{3d^N}$ be introduced as a summand in those cases only, where the respective thermodynamic function, as Gibbs free energy ΔG^o, enthalpy ΔH^o, or inner energy ΔU^o of a reaction, is defined in the same way. This is the case for the introductory example of enthalpies of hydration $\Delta H_{h,m}$, because this quantity is defined by the reaction of the free ion in the gaseous state M_g^{n+}

$$M_g^{n+} + ne_g^- + nH_{aq}^+ = \frac{n}{2}(H_2)_g + M_{aq}^{n+} \qquad (19)$$

On the other hand, in reactions not involving the state of free, gaseous $3d^N$-ions, absolute values of $CFSE_{3d^N}$ cannot directly be used. However, the changes in $CFSE_{3d^N}$, involved in reactions, $\Delta CFSE3d^N$, may indeed be introduced in the changes of thermodynamic functions typical for the reaction. Changes in the crystal field stabilization energy on reaction occure as a consequence of different crystal fields in the various sites of the phases participating in the reaction.

Because the mean M-O distance R, which influences 10Dq strongly, is a function of pressure and temperature, the same is true for $CFSE_{3d^N}$ and $\Delta CFSE_{3d^N}$. However, thermal expansion and compressibility

$$\alpha = \frac{1}{V}\left(\frac{\delta V}{\delta T}\right)_p \quad \text{and} \quad \chi = -\frac{1}{V}\left(\frac{\delta V}{\delta p}\right)_T \qquad (20)$$

beeing bulk properties of the phases involved, cannot be used to calculate $CFSE_{3d^N} = f(T,p)$ or $\Delta CFSE_{3d^N} = f(T,p)$. For this purpose local thermal expansion and compressibility, i.e. that of the $3d^N$-ion centered polyhedron or polyhedra must be known. High pressure-high temperature X-ray diffraction yields averaged data on $[M_{1-x}M(3d^N)]O_n$ polyhedra in case of solid solutions. Therefore, the pressure-temperature dependence of $CFSE_{3d^N}$ must be determined from high pressure-high temperature spectra. Although there is good progress in this field (Burns 1982), we are far away from having systematic data on $CFSE_{3d^N} = f(T,p)$ in silicate minerals.

An example for the influence of crystal field stabilization energy on a phase transition is presented by Burns and Sung (1978) by searching the olivine (α) / spinel (β) transition for some $M_2[TO_4]$ compounds. At equilibrium pressure of transformation, $p_t^{\alpha/\gamma}$

$$\Delta G^{\alpha/\gamma} = \Delta U^{\alpha/\gamma} + p_t^{\alpha/\gamma}\Delta V^{\alpha/\gamma} - T\Delta S^{\alpha/\gamma} = 0 \tag{21}$$

and

$$p_t = \frac{T_t\Delta S^{\alpha/\gamma} - \Delta U^{\alpha/\gamma}}{\Delta V^{\alpha/\gamma}} \tag{22}$$

Syono et al. (1971) assumed the changes in entropy and volume on transition to be independent in a first approximation on the type of M and T in $M_2[TO_4]$. Then, the equilibrium pressure is a function of the change of the inner energy U of the reaction only (T = const) and on the basis of simple ionic model $\Delta U^{\alpha/\gamma} = const(R_M/R_T)$. With these assumptions, Syono et al. (1971) obtained indeed a linear relation $(R_M/R_T) = const'\cdot p_t^{\alpha/\gamma}$ for nontransition metal silicates and germanates. However, if M was a $3d^N$-ion, then strong deviations of $p_t^{\alpha/\gamma}$ towards lower pressure occurred, indicating additional stabilization of the γ-phases in case of $M = 3d^N$. Burns and Sung (1978) pointed out that the overall Gibbs free energy of reaction of equn. (21) is to be separated into two components

$$\Delta G^{\alpha/\gamma} = \Delta G'^{\alpha/\gamma} + \Delta G^{\alpha/\gamma}_{CFS} \tag{23}$$

where $\Delta G'^{\alpha/\gamma}$ is the overall Gibbs free energy change for the $\alpha\to\gamma$ transition for the hypothetical condition of zero crystal field splitting. $\Delta G^{\alpha/\gamma}_{CFS}$ is the overall change of Gibbs free energy caused by crystal field splitting

$$\Delta G^{\alpha/\gamma}_{CFS} = \Delta CFSE^{\alpha/\gamma}_{3d^N} - T\Delta S^{\alpha/\gamma}_{CFS} \tag{24}$$

$\Delta S^{\alpha/\gamma}_{CFS}$ is the change in electronic configurational entropy on transition

$$\Delta S^{\alpha/\gamma}_{CFS} = S_e^{\gamma} - S_e^{\alpha} \tag{25}$$

S_e is the electronic configurational entropy in the respective phase

$$S_e = k\cdot \ln W \tag{26}$$

$$W = N^{2N_L} \tag{27}$$

with k Boltzmann constant, W number of different ways to arrange the N electrons (d-electrons in the present case), and N_L Avogadro's number. In the hypothetical case of zero or spherical field N = 5. In cubic field of the spinel or γ-phase , the d-electron arrangement in the $^5T_{2g}$ground state is triply degenerate, hence N = 3 in spinel, whereas in the distorted octahedra of the olivine or α-phase (M1 point symmetry C_i, M2 point symmetry C_s), the ground state degeneracy is totally lifted and hence N = 0. Therefore, $\Delta S^{\alpha/\gamma}_{CFS}$ is always negative for α→γ. By evaluating spectra of α- and γ-$Fe_2[SiO_4]$ obtained at various pressures and temperatures, Burns and Sung (1978) were able to show that $\Delta CFSE^{\alpha/\gamma}_{Fe2+}$ is always negative for α→γ, its value decreasing on raising temperature. Furthermore, the authors stated that the change of electronic configurational entropy has comparable value as the molar entropy change ($\Delta S^{\alpha/\gamma} \cong -11.3$ $J \cdot mole^{-1} \cdot K^{-1}$) and that $\Delta S^{\alpha/\gamma}_{CFS}$ does not strongly depend on pressure and temperature, whereas the absolute values of S^{α}_{e} and S^{γ}_{e} increase with temperature. By means of this crystal field based formalism and on the basis of spectroscopic data available, Burns and Sung (1978) calculated, for three models of the value of ground state splitting, δ, in the γ-phase, the univariant curve of the α/γ-transformation in the pT-field. Though the best fit was found at appreciably higher pressures than experimentally determined (Akimoto et al. 1965), it is much closer to the experimental reality than the p_t predicted by the simple spheric ion model (see above). Furthermore, the slope of the experimental curve is closely fitted. Both observations show that the approach of equns. (23) and (24) is principally correct, and that there is a strong contribition from $\Delta CFSE_{3d}N$ to $\Delta G^{\alpha/\gamma}$ or $\Delta U^{\alpha/\gamma}$, respectively.

Another attempt to introduce changes of crystal field stabilization energy into thermodynamic considerations is made by Ganguly (1976) in a theoretical treatment of the binary solid solutions uwarovite-grossularite, uvarovite-andradite, and grossularite-andradite and the ternary ugrandite system. Using the "simple mixture" formalism of Guggenheim (1967), which gives the excess free energy of mixing symmetrical with molar fractions of components 1 and 2

$$\Delta G^{ex} = W_{12}X_1X_2 \qquad (28)$$

Ganguly (1976) argues that the pressure and temperature dependencies of the adjustable interaction parameter W_{12}

$$(\delta W/\delta p)_T = \Delta V^{ex}/X_1X_2 \quad \text{and} \quad (\delta W/\delta T)_p = -\Delta S^{ex}/X_1X_2 \qquad (29)$$

may be neglected ($\Delta V^{ex} \cong 0$ for the Ca-garnets, $\Delta S^{ex} \cong 0$ assuming random distribution) and that

$$\Delta H^{mix} \cong \Delta G^{ex} = W_{12}X_1X_2 \qquad (30)$$

Furthermore, the enthalpy of mixing was interpreted as

$$\Delta H^{mix} \cong \Delta H^{mix}_{est} + \Delta H^{mix}_{cov} + \Delta H^{mix}_{dis} + \Delta CFSE^{mix} \qquad (31)$$

In this interpretation, ΔH^{mix}_{est} represents contributions from changes in the electrostatic interactions between cation and coordinating anions to the heat of mixing, ΔH^{mix}_{cov} represents contributions of changes in covalent character of the cation-anion bond, ΔH^{mix}_{dis} is a contribution arising from structure distortions caused by mutual substitution of ions different in size, and, finally, $\Delta CFSE^{mix}$ is the change of crystal field stabilization energy on formation of the $3d^N$-ion bearing solid solution.

As the $3d^5$-configurated ion Fe^{3+} experiences no crystal field splitting of the ^6S spectroscopic state (Figure 3), ΔH^{mix} of the grossularite-andradite solid solution is completely due to non-crystal field effects

$$\Delta H^{mix}_{(Gross\text{-}Andr)} \cong \Delta H^{mix}_{est} + \Delta H^{mix}_{cov} + \Delta H^{mix}_{dis} \quad (32)$$

Values of $\Delta H^{mix}_{(Gross\text{-}Andr)}$ for the series were obtained from a theoretical evaluation of data on $Y_3Al_tFe^{3+}_{5-t}O_{12}$ garnets. $\Delta CFSE^{mix}$ of the uwarovite-grossularite and uvarovite-andradite solid solutions series were calculated as

$$\Delta CFSE^{mix}(X) = X\cdot[4n - 6(N - n)]\cdot(Dq - Dq_{ss})\cdot 11.966 \ J\cdot mole^{-1} \quad (33)^*$$

where Dq and Dq_{ss} are the crystal field splittings of Cr^{3+} in end member uvarovite and in the solid solution with molar fraction X, respectively. Dq_{ss} values for various X were calculated by using the $1/R^5$-relation given in equn. (11) from Dq in uvarovite end member with the assumption, that the decrease (increase) of the mean $Y^{[6]}$-O distance is given by a linear interpolation between the respective distances in uvarovite and grossular (andradite) as the uwarovite molar fraction decreases.

The values of $\Delta H^{mix}_{(Gross\text{-}Andr)}$, representing the non-crystal field contributions, were found to be much smaller than $\Delta CFSE^{mix}$ (ca. 15 o/o only in the extreme point of the uwar-gross$_{ss}$). This is in qualitative accordance with a result obtained in comparing the change in lattice energy ΔU on substitution of an ion with radius r by another one with radius r_o. When the crystal is, as a first approximation, considered as a continous, isotropic medium with compressibility χ and Poisson's ratio δ, then

$$\Delta U = \frac{6\pi r_o^3(\gamma - 1)}{\chi\cdot\gamma}\,\theta^2\left(1 - \frac{\theta}{\gamma}\right) \quad (34)$$

with $r = r_o(1 + \theta)$ and $\gamma = 3(1 - \delta)/(1 + \delta)$ (Nagasawa 1966). Using equn. (34), Schwarcz (1967) calculated $\Delta U \cong 4.6$ kJ/g-atom$_{Cr^{3+}}$ for corundum-eskolaite solid solutions, whereas the crystal field stabilization energy is around 245 kJ/g-atom$_{Cr^{3+}}$.

Therefore, it is obvious that $(\Delta H^{mix}_{est} + \Delta H^{mix}_{cov} + \Delta H^{mix}_{dis})$ is of minor importance in ΔH^{mix}, such that the respective values of (Andr-Gross)$_{ss}$ may be used along with the respective $\Delta CFSE^{mix}$ values to obtain ΔH^{mix} of the (Uwar-Gross)$_{ss}$- and (Uvar-Andr)$_{ss}$-series (Ganguly 1976).

The only problem in this approach may be assumption of a linear

* Ganguly (1976) gave a slightly different equation using $\Delta = 10Dq$, which, however, is equivalent (33). This equn. is given here for consistency with equn. (10).

interpolation of the mean $Y^{[6]}$-O distances, because there are indications from piemontite spectra (Abu-Eid and Langer unpubl.) and from spectra of manganian andalusites (see below) that there may exist much smaller gradients $\pm(dR/dX)$ for $3d^N$-ion centered octahedra than predicted from a linear interpolation of end member R-values and as obtained from X-ray structure refinements. This implies that the mean octahedral distance $R_{M(3d^N)}$ of a larger transition metal ion decreases on substitution by Al to a smaller extent than predicted by the assumption that

$$R(X) = [(1 - X)R_{M(3d^n)} + XR_{Al}] \qquad (35)$$

Such an observation is in accordance with the assumption of a contribution of ΔH^{mix}_{dis} (equn. 31), because in case of validity of equn. (35) there would be no structural distortion causedby substitution of ions different in size.

A final example for the incorporation of $\Delta CFSE_{3d^N}$ in thermodynamic functions may be briefly discussed here, which is concerned with the distribution of $3d^N$-ions between coexisting phases or different sites of one phase. The equilibrium constant of the exchange reaction

$$Y^A + E^B = E^A + Y^B \qquad (36)$$

where E and Y are two elements substituting for each other and A and B either two sites in one phase (intracrystalline distribution) or two phases both with one site for E or Y (intercrystalline distribution),is given by

$$K_E^{A/B}(p,T) = \left(\frac{X_E^A X_Y^B}{X_Y^A X_E^B}\right)\left(\frac{\gamma_E^A \gamma_Y^B}{\gamma_Y^A \gamma_E^B}\right) \qquad (37)$$

Here, X are site fractions or mole fractions, respectively, of the elements E and Y in the respective sites or phases, respectively, and γ are the corresponding activity coefficients. Furthermore, the partition coefficient K_D is given by

$$K_E^{A/B}/K_\gamma = K_D = f(p,T,X) \qquad (38)$$

Assuming $K_\gamma = 1$, one may obtain

$$K_D^{A/B} = \left(\frac{X_E^A}{1 - X_E^A} / \frac{X_E^B}{1 - X_E^B}\right) = \exp(-\Delta G^o/RT) \qquad (39)$$

Schwarcz (1967) interprets the Gibbs free energy for the exchange of one g-atom E and Y between the two sites or phases, respectively, in a similar way as just discussed for the enthalpy of mixing

$$\Delta G^o = \Delta G^o_S + \Delta G^o_{EZ} + \Delta G^o_{EF} \qquad (40)$$

where the functions characterized by subscripts S, EZ, and EF are contributions from differences in structural properties, i.e. geometry and ligands, of the two sites (phases) A and B, from differences of the effective ionic charges or covalency of the E-O and Y-O bond in A and B, and from differences in the crystal field stabilization in A and B, resoectively.

For dilute solid solutions with X_E^A, $X_E^B \ll 1$, i.e. for the case of trace element distribution, the approximation

$$K_{D,E}^{A/B} \cong K_E^{A/B} = X_E^A/X_E^B \quad (41)$$

is valid, in which $K_E^{A/B}$ is the Nernst distribution coefficient. Schwarcz (1967) was able to demonstrate that the Nernst distribution coefficients of divalent and trivalent $3d^N$-ions in coexisting pairs of biotite/hornblende, olivine/orthopyroxene, and garnet/biotite follows the sequence of $CFSE_{3d^N}$ in the respective phases and that $\Delta CFSE_{3d^N}^{Bio/Hbl} > 0$, $\Delta CFSE_{3d^N}^{Ol/Opx} > 0$, and $\Delta CFSE_{3d^N}^{Bio/Gt} > 0$. These results demonstrate clearly, that the interpretation of the Gibbs free energy of distribution ΔG^o of equn. (39) suggested in equn. (40) is at least qualitatively correct.

3. METHODS OF MICROSCOPE-SPECTROMETRY

Polarized spectra of small natural and synthetic single crystals of $3d^N$-ion bearing silicate minerals are measured in our laboratory with an automated single-beam microscope-spectrometer (Langer and Frentrup 1979) schematically outlined in Figure 5A.

The instrument is operated in a wavenumber step-scan procedure as indicated in Figure 5B and C. This produces high relative accuracies of absorbance values, typically ± 0.001 to ± 0.002 in the spectral range 35,000 - 12,000 cm^{-1} and ± 0.01 to ± 0.02 in the near infrared. The instrument is equipped with UV-transparent condenser and objectives with proper chromatic correction down to about 12,000 cm^{-1} (Zeiss Ultrafluars). This allows for measuring spectra up to 40,000 cm^{-1} (250 nm) in the UV.

The scheme of the instrument in Figure 5A shows that an image of the prisms is produced in the plane of the entrance diaphragm, in the object plane, and in the plane of the exit diaphragm. Therefore, the absolute diameter of entrance aperture and the magnification scale of the condenser define the effective diameter, d_a, of the illuminated area in the object plane. The smallest d_a may be about 8 μm. Furthermore, from the divergent beam produced by the objective, the exit diaphragm can be set to cut an even smaller measuring area, the diameter, d_m, of which is defined by the magnification scale of the objective and the absolute diameter of the exit diaphragm. The smallest d_m may be about 3 μm. However, such small spots can only be used in spectral rages, where the response of the detectors is high.

The spectral resolution depends on the spectral band width, $\Delta\tilde{\nu}_{1/2,m}$, of the measuring radiation. To obtain sufficient measuring intensity, $\Delta\tilde{\nu}_{1/2,m}$ cannot be smaller than 50 cm^{-1}. Thus, all bands with true half band widths $\Delta\tilde{\nu}_{1/2,t} \gtrsim 500$ cm^{-1} are not subject to finite spectral resolution (Ramsey 1952). Most electronic transitions of minerals produce bands with true half widths larger than 500 cm^{-1} (Table 1).

A problem may arise in microscope-spectrometry when band intensities

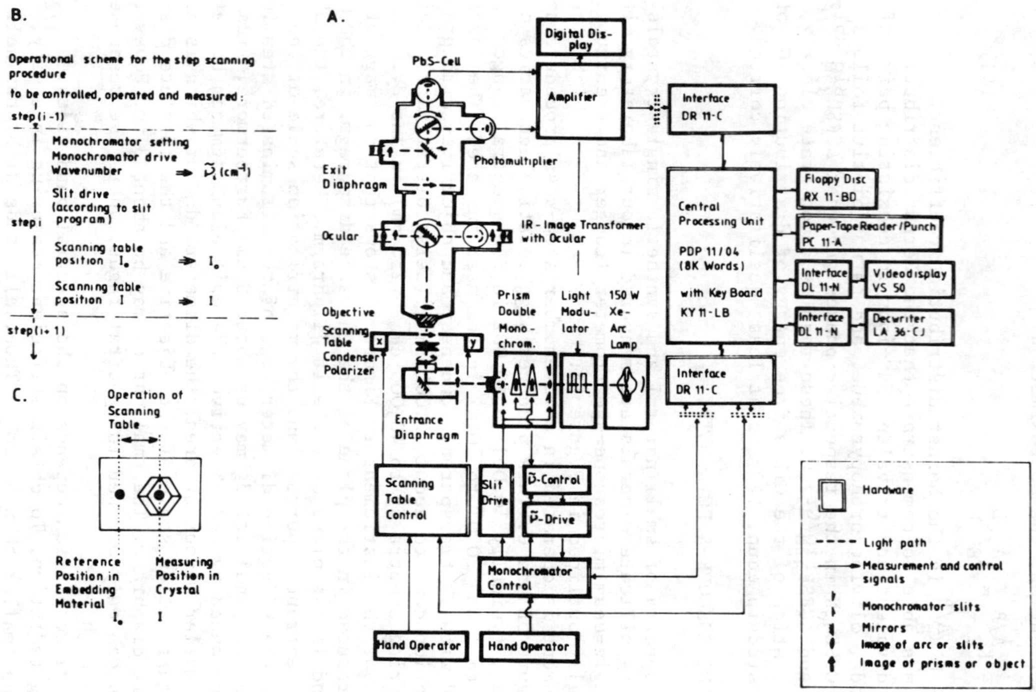

Figure 5. Schematical sketch of an automated microscope-spectrometer (Langer and Frentrup 1979), A, and the wavenumber step scanning procedure adopted to record the spectra, B and C. In favourable spectral ranges, the smallest measuring spot may have diameter of about 3 μm.

are to be evaluated from polarized spectra obtained with strongly divergent measuring radiation (high magnification of the objective). In such cases, maximum intensities may be reduced compared to the true values, an effect caused by mixing of polarizations, as pointed out by Goldman and Rossman (1977). In such cases only relative α_{max}-values may be obtained from spectra scanned with the same measuring conditions.

The preparation of noncubic crystals for measuring polarized spectra with $\vec{E}$ parallel to the main optical directions X, Y and Z (n_α, n_β n_γ) requires careful work. Two methods are used:

(i) In case of crystals, which are too small to be handled as individual single crystals, i.e. with longest dimension below about 100 μm, and/or in cases where the UV spectra are to be scanned above about 30,000 cm^{-1}, the crystals are spread on quartz glass slide, embedded in glycerole and covered with a thin quartz glass covering slide, which is fixed by a glue. Crystals in the appropriate orientation, to allow for measurements with $\vec{E}\parallel$ to two of the main optical directions, are selected by means of orthoscopic as well as conoscopic observations.

(ii) When crystals are large enough to be manipulated individually and when spectra are to be scanned only up to about 30,000 cm^{-1}, then following procedure is adopted: The crystals are oriented by X-ray oscillation photographs such that the rotation axis is parallel to either a, b, or c crystallographic axis. In this orientation, the crystals are embedded on a glass slide in two component epoxy resin (e.g. Araldite E) showing no absorption, at least for thin films, in the NIR/VIS/UV up to about 30,000 cm^{-1}. On two opposite sides of the crystal, quartz plates parallel c are also embedded and on one of the two other sides a diaspore (010) platelet is mounted. The mount is then ground and, if necessary, polished from both sides until the desired thickness t is obtained, usually 20 to 50 μm. The thickness t of the mount, i.e.also that of the crystal to be studied, can be accurately determined from the retardation $\Gamma = t(n_\gamma - n_\alpha)$, measured on the diaspore plate by means of a comparator. Due to the high birefringence of diaspore, $(n_\gamma - n_\alpha) = 0.054$, thicknesses may be determined to within ±0.2 μm. The success of the whole procedure is checked by conoscopic observations.

The second procedure has the advantage of accurate orientation to within about ±2° of $\vec{E}$ with respect to X, Y, or Z and of accurate thickness measurements. The latter are indispensible for obtaining $\alpha_{\tilde{\nu}}$-values needed for any quantitative evaluation of the spectra in terms of concentration of the absorbing species (equns. 8, 9).

For recording spectra at low temperature, a liquid nitrogen cell has been developed (Smith et al. 1982) allowing for measurements at 100±10 K.

4. APPLICATIONS

4.1. Cr^{3+}, an ion with nondegenerate ground state, in garnet, clinopyroxene, and kyanite

It was already pointed out that Cr^{3+} does not gain additional stabilization energy in distorted octahedra compared to octahedra with point group O_h because of its nondegenerate crystal field ground state.

Therfore, differences in the mean Cr^{3+}-O distances, $R_{Cr^{3+}-O}$, are of greatest importance for v lues of 10Dq and, hence, also of $CFSE_{Cr^{3+}}$ in the octahedra of garnet, clinopyroxene, and kyanite with their site symmetries $\bar{3}$ (Y-site), 2 (M1 site), and 1 (M1 and M2 sites, where Cr^{3+} for Al substitution takes place, Langer and Seifert 1971), respectively. Differences in term-term interactions and in the degree of spin-orbit coupling may contribute.

On the basis of the discussion in section 2.2 it may obviously be expected that an evaluation of $CFSE_{Cr^{3+}}$ in garnet, clinopyroxene, and kyanite might provide an understanding of the partitioning behaviour of Cr^{3+} in kyanite-bearing eclogite xenoliths from the mantle or in metamorphic kyanite-bearing eclogites.

Examples of spectra of Cr^{3+} in garnets were already presented in Figure 1. For a detailed study the reader is referred to the paper by Amthauer (1976). It is obvious from the data of this author, that 10Dq and, hence, $CFSE^{Y}_{Cr^{3+}}$ depends on site fractions $x^{Y}_{Cr^{3+}}$ and $x^{X}_{Mg,Ca,Fe^{2+}}$ as well, quantitative relations still being to be established on the basis of spectra of synthetic garnets with systematic variations of the site fractions mentioned. In any case, the above observation is in accordance with the X-ray refinement results of Novak and Gibbs (1971), who found the Y-O distance to depend on the population of both sites, octahedral Y and triangular dodecahedral X.

Polarized single crystal spectra on synthetically grown crystals of the clinopyroxene kosmochlore, $NaCr^{3+}[Si_2O_6]$, were measured by Abs-Wurmbach et al. (1985). Also, spectra of natural, Ti^{3+}-bearing kosmochlore from the Toluca meteorite (original sample, British Museum), and of a Cr^{3+}-aegirine-augite from St. Marcel, Italy, were obtained. Spectra of crystals from the first two samples are shown in Figure 6. The spectra of the synthetic end member kosmochlore were recorded on idiomorphic, prismatic crystals approximately 60 μm long and 20 μm in width, whereas crystals of the natural material were somewhat larger, but irregularely shaped fragments. Their extinction was uniform.

When comparing the spectra of the chemically complex natural material with those of the pure end member crystals (Figure 6), it is evident that the major band systems of the natural crystals are caused by dd transitions of Cr^{3+}, except for the 27,100 cm^{-1} band in the X-spectrum. This band was interpreted as caused by $Fe^{2+}(M1)Ti^{4+}(M1)$ charge transfer on the basis of the polarization behaviour and of the observed energy compared to that calculated (27,400 cm^{-1}, Abs-Wurmbach et al. 1985) for the M1-M1 distance in kosmochlore, 3.084 Å (Clarke et al. 1969). Furthermore, it was concluded, that the apparent intensity enhancement of the band system at 21,900 cm^{-1} in natural kosmochlore is caused by dd transitions of Ti^{3+}, overlapping with the second spin-allowed dd band of Cr^{3+}.

The site symmetry of the M1 octahedron in clinopyroxene is 2. Therfore, Figure 7 shows term diagrams of Cr^{3+} on descending symmetry from O_h to C_2 (C_2'', diagonal twofold axis of D_{4h} retained). Also shown in this diagram are the symmetry allowed transitions with their polarizatons along the octahedral axes x, y, and z. In addition, a part of the octahedral M1 zig-zag chain (a.), a single octahedron with axes set for C_{2v} (C_2'') pseudosymmerty (b. and d.) as well as orientation of the main optical directions n_X, n_Y, and n_Z, with respect to the octahedral axes x_o, y_o, and z_o.

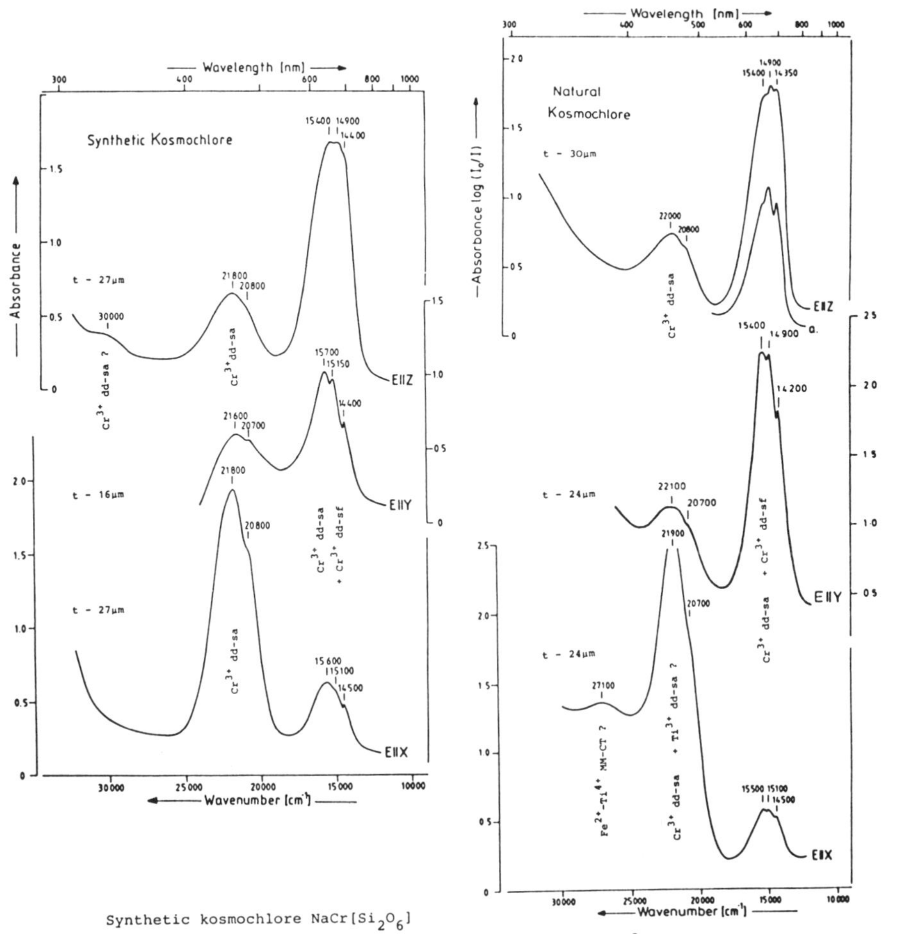

Synthetic kosmochlore $NaCr[Si_2O_6]$

Natural kosmochlore $(Na_{0.93}Ca_{0.06})(Mg_{0.08}Fe^{2+}_{0.02}Al_{0.04}Ti^{3+}_{0.09}Cr_{0.77})$
$[Si_{2.02}O_6]$ (Toluca Meteorite, BM 81870)

Figure 6. Polarized absorption spectra of micro-crystals of synthetically grown end member kosmochlore, $NaCr^{3+}[Si_2O_6]$ (left) and natural, meteoritic kosmochlore (right) (Abs-Wurmbach et al. 1985).

It is obvious from inspe tion of Figure 7 and comparison with the obseved band polarizations in Figure 6, that neither $C_2(C_2'')$ symmetry nor $C_{2v}(C_2'')$ pseudosymmetry do explain the pleochroitic behaviour of dd bands of Cr^{3+} in clinopyroxene M1 sites. It was shown, instead, that cyclic pseudosymmetry D_{2d} could best explain the polarization behaviour (Abs-

Wurmbach et al. 1985). In D_{2d}, x_o is parallel to optical n_Y, and y_o or z_o are approximately parallel to optical n_Z or n_X, respectively. Symmetry selection rules for Cr^{3+} in D_{2d} (Figure 7) are, thus in accordance with the observed pleochroism. This assignment is also in accordance with subgroup-supergroup relations between C_2 and D_{2d}.

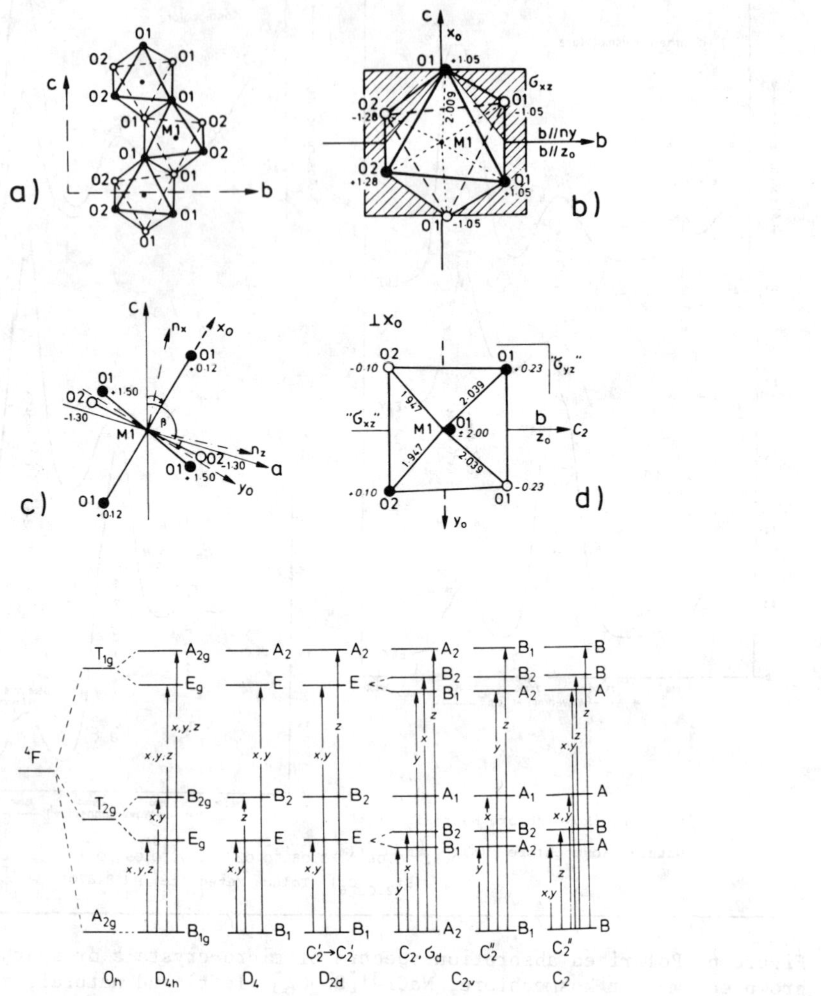

Figur 7. An attempt to analyze the observed pleochroism of dd bands of Cr^{3+} in clinopyroxene M1 octahedra on the basis of term diagrams on descending symmetry from O_h to C_2, the site symmetry of M1 (lower part). The upper part shows the M1 zig-zag chain (a), the M1 octahedron taken as a C_{2v} polyhedron (b, d), and the orientation of the optical directions n_X, n_Y, n_Z with respect to the octahedral axes x_o, y_o, z_o in C_{2v} and C_2 (c) (Abs-Wurmbach et al. 1985).

Spectra of synthetic crystals of two chromium kyanites with 15 or with 50 mole o/o of the theoretical end member "Cr_2SiO_5" are shown in Figure 8. These spectra were recorded on prismatic crystals approximately 50 μm long and 15 μm in width.

Apart from absorption edges with position relatively far up in the UV, three main band systems show up which are centered near 16,500, 24,200, and 29,000 cm^{-1} in the spectra of low chromium kyanite (Figure 8).

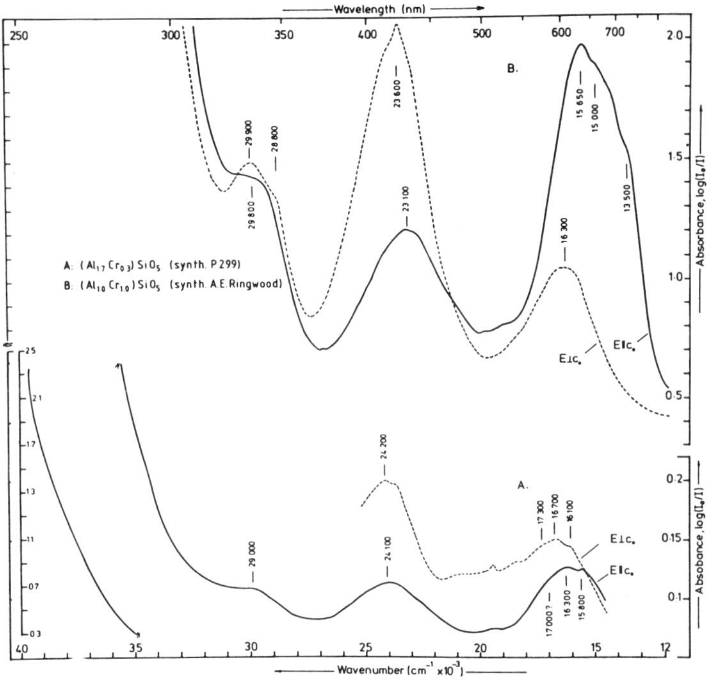

Figure 8. Polarized single crystal spectra of two different Cr^{3+}-containing kyanites (crystals of the high-chromium kyanite were obtained by courtesy of A.E.Ringwood, Canberra). Due to small crystal sizes (prismatic crystals about 50 μm long and 15 μm in width) spectra could only be recorded with $\vec{E}//c$ and $\vec{E}$ c

In the high chromium kyanite, the two first mentioned band systems occur at lower energies compared to the low chromium kyanite. The three band systems originate from low symmetry split transitions of Cr^{3+} in O_h field $^4A_{2g} \rightarrow {}^4T_{2g}(F)$, $\rightarrow {}^4T_{1g}(F)$, and $\rightarrow {}^4T_{1g}(P)$, respectively (cf. Figure 4). In the kyanite structure, the M1 and M2 octahedra, in which the chromium for aluminum substitution predominantly occurs (Langer nad Seifert 1971), have point symmetry C1 and, hence, all degeneracies are lifted. It is likely that the fine structure of the band systems near 24,200 and 16,500 cm^{-1} is caused by this lifting of degeneracy of the $^4T_{1g}(F)$ and $^4T_{2g}(F)$ excited states. On the other hand, the two low energy band systems are polarized in a way indicating again a cyclic pseudosymmetry. However, due to their small size, the crystals could not be manipulated as individual crystals. Therfore, sections allowing for fully establishing the pleochroic scheme could not be prepared. Nevertheless, from the barycenter of the $^4A_{2g} \rightarrow {}^4T_{2g}$ band system, $CFSE_{Cr}3+$ in kyanite could be determined.

Table 3 presents the chromium distribution for kyanite bearing eclogites as well as $CFSE_{Cr}3+$ in several clinopyroxenes, garnets, and kyanite. The stabilization energy does not depend strongly on composition in case of the clinopyroxenes and has always lowest value in this phase compared to garnet and kyanite. This is in accordance with the distribution behaviour of Cr^{3+} in the grosspydite xenolite. On the other hand, in garnet $CFSE_{Cr}3+$ strongly depends on the site fractions of Cr^{3+} and those of Ca, Mg, and Fe^{2+}. Therefore, $CFSE_{Cr}3+$ is to be determined on garnets from the actual paragenesis. If, however, $CFSE_{Cr}3+$ in uwarovite does apply in a first approximation in case of the grosspydite, and this is insofar a reasonable assumption as the garnet in this eclogite is rich in Ca, then

Table 3. Chromium contents and $CFSE_{Cr}3+$ in eclogite minerals.

(i)	Grosspydite-xenolite from Zagadochnaya kimberlite, Yakutia (Sobolev et al. 1968) p 15 kbar/ T 900°C (Seifert and Langer 1970)							
		Rock	Cpx		Gt		Ky	
	wt% Cr_2O_3	1.5	0.92	<	1.50	<	4.80	(sample Z13)
(ii)	Eclogite from Oberkotzau, Münchberger Gneiss Massiv p 13-17 kbar/T 620°C (Franz et al. 1986)							
		Rock	Cpx		Gt		Ky	
	wt% Cr_2O_3	0.084	0.126±0.084 (N=17)	>	0.066±0.052 (N=11)	<	0.284±0.145 (N=23)	(sample G262-a)
(iii)	$CFSE_{Cr}3+$		18 544 (a)		19 500 (d)		19 800 (f)	[cm^{-1}]
			222		233		240	[kJ·g-atom$_{Cr}$3+]
			18 470 (b)		21 050 (e)			
			221		251			
			18 650 (c)					
			223					

N number of point analyses by EMP on paragenetic grains (Kersten and Langer unpubl.)
(a) synthetic kosmochlore, (b) Cr-aegirine-augite (Abs-Wurmbach et al. 1985)
(c) Cr-omphacite, Yakutia (Khomenko and Platonov 1985)
(d) uwarowite, Outukumpu, Finland, 16.9 wt% Cr_2O_3 (Amthauer 1976, Langer 1984)
(e) pyrope, Meronitz, Bohemia, 5.5wt% Cr_2O_3 (Amthauer 1976)
(f) synthetic kyanite, 15 mole% "Cr_2SiO_5" (Langer 1984)

the sequence of the stabilization energies of chromium in clinopyroxene, garnet, and kyanite can indeed explain the distribution of chromium between these phases in the eclogite xenolite but not in the metamorphic eclogite, where garnet is lowest in Cr^{3+}. The consequence is that a strict proof of the governing influence of $CFSE_{Cr^{3+}}$ on Cr^{3+} distribution and a quantitative relation between these two properties can only be obtained by measuring polarized spectra on the paragenetic minerals themselves after their composition has been established by microprobe analyses. Such work is currently under way.

4.2. Mn^{3+}, an ion with degenerate ground state, in various silicate structures

Trivalent manganese requires high oxygen fugacities to be stable with respect to reduction. At least the oxygen fugacities of the Mn_3O_4/Mn_2O_3 (hausmannite/partridgeite) equilibrium or of other redox equilibria involving $Mn^{2+}Mn_6^{3+}SiO_{12}$ (braunite) in silicious systems are required. Provided such conditions prevailed during rock formation, then it is expected from the theoretical considerations in section 2.1. that Mn^{3+} should fractionate into those phases, whose structures contain octahedral sites with geometry close to D_{4h} symmetry, elongated or compressed, thus allowing for lifting of the ^5E_g ground state degeneracy.

The system Al_2SiO_5 - "Mn_2SiO_5" (apostrophies mark a theoretical end member) has been extensively studied by both, phase equilibrium experiments (Abs-Wurmbach et al. 1983) pertaining on the stability relations and Mn^{3+} distribution between the Al_2SiO_5 polymorphs, andalusite, sillimanite, and kyanite, and by single crystal structure refinements and spectroscopic studies of Mn^{3+}-bearing andalusites (Abs-Wurmbach et al. 1981) in order to evaluate the behaviour of the $3d^4$-configurated ion Mn^{3+} in this structure type.

Polarized spectra of synthetic and natural manganian andalusites are presented in Figure 9. Natural solid solutions may be represented by the formula $(Al_{1-x-y}Mn_x^{3+}Fe_y^{3+})_2[O/SiO_4]$ and pure synthetic ones by $(Al_{1-x}Mn_x^{3+})_2[O/SiO_4]$.

All spectra are dominated by two, may be three band systems centered near 15,000 (I), 22,000 (II), and near 23,500 (III) cm^{-1}. The last strong band shows up in spectra of natural viridines from Ultevis and Darmstadt, may form a high energy shoulder to band II in case of the material from Kanona, and could be contained as a component band in the spectra of synthetic crystals (Figure 9, lower part right). These bands are caused by spin-allowed dd transitions of Mn^{3+}, as they occur in natural and pure synthetic crystals as well.

The structure refinements prove, within the limits of experimental error, that Mn^{3+} substitutes for Al in the octahedral M1 sites of the structure only (Abs-Wurmbach et al. 1981). Therefore, the just mentioned bands were interpreted, using D_{4h} approximation of these octahedra (cf. Figure 10). The energy of band I represents the ground state splitting δ_2 of Mn^{3+} in the strongly eleongated M1 octahedra (Jahn-Teller splitting).

The octahedral elongation grows strongly on increasing substitution (x + y) of the Al(1) ions in the andalusite solid solution (Figure 11, left part). This is also obvious from the concomitant increase in ground

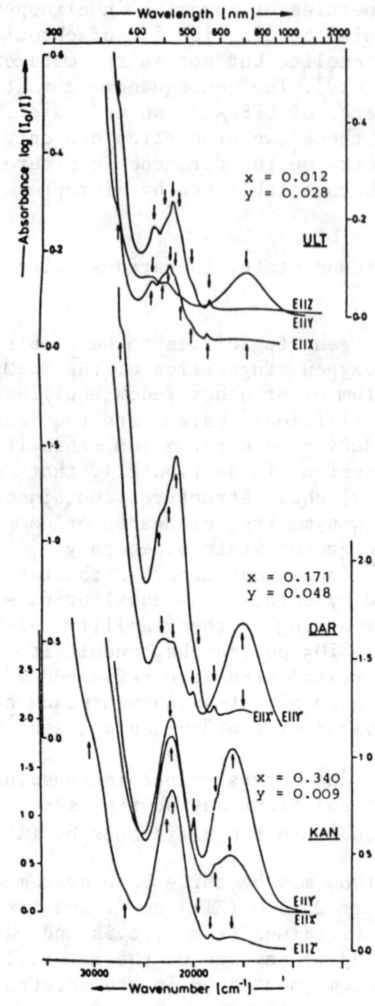

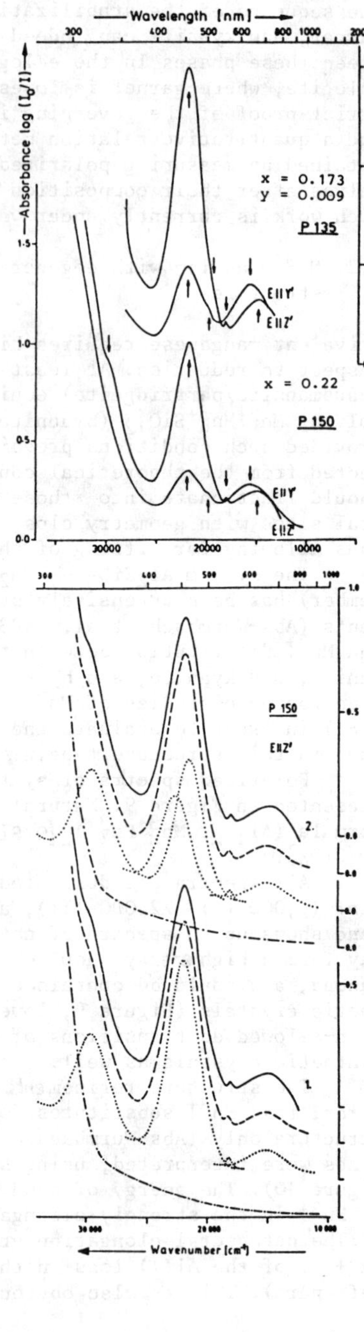

Figure 9. Polarized single crystal spectra of natural manganian andalusites, $(Al_{1-x-y}Mn_x^{3+}Fe_y^{3+})_2[O/SiO_4]$ (left) and of synthetic crystals, $(Al_{1-x}Mn_x^{3+})_2[O/SiO_4]$, (right) grown at 15 kbar/900°C/fO_2 of the Mn_2O_3/MnO_2 buffer. ULT, DAR = viridines from Ultevis and Darmstadt, resp., KAN = kanonaite from type locality, Zambia (from: Abs-Wurmbach et al. 1981).

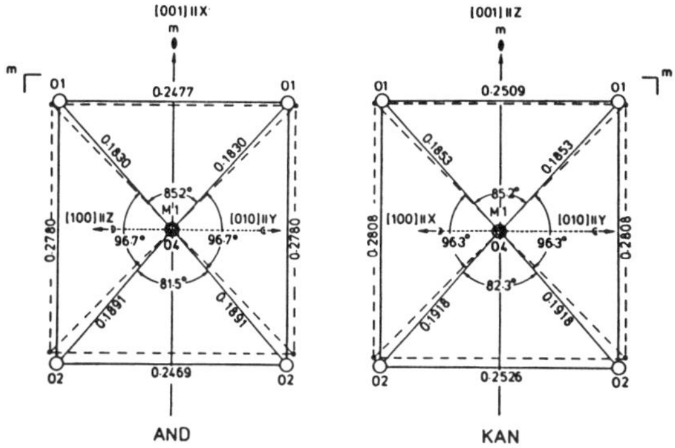

M(1) octahedra in the andalusite type structure.
The octahedra are viewed along the O(4)-M(1)-O(4) pseudoaxis C_4.
AND, andalusite M(1)-O(4) 0.2085 nm
KAN, kanonaite M(1)-O(4) 0.2245 nm

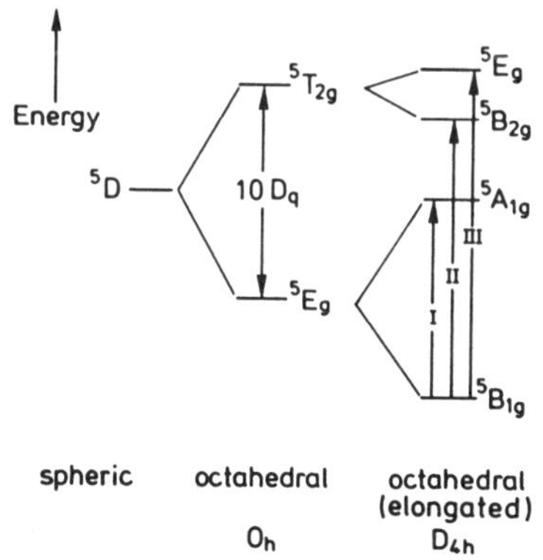

Figure 10. Geometry of the M1 octahedron in andalusite type structures (upper part). AND = andalusite (Burnham and Buerger 1961), KAN = kanonaite. The dashed lines represent an undistorted octahedral plane with same area as the actual one. Note that the geometry of the site is close to D_{4h} symmetry, when we neglect the small deviation of the O4-M1-O4 angle from 180° (deviation is 1.0 to 1.2° depending on composition). Also given is the energy level scheme of $3d^4$-configurated Mn^{3+} in octahedral coordination with O_h and D_{4h} symmetry (lower part) (from: Abs-Wurmbach et al. 1981).

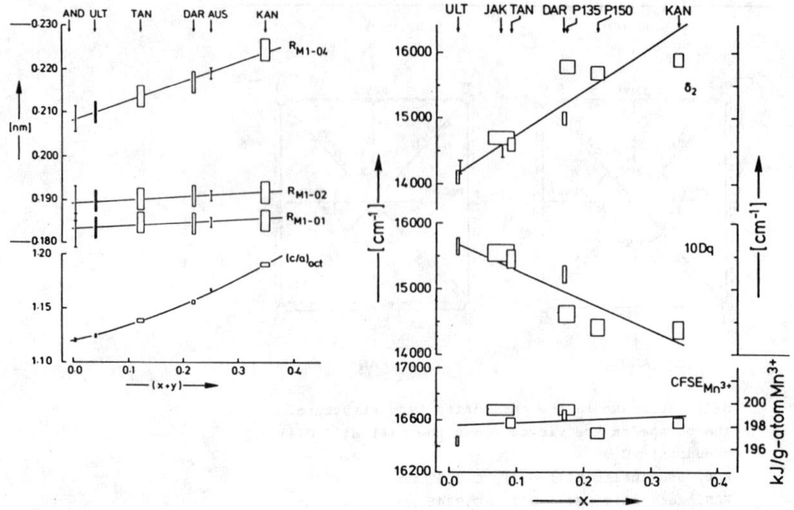

Figure 11. M1 octahedral geometry as obtained from X-ray structure refinements (left) and crystal field parameters of Mn^{3+}(M1) in the andalusite type solid solutions. δ_2 is the ^5E_g ground state splitting of Mn^{3+} (Jahn-Teller splitting) (from: Abs-Würmbach et al. 1981).

state splitting δ_2 (Figure 11, right part). An increase of the mean distance $R_{O-Mn^{3+}}$ within the solid solution series is indicated by the decrease of the crystal field parameter 10Dq (Figure 11, right part). Howevr, because the crystal field stabilization energy includes the ground state splitting

$$-CFSE_{Mn^{3+}} = 6Dq + \frac{1}{2}\delta_2 \tag{42}$$

the increase in δ_2 on raising Mn^{3+} contents counterbalances the decreasing crystal field splitting parameter 10Dq, such that $CFSE_{Mn^{3+}}$ is nearly constant within the solid solution series (Figure 11, right part).

We may calculate the mean $R_{Mn^{3+}-O}$ distance for the Mn^{3+} centered octahedra from the slope of the 10Dq vs. x plot of Figure 11 (right) using equn. (11)

$$R_{(Mn^{3+}-O)_2} = R_{(Mn^{3+}-O)_1}(\Delta_1/\Delta_2)^{1/5} \tag{43}$$

and $R_{(Mn^{3+}-O)_1} = 2.005$ Å for kanonaite with x = 0.34. By this, we obtain $R_{(Mn^{3+}-O)_2} = 1.971$ Å for the Ultevis viridine, appreciably higher than the mean M1-O distance 1.943 Å found in the X-ray structure refinement, which averages over both, Al-centered and Mn^{3+} centered M1 octahedra.

The equilibrium experiments in the system Al_2SiO_5-"Mn_2SiO_5" showed Mn^{3+} to strongly fractionate into the andalusite structure phase. This creates wide fields of andalusite$_{ss}$-sillimanite$_{ss}$ and andalusitesskyanite$_{ss}$ coexistencies. Furthermore, the incorporation of large amounts of Mn^{3+} in the andalusite solid solution enlarges the p,T-stability range of this structure type as compared with the pure, Mn^{3+}-free system Al_2SiO_5. This effect is shown in Figure 12, where breakdown curves for various contents of the theoretical end member "Mn_2SiO_5" in the andalusite solid solution are plotted. An andalusite$_{ss}$ with 20 mole o/o of the theoretical end member is stabilized by more than 8 kbar (T = 500°C) and by more than 500°C (p = 3.8 kbar) compared to the triple point conditions of 500°C/3.8 kbar in the pure Al_2SiO_5 system (Holdaway 1971).

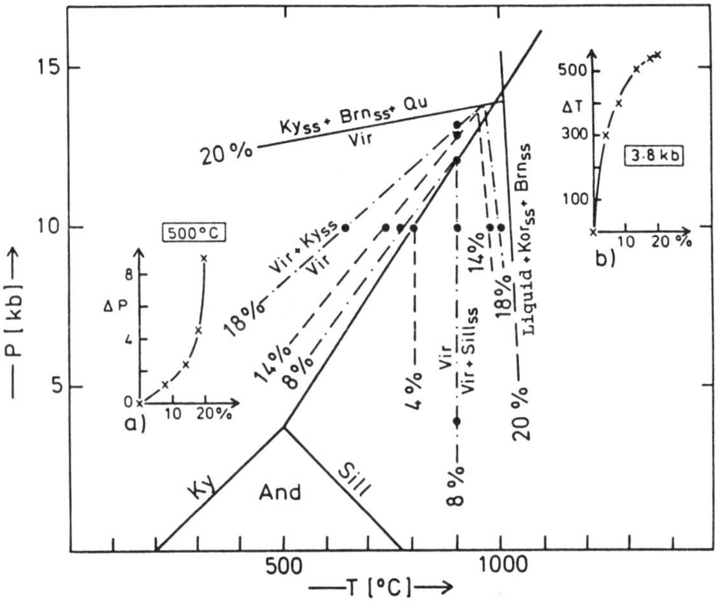

Figure 12. Phase relations of the system Al_2SiO_5-"Mn_2SiO_5" in the p,T, X-space, projected onto the p,T-plane. Equilibrium experiments were performed with the fO_2 of the Mn_2O_3/MnO_2-buffer. Note the large expansion of the and$_{ss}$ field on increasing Mn^{3+} contents (given in mole o/o of the theoretical "Mn_2SiO_5" end member) into the stability fields of sillimanite and kyanite typical for the pure system Al_2SiO_5. Inserts left and right show pressure and temperature increase of andalusite$_{ss}$ as compared with triple point conditions in the pure Al-system (Holdaway 1971)(from: Abs-Wurmbach et al. 1983).

This enormous effect must in some way be due to the exceptionally high crystal field stabilization energy of Mn^{3+} in the andalusite matrix (for comparison with other silicates cf. Table 4). The effect might be worked out by an interpretation, analogous to those given in section 2.2, of the Gibbs free energy of the reactions r

$$Al_2SiO_5(\text{and,ky,sill}) + \frac{2}{7}xMn^{2+}Mn_6^{3+}SiO_{12}(\text{brn}) + \frac{5}{7}xSiO_2 + \frac{1}{14}xO_2$$
$$= (Al_{1-x}Mn_x^{3+})_2SiO_5(\text{and}_{ss},\text{sill}_{ss},\text{ky}_{ss}) \quad (44)$$

For these reactions, r, one may then write

$$\Delta G^{o,r} = \Delta G^{o',r} + \Delta G^{o,r}_{CFS}$$
$$= 2x\Delta H^{o',r} - T\Delta S^o_c + 2x\Delta CFSE^{(\text{and,ky,sill})/\text{brn}}_{Mn^{3+}}$$
$$- T\Delta S^{o(\text{and,ky,sill})}_{CFS} \quad (45)$$

(cf. section 2.2, $\Delta S^o_c = R\,\Sigma(2x_i \ln 2x_i)$ = configurational entropy of mixing)

Not all data required are available up to now. Whereas x-values in the there structures are principally known for various p and T (Abs-Wurmbach et al. 1981), $CFSE_{Mn^{3+}}$ in kyanite$_{ss}$ and sillimanite$_{ss}$ as well as in braunite is still unknown. Polarized spectra on a Mn^{3+}-bearing natural kyanite, intergrown with purple yoderite from Mautia Hill, Tanzania, measured down to 12,000 cm^{-1} exhibit a band system near 21,000 cm^{-1} caused by Mn^{3+} and corresponding to band system II in manganian andalusites. However, there is no sign down to 12,000 cm^{-1} of a band caused by Mn^{3+} ground state splitting. It may occur at lower energies, not measured so far, but in any case the result is $CFSE^{ky}_{Mn^{3+}}$ $CFSE^{and}_{Mn^{3+}}$, in agreement with the observed partition behaviour of Mn^{3+}. The greatest problem in an evaluation of equn. (45) poses the opaque phase braunite*.

For these reasons, it is impossible to prove the stabilizing effect of Mn^{3+} in the andalusite matrix by numerical evaluation of equn. (45). However, this effect is so strong that differences in the configurational entropy in and$_{ss}$, ky$_{ss}$ and sill$_{ss}$, simply caused by different maximum x-values at given p,T in the three phases, cannot account for it alone.

Another extreme for the crystallochemical behaviour of Mn^{3+} is posed by the clinopyroxene-, the clinoamphibole- and by the garnet-structure. From natural systems it is known that only small amounts, if any, of Mn^{3+} are incorporated in these phases (e.g. Roy 1966) even in cases where the requisite high oxygene fugacities have prevailed. Then, it may be predicted that $CFSE_{Mn^{3+}}$ in these phases has small value, at least smaller than in the Mn^{3+}-bearing excess phase formed beside the just mentioned phases, i.e. mostly braunite.

An example is blanfordite, a Mn^{3+}-containing clinopyroxene of the acmite-diopside series. The mineral occurs in highly oxidized manganese rich layers of Indian gondites (Roy 1966, Lahiri 1971). This clinopyroxene with 0.09 Mn^{3+}_{tot} per formula unit and two clinomphiboles equally

---

* To the authors knowledge, nobody showed so far how to extract weak crystal field bands from mostly rather uncontoured reflection spectra caused by O-Mn^{3+} charge transfer or band transitions with their enormous oscillator strengths.

low or even lower in Mn^{3+} contents, although coexisting with Mn^{3+}-bearing excess phases, were studied by X-ray structure refinements, Moessbauer and polarized absorption spectroscopy (Ghose et al. 1986). From the refinements, Mn^{3+} was found to be contained in the clinopyroxene M1 octahedra and in the clinoamphibole M2 octahedra. As an example, the polarized spectra and their interpretation of blanfordite are shown in Figure 13.

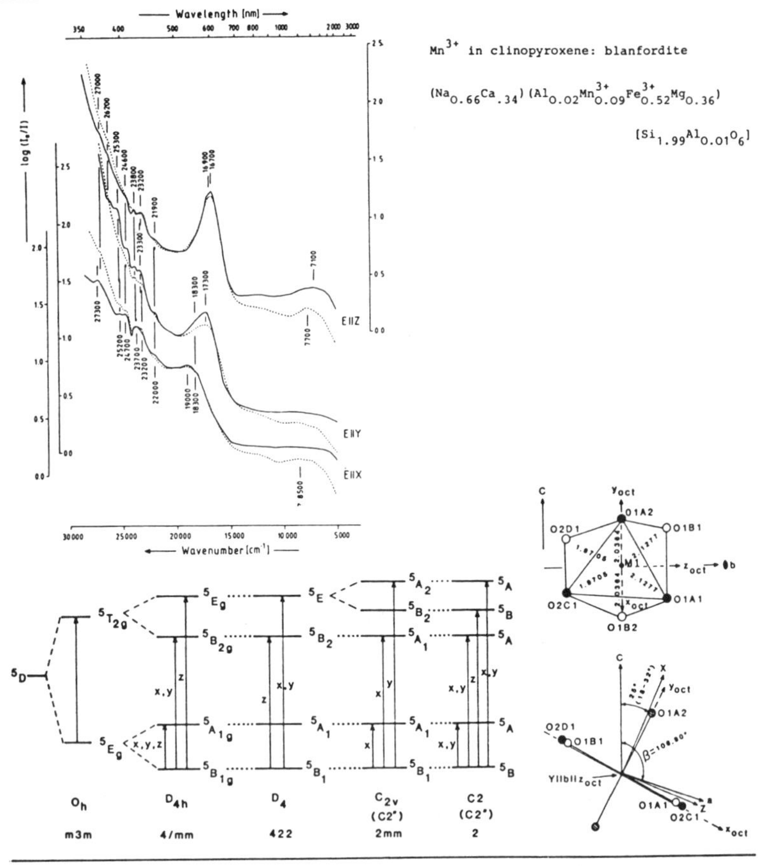

Figure 13. Polarized spectra of the Mn^{3+}-bearing clinopyroxene blanfordite (above). Dashed and solid lines: spectra at room or liq.N_2 temperature, respectively. Also shown is the energy level scheme for Mn^{3+} on descending symmetry and the M1 octahedron for C_{2v} and C_2 (both C_2''), as well as the orientation of optical directions with rescpect to the octahedra axes (from: Ghose et al. 1986).

The spectra are dominated by a weakly polarized absorption edge, to which minor absorption bands are attached attributed by Ghose et al. (1986) to spin-forbidden transitions of Fe^{3+} and Mn^{2+}. Thus, parts of total manganese are divalent. Furthermore, a strong band system between 20,000 and 15,000 cm^{-1} shows up, in addition to weak broad band in the near infrared with $\vec{E}//Z$ and possibly also $\vec{E}//X$. The band system between 20,000 and 15,000 cm^{-1} was found to consist of components $\tilde{\nu}_4$ at 19,000 cm^{-1} with $\vec{E}//X$, $\tilde{\nu}_3$ at 17,300 cm^{-1} with $\vec{E}//Y$, and $\tilde{\nu}_2$ at 16,800 cm^{-1} with $\vec{E}//Z$, whereas $\tilde{\nu}_1$ is at ca. 7,700 cm^{-1} with $\vec{E}//Z$ (possibly ca. 8,500 cm^{-1} with $\vec{E}//X$). The energy level diagrams (Figure 13) show that these results are best interpreted by the scheme for $C_2(C_2'')$ symmetry. Furthermore, the ground state splitting of Mn^{3+}, i.e. the Jahn-Teller effect, is much smaller in clinopyroxene M1 octahedra than in andalusite. Therefore, the crystal field stabilization energy of Mn^{3+} in clinopyroxene is indeed small (Table 4).

An even more extreme example in this respect is Mn^{3+} in the garnet structure as studied on garnet crystals grown under high pressure and temperatures and under defined oxygen fugacities in the systems Ca-Mn-Al-Si-O (Frentrup and Langer 1981) and Mn-Fe-Si-O (Lattard and Schreyer 1983). Single crystal spectra of such garnets are presented in Figure 14.

The spectra show strong absorption edges in the visible range, thus explaining for the orange colour of the garnet crystals. Attached to these edges, there occur minor sharp features assi gned to spin-forbidden transitions of Mn^{2+} (left spectrum) and of Mn^{2+} plus Fe^{3+} (right spectrum, lower part).In addition, the spectral envelope of the Ca-Mn^{3+}-garnet could only be fitted by assuming two additional bands near 20,400 and 17,000 cm^{-1} (Figure 14, left part). These component bands correspond to those indicated by the clearly discernible inflections near 20,000 and 19,200 cm^{-1} in the calderite-blythite solid solution. These two bands are caused by Mn^{3+} with certainty, because their intensity increases on increasing blythite content(Figure 14, right part). The $\varepsilon_{Mn^{3+}}$-values calculated for the two bands under discussion correspond to those expected for spin-allowed dd transitions in case of both garnets. The band energies were found to be in accordance (Frentrup and Langer 1981) with the ligand field energy diagram calculated for trigonal fields (Koenig and Kremer 1977), such as the octahedral site symmetry $\bar{3}$ in the garnet structure.

In the spectra of the calderite-blythite solid solutions (Figure 14, right part), no band indicative for ground state splitting of Mn^{3+} is observed. Hence, there is no Jahn-Teller effect of Mn^{3+} in this garnet. In the case of the Ca-Mn^{3+}-garnet (Figure 14, left part), such band could be present near 8,000 cm^{-1}. If the $(Mn^{3+}O_6)$ octahedra in both the garnet phases studied have really site symmetry $\bar{3}$, as the Y-sites in other garnets, then no such band is expected (Koenig and Kremer 1977). A structure refinement on the Ca-Mn^{3+}-garnet, whose spectrum is shown in Figure 14 (left part) confirmed this result (Arni et al. 1985). There occurred no deviations from Ia3d symmetry, and the octahedral geometry proved to be even closer to regular than in case of the Fe^{3+}-bearing garnet andradite (Figure 15). Therefore, it might be concluded that there exists no ground state splitting δ_2 of Mn^{3+}, comparable to that in other phases, in the garnet structural matrix and, as a consequence, $CFSE_{Mn^{3+}}$ in garnet is much lower than in other silicate structure.

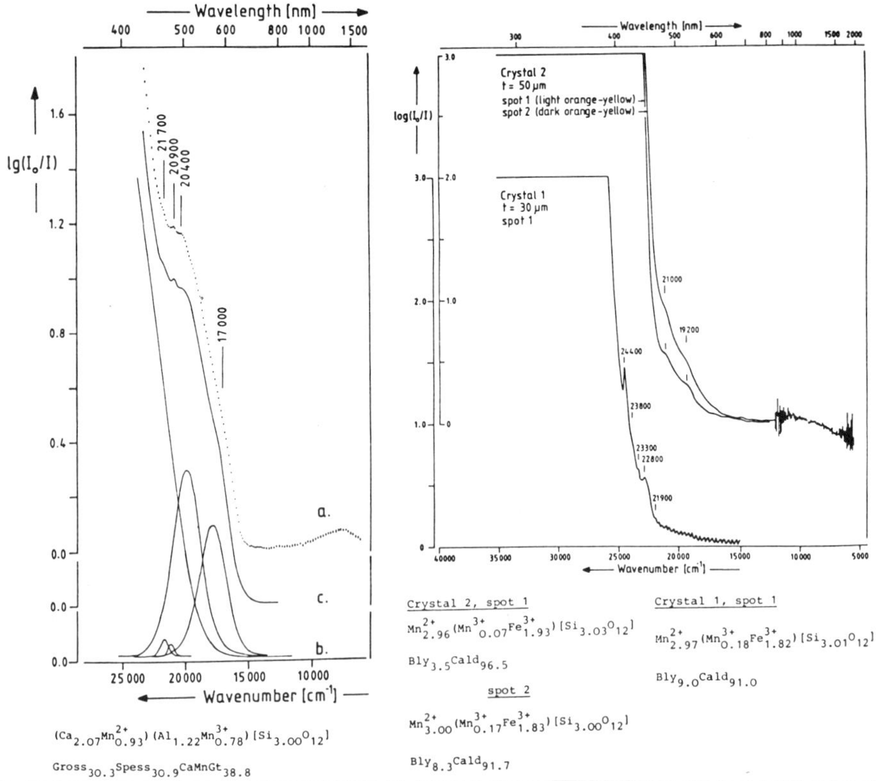

Figure 14. Spectra of a $Ca_3Mn_2^{3+}Si_3O_{12}$-bearing garnet solid solution (left, Frentrup and Langer 1981) and of a calderite-blythite solid solution (right, Langer and Lattard 1984). Abbreviations: Gross = grossularite ($Ca_3Al_2Si_3O_{12}$), Spess = spessartine ($Mn_3^{2+}Al_2Si_3O_{12}$), CaMnGt = $Ca_3Mn_2^{3+}Si_3O_{12}$ garnet Nishizawa and and Koizumi 1975), Bly = blythite ($Mn_3^{2+}Mn_2^{3+}Si_3O_{12}$), Cald = calderite ($Mn_3^{2+}Fe_2^{3+}Si_3O_{12}$).

Table 4 presents data on the crystal field splitting parameter 10Dq and crystal field stabilisation energy of trivalent manganese in various silicate structures. The sites, and site symmetries of the Mn^{3+}ontaining octahedra are also given along with the distortion parameters

$$\lambda_{oct} = \frac{1}{6}\sum_{i=1}^{6}(l_i/l_o)^2 \quad \text{(mean quadratic elongation, } l_o = \text{bond length in undistorted octahedron with same volume)} \qquad (46)$$

$$\sigma^2_{oct} = \frac{1}{11}\sum_{i=1}^{12}(90 - \theta_i)^2 \quad \text{(bond angle variance)} \qquad (47)$$

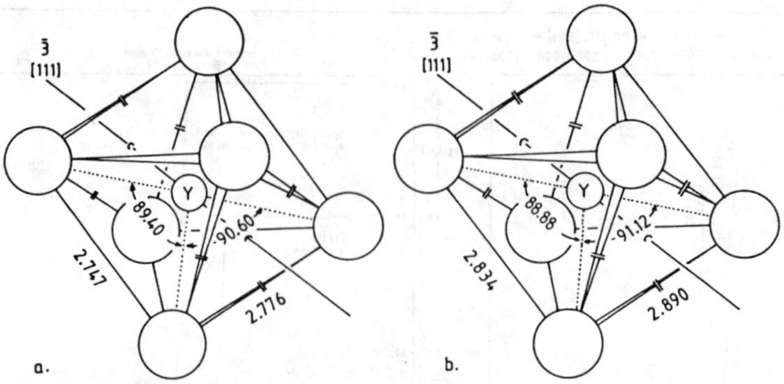

Figure 15. Octahedral geometry of a Mn^{3+}-bearing garnet with Y = $(Mn^{3+}_{0.36}Al_{0.64})$ of the system Ca-Mn-Al-Si-O (Arni et al. 1985), a., and of a natural andradite with Y = $(Fe^{3+}_{1.99}Al_{0.01})$ (Novak and Gibbs 1971), b. The Y-O distances are 1.953(2) for a. and 2.024(2) Å for b.

as introduced by Robinson et al. (1971).

The values of $CFSE_{Mn^{3+}}$ in the various mineral structures can indeed contribute to explain, in a qualitative way, the abundances of Mn^{3+} in the respective silicates apart from the conditions of excess manganese and of sufficiently high oxygen fugacities. One problem is posed by muscovite, where $CFSE_{Mn^{3+}}$ is found to be higher than in piemontite, whereas the Mn^{3+}-contents of alurgites are always low, even in parageneses with Mn^{3+}-bearing excess phases.

Any further quantitative evaluation of the influence of $CFSE_{Mn^{3+}}$ on formation of silicate solid solutions or on Mn^{3+}-partitioning between coexisting phases, according to formalims as in equn. (45), requires measurements of the composition and polarized spectra of the coexisting Mn^{3+}-containing minerals in the respective parageneses, except for those mineral phases in which $CFSE_{Mn^{3+}}$ is nearly independent on composition.

4.3. Fe^{3+}-bearing point defects in fayalite, an attempt to solve a kinetic problem

Nitsan (1971) calculated the stability of fayalite, $Fe_2[SiO_4]$, and of olivine solid solutions, $(Mg_{1-x}Fe^{2+}_x)SiO_4$, with respect to reduction and oxidation in the $\log fO_2$ - $1/T$ field. Within the fO_2-stability range at a fixed temperature, e.g. a range of fO_2 from $10^{-19.7}$ to $10^{-15.6}$ for fayalite at 800°C, there exist Fe^{3+}-bearing point defects, whose concentration increases with raising fO_2. As a consequence, fayalite has a ternary compositional field in the system Fe-Si-O, the extension of which is described in terms of the nonstoichiometry parameters $\bar{\eta}$ and $\bar{\xi}$ as shown in Figure 16 (Nakamura and Schmalzried 1983). The majority defects are:

$Fe^{\bullet}_{Fe}$ = Fe^{3+} on octahedral sites ("defect electrons")

V''_{Fe} = vacancies on octahedral sites (two noncompensated charges)

Table 4. Trivalent manganese in some silicate structures.

Phase	Mn^{3+} site	λ_{oct}	σ^2_{oct}	10Dq $[cm^{-1}]$	δ $[cm^{-1}]$	$CFSE_{Mn^{3+}}$ $[kJ \cdot g\text{-}atom^{-1}]$	Literature
Garnet $(Ca_{2.21}Mn^{2+}_{0.79})^{[8]}(Al_{1.20}Mn^{3+}_{0.80})^{[6]}[SiO_4]_3$ synthetic garnet	Y $\bar{3}$	1.0037	0.4	15,000	5,000	92	Frentrup and Langer (1981), Arni et al. (1985) [(a)]
Clinoamphibole $(Na_{0.08}K_{0.31})^{[A]}(Na_{0.66}Ca_{0.34})^{[8]}(Fe^{3+}_{0.87}Mn^{3+}_{0.48}Ti_{0.10}Mg_{3.51})^{[6]}[Si_{7.90}Al_{0.09}O_{22}(OH)_2]$ natural juddite	M2 2	1.0130	38.4	11,925⁺	ca.6,100⁺	142	Ghose et al. (1986)
Clinopyroxene $(Na_{0.66}Ca_{0.34})^{[8]}(Fe^{3+}_{0.52}Al_{0.02}Mn^{3+}_{0.09}Mg_{0.36})^{[6]}[Si_{1.99}O_6]$ natural blanfordite	M1 2	1.0221	32.0	13,650⁺	ca.8,100⁺	146	Ghose et al. (1986)
Clinozoisite $Ca_2(Al_{2.05}Mn^{3+}_{0.63}Fe^{3+}_{0.33})^{[6]}[OH/O/SiO_4/Si_2O_7]$ natural piemontite	M3 m	1.0344	85.4	13,450⁺	12,330	168	Burns and Strens (1968), Dollase (1969) [(a)]
$Ca_2(Al_{2.09}Mn^{3+}_{0.91})^{[6]}[HO/O/SiO_4/Si_2O_7]$ synthetic piemontite	M3 m	1.0349	83.7	13,460⁺	12,500	171	Kersten et al. (1987)
Zosite $(Ca_{1.97}Zn_{0.10})(Al_{2.86}Mn^{3+}_{0.03}Fe^{3+}_{0.10})^{[6]}[OH/O/SiO_4/Si_{2.02}O_7]$ natural thulite	M3 m	1.0211	29.1	14,000⁺	12,900	178	Kersten and Langer unpubl.), Tillmanns et al. (1984) [(a)]
Muscovite natural alurgite with ca. 0.6 wt% Mn_2O_3	M2 1	1.0425		14.050⁺	13,200	180	Annersten and Halenius (1976), Güven (1971) [(a)]
Andalusite $(Al_{0.960}Mn^{3+}_{0.012}Fe^{3+}_{0.028})_2[O/SiO_4]$ natural viridine (Ultevis)	M1 2	1.0131	17.2	15,700⁺	13,200	197	Abs-Wurmbach et al. (1981)
$(Al_{0.651}Mn^{3+}_{0.340}Fe^{3+}_{0.009})_2[O/SiO_4]$	M1 2	0.9678	16.0	14,400⁺	15,900	198	Abs-Wurmbach et al. (1981)

⁺ derived from mean values of the respective X,Y,Z polarized bands

(a) structural data, if not in the first reference

Fe'_{Si} = Fe^{3+} on tetrahedral Si-sites (one noncompensated negative charge)

$\{Fe^{\bullet}_{Fe}Fe'_{Si}\}$ = associates of Fe^{3+} on octahedral Fe-sites with Fe^{3+} on tetrahedral Si-sites

These defects are formed by the following reactions (Nakamura and Schmalzried 1983)

$$6Fe^{X}_{Fe} + SiO_2 + O_2(g) = 2V''_{Fe} + 4Fe^{\bullet}_{Fe} + Fe_2SiO_4 \quad (48)$$

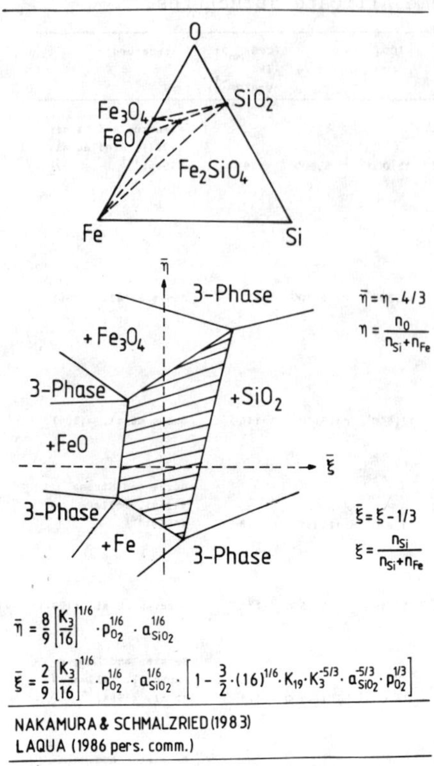

Figure 16. Ternary compositional field of fayalite caused by fO_2-T-dependent concentration of point defects and nonstoichiometry parameters, $\bar{\eta}$ and $\bar{\xi}$, which characterize the extension of the field.

$$Fe^{x}_{Fe} + Si^{x}_{Si} + \tfrac{1}{2}Fe_2SiO_4 + \tfrac{1}{2}O_2(g) = Fe'_{Si} + Fe^{\cdot}_{Fe} + \tfrac{3}{2}SiO_2 \quad (49)$$

$$Fe'_{Si} + Fe^{\cdot}_{Fe} = \{Fe'_{Si}Fe^{\cdot}_{Fe}\} \quad (50)$$

In the temperature range 1,000 - 1,280°C, reaction (50) is completely on the right side and, therefore, reactions (48) to (50) can be combined

$$10Fe^{x}_{Fe} + Si^{x}_{Si} + 2O_2(g) = 3V''_{Fe} + 6Fe^{\cdot}_{Fe} + \{Fe'_{Si}Fe^{\cdot}_{Fe}\} + Fe_2SiO_4 \quad (51)$$

Nakamura and Schmalzried (1983) carried out accurate thermogravimetric measurements on fayalite and olivine solid solutions heated under well defined, different oxygen fugacities and were able to evaluate the equilibrium constants K_3, K_2, and K_{11} for reactions (48), (49), and (50), respectively, from measured weight changes δ. The equilibrium constant $K_{19} = K_3^{3/2} \cdot K_2 \cdot K_{11}$ was obtained for reaction (51).

Using these equilibrium constants and their temperature dependence, one may calculate the nonstoichiometry parameters (Figure 16) and, hence, the total site fraction of Fe^{3+} in fayalite may be obtained as

$$x_{Fe^{\cdot}} = [Fe^{\cdot}_{Fe}] + [\{Fe^{\cdot}_{Fe}Fe'_{Si}\}] \qquad (52)$$

where $$[Fe^{\cdot}_{Fe}] = 3\delta, \quad [\{Fe^{\cdot}_{Fe}Fe'_{Si}\}] = 1\delta \qquad (53)$$

and $$\delta = \frac{4}{3}\eta = \frac{2}{3}\left(\frac{K_3}{16}\right)^{1/6} \cdot a_{SiO_2}^{1/6} \cdot p_{O_2}^{1/6} \qquad (54)$$

(Nakamura and Schmalzried 1983). From such results, we can calculate the concentration of Fe^{3+} in terms of site fractions $x_{Fe^{\cdot}}$ for temperatures and oxygen fugacities within the stability range of fayalite.

Point defects in olivines play a prominent role in electrical conductivity and deformation behaviour of these minerals. Furthermore, defect concentrations in natural olivines may contain information on the intrinsic oxygen fugacity during their formation in the earths mantle, provided the defect equilibria can be quenched. Therefore, information is needed on the time dependence of the above Fe^{3+}-forming reactions at various temperatures.

For this purpose, we designed following diffusion experiment (Cemic et al. 1986, 1987): A fayalite single crystal, grown or equilibrated at a given high oxygene fugacity fO_2', is oriented crystallographically, and is ground to a cube with edges parallel to the crystallographic a, b, and c axes (Figure 17). Then, the cube is reheated at a lower oxygene fugacity fO_2'' for a time shorter than required for complete re-equilibration. This reheating will cause a diffusional stream of defect

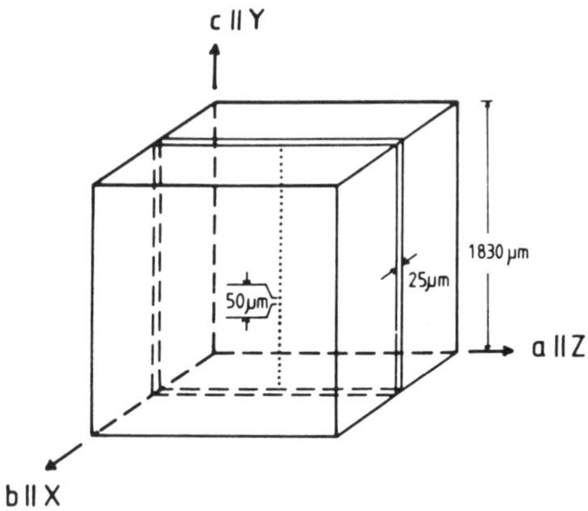

Figure 17. Schematic sketch of a crystallographically oriented cube ground from a fayalite single crystal, which was grown or equilibrated at relatively high oxygen fugacity fO_2' (in our experiment: 1,200°C, fO_2' defined by H_2/CO_2 = 1:20, yielding $fO_2' = 4.0 \cdot 10^{-9}$atm). The cube is now ready for a diffusion experiment at low fO_2'', whereafter an oriented slice parallel (010) is cut,to evaluate one dimensional Fe^{3+}-concentration profiles, e.g. along the c axis (dotted line). fO_2' and fO_2'' are to be kept within the stability range at the respective temperature(s).

electrons, $Fe_{Fe}^{\cdot}$, and of vacancies, V_{Fe}'', towards the surface of the cube, which is exposed to fO_2'', and a stream of Fe_{Fe}^{x}, i.e of Fe^{2+}, in the opposite direction, is equivalent to it. Diffusion on orthorhombic fayalite is a three dimensional, anisotropic process involving diffusion coefficients D_{xx}, D_{yy}, and D_{zz}., Thus, in the experiment a continuum of iso-$x_{Fe^{\cdot}}$-surfaces will be created inside the cube, whose form depend on the values of the just mentioned diffusion coefficients. When cutting a thin, oriented slice from the cube after the experiment (Figure 17), then the evaluation of three dimensional diffusion process can be reduced to two one dimensional cases by measuring Fe^{3+}-concentration profiles, i.e. $x_{Fe^{\cdot}}$-profiles, e.g. along Y//c and Z//a in the (010) slice of Figure 17. Such profiles may then be evaluated for mean, concentration independent, chemical diffusion coefficients, e.g. $\tilde{D}_{zz}$ in case of the profile along Y//c.

Obviously, the concentration profile cannot be analyzed in terms of the Boltzmann-Matano formalism, because this applies for the case of diffusion in two indefinite half rooms. Our experiment corresponds rather to the model of diffusion into a plate with thickness 2h, which is the cube's thickness along the profile (Figure 17). For this case, we obtain according to Crank (1975)

$$c_{Fe^{2+}} = c_{o,Fe^{2+}} + (c_{\infty,Fe^{2+}} - c_{o,Fe^{2+}})\{1 - \frac{4}{\pi}\sum_{n=0}^{\infty}\frac{(-1)^n}{2n+1} \left[\exp-(2n+1)^2\frac{\pi^2}{4}\cdot\frac{\tilde{D}t}{h^2}\right]\cos\left(\frac{2n+1}{2}\pi\frac{x}{h}\right) \qquad (55)$$

with $c_{Fe^{2+}}$ = concentration at point x in the crystal (profile), $c_{o,Fe^{2+}}$ = starting concentration (equilibrated with fO_2'), $c_{\infty,Fe^{2+}}$ = final concentration (after reequilibration with fO_2''), h = half crystal thickness in cm, t = time in sec, x = point in the c ystal (profile), and n = integer n = 0, 1, 2, 3, ...∞. For simplicity, equn. (55) is formulated for the diffusional stream of Fe^{2+}, i.e. Fe_{Fe}^{x}, into the crystal, which is equivalent to the diffusional stream of defect electrons, $Fe_{Fe}^{\cdot}$ or Fe^{3+}, and vacancies, V_{Fe}'', to the surface.

The problem arises then, how to measure small site fractions of Fe^{3+}, $x_{Fe^{\cdot}}$, in fayalite, which are in the order of 10^{-4} to 10^{-3} (corresponding to 0.0X - 0.X wt o/o Fe_2O_3) with an areal resolution in the μm range. Microscope-spectrometry might help, however dd bands of Fe^{3+} cannot be used, because these are all spin-forbidden and, hence, too weak in intensity to be measured on slices, about 20 μm thick, of crystals containing the just mentioned low Fe^{3+}-concentrations.

On the other hand, charge transfer transitions (Table 1) of $(Fe^{3+}O_6)$-clusters could be useful. Since these occur at lower energies than charge transfer transitions in $(Fe^{2+}O_6)$-clusters (Tossell et al. 1974), they should cause a shift of the UV absorption edge, caused by charge transfer in the $(Fe^{2+}O_6)$-clusters, towards lower energies (Figure 18). In other words, at a given wavenumber $\tilde{\nu}$ in the edge, the linear absorption coefficient $\alpha_{\tilde{\nu}}$ will be enhanced by increasing $x_{Fe^{\cdot}}$.

To check for such an effect, a series of fayalite crystal growth experiments were performed at 800°C, 10 kbar, and the oxygen fugacities

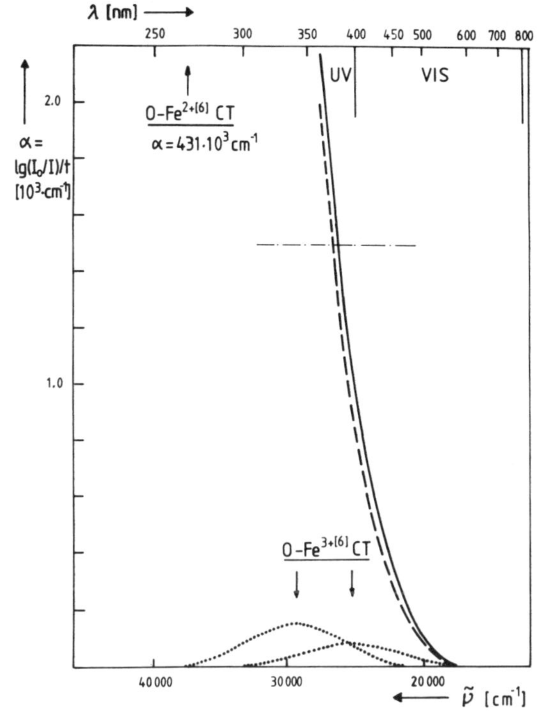

Figure 18. Theoretically calculated influence of O-Fe^{3+}-charge transfer transitions, caused by traces of Fe^{3+}, on the UV absorption edge of fayalite, i.e. the low energy wing of a band caused by O-Fe^{2+} charge transfer (polarization neglected). Dashed line: ideal fayalite with Fe^{2+} only, $x_{Fe^{2+}} = 1$, calculated for $\tilde{\nu}_{max} = 37,500\ cm^{-1}$ (Tossell et al. 1974), $\varepsilon_{Fe^{2+}} = 10^4$ [l·g-atom^{-1}·cm^{-1}], and $\Delta\tilde{\nu}_{1/2} = 8,000\ cm^{-1}$. The O-$Fe^{3+}$-charge transfer bands (dotted curves) are calculated for $x_{Fe^{3+}} = 1.7 \cdot 10^{-4}$ in fayalite (800°C/$fO_2 = 1.86 \cdot 10^{-12}$) obtained from data of Nakamura and Schmalzried (1983). O-Fe^{3+} charge transfer was assumed to occur at 29,400 and 25,300 cm^{-1} (Tossell et al. 1974) with $\varepsilon_{Fe^{3+}} = 5 \cdot 10^3$ [l·g-atom^{-1}·cm^{-1}] and $\Delta\tilde{\nu}_{1/2} = 8,000\ cm^{-1}$ (Lehmann 1970). Full line: sum of O-Fe^{2+} and O-Fe^{3+} charge transfer. Dot-dashed line: upper limit of $\alpha = f(\tilde{\nu})$ measurements for a plate with 20 μm thickness.

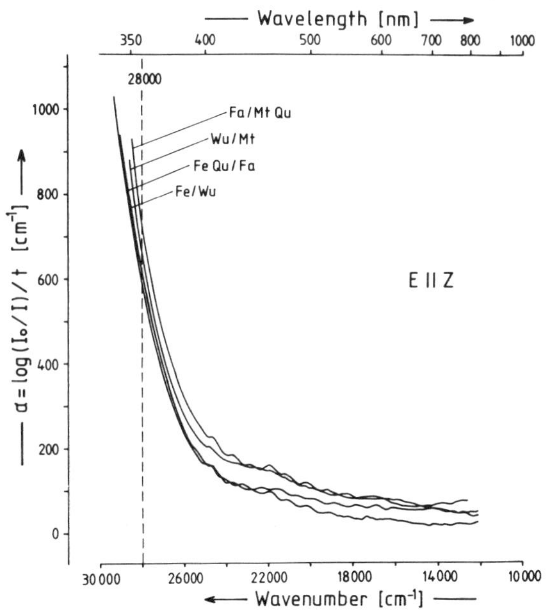

Figure 19. Ultraviolet spectra with $\vec{E}$//Z of fayalite crystals grown under different oxygen fugacities produced by the buffers indicated. Note the shift of the absorption edge with increasing oxygen fugacity (Cemic et al. 1986).

of the FeQu/Fa ($fO_2 = 4.68 \cdot 10^{-20}$atm), Fe/Wu ($fO_2 = 4.90 \cdot 10^{-19}$ atm), Wu/Mt ($fO_2 = 1.86 . 10^{-17}$ atm), and Fa/MtQu ($fO_2 = 1.86 \cdot 10^{-14}$ atm) buffers which produce $x_{Fe^{\cdot}} = 0.62 \cdot 10^{-4}$, $0.91 \cdot 10^{-4}$, $1,67 \cdot 10^{-4}$, and $5.27 \cdot 10^{-4}$, respectively.

Spectra with $\vec{E}//Z$ of oriented slices of such crystals are presented in Figure 19., which really shows the effect theoretically deduced in Figure 18. The $\vec{E}//X,Y$ spectra show similar results. Obviously, the red shift of the edge, or the increase in α at a given wavenumber, can be used to determine trace amount of Fe^{3+} in fayalite, $x_{Fe^{\cdot}}$, provided the required calibration is available. Such a calibration is shown in Figure 20. The correlations are $\alpha_{X,Y,Z} = a \cdot x_{Fe^{\cdot}} + b$ with a = 107,901, b = 647 for $\vec{E}//X$ (correlation coefficient r = 0.999), a = 69,468, b = 505.5 for $\vec{E}//Y$ (r = 0.93), and a = 165,804, b = 640,6 for $\vec{E}//Z$ (r = 0.993).

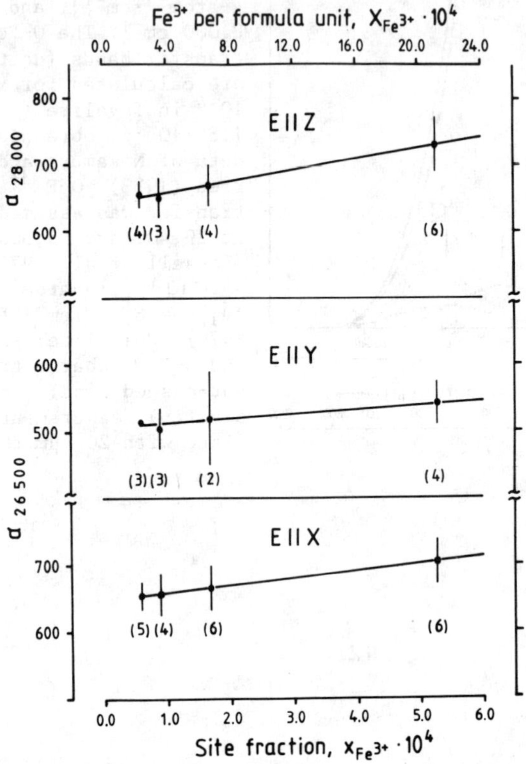

Figure 20. Calibration curves for α-values over background in the UV spectra of fayalite vs. Fe^{3+}-concentration, $x_{Fe^{\cdot}}$ (site fraction) as calculated on the basis of the results of Nakamura and Schmalzried (1983). Numbers in brackets indicate the number of different crystal mounts used (mounts prepared following procedure (ii) of section 3, each mount was measured four times) (Cemic et al. 1986).

From the slopes of the calibration curves, molar absorptivities ε close to those of other Fe^{3+}-oxygen complexes (Lehmann 1970) were obtained, $\varepsilon_{X,26,500} = 2.48\cdot10^3$, $\varepsilon_{Y,26,500} = 1.61\cdot10^3$, and $\varepsilon_{Z,28,000} = 3.84\cdot10^3$ $l\cdot g\text{-}atom^{-1}\cdot cm^{-1}$. Therefore, the calibrations are likely to be correct despite the large error bars. The correctness of the calibrations is further substantiated by the fact, that it was possible, by using the above regressions, to accurately calculate from spectral measurements the fO_2' at which the fayalite has grown (1,200°C, H_2/CO_2 = 1:20) serving in the diffusion experiment, despite long extrapolation of the regressions in Figure 20 up to about $x_{Fe^{\bullet}} = 40\cdot10^{-4}$ (fO_2' calculated from spectral data $8.0\cdot10^{-9}$ atm and from the H_2/CO_2 mixture at 1,200°C $fO_2'=4.0\cdot10^{-9}$) (Cemic et al. 1986). It was concluded, then, that the spectral method and the calibrations obtained could be used to analyze diffusion profiles obtained in point defect diffusion experiments as outlined above.

Figure 21 shows the results of such an experiment. It was performed on the fayalite cube, schematically shown in Figure 17 and ground from a crystal grown at $fO_2' = 4.0\cdot10^{-9}$ atm (1,200°C). The cube was reheated at $fO_2'' = 9.7\cdot10^{-18}$ atm, fixed by CO/CO_2, for 13.3 hrs and then quenched. A slice parallel to (010) and 24.0 µm thick (from the retardation in diaspore, cf. section 3), was ground as shown in Figure 17. Values of $\alpha_{Y,26,500}$ were obtained from $\vec{E}//Y$ spectra, measured with 3 µm spot in steps of 50 µm. By means of the calibration (Figure 20), site fractions $x_{Fe^{\bullet}}$ were obtained from the $\alpha_{Y,26,500}$-values, and were recalculated to vacancy concentrations $c_{V''_{Fe}}$. These are shown along two profiles measured parallel to the cryallographic c axis (Cemic et al. 1987).

The large scattering of the values, due to the scattering of the α-values measured, is at least in part caused by scattering in spectral I_o- and/or I-values, originating from submicroscopic inclusions in the sample mount and by dislocations in the fayalite plate. Despite their scattering, the values, as a total, show the right overall slope when considering that $fO_2'' < fO_2'$. Therefore, the data were evaluated using equn. (55) (Cemic et al. 1987). This produced the fitted curves in Figure 21 with

$$\tilde{D}_{zz,V''_{Fe}} = 5.7\cdot10^{-8}\ [cm^2\cdot sec^{-1}] \quad \text{(upper profile)}$$

$$\tilde{D}_{zz,V''_{Fe}} = 4.0\cdot10^{-8}\ [cm^2\cdot sec^{-1}] \quad \text{(lower profile)}$$

(Cemic et al. 1987). These values are in fair agreement with vacancy diffusion coefficient in fayalite obtained by Nakamura and Schmalzried (1983, 1984) from Mg-Fe-interdiffusion experiments

$$\tilde{D}_{V''_{Fe}} = 6.48\cdot10^{-8}\ [cm^2\cdot sec^{-1}]$$

(900°C, $fO_2 = 10^{17}$ atm, $a_{SiO_2} = 1$, orientation dependence not measured).

5. CONCLUSIONS

Using the microscope-spectrometric methods discussed, polarized spectra may be recorded on single crystals, whose smallest dimension is in the

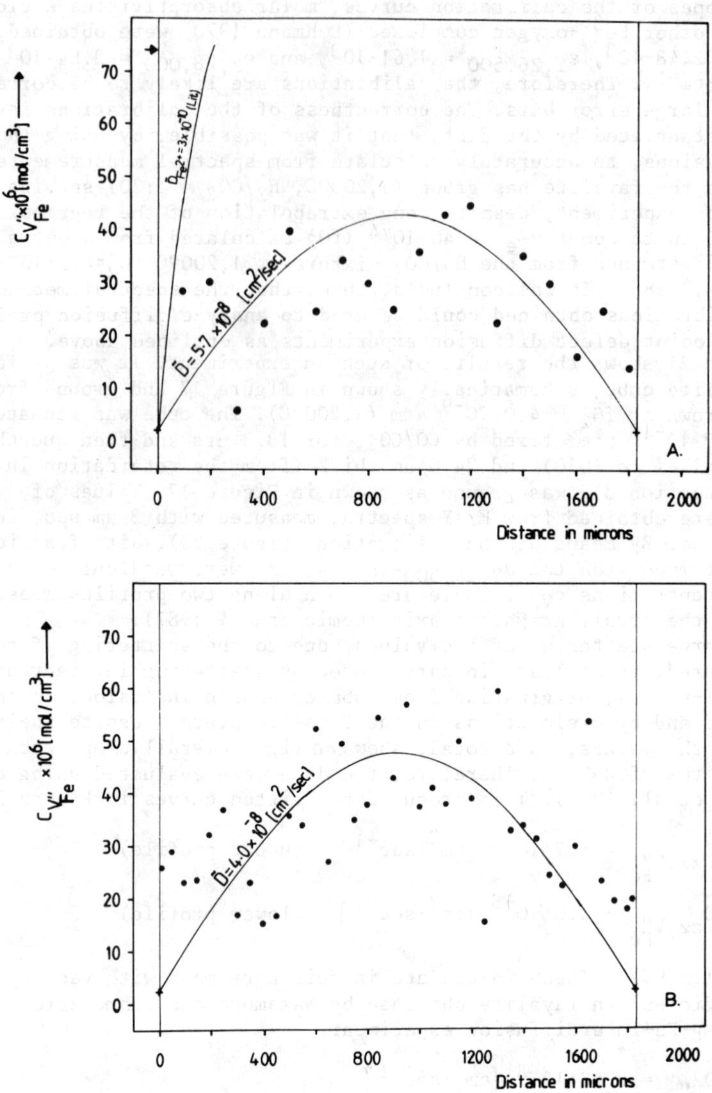

Figure 21. Two profiles, along the crystallographic c axis in fayalite, of vacancy concentrations $c_{V''_{Fe}}$, as determined by microscope-spectrometry. The V''_{Fe} (corresponding to $x_{Fe^{\cdot}}$, after recalculation) profile was generated by a diffusion experiment at $fO''_2 = 9.7 \cdot 10^{-18}$ atm (895°C, 13.3 hrs) in fayalite grown at $fO'_2 = 4.0 \cdot 10^{-9}$ atm (1,200°C). Upper and lower profiles are measured 400 μm from the rim of the (010) plate and in its center, respectively (cf. Figure 17).

order of 15 μm. This allows single crystal spectra to be recorded on silicate minerals obtained in routine high-pressure high-temperature experiments under defined thermodynamic conditions, thus producing chemically well defined silicate phases. Such spectra contribute to solve problems of band assign ments in the case of chemically more complex natural minerals.

From the evaluation of dd transitions of $3d^N$-ions, the geometry of the $3d^N$-centered sites (site symmetry or "pseudosymmetry"), the crystal field parameters 10Dq and B, as well as the crystal field stabilization energy of the respective $3d^N$-ion in the respective site or sites may be obtained. Changes of the lattest quantity, $\Delta CFSE^{r}_{3d}N$, during phase transitions, formation of solid solutions, and intra- or intercrystalline $3d^N$-ion distribution reactions occur as a consequence of differences of the crystal field at the $3d^N$-ion centered sites in the structures of the phases involved.

Changes in crystal field stabilization energy on reaction, $\Delta CFSE^{r}_{3d}N$, contribute as a summand to calorimetrically measured values of $\Delta G^{o,r}$, the change of Gibbs free energy on reaction r. However, to evaluate this, we need accurate values for the zero field contributions to $\Delta G^{o,r}$. Furthermore, we need a complete set of $CFSE_{3d}N$-values for all phases (sites) involved in the reaction under study. In the case of a quantitative analysis of $3d^N$-ion intercrystalline partitioning in terms of $\Delta CFSE_{3d}N$, paragenetic mineral crystals have to be studied by microprobe and by microscope-spectrometry.

Microscope-spectrometry may also be useful to measure concentration profiles obtained in diffusion experiments with $3d^N$-ions, such as Fe^{3+}, Mn^{3+}, Ti^{3+}, Cr^{2+}, which cannot be analyzed for by other techniques with high areal resolution, e.g. electron microprobe, due to the fact that such techniques cannot detect valencies.

Further careful studies of all thes aspects is needed combining microprobe work, structure refinements , Moessbauer spectroscopy and transmission electron microscopy with microscope-spectrometry on $3d^N$-ion containing silicate and oxide minerals.

ACKNOWLEDGEMENTS

To all coworkers, collegues, and friends, who contributed by their work to this paper my sincere thanks are due. These are: I. Abs-Wurmbach, Bochum, R. M. Abu-Eid, Kuwait, W. Assmus, Frankfurt, R. Arni, Berlin, L. Cemic, Berlin, G. Franz, Berlin, K. R. Frentrup, Holderbank, S. Ghose, Seattle, St. Grammenopoulou-Bilal, Berlin, U. Hålenius, Luleå, M. Kersten, Berlin, W. Laqua, Giessen, G. Rossi, Pavia, F. Seifert, Bayreuth, G. Smith, Stonehaven, E. Tillmanns, Wuerzburg, L. Ungaretti, Pavia. Help was also obtained from discussions with R. G. Burns, Cambridge/Mass., A. . Marfunin, Moscow, A. N. Platonov, Kiev, and G. R. Rossman, Pasadena.
Mrs. S. Freitag and H. Reuff, Berlin, contributed by their valuable help in crystal orientation and preparation, respectively.
Generous financial support was provided by the Deutsche Forschungsgemeinschaft, Bonn-Bad Godesberg.

LITERATURE REFERENCES

Abs-Wurmbach, I., Langer, K., Seifert, F., Tillmanns, E. (1981) 'The crystal chemistry of (Mn^{3+},Fe^{3+})-substituted andalusites (viridines and kanonaite), $(Al_{1-x-y}Mn^{3+}_xFe^{3+}_y)_2[O/SiO_4]$: crystal structure refinements, Moessbauer, and polarized optical absorption spectra.' Z. Kristallogr. 155 81 - 113

Abs-Wurmbach, I., Langer, K., Schreyer, W. (1983) 'The influence of Mn^{3+} on the stability relations of the Al_2SiO_5 polymorphs with special emphasis on manganian andalusites (viridines), $(Al_{1-x}Mn^{3+}_x)_2(O/SiO_4)$: An experimental investigation.' J. Petrology 24 48 - 75

Abs-Wurmbach, I. Langer, K., Oberhaensli, R. (1985) 'Polarized absorption spectra of single crystals of the chromium bearing clinopyroxenes kosmochlore and Cr-aegirine-augite.' N. Jb. Miner. Abh. 152 293 - 319

Akimoto, S., Fujisawa, H., Katsura, T. (1965) 'The olivine spinel transition in Fe_2SiO_4 and Ni_2SiO_4.' J. Geophys. Res. 70 1969 - 1977

Amthauer, G. (1976) 'Kristallchemie und Farbe chromhaltiger Granate.' N. Jb. Miner. Abh. 126 158 - 186

Annersten, H., Halenius, U. (1976) 'Ion distribution in pink muscovite: a discussion.' Amer. Mineral. 61 1045 - 1050

Arni, R., Langer, K., Tillmanns, E. (1985) ' Mn^{3+} in garnets III. Absence of Jahn-Teller distortion in synthetic Mn^{3+}-bearing garnet.' Phys. Chem. Minerals 12 279 - 282

Ballhausen, C. J. (1954) 'Studies of absorption spectra. II. Theory of copper (II) spectra.' Kgl. Danske Videnskab Selskab Mat. Fys. Medd. 29 No. 4, 3 - 18

Ballhausen, C. J. (1962) ' An introduction to ligand field theory.' McGraw-Hill, New York

Burnham, Ch. W., Buerger, M. J. (1961) 'Refinement of the crystal structure of andalusite.' Z. Kristallogr. 115 269 - 290

Burns, R. G., Fyfe, W. S. (1964) 'Site of preference energy and selective uptake of transition-metal ions from a magma.' Science 144 1001 - 1003

Burns, R. G. (1966) ' Apparatus for measuring polarized absorption spectra of small crystals.' Journ. Sci. Instr. 43 58 - 60

Burns, R. G., Strens, R. G. J. (1967) 'Structural interpretation of polarized absorption spectra of the Al-Fe-Mn-Cr-epidotes.' Mineralog. Mag. 36 204 - 226

Burns, R. G. (1970a) 'Crystal field spectra and evidence for cation orderingin olivine minerals.' Amer. Mineral. 55 1608 - 1632

Burns, R. G. (1970b) ' Mineralogical applications of crystal field theory.' Cambridge University Press, Cambridge

Burns, R. G., Sung, C.-M. (1978) 'The effect of crystal field stabilization on the olivine → spinel transition in the system Mg_2SiO_4 - Fe_2SiO_4.' Phys. Chem. Minerals 2 349 - 364

Burns, R. G. (1982) 'Electronic spectra of minerals at high pressures: how the mantle excites electrons.' In: Schreyer, W. ed. 'High pressure researches in geoscience.' Schweizerbart, Stuttgart, 223 - 246

Cemic, L., Grammenopoulou-Bilal, St., Langer, K. (1986) 'A microscope-spectrometric method for determining small Fe^{3+}-concentrations due to Fe^{3+}-bearing defects in fayalite.' Ber. Bunsenges. Phys. Chem. 90 654 - 661

Cemic, L., Grammenopoulou-Bilal, St., Langer, K. (1987) 'Determination of the chemical diffusion coefficient $\tilde{D}$ in fayalite by a microscope-spectrometric method.' Terra Cognita 7 285

Clark, J. R., Appleman, D. E., Papike, J. J. (1969) 'Crystal-chemical characterization of clinopyroxenes based on eight new structure refinements.' Mineral. Soc. Amer. Spec. Pub. 2 31 - 50

Crank, J. (1975) 'Mathematics of diffusion.' Clarendon Press, Oxford, 2nd ed.

Curtis, C. D. (1964) 'Applications of the crystal-field theory to the inclusion of trace transition elements in minerals during magmatic differentiation.' Geochim. Cosmochim.Acta 28 389 - 403

Dollase, W. A. (1969) 'Crystal structure and cation ordering of piemontite.' Amer. Mineral. 54 710 - 717

Dunn, T. M. McClure, D. S., Pearson, R. G. (1965) 'Some aspects of crystal field theory.' Harper and Row, New York

Figgis, B. N. (1966) 'Introduction to ligand fields.' Wiley and Sons, New York

Frentrup, K. R., Langer, K. (1981) 'Mn^{3+} in garnets: optical absorption spectrum of a synthetic Mn^{3+}-bearing silicate garnet.' N. Jb. Miner. Mh. 245 - 256

Ganguly, J. (1976) 'The energetics of natural garnet solid solution. II. Mixing of the calcium silicate end members.' Contrib. Mineral. Petrol. 55 81 - 90

George, P., Mc.Clure, D. S. (1059) 'The effect of inner orbital splitting on the thermodynamic properties of transition metal compounds and coordination complexes.' Progr. Inorg. Chem. 1 381 - 463

Ghose, S., Kersten, M., Langer, K., Rossi, G., Ungaretti, L. (1986) 'Crystal field spectra and Jahn-Teller effect of Mn^{3+} in clinopyroxene and clinoamphiboles from India.' Phys. Chem. Minerals 13 291 - 305

Goldman, D. S., Rossman, G. R. (1977) 'Determination of quantitative cation distribution in orthopyroxenes from electronic absorption spectra.' Phys. Chem. Minerals 4 43 - 53

Gueven, N. (1971) 'The crystal structures of $2M_1$ phengite and $2M_1$ muscovite.' Z. Kristallogr. 134 196 - 212

Guggenheim, E. A. (1967) 'Thermodynamics' North Holland Publ. Comp., Amsterdam

Hålenius, U., Langer, K. (1980) 'Microscope-photometric methods for non-destructive Fe^{2+}-Fe^{3+} determination in chloritoids $(Fe^{2+},Mn^{2+},Mg)_2(Al,Fe^{3+})_4Si_2O_{10}(OH)_4$.' Lithos 13 291 - 294

Hålenius, U.,Annersten, H., Langer, K. (1981) 'Spectroscopic studies on natural chloritoids.' Phys. Chem. Minerals 7 117 - 123

Holdaway, M. J. (1971) 'Stability of andalusite and the aluminum silicate phase diagram.' Amer. J. Sci. 269 97 - 131

Holmes, O. G., McClure, D. S. (1957) 'Optical spectra of hydrated ions of the transition metals.' J. Chem. Phys. 26 1686 - 1694

Hutchings, T. M. (1964) 'Point-charge calculations of energy levels of magnetic ins in crystalline electric fields.' Solid State Physics 16 227 - 273

Jahn, H. A., Teller, E. (1937) 'Stability of polyatomic molecules in degenerate electronic states. I. Orbital degeneracy.' Proc. Roy. Phys. Soc. (London) A161 220 - 235

Joergensen, C. K. (1963) 'Inorganic complexes' Academic Press, New York - London
Kersten, M., Langer, K., Almen, H., Tillmanns, E. (1987) 'Kristallchemie von Piemontiten: Strukturverfeinerungen und polarisierte Einkristallspektren.' Ref. 26 Diskusstag. AGKr., Vorabdr. Z. Kristallogr. 120 121
Khomenko, V. M., Platonov, A. N. (1985) 'Electronic absorption spectra of Cr^{3+} ions in natural clinopyroxenes.' Phys. Chem. Minerals 11 261 - 265
Koenig, E. (1971) 'The nephelauxetic effect: calculation and accuracy of interatomic repulsion parameters in cubic high-spin d^2, d^3, d^7, and d^8 systems.' Structure and Bonding 9 175 - 211
Koenig, E., Kremer, S. (1977) 'Ligand field diagrams.' Plenum Press, New York
Lahiri, D. (1971) 'Mineralogy and genesis of the manganese oxide and silicate rocks in Kajlidongri and surrounding areas, Jabuha District, Madhya Pradesh, India.' Econ. Geol. 66 1176 - 1185
Langer, K., Seifert, F. (1971) 'High pressure-high temperature synthesis and properties of chromium kyanite, $(Al,Cr)_2SiO_5$.' Z. anorg. allg. Chem. 383 29 - 39
Langer, K., Abu-Eid, R. M. (1977) 'Measurement of the polarized absorption spectra of synthetic transition metal-bearing silicate microcrystals in the spectral range 44,000 - 4,000 cm^{-1}.' Phys. Chem. Minerals 1 273 - 299
Langer, K., Frentrup, K. R. (1979) 'Automated microscope-absorption-spectrophotometry of rock-forming minerals in the range 40,000 - 5,000 cm^{-1} (250 - 2,000 nm).' Journ. Microscopy 116 311 - 320
Langer, K. (1984) 'Die Farbe von Mineralen und ihre Aussagefaehigkeit fuer die Kristallchemie.' Rhein.-Westfael. Akad. Wiss., Vortraege N332 7 - 60
Langer, K,, Lattard, D. (1984) 'Mn^{3+} in garnets II: optical absorption spectra of blythite bearing synthetic calderites, $Mn_3^{2+[8]}(Fe_{1-x}^{3+}Mn_x^{3+})^{[6]}[SiO_4]_3$.' N. Jb. Miner. Abh. 149 129 - 141
Lattard, D., Schreyer, W. (1983) 'Synthesis and stability of the garnet calderite in the system Fe-Mn-Si-O.' Contrib. Mineral. Petrol. 84 199 - 214
Lehmann, G. (1970) 'Ligand field and charge transfer spectra of Fe(III)-O complexes.' Z. Phys. Chem. N. F. 72 279 - 297
Lever, A. B. P. (1968) 'Inorganic electronic spectroscopy.' Elsevier, Amsterdam
Mao, H. K., Bell, P. M. (1971) 'Apparatus for the measurement of crystal field spectra of single crystals.' Ann. Rep. Dir. Geophys. Lab., Carnegie Inst., Washington 608 - 611
Mao, H. K,, Bell, P. M. (1973) 'Polarized crystal field spectra of the moon.' In: Analytical Methods Developed for Application in Lunar-Sample Analysis. p. 100, ASTM, STP 539, Amer. Soc. Testing Materials
Marfunin, A. S. (1977) 'Physics of minerals and inorganic materials.' Springer, Berlin - Heidelberg - New York
McClure, D. S. (1059) 'Electronic spectra of molecules and ions in crystals.' Solid State Physics 9 399 - 525
Nagasawa, H. (1966) 'Trace element partition coefficients in ionic crystals.' Science 152 767 768
Nakamura, A., Schmalzried, H. (1983) 'On the nonstoichiometry and point defects of olivine.' Phys. Chem. Minerals 10 27 - 37

Nakamura, A., Schmalzried, H. (1984)'On the Fe^{2+}-Mg^{2+}-interdiffusion in olivine (II).' Ber. Bunsenges. Phys. Chem. 88 140 - 145
Nishizawa, H., Koizumi, M. (1975) 'Synthesis and infrared spectra of $Ca_3Mn_2Si_3O_{12}$ and $Cd_3B_2Si_3O_{12}$ (B: Al,Cr,V,Fe,Mn) garnets.' Amer. Mineral. 60 84 - 87
Nitsan, U. (1974) 'Stability of olivine with respect to oxidation and reduction.' J. Geophys. Res. 79 706 - 711
Novak, G. A., Gibbs, G. V. (1971) 'The crystal chemistry of the silicate garnets.' Amer. Mineral. 56 791 825
Orgel, L. (1966) 'An introduction to transition-metal chemistry: Ligand field theory.' Methuen, London
Poole, C. P. jr. (1964) 'The optical spectra and color of chromium containing solids.' J. Phys. Chem. Solids 25 1169 - 1182
Ramsey, D. A. (1952) 'Intensities and shapes of infrared absorption bands of substances in the liquid phase.' J. Amer. Chem. Soc. 74 72 - 80
Reinen, D. (1969) 'Ligand field spectroscopy and chemical bonding in Cr^{3+}-containing oxidic solids.' Structure and Bonding 6 30 - 51
Robinson, K., Gibbs, G. V., Ribbe, P. H. (1971) 'Quadratic elongation: A quantitative measure of distortion in coordination polyhedra.' Science 172 567 - 570
Roy, S. (1966) 'Syngenetic manganese formations of India.' Jadavpur Univ., Calcutta
Schlaefer, H. L., Gliemann, G. (1961) 'Einfuehrung in die Ligandenfeldtheorie.' Akademische Verlagsgesellschaft, Frankfurt/M.
Schwarcz, H. P. (1967) 'The effect of crystal field stabilization on the distribution of transition metals between metamorphic minerals.' Geochim. Cosmochim. Acta 31 503 - 517
Smith, G., Halenius, U., Langer, K. (1982) 'Low temperature spectral studies of Mn^{3+}-bearing andalusite and epidote type minerals in the range 30000 -5000 cm^{-1}.' Phys. Chem. Minerals 8 136 - 147
Steffen, G., Seifert, F., Langer, K. (1987)'Quantative determination of Fe^{2+} distribution in orthopyroxenes by polarized electronic absorption spectra.' Terra Cognita 7 391
Strens, R. G. J. (1968) 'Stability of the Al_2SiO_5 solid solutions.' Mineralog. Mag. 36 839 - 849
Sugano, S., Tanabe, Y., Kamimura, H. (1970) 'Multiplets of transition-metal ions in crystals.' Academic Press, New York - London
Syono, T., Tokonami, M., Matsui, Y. (1971) 'Crystal field effects on the olivine - spinel transformation.' Phys. Earth Planet. Interiors 4 347-352
Tanabe, Y., Sugano, S. (1954 a, b) 'On the absorption spectra of complex ions. I, II' Journ. Phys. Soc. Japan 9 753 - 766 and 767 - 779
Tillmanns, E., Langer, K., Arni, R. (1984) 'Crystal structure refinements of coexisting thulite and piemontite, $Ca_2Al_{3-p}M_p^{3+}[OH/O/SiO_4/Si_2O_7]$ (M^{3+} = Mn^{3+}+Fe^{3+}).' 13th Int. Congr. Crystallogr., Coll Abstr. C258
Tossell, J. A., Vaughan, D. J., Johnson, K. H. (1974) 'The electronic structure of rutile, wustite, and hematite from molecular orbital calculations.' Amer. Mineral. 59 319 - 334
Wilson, E. G., Decius, J. C., Cross, P. C. (1955) 'Molecular vibrations' McGraw-Hill, New York
Wood, B. J., Strens, R. G. J. (1972) 'Calculation of crystal field splittings in distorted coordination polyhedra: spectra and thermodynamic properties of minerals.' Mineralog. Mag. 38 909 - 917

Manning, [illegible] (1970) [illegible] the [illegible] interpretation [illegible] olivine [illegible]. Journal of Phys. Chem. 53 [illegible]

[illegible] Mesozoic [illegible] (1977) Hypotheses and [illegible] Ca [illegible] and Mg [illegible] Cr [illegible] [illegible] Rev. Mineral. [illegible]

Mitchell, [illegible] (1964) [illegible] olivine with respect to [illegible] and [illegible] Geophys. Res. 69 [illegible]

Moore, [illegible] A., Gibbs, G. V. (19[illegible]) The crystal chemistry [illegible] garnets. Amer. Mineral. 56 [illegible]

Orgel, L. E. (1960) An introduction to transition-metal chemistry. Ligand-field theory. Methuen, London.

[illegible] (1964) The spectral [illegible] of [illegible] complexes. J. Chem. Phys. 30 [illegible]

[illegible] (1952) [illegible] and [illegible] of [illegible] absorption [illegible] in the [illegible] phase. J. Am. Chem. Soc. 74 [illegible] 80

[illegible] (1969) [illegible] spectroscopy [illegible] Proc. [illegible] and Space [illegible]

Robinson, K., Gibbs, G. V., Ribbe, P. H. (1971) Quadratic elongation: A quantitative measure of distortion in coordination polyhedra. Science 172 [illegible]

Rose, S. (1960) Symmetrie und [illegible] in [illegible] Univ. [illegible]

[illegible] (1964) [illegible] Gallertyi [illegible]

[illegible] (1967) The effect of crystal field stabilization on the distribution of transition metal [illegible] in minerals. Geochim. Cosmochim. Acta [illegible]

[illegible] (1972) [illegible] temperature spectra [illegible] in minerals [illegible] Geochim. Cosmochim. Acta [illegible] 117

[illegible] (1976) [illegible] distribution [illegible] minerals [illegible] spectra. Terra Cognita 1 [illegible]

[illegible] (1976) [illegible] solid solutions [illegible]

Sugano, S., Tanabe, Y., Kamimura, H. (1970) Multiplets of transition-metal ions in crystals. Academic Press, New York-London.

[illegible] (1977) [illegible] spectra [illegible] in minerals [illegible] J. Chem. Phys. [illegible] 55

[illegible] structure replacements [illegible] [illegible] Fe^{3+} [illegible] Mg [illegible] J. Chem. Soc. [illegible] 58

[illegible] (1956) Molecular [illegible] [illegible]

Wood, B. J., Strens, R. G. J. (1972) Calculation of crystal field splittings in distorted coordination polyhedra: spectra and thermodynamic properties of minerals. Mineral. Mag. 38 [illegible]

Recent Advances in the Mineralogical Applications of the ^{57}Fe Mössbauer Effect

F. Seifert
Bavarian Research Institute for Experimental Geochemistry
and Geophysics, University Bayreuth, P.O. Box 101251,
D-8580 Bayreuth, FRG

ABSTRACT Application of the ^{57}Fe Mössbauer effect to mineralogical problems can contribute to an understanding of the structural role of Fe in terms of valence, electronic equilibria, bonding and site geometry, site occupancy, magnetic properties, diffusion and as a probe for phase transformations. Studies in the fields of short-range order behaviour, amorphous states, magnetic structure and phase transitions are identified as "areas of growth" for Mössbauer spectroscopy and specific examples are described. However, Mössbauer spectroscopic results often need to be complemented by independent techniques, as demonstrated by two examples of potential pitfalls.

1. INTRODUCTION

The Mössbauer effect, or recoilless nuclear gamma ray resonance, has been discovered only in 1958 and been introduced into earth sciences as a structural tool in the late sixties (cf. Bancroft et al. 1967, 1968). Since then, it has become a more or less standard technique for characterizing crystalline and amorphous compounds, mostly with respect to the role of Fe which is the geochemically most useful Mössbauer active nucleus and particularly easy to measure.

Mössbauer spectroscopy is an element-specific technique such as nuclear magnetic resonance (NMR) or extended x-ray absorption fine structure (EXAFS). This presents an advantage (no interference by other atomic species) but at the same time also a drawback and possible source of errors, since only part of the structural information can be seen in the spectra, and their interpretation might not be straightforward. Mössbauer studies, particularly of chemically and structurally complex phases such as most minerals, should therefore generally be supplemented by independent techniques.

It is the purpose of this review to illustrate the general use and, in a few specific examples, recent applications of the ^{57}Fe Mössbauer effect in mineral physics and chemistry. We will not discuss in detail the theory behind it nor experimental techniques because of the wealth of existing adequate textbooks and summaries (cf. Bancroft 1973, Gonser (ed.) 1975, Gütlich et al. 1978, Maddock 1985). It may, however, be useful to discuss in brief the most important parameters and their physical background, as far as they are relevant for the following review.

E. K. H. Salje (ed.), Physical Properties and Thermodynamic Behaviour of Minerals, 687–703.

2. MÖSSBAUER PARAMETERS

The Mössbauer effect is generally described by a number of parameters (some of them called hyperfine parameters) having different physical sources and, therefore, yielding different kinds of information.

2.1. Debye-Waller factor

The Debye-Waller factor describes the fraction of gamma-ray emissions or absorptions that occurs without recoil and thus controls the intensity of a Mössbauer line - besides the concentration of the Mössbauer-active nucleus. The Debye-Waller factor f can be expressed as

$$f = \exp -k^2 < x^2 > \qquad (1)$$

where k is the magnitude of the wave vector of the gamma ray (equal to $2\pi / \lambda$, λ = wavelength) and $< x^2 >$ is the mean square vibrational amplitude of the emitting (or absorbing) nucleus. It follows from (1) that f is, besides its energy dependence, also affected by the vibrational properties of the nucleus, which again depend on temperature, pressure, site geometry, bonding, and, for sites with lower than cubic point symmetry, also on the direction. For any site, however, f approaches 1 as temperature goes towards absolute zero. This temperature dependence can be described by a Debye approximation (Heberle 1971).

2.2. Line shape

The energy distribution of a Mössbauer line around a central energy E value is, in theory, a Lorentzian distribution that can be described by E, the half width Γ and the intensity. This is strictly valid only if an absorption is caused by a nucleus in a single, well-constrained crystallographic site and if the sample is infinitesimally thin. However, most authors choose to describe a line that might deviate slightly in shape still by a (broadened) Lorentzian shape, whereas others use a Gaussian component (e.g. Dowty and Lindsley 1973) or a distribution of Lorentzian lines (Wivel and Mörup 1980).

2.3. Full width at half maximum intensity (half width)

The "natural" (Heisenberg) half width Γ_H of a transition (i.e. the spread in energy around a mean value E_o) is given by

$$\Gamma_H = \hbar / \tau \qquad (2)$$

where $\hbar = h/2\pi$, h = Planck's constant, τ = mean lifetime of the excited state. For the 14.4 keV transition in ^{57}Fe, τ = 141 ns and $\Gamma_H = 5.67 \times 10^{-9}$ eV (equivalent to a "sharpness" of the transition $\Gamma_H/E_o = 3 \times 10^{-13}$). In terms of the conventional Mössbauer energy scale this half width corresponds to 0.097 mm s^{-1}.

In any experiment with an emitting source and absorbing sample the minimum half width that can be expected is $2 \times \Gamma_H$ = 0.194 mm s^{-1}. In practice, 0.24 mm s^{-1} presents about the lowermost limit attainable and line widths of 0.30 - 0.35 mm s^{-1} are still considered satisfactory.

A number of effects tend to increase this halfwidth, even if there is no instrumental line broadening: Most important are saturation effects (due to the non-zero concentration of the nucleus in a real sample), relaxation processes, fluctuating environments due to phase transitions, association of defects etc. Also, movement of atoms in jump processes (diffusion) leads to an apparent motion of the nucleus relative to the source and thus to a line broadening, when the jump frequency of the nucleus is of the same magnitude as $1/\tau$. Finally, line broadening becomes apparent whenever a site is not strictly defined (in terms of geometry, bonding etc.) but actually a distribution of sites is present (such as in glasses or by next-nearest neighbours interaction) where the resolution of individual Lorentzian contributions generally is insufficient.

2.4. Isomer shift

The mean energy of the absorbed or emitted γ-radiation is affected by the interaction of the nuclear charge distribution with the electron density at the absorbing or emitting nucleus. Therefore, the maximum absorption in a sample A occurs generally at a different energy than emission of the source S:

$$\delta = C\, \partial R/R\, [\,|\psi_A(0)|^2 - |\psi_S(0)|^2] \qquad (3)$$

where δ = isomer shift, C = constant for a given nucleus, $\partial R/R$ is the change of nuclear radius between excited and ground state, and the term in brackets is the difference in electron density at the nucleus between absorber and source. With $\psi_S(0)$ generally constant, δ thus presents a direct though only relative measure of the electron density at the nucleus of the sample. It is affected by the valence of the nucleus and its bonding (e.g. coordination number, electronegativity of the surrounding atoms, covalency effects, metal-anion distances). In the following, isomer shifts will always be reported relative to metallic iron.

2.5. Quadrupole splitting

Whenever a Mössbauer active nucleus is surrounded by a non-cubic field, the nuclear excited states split into sublevels, and the magnitude of this quadrupole splitting Δ is proportional to the nuclear electric quadrupole moment eQ and the Z component of the electric field gradient (EFG) tensor. For ^{57}Fe,

$$\Delta = -\tfrac{1}{2} V_{zz}\, e\, Q\, (1+\eta^2/3)^{1/2} \qquad (4)$$

with $\eta = (V_{xx}-V_{yy})/V_{zz}$ (asymmetry parameter). Note that for a given nucleus this equation still contains two unknowns (V_{zz} and η). These depend on the contribution by the external ligands (lattice contribution) and a valence contribution that may - as in the case of Fe^{2+} - also contribute to a distortion of the inner electrons and thus to an EFG. It is often difficult to separate these effects, and in complex structures such as most silicates an empirical approach for the correlation between quadrupole splitting and site distortion is mostly employed (cf. Steffen et al. 1984).

2.6. Nuclear Zeemann Effect

By the interaction of the nuclear magnetic dipole moment with a magnetic field the nuclear levels are split further (into four levels for the I = 3/2 excited state of Fe and two levels of the I = 1/2 ground state of Fe) producing six-line magnetic spectra. This splitting E_m is proportional to the magnetic quantum number m_i, the nuclear Bohr magneton β_N, the nuclear Landé splitting factor g_N, and the magnetic field H at the site of the nucleus:

$$E_m = -g_N \beta_N H m_i \qquad (5)$$

3. VALENCE

Addition or removal of p or d electrons, which do not interact directly with the nuclear charge density, nevertheless affects the charge at the site of the nucleus by expanding or contracting the 3s electrons. Because for ^{57}Fe $\partial R/R$ is negative, the d^6 ion (Fe^{2+}) has an appreciably larger isomer shift than a d^5 ion (Fe^{3+}). This leads to a wide separation of the Fe^{3+} and Fe^{2+} lines in the Mössbauer spectrum (typically $\delta =$ 0.15-0.50 mm s^{-1} for Fe^{3+}, 0.9-1.25 mm s^{-1} for Fe^{2+} at room temperature) and thus allows even quantitative determinations of ferrous/ferric ratios. Because bonding and vibrational characteristics are grossly different for Fe^{2+} and Fe^{3+}, the different Debye-Waller-factors for the two species must be taken into account (cf. Andersen et al. 1975, Amthauer et al. 1976) and measurements of area ratios should be performed as a function of temperature and then be extrapolated to 0K.

Mössbauer spectroscopic valence determination offers advantages over conventional techniques:

a. None of the valence states is determined by difference (as for instance in wet chemical techniques generally yielding total iron and Fe^{2+} only). This is particularly useful in refractory phases with high Fe^{2+} and low Fe^{3+} such as many natural garnets.

b. In oxidation studies it can be expected that Fe^{2+} in different sites of the crystal structure is oxidised by different mechanisms and at different rates. Mössbauer spectroscopy may thus be used to follow the oxidation process in greater structural detail. For instance, Sanz et al. (1983) where able to show by a combination of Mössbauer and infrared spectroscopic techniques that upon dehydroxylation of biotite the mechanisms by which OH linked to Fe^{2+} and OH linked to a vacancy are lost are completely independent. Ferrow (1987) demonstrated, on the other hand, that oxidation of Fe^{2+} in the M2 position of biotite is relatively easier that of that of Fe^{2+} in M1.

c. When more than one transition metal is present in a sample, the valence state of Fe may not be uniquely be determined by purely chemical techniques and Mössbauer spectroscopy in addition allows a characterization of fluctuating charges between atoms of the same kind (see following section).

4. ELECTRON DELOCALIZATION AND HOMOGENEOUS ELECTRONIC EQUILIBRIA

In a crystal structure containing both ferrous and ferric iron, the electrons may be delocalized, i.e. exchange their positions between different Fe atoms. If the mean time of residence of an electron at a given nucleus is larger than the mean life time of the excited state, the Mössbauer experiment will "see" either a trivalent or a divalent Fe. If the exchange processes are faster, however, an average valence state of 2.5 will be observed, with hyperfine parameters intermediate between those of Fe^{2+} and Fe^{3+}. These processes are thermally activated, i.e. faster and more frequent at higher temperatures. Therefore, in the Mössbauer experiments localized electrons (or integer oxidation numbers) will be observed at low temperatures but mixed valence states at higher temperatures.

In a comprehensive Mössbauer and optical absorption study of iron-bearing minerals Amthauer and Rossmann (1984) were able to show that this thermally activated electron delocalization occurs only if Fe^{2+} and Fe^{3+} occupy crystallographically equivalent or geometrically very similar sites which share common edges and form extended structural units. Well-studied examples are deerite (Amthauer et al. 1980), ilvaite (Litterst and Amthauer 1984) and vonsenite (Swinnea and Steinfink 1983). The complex spectra of schorlomite garnets have recently also been interpreted in terms of electron delocalization processes (Wu and Mu 1986) between two different sets of sites.

Similar electronic exchange processes can also be expected if two different transition metals are present in a crystal structure, e.g. Fe + Mn or Fe + Ti, for which an exchange reaction may be written such as

$$Fe^{2+} + Mn^{3+} \rightleftharpoons Fe^{3+} + Mn^{2+} \quad (6)$$

or

$$Fe^{2+} + Ti^{4+} \rightleftharpoons Fe^{3+} + Ti^{3+}. \quad (7)$$

Mössbauer spectroscopy can locate the equilibrium state. It is well known that e.g. in ilmenite $FeTiO_3$ this equilibrium is shifted (at least at low temperatures) entirely to the left whereas in braunite and phlogopite it is shifted to the right hand side (Seifert and Dasgupta 1982, Smith et al. 1983). In yoderite, Abu-Eid et al. (1978) inferred the coexistence of Fe^{2+}, Fe^{3+}, Mn^{2+} and Mn^{3+} from a combined chemical, Mössbauer spectroscopic and optical absorption study.

A theoretical treatment of such exchange processes has recently been given by Sherman (1987 a,b) on the basis of molecular orbital theory. Electron delocalization between adjacent Fe cations requires them to be ferromagnetically coupled. High-temperature Mössbauer spectroscopy on samples with more than one transition metal could certainly improve our understanding of such equilibria at elevated temperatures.

5. SITE CHARACTERIZATION

Since the hyperfine parameters are affected by the charge density at the site of the nucleus and the EFG, they can serve as a structural tool to characterize a site and, by comparison with the crystal structure, make the right band assignments. In general, only the nearest neighbours (i.e. the surrounding anions) are considered in such studies.

5.1. Ferric iron

The isomer shift increases with increasing coordination number (typically 0.15 to 0.35 mm s^{-1} for fourfold coordination and 0.35-0.50 mm s^{-1} for sixfold coordination by oxygen). Increasing average Fe-O bond length increases the isomer shift of tetrahedrally coordinated Fe^{3+}, and increasing distortion (either defined as bond length or bond angle variation) increases the quadrupole splitting (e.g. Steffen et al. 1984).

5.2. Ferrous iron

Isomer shifts are indicative of the coordination number with respect to oxygen (typically 0.9-1.05 for $Fe^{[4]}$, 1.10-1.20 for $Fe^{[6]}$ and 1.25 mm s^{-1} for $Fe^{[8]}$ at room temperature). Because of the d^6 nature of the ion, both the lattice and the valence contribution to the EFG become effective. This leads first to an increase of the quadrupole splitting with increasing distortion and then to a decrease (e.g. Dowty and Lindsley 1973). Most Fe^{2+} polyhedra in silicate structures seem to be highly distorted, so that a smaller quadrupole splitting is in general associated with a larger distortion, and vice versa.

Thus, for a known crystal structure, band assignments are usually straightforward (unless the structure contains too many distinct Fe-bearing positions). Compared to X-ray diffraction studies, the main advantage of Mössbauer spectroscopy is the localization of even small amounts of Fe in individual sites, and in the detection of small relative changes in distortion as a function of composition (e.g. Seifert 1977).

5.3. Next nearest neighbours interaction

Since the entire structure contributes to ψ_A and the EFG, effects of next-nearest neighbours (i.e. the next cations or cation vacancies) interactions can be expected. As a consequence, a doublet due to Fe in a given site may split into a number of doublets representing the number of different environments in terms of next cations. This effect often leads to a great complexity of the spectra with only marginal resolution.

The magnitude of next-nearest neighbours interaction will depend on the differences in size and charge of substituting cations in the individual environments. Therefore, in solid solution systems with isovalent substitutions involving similar sized cations (e.g. Mg-Fe^{2+}) the effects will generally be small. Examples are the orthopyroxene series (Virgo and Hafner 1970) and the system (Mg, Fe)S (Osborne and Fleet 1984). Larger differences in ionic size (e.g. Fe^{2+}-Ca such as in the system clinoferrosilite-hedenbergite, Dowty and Lindsley 1973, or Al-Cr in the system $FeAl_2O_4$-$FeCr_2O_4$, Bancroft et al. 1983) lead to significant line broadening which can be attributed to a splitting of sites due to next-nearest neigbours interactions. Even stronger effects may occur at heterovalent substitutions of e.g. the Tschermak type (Mg, Fe^{2+}) + Si = 2(Al, Fe^{3+}) or edenite type (Mg, Fe^{2+}) + Ca = (Fe^{3+}, Al) + Na (e.g. Dollase and Gustafson 1982, Seifert 1983) but it is often not clear whether the line broadening or splitting is due to true next-nearest neighbours interactions or inhomogeneities on a microscale combined with changes in the Fe^{2+} d-orbital configuration (Dollase and Gustafson 1982).

On the other hand, if line splitting can be related to next-nearest neigbours interaction, the area ratios could be used to determine short range order SRO. From the known chemical composition the probability of the individual environments in a random structure without SRO may be calculated from combination theory (Dowty and Lindsley 1973, Seifert 1983). Any deviations of the observed intensities from the spectrum thus predicted can then be attributed to SRO. Such an interpretation of the broadened spectra of aluminous orthopyroxenes has been attempted by Seifert (1983). Although the spectra suffer from low resolution (even at liquid nitrogen temperatures), indications are that Fe^{2+} preferentially orders into those M2 positions that are surrounded by at least one (Al, Fe^{3+}) in a neighbouring M1 site.

A new approach for the evaluation of mineral spectra split by next-nearest neigbours interaction has been introduced by Bancroft et al. (1983) by the use of partial quadrupole splittings. Studying spinel of composition $FeCr_{1.9}Al_{0.1}O_4$ they argued that the extra doublet in this spectrum appearing in addition to the singlet line of the $FeCr_2O_4$ spectrum is due to an environment where three oxygens of the Fe^{2+} tetrahedron are linked to 2 Cr,1 Al and the other one to 3 Cr. From the quadrupole splitting observed for this configuration they calculated, by linear combination, quadrupole splittings for other configurations and used these as a constraint in the fitting of the complex spectra. The results indicated large random distribution of Cr and Al over the octahedral sites, i.e. no clustering.

Compared to the unconstrained approach as used by e.g. Dowty and Lindsley (1973) or Seifert (1983) this method avoids one severe problem: In unconstrained fits the lines associated with one particular configuration often show a compositional dependence of the quadrupole splittings which is of a similar magnitude as the difference between individual configurations. Future experiments should therefore be performed on samples characterized by independent techniques in order to test the basic and crucial assumption of the partial quadrupole splitting approach that the increase in the valence contribution of V_{zz} is linear with increase of distortion from e.g. ideal tetrahedral symmetry.

Particularly strong effects of next-nearest neighbours interactions are present in wüstite, $Fe_{1-x}O$. Here the singlet of Fe^{2+} in the undistorted NaCl structure is split into one or several doublet by next-nearest Fe^{3+} and vacancies. There concentration and configuration around Fe^{2+} depends on x and the type of clustering (Simons and Seifert 1979, McCammon and Price 1985).

6. SITE OCCUPANCY

A comparison of different techniques for studying site occupancies has recently been given by Hawthorne (1983). If a clear-cut assignment of doublets to structural sites can be made, then Mössbauer spectroscopy may be used as a rapid and precise technique to study the relative concentrations of Fe in these sites, particularly if the spectrum and the structure is simple, so that rather few, well-resolved bands exist, and next-nearest-neighbours interactions are virtually absent. A further prerequisite is that spectra should be taken at low temperatures so that the individual Debye-Waller factors become nearly equal. Since Mössbauer spectroscopy is element-specific, assumptions on the distribution of the other elements must be made if more than two different cations may be involved in ordering processes, or independent techniques have to be used in addition.

The classical example of site occupancy determinations by Mössbauer spectroscopy is the quasibinary system enstatite-ferrosilite in which the Mg occupancy of the octahedral sites can be calculated by difference (Virgo and Hafner 1969, 1970, Saxena and Ghose 1971, Besancon 1981) although recently automated x-ray methods seem to compete successfully even in this system and give more detailed information (Saxena et al. 1987, Tazzoli and Domeneghetti 1987) e.g. on the role of minor additional components. The problem of the interpretation of Mössbauer spectra of aluminous orthopyroxenes has already been addressed above.

Spectra of ^{57}Fe in slightly more complex structures can still be interpreted quantitatively. For instance in amphiboles, Fe^{2+} may partition over four (M1,M2,M3,M4) different octahedral sites, but in Al-poor orthoamphiboles and cummingtonites the hyperfine parameters of Fe in M1, M2 and M3 are so similar (and distinct from those of Fe in M4) that these may be treated as a single site. Since it is known from single-crystal work that fractionation of Mg and Fe between M1, M2 and M3 is only slight in these amphiboles, the Mössbauer spectra can be interpreted on the basis of a pseudo-two-site (M123, M4) model, and this has been used extensively in the past (e.g. Hafner and Ghose 1971, Seifert 1977). In Ca-bearing, Al-poor amphiboles the M2 site can be resolved in addition (Goldman 1979, Skogby and Annersten 1985).

The main use of such data is in kinetics, where a large number of samples annealed a different temperatures and times has to be studied (e.g. Besancon 1981, Skogby 1987).

As discussed above, the Mössbauer spectra get blurred in samples of more complex compositions, particularly if heterovalent substitutions are involved, and they are then of little use for quantitative purposes.

We will return to the problem of site occupancies in the section on pitfalls.

7. MAGNETIC PROPERTIES

By virtue of the nuclear Zeemann effect Mössbauer spectroscopy may yield information on the internal magnetic field at the site of the nucleus. Such studies - in general performed as a function of sample temperature and with or without an applied magnetic field - have recently successfully been employed to characterize mineralogical samples in the following fields:

a. By measuring the temperature variation of the internal field, the ordering (Néel) temperature T_N can be determined, and, below T_N, the magnitude of the internal field and its polar angle with V_{zz}. For instance, Guillen and Regnard (1985) studied these parameters in the system $FeWO_4$-$MnWO_4$ and correlated their variation with composition. Hayashi et al. (1987) investigated the pressure dependence of T_N in fayalite in a diamond-anvil cell up to 16 GPa and interpreted the observed dT_N/dP (2.2 ± 0.2 K/GPa) in terms of a superexchange

interaction that varies as the -10/3 power of the volume. Transitions of the magnetic structure may be detected in addition: Linares et al. (1983) observed marked changes in the magnetic spectra of grunerite in the range 5 to 8 K and a maximum of the susceptibility of grunerite at 7 K and they related it to a spin canting transition, which was later confirmed by Ghose et al. (1987) using neutron diffraction.
b. If a crystal structure contains Fe in two crystallographically distinct, but very similar sites, resolution in the paramagnetic region may be impossible due to the very similar isomer shifts and quadrupole splittings. However, in the magnetically ordered region at low temperatures, the differences in H at the two sites and differences in T_N might be sufficient to resolve the two types of Fe. This case has recently been reported for almandine by Murad and Wagner (1987) and was attributed to either the existence of two different polar angles between V_{zz} and the direction of the magnetic field, or the presence of two distinctly different EFG's.
c. Similarly, the Néel transition can also be used as a probe for characterizing fine-grained materials (for instance Al-rich goethites, Murad and Bowen 1987, and following section).
d. By applying an external field to the sample, the sign of the exchange coupling between two neighbouring Fe ions can be studied, i.e. ferromagnetic coupling can be distinguished from antiferromagnetic coupling (e.g. Ballet and Coey 1982, Ballet et al. 1985, Townsend and Longworth 1985, Ballet 1986).
e. At small grain sizes superparamagnetic relaxation is observed: The magnetic moments of small single domain crystallites fluctuate rapidly and thus the time-averaged field at the Fe nuclears is reduced or vanishes. Because these fluctuations are thermally activated, the superparamagnetic fraction of a sample increases rapidly with temperature, and its temperature dependence may allow a determination of the crystallite size distribution (e.g. Fysh and Ostwald 1983).

8. CRYPTOCRYSTALLINE AND AMORPHOUS PHASES

Whereas in crystalline phases much of the information on Fe (e.g. site geometry, vibrational properties, site occupancy) can also be obtained by diffraction techniques, Mössbauer spectroscopy can become superior in the study of cryptocrystalline or amorphous phases.

8.1. Poorly crystalline lepidocrocite, γ-FeOOH

De Grave et al. (1986) studied Mössbauer spectra of synthetic lepidocrocite samples of different degrees of crystallinity. They observed a broad distribution of Néel temperatures which, by analogy to β-FeOOH (Chambaere and De Grave 1984), may be related to a nonuniform content of structural water. In the paramagnetic range the quadrupole splitting distribution curve is bimodal. The relative probabilities of these two components correlate with the surface area of the samples and may thus be related to "bulk" versus "surface" Fe^{3+}. This result can be supported by data from (surface sensitive) conversion electron Mössbauer spectroscopy.

8.2. Formation of an amorphous intermediate phase on dehydroxylation of akageneite, β-FeOOH

Chambaere and De Grave (1985) monitored the formation of a "quasi-amorphous intermediate phase" (QUAIPH) in the heating process of akageneite by a combination of XRD, TEM, TGA and Mössbauer spectroscopy. In the dehydroxilation process a new Mössbauer doublet of ferric iron with an unusually high quadrupole splitting appears first at heating temperatures in the vicinity of 180°C. This is associated with a surface site. As the next step in the evolution of this paramagnetic QUAIPH phase upon heating, the Néel temperature increases and two sextets appear in the room temperature spectra. Only at temperatures between 310° and 420°C (depending on the water content of the starting material) the transformation of QUAIPH into hematite is complete.

Although such studies probably tell us little about thermodynamic equilibrium states, they are very helpful in characterizing the mechanisms and the participation of amorphous intermediate phases in reaction progress.

8.3. Silicate glasses

The valence, coordination and speciation of iron in silicate melts is a decisive factor for physical properties such as viscosity, density or crystallization behaviour, and many studies have therefore been devoted to this subject in recent years. Because Mössbauer spectra at high temperatures (in the melt region) are difficult to collect and to interpret, the quenched equivalents (glasses) are generally studied at low temperatures and it is assumed that the melt structure is frozen in during quenching at the glass transition temperature without significant changes.

Contrary to crystalline substances glasses exhibit broad distributions of site geometries, leading to broadened lines or, more exactly, a distribution of hyperfine parameters. The interpretation of the spectra is then often not straightforward and additional methods have often be used such as Raman spectroscopy, EXAFS or luminescence spectroscopy.

The determination of ferrous/ferric ratios in glasses by Mössbauer spectroscopy may be influenced by the fitting model and by unequal recoilless fractions of different Fe species. Mysen et al. (1985a) therefore compared Fe^{2+}/Fe^{3+} ratios obtained by wet chemical analysis to those derived from Mössbauer spectra fitted to one ferrous, one ferric Lorentzian doublet. This comparison exhibited no systematic difference between the two methods and similar uncertainties. An advantage of the Mössbauer technique is that, working with enriched ^{57}Fe, smaller sample quantities are sufficient and that structural information e.g. on the coordination number can be extracted in addition to the ferrous/ferric ratio.

There is general consensus (see Virgo and Mysen 1985), based on a variety of methods, that ferrous iron generally is largely in octahedral coordination, although the isomer shifts (about 0.90 to 1.05 mm s^{-1} at room temperature) are systematically lower than for $Fe^{2+[6]}$ in crystalline compounds (1.10 - 1.20 mm s^{-1}). The hyperfine parameters exhibit broad distribution functions (Virgo and Mysen 1985) and sometimes there are indications that this distribution is bimodal.

The structural role of Fe^{3+} in silicate glasses is more complex and has been the subject of much controversary until Virgo and Mysen (1985) and Mysen et al. (1985b) studied systematically the Fe^{3+} isomer shifts as a function of Fe^{2+}/Fe^{3+} ratio in a number of synthetic systems. In oxidized samples, the Fe^{3+} isomer shift is rather constant and at values of about 0.25 - 0.30 mm s^{-1}, indicative of tetrahedral coordination. In reduced samples, on the other hand, it is close to 0.5 mm s^{-1} and octahedral coordiantion can be inferred. At intermediate redox ratios, a continous change in isomer shifts with redox ratio is observed, and this may be interpreted as coexistence of the two coordination polyhedra of Fe^{3+} in changing proportions. Because the coordination change is coupled with the Fe^{2+}/Fe^{3+} ratio, these two species may be associated in the form of a complex (e.g. Fe_3O_4).

Mössbauer spectroscopy has also been applied to natural glasses (e.g. obsidian, Schmidbauer et al. 1986, Spiering and Seifert 1985) for ferrous/ferric determination and site characterization. A handicap of such studies is, however, that chemistry and structure are complex and that the chemical composition cannot be varied systematically.

9. DIFFUSION STUDIES

For the Mössbauer experiment itself it is irrelevant whether the source or the sample is moved. If both the source and the Mössbauer active nuclei exhibit movement (the latter by diffusion within a fixed reference frame) then line broadening will occur. The theory has already been developed by Singwi and Sjölander (1960) for continuous movements as well as for discrete jumps.

9.1. Selfdiffusion of ferric iron in rutile

Drissen (1980) studied the line broadening in the Mössbauer spectra of Fe^{3+}-doped rutile as a function of temperature. Fig. 1 gives the data in an Arrhenius plot. Assuming a jump distance of $c_o/4$ (Stampfl et al. 1973) a preexponential factor D_o of 0.6×10^{-7} cm^2s^{-1} and an activation energy of 75 ± 8 kJ $mole^{-1}$ can be calculated. It is clear that other methods for the determination of the diffusion coefficient are more precise and sensitive: the minimum diffusion coefficient detectable by line broadening is on the order of 10^{-9} cm^2s^{-1} and in most minerals such values are only achieved at temperatures above some 1200°C where the measurement of Mössbauer spectra becomes difficult.

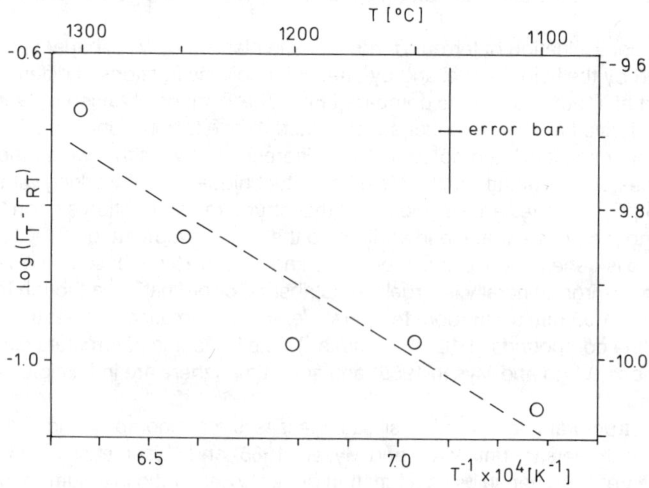

Fig. 1 Line broadening (expressed as the logarithm of the difference in line width at temperature T and room temperature) in Fe^{3+}-doped rutile as a function of reciprocal absolute temperature. On the right hand side, the logartihm of the equivalent diffusion coefficient (in cm^2 s^{-1}) is given. Data from Drissen (1980).

9.2. Vacancy-related selfdiffusion in magnetite

The approach described above suffers from a poorly defined thermodynamic state: As temperature changes, the defect concentration (and type ?) will also change, and both will affect the diffusion process. Therefore, Becker and v. Wurmb (1986) studied Mössbauer spectra of magnetite at constant temperature as a function of a_{O2} and were thus able to isolate the effects of vacancy concentration through the entire range of homogeneity of magnetite at this temperature.

In the paramagnetic region (i.e. above $T_N = 566^oC$) one octahedral doublet and one tetrahedral singulet are observed, representing the two Fe sublattices. The a_{O2}-depending spectra can be interpreted on the basis of two different models:

a. The sublattices are decoupled to a large extent and the cations migrate predominantly in the one or other sublattice. In this case the Mössbauer jump frequency (related to the diffusion coefficient) can be calculated from the line broadening as in section 9.1. Becker and v. Wurmb (1986) showed in the iron-oversaturated magnetites (i.e. those lying between stoichiometric magnetite and the phase boundary towards the magnetite + wüstite field) the line widths corresponding to the two sublattices are independent of a_{O2}. In Fe-deficient magnetites (those between stoichiometric Fe_3O_4 and the phase boundary towards the magnetite + hematite field), however, the line widths due to the Fe in octahedral sites broaden by a factor of more than two, whereas the singulet due to the tetrahedral sites is not affected. In terms of this model these data indicate that jumps occur preferentially on the octahedral sublattice and that increased concentration of defects greatly enhances cation mobility on this sublattice (vacancy-type migration mechanism).

b. Cation motion occurs mostly by exchange jumps between the sublattices. Line shapes can then be treated by relaxation theory, i.e. an immobile nucleus experiences fluctuations of interactions, provided that diffusion of the Mössbauer nuclei is slow. By computing relaxation spectra as a function of exchange rate Becker and v. Wurmb (1986) were able to show the collapse of the three-line (one octahedral doublet, one tetrahedral singulet) spectrum into a two-line spectrum with increasing exchange rate. Applying this concept to the a_{O2}-dependent spectra it has been deduced that the exchange rate increases with increasing oxygen activity, i.e. it is contolled by the vacancy concentration. The a_{O2}-dependence is only consistent with jumps from one sublattice to an empty neighbouring site of the other sublattice, but not with a mechanism involving ions on interstitial sites.

Mössbauer spectroscopy under in-situ conditions can thus yield constraints on atomistic transport processes.

10. PHASE TRANSITIONS

Through the response of hyperfine parameters to changes of the geometry and bonding characteristics of a site phase transitions of a crystal structure may be detected by taking Mössbauer spectra at variable temperatures (or pressures). Such studies are not confined to phases containing Fe in large (stoichiometric) amounts because working with enriched ^{57}Fe instead of natural iron about 0.1 wt.% Fe are sufficient to measure the spectra. Therefore, Fe may be implanted as a probe into the structure of a Fe-free phase and the phase transition may be monitored without too much chemical change.

Magnetic and electronic transitions as well as the study of intracrystalline order-disorder reactions has already been described above, and we will only discuss an example for a commensurate-incommensurate phase transition.

Seifert et al. (1987) studied Mössbauer spectra of melilites in the sytem $Ca_2MgSi_2O_7$ - $Ca_2FeSi_2O_7$. Whereas the high-temperature commensurate phase contains only one Fe-bearing tetrahedral site T1 and consequently exhibits only one doublet with narrow line widths, this doublet broadens at temperatures below 85 to 250°C (depending on composition) and on further decrease of temperature eventually splits into two well-separated and sharp doublets. This phase transition can independently be followed by electron diffraction and transmission electron microscopy, and the appearance of the Mössbauer doublets correlates with the formation of an incommensurate phase exhibiting satellite reflections and a modulated structure.

Mössbauer spectroscopy gives us further insight into the nature of the incommensurate structure: The hyperfine parameters prove that Fe still resides in tetrahedral coordination (although in two distinct sites). The sharp peaks of the two doublets indicate that two well-defined sites are present and not a continuum of sites which would produce only one broadened doublet. Therefore, the modulation cannot be of the more common sinusoidal type but must be more or less rectangular. Although resolution of the two doublets goes towards zero as the phase transition temperature is approached, it can be inferred that the ratio of the number of the two iron-bearing tetrahedral sites is close to 1 when the commensurate high-temperature phase is formed. Similar results can also be obtained for Si in the T2 site by NMR spectroscopy at elevated temperatures.

11. TWO EXAMPLES FOR PITFALLS

The two following examples are rather trivial, but emphasize the importance of careful consideration of error propagation and of independent information in the interpretation of Mössbauer spectra.

11.1 Extraction of thermodynamic data from Mössbauer-derived site occupancies

Saxena and Ghose (1971) noted that the Fe-Mg distribution over the M1 and M2 sites in orthopyroxenes (both heated and natural) deviated from ideality in the vicinity of the Fe-rich end member. Compared to a hyperbola in a Rooseboom diagram that would hold for a constant distribution coefficient K_D, the actual data points are displaced towards lower X_{M2}^{Fe} values. Saxena and Ghose (1971) and Saxena (1973) evaluated these apparent deviations from ideality in terms of a regular solution model and calculated temperature-dependent Margules interaction parameters W_{M1} and W_{M2}. As in most Mössbauer studies of this kind it was assumed that the Debye-Waller factors for Fe in the two positions (or their Debye temperatures) are sufficiently similar and that slight differences, if present, would not affect the results.

Let us consider what the effect on the apparent distribution curve would be if the ratio of the two Debye-Waller factors f_{M1}/f_{M2} were not equal to 1 but would deviate by 10%. This is the limit given by Virgo and Hafner (1968) for ferrosilite. Fig. 2 shows the distribution curve for a constant $K_D = 0.08$ (i.e. corresponding to ideality) and a curve that would be found experimentally with the assumption $f_{M1}/f_{M2} = 1.0$, if the true distribution would lie along the $K_D = 0.08$ curve, and the true f_{M1}/f_{M2} were 1.10. Obviously much of the "nonideality" postulated vanishes. f_{M1} larger than f_{M2} does not seem to be an unreasonable assumption since Fe in M1 can be expected to be bonded more tightly than Fe in M2 and therefore to exhibit a higher recoil-free fraction.

We do not want to imply that Fe-Mg distribution in orthopyroxenes is necessarily ideal, but it should be pointed out that due to the calculation of x^{Mg} by difference, uncertainties in K_D become very large as x^{Fe} approaches 1, and that more work is required before thermodynamic data can be extracted. A crucial experiment would be careful measurement of temperature - dependent M1/M2 area ratios in a pure and well-characterized ferrosilite end member sample where the two site occupanices are fixed by stochiometry and the ratio of their recoilless fractions can be determined.

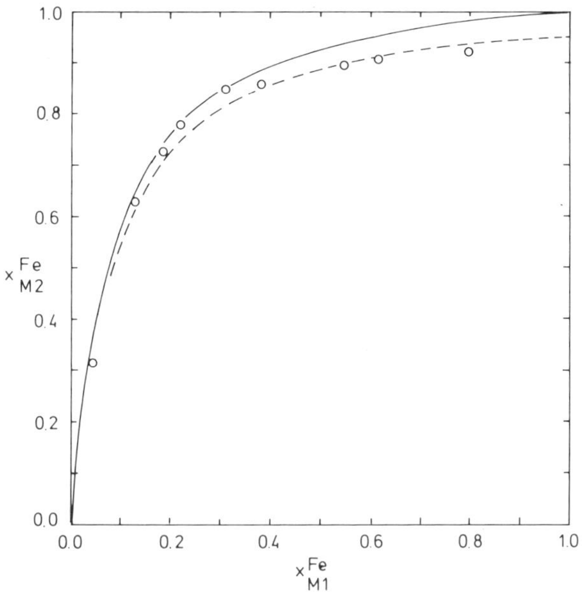

Fig. 2 Effect of $f_{M1}/f_{M2} \neq 1.0$ on the apparent Fe-Mg distribution in orthopyroxene. The full curve gives an assumed true distribution with a $K_D = x_{M1}^{Fe}(1-x_{M2}^{Fe})/x_{M2}^{Fe}(1-x_{M1}^{Fe})$ of 0.08, and the dashed curve represents a distribution that would be obtained from such phases if f_{M1}/f_{M2} were arbitrarily set to 1.0, but the true ratio would be 1.10. Circles represent the 600°C data obtained by Saxena and Ghose (1971). For discussion see text.

11.2 Ferric iron in melilite

Akasaka and Ohashi (1985) described Mössbauer spectra of synthetic ferriferous melilite solid solutions. In most spectra, two rather well-resolved doublets were present with equal area (within the limits of error) but differing in isomer shift and quadrupole splitting. The authors compared this to the melilite structure and assigned one doublet to the T1, the other to the T2 position.

These results are surprizing in several respects: First, the hyperfine parameters deviate significantly from the well-established correlation between bond length and isomer shift of Fe^{3+} in tetrahedral coordination (e.g. Steffen et al. 1984). Second, it would be very unusual that Fe^{3+} and Al should not be fractionated between T1 and T2, because these sites have very different metal-oxygen bond lengths (mean T1-O = 1.915 Å vs. mean T2-O = 1.619 Å in akermanite, Kimata and Ii 1981).

It has been demonstrated in the meantime that ferriferous melilites exhibit incommensurate structures at room temperature, similar to those described in the section for ferrous-bearing melilites (Roethlisberger unpublished 1987). An alternate, presently at least as reasonable assignment of the two doublets would be to two T1 positions distorted to a different degree in the incommensurate phase. The critical test will be to measure temperature-dependent Mössbauer spectra and to compare the results to other structural information obtained by e.g. transmission electron microscopy.

REFERENCES

Abu-Eid R.M., Langer K., Seifert F. (1978) 'Optical absorption and Mössbauer spectra of purple and green yoderite, a kyanite-related mineral'. Phys. Chem. Minerals **3**:271-289.

Akasaka M., Ohashi H. (1985) '^{57}Fe Mössbauer study of synthetic Fe^{3+}-melilites.' Phys. Chem. Minerals **12**:13-18.

Amthauer G., Annersten H., Hafner S.S. (1976) 'The Mössbauer spectrum of ^{57}Fe in silicate garnets.' Z. Kristall. **143**:14-55.

Amthauer G., Langer K., Schliestedt M. (1980) 'Thermally activated electron delocalization in deerite.' Phys. Chem. Minerals **6**:19-30.

Amthauer G., Rossman G.R. (1984) 'Mixed valence of iron in minerals with cation clusters.' Phys. Chem. Minerals **11**:37-51.

Andersen E.B., Fenger J., Rose-Hansen J. (1975) 'Determination of Fe^{2+}/Fe^{3+} ratios in arfvedsonite by Mössbauer spectroscopy.' Lithos **8**:237-246.

Ballet O. (1986) 'Comment on magnetism in biotites.' Phys. Chem. Minerals **13**:281.

Ballet O., Coey J.M.D. (1982) 'Magnetic properties of sheet silicates; 2:1 layer minerals.' Phys. Chem. Minerals **8**:218-229.

Ballet O., Coey J.M.D., Burke K.J. (1985) 'Magnetic properties of sheet silicates; 2:1:1 layer minerals.' Phys. Chem. Minerals **12**:370-378.

Bancroft G.M. (1973) Mössbauer spectroscopy. An Introduction for Inorganic Chemists and Geochemists. McGraw Hill 252 p.

Bancroft M., Maddock, A.G., Burns R.G. (1967) 'Application of the Mössbauer effect to silicate mineralogy I. Iron silicates of known crystal structure.' Geochim. Cosmochim. Acta **31**:2219-2246.

Bancroft G.M., Burns R.G., Stone A.J. (1968) 'Applications of the Mössbauer effect to silicate mineralogy II. Iron silicates of unknown and complex crystal structures.' Geochim. Cosmochim. Acta **32**:547-559.

Bancroft G.M., Osborne M.D., Fleet M.E. (1983) 'Next-nearest neigbour effects in the Mössbauer spectra of Cr-spinels: an application of partial quadrupole splittings.' Solid State Comm. **47**:623-625

Becker K.D., v. Wurmb V. (1986) 'Oxygen activity dependence of high temperature Mössbauer spectra in the iron oxides. I. Magnetite ($Fe_{3-\delta}O_4$).' Zeitschrift Phys. Chemie NF **149**:77-89.

Besancon J.R. (1981) 'Rate of cation ordering in orthopyroxenes.' Am. Mineral. **66**:965-973.

Chambaere D.G., De Grave E. (1984) 'On the Néel temperature of β-FeOOH: structural dependence and its implications.' J. Magn. Mat. **42**:263-268.

Chambaere D.G., De Grave E. (1985) 'The β-FeOOH to α-Fe_2O_3 phase transformation: Structural and magnetic phenomena.' Phys. Chem. Minerals **12**:176-184.

De Grave E., Persoons R.M., Chambaere D.G., Vandenberghe R.E., Bowen L.H. (1986) 'An ^{57}Fe Mössbauer effect study of poorly crystalline γ-FeOOH.' Phys. Chem. Minerals **13**:61-67.

Dowty E., Lindsley D. (1973) 'Mössbauer spectra of synthetic hedenbergite-ferrosilite pyroxenes.' Am. Mineral. **58**:850-868.

Drissen P.M (1980) 'Versuche zur Bestimmung von Diffusionskoeffizienten des Fe^{3+} in Silikaten und Oxiden über die Linienverbreiterung in Mössbauerspektren.' Diploma Thesis Kiel, 73 p.

Ferrow E. (1987) 'Mössbauer and X-ray studies on the oxidation of annite and ferriannite.' Phys. Chem. Minerals **14**:270-275.
Fysh S.A., Ostwald J. (1983) 'A Mössbauer study of some Australian iron ore minerals.' Mineral. Mag. **47**:209-217.
Ghose S., Cox D.E., Van Dang N. (1987) 'Magnetic order in grunerite, $Fe_7Si_8O_{22}(OH)_2$ A quasi-one dimensional antiferromagnet with a spin canting transition.' Phys. Chem. Minerals **14**:36-44.
Goldman D.S. (1979) 'A reevaluation of the Mössbauer spectroscopy of calcic amphiboles.' Am. Mineral. **64**:109-118.
Gonser U. (ed.) (1975) Mössbauer Spectroscopy. Topics in Applied Physics **5**, Springer Verlag, 241 p.
Gütlich P., Link R., Trautwein A. (1978) Mössbauer Spectroscopy and Transition Metal Chemistry. Springer Verlag, 280 p.
Guillen R., Regnard J.R. (1985) 'Magnetic properties of natural and synthetic wolframites $Fe_xMn_{1-x}WO_4$.' Phys. Chem. Minerals **12**:246-254.
Hafner S.S., Ghose S. (1971) 'Iron and magnesium distribution in cummingtonites $(Fe,Mg)_7Si_8O_{22}(OH)_2$.' Z. Krist. **133**:301-326.
Hawthorne F.C. (1983) 'Quantitative characterization of site-occupancies in minerals. 'Am. Mineral. **68**:287-306.
Hayashi M., Tamura I., Shimomura O., Sawamoto H., Kawamura H. (1987) 'Antiferromagnetic transition of fayalite under high pressure studied by Mössbauer spectroscopy.' Phys. Chem. Minerals **14**:341-344.
Heberle J. (1971) 'The Debye integrals, the thermal shift and the Mössbauer fraction' Mössbauer Effect Methodology **7**:299-308.
Kimata M., Ii N. (1981) 'The crystal structure of synthetic akermanite $Ca_2MgSi_2O_7$.' N. Jb. Mineral. Mh. 1981:1-10.
Linares L., Regnard J.R., Van Dang N. (1983) 'Magnetic behaviour of grunerite from Mössbauer spectroscopy.' J. Magnetism Magnetic Mat. **31-34**:715-716.
Litterst F.J. Amthauer G. (1984). 'Electron delocalization in ilvaite, a reinterpretation of its ^{57}Fe Mössbauer spectrum.' Phys. Chem. Minerals **10**:250-255.
Maddock A.G. (1985) 'Mössbauer spectroscopy in mineral chemistry.' In: Chemical Bonding and Spectroscopy in Mineral Chemistry (F.J. Berry and D.J. Vaughan ed.) Chapman and Hall, 141-208.
McCammon C.A., Price D.C. (1985) 'Mössbauer spectra of Fe_xO (> 0.95).' Phys. Chem. Minerals **11**:250-254.
Murad E., Bowen L.H. (1987) 'Magnetic ordering in Al-rich goethites: Influence of crystallinity.' Am. Mineral. **72**:194-200.
Murad E., Wagner F.E. (1987) 'The Mössbauer spectrum of almandine.' Phys. Chem. Minerals **14**:264-269.
Mysen B.O., Carmichael I.S.E., Virgo D. (1985a) 'A comparison of iron redox ratios in silicate glasses determined by wet-chemical and ^{57}Fe Mössbauer resonant absorption methods.' Contr. Mineral. Petrol. **90**:101-106.
Mysen B.O., Virgo D., Neumann E.-R., Seifert F. (1985b) 'Redox equilibria and the structural states of ferric and ferrous iron in melts in the system CaO-MgO-Al_2O_3-SiO_2-Fe -O: relationships between redox equilibria, melt structure and liquidus phase equilibria.' Am. Mineral. **70**:317-331.
Osborne M.D., Fleet M.E. (1984) 'Mössbauer investigation of niningerite solid solutions (Mg, Fe)S.' Phys. Chem. Minerals 10:245-249.
Sanz J., Gonzalez-Carreno T., Gancedo R. (1983) 'On dehydroxylation mechanisms of a biotite in vacuo and in oxygen.' Phys. Chem. Minerals **9**:14-18.
Saxena S.K. (1973) Thermodynamics of rock-forming crystalline solutions. Springer-Verlag

Berlin, 188 p.
Saxena S.K., Ghose S. (1971) 'Mg^{2+}-Fe^{2+} order-disorder and the thermodynamics of the orthopyroxene crystalline solution.' Am. Mineral. **56**:532-559.
Saxena S.K., Tazzoli F., Domeneghetti M.C. (1987) 'Kinetics of Fe-Mg distribution in aluminous orthopyroxenes.' Phys. Chem. Minerals in the press.
Schmidbauer E., Mosheim E., Semioschkina N (1986) 'Magnetization and ^{57}Fe Mössbauer study of obsidians.' Phys. Chem. Minerals **13**:256-261.
Seifert F. (1977) 'Compositional dependence of the hyperfine interaction of ^{57}Fe in anthophyllite.' Phys. Chem. Minerals **1**:43-52.
Seifert F. (1978) 'Equilibrium Mg-Fe^{2+} cation distribution in anthophyllite.' Am. J. Science **278**:1323-1333.
Seifert F. (1983) 'Mössbauer line broadening in aluminous orthopyroxenes: Evidence for next nearest neigbors interactions and short-range order.' N. Jahrb. Mineral. Abh. **148**:141-162.
Seifert F., Dasgupta H.C. (1982) 'A note on the Mössbauer spectrum of ^{57}Fe in braunite.' N. Jahrb. Mineral. Mh. **1982**:11-15.
Sherman D.M. (1987a) 'Molecular orbital (SCF-X_δ-SW) theory of metal-metal charge transfer processes in minerals. I. Application to $Fe^{2+} \rightarrow Fe^{3+}$ charge transfer and "electron delocalization" in mixed-valence iron oxides and silicates.' Phys. Chem. Minerals **14**:355-363.
Sherman D.M. (1987b) 'Molecular orbital (SCF-X_δ-SW) theory of metal-metal charge transfer processes in minerals. II. Application to $Fe^{2+} \rightarrow Ti^{4+}$ charge transfer transitions in oxides and silicates.' Phys. Chem. Minerals **14**:364-367.
Singwi K.S., Sjölander A. (1960) 'Resonance absorption of nuclear gamma rays and the dynamics of atomic motions.' Phys. Rev. **120**:1093-1099.
Simons B., Seifert F. (1979) 'High-pressure wüstite: Cell parameters and Mössbauer spectra.' Carnegie Inst. Wash. Year Book **78**:625-626.
Skogby H., Annersten H. (1985) 'Temperature dependent Mg-Fe cation distribution in actinolite-tremolite.' N. Jahrb. Mineral. Mh. 1985:193-203.
Smith G., Halenius U., Annersten H., Ackermann L. (1983) 'Optical and Mössbauer spectra of manganese-bearing phlogopites: $Fe_{IV}{}^{3+}$-$Mn_{VI}{}^{2+}$ pair absorption as the origin of reverse pleochroism.' Am. Mineral. **68**:759-768.
Spiering B., Seifert F. (1985) 'Iron in silicate glasses of granitic composition: a Mössbauer spectroscopic study.' Contrib. Mineral. Petrol. **90**:63-73.
Stampfl P.P., Travis J.C., Bielefeld M.J. (1973) 'Mössbauer spectroscopic studies of iron doped rutile.' Phys. Stat. Sol. (a) **15**:181-185.
Steffen G., Seifert F., Amthauer G. (1984) 'Ferric iron in sapphirine: a Mössbauer spectroscopic study.' Am. Mineral. **69**:339-348.
Swinnea J.S., Steinfink H. (1983) 'Crystal structure and Mössbauer spectrum of vonsenite, $2FeO.FeBO_3$.' Am. Mineral. **68**:827-832.
Tazzoli V., Domeneghetti M.C. (1987) 'Crystal chemistry of natural and heated aluminous orthopyroxenes.' Phys. Chem. Mineral in the press.
Townsend M.G., Longworth G. (1985) 'Sign of the magnetic coupling of Fe^{2+} and Fe^{3+} in biotite.' Phys. Chem. Minerals **12**:141-144.
Virgo D., Hafner S.S. (1968) 'Re-evaluation of the cation distribution in orthopyroxenes by the Mössbauer effect.' Earth Planet. Sci. Letters **4**:265-269.
Virgo D., Hafner S.S. (1969) 'Fe^{2+}-Mg order-disorder in heated orthopyroxenes.' Min. Soc. Am. Special Paper **2**:67-81.
Virgo D., Hafner S.S. (1970) 'Fe^{2+}-Mg order-disorder in natural orthopyroxenes. 'Am. Mineral. **55**:201-233.
Virgo D., Mysen B.O. (1985) 'The structural state of iron in oxdized vs. reduced glasses at 1

atm: A ^{57}Fe Mössbauer spectroscopic study.' Phys. Chem. Minerals **12**:65-76.

Wivel C., Mörup S. (1980) 'Improved computational procedure for evaluation of overlapping hyperfine field distributions in Mössbauer spectra.' J. Phys. **E 14**:605-610.

Wu G., Mu B. (1986) 'The crystal chemistry and Mössbauer study of schorlomite.' Phys. Chem. Minerals **13**:198-205.

SUBJECT INDEX